NEAR-EARTH OBJECTS

THE UNITED NATIONS INTERNATIONAL CONFERENCE

UNITED NATIONS OFFICE AT VIENNA   OFFICE DES NATIONS UNIES A VIENNE

THE EXPLORERS CLUB

ANNALS OF THE NEW YORK ACADEMY OF SCIENCES
Volume 822

NEAR-EARTH OBJECTS

THE UNITED NATIONS INTERNATIONAL CONFERENCE

Edited by John L. Remo

The New York Academy of Sciences
New York, New York
1997

∞ *The paper used in this publication meets the minimum requirements of American National Standard for Information Sciences—Permanence of Paper for Printed Library Materials, ANSI Z39.48-1984.*

Cover: *The art on the softcover version of this book shows a Russian stamp issued in 1957 commemorating the 1947 Sikhote-Alin fireball. The picture on the stamp was painted by P.I. Medvedev, who witnessed the towering multicolored cloud of debris produced by the impact of this meteorite in Siberia. This image, which was published in* Rocks from Space: Meteorites and Meteorite Hunters *(Missoula, MT: Mountain Press Publishing Company, 1996) was generously lent by the author of the book, Dr. O. Richard Norton. The image at the bottom of the cover is a 1994 supercomputer simulation of the ballistic fireball generated by the collision of comet Shoemaker-Levy 9 with Jupiter, which shows a stricking similarity to the picture above. These pictures are discussed in greater detail on page 262.*

GYAT / PCP
Printed in the United States of America
ISBN 1-57331-040-9 (cloth)
ISBN 1-57331-041-7 (paper)
ISSN 0077-8923

ANNALS OF THE NEW YORK ACADEMY OF SCIENCES

Volume 822
May 30, 1997

NEAR-EARTH OBJECTS
THE UNITED NATIONS INTERNATIONAL CONFERENCE[a]

Editor
JOHN L. REMO

Organizing Committee
C. BLACK, J. LORET, AND J.L. REMO, *Co-Chairs*
L. FRIEDMAN, E. NADLER, J. RAHE, J. TANCREDI, W. TEDESCHI, AND N. TYSON

Program Review Committee
J.L. REMO, *Chair*
C. BLACK, C. CHAPMAN, T. GEHRELS, B. MARSDEN, AND P. SFORZA

CONTENTS

[a]The papers in this volume were presented at the **International Conference on Near-Earth Objects** held on April 24–26, 1995 in New York, New York and sponsored by the American Museum of Natural History, the Explorers Club, NASA, the Planetary Society, Quantametrics, Inc., Sandia National Laboratories, and the United Nations Office for Outer Space Affairs.

II. Earth and Planetary Sciences

III. Astronautics

IV. Detection and Mitigation

Financial assistance was received from:

- **AMERICAN MUSEUM OF NATURAL HISTORY**
- **THE EXPLORERS CLUB**
- **NATIONAL AERONAUTICS AND SPACE ADMINISTRATION**
- **THE PLANETARY SOCIETY**
- **QUANTAMETRICS, INC.**
- **SANDIA NATIONAL LABORATORIES**
- **UNITED NATIONS OFFICE FOR OUTER SPACE AFFAIRS**

Preface

JOHN L. REMO

Quantametrics, Inc.
1 Brackenwood Path, Head of Harbor
St. James, New York 11780

This volume is a compilation of the proceedings of a meeting from April 24 through 26, 1995, when scientists from around the world gathered at the United Nations World Headquarters and at the Explorers Club in New York City to discuss a broad range of scientific issues associated with near-Earth objects (NEOs). The primary focus of the topics was directed toward the need for gathering and interpreting scientific data to understand the Earth's past history, evaluate current astronomical observations, and outline future exploration missions to NEOs. Of critical importance is the need to support astronomical observation programs in both the northern and southern hemispheres.

The topics presented in this volume can be thought of as a continuation of those contained within the foundation volume, *Hazards Due to Comets and Asteroids* [1], which summarizes most of the preceding work in NEOs and serves as the ideal starting place for those seriously interested in NEO studies.

At the conclusion of the conference, there was a general consensus that the study of NEOs and their possible interaction with the Earth was of both international as well as interdisciplinary scientific significance. Perspectives on this consensus have been published elsewhere [2].

In addition to the conference supporters, we are all indebted to the staff of the United Nations and especially to both N. Jasentuliyana, director of the United Nations Office for Outer Space Affairs and to M. Whaba, Executive Office of the Secretary General, as well as to the U.S. Mission to the United Nations, without whom this conference would not have been possible. I also wish to personally thank L. Friedmen, T. Gehrels, D. Goldin, R. Hagengruber, H.J. Haubold, and B. Marsden, for their encouragement, support, and guidance, and John C. Remo for his stewardship with the audiovisuals and media materials. Without the support and assistance of B. Boland and his staff at the New York Academy of Sciences, especially Sheila Kane and Stephanie J. Bludau, this volume would not have been possible. Finally, I wish to personally thank all of the authors for providing an exciting conference and the fine papers contained in this volume.

REFERENCES

1. GEHRELS, T. 1994. Hazards Due to Comets and Asteroids. University of Arizona, Tuscon, AZ.
2. REMO, J.L. 1996. Policy perspectives from the UN international conference on near-Earth objects. Space Policy. **12:** 13–17.

Development NEO Research: A Chronological Outline

JOHN L. REMO

Quantametrics, Inc.
1 Brackenwood Path, Head of Harbor
St. James, New York 11780

ABSTRACT: Milestones in the historical development of NEO studies are briefly outlined.

INTRODUCTION

Milestones of near-Earth object (NEO) research are listed in chronological order. A suitable scientific beginning point for the time-line is mid to late 1700's which are marked by the recognition of the periodic nature of Halley's comet in 1758 and Chladni's work in the late 1700's that led to the acceptance of meteorites as extraterrestrial material. The chronology continues through the mid-twentieth century when the field of planetary science emerged with assistance from the tools of nuclear and quantum physics, chemistry, and astronomy. These tools provided laboratory methodologies which made possible the quantitative analyses that have paved the way toward understanding the important evolutionary relationships between stars, the interstellar medium, and the systems of planets and minor objects that surround them. During this same period, space missions became possible and opened the way to vigorous planetary exploration programs. Recovered samples and other (imaging and spectral) data from these missions have fundamentally shaped planetary science and they promise to influence the study of minor planets in the same way. Today, planetary science stands at the threshold of a new era of discovery with the Hubble Space Telescope and the ground-based NEAT (Near-Earth Asteroid Tracking) and Spacewatch types of telescopes, and the NEAR and Rosetta spacecraft rendezvous, which will extend the time line of discovery far into the future.

The importance of NEO studies received an important impetus with the discovery of the iridium anomaly at the K/T (Cretaceous-Tertiary) boundary. No other event has so clearly demonstrated the influence of minor objects on the evolution of terrestrial life. The introduction of studies based on the fossil record has broadened the interdisciplinary nature of planetary science, encompassing fundamental concepts of Earth history, mammalian evolution, and contemporary natural hazards both on Earth and in space.

Giotto's representation of comet Halley as a sign of hope for mankind must be tempered with awareness about the hazard they may present.

Toward the end of the time line is the impact of SL-9 fragments into Jupiter which presented both a majestic display and an unequivocal warning of our vulnerability. Both space and ground-based instruments were able to monitor and analyze this event because of prior commitments to basic research. Hopefully, we will have the assets in place to observe a NEO that threatens the Earth in the future. The future progress of NEO research and hazard mitigation will depend on the steadfast commitment of world leaders to support both pure and applied research in general, and a sensible NEO detection and exploration program in particular. What makes science so exciting is that one cannot predict what discoveries are yet to come.

ASTEROID AND COMET DISCOVERY TIME LINE

1304: Halley's comet (1301) tail and coma, while a fright at the time, depicted in Giotto's "Adoration of the Magi," as the "Star of Bethlehem," a sign of hope.

1758: Reappearance of comet Halley as predicted by E. Halley over half a century earlier (1705), represented a triumph for Newtonian dynamics. Halley also noted Jupiter undoubtedly had serious effects on the comet's motion as well as the possibility of a comet planet collision.

1776: Titius-Bode algorithm predicts planetary distances.

1770: A comet passes less than 3 million km from the Earth. Orbit studied by A. Lexell, who found it to have a period of only six years and to have made very close encounters with Jupiter in 1767 and 1779.

1781: Uranus discovered, W. Herschel.

1794: E. Chladni concludes that, "meteorites do indeed fall from the sky."

1803: J. Biot methodically demonstrates from eyewitnesses that meteorites can fall from the sky.

1801: G. Piazzi discovers asteroid 1, Ceres. K.F. Gauss computes semi-major axis of the elliptical orbit to be 2.8 AU as predicted by Titius-Bode's algorithm.

1819: J.F. Encke's comet recognized at four apparitions back to 1786, found that orbit is affected by forces other than gravitational.

1866: Gaps at 3.28 and 2.50 AU correspond to orbital periods that are fractions of Jupiter's period, D. Kirkwood.

1866: Perseid meteor shower occurs when comet (1862 III) crosses the Earth's orbit, G. Schiaparelli.

1868: Comet spectroscopy reveals carbon emission bands, W. Huggins.

1891: Photography introduced to asteroid observations, M. Wolf.

1898: The first discovered Earth-approaching asteroid, (433) Eros, discovered, G. Witt.
1898: Asteroid light variation interpreted as rotation of an irregular object.
1906: The first Trojan, (588) Achilles, discovered, M. Wolf. 1908: Tunguska explosion; 10–20 megaton (TNT equivalent) airburst.
1908: The Tunguska event.
1918: K. Hirayama introduces concept of asteroid families.
1920: 944 Hidalgo discovered, W. Baade.
1921: L.A. Kulik expedition to Tunguska.
1932: 1862 Apollo, first recognized Earth-crossing asteroid discovered by K. Reinmuth. However, (887) Alinda found by M. Wolf in 1918 is now recognized as an Earth-crossing asteroid.
1944: O. Schmidt postulates that incipient planet formation near 2.8 AU by planetesimals is interrupted by Jupiter's perturbation, resulting in asteroids.
1947: IAU established Minor Planet Center in Cincinnati, Ohio; moved to Cambridge, Massachusetts in 1978.
1949: Yerkes-McDonald photometric asteroid survey, G. Kuiper.
1949: R. Baldwin publishes "The Face of the Moon."
1950: F. Whipple suggests dirty snowball model for comet nuclei.
1950: Oort cloud hypothesized for ~10^{12} (P. Weissman, 1993), inactive long period (~50×10^6 y) comets in a spherical cloud about 50,000–150,000 AU from the Sun.
1951: Kuiper belt proposed for ~10^9 short period (~100 y) comets in a planar disc population 30–100 AU from the Sun.
1951: E. Opik calculates probability of Earth-crossing asteroid (ECA) collision with Earth.
1951: K-Ar meteorite formation ages measured, E.K. Gerling.
1953: Asteroid collision study, S.L. Poitowski.
1957: Cosmic ray exposure ages of meteorites measured, F.Begemann.
1959: First precise record of a meteorite fall (Pribram), Z. Ceplecha.
1960: Analysis of impact mechanics at Meteor Crater, Arizona, G. Shoemaker.
1964: Solar wind He implantation in certain meteorites, H. Wanke.
1964: Theory of the origin of meteorites in asteroids, E. Anders.
1964: Iron meteorite cooling rates imply a parent asteroid size of ~100–500 km in diameter, J. Wood.
1967: Dynamics of meteorite delivery from asteroid belt, G. Wetherill.
1968: Radar observation of the Apollo asteroid (1566) Icarus from Goldstone and Haystack.
1969: Process for studying and evaluating cometary nongravitational forces initiated, B. Marsden.

1969: Spectrophotometry of asteroids by T. McCord; applied to many asteroids by C. Chapman.
1970: Polarimetric reflectivity of (1566) Icarus by J. Veverka; applied to many asteroids by B. Zellner.
1970: Palomar-Leiden Survey of faint minor planets published, E. J. van Houten et al.
1970: Infrared diameter of asteroid (4) Vesta, D. Allen. Infrared radiometry applied to many asteroids.
1971: Modern ECA wide field search initiated; Daedalus discovered, T. Gehrels.
1971: Exploitation of asteroids proposed, S. Herrick.
1972: $\sim 10^6$ kg object skips off the Earth's atmosphere.
1972: Protoplanets scatter icy planetesimals to Oort cloud, V. Safronov.
1973: Planet-crossing asteroid survey (PCAS) initiated using 0.46 m Schmidt telescope, E. Helin, G. Shoemaker, et al.
1973: Energy distributions for (dynamically new) observed long-period comets from the Oort cloud, B. Marsden and Z. Sekanina.
1975: Survey of UBV photometry of asteroids, T. Bowell.
1975: Combined results of polarimetry, radiometry, and spectrophotometry used to define the broadband system of classification in terms of C, S, and other asteroid types which resemble (meteorite) mineral spectra, C. Chapman, D. Morrison, and B. Zellner.
1975: Fourier transform spectroscopy of (4) Vesta, S.M. Larsen and U. Fink.
1976: Asteroid (2062) Aten found with semimajor axis smaller than Earth's orbit, E. Helin.
1977: Icy mantles on silicate grain cores radiated prior to their integration into comet nuclei, J.M. Greenberg.
1977: Discovery of the first "centaur," Chiron, a large asteroid -comet between the orbits of Saturn and Uranus, C. Kowal.
1980: L. and W. Alvarez, F. Asaro, and H. Michel discover that the rare element iridium is anomalously enhanced in a thin clay layer near Gubbio, Italy. Same enhancement found to occur worldwide in 2–3 cm layer, both on land and sea. Suggest huge asteroid impact which resulted in mass extinctions of life forms at the K/T boundary, 65 million years ago.
1981: Spacewatch CCD scanning program initiated at Kitt Peak, Arizona, T. Gehrels.
1983: IRAS — distinct dust rings circle sun at asteroid belt.
1986: IRAS — dust trails spread along orbits of active comets.
1986: Giotto, Vega I & II, and Sakigake and Suisei missions to Halley. Giotto observes nucleus 15 km long and 7–10 km wide, ~80% water

vapor, carbon based particles, and gas and dust jets, and abundance of C, H, O, N.

1989: Radar (double lobe) image of (4769) Castalia, S. Ostro.

1980s: Asteroid compositional classes concentrated at different solar distances.

1990: Anglo-Australian Near-Earth Asteroid Survey (ANEAS) using a 1.2 m Schmidt at Siding Spring Observatory, New South Wales, D. Steel, R. Mc Naught, and K. Russell.

1990: Chicxulub 180 km crater appears to fit requirements for K/T impact site, A. Hildebrand.

1991: Galileo spacecraft flyby of (951) Gaspra.

1992: Discovery of the "first" member of the Kuiper belt, 1992 QB1, with the conclusion that the "zeroth" member was discovered in 1930 as Pluto, D. Jewitt and J. Luu.

1993: Galileo spacecraft flyby to Ida and satellite Dactyl. Galileo imaging team confirms steepening toward smaller sizes of the inner solar system crater size distribution.

1993: Chicxulub crater gravitational anomaly work shows evidence of impact, V. Sharpton.

1993: Comet Shoemaker-Levy 9 discovered, C. and G. Shoemaker and D. Levy. B. Marsden computes SL-9 orbit and notes that it would impact Jupiter.

1994: Shoemaker-Levy 9 fragments and impacts on Jupiter. Observed by Hubble space telescope, the Galileo satellite, and numerous ground observatories.

1994: Comprehensive review of NEOs in "Hazards Due to Comets and Asteroids," T. Gehrels, editor and 120 collaborators.

1995: United Nations Conference on near-Earth objects.

1995: Near-Earth Asteroid Tracking (NEAT) becomes operational, E. Helin.

1996: Discovery of 1996 JA1, the largest object (diameter 150 m) known to have approached the Earth to within a moon's apogee distance, T. Spahr and C. Hergenrother.

1996: Discovery of 1996 PW, asteroidal in appearance, but with an orbit similar to that of a long-period comet, blurs distinctions between asteroids and comets, E. Helin and G. Williams.

Near-Earth Asteroid Tracking (NEAT) Program

ELEANOR F. HELIN, STEVEN H. PRAVDO, DAVID L. RABINOWITZ, AND KENNETH J. LAWRENCE

Jet Propulsion Laboratory
California Institute of Technology
4800 Oak Grove Drive, MS 183-501
Pasadena, California 91109

ABSTRACT: The discoveries of near-Earth asteroids (NEAs) and comets have increased enormously over the last 10–20 years. This is a consequence, in large part, of the success of programs that have systematically searched for these objects. These programs have been motivated by the relationships of NEAs to terrestrial impacts, meteorites, comets, and their relative accessibility to spacecraft missions. This paper will review the long-term Palomar Planet-Crossing Asteroid Survey (PCAS), a photographic program, and the current Near-Earth Asteroid Tracking (NEAT) system, NASA's new electronic detection program. The primary goal of NEAT is to discover and inventory near-Earth asteroids and comets, collectively called near-Earth objects or NEOs, larger than 1 km in size. Details of the NEAT system and program results are presented and discussed.

HISTORICAL BACKGROUND

The predecessor to the Near-Earth Asteroid Tracking (NEAT) program was the Palomar Planet-Crossing Asteroid Survey (PCAS) [5], [6] a photographic program conducted at Palomar Observatory using the 0.46 m Schmidt telescope. It was the first and longest running program dedicated to the detection of NEOs. It was discontinued in June 1995, after 23 years of operation.

The dedicated, systematic PCAS program searched the sky six nights monthly, typically covering about 4,000 square degrees of sky, each observing run centered near opposition and along the ecliptic. This coverage led to the discovery of 1 to 2 NEOs each month. Each exposure taken with this wide-angle telescope covers about eight degrees of sky in diameter or about 60 square degrees, almost the size of the bowl of the Big Dipper. The venerable old Schmidt camera telescope is a fast photographic system (focal ratio of 2). Film pairs of preselected centers (standard fields) were taken during the search program producing important discoveries and an extensive archive of films.

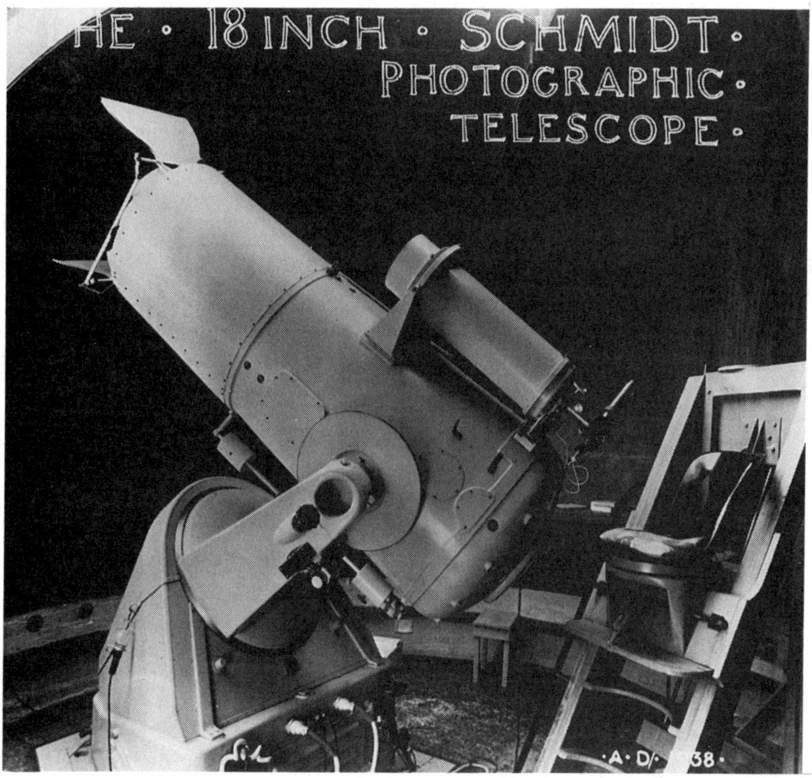

FIGURE 1. For over 60 years the Palomar 0.46 m (18-inch) telescope has been operational. It resembles a large mortar cannon ready for firing and has been responsible for two-thirds of the near-Earth asteroids that have been found in the last 20 years before electronic search systems became active.

METHODS AND PROCEDURES USED DURING THE PROGRAM

Methods used were standardized as soon as the system was working satisfactorily given the constraints of the telescope and photographic material available. The observing part of the program remained essentially the same in terms of acquiring the data. Eight films, or four pairs of films, were exposed for 8–10 minutes and then developed. Film pairs were exposed at the same position in the sky separated by approximately 30 minutes in time. Once the films were dry, they were scanned on a custom-designed stereomicroscope. This inspection was done promptly so discoveries could be made close to real time and new, fast-moving objects (fmo) could be followed up the same night if possible. These additional observations were acquired as quickly as possible in order to assure sufficient positions to determine a preliminary orbit (for example, the independent discovery of the bright asteroid Toutatis).

With the introduction of improved Kodak 4415 film, and new hypersensitization procedures, it was possible to reduce the exposure time to 6 minutes. The shorter exposure time allowed more photographs to be acquired resulting in greater sky coverage. With the "hypered" film and optimum exposure time, we were able to extend the limiting magnitude to approximately 18.5 for stellar objects. With greater sky coverage and limiting magnitude, the PCAS discovery rate increased, achieving the primary objective of our search program and maximizing the discovery of NEOs.

The typical observing team was comprised of three, sometimes four members alternating tasks throughout the night. Telescope activities included the telescope observing, developing film, scanning and inspection of films, labeling films, interacting with the Minor Planet Center and other observers via e-mail. During the first few hours of the night, fields for follow-up observations were taken. Usually, this was for objects recently discovered on prior observing runs. As soon as two hours postopposition was high enough in the sky, we concentrated heavily upon the anti-Sun or opposition area of the sky for the majority of the night. Our main concentration was from two hours pre- through postopposition and along the ecliptic. In addition, we always tried to obtain follow-up fields on a second night two or three nights later in the run. This would allow for confirmatory positions of any new objects discovered.

RESULTS FROM THE SYSTEMATIC PHOTOGRAPHIC SURVEY: PCAS

Overall, PCAS' contributions to the scientific community have been substantial and multifaceted: a total of 95 NEA discoveries (plus many independent discoveries), 17 comets, plus dozens of crucial confirmatory and prediscovery observations. Also, PCAS was instrumental in attracting worldwide NEO participation through its on-going program.

The majority of NEOs in the last twenty years have been discovered with wide-field photographic Schmidt telescopes. This growing number of discoveries gained the attention of the scientific community and in time, sparked the interest of the press and other media. From the small Schmidt telescope at Palomar Observatory, the long-term PCAS program had been productive in the discovery of a significant number of the NEOs.

SOME NOTABLE PCAS DISCOVERIES

The first NEA discovered was 5496 1973 NA, a rapidly moving asteroid found on the July 4, 1973. It was a bright object (magnitude 10–11) moving

11° per day into the southern hemisphere. It has a very high inclination of 68°, still holding the record for highest inclination among the known NEAs. It is probably a degassed comet.

2062 Aten, the first asteroid with an orbit smaller than the Earth's, was found January 7, 1976. It became the prototype of a new class of rare asteroids which, because of their small orbits, never venture far from the orbit of the Earth. Ultimately, these asteroids with relatively short dynamical lifetimes have a high potential for colliding with the Earth [4].

In 1979, a bright asteroid, 4015 1979 VA, was found moving in a cometary-like orbit [6]. This asteroid, 1979 VA, was later linked with an image photographed in 1949 during the Palomar Observatory Sky Survey. In the 1949 photo, it had a "tail," providing evidence that 30 years prior to its discovery as an asteroidal object, it was a comet with a tail [1]. This was the first direct proof of a comet evolving into an asteroid after its supply of ices and volatiles were depleted. As the discoverer, Helin had not yet named the numbered asteroid when it was linked to comet 107P/Wilson-Harrington. Henceforth, it was called by that name. A numbered asteroid with a comet name, P/107 Wilson-Harrington = 4015 1979VA [2].

Several asteroids, 3757 1982 XB, 6489 Golevka, 1989 ML are excellent candidates for spacecraft missions because of the low-energy expenditures required for rendezvous. In 1996, 6489 Golevka was imaged by radar (from Goldstone in the US, Evpatoria in Russia, and Kashima in Japan) greatly refining its orbit/ephemeris for future mission consideration [14].

BACKGROUND ON THE UNITED STATES AIR FORCE "GROUND-BASED ELECTRO-OPTICAL DEEP SPACE SURVEILLANCE SYSTEM (GEODSS)

The GEODSS satellite surveillance system has been in existence since the early 1980s. They utilize two "main" telescopes of 1 m aperture and one auxilary telescope, a 0.35 m (14-inch) folded Schmidt, at each site distributed around the globe. GEODSS sites provide "deep" space surveillance in detecting, tracking, and identifying manmade objects. This network of telescopes have always been attractive to astronomers. Discussions began in midsummer of 1992 to explore the possibility of using the 1 m telescope by placing a state-of-the-art charge-coupled device (CCD) camera at the focal plane, fully utilizing the large-field of view of their instrument. Some observing time became available in the aftermath of the Cold War and consideration was given to using this powerful, fast detection system for natural, near-Earth objects. This collaboration was formalized and the NEAT program began in December 1995. It has served as an unparalleled opportunity for NASA/JPL to collaborate with the Air Force in searching for NEAs. A highly successful detection program has been developed using a military telescope for scientific purposes, producing a economical and efficient system through an innovative and mutually beneficial partnership.

THE NEAR-EARTH ASTEROID TRACKING (NEAT) PROGRAM

Introduction

The Near-Earth Asteroid Tracking (NEAT) program is an autonomous observational system located at the United States Air Force/Ground-based Electro-Optical Deep Space Surveillance (GEODSS) site on Haleakala, Maui, Hawaii. It is a cooperative effort between the National Aeronautics and Space Administration/Jet Propulsion Laboratory and the United States Air Force. It is an innovative detection system using a charge-coupled device (CCD) system and pattern recognition software and has been in operation since December, 1995. It is the first fully-automated detection program, operated remotely without JPL personnel on site. For the first year, NEAT observations were scheduled for a 12-night period centered near the new moon. Beginning in January 1997, NEAT began observing the 6 nights prior to the new moon each month. An observing script, consisting of a list of software instructions, is transmitted to a JPL Sparc Station computer located at the GEODSS facility each afternoon. The observations are taken as scheduled through the night until morning twilight. The NEAT observing file is "clicked" off at dawn and the software-identified images/observations are transmitted to JPL. The detection data are in the form of small snap shot images (patches) which are visually inspected. Shortly thereafter, 3 positions of each object are electronically mailed to the Smithsonian Astrophysical Observatory (Minor Planet Center) for recording, identification, and designation. This program has now produced over 10,000 detections, 5637 new asteroids, 886 new designations. NEAT has discovered 14 near-Earth asteroids, 5 Earth-crossers including 1 Aten, 8 Earth-approachers; (see Table of Significant Discoveries), 2 long-period comets, and the unique asteroid, 1996 PW, which moves in a comet-like orbit and has the highest eccentricity known for an asteroid. In addition, dozens of high inclination asteroids and numerous Mars-crossers have been found.

Purpose

The fundamental goal of the NEAT program is to significantly increase the known population of NEOs to limiting magnitude 20.0 with the current CCD/telescope system. It was designed to complete a comprehensive search of the sky for near-Earth asteroids and comets, collectively called near-Earth objects or NEOs, as defined within the Spaceguard Survey report. This forms the foundation for addressing all of the scientific and technical objectives under consideration, including which NEOs, if any, could be a threat to the Earth capable of "global devastation," or those objects with diameters greater than 1 km. With a more complete inventory of objects, it will be possible to obtain more incisive information concerning their orbital distributions. This, in turn, will lead to a better understanding of families of objects and their interactions with one another. It may also lead to identification of entirely new families and groups. In addition, objects suitable for future space missions will be identified. After only 15 months, and the discovery of many unusual objects, these goals are being realized.

Telescope Site

The Air Force 1 m GEODSS telescope is located at 10,500 ft at the summit of the Haleakala extinct volcano where a complex of Air Force telescopes are used for satellite surveillance. In conjunction with the GEODSS/ NEAT program, and co-located at the site, the Air Force Maui Optical Site (AMOS) 1.2 m telescope is utilized for follow-up purposes, producing a very effective and valuable supplement to the discovery program.

FIGURE 2. Aerial photograph of Air Force telescope facility in Mt. Haleakala, Maui, HI.

NEAT CAMERA

Version I

JPL designed and fabricated the NEAT camera and computer system. Version I of the camera employed a 2048 × 2048 Loral CCD chip with 15 micron pixels. This gave a scale of 1.43″/pixel and covered 0.64 square degrees per image when coupled to the 1 m f/2.15 GEODSS telescope. This smaller CCD was used from December, 1995 through March, 1996.

Version II

In April 1996, NEAT Version II was installed on GEODSS which is a 4096 × 4096 element Loral CCD chip, also with 15 micron pixels. The scale therefore remains the same, but the camera now covers more than 2.5 square degrees of sky in each frame and uses the full field of the GEODSS telescope. This CCD has four amplifiers on the chip and can be read out in 80 sec. With each image size at 33 Mbytes, an entire night can produce 4–5 Gbytes of data.

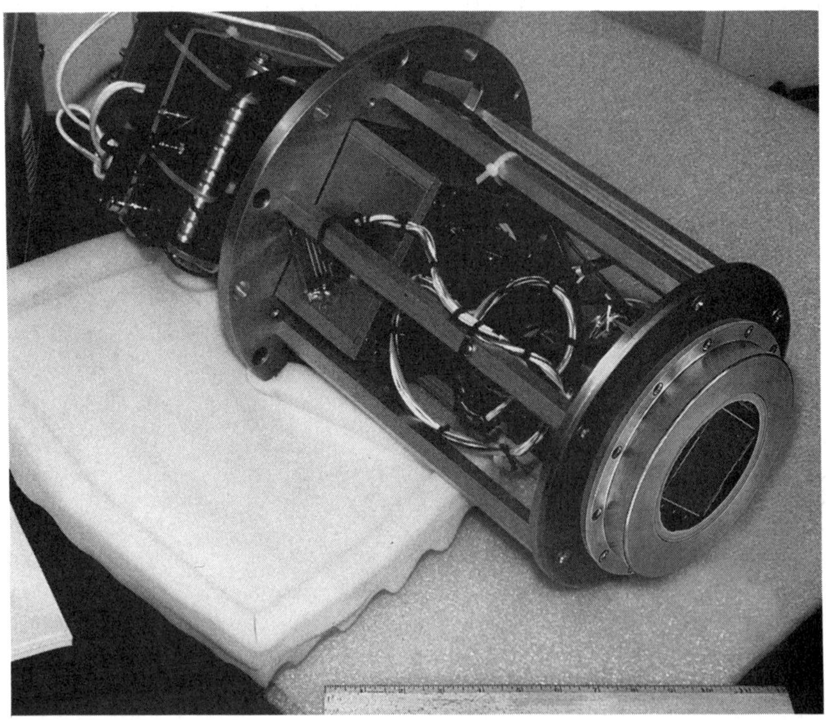

FIGURE 3. Photograph of NEAT camera in frame (housing) ready for telescope installation after the Ebsicon detector has been removed.

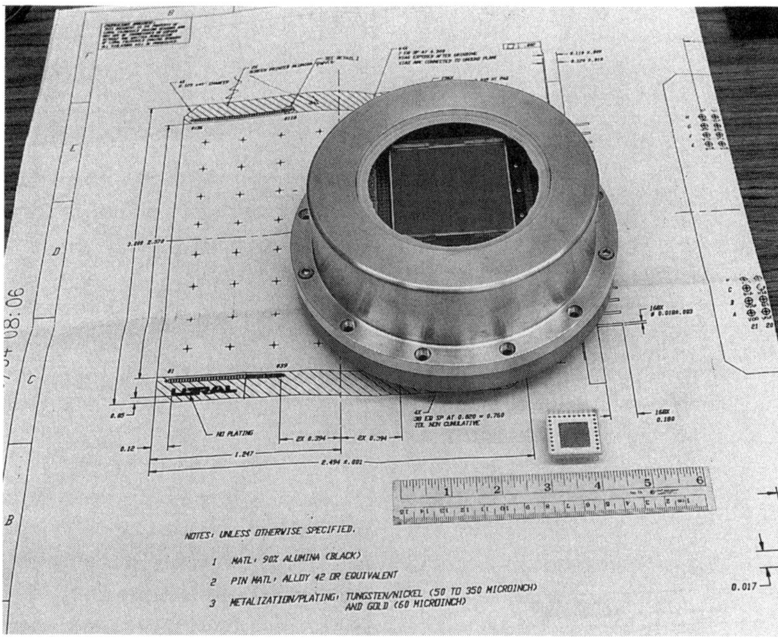

FIGURE 4. Photograph 4k × 4k CCD chip.

Version III

The NEAT Version III will be using new analog chain electronics from San Diego State University scheduled to be on-line in May, 1997. This will provide faster CCD chip read-out time which will result in an increase of sky coverage by 4 to 10 times. NEAT Version III will cover about 1000 square degrees per night.

AUTOMATIC DATA ANALYSIS SYSTEM

The automatic data analysis software examines the three images taken at the same sky positions. After the dark frames are subtracted, the software registers the images and removes any stationary objects, e.g., stars, galaxies, and/or permanent artifacts. Bright stars are fitted to the Guide Star Catalog to determine the conversion between *x–y* pixels and RA and Dec. Small areas, 25 pixels squared, around the objects that have been determined to have moved linearly and consistently over the time period of the three exposures are saved. These "patches" are transmitted to JPL in the morning. The entire image, however, is also saved to tape for shipment to JPL after the observing run. During the morning, the patches are reviewed as to which objects are real. The positions of the real objects are then reported to the Minor Planet Center. Follow-up observations of any interesting objects are made the next night. A representation of the image patches are objects given in Fig. 5.

The candidate object appears to move from the upper left to the lower right. Each small panel is a 25 × 25 pixel subimage extracted from the 4096 × 4096 pixel raw images. The moving object is found at 3 different positions during the 3 observation times. The top 3 panels are centered at position 1 during times 1–3. The moving object is centered in panel 1 only. The middle 3 panels are centered at positions 2 during times 1–3. The moving object is centered in panel 2 only. The bottom 3 panels are centered at position 3 during times 1–3. The moving object is centered in panel 3 only. Notice that fixed stars, if any, remain in the same relative position in all 3 horizontal panels. To review, the moving objects can only appear in the center of the panels in the upper-left to lower-right diagonal. All of the other panels are "anticoincidence" tests. The object can not be centered in any of them.

Another feature of the NEAT software analysis is each object's motion plot. The diagram to the right of the images depicts objects' motion in degrees per day, along the ecliptic longitude (horizontal axis) and latitude (vertical axis) directions. In Fig. 5, the object, 1996 TO5, is moving at −0.42° per day, mostly in ecliptic latitude. The polygon shows the boundaries for the expected motions of main-belt asteroids. 1996 TO5 is a near-Earth asteroid, an Earth-approaching Amor, consistent with the fact that its motion falls outside the polygon. In some cases the polygon delineating the main-belt motions will not be closed because the boundaries are not known precisely.

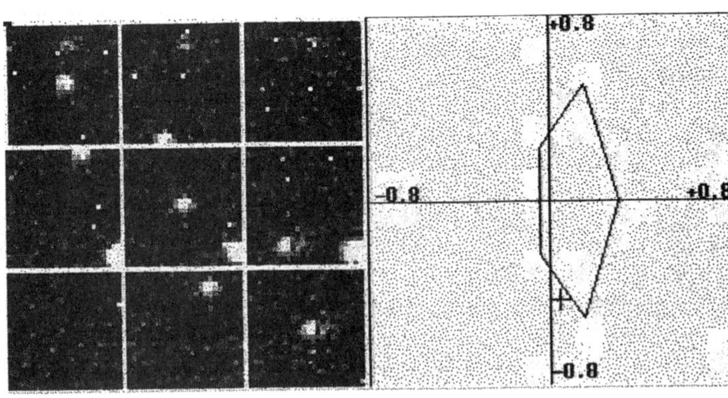

FIGURE 5. Transmitted "patches" with motion plot of 1996 TO5.

TABLE OF SIGNIFICANT DISCOVERIES: NEAT Near-Earth Asteroid Discoveries

Designation	*a*	*e*	*i*	*q*	Period	Arc	H-Mag	MPC Ref.
1996 EN	1.507	0.431	37.95	0.857	1.85	2 opp	16.50	28618
1996 EO	1.341	0.401	21.60	0.804	1.55	39	19.00	27328
1996 FQ3	2.031	0.471	1.07	1.075	2.89	122	21.00	27731
1996 FR3	2.165	0.795	8.31	0.443	3.19	87	17.00	27457
1996 KE	2.565	0.537	24.31	1.187	4.11	60	19.00	27569
1996 PH2	2.115	0.380	13.91	1.311	3.08	25	20.00	27884
1996 RY3	1.211	0.139	37.42	1.042	1.33	16	21.00	28053
1996 RC4	1.704	0.227	29.52	1.317	2.22	33	18.00	28619
1996 SK	2.428	0.797	1.96	0.494	3.78	82	17.00	28890
1996 TO5	2.382	0.517	21.00	1.152	3.68	96	17.00	28864
1996 TD9	1.333	0.404	5.02	0.795	1.54	4	24.00	28055
1996 TE9	1.793	0.326	21.64	1.209	2.40	54	19.00	28597
1997 AC11	0.914	0.369	31.80	0.577	0.87	9	20.80	28813
1997 GH3	2.439	0.565	2.96	1.061	3.81	3	17.00	MPEC 97-G03

NOTABLE NEAT DISCOVERIES

NEAT has discovered bright (large) asteroids, as well as faint objects, and many of moderate faintness, demonstrating the excellent capability of the NEAT camera combined with the 1 m Air Force telescope.

Diffuse and elongated images also have been detected and identified as comets, new or independently found, as well as asteroids moving rapidly during the exposure. Both the hardware and software were brought on line in a relatively short period of time from concept to operation and at a modest cost.

"Potentially hazardous asteroids" (PHAs) are defined as asteroids which can come within 5 million km to the Earth or about 20 times the distance of the Moon [10]. These asteroids could be perturbed into a collision trajectory within about 100 years. None of the 88 currently known PHAs greater than 1 km in diameter, nine discovered in 1996 (3 by NEAT), will actually be a threat in less than 1000 years. Estimates indicate that there are 1000–2000 undiscovered PHAs. The odds of an impact from one of these is about 1 in 1000 within 100 years.

FIGURE 6. NEAT's first Earth-crossing asteroid was 1996 EN, a "PHA," found in March 1996. However, with a high inclination to the ecliptic, the chance of impacting the Earth is less likely.

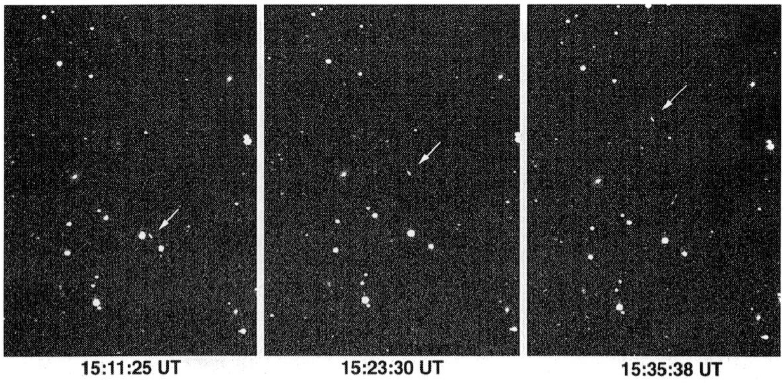

FIGURE 7. NEAT discovered its first NEA of 1997 on 1/10/97. Aten class asteroid, 1997 AC11 is imaged here five days after its discovery. It is the first Aten discovered with NEAT and is one of only 24 Atens, i.e., asteroids with semimajor axes less than 1 astronomical unit, currently known.

NEAT's first Earth-crossing asteroid was 1996 EN, a "PHA," found in March 1996. It is a large body of 3 km in diameter and is a deep Earth-crosser. However, it has a high inclination of 31° to the ecliptic which makes the chance of impacting the Earth less likely (see Fig. 6).

1997 AC11, the first Aten discovered by NEAT, has an high inclination and a small orbit. It is the 24th Aten asteroid discovered since 1976, when 2062 Aten was found by Helin at Palomar, the first asteroid with an orbit smaller than the Earth's and a period less than one year. On January 10, 1997 when it was discovered it was moving 1.8° per day and accelerated relative to the Earth during NEAT's observing period so that on January 15, it was traveling at 3.9° per day (see Fig. 7).

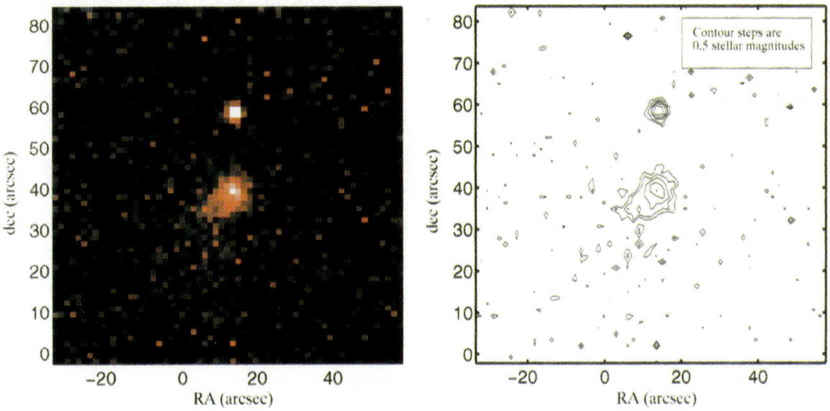

FIGURE 8. Discovery image of Comet C/1996 E1 (NEAT) as observed with NEAT. Image contours on right are in 0.5 magnitude steps. Object to the north of the comet is a star. (Image analysis by S. Shaklan/S. Pravdo) .

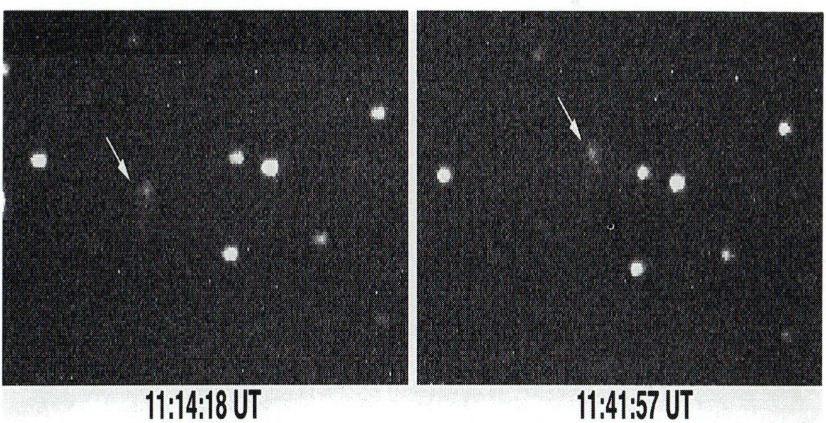

FIGURE 9. Parabolic Comet C/1997 A1 (NEAT) as observed with NEAT on January 15, 1997.

TWO LONG-PERIOD COMETS

Comet 1996 E1 (NEAT) was discovered March 15, the first major discovery of the new program (see Fig. 8). The comet, which has a parabolic orbit, is highly inclined to the ecliptic plane and made its closest approach to the Sun on 27 July 1996 at 1.36 AU. Upon discovery, it was diffuse with strong central condensation and a 15 arcsec tail.

Comet 1997 A1 (NEAT) was the first comet discovered in the new year on January 7, 1997 (see Fig. 9). It has a parabolic orbit and had a long observing window allowing observations from around the world. When discovered, it was at 3.5 AU and moved closer towards perihelion at 3.1 AU on June 19, 1997.

UNIQUE OBJECT

1996 PW has remained asteroidal in appearance throughout its observing window, but it's orbit is similar to that of a long-period comet. It has an orbit of high inclination (29.8°) and a period of ≈5900 years. Assuming a stony asteroid, its diameter would be about 8 km. If a typical cometary albedo is assumed, it could be about 15 km in diameter. Gareth Williams (Minor Planet Center) on reviewing the submitted NEAT data, noted the unusual motion of this unique, slow-moving object.

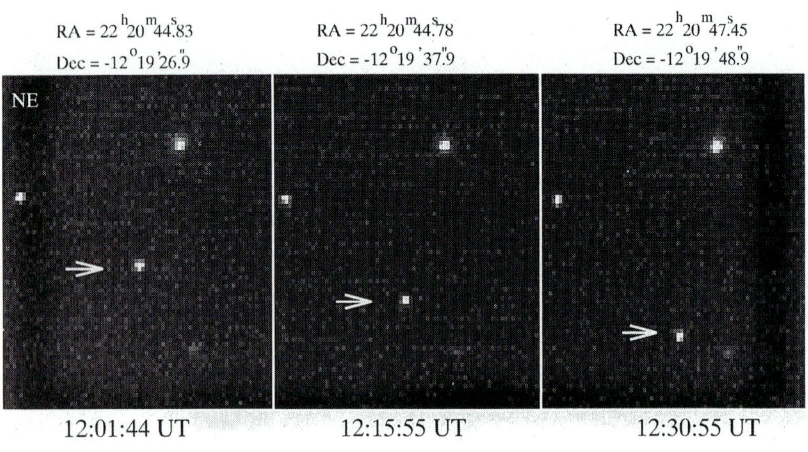

FIGURE 10. NEAT discovery image of 1996 PW obtained on August 9, 1996. (See also M.P.E.C. 1996-P03, -Q03, IAUC 6452). (Imagery by S. Shaklan/S. Pravdo.)

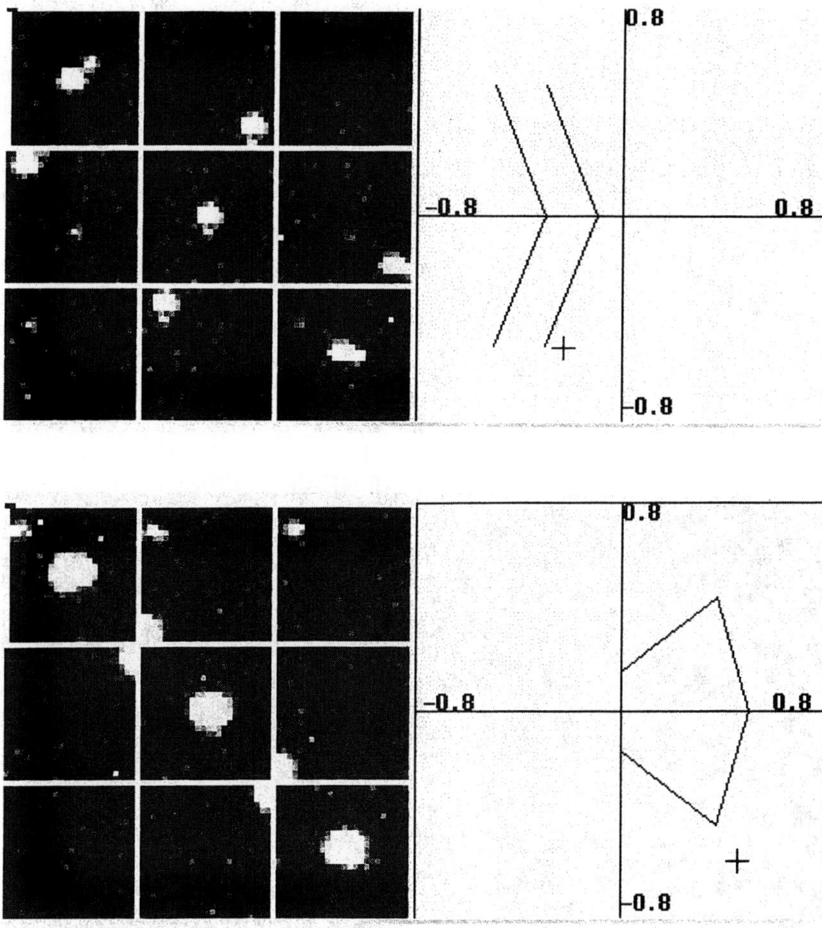

FIGURE 11. Mars-crossing asteroids 1997 CO5 (top) and 1997 CZ5 (bottom). Both discovered by NEAT during its February 1997 observing run.

Figure 10 is the discovery image of 1996 PW, discovered with NEAT on August 9, 1996. It is a unique object with perihelion between Jupiter and Mars — about 2.5 AU — and an aphelion estimated at 652 astronomical units (see also M.P.E.C. 1996-P03, -Q03, IAUC 6452, MPC 28618). With discoveries like 1996 PW, distinctions between asteroids and comets become more blurred and challenges us to understand the history and nature of this puzzling object.

OTHER UNUSUAL DISCOVERIES: MARS-CROSSERS

NEAT discovered two new high-eccentricity Mars-crossing asteroids whose orbital elements are given in Minor Planet Electronic Circulars. They are classified among 106 unusual objects. 1997 CZ5 (Fig. 11, bottom) is one of the larger ones — 5 to 12 km (3 to 8 miles) diameter — ranking 16th in size. It was discovered with NEAT on 2/7/97 moving 0.75° per day and recovered on subsequent days by NEAT and other observers worldwide. 1997 CO5 (Fig. 11, top) was discovered with NEAT on 2/6/97 moving 0.6° per day and recovered on subsequent days by NEAT and other observers worldwide. It is considerably smaller — 1 to 2 km (about 1 mile) diameter. Both have perihelia between 1.3 and 1.4 astronomical units, just beyond the 1.30 AU limit required for classification as Amor, near-Earth asteroids. As such, they may be precursors to NEAs, transitioning after gravitational perturbations move their orbits closer to the Earth.

FUTURE OF NEAT

The future plans for the NEAT program are to continue to increase the discovery rate, to essentially 90 percent completeness of NEAs larger than 1 km in diameter in the next decade and provide partial inventory of other smaller objects of interest to the science community. In principle, additional NEAT cameras can be placed on other GEODSS telescopes located throughout the world to dramatically increase NEO discoveries. A NEAT/GEODSS Network would collaborate with the International Spaceguard Survey.

COORDINATION AND THE SPACEGUARD CONCEPT

At present only two modern NEO discovery programs are fully in operation in the U.S., NEAT and Spacewatch (Gehrels, 1996). NEAT covers about 1000 square degrees/night to about magnitude 20.0 while Spacewatch observes about ten percent of the NEAT coverage, to about magnitude 22. Both systems are soon to enjoy upgrades, thereby increasing their sky coverage dramatically. Other systems from around the world are expected to become operational in coming months and years. NEAT and Spacewatch are the first U.S. participants of the International Spaceguard Survey.

An important aspect of the NEAT/GEODSS program, is the need for follow-up of the many new discoveries. The response from the international astronomical community has been outstanding (see the following discovery announcement and observations of Comet NEAT, C/1997 A1; IAUC 6532, 6534, MPECs 1997 A03) illustrating the wide representation of countries and astronomers contributing observations. Planetary defense from potentially dangerous NEOs is an international concern, requiring participation from all countries of the world [15], [16].

Circular No. 6532

Central Bureau for Astronomical Telegrams
INTERNATIONAL ASTRONOMICAL UNION
Mailstop 18, Smithsonian Astrophysical Observatory, Cambridge, MA 02138, U.S.A.
IAUSUBS@CFA.HARVARD.EDU or FAX 617-495-7231 (subscriptions)
BMARSDEN@CFA.HARVARD.EDU or DGREEN@CFA.HARVARD.EDU (science)
URL http://cfa-www.harvard.edu/cfa/ps/cbat.html
Phone 617-495-7244/7440/7444 (for emergency use only)

COMET 1997 A1

E. F. Helin, Jet Propulsion Laboratory, on behalf of the Near-Earth-Asteroid Tracking Team (cf. IAUC 6341), reports the discovery of a comet on images obtained by the on-site USAF/Ground-based Electro-Optical Deep Space Surveillance Team in Hawaii. Available precise positions:

1997 UT	R.A. (2000)	Decl.	m1	Observer
Jan. 10.51495	9 39 18.96	+16 26 18.9	18.6	Helin
10.52334	9 39 17.97	+16 26 28.9	18.8	"
10.53174	9 39 17.04	+16 26 38.6	18.9	"
11.33935	9 37 46.18	+16 42 31.0	18.9	Rabinowitz
11.34959	9 37 45.11	+16 42 43.9		"

E. F. Helin, K. J. Lawrence and S. Pravdo (Haleakala-NEAT/GEODSS). 1-m f/2.2 Ritchey-Chretien + CCD. Images clearly diffuse.
D. L. Rabinowitz (Table Mountain Observatory). 0.6-m reflector + CCD. Near limiting magnitude of exposure. 15" coma, extending asymmetrically to the south.

1997 January 11 (6532) Brian G. Marsden

Circular No. 6534

Central Bureau for Astronomical Telegrams
INTERNATIONAL ASTRONOMICAL UNION
Mailstop 18, Smithsonian Astrophysical Observatory, Cambridge, MA 02138, U.S.A
IAUSUBS@CFA.HARVARD.EDU or FAX 617-495-7231 (subscriptions)
BMARSDEN@CFA.HARVARD.EDU or DGREEN@CFA.HARVARD.EDU (science)
URL http://cfa-www.harvard.edu/cfa/ps/cbat.html
Phone 617-495-7244/7440/7444 (for emergency use only)

COMET C/1997 A1 (NEAT)

Further precise CCD positions have been reported as follows:

1997 UT	R.A. (2000)	Decl.	m1	Observer
Jan. 9.65247	9 40 54.40	+16 09 29.4	18.5	Kobayashi
9.67270	9 40 51.96	+16 09 52.0		"
11.38658	9 37 40.78	+16 43 27.4	18.9	Helin
11.40340	9 37 38.85	+16 43 46.6	18.7	"
11.61977	9 37 13.96	+16 48 03.6	18.2	"
11.63868	9 37 11.67	+16 48 25.7	18.2	"
11.79751	9 36 53.65	+16 51 34.7	19.4	Zhu
11.80222	9 36 53.07	+16 51 40.8		"
12.44963	9 35 38.41	+17 04 33.5	18.6	Helin
12.47042	9 35 35.91	+17 04 58.7	18.7	"
12.58014	9 35 23.25	+17 07 08.9		Abe
12.58494	9 35 22.68	+17 07 15.0		"
12.58889	9 35 22.21	+17 07 19.8	18.5	Sugie
12.59722	9 35 21.20	+17 07 30.0		"

T. Kobayashi (Oizumi). 0.25-m f/4.4 reflector. Prediscovery images.
E. Helin, K. J. Lawrence, S. Pravdo and D. L. Rabinowitz (Haleakala-NEAT/GEODSS). 1-m f/2.2 Ritchey-Chretien.
J. Zhu and L. J. Ma (Xinglong). 0.60-m Schmidt. Measurer Zhu.
H. Abe (Yatsuka). 0.26-m f/6.0 reflector. Communicated by S. Nakano.
A. Sugie (Dynic). 0.60-m f/3.7 reflector. Communicated by Nakano.

Parabolic orbital elements from 16 observations Jan. 9-12:

T = 1997 June 17.208 TT   Peri. = 39.270
                           Node = 135.774   2000.0
q = 3.17375 AU   Incl. = 145.069

1997 TT	R. A. (2000)	Decl.	Delta	r	Elong.	Phase	m1
Jan. 12	9 36.51	+16 55.6	2.616	3.508	150.9	7.8	18.5
17	9 26.41	+18 37.0	2.554	3.489	158.6	5.9	18.5
22	9 15.40	+20 21.3	2.505	3.469	166.3	3.8	18.4
27	9 03.57	+22 06.3	2.471	3.450	173.0	2.0	18.3
Feb. 1	8 51.11	+23 49.5	2.452	3.432	173.1	2.0	18.3
6	8 38.22	+25 28.6	2.448	3.414	166.3	3.9	18.3
11	8 25.16	+27 01.3	2.460	3.397	158.5	6.1	18.3

1997 January 12 (6534) Brian G. Marsden

The Minor Planet Electronic Circulars contain information on unusual minor planets and routine data on comets. They are published on behalf of Commission 20 of the International Astronomical Union by the Minor Planet Center, Smithsonian Astrophysical Observatory, Cambridge, MA 02138, U.S.A.

BMARSDEN@CFA.HARVARD.EDU or GWILLIAMS@CFA.HARVARD.EDU
URL http://cfa-www.harvard.edu/cfa/ps/mpc.html

COMET C/1997 A1 (NEAT)

Observations:

CJ97A010	C1997 01 09.65247	09 40 54.40	+16 09 29.4	18.5 T	411
CJ97A010	C1997 01 09.67270	09 40 51.96	+16 09 52.0		411
CJ97A010	C1997 01 10.51495	09 39 18.96	+16 26 18.9	18.6 T	566
CJ97A010	C1997 01 10.52334	09 39 17.97	+16 26 28.9	18.8 T	566
CJ97A010	C1997 01 10.53174	09 39 17.04	+16 26 38.6	18.9 T	566
CJ97A010	C1997 01 11.33935	09 37 46.18	+16 42 31.0	18.9 T	673
CJ97A010	C1997 01 11.34959	09 37 45.11	+16 42 43.9		673
CJ97A010	C1997 01 11.38658	09 37 40.78	+16 43 27.4	18.9 T	566
CJ97A010	C1997 01 11.39499	09 37 39.83	+16 43 37.3	18.7 T	566
CJ97A010	C1997 01 11.40340	09 37 38.85	+16 43 46.6	18.7 T	566
CJ97A010	C1997 01 11.61977	09 37 13.96	+16 48 03.6	18.2 T	566
CJ97A010	C1997 01 11.63025	09 37 12.75	+16 48 16.3	18.5 T	566
CJ97A010	C1997 01 11.63868	09 37 11.67	+16 48 25.7	18.2 T	566
CJ97A010	C1997 01 11.79278	09 36 54.17	+16 51 30.2		327
CJ97A010	C1997 01 11.79751	09 36 53.65	+16 51 34.7	19.4 T	327
CJ97A010	C1997 01 11.80222	09 36 53.07	+16 51 40.8		327
CJ97A010	C1997 01 11.80696	09 36 52.48	+16 51 46.3		327
CJ97A010	C1997 01 12.44963	09 35 38.41	+17 04 33.5	18.6 T	566
CJ97A010	C1997 01 12.45994	09 35 37.05	+17 04 47.0	19.2 T	566
CJ97A010	C1997 01 12.47042	09 35 35.91	+17 04 58.7	18.7 T	566
CJ97A010	C1997 01 12.57534	09 35 23.79	+17 07 02.1	18.4 T	367
CJ97A010	C1997 01 12.58014	09 35 23.25	+17 07 08.9		367
CJ97A010	C1997 01 12.58494	09 35 22.68	+17 07 15.0		367
CJ97A010	C1997 01 12.58889	09 35 22.21	+17 07 19.8	18.5 T	402
CJ97A010	C1997 01 12.59306	09 35 21.70	+17 07 24.5		402
CJ97A010	C1997 01 12.59722	09 35 21.20	+17 07 30.0		402
CJ97A010	C1997 01 12.91464	09 34 44.08	+17 13 49.2	18.5 T	046
CJ97A010	C1997 01 12.91691	09 34 43.72	+17 13 52.7		046
CJ97A010	C1997 01 12.91902	09 34 43.40	+17 13 55.9		046
CJ97A010	C1997 01 12.92096	09 34 43.33	+17 13 57.7		046
CJ97A010	C1997 01 12.92786	09 34 42.36	+17 14 07.2		046
CJ97A010	C1997 01 13.46442	09 33 39.01	+17 24 54.4	18.2 T	566
CJ97A010	C1997 01 13.47283	09 33 38.00	+17 25 04.2	18.1 T	566
CJ97A010	C1997 01 13.48124	09 33 36.98	+17 25 14.1	18.3 T	566
CJ97A010	C1997 01 13.55558	09 33 28.35	+17 26 42.6		867
CJ97A010	C1997 01 13.56218	09 33 27.50	+17 26 51.0		867
CJ97A010	C1997 01 13.56773	09 33 26.79	+17 26 58.2		867
CJ97A010	C1997 01 13.64462	09 33 17.62	+17 28 32.0		900
CJ97A010	C1997 01 13.66438	09 33 15.17	+17 28 55.9		900
CJ97A010	C1997 01 13.67892	09 33 13.38	+17 29 14.3		897
CJ97A010	C1997 01 13.69010	09 33 12.14	+17 29 27.3	17.2 T	360
CJ97A010	C1997 01 13.69300	09 33 11.81	+17 29 29.9		897
CJ97A010	C1997 01 13.69368	09 33 11.64	+17 29 34.3	18.4 T	422
CJ97A010	C1997 01 13.69705	09 33 11.30	+17 29 35.9		360

—continued

Observations — continued

CJ97A010	C1997	01	13.70028	09	33	10.88	+17	29	42.9	18.5 T	422
CJ97A010	C1997	01	13.70093	09	33	10.81	+17	29	40.9	17.7 T	897
CJ97A010	C1997	01	13.70330	09	33	10.52	+17	29	43.0		360
CJ97A010	C1997	01	13.70562	09	33	10.22	+17	29	48.1	18.5 T	422
CJ97A010	C1997	01	14.08245	09	32	25.10	+17	37	22.2	17.8 T	046
CJ97A010	C1997	01	14.08608	09	32	24.67	+17	37	26.6		046
CJ97A010	C1997	01	14.08823	09	32	24.38	+17	37	29.2		046
CJ97A010	C1997	01	14.10635	09	32	22.17	+17	37	51.1		557
CJ97A010	C1997	01	14.10951	09	32	21.80	+17	37	55.0	17.9 N	557
CJ97A010	C1997	01	14.11243	09	32	21.43	+17	37	58.5	16.9 T	557
CJ97A010	C1997	01	14.54427	09	31	29.24	+17	46	44.8	18.2 T	566
CJ97A010	C1997	01	14.55267	09	31	28.25	+17	46	55.0	18.2 T	566
CJ97A010	C1997	01	14.56109	09	31	27.19	+17	47	06.0	18.1 T	566
CJ97A010	rC1997	01	15.05873	09	30	26.61	+17	57	11.8	17.8 T	118
CJ97A010	rC1997	01	15.06181	09	30	26.21	+17	57	15.4		118
CJ97A010	rC1997	01	15.06903	09	30	25.34	+17	57	24.8		118
CJ97A010	C1997	01	15.08287	09	30	23.64	+17	57	40.5	17.0 T	966
CJ97A010	C1997	01	15.08780	09	30	23.00	+17	57	46.8		966
CJ97A010	C1997	01	15.09020	09	30	22.82	+17	57	50.6		966
CJ97A010	C1997	01	15.09236	09	30	22.53	+17	57	53.8		966
CJ97A010	C1997	01	15.16380	09	30	13.64	+17	59	20.4	17.9 T	046
CJ97A010	C1997	01	15.16507	09	30	13.50	+17	59	22.0		046
CJ97A010	C1997	01	15.16728	09	30	13.23	+17	59	25.1		046
CJ97A010	C1997	01	15.17091	09	30	12.74	+17	59	29.3		046
CJ97A010	C1997	01	15.41022	09	29	43.42	+18	04	22.5		658
CJ97A010	C1997	01	15.41623	09	29	42.65	+18	04	29.3		658
CJ97A010	C1997	01	15.42234	09	29	41.88	+18	04	36.8		658

Observer details:

046 Klet. Observers J. Ticha, M. Tichy, Z. Moravec. 0.57-m f/5.2 reflector + CCD.

118 Modra. Observers P. Koleny, L. Kornos. 0.6-m f/5.5 reflector + CCD.

327 Peking Observatory, Xinglong Station. Observers J. Zhu, L. J. Ma. 0.6-m Schmidt + CCD.

360 Kuma Kogen. Observer A. Nakamura. 0.60-m f/6.0 Ritchey-Chretien + CCD.

367 Yatsuka. Observer H. Abe. 0.26-m f/6.0 reflector + CCD.

402 Dynic Astronomical Observatory. Observer A. Sugie. 0.60-m f/3.7 reflector + CCD.

411 Oizumi. Observer T. Kobayashi. 0.41-m f/4.3 reflector + CCD. Prediscovery images.

422 Loomberah. Observer G. J. Garradd. 0.25-m f/4.1 reflector + CCD.

557 Ondrejov. Observers P. Pravec, L. Sarounova. 0.65-m f/3.6 reflector + CCD.

566 Haleakala-NEAT/GEODSS. Observers E. F. Helin, S. Pravdo, K. J. Lawrence, D. L. Rabinowitz for the NEAT team. 1.0-m f/2.2 GEODSS telescope + CCD.

658 Dominion Astrophysical Observatory. Observer G. C. L. Aikman. 1.82-m reflector + CCD.

673 Table Mountain Observatory, Wrightwood. Observer D. L. Rabinowitz. 0.6-m reflector + CCD.

Observations — continued

```
867 Saji Observatory.  Observers M. Yamanishi, A. Miyamoto, M. Aimoto, T.
    Oribe.  1.03-m f/4.8 reflector + CCD.
897 Chiyoda.  Observer T. Kojima.  0.25-m f/6.0 reflector + CCD.
900 Moriyama.  Observer Y. Ikari.  0.25-m f/6.3 reflector + CCD.
966 Church Stretton.  Observer S. P. Laurie.  0.25-m Schmidt-Cassegrain + CCD.

Orbital elements:
    C/1997 A1 (NEAT)
T 1997 June 19.45397 TT                                      Marsden
q   3.1587095             (2000.0)          P                  Q
                 Peri.    39.94749     -0.18209130        +0.89846855
                 Node    135.76550     +0.69042376        -0.17245471
e   1.0          Incl.   145.06305     +0.70011270        +0.40374948
From 65 observations 1997 Jan. 9-15.

 Brian G. Marsden           (C) Copyright 1997 MPC           M.P.E.C. 1997-A03
```

ACKNOWLEDGMENT

The research described in this paper was carried out by the Jet Propulsion Laboratory, California Institute of Technology, under a contract with the National Aeronautics and Space Administration.

REFERENCES

1. BOWELL, E.L.G., B.A. SKIFF, & R.M. WEST. 1992. (4015) 1979 VA = P/Comet Wilson-Harrington (1949 III), IAU Circulars 5585 and 5586, Aug. 13, 1992.
2. FERNANDEZ, YANGA R., L. MCFADDEN, C.M. LISSE, E.F. HELIN, A.B. CHAMBERLIN. Analysis of POSS images of comet-asteroid transition object 107P/1949 W1 (Wilson-Harrington). Icarus, in press.
3. GEHRELS, T. 1991. Scanning with charge-couple devices. Space Science Reviews. **58:** 347–375.
4. HELIN, E.F. & E.M. SHOEMAKER. 1977. Discovery of Asteroid 1976 AA. Icarus. **32:** 415–419.
5. HELIN, E.F. & E.M. SHOEMAKER. 1979. Palomar planet-crossing asteroid survey 1973–1978. Icarus **40:** 321–328.
6. HELIN, E.F. & R.S. DUNBAR. 1985. The Palomar planet-crossing search: 1979–1984. Bull. Amer. Astron. Soc. **17:** (3) 32.
7. HELIN, E.F. & R.S. DUNBAR. 1990. Search techniques for near-Earth asteroids. Vista in Astronomy. **33:** 21–37.
8. HUDSON, R.S., & S.J. OSTRO. 1994. Shape of asteroid 4769 Castalia (1989 PB) from inversion of radar images. Science. **263:** 940–943.

9. LAU, C.O. & N.D. HULKOWER. 1985. On the Accessibility of Near-Earth Asteroids. AAS/AIAA Astrodynamics Specialist Conference, Vail, paper no. AAS 85-352, 1985.
10. MARSDEN, B.G. 1997. Overview of Orbits. In this volume.
11. MARSDEN, B.G. 1997. Discovery Announcement and Positional Data of Comet NEAT, C/1997 A1 IAU Circular Nos. 6532, 6534; Minor Planet Electronic Circular, MPEC, 1997 A03.
12. MCADAMS, J.V. 1991. Mission Options for Rendezvous with the Most Accessible Near-Earth Asteroid — 1090 ML. AAS/AIAA Astrodynamics Specialist Conference, Paper no. AAS 91-397, 1991.
13. MORRISON, D., Chair. 1992. The Spaceguard Study: Report of the NASA International Near-Earth Object Detection Workshop. Washington D.C.: NASA, January 1992.
14. OSTRO, S.J., J.F. CHANDLER, A.A. HINE, I.I. SHAPIRO, K.D. ROSEMA, & D.K. YEOMANS. 1990. Radar images of asteroid 1989 PB. Science. **248:** 1523–1528, abstract.
15. REMO, JOHN L. 1996. Policy Perspectives from the United Nations international conference on near-Earth objects. Space Policy. **12:** 13–17.
16. REMO, JOHN L. 1996. United Nations conference on near-Earth objects. Earth and Space Review. **5:** 16–23.

The Uncertain Nature of Cometary Motions

DONALD K. YEOMANS
California Institute of Technology
301-150G, Jet Propulsion Laboratory
4800 Oak Grove Drive
Pasedena, California 91109

The number of active short- and long-periodic comets crossing the Earth's orbit each year is less than 10 percent of the corresponding number of asteroids crossing the Earth's orbit. However, the higher relative velocities of comets with respect to the Earth and the uncertainties associated with accurately computing their future trajectories can cause considerable problems when assessing the risks of Earth-crossing objects. Unlike asteroids, the motions of active comets are often affected by so-called nongravitational (outgassing) forces that are imperfectly modeled. In addition, the astrometric optical observations that are used to refine a comet's orbit are often imprecise because a comet's center of mass can be hidden by atmospheric gas and dust. For long-period comets, there is the additional problem of having to base orbital solutions on relatively short observational data intervals.

Long-term numerical integrations extending two centuries into the future have been carried out to investigate upcoming Earth-close approaches by known periodic comets. Error analyses and impact probabilities have been computed for those comets that will pass closest to the Earth. Although there are no known comets that will make dangerously close Earth approaches in the next two centuries, there are a few objects that warrant future monitoring.

Because the cometary atmosphere prevents precision astrometic data and cometary nongravitational forces are imperfectly modeled, the uncertainty in the future motion of an active short-periodic comet is substantially larger than the corresponding uncertainty for an asteroid with a comparable observational history. In addition, comets can split, enter into temporary orbits about Jupiter, and collide with the Sun, Jupiter, or any of the other planets, including Earth. Using comets such as Swift-Tuttle, Finlay, and Comet Shoemaker-Levy 9 as examples, the problems associated with cometary orbital dynamics and close approaches will be demonstrated.

The EUNEASO Project: A European NEO Search, Follow-up, and Physical Observation Programme

G. HAHN,[a] S. MOTTOLA,[a] A. ERIKSON,[a] A.W. HARRIS,[a] G. NEUKUM,[a]
A. MAURY,[b] R. SAVALLE,[b] H. SCHOLL,[b] A. BIJAOUI,[b]
M. DIMARTINO,[c] C. BARBIERI,[d] M. LAZZARIN,[d] F. MIGLIORINI,[d]
P. PRAVEC,[e] M. WOLF,[e] L. SAROUNOVA,[e] F.P. VELICHKO,[f]
YU.N. KRUGLY,[f] I.N. BELSKAYA,[f] AND V.G. SHEVCHENKO[f]

[a]*Institute for Planetary Exploration, DLR*
Berlin, Germany

[b]*Observatoire de la Côte d'Azur*
Nice, France

[c]*Osservatorio Astronomico di Torino*
Pino Torinese, Italy

[d]*Istituto di Astronomia*
Università di Padova
Padova, Italy

[e]*Ondrejov Observatory*
Czech Republic

[f]*Astronomical Observatory*
Kharkiv State University
Kharkiv, Ukraine

EUNEASO, which stands for EUropean Near-Earth Asteroids Search Observatories, is a cooperation to organize a large-scale NEO search programme. The participating groups do astrometric follow-up observations as well as photometric and spectroscopic studies of NEOs. Currently, the development of CCD cameras for the search programme is done at the DLR in Berlin and the OCA in Nice, where tests have been performed with the 90 cm Schmidt telescope at Caussols.

Scanning photographic films taken by a Schmidt telescope appears to be a promising tool to detect NEOs. Corresponding tests with commercial scanners are carried out at OCA.

Plans exist to install a 1.5k × lk CCD at the 67 cm Schmidt telescope in Asiago, Italy.

Astrometric follow-up and photometric observations of NEOs are currently done on a routine basis at Ondrejov Observatory, Czech Republic, using a 60 cm telescope equipped with a CCD camera. A similar system is available at Kharkiv Observatory, Ukraine, attached to a 70 cm telescope.

Photometric CCD observations of NEOs are performed as part of the asteroid observing programme of the DLR with the 60 cm Bochum telescope at ESO, which is available for some 100 nights per year.

The European NEO Search Project within EUNEASO

ALAIN MAURY,[a] RENAUD SAVALLE,[a] GERHARD HAHN,[b]
STEFANO MOTTOLA,[b] AND GERHARD NEUKUM[b]

*[a]Observatoire de la Côte D'Azur
Nice, France
and
[b]Institute of Planetary Exploration, DLR
Berlin, Germany*

EUNEASO is a European program aimed at discovering and observing near-Earth objects. It was started in 1992, and is an informal partnership to promote and coordinate research on NEO in Europe (see poster by Hahn et al. in this volume).

Efforts are made to develop and operate a CCD array camera system in order to participate in a large scale NEO search programme. Work is progressing on the following steps:

(1) Construction of hardware and software required to automatically detect near-Earth objects using a large CCD camera. Our aim is to install a 9 2K CCD camera at the focus of the Observatoire de Cote d'Azur 90/150/315 cm Schmidt telescope, and if possible a similar telescope in the southern hemisphere, i.e., the ESO Schmidt.

(2) Integration of this hardware in the worldwide asteroid observing network. An European asteroid web server has been installed at DLR. It allows to connect to several databases of interest, but also, on the OCA Schmidt telescope where, if permission is granted, observing requests may be posted.

(3) Automated detection software has been written at OCA (detection part) and DLR (reduction part) and tested on images obtained with a single 2K camera in 1994. It allows a complete image detection down to 3 sigmas with 200 kpixels/second speed using a 12 megaflops Unix workstation. It will be ported on a network of Digital Signal Processors during the 1995 summer. This should allow real-time detection using 9 CCDs (76 megabytes per image).

(4) The construction of the multi CCD camera has started. The CCD controller is currently being tested. It will be able to read out a 2K CCD in 25 seconds, and will use dynamic antiblooming mode in order to be able to use Hipparcos reference stars for very high precision astrometry.

(5) The mechanical assembly has also begun. Tests will be done at the telescope during the 1995 summer. We hope to start regular observations with a single CCD camera at the OCA as soon as possible.

Meteoroid Orbits: Implications for Near-Earth Object Search Programs

DUNCAN STEEL[a,b]

Spaceguard Australia P/L
P.O. Box 3303, Rundle Mal
Adelaide, SA 5000
Australia

ABSTRACT: The available orbital database on macroscopic potential impactors of our planet (asteroids and comets, collectively near-Earth objects or NEOs) numbers less than one thousand, whereas there have been some hundreds of thousands of orbits of Earth-impacting meteoroids determined in various surveys, mostly using meteor radars. If one assumes that NEOs have orbital characteristics broadly similar to meteoroids, then the orbits of the latter can give important indications concerning the conduct of search programs designed to discover large NEOs well ahead of any catastrophic impact, allowing ameliorative action to be taken. For smaller NEOs that cannot be telescopically detected until the day or so before impact, the radiant distribution of observed meteors shows the regions of the sky from which impactors are most likely to emanate. It is shown that the vast majority of meteoroids striking the Earth have geocentric (apparent) radiants within two near-ecliptic regions a few tens of degrees wide and centered on longitudes ±90° from the apex of the Earth's way (the so-called helion and anti-helion sources). These are bodies with low inclinations, large eccentricities ($e = 0.7 - 0.9$) and quite small semimajor axes (mostly $a = 1.3 - 2.5$ AU). After allowing for the terrestrial motion about the Sun (conversion to the true radiant), the longitudes are around ±120° from the apex. For a ground-based search on the nightside, the best search region is that within ~20° of the arc joining the geocentric and true radiants (longitudes 90° and 120°). On the dayside, proximity to the solar direction argues for a space-based surveillance program, if small NEOs are to be found just prior to impact.

1. INTRODUCTION

This paper is predicated on a very simple idea: if you want to know about the large objects that strike the Earth occasionally, a useful source of information is the smaller ones that are hitting our planet all the time. For example, few of the known Earth-approaching asteroids and comets (collectively, near-Earth objects or NEOs) have orbits that make a terrestrial impact possible in the present epoch, so that orbital evolution studies are needed in order to determine with any confidence the hazard that they pose to us, whereas all observed meteoroids certainly *did* have Earth-intersecting orbits.

[a]E-mail: dis@a011.aone.net.au.
[b]Also with *The Spaceguard Foundation*, Rome, Italy. Previously with the University of Adelaide and the Anglo-Australian Observatory.

The available database on the orbital parameters of Earth-crossing macroscopic objects in late 1996 is limited to about 225 asteroids, around 30 periodic comets, and almost 450 near-parabolic comets [1]. This compares with the meteoroid database of about 69,000 orbits from various photographic, TV system, and radar surveys held in the International Astronomical Union Meteor Data Center at Lund in Sweden [2],[3] and upwards of 400,000 orbits derived from a new meteor orbit radar operating at the University of Canterbury in New Zealand [4]. Numerically, then, the meteoroid orbits represent a rich data resource. Although on an individual basis such orbits are by no means as precise as the telescopically-observed orbits of asteroids and comets — meteors can be observed only during their fleeting lifetimes, mostly less than a second, whereas the larger bodies in space may be tracked astrometrically for months, years, and decades — they are statistically useful, for example, in linking specific meteor showers to parent objects.

Meteoroids are defined as being solid particles in space larger than 100 μm in size (smaller bodies comprise the interplanetary dust cloud) but smaller than about 10 meters [5]. Larger inert bodies are termed asteroids. Although meter-plus meteoroids do enter the atmosphere from time to time, producing high altitude detonations, releasing energies equivalent to some kilotons of TNT, these are not frequent enough to make possible observational programs rendering statistically-useful numbers of pre-encounter orbits. Our database for the orbits of the largest meteoroids/smallest asteroids is now being built up through the efforts of the *Spacewatch* team at the University of Arizona [6], and in the future it is to be expected that a program such as that recommended in the *Spaceguard* report [7],[8] will lead to a rapid increase in our knowledge of the orbits of small asteroids.

Going down in size, meteoroids in the decimeter to meter size range produce bright meteors or fireballs. Three photographic fireball surveys conducted in the USA, Canada, and central Europe have produced about a thousand good orbits [2], allowing detailed study of the physics of the interaction between meteoroids and the atmosphere [9], and in some cases rendering sufficiently-precise orbit determinations that meaningful backwards integrations may be carried out to allow investigations of the general orbital behavior of such bodies [10].

Photographically-detected meteors result from meteoroids typically in the centimeter to decimeter size range, and about 6,000 orbits have been determined from observations of this type [2]. Fainter meteors may be detected optically using TV techniques, and about 500 orbits were determined in this way in a Canadian program in the mid-1980s (again, see [2]).

All of the above observation techniques suffer from the limitation that they cannot be applied on the dayside of the Earth (except for IR observations from space of some very bright fireballs/detonations). This is not the case for radar observations of meteoric ionization trains, which allow meteors to be detected 24 hours a day (except in the presence of electrical noise such as that produced by intense lightning storms). The bulk of the meteoroid orbits archived at the IAU Meteor Data Center (numbering, in all, about 63,000) are derived from radar observations [2],[3], the original particles being mostly in the 100 μm – 3 mm size range. The limiting size/mass attained is a function of the power of the radar, the antenna gains, and other features of the particular equipment used.

The new orbit radar in New Zealand [4] is producing data at a prolific rate, upwards of a thousand meteoroid orbits (i.e., more than the total NEO database) resulting from each day of operation. Although that radar has now been running for several years, there are some weeks during the year (some solar longitudes) in which observations have yet to be made. For this reason, and several other considerations (such as the familiarity of other researchers with the data sets archived at the IAU Meteor Data Center), the present paper will make use only of meteoroid orbit data collected in earlier programs. Because this area of astronomical science has been substantially moribund in recent decades, this means that the data are derived from campaigns carried out in the 1960s. A later analysis of the type described here employing the new data from New Zealand will be useful on account of: (i) the larger number of orbits available; (ii) the greater individual precision of those data, due to the modern signal processing techniques used; and (iii) the fact that it has only been in the past few years that various radar meteor selection effects (in particular the echo ceiling and Faraday rotation effects described in Section 3) have been understood, allowing compensation to be made in the analysis of the meteoroidal influx detected using currently-functioning radars with known operational parameters.

In this paper, I concentrate upon two great fallacies of meteroid science, and show how a consideration of them can lead to some clues — at least — being derived about the range of dynamical characteristics of NEOs, and hence help us in designing ways to defend humankind against catastrophic impacts. These two great fallacies are:

(a) meteoroids come from comets and have comet-like orbits; and

(b) sporadic meteoroids come from random directions.

Point (a) will be considered in the next section, dealing with the general orbital characteristics of meteoroids; we will see that although meteoroids have orbits covering the range of orbital parameters displayed by NEOs, there are broad meteoroid classes that are not represented in the known NEO database, and this might indicate that we are currently in ignorance of a significant part of the NEO population having orbits quite different in character to the discovered fraction. In the following section I will turn to point (b) and discuss the radiant distribution of sporadic (nonshower) meteors, and show that these predominantly approach the Earth from two distinct directions, known to meteor researchers for many years as the *helion* (sunward) and *antihelion* (near-opposition) ecliptic sources; this might be interpreted as implying that *if* one were in the business of attempting to spot small NEOs (say, 20–200 meters in size) in the day or less before impact, to enable short-term ameliorative steps to be taken, then one would do well to concentrate one's efforts on a quite restricted fraction of the celestial sphere.

2. GENERAL ORBITAL DISTRIBUTIONS OF METEOROIDS

One recent book on comets (which will remain uncited here so as not to embarrass the authors) confidently stated that meteoroids observed in the Earth's atmosphere originate from comets, to at least the 99% level. Although there are many spectacular meteor showers resulting from meteoroid streams that definitely can be associated with specific comets (see [11] for a discussion and suitable citations), the observed orbital characteristics of meteoroids do not support the idea that they are in the main the products of comets observed in the present epoch. I have previously published [12] a discussion of this point, using for illustrative purposes the radar meteor orbits determined in the Obninsk (former Soviet Union) survey [13] of 1967–68, and showed that whilst specific showers are well-represented in the meteoroid orbit catalogs, the gross distribution is in many respects more like that of Earth-crossing asteroids than that of known comets. Here I will use the meteoroid orbits from the two Adelaide (South Australia) surveys [14]–[16] of the 1960s to take this point further.

Two distinct surveys were carried out, the first [14] in 1960–61 rendering 2092 orbits to a limiting radar magnitude of about +6 (typical particle sizes ~3 mm), and the second [15] in 1968–69 producing 1667 orbits to magnitude +8 (particle sizes ~1 mm). For the combined set of 3759 orbits I have presented elsewhere [16] the distributions of individual orbital elements as histogram plots. These show, for example, that although most orbits are of low inclination ($i < 15°$) and high eccentricity ($e > 0.7$), similar to the known Jupiter-family comets [1], their semimajor axis distribution peaks at $a = 1.5 - 3.0$ AU, so that most are in cis-jovian orbits (aphelia $Q < 5$ AU).

One might consider that such a distribution, which is similar to that of Earth-crossing asteroids but quite unlike the known comets, might be diagnostic of meteoroids released by comets into Jupiter-crossing orbits, most of those meteoroids then having been ejected from the solar system by that giant planet, leaving only the meteoroids having orbits shrunken by the Poynting-Robertson (P–R) effect [17] to $Q < 4.5$ AU. However, the time-scales for ejection by Jupiter and orbital shrinkage under the P–R effect are very similar to each other for ~1 mm meteoroids in orbits like Jupiter-family comets [18], implying that we might expect the meteoroid Q-distribution to reflect their source (i.e., a Q-distribution peaking at 5–6 AU if that source is similar to the gross parameters of the Jupiter-family comets observed today). In fact the Q-distribution peaks at ~3 AU for the Adelaide orbits [16, Fig. 5], and at 1–2 AU for the Obninsk orbits ([12, Fig. 1]; limiting radar magnitude +7.5, hence again ~1 mm meteoroids), which might be interpreted as implying that these meteoroids were *released* on cis-jovian orbits. The orbits determined in other surveys support this general picture: see [2] for suitable plots.

A better insight into the distribution of orbits of these meteoroids may be gained from two-element plots such as those presented here in Fig. 1. At upper left the plot of inverse semimajor axis (orbital energy) against inclination shows that, indeed, there is a preponderance of orbits with $a = 1.5 - 3.0$ AU and $i < 15°$. Of these, the majority have $e > 0.7$ (upper right plot), and thus perihelion distance $q = 0.2 - 0.6$ AU (lower left plot) and aphelion distance $Q < 5$ AU (lower right plot). Bearing in mind that it has been known to be the parent of the Taurid meteor showers for some decades [19], [20], the object that immediately springs to mind as having an orbit similar to this concentration of meteoroid orbits is Comet Encke (new designation [1] 2P/Encke).

In a seminal paper [21] in which he considered the mass balance of the zodiacal dust cloud (the dust being produced largely through the collisional comminution of larger meteoroids [22]–[24]), Fred Whipple showed that the present complex requires an exceptional recent input ("recent" meaning at most ~10^5 yr ago because this is the collisional lifetime for small meteoroids [18], [24]), and he identified 2P/Encke as the likely source of this input. Other studies have shown that the zodiacal cloud is not presently in equilibrium, with an over-supply of meteoroids [23], and other researchers [25]–[27] have suggested that 2P/Encke has been a major contributor to that complex. That comet is known to exhibit anomalous dust/meteoroid production in the present epoch [28], but of greater significance here is its behavior over the past 10^4–10^5 yr: its current activity phase began in AD 1786, and it has not been identified in earlier historical records [29], but the number of precession cycles of its argument of perihelion needed for it to have produced the meteor showers with which it is associated requires meteoroid ejection continuing back over ~2×10^4 yr [30], [31]. Studies of the orbital trends within the meteoroid stream lead to a similar time-scale [32].

The above might be interpreted as implying that the main orbital concentration of meteoroids is *not* of significance in arriving at any prediction of the orbits of unknown NEOs, with solely 2P/Encke being responsible for that concentration's genesis, were it not for the following: it has now been shown that there are at least a dozen large asteroids (and several smaller asteroids/large meteoroids) in broadly the same orbit, and this has been interpreted as being evidence for the break-up of a massive proto-Encke in a cis-jovian orbit ~2–3 × 10^4 yr ago [33]–[35]. Further, it has been shown [36] that the arrival of 2P/Encke in its orbit de-coupled from Jupiter, which has been a long-term problem [29], is explicable in terms of a random walk under sensible nongravitational forces on such a time-scale, and I have argued [37] that one should regard asteroids apparently-associated with meteoroid streams as being merely the larger inert (or possibly just dormant) cometary fragments within those streams. If this is the case then, especially in view of the high impact speed of objects in the 2P/Encke-related complex, it would be wise to optimise any NEO search for objects in such orbits. Additional evidence for this is presented in Section 3.

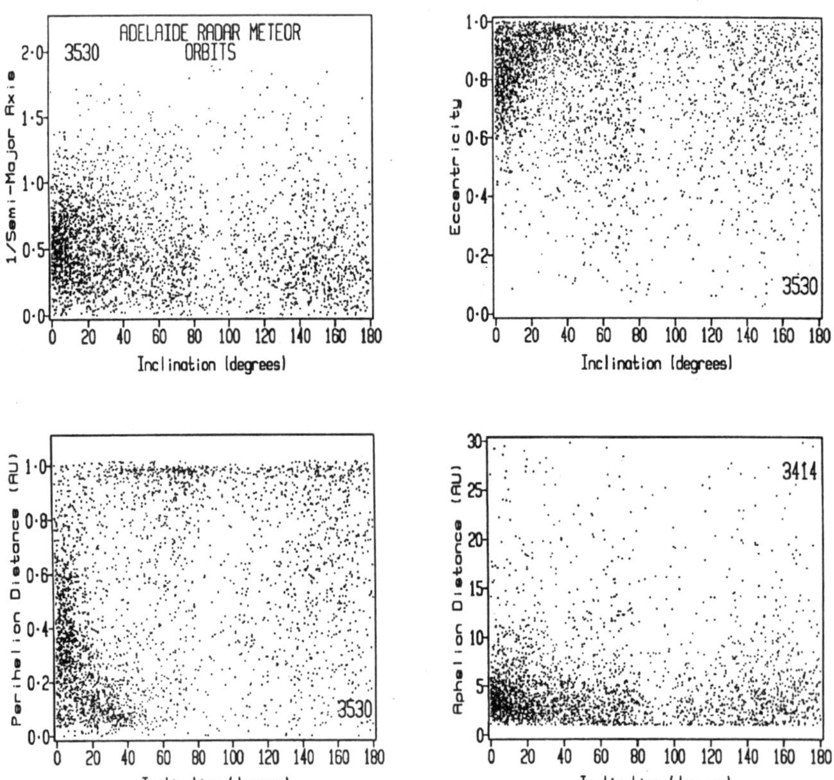

FIGURE 1. Two-element distribution plots for the meteoroid orbits determined in the Adelaide radar surveys. The number given within each plot indicates how many individual orbits are plotted. A total of 3759 orbits are available, but 229 of those are hyperbolic, leaving 3530 elliptical/parabolic orbits. Of those 116 have $Q > 30$ AU, leaving 3414 orbits in the aphelion/inclination plot at lower right.

Having discussed at length the main concentration of orbits in Fig. 1, before passing on to a consideration of the radiant distribution of meteors I should mention some other pertinent points to be gained from that figure. Firstly, although there are many meteoroids with retrograde paths combined with large semimajor axes and high eccentricities (i.e., like 1P/Halley, 109P/Swift-Tuttle, and other known, or dormant, or extinct, or yet-to-be-discovered, intermediate-period comets), there are also large numbers of retrograde cis-jovian orbits, some even of the Aten-type ($a < 1$ AU); see also [12, Fig. 4] for the equivalent plot for the Obninsk orbits. I have previously discussed [38] the presence of such orbits in the data collected in the Harvard Radio Meteor Project (HRMP, those orbits being used later in the present paper). It is clear that there would be great difficulties in explaining how retrograde parent objects could enter cis-jovian orbits, which argues in favor of these meteoroids being the result of release from intermediate-period comets with gradual orbital shrinkage under the P–R effect [17]; on the other hand, one could point out that the origin of Aten-type or Phaethon-type asteroid orbits still presents a problem in celestial mechanics. The existence of NEOs in previously unrecognized orbital classes has significant consequences for both estimations of the Earth-impact rate, and also the design of search programs, as is demonstrated [39], [40] by the discovery of (5335) Damocles (formerly 1991 DA). Retrograde NEOs have impact probabilities and speeds upon the Earth which are much higher than the values for prograde objects [41], [42].

Secondly, the lower left plot in Fig. 1 shows a substantial number of orbits with $q < 0.2$ AU and $i = 10°–40°$. Although some of these are members of the Geminid shower — which shows up vividly in the Obninsk data [12, Fig. 4], that radar having been better-placed for Geminid observations — the detection of such meteoroids at other times of year (than the Geminid epoch in mid-December) indicates that not all these orbits are associated with the stream producing that shower. Following the line of reasoning in this discussion (that the meteoroid-orbit distribution can give clues as to the orbital parameters of undiscovered NEOs), one could have predicted 25 years ago on the basis of these meteoroid data that there are NEOs with such small perihelion distances, moderate inclinations, large eccentricities (else they would not cross the terrestrial orbit: meteor observations cannot inform us about possible Cytherian or Hermian asteriods), and perhaps small semimajor axes (from the upper left plot in Fig. 1). Whilst one such asteroid [(1566) Icarus, discovered in 1949] was known at that time, the past decade or so has seen a number of other asteroids discovered which fall into the same category, including (3200) Phaethon, (5786) Talos (formerly 1991RC; see [43]), 1989VA, 1990UO, 1994XL$_1$, 1995LG, and 1996FS$_1$. (3200) Phaethon was recognized as being the parent of the Geminid stream soon after its discovery in 1983, and its existence was long-suspected (Cook [44] recommended that "Small asteroids might be searched for along the orbits of the Geminids and the Sextantids," these two showers having since been quite securely linked

[31]). The question now is whether the broad concentration of orbits in the lower left of the q–i plot is indicative of many more NEOs with similar orbits which await discovery.

Thirdly, I note briefly that there are many orbits with perihelia near 1 AU (see also [16, Fig. 4]). Caution must be applied in interpreting these as a real concentration of orbits in heliocentric space, because the collision probability with the Earth is elevated if $q \simeq 1$ AU [41], [42]. The derivation of the true distribution of Earth-crossing meteoroid orbits in space from the raw distribution shown in Fig. 1 requires that such weighting factors be allowed for, along with other selection effects described in Section 3. However, I note that the $q \simeq 1$ AU orbits show concentrations at i = 40°–80° and $i > 120°$, which again may be indicative of yet-to-be-discovered parent NEOs, a suggestion that is supported by broadly-similar concentrations being seen in the Obninsk orbits [12, Fig. 4]. Having written all of the above, a cautionary note must be added: only the raw orbital element distributions have so far been considered, without allowance for any selection effects, and in the next section it will be seen that the selection effects are severe and may been sufficient to negate the first (retrograde cis-jovian orbits) and third (high-inclination orbits with perihelia near 1 AU) suggestions made above.

3. THE RADIANT DISTRIBUTION OF SPORADIC METEORS

Using fictitious asteroid orbits based upon the observed NEO population, Hills and Leonard [45] have investigated the arrival directions of such bodies, and suggested suitable observational programs that might permit their identification in the last few days before an impact. This could allow such objects to be deflected if suitable ready-on-station interceptors were maintained [8], [46], or enable the regions likely to be affected to be evacuated (for example, coastal areas if an entry above an ocean, producing a giant tsunami, were predicted). Although I am not an advocate of the maintenance of the sort of missile arsenal required to allow impending impactors to be deflected or destroyed, the modest search programs described by Hills and Leonard, making evacuations of at-risk regions feasible, do seem to be judicious and advisable on both humanitarian and economic grounds. Personal views aside, as I stated in the Introduction, *if* one were in the business of attempting to spot small NEOs just prior to impact — as opposed to finding larger NEOs many years ahead of any potential impact as described in such studies as the *Spaceguard* report [7] — then it would be useful to give some thought as to where one should concentrate one's efforts. In this section, I discuss how observed meteoroid orbits/meteor radiants can assist in this regard.

In the Introduction, I mentioned that one of the two great fallacies of meteoroid science is that "sporadic meteoroids come from random directions." Recently there has been an excellent discussion of this point by Jones and

Brown [47], who outlined the history of the idea, discussed the recognition that the nonshower (hence "sporadic," by definition) meteors have radiants concentrated towards certain regions of the sky, and presented raw (i.e., not corrected for selection effects) radiant distributions for various meteoroid orbit surveys. Their discussion will not be repeated here, but the reader is encouraged to consult that paper [47] and a subsequent publication [48] by the same authors.

Broadly speaking, the sporadic meteors having radiants close to the ecliptic may be divided into three sources: in *geocentric* coordinates (i.e., the observed or *apparent* radiant directions, without allowing for the Earth's motion) usually termed the *helion*, *apex,* and *antihelion* sources, the names being self-explanatory. The apex source may be sub-divided into two components with mean radiants 10°–20° north and south of the ecliptic (see later). A fourth source is termed the *toroidal* component, with a radiant direction at the same longitude as the apex but centered at an ecliptic latitude of ~60°, and again north and south branches are recognized. Various meteoroid orbit surveys have resulted in the detection of these sources, but with differing relative fluxes (see the figures in [47] and [48], and also Figs. 2–4 below). The sources, as expected, show quite different orbital characteristics: some examples are presented by Jones and Brown [47] (e.g., their Fig. 9 shows that the north toroidal source has $a \approx 1$ AU and small values of e, resulting in its name). It should also be mentioned that the sources — in particular the helion and antihelion components — show quite marked seasonal variations in relative strength (e.g., see [49], [50]), and this argues for an unrelaxed population resulting from an exceptional cometary disintegration event some time in the astronomically-recent past, as discussed in Section 2. Similar arguments, for a broad "sporadic stream," have been advanced by Štohl [27], [51]–[53] on the basis of the helion and antihelion sources' radiants and orbits.

Displayed in Fig. 2 is the radiant distribution of meteors with orbits determined in the Adelaide radar surveys. This is the raw distribution: no allowance has been made for the orbital dependence of the probability of meeting the Earth, station seeing effects, or other weighting factors discussed below in connection with the HRMP. The coordinate system used is the ecliptic system, with the apex being at ecliptic longitude $\lambda = 0°$, the sunward direction being $\lambda = -90°$ and the opposition direction $\lambda = +90°$. Because Adelaide is in the southern hemisphere, most radiants are at ecliptic latitudes $\beta < 0°$, the furthest north being at $\beta \simeq +30°$. All four sources may been seen in this contour plot, the strongest by far being the antihelion source, followed by the helion source. These are, as aforesaid, broad near-ecliptic sources seen here to radiate from regions a few tens of degrees in extent and centered near $\lambda \simeq \pm 70°$. The (southern) apex source is seen to be a diffuse source around $\lambda = 0°$, $\beta = -15°$, whilst the (southern) toroidal source is close to $\lambda = 0°$, $\beta = -60°$.

Not all surveys show the same relative strengths of these sources, despite perhaps having similar limiting radar magnitudes (similar particle sizes detected), as can immediately be seen from Fig. 3 in which are plotted the radiants from the Obninsk program [2], [3], [12], [13]. Note that the available Obninsk data consist solely of orbits observed at their descending nodes, and hence producing radiants at $\beta \geq 0°$; the apparent presence of radiants at $\beta < 0°$, and the shift northwards of the center of the antihelion source to $\beta \simeq +10°$, is an artifact of the contouring procedure used in producing Fig. 3, in conjunction with the artificial cut-off applied by the original experimenters at Obninsk. Here the antihelion source is strong, but only about the same strength as the toroidal source and it is weaker than the apex source. The helion source is evidenced by a contour plot lobe in the appropriate position but it is much less prominent than in the Adelaide data (Fig. 2). I note now, in conjunction with Fig. 2, that the apex source is clearly split into northern and southern components, as was first recognized by Elford and Hawkins [54], and discussed by Jones and Brown [47], [48].

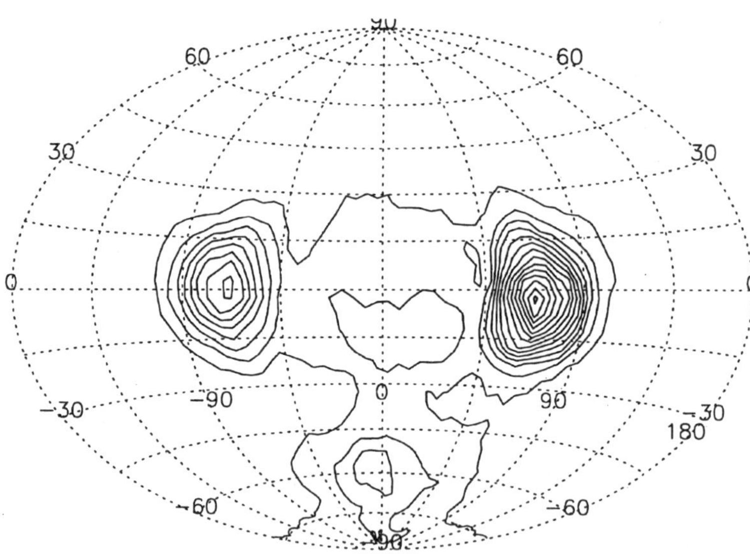

FIGURE 2. Contour plot of the geocentric (apparent) radiant distribution for the radar meteors from the Adelaide surveys, the orbital parameters of these having been plotted in Fig. 1. The plot is in ecliptic coordinates, using the apex of the Earth's way as the origin for longitude. Thus the Sun is located at longitude $\lambda = -90°$, latitude $\beta = 0°$, whilst the opposition point is at $\lambda = +90°$. Four distinct sources are evidenced: the helion and antihelion sources which are near-ecliptic and close to $\lambda = \pm 70°$, the southern apex source around ($\lambda = 0°$, $\beta = -15°$), and the southern toroidal source around ($\lambda = 0°$, $\beta = -60°$).

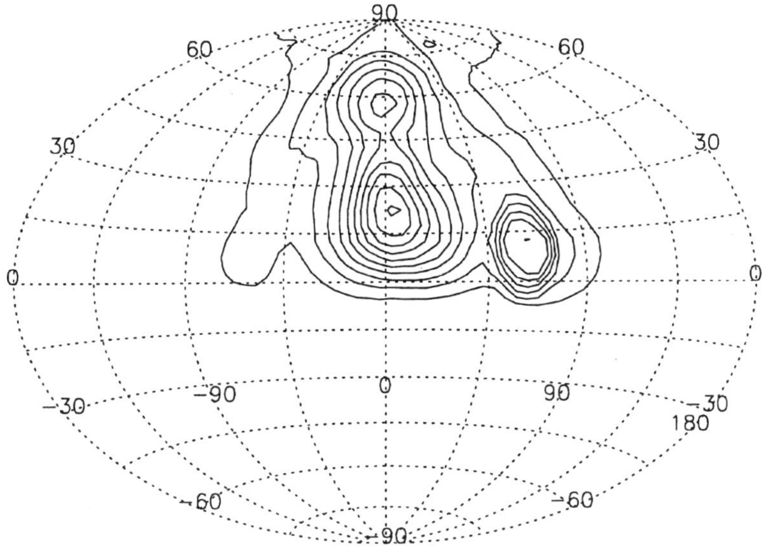

FIGURE 3. As Fig. 2, except for the Obninsk radar meteors. This plot contrasts strongly with the Adelaide distribution, but the following analysis of the Harvard Radio Meteor Project orbits indicates that this is due to selection effects, meaning that the source strengths shown in these raw distributions can tell one little about the real distribution of incoming meteoroids unless those selection effects can be properly accommodated in the analysis.

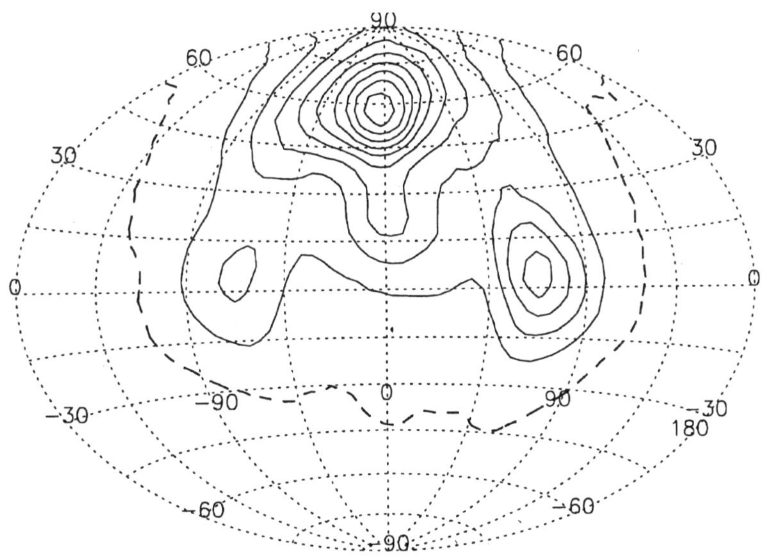

FIGURE 4. As Figs. 2 and 3, except for the almost 20,000 orbits determined in the Synoptic Year program of the Harvard Radio Meteor Project. The contours here show steps of one meteor radiant per square degree. The dashed line shows the limit of the region where at least one meteor per ten square degrees was detected. This is the raw radiant distribution before any corrections were applied for selection effects.

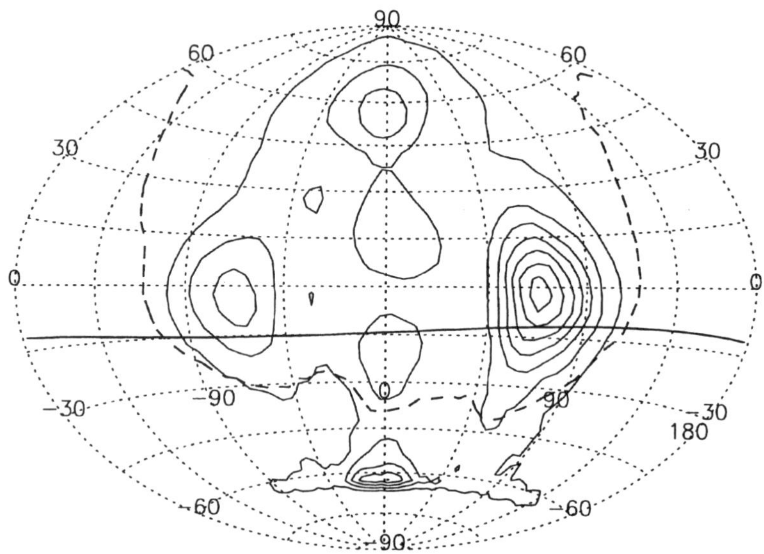

FIGURE 5. This is the distribution from Fig. 4 after a correction was applied for the amount of time that each radiant position was observed during the year in question, taking into account the response function of the HRMP radar. The northern toroidal source, being circumpolar, was observable continuously, and the resultant low weight causes it to be much less prominent here than in Fig. 4. The solid line crossing the plot near $\beta = -15°$ indicates the limit at which radiants were observable for at least 10% of the time over which the most-observed point (near the north celestial pole) was observed. Note that both the north and south apex sources *and* the north and south toroidal sources are evidenced here.

The apparent inter-survey variations in these source strengths should not be imagined to be real, but should be understood as being due to the observational biases inherent in designing, building, and operating radar systems such as those utilized in collecting the data plotted in Figs. 2 and 3. In order to derive an idea of the real radiant distribution, it is necessary to make a number of corrections to the raw distributions plotted in those figures. One critical point in this respect is the *response function* of the radar, this being a measure of its sensitivity to different meteor radiants at a specified time: in essence the rotation of the Earth means that the radar response function is diurnally swept across the sky. This concept was introduced [55] by my colleague W.G. Elford in connection with the HRMP. For more recent implementations of the response function concept with other radars see [56] and [57]; in the latter paper is given a discussion of alternative approaches which might be applied. Because I do not have available the required information regarding the Obninsk radar equipment, it is not possible to make the necessary corrections. In principle the Adelaide data could be corrected for station seeing effects, but: (i) the HRMP orbits are more plentiful and, as will be seen below, even those are not sufficiently numerous to allow the final plot derived to be as lacking in noise and great in sky coverage as might be desired; and (ii) Elford's derivation [55] of the HRMP response function is available, and so easily applied. Thus the data set used in this analysis will be restricted to that resulting from the HRMP; specifically, the 19,818 orbits measured in the Synoptic Year program of 1968–69 will be used [2], [3], [58]–[61]. I should make clear here that the analysis presented below is predominantly due to my colleague A.D. Taylor.

In Fig. 4 is shown a contour plot of the same type as that given in Figs. 2 and 3 for the Adelaide and Obninsk surveys; that is, this is the raw radiant distribution of the HRMP data, with no corrections yet having been applied. The contours go up in steps of one meteor per square degree; the dashed line shows the contour outside of which there is less than one meteor per ten square degrees. The overall distribution may be compared to the same data plotted with a dot for each meteor [61, Fig. 3].

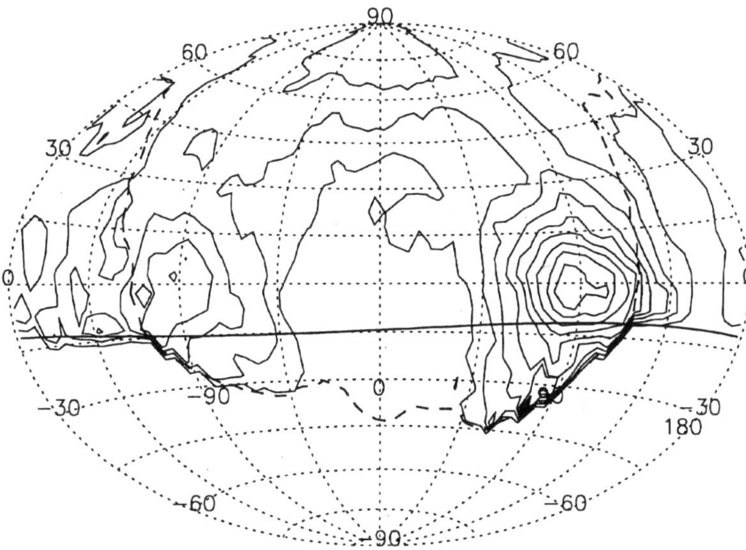

FIGURE 6. Here the distribution from Fig. 5 has been reduced to a constant mass, with radiants which are both south of the solid line and outside of the dashed line having been excluded else the combination of two large weighting factors can give unreasonable prominence to singular meteors within that suppressed region. Knowing the limiting magnitude (equivalent to an electron line density) of the HRMP, and the approximate mass of each observed meteoroid, and using standard ionizing efficiencies and an assumed mass distribution exponent of 1.1, each meteoroid was given a weighting so as to reflect the number of meteoroids to a limiting mass of ~10^{-4} g at a standard speed of 30 km/sec that would be detected by an ideal system with no selection effects. This plot, then, is our best estimate of the incoming flux of meteoroids in terms of the distribution of their arrival directions. The helion and antihelion sources are the most prominent, the latter especially so; this is discussed in the text.

The first correction to be applied is that for station seeing. The response function calculated by Elford [55] was that for the HRMP installation in 1961–65, which produced 19,327 orbits [2], [3], [58], [62]. That radar equipment was altered to some extent for the Synodic Year study, and an aircraft was used to measure the antenna patterns. The major alteration from the earlier evaluation was to put the elevation response about 6° lower, and this later version of the response function was used in deriving the results presented here in Fig. 5 (and thence 6). The most-covered area of the sky is, of course, the circumpolar region, and this leads to more meteors being detected coming from northerly radiants than regions further south. By allowing for the different amounts of time over which different regions were observed, Fig. 5 results: the effective observing time was estimated from the published observing schedule, and then the response function was integrated over the Synodic Year observations to derive a year-averaged weighting distribution for each ecliptic coordinate cell. The solid slightly-wavy line indicates the southerly extent at which the radiants were covered for at least 10% of the time with which the most-observed radiant cell was observed. The northern toroidal source, which dominates in the raw radiants (Fig. 4), is now much reduced in its prominence, whilst the helion and antihelion (in particular) sources start to show their significance.

However, there are still important selection effects to be considered. In particular, the differing sources have different characteristic meteor speeds (V), and these affect the meteor detectability through: (i) the ionizing efficiency of the meteoroid varying as $\sim V^{3.5}$; and (ii) the higher-speed meteoroids ablating at greater heights, which are selected against by VHF radars due to the echo-ceiling effect [63], [64]. Note that Taylor [65] has recently reanalyzed the HRMP results [59], [60] for the mean meteor speed and corrected speed distribution, and shown these to be flawed due to a simple numerical error (see also [66]). Leaving this aside, here the echo-ceiling effect has been accommodated using the correction applied by Sekanina and Southworth [59], [60], which almost certainly underestimates the effect, according to more recent HF radar observations [56], [64], and thus fails to apply a sufficiently-large weighting to the higher-speed meteors [66]. The flux for each radiant (each square degree in ecliptic coordinates) was then reduced to a "constant mass meteoroid" value by scaling each observed meteoroid (mass estimates are available for each) to the $\sim 10^{-4}$ g mass expected for ~30 km/sec meteors, this producing the limiting ionization line density for the HRMP; a mass distribution index of 1.1 was used [56], with the ionization production parameters given by Verniani [67]. The result is shown in Fig. 6.

The effect of these correction factors is seen to be quite startling, when one compares Fig. 6 with the raw distribution in Fig. 4. The apex and toroidal sources have now all but disappeared, so that they are understood to be due largely to selection effects rather than the presence of any significant distinct meteoroid populations (cf. discussion at the end of Section 2). This is heartening in that the orbits that characterize them are dissimilar to any known NEOs which might be regarded as potential parents (the apex source consists mainly of retrograde meteoroids of small semimajor axis, whilst the toroidal source comprises high-inclination meteoroids in similarly-small orbits: see [47, Figs. 7 and 9], respectively).

Turning to the most important point of this analysis, the dominant sources are seen to be the antihelion and helion components, with the latter appearing to be rather weaker than the former (as is found in other surveys: Figs. 2 and 3). Because one might expect these sources to be of equal strength — since they might be thought of as being just the pre- and postperihelion intersections with the Earth of the same basic types of orbit — this contrast poses a puzzle which requires further investigation. The likely source of this inequality between the *detected* helion and antihelion fluxes has only just been identified [68], [69]; it appears to be as follows. In the dayside atmosphere, D-region ionization caused by the solar high-energy photon flux causes Faraday rotation of the plane-polarized radio waves used in most radars, such that the echoes are attentuated, reducing the proportions of the helion source meteor echoes which have amplitudes above the detection limits of the various systems used. Full modelling in which this effect is accommodated has yet to be attempted; it will be of interest to apply *all* the above selection effects, including the Faraday rotation effect, to the data being collected with the radar in New Zealand [4] in order to see whether the helion and antihelion sources are indeed of equal strength in terms of the actual influx to the atmosphere.

It is notable that the effect of applying the various correction factors (Fig. 6: no correction for Faraday rotation possible) is to move the locations of the helion and antihelion sources out to longitudes of ±90°. This might suggest to the reader that the optimal position to search for NEOs in the last day before impact would be in the directly sunward and opposition locations (and of course real telescopic NEO search programs do concentrate upon the opposition position [6]–[8]), but it should be recalled that the positions plotted in Figs. 2–6 are the *geocentric* (or *apparent*) radiants (i.e., compounded with the terrestrial motion about the Sun). The transformation from an observed geocentric radiant to the *true radiant*, as used in calculating meteoroid orbits [70] is quite straightforward; the converse is used in deriving theoretical shower radiants for putative parent objects [37], [70]. Here, however, the low inclinations of the meteoroid orbits allow a very simple transformation to be used, as follows.

The angle γ with which an orbit with semimajor axis a and eccentricity e cuts a circular, coplanar orbit at a heliocentric distance R is given by:

$$\cos^2\gamma = a^2(1 - e^2)/R(2a - R). \tag{1}$$

Because the helion and antihelion sources are of low inclination we can assume, for these purposes, that they are coplanar with the Earth, which is taken to have a circular orbit at $R = 1$ AU. For typical orbits in these sources ($a \approx 1.3 - 2.5$ AU, $e \approx 0.7 - 0.9$; [47, Fig. 8]) one derives $\gamma \approx 60°$, indicating true radiants centered on longitudes of ±120° from the apex. (In passing, I note that the only other radar meteor survey to reach a similar limiting magnitude as the HRMP and present New Zealand program, that at Kharkov in Ukraine, produced similar orbital elements for the helion and antihelion sources as those given here: see [52, Fig. 5] and [2]).

Where, then, should the search for small NEOs in the day or two before impact be directed? If one were examining a stream of objects passing by the Earth (for example, looking for backscatter from particles in a meteoroid stream as the Earth moved through that stream) then the appropriate region of the sky to scan is that around the *true* radiant [71], [72]. On the other hand, particles due to actually strike the Earth would appear to emanate, shortly before impact, from the *geocentric* radiant. Kessler and Zook commented, in the context of looking for a concentration of meteoroids within a stream, that the region to search is along the arc connecting the geocentric and the true radiants [72]. For the antihelion source, the region to concentrate upon is therefore that within about 20° of a line along the ecliptic between longitudes (from the apex) of about 90° (the geocentric or apparent radiant) and 120° (the true radiant). The helion source, of course, would be more problematical, due to the glare of the Sun; a space-based search system would be necessary [73].

A note on telescopic NEO identification is necessary. So close to the Earth/to impact the NEO angular speed in the sky plane is very small, making detection through the usual techniques which rely on angular displacements between exposures with a monostatic system very difficult: an NEO which is going to hit the Earth is moving directly towards the planet, radiating from a point (cf. meteor terminology). Identification through the significant parallax obtained from bistatic detection systems when the target is at a small range is a possibility [45].

Finally, in this section I make clear that no attempt has been made to take out the selection effect due to the collision probability with the Earth, which is a strong function of the impactor's orbital elements [41], [42], [74], as was mentioned earlier. This is quite deliberate. In searching for NEOs which might strike our planet, those of greatest interest are the ones which are most likely to hit us. If one wanted to derive the distribution in space of meteoroids, then it would be necessary to accommodate such collision probabilities, but from the present perspective no such correction is desired: if NEOs have similar orbital characteristics to the smaller meteoroids, then their most likely arrival directions are indicated by Fig. 6.

4. CONCLUSIONS

There is no doubt in my mind that the best technique to be used by humankind to protect itself against NEO impacts of sufficient magnitude to cause a global catastrophe is a *Spaceguard*-type program aimed at discovering all potentially-dangerous asteroids and comets some years ahead of any impact in order that the impending impact be forecast and then obviated. However, it is also clear from the *Spaceguard* study [7] and similar research [8] that it is not feasible to detect and track all small (larger than ~50 meter) NEOs that could cause at least major local, and possibly continental, damage. The population and hence influx of these is much greater than that of the larger (> 1–2 kilometer) NEOs capable of causing global effects. In this paper I have taken the view that the plentiful — and rapidly growing — meteoroid orbit database might be used as a *prima facie* indicator of the orbital distribution of small NEOs. If this is actually the case then, as shown above, the region of the sky that needs to be scoured most intensely is quite a small fraction — less than 10 % — of the whole celestial sphere.

That conclusion in itself may also be applied to the design of search programs for the larger NEOs. If the observed meteoroid orbits are diagnostic of the orbits of NEOs, then they provide an important information source which could allow NEO search strategies to be tuned so as to produce optimal discovery rates for the objects most likely to collide with our planet.

ACKNOWLEDGMENTS

Figures 2 through 6 were kindly prepared for me by Andrew Taylor. Discussions concerning radar meteor detection with Graham Elford and Jack Baggaley, and concerning theoretical radiants with Taylor and Brian Marsden, are also gratefully acknowledged.

REFERENCES

1. MARSDEN, B.G. & G.V. WILLIAMS. 1996. Catalogue of Cometary Orbits. 11th edit. Minor Planet Center. Cambridge, Massachusetts.
2. STEEL, D.I. 1997. Meteoroid orbits. Space Sci. Rev. **78:** 507–553.
3. LINDBLAD, B.A. & D.I. STEEL. 1994. Meteoroid orbits available from the IAU Meteor Data Center. *In* IAU Symp. 160: Asteroids, Comets, Meteors 1993. A. Milani, M. Di Martino, & A. Cellino, Eds.: 497–501. Kluwer Publishing. Dordrecht, Holland.
4. BAGGALEY, W.J., R.G.T. BENNETT, D.I. STEEL, & A.D. TAYLOR. 1994. The Advanced Meteor Orbit Radar Facility: AMOR. Quart. J. Roy. Astron.Soc. **35:** 293–320.
5. BEECH, M. & D.I. STEEL. 1995. On the definition of the term "meteoroid." Quart. J. Roy. Astron. Soc. **36:** 281–284.

6. GEHRELS, T. & R. JEDICKE. 1996. The population of near-Earth objects discovered by Spacewatch. Earth, Moon & Planets. **72:** 233–242.
7. MORRISON, D., Ed. 1992. The Spaceguard Survey: Report of the NASA International Near-Earth-Object Detection Workshop. Jet Propulsion Laboratory. Pasadena, California.
8. GEHRELS, T., Ed. 1994. Hazards Due to Comets and Asteroids. University of Arizona Press. Tucson, Arizona.
9. CEPLECHA, Z. 1994. Meteoroid properties from photographic records of meteors and fireballs. *In* IAU Symp. 160: Asteroids, Comets, Meteors 1993. A. Milani, M. Di Martino, & A. Cellino, Eds.: 343–356. Kluwer Publishing. Dordrecht, Holland.
10. JOPEK, T.J., P. FARINELLA, CH. FROESCHLÉ, & R. GONZI. 1995. Long-term dynamical evolution of the brightest bolides. Astron. Astrophys. **302:** 290–300.
11. STEEL, D.I. 1994. Meteoroid streams. *In* IAU Symp. 160: Asteroids, Comets, Meteors 1993. A. Milani, M. Di Martino, & A. Cellino, Eds.: 111–126. Kluwer Publishing. Dordrecht, Holland.
12. STEEL, D.I. 1991. The orbital distribution and origin of meteoroids. *In* IAU Colloq. 128: Origin and Evolution of Interplanetary Dust. A.C. Levasseur-Regourd & H. Hasegawa, Eds.: 291–298. Kluwer Publishing. Dordrecht, Holland.
13. LEBEDINETS, V.N., V.N. KORPUSOV, & A.V. MANOCHINA. 1981, 1982. Radio Meteor Investigations in Obninsk: Catalogue of Orbits. Materials of the World Data Center B. Moscow, USSR.
14. NILSSON, C. 1964. A southern hemisphere radio survey of meteor streams. Aust. J. Phys. **17:** 205–256.
15. GARTRELL, G. & W.G. ELFORD. 1975. Southern hemisphere meteor stream determinations. Aust. J. Phys. **28:** 591–620.
16. OLSSON-STEEL, D. 1988. Identification of Meteoroid Streams from Apollo Asteroids in the Adelaide Radar Orbit Surveys. Icarus. **75:** 64–96.
17. BURNS, J.A., PH. L. LAMY, & S. SOTER. 1979. Radiation Forces on Small Bodies in the Solar System. Icarus. **40:** 1–48.
18. OLSSON-STEEL, D. 1986.The origin of the sporadic meteoroid component. Mon. Not. Roy. Astron. Soc. **219:** 47–73.
19. WHIPPLE, F.L. 1940. Photographic meteor studies. III. The Taurid shower. Proc. Amer. Phil. Soc. **83:** 711–745.
20. WHIPPLE, F.L. & S.E. HAMID. 1952. On the origin of the Taurid meteor streams. Helwan Obs. Bull. **41:** 1–30.
21. WHIPPLE, F.L. 1967. On maintaining the meteoritic complex. *In* NASA SP–150, The Zodiacal Light and the Interplanetary Medium, J.L. Weinberg, Ed.: 409–426. NASA, Washington, DC.
22. LEINERT, C., S. RÖSER, & J. BUITRAGO. 1983. How to maintain the spatial distribution of interplanetary dust. Astron. Astrophys. **118:** 345–357.
23. GRÜN, E., H.A. ZOOK, H. FECHTIG, & R.H. GIESE. 1985. Collisional Balance of the Meteoritic Complex. Icarus. **62**: 244–272.
24. STEEL, D. & W.G. ELFORD. 1986. Collisions in the Solar System—III. Meteoroid survival times. Mon. Not. Roy. Astron. Soc. **218:** 185–199.
25. KRESÁK, Ľ. 1980. Sources of interplanetary dust. *In* IAU Symp. 90, Solid Particles in the Solar System, I. Halliday & B.A. McIntosh, Eds.: 211–222. Reidel, Dordrecht, Holland.
26. KRESÁK, Ľ. & M. Kresáková. 1987. The mass loss rates of periodic comets. *In* ESA SP–278, Symp. on the Diversity and Similarity of Comets, E.J. Rolfe & B. Battrick, Eds.: 739–744. European Space Agency, Paris, France.
27. ŠTOHL, J. 1987. Meteor contribution by short-period comets. Astron. Astrophys. **187:** 933–934.

28. CAMPINS, H. 1988. The Anomalous Dust Production in Periodic Comet Encke. Icarus. **73:** 508–515.
29. SEKANINA, Z. 1991. Encke, the comet. J. Roy. Astron. Soc. Canada. **85:** 324–376.
30. BABADZHANOV, P.B., YU.V. OBRUBOV, & N. MAKHMUDOV. 1990. Meteor streams of Comet Encke. Solar System Res. **24**: 12–19.
31. BABADZHANOV, P.B. & YU.V. OBRUBOV. 1992. Evolution of short-period meteoroid streams. Cel. Mech. Dyn. Astron. **54:** 111–127.
32. STEEL, D.I., D.J. ASHER, & S.V.M. CLUBE. 1991. The structure and evolution of the Taurid Complex. Mon. Not. Roy. Astron. Soc. **251:** 632–648.
33. ASHER, D.J., S.V.M. CLUBE, W.M. NAPIER, & D.I. STEEL. 1994. Coherent Catastrophism. Vistas Astron. **38:** 1–27.
34. STEEL, D.I. & D.J. ASHER. 1995. The orbital dispersion of the macroscopic Taurid objects. Mon. Not. Roy. Astron. Soc. **280:** 806–822.
35. ASHER, D.J. & D.I. STEEL. 1995. Theoretical meteor radiants for macroscopic Taurid Complex objects. Earth, Moon & Planets. **68:** 155–164.
36. STEEL, D.I. & D.J. ASHER. 1996. On the origin of Comet Encke. Mon. Not. Roy. Astron. Soc. **281:** 937–944.
37. STEEL, D.I. 1995. The association of Earth-crossing asteroids with meteoroid streams. Earth, Moon & Planets. **68:** 17–30.
38. OLSSON-STEEL, D. 1989. Meteoroid streams and the zodiacal dust cloud. *In* Catastrophes and Evolution: Astronomical Foundations. S.V.M. Clube, Ed.: 169–193. Cambridge University Press, Cambridge, United Kingdom.
39. STEEL, D., R.H. MCNAUGHT, & D. ASHER. 1992. 1991 DA: An asteroid in a bizarre orbit. *In* Asteroids, Comets, Meteors 1991, A. Harris & E. Bowell, Eds.: 573–576. Lunar & Planetary Institute. Houston, Texas.
40. STEEL, D.I. & D.J. ASHER. 1992. The past and future orbit of (extinct comet?) 1991 DA. *In* Periodic Comets, J.A. Fernández & H. Rickman, Eds.: 65–73. Universidad de la República, Montevideo, Uruguay.
41. STEEL, D. 1992. Cometary supply of terrestrial organics: lessons from the K/T and the present epoch. Orig. Life & Evol. Biosphere. **21:** 339–357.
42. STEEL, D. 1993. Collisions in the solar system—V. Terrestrial impact probabilities for parabolic comets. Mon. Not. Roy. Astron. Soc. **264:** 813–817.
43. STEEL, D., R.H. MCNAUGHT, & D. ASHER. 1992. Did Icarus have a twin brother? Minor Planet Bull. **19:** 9–11.
44. COOK, A.F. 1973. A Working List of Meteor Streams. *In* NASA SP-319, Evolutionary and Physical Properties of Meteoroids. C.L. Hemenway, P.L. Millman, & A.F. Cook, Eds.: 183–191. NASA, Washington, DC.
45. HILLS, J.G. & P.J.T. LEONARD. 1995. Earth-crossing Asteroids: The Last Days Before Earth Impact. Astron. J. **109:** 401–417.
46. CANAVAN, G.H., J.C. SOLEM, & J.D.G. RATHER, Eds. 1993. Proceedings of the Near-Earth-Object Interception Workshop. LA-12476-C. Los Alamos National Laboratory. Los Alamos, New Mexico.
47. JONES, J. & P. BROWN. 1993. Sporadic meteor radiant distributions: Orbital survey results. Mon. Not. Roy. Astron. Soc. **265:** 524–532.
48. JONES, J. & P. BROWN. 1994. The radiant distribution of sporadic meteors. Planet. Space Sci. **42:** 123–126.
49. KEAY, C.S.L. 1963. The annual variation in the radiant distribution of sporadic meteors. J. Atmos. Terr. Phys. **25:** 507–513.
50. ELFORD, W.G. 1967. Incidence of Meteors on the Earth Derived from Radio Observations. Smithson. Contrib. Astrophys. **11:** 121–131.
51. ŠTOHL, J. 1983. On the distribution of sporadic meteor orbits. *In* Asteroids, Comets, Meteors, C.-I. Lagerkvist & H. Rickman, Eds.: 419–424. University of Uppsala, Sweden.

52. ŠTOHL, J. 1986.The distribution of sporadic meteor radiants and orbits. *In* Asteroids, Comets, Meteors II. C.-I. Lagerkvist, B.A. Lindblad, H. Lundstedt, & H. Rickman, Eds.: 565–574. University of Uppsala, Sweden.
53. ŠTOHL, J. & V. PORUBČAN. 1992. Dynamical aspects of the Taurid meteor complex. *In* IAU Symp. 152, Chaos, Resonance and Collective Dynamical Phenomena in the Solar System, S. Ferraz-Mello, Ed.: 315–324. Kluwer. Dordrecht, Holland.
54. ELFORD, W.G. & G.S. HAWKINS. 1964. Meteor Echo Rates and the Flux of Sporadic Meteors. Radio Meteor Project Res. Rep. **9**, Harvard College Observatory & Smithsonian Astrophysical Observatory, Cambridge, Massachusetts.
55. ELFORD, W.G. 1964. Calculation of the Response Function of the Harvard Radio Meteor Project Radar System. Radio Meteor Project Res. Rep. **8,** Harvard College Observatory & Smithsonian Astrophysical Observatory. Cambridge, Massachusetts.
56. THOMAS, R.M., P.S. WHITHAM, & W.G. ELFORD. 1988. Response of high frequency radar to meteor backscatter. J. Atmos. Terr. Phys. **50:** 703–724.
57. ELFORD, W.G., M.A. CERVERA, & D.I. STEEL. 1994. Single station identification of radar meteor shower activity: The June Librids in 1992. Mon. Not. Roy. Astron. Soc. **270:** 401–408.
58. COOK, A.F., M.R. FLANNERY, H. LEVY, R.E. MCCROSKY, Z. SEKANINA, C.-Y. SHAO, R.B. SOUTHWORTH, & J.T. WILLIAMS. 1972. Meteor Research Program. NASA Contractor Report CR–2109, Washington, DC.
59. SOUTHWORTH, R.B. & Z. SEKANINA. 1973. Physical and Dynamical Studies of Meteors. NASA Contractor Report CR–2316, Smithsonian Astrophysical Observatory, Cambridge, Massachusetts
60. SEKANINA, Z. & R.B. SOUTHWORTH. 1975. Physical and Dynamical Studies of Meteors: Meteor-Fragmentation and Stream Distribution Studies. NASA Contractor Report CR–2615, Smithsonian Astrophysical Observatory. Cambridge, Massachusetts.
61. SEKANINA, Z. 1976. Statistical model of meteor streams. IV. A study of radio streams from the synoptic year. Icarus. **27:** 265–321.
62. SEKANINA, Z. 1973. Statistical Model of Meteor Streams. III. Stream Search Among 19303 Radio Meteors. Icarus. **18:** 253–284.
63. MCKINLEY, D.W.R. 1961. Meteor Science and Engineering. McGraw-Hill, New York.
64. STEEL, D.I. & W.G. ELFORD. 1991. The height distribution of radio meteors: Comparison of observations at different frequencies on the basis of standard echo theory. J. Atmos. Terr. Phys. **53:** 409–417.
65. TAYLOR, A.D. 1995. The Harvard Radio Meteor Project meteor velocity distribution reappraised. Icarus. **116:** 154–158.
66. ELFORD, W.G., M.A. CERVERA, & D.I. STEEL. 1995. Meteor velocities: A new look at an old problem. Earth, Moon, & Planets. **68:** 257–266.
67. VERNIANI, F. 1973. An analysis of the physical parameters of 5759 faint radio meteors. J. Geophys. Res. **78:** 8429–8462.
68. ELFORD, W.G. & A.D. TAYLOR. 1997. Measurement of Faraday rotation of radar meteor echoes for the modelingof electron densities in the lower ionosphere. J. Atmos. Terr. Phys. In press.
69. TAYLOR, A.D., W.G. ELFORD, & D.I. STEEL. 1997. New results in radar meteor science. COSPAR Conference Proc. **10**. In press.
70. PORTER, J.G. 1952. Comets and Meteor Streams. Chapman & Hall. London, United Kingdom.
71. BAGGALEY, W.J. 1977. The possibility of the detection of meteor streams in interplanetary space. The Observatory. **97:** 123–129.
72. MARSDEN, B.G. 1993. Perseid meteors. IAU Circ. 5840.

73. ISOBE, S. & M. YOSHIKAWA. 1996. Asteroids approaching the Earth from directions around the Sun. Earth, Moon & Planets. **72:** 263–266.

74. STEEL, D. 1995. Collisions in the solar system—VI. Terrestrial impact probabilities for the known asteroid population. Mon. Not. Roy. Astron. Soc. **273:** 1091–1096.

Overview of Orbits

BRIAN G. MARSDEN

Harvard-Smithsonian Center for Astrophysics
60 Garden Street
Cambridge, Massachusetts 02138

ABSTRACT: After a brief summary of the distribution of both large and small bodies in the solar system, there follows a categorization of the various hazards from comets and asteroids, the concept of PHA, or "potentially hazardous asteroid," being introduced to include one that is large enough and can pass close enough to the Earth's orbit that it might conceivably yield a globally-damaging impact in the forseeable furture. Future search stragtegies are discussed, as they involve increasing both sky coverage and faintness limit, arguments being given for ensuring much fainter detections in the general opposition region than at much smaller elongations. It is concluded that a proposed "impact hazard scale" is generally less useful than recognition of the PHA status. The current situation with regard to follow-up astrometric activity, particularly by amateur astronomers, is viewed favorably, but professional search activity is at a depressingly low level.

1. INTRODUCTION

The solar system is a fascinating place. The Earth-Moon system and three other terrestrial planets that travel around the Sun in more-or-less circular, coplanar, codirectional orbits are separated from four more distant, much larger, similarly-orbiting planets by a broad gap that contains the Main Belt of minor planets. The Main Belt, the first members of which were discovered in the early years of the nineteenth century, includes a couple of hundred bodies in the size range 100–1000 km. These bodies, and numerous smaller companions, also orbit the Sun in much the same way. During the past few years it has been established that there is *another* set of bodies orbiting just beyond the giant planets, in what is termed the Kuiper Belt. With a diameter of more than 2000 km, Pluto is the largest member of the Kuiper Belt. Although only a couple of dozen bodies in the Kuiper Belt are currently known, the number larger than 100 km must surely be counted in the tens of thousands, and there are again innumerable smaller objects [1].

In the main, the bodies in the solar system that are 100 km and more across have orbits that are currently quite stable, with very little chance that they will collide with, or be gravitationally deflected by, any of the other large bodies. Certainly, this was not always so. In the early days of the solar system there must have been many collisions, some of which allowed bodies to increase in size, while many others caused the production of smaller fragments; and the chaos associated with the gravitational interplay among the major and minor planets largely cleaned the minor bodies from unstable regions of the Main Belt after only millions of years. At the larger heliocentric distance of the Kuiper Belt this orbital rearrangement takes place much more slowly. Moderately large bodies from the extensive reservoir of the Kuiper Belt are still occasionally being deflected so that a passage close to Neptune becomes possible. After an initial encounter with Neptune, further encounters will occur within a matter of millennia, and the orbital modifications can also rapidly involve encounters with Uranus, Saturn, and Jupiter. The half-dozen or so "centaur" 100 km objects currently observed in unstable orbits in the region of the giant planets, in some instances extending out beyond Neptune at aphelion, rather clearly illustrate this evolution.

Whereas the Main Belt consists essentially of objects of a stony or a metallic composition, in many respects similar to the terrestrial planets, the great heliocentric distance of the Kuiper Belt suggests that its components have the icy composition one associates with comets. Indeed, Chiron, the first of the centaurs to be recognized, is observed to have a coma or tail and is by far the largest body actually seen to show such cometary activity.

That there are very many smaller cometary objects in the Kuiper Belt has recently been claimed as the result of observations with the Hubble Space Telescope [2]. One likewise expects that there must at any time be an extensive transient population of smaller objects among the centaurs. Many of the latter must sooner or later be able to make close approaches to Jupiter — occasionally as disastrously as comet D/1993 F2 (Shoemaker-Levy 9) in 1992 and 1994 — and become typical short-period comets of the Jupiter family, with their aphelia generally near Jupiter and their perihelia sufficiently closer to the Sun that water-ice vaporization can occur, and the comets become active. This activity causes the physical dissipation of such comets, although current thought is that this can be a complex process, with the possibility of a complete mantle of particulate material leaching out to insulate the whole nuclear surface for extended periods of time. Possibly, there is some minimum size for an active comet. That size may depend on the distance from the Sun and is perhaps a few hundred meters at the Earth's distance.

In the solar system's early stages, there must also have been extensive migration of cometary material from the vicinity of Uranus and Neptune toward the *outside*, with objects boosted to aphelia much farther from the Sun. At these distances they are largely free from the physical dissipation that severely afflicts the Jupiter-family comets. There arises the possibility of perturbations by passing stars and giant molecular clouds, as well as by the general gravitational field of the Galaxy. These long-period comets are thus drawn into the three-dimensional distribution of the Oort Cloud, a small fraction of them coming back close enough to the Sun that they, too, undergo water-ice vaporization and physical dissipation.

At some level, orbital evolution of small minor planets in the Main Belt is also still going on. Such objects can therefore also depart from the Main Belt on tracks that take them both closer to and farther from the Sun.

2. CATEGORIZATION OF EARTH-THREATENING BODIES

The outcome of all this activity is that there is a mixed population of small objects that have the capability of making close approaches to — even colliding with — the Earth. This population contains live comets, of both long period and short period; inactive comets, both temporarily deactivated with insulating mantles and completely stripped of volatiles; and fragments of minor planets, mainly stony objects but in a few instances composed of iron. If the objects are small enough, collisions with the Earth take place in which the resistive effect of the atmosphere prevents significant damage. Such collisions occur quite frequently, the principal danger to the Earth being from roughly centennial localized "Tunguska" events involving any of a million or more stony Earth-crossing minor planets 50–100 meters across.

It is the collisions with roughly kilometer-sized objects that are of global concern [3]. Such events are so potentially devastating and are statistically so infrequent that it is in fact inappropriate to consider them statistically. The end of the human race could be brought about by means of a single collision at any time. It is essential to try and guard against such a possibility. By a convenient coincidence, the estimated number of dangerous objects in short-period orbits in this size range is small enough — a few thousand — that the process of surveillance for them can be carried out rather expeditiously.

The term NEO, or "near-Earth object," tends to be applied, indiscriminately and rather confusingly, to the population of objects that cross the orbit of the Earth; if the object appears to be a minor planet — or, at least, not an active comet — it may be called an NEA, or "near-Earth asteroid." Many such objects have absolutely no possibility of actually passing close to the Earth, by virtue either of their current orbital configurations or the configurations resulting from the perturbations the orbits are likely to undergo over tens of thousands of years. This has led to the concept of an ECA, or "Earth-crossing asteroid" (the orbits of comets being too chaotic to permit meaningful long-term computations), to denote an object that might have the possibility of actually colliding with the Earth in the long run.

From a purely practical point of view — and surveillance and hazard assessment are very practical matters — it therefore seems appropriate to define a category of PHA, or "potentially hazardous asteroid," a subset of the ECA or NEA population that could in fact make particularly close approaches to the Earth during the foreseeable future, and that are large enough that impacts would cause damage on a global scale. This requires a computation of the minimum distance between the orbits of the objects and the Earth. The general computation of the minimum distance between two ellipses (confocal or otherwise) is actually quite complicated. Furthermore, since the ellipses do change with time, one should monitor those changes. Nevertheless, it is quite sufficient to simplify the computation by treating the Earth's orbit as a circle of radius 1 AU in the plane of the ecliptic and otherwise use principles well known to astronomers who study meteors.

One therefore needs to examine the departure of the nodal distances of the orbit of the object of interest from 1 AU, i.e., the two quantities

$$N = \left|\frac{q(1+e)}{1 \pm e\cos\omega} - 1\right|,$$

q, e, and ω being the perihelion distance (in AU), eccentricity and argument of perihelion; and also the distance of the object from the Earth's orbit when it is 1 AU from the Sun, i.e., the two quantities

$$M = |\sin(\omega \pm \nu)|\sin i,$$

where i is the object's orbital inclination to the ecliptic, and the values of the true anomaly $\pm\nu$ are given by

$$\pm\nu = \arccos\left[\frac{q(1+e)-1}{e}\right].$$

The minimum m of the four N and M values is noted and the object considered for PHA status if $m \leq 0.05$. As a distance in AU, this number is large enough to contain the error arising from the neglected eccentricity of the Earth's orbit and, generally, also the effect of perturbations over a few centuries; the latter point will be discussed further in the next Section. It is necessary also to consider that objects with perihelia up to $q = 1.05$ AU might approach within 0.05 AU of the circle representing the Earth's orbit. In such a case, the above values of $\pm\nu$ are imaginary, but one can suppose the object to be at perihelion and compute the single value of M corresponding to $\nu = 0$. In determining the minimum m, however, it is then necessary to replace M by $[M^2 + (q-1)^2]^{1/2}$, a point that was missed in an earlier discussion of this computation [4].

PHA status also requires the object to be at least 1 km in diameter. In the probable absence of specific information about albedo, it is thus more than sufficient to consider that the object's absolute visual magnitude $H < 20$.

3. PHAS AND PHCS

Table 1 lists, in chronological order of discovery, the 64 currently known PHAs. After the columns giving number, name and provisional designation, there follow columns with H, i (in degrees), q and the aphelion distance (in AU) $Q = q(1+e)/(1-e)$. The next column gives the minimum m, an A or D following the number indicating that this is an N value that occurs at the ascending or descending node, respectively; if no letter is appended, *m* is one of the values of *M*. The column Δm gives the difference of *m* from the value obtained using orbital elements osculating a century ago. The quantity $V_Q = H + 5\log Q(Q - 1)$ is the apparent visual magnitude at an aphelic opposition, and "Arc" denotes the time span, in days, covered by the observations of those objects observed at only a single opposition.

TABLE 1. PHAs

Object			H	i	q	Q	m	Δ_m	V_Q	Arc
(1862)	Apollo	1932 HA	16.2	6.4	0.647	2.29	0.028	+0.003	18.6	
(2101)	Adonis	1936 CA	18.7	1.4	0.441	3.31	0.012	−0.003	23.1	
	Hermes	1937 UB	18.0	6.1	0.619	2.67	0.003	+0.002	21.2	4
(2201)	Oljato	1947 XC	15.2	2.5	0.630	3.72	0.000	−0.003	20.2	
(1566)	Icarus	1949 MA	16.4	22.9	0.187	1.97	0.040	−0.005	17.8	
(1620)	Geographos	1951 RA	15.6	13.3	0.828	1.66	0.046	−0.003	15.8	
		1954 XA	19.0	3.9	0.509	1.05	0.038		12.6	6
(4183)	Cuno	1959 LM	14.5	6.8	0.719	3.24	0.038	−0.009	18.8	
(5011)	Ptah	6743 P-L	17.0	7.4	0.817	2.45	0.026	0.003	19.8	
(1981)	Midas	1973 EA	15.0	39.8	0.622	2.93	0.001	−0.003	18.8	
(2102)	Tantalus	1975 YA	15.3	64.0	0.905	1.68	0.029A	−0.004	15.6	
(2135)	Aristaeus	1977 HA	17.9	23.0	0.795	2.40	0.015	−0.005	20.5	
		1978 CA	18.0	26.1	0.883	1.37	0.026D	−0.002	16.5	32
(4015)	Wilson-Har.	1979 VA	16.0	2.8	0.999	4.29	0.048	+0.002	21.7	
		1979 XB	19.0	24.9	0.649	3.88	0.040		24.2	4
(4660)	Nereus	1982 DB	18.3	1.4	0.953	2.03	0.005	+0.001	19.9	
(3361)	Orpheus	1982 HR	19.0	2.7	0.819	1.60	0.013	+0.001	18.9	
(3757)		1982 XB	19.0	3.9	1.016	2.65	0.025	+0.001	22.2	
		1983 LC	19.0	1.5	0.766	4.50	0.023	+0.006	25.0	19
(3200)	Phaethon	1983 TB	14.6	22.1	0.140	2.40	0.026	−0.010	17.2	
(3671)	Dionysus	1984 KD	16.3	13.6	1.003	3.39	0.034D	+0.001	20.8	
(3362)	Khufu	1984 QA	18.1	9.9	0.526	1.45	0.018	−0.001	17.2	
		1986 JK	19.0	2.1	0.893	4.70	0.006	0.000	25.2	179
(4034)		1986 PA	18.1	11.2	0.589	1.53	0.023	+0.002	17.6	
(4486)	Mithra	1987 SB	15.4	3.0	0.743	3.66	0.045	+0.001	20.3	
(4450)	Pan	1987 SY	17.1	5.5	0.596	2.29	0.027	+0.004	19.5	
(6037)		1988 EG	18.7	3.5	0.635	1.90	0.024	−0.001	19.9	
		1988 XB	17.5	3.1	0.761	2.17	0.005	−0.001	19.5	
(4179)	Toutatis	1989 AC	14.0	0.5	0.919	4.11	0.006	+0.004	19.5	
		1989 DA	18.0	6.4	0.988	3.34	0.049	−0.004	22.5	89
		1989 JA	17.0	15.2	0.913	2.63	0.041	+0.008	20.2	237

—*continued*

TABLE 1. PHAs

Object		H	i	q	Q	m	Δ_m	V_Q	Arc
(4769) Castalia	1989 PB	16.9	8.9	0.549	1.58	0.023	+0.002	16.7	
(6239) Minos	1989 QF	18.1	3.9	0.676	1.63	0.028	–0.001	18.2	
	1989 UQ	19.0	1.3	0.673	1.16	0.015	+0.001	15.3	
	1989 UR	18.0	10.3	0.695	1.47	0.031	+0.002	17.2	34
	1990 HA	17.0	3.9	0.790	4.36	0.010	–0.009	22.8	212
	1990 MF	18.5	1.9	0.951	2.54	0.018	–0.002	21.5	175
(4953)	1990 MU	14.3	24.4	0.556	2.69	0.041	–0.012	17.6	
	1990 SM	16.5	11.5	0.486	3.76	0.019		21.6	24
	1990 UA	19.5	1.0	0.770	2.67	0.012		22.7	7
(5189)	1990 UQ	17.5	3.6	0.810	2.29	0.044	+0.001	19.9	
	1991 AQ	17.0	3.2	0.495	3.95	0.021	–0.018	22.3	
	1991 CS	17.5	37.1	0.938	1.31	0.032D	0.000	15.5	60
	1991 DG	19.0	11.2	0.909	1.95	0.048	+0.003	20.3	238
	1991 EE	17.5	9.8	0.844	3.65	0.033	–0.021	22.4	209
	1991 GO	19.0	9.7	0.663	3.26	0.022		23.3	9
	1991 JW	19.5	8.7	0.916	1.16	0.036A	–0.001	15.8	
(6489)	1991 JX	19.0	2.3	1.010	4.02	0.038	–0.020	24.4	
	1991 RB	19.0	19.5	0.749	2.15	0.047		21.0	23
	1991 VH	17.0	13.9	0.973	1.30	0.013D	–0.001	15.0	66
(5604)	1992 FE	17.0	4.8	0.551	1.30	0.037	+0.001	15.0	
	1992 UY4	17.5	2.8	1.009	4.30	0.031	+0.004	23.3	155
	1993 DQ1	17.0	10.0	1.038	3.04	0.050A		21.0	28
(5693)	1993 EA	17.0	5.1	0.527	2.02	0.008	+0.002	18.6	
	1993 KH	19.0	12.8	0.850	1.62	0.008	+0.003	19.0	226
	1993 VB	19.5	5.1	0.916	2.90	0.006	+0.004	23.2	
	1994 AW1	17.5	24.1	1.022	1.19	0.036A	0.000	14.3	269
	1994 CC	18.0	4.6	0.954	2.32	0.011	–0.002	20.4	43
	1994 CN2	16.5	1.4	0.952	2.19	0.014	0.000	18.6	76
	1994 PM	17.5	18.0	0.366	2.59	0.026	+0.010	20.6	
	1994 PC1	16.0	33.5	0.905	1.79	0.016A	0.000	16.8	
	1994 RC	19.0	4.7	0.903	3.63	0.046	–0.006	23.9	33
	1994 XD	19.0	4.3	0.639	4.08	0.018		24.5	27
	1995 EK1	18.0	8.8	0.507	4.02	0.050	+0.024	23.4	41

Examination of the table shows that just half of the PHAs have been discovered during the past half-dozen years, at average intervals of somewhat more than two months. Fully three-quarters of these objects were found rather equally in just three observing programs — the AANEAS photographic program with the U.K. Schmidt at Siding Spring, the PCAS program conducted by Eleanor Helin with the Palomar small Schmidt, and the Spacewatch scanning CCD program at Kitt Peak. Understandably, it is the recent discoveries that are most likely to have been observed at a single opposition, for there may not yet have been time to make a recovery. Many of the observed arcs are long enough to ensure that eventual recovery will be possible, however, and a few of the orbit computations also have the benefit of radar observations.

Uncertainties in orbital determinations obviously affect the computed values of m for the short-arc orbits. Nevertheless, particularly when an object is relatively close to the Earth at discovery, a reliable value can sometimes be determined surprisingly quickly. Even 2-day arcs often give values of m that are correct to 0.01.

The quantity Δm is not available for some of the short-arc orbits. The median value without regard to sign is less than 0.003, confirming that the limit of 0.05 is fully sufficient to identify the objects that may threaten the Earth over the course of centuries. Fewer than 10 percent of the $|\Delta m|$ values are greater than 0.01. These essentially involve cases where $Q \gtrsim 3.5$ AU. Very specifically, they may be cases where the objects are close to being in 3:1 mean-motion resonance with Jupiter. The two objects (1915) Quetzálcoatl and (2608) Seneca are not shown in Table 1 because they currently have $m = 0.089$ and $m = 0.112$, respectively. However, an effect of the resonance has been steadily to increase q through the value 1.0 AU over the course of the century, the Δm values for these objects being +0.046 and +0.079, respectively. Clearly, some caution is required in using m as a diagnostic for PHA status if an orbit might be near the 3:1 resonance.

For much the same reason, m is not a good diagnostic for PHCs ("potentially hazardous comets"), unless it can be reevaluated from osculating elements specified at frequent epochs. Nevertheless, one should note that there are currently seven comets of short and intermediate period (ranging from the 5-year period of 26P/Grigg-Skjellerup to the 130-year period of 109P/Swift-Tuttle) with $m \lesssim 0.05$, suggesting that the relative threat of active short-period comets and minor planets (which presumably include some passive, i.e., inactive, short-period comets) is around 1:10.

There is little point in computing m for long-period comets, which — by definition — have at most one opportunity to hit the Earth over the course of two centuries. Object for object, the relative threat of long-period comets over short-period objects has been estimated at perhaps 2 percent [4], although the lack of further opportunities in reality makes the relative threat very much less. Nevertheless, some authorities have remarked that the relative contribution of long-period and short-period bodies to be 10 percent and more [5]. This is evidently because they have counted it in terms of the *kinetic energies* of the collisions one wishes to avoid.

In the case of a discovery of a very distant comet, hazard assessment involves the assumption of parabolic motion and the establishment of whether q is or is not significantly greater than 1 AU. The parabolic assumption can be wrong and q difficult to determine, and there is likely to be ambiguity as to whether the comet is approaching or receding from the Sun.

4. FUTURE SEARCH STRATEGIES AND ORBIT DETERMINATION

The entries in Table 1 represent perhaps 5 percent of the total threat from PHAs, most of which have not yet been discovered. With a new discovery only every two months, it would take several centuries to find them all. Clearly, the main task is substantially to increase the sky search coverage, which for the single Spacewatch instrument is only about one-tenth that of one of the photographic programs at Palomar, even though it goes to fainter levels.

An increase in CCD sky coverage to the level of the Palomar programs was the most important recommendation of the Spaceguard Survey workshops [3]. The recommended sky coverage was to the near-opposition third of the available night sky each month, the images being obtained in such a way as to secure automatic follow-up of a substantial fraction of the objects found. It was estimated that, if such a survey were made for 25–30 years with large-enough telescopes that NEAs (or PHAs) could be detected at apparent visual magnitude $V = 22$, it would be possible to find and determine orbits for more than 90-percent of the 1 km objects and perhaps 65-percent of the 0.5 km objects that could be a hazard. There was some concern that Spaceguard as defined in this way would miss objects with their aphelia near the Earth, and thus, at best a very rare passage through opposition, that the magnitude limit would not be good enough for short-period comets, and that both sky coverage and magnitude limit would *each* have to be improved if there were to be any real hope of finding threatening long-period comets with notice of more than a month or two [6].

The pay-off between sky coverage and limiting magnitude is a useful topic for discussion. Recently, arguments have been made in favor of complete night-sky coverage to a relatively bright limit. Specifically, there has been some consideration that the ready availability of the U.S. Air Force 1-m GEODSS cameras and fast-readout CCDs allows one to detect 80 percent of the objects of interest, and in only ten years, by covering the whole sky down only to magnitude 20 [7]. Before one completely embraces this concept, there are some points to consider. The most obvious one is that not a single minor planet was ever discovered with the naked eye — admirable though the naked eye may be for complete sky coverage — yet efforts during the past few years have shown that systems capable of reaching magnitude 23 or so yield several Kuiper-Belt candidates per square degree!

Often overlooked by proponents of the GEODSS all-sky search is that it strives only to make initial detections. Confirmation of an object on a second night doubles the number of telescopes needed. Additional telescopes are then needed to secure the subsequent follow-up observations necessary for orbit determination. Since this follow-up is to be targeted toward specific discoveries, considerable logistics are required. This must be compared with the Spaceguard concept, which automatically provides its own confirmation and follow-up observations.

An important point about concentration near opposition is that any objects found can be observed for most of the night. Not only can an object's apparent motion at opposition be a fair indicator of distance (unless the orbit happens to be unduly eccentric or inclined), but there exists the possibility of refining the distance by means of parallactic observations from a single site during the course of the night. For greatest efficiency the observing could be organized so that a particular object is recorded near midnight on either the detection or confirmation night and a few hours both earlier and later on the other. Because there will likely be a greater departure from an apparent great-circle track across the sky, three sets of observations like this on two nights will generally provide more useful information about the object's orbit than midnight observations on three consecutive nights. In any case, two-point "Väisälä-orbits," computed for assumed distances and taking the object to be at perihelion or aphelion, can be at least as meaningful as three-point orbits computed by rote. By considering similar sequences of observations early and late in the dark run, or even from one month to the next, linkages can be made automatically. If the observations straddle opposition, variations in an object's brightness from one sequence to the next is minimized. In an automatic operation like this, the burden is appropriately put on the computations. Time-consuming and unproductive communications are minimized.

If there is extensive observational effort at smaller solar elongations, both the second night confirmation and the later follow-up can be significantly compromised. The time span available for observation on any one night will likely be too short to establish distance by parallax. It may not help to make further observations from a different site. If that site differs significantly in latitude, the object may be above the horizon only in the daytime sky. To observe from the same latitude but a different longitude would not help because the observations would have to be made when the object was at the same local hour angle, and the parallactic effect would be the same. Month-to-month follow-up can be rendered difficult by a probable considerable change in an object's brightness and would be impossible if its solar elongation becomes too small.

Another disadvantage of observations made far from opposition is that the orbital characteristics of an object discovered are generally less immediately apparent. There can also be two entirely distinct orbit solutions, even for solar elongations as high as 116°. Table 2 illustrates an example of this, taken from a recent study of difficult orbit solutions [8]. From observations on five nights over the course of a week, there was no way of telling whether the object, at solar elongation 62°, was a PHA with m = 0.03 in an orbit ranging from q = 0.17 to Q = 1.30 AU (Solution 1), or whether it was simply a minor planet in the Main Belt on the far side of the Sun (Solution 2). The (O–C) residuals and heliocentric and topocentric distances for the two solutions are shown. There are also three fictitious observations, denoted with asterisks, for times 0.1 day after real observations. Quite apart from the fact that the object's low altitude would have made these observations impossible, it is clear that they would not have helped resolve the double-solution problem. It turns out that, as one might statistically expect, the Main-Belt Solution 2 is the one that is physically valid, in this case because the object can in fact be identified with (834) Burnhamia.

Finally, reliable orbit solutions that allow long-term predictability require (in the absence of radar detections) the linkage of observations in different years. From Kepler's second law it follows that in most of those years the object will be in the vicinity of aphelion. It is therefore desirable to be able to detect the object at an aphelic opposition. For the known PHAs Table 1 shows a rather uniform distribution of Q in the range from 5 AU down to 3 AU, with then a rather sudden change to about twice the frequency down to 1.2 AU. Possibly, the larger values statistically refer to comets that have become inactive and the smaller values to fragments that originated in the Main Belt [9]. There is an obvious underrepresentation of objects with $Q < 1.2$ AU, but there is little indication that the concentration of observations near opposition impedes either discovery or follow-up for $Q > 1.2$ AU. At $Q = 1.2$ AU the loss of objects is perhaps 10 percent.

Now look at the column V_Q. A limiting magnitude $V = 20$ would allow only 52 percent of these representative objects to be observed at aphelion, the fraction increasing to 73 percent for limiting $V = 22$. It is not a coincidence that the earlier Spaceguard was designed to have a limit of $V = 22$. One gets a better orbital solution from optical observations when they are distributed all around the orbit, not just near perihelion when the object is bright.

The difference between limiting magnitudes of 21 and 22 is significant because it is here that there is an increased number of single-opposition objects — something that one should attempt to reduce. Limiting magnitude 24 would be even better, reaching 92 percent of the PHAs at aphelion, as well as half of the short-period comets. On the other hand, very far from opposition, it is fully appropriate, and confusion with unknown Main-Belt objects is minimized, if the limiting magnitude is 17.

TABLE 2. Double Solution

UT	R.A. Residuals arcsec	Decl. Residuals arcsec	Helio. Distances AU	Topoc. Distances AU	R.A. Residuals arcsec	Decl. Residuals arcsec	Helio. Distances AU	Topoc. Distances AU
1966 November								
9.024	−0.4	+0.4	1.165	1.182	−0.2	−0.3	2.541	2.812
9.124*	0.0	0.0			−0.4	+0.1		
13.024	−0.1	−0.1	1.138	1.201	+0.3	0.0	2.547	2.862
13.124*	+0.2	−0.5			+0.1	+0.3		
14.017	+0.5	+0.8			+0.9	+0.7		
15.010	−1.6	+1.6			−1.2	+1.1		
16.009	0.0	−0.2	1.116	1.212	+0.2	−1.4	2.551	2.899
16.109*	+0.4	−0.4			−0.1	−0.9		
Elements for Epoch 1966 Nov. 11.0:								
ω, Ω, i	234.66	270.82	3.07		89.76	184.46	3.93	
M, e, a	252.94	0.7706	0.7361		23.79	0.2244	3.1556	

5. CURRENT ASTROMETRIC ACTIVITY

Table 3 summarizes the total astrometric activity on comets and minor planets during the past twelve months. A total of 72,438 observations were published in the *Minor Planet Circulars*, referring to some 8400 minor planets — about 100 of them "unusual" — and 56 comets. Some 6 percent of the observations were of unusual minor planets, and 8 percent were of comets (notably D/1993 F2). The observations were made at 126 different observatories in 24 different nations, although only seven observatories were responsible for fully 70 percent of the observations, and observatories in just six nations made 92 percent of the observations. By far the dominant contributor was Spacewatch, with some 30 percent of the observations; furthermore, these observations should be supplemented with some 61,700 more single-night Spacewatch observations that have been filed but not published. It is also significant that 65 percent of the observatories were owned and operated by amateur astronomers, who were collectively responsible for making 26 percent of the observations. Most of these amateurs are in Japan and Italy, the third and fourth most productive nations for observations in the past year, and amateurs were in fact responsible for almost all the observations made in those two nations.

TABLE 3. Relative Contribution of Observations

Percentage by observatory:	
Spacewatch	30
ESO	19
Oizumi (amateur)	5
Oak Ridge	4
Klet	4
Siding Spring	4
Palomar	4
Others	30
Percentage by nation:	
United States	42*
Chile**	19
Japan	16
Italy	6
Czech Republic	5
Australia	4
Others***	8

*Arizona 32, Massachusetts 5, California 4.
**European Southern Observatory.
***Germany, France, Canada, Austria, Spain, United Kingdom, New Zealand, Russia, Slovakia, India, Sweden, Switzerland, China, Belgium, Argentina, Poland, Slovenia, Tajikistan.

If one includes the single-night Spacewatch data, he can see that observations were being made by an observer somewhere in the world and later processed by the Minor Planet Center at an average rate of one every four minutes. If Spaceguard comes to fruition, it has been estimated that this will be the approximate rate at which observations are made of NEAs, the total rate at which observations are made being perhaps 100 times greater, or one every 2 seconds. Thanks to the recent acquisition of new computers and the availability of a very small but dedicated staff, the Minor Planet Center is currently processing observations more completely and efficiently than ever before. As noted, that processing includes preparation for the *printed* publication of all but the single-night observations. If this printing and much of the back-and-forth discussion with observers about errors in their data were substantially curtailed, the throughput could probably be improved to one observation every 20 seconds, without even the addition of another staff member. For the moment, however, it should be noted that *ten times* (i.e., 240) as many subscribers want to receive the printed *Minor Planet Circulars* as want the electronic version. To reach the estimated 2-second interval between observations would clearly require some redeployment and augmentation of staff and computing capability, the amount of activity on communication alone depending critically on whether or not follow-up observations are made in a targeted manner.

6. IMPACT HAZARD SCALE

In order to guard against those unfortunate fiascos in the press that arise from time to time as to whether a particular object may or may not collide with the Earth, an "impact hazard scale" has been proposed [10] for the characterization of the probability that a particular NEO will hit the Earth around a particular date. The scale is logarithmic, with values ranging from 0 for "no chance of collision" to 5 for "certain collision," and a second parameter characterizes a possible impact in terms of kinetic energy.

In practice, of course, except perhaps for a small but finite time following its discovery, almost every object will score 0 all the time, and if that is the case, the second parameter is irrelevant. Furthermore, the specification of an impact probability depends heavily on the ground rules adopted for the orbit computation [11]. For example, it makes a difference whether or not one assumes the orbit of a newly-discovered comet to be a parabola; chances are that the orbit is close to parabolic, but a general solution can give a very different result. Then there is the possibility that some solutions would send the object close to a planet — major or minor — that could perturb it in a significant but unknowable way. Likewise, the calculation can depend on the reliability or otherwise of a critical single observation, something that cannot be independently assessed.

The absence of parallax between early observations from Australia and Japan immediately made it clear that comet C/1995 O1 (Hale-Bopp) was then more than 2 AU from the Earth. There was a smattering of interest in the press, some of it even hinting at a hit, but it was surely more productive actually to continue to make observations and to refine the orbit computation than to attempt to compute impact probabilities. After a day or two, the good continuing supply of observations of the bright object made it possible to discount that it was some 3–4 AU away, in a low-inclination elliptical orbit typical of a short-period comet. Assumption of a parabolic orbit increased the distance to beyond 5 AU, but for several days it was unclear whether the distance was 5 AU, 9 AU, or in between, smaller values indicating $q < 1$ AU and larger values $q \gg 1$ AU. Furthermore, to assume $q < 1$ AU gave an uncertainty of several months in perihelion time (all in the future), some values indicating close approaches to the Earth. Coincidentally, there next seemed to be the possibility of a close approach to Jupiter, a circumstance that could alter subsequent developments in a highly unpredictable manner. The possibility of a Jupiter approach then receded, and by the time the observations covered some eight days, it was clear that the orbit was at least almost parabolic, that $q = 0.9$ AU and that the minimum distance from the Earth (on this pass) would be 1.3 AU. From the hazard point of view, it did not even matter that an image of the comet could then be recognized on a single photographic plate taken two years earlier — although that observation was crucial in attempting to show that the orbit did in fact differ from a parabola. But the most relevant point about this comet is that it should have been bright enough to be *discovered visually by amateur astronomers* only 22° from opposition. Despite the interest — even obsession — nowadays with NEO searches and the argument about whether or not to cover the whole sky, the fact is that the curtailment of the Palomar photographic programs and failure so far to produce a replacement means that NEO surveillance is worse in 1995 than at any time during the past 20 years or more!

It is interesting to note that C/1995 O1, the nucleus of which appears to be the largest body (perhaps 40 km across) to approach the orbit of the Earth in two centuries (its immediate predecessors being the great comets of 1811 and 1577), happens to have its orbital nodes near the orbits of the Earth and Jupiter. C/1995 O1 comes to perihelion on 1997 April 1, some 4200 years after its previous approach. If it had come just four months earlier, the comet would have passed only 0.10 AU from the Earth and been a truly awe-inspiring sight. If it had come just four months later, the comet would have met that earlier speculation and passed extremely close to Jupiter in June 1996, with unpredictable results. Presumably, there was such a Jovian encounter as it approached the Sun on an earlier trip — perhaps even its last trip.

Neverthless, the principal threat to the Earth is clearly from objects in short-period orbits. Over the course of a couple of centuries, such an object may have several opportunities to strike the Earth, but the orbital inaccuracy can be such that one does not know *when* the Earth is actually vulnerable. Yet there are also many objects, even NEAs, that cannot conceivably be a threat during this time, because their orbits come nowhere near that of the Earth. The important thing is to establish, as soon as possible after discovery, whether or not a specific NEA should be added to the list of PHAs. Unless the resources for support were to become much more widespread, further studies (astrometric and physical observations, detailed orbital calculations) can then be concentrated on the PHAs and the small fraction of comets that might conceivably be a threat.

REFERENCES

1. Jewitt, D. C. & J. Luu. 1995. The solar system beyond Neptune. Astron. J. **109:** (6) 1867– 1876.
2. Cochran, A., H. Levison, S.A. Stern, & M.J. Duncan. 1995. The discovery of Halleysized Kuiper belt objects using HST. Astrophys. J. **455:** (1) 342–346.

3. MORRISON, D. Ed. 1992. The Spaceguard Survey. National Aeronautics and Space Administration. Washington, DC.
4. MARSDEN, B.G. 1993. To hit or not to hit. *In* Proc. Near-Earth Object Interception Workshop, G.H. Canavan, J. Solem & J.D. G. Rather, Eds.: 67–71. Los Alamos National Laboratory. Los Alamos, New Mexico.
5. SHOEMAKER, E.M., R.F. WOLFE, & C.S. SHOEMAKER. 1990. Asteroid and comet flux in the neighborhood of Earth. *In* Global Catastrophes in Earth History, V.L. Sharpton & P. D. Ward, Eds.: 155–170. Geological Society of America, Boulder, Colorado.
6. MARSDEN, B.G. & D.I. STEEL. 1994. Warning times and impact probabilities for long period comets. *In* Hazards due to Comets and Asteroids, T. Gehrels, Ed.: 221–239. University of Arizona Press. Tucson, Arizona.
7. SHOEMAKER, E.M., ED. 1995. Report of the near-Earth object survey working group. National Aeronautics and Space Administration. Washington, DC.
8. MARSDEN, B.G. 1991. The computation of orbits in indeterminate and uncertain cases. Astron. J. **102:** (4) 1539–1552.
9. MARSDEN, B.G. 1972. Evolution of comets into asteroids? In Physical Studies of Minor Planets, T. Gehrels, Ed.: 413–421. National Aeronautics and Space Administration. Washington, DC.
10. BINZEL, R.P. 1995. An impact hazard scale for near-Earth objects. *In* Proceedings of the United Nations International Conference on Near-Earth Objects, this volume. Annals of the New York Academy of Sciences. New York.
11. MARSDEN, B.G. 1995. Orbits of near-Earth objects: the irrelevance of probability. *In* Erice seminar: The Collision of an Asteroid or Comet with the Earth. R. Coppola, Ed.: In press. Saginaw Valley State University, University Center, Michigan.

Long-period Comets and the Oort Cloud

PAUL R. WEISSMAN[a,b]

Earth and Space Sciences Division
Jet Propulsion Laboratory, MS 183-601
4800 Oak Grove Drive
Pasadena, California 91109

ABSTRACT: The long-period comets pose a unique problem for the impact hazard problem. Because of their very long orbital periods and generally large distances from the Sun, they cannot be surveyed and catalogued in the same manner as the near-Earth asteroids and short-period comets. They appear at random, uniformly distributed on the celestial sphere. Current technologies can detect long-period comets at distances of ~5 AU, giving somewhat less than a one year warning time for potential Earth impactors. The mean impact probability for a long-period comet crossing the Earth's orbit is 2.2 to 2.5×10^{-9} per perihelion passage. The mean impact velocity is ~52 km sec^{-1}, but the most probable impact energy is characterized by a velocity of 56 to 58 km sec^{-1}. The estimated current impact rate for cometary nuclei large enough to create 10 km diameter (or larger) craters on the Earth is between 5×10^{-7} and 2.8×10^{-6} per year, with a best estimated value of 1.0×10^{-6} per year. Nuclei large enough to initiate global climatic disturbances strike the Earth, on average, every 16 Myr. The impact frequency may be increased substantially for brief periods of time during cometary showers, initiated by major perturbations of the Oort cloud. Improved technologies are needed to detect approaching long-period comets at large heliocentric distances so as to increase the warning time for potential impactors.

INTRODUCTION

The long-period comets are the least predictable component of the flux of small bodies in Earth-crossing orbits. Their orbital periods, typically 10^3 to 10^7 years, are in general so long that they have not been observed on previous passages through the planetary system. Although human societies have recorded appearances of comets for over 2,000 years [22], [66], [42], most of the older observations are of the few comets that were bright enough or came close enough to the Earth to be seen by naked-eye observers, and the positions are not sufficiently accurate to determine precise orbits. Current observational programs are not designed to search for long-period comets; a significant fraction of current discoveries are still made visually by amateur astronomers. At present, approaching long-period comets are rarely found at

[a]This work was supported by the NASA Planetary Geology and Geophysics Program. The research reported herein was performed at the Jet Propulsion Laboratory under contract with the National Aeronautics and Space Administration.

[b]E-mail: pweissman@issac.jpl.nasa.gov.

distances beyond the orbit of Jupiter, giving at best a one year warning time for Earth-crossing objects.

The long-period comets thus provide a significant dilemma for those concerned with the hazard posed by comet and asteroid impacts on the Earth. Although the majority of Earth impacts will come from asteroids and defunct short-period comets [63], [77], [64], the long-period comets still contribute between 10 and 30% of the total impactor flux (measured in terms of impact energy), and thus must be considered as a significant component of the impact hazard.

This paper is designed to serve as an introduction to the topic of long-period comets, in particular for individuals who are not planetary astronomers. Section 2 describes the existing observational record for the long-period comets. Section 3 discusses the Oort cloud, which is the source of the long-period comets. Section 4 reviews the dynamics of long-period comets and their residence lifetimes in the planetary system. Section 5 discusses possible temporal variations in the cometary flux. Section 6 describes estimates of the impact probability of long-period comets on the Earth, and the expected flux of impacting objects. Section 7 summarizes the material in this paper and points out areas for future work.

The formal definition of a long-period comet is a comet with an orbital period of 200 years or more. This definition is somewhat arbitrary and is based on the fact that the good observational positions needed to determine accurate orbits exist only over the last 200–300 years, corresponding in large part with the invention of the telescope. Despite the fact that the actual observational record extends over two millennia, no verified return of a long-period comet has ever been identified. The orbital periods of most long-period comets are well in excess of 10^3 years.

More detailed reviews on the dynamics of long-period comets and the Oort cloud can be found in Marsden and Roemer [37], Weissman [78], and Fernández [17]. A good review on the origin of comets is that by Mumma et al. [46], and useful reviews on short-period comets are by Weissman and Campins [81] and Shoemaker et al. [64].

Long-period comets are usually named after their discoverer(s). Official catalog designations give the year the comet was discovered and a letter-number code that designates the approximate time of year and order in which each comet was found. For convenience in the following text, long- and short-period comets will be referred to as LP and SP comets, respectively.

THE OBSERVATIONAL RECORD

The most recent Catalogue of Cometary Orbits [42] contains 698 orbits of long-period comets observed through the end of 1995. The earliest record is for comet C/146 P1, observed in 147 B.C., one of many LP comets recovered from ancient Japanese and Chinese records [22]. Only 67 of the orbits are for comets observed earlier than 1610, prior to the invention of the telescope. All of those are parabolic orbits only; no orbital period can be assigned. Because of the diffuse nature of cometary comae, pre-telescopic discoveries required the objects to reach a visual magnitude between 3.5 and 4.0, considerably brighter than the limit of $m_V = 6$ for objects that are stellar in appearance, such as supernovae [66], [89], [90].

Seventy-three orbits in the catalog are for LP comets observed between 1610 and 1800, with only 3 of those having non-parabolic elements. The quality of orbits gets progressively better with time. For the 202 LP comets which were discovered in the 19th century, 86 have non-parabolic orbits, or 43% of the total. For the 20th century the numbers are 262 non-parabolic orbits out of a total of 356 LP comets discovered, or 74%.

Of the 698 LP comets in the catalog, 411 have perihelion distances less than 1.0 AU and are thus in Earth-crossing orbits (an additional 10 LP comets have perihelia between 1.0 and 1.0167 AU, the Earth's aphelion distance). Observational selection effects bias discovery statistics to comets with perihelia near the Earth's orbit [11]. Prior to 1900, 245 of 333 discovered LP comets or 74% were in Earth-crossing orbits (not including 5 Sun-grazing comets, most of which are fragments of a single disrupted comet: see below). Since 1900, only 142 of 341 observed LP comets, or 42% are Earth-crossing (again, not counting 19 Sun-grazers), due to the increasing number of photographic and CCD discoveries of LP comets with larger perihelion distances.

Nine of the LP comet orbits in the catalog may be short-period comets, but were too poorly observed to obtain a non-parabolic orbital solution. Short-period orbits are characterized by low inclinations and perihelion directions near the ecliptic. Three of these nine comets were pre-telescopic discoveries, while three have been found since 1900. Six of these objects were in Earth-crossing orbits, so they may present a hazard to the Earth.

The current discovery rate for LP comets, averaged over the 10-year period 1985–1994 is 7 comets per year (excluding Sun-grazing comets; see below). In the same 10-year period an average of 2.4 discovered LP comets per year had Earth-crossing orbits, i.e., they had perihelion distances $\leq$ 1 AU. However, many LP comets are likely missed due to observational selection effects. For example, comets can be easily missed if they pass perihelion on the opposite side of the Sun from the Earth. Everhart [11], [12] studied observational selection effects in detail and estimated that ~8,000 LP comets had passed perihelion within 4 AU of the Sun over a 127-year period. Everhart also derived corrected absolute magnitude and perihelion distributions for the LP comets; the perihelion distribution is shown in Fig. 1. The number of Earth-crossing comets increases linearly with increasing perihelion distance between 0 and 1 AU. Using Everhart's estimates, one finds that there are ~10 Earth-crossing LP comets per year brighter than absolute magnitude, H_{10} = 11.[1] Comparing this estimate with the observed statistics above, it appears that three-fourths of all Earth-crossing LP comets are currently missed.

The orbits of the LP comets are randomly oriented on the celestial sphere. This is reflected in the orbit element distributions. For random orientations, both the argument of perihelion and the longitude of the ascending node are expected to be uniformly distributed between 0 and 360°, while the distribution of inclinations is such that the number of comets at any inclination, i, is proportional to sin(i). Although the observed distributions do not match the expected distributions perfectly, Everhart [11] showed that the differences between them could readily be accounted for by observational selection effects. The most notable selection effect is caused by the preponderance of northern hemisphere observers, which leads to an excess of comets being discovered if they pass perihelion north of the celestial equator. A second notable effect is an excess in the number of retrograde comets discovered, because their retrograde motion is more likely to bring them close to the Earth during the limited time when they are bright enough to be discovered. The preponderance of northern hemisphere observers also results in seasonal variations in the discovery rate of comets, fewer comets being found in northern hemisphere winter when inclement weather and cold are likely to keep some observers away from telescopes.

Some researchers have identified possible asymmetries in the orbital element distributions of the LP comets, which they associate with particular theories of the origin of comets. Most notable of these is an alleged clustering of perihelia near the solar apex (e.g., [69], [4]). However, Neslušan [47] showed that the asymmetry and apex clustering is largely a result of the north-south asymmetry in observers. It has also been proposed that a contribution to the asymmetries may come from the most recent stellar perturbation of comets in the Oort cloud ([3]; see Section 3).

A special case of Earth-crossing LP comets is the Sun-grazing comets [34], [35]. These 24 observed comets have perihelia < 0.01 AU, within one solar radius of the Sun's photosphere. Thus, they are within the Roche limit of the Sun where tidal forces can disrupt the nucleus of a comet. All but one of the 24 Sun-grazing comets have very similar orbital elements, though some orbits are assumed because of limited observations. They are collec-

[1]The absolute magnitude of a comet, H_{10}, is the total brightness a comet would have, with coma, if it was 1 AU from both the Sun and Earth, and its brightness varied as $1/r^2$ versus geocentric distance, and $1/r^4$ versus heliocentric distance. This brightness law generally works reasonably well for LP comets when they are less than ~3 AU from the Sun. However, the heliocentric exponent can vary between ~2.5 and 6 or more, generally being less for dynamically new comets, and larger for older, more physically evolved comets. Based on modeling by Weissman [77], an H_{10} value of 11 corresponds to a nucleus mass of 4×10^{15} g, or a nucleus radius of 1.2 km assuming a bulk density of 0.6 g cm^{-3}.

tively referred to as the Kreutz group. The similarity of the orbits of the Kreutz group suggest that they are fragments of a larger parent comet that disrupted on one or more previous returns. In recent years, discovery of Sun-grazing comets has been greatly supplemented by spaceborne instruments, including the SOLWIND experiment and the Solar Maximum Mission, and some fragments have been observed to impact the Sun. In addition to tidal forces, the intense heating of the nucleus and the resulting thermal stresses probably play a role in the disruption of these nuclei.

The Kreutz group comets have semimajor axes on the order of 60–100 AU, yielding orbital periods between ~500 and 10^3 years. However, the orbits of the Kreutz group are inclined ~139–144° to the ecliptic and are oriented such that the fragments cannot make close approaches to the Earth. Repeated passages within the Sun's Roche limit will likely destroy these objects before their orbits can evolve very far. Thus, they likely play no significant role in the impact hazard on the Earth.

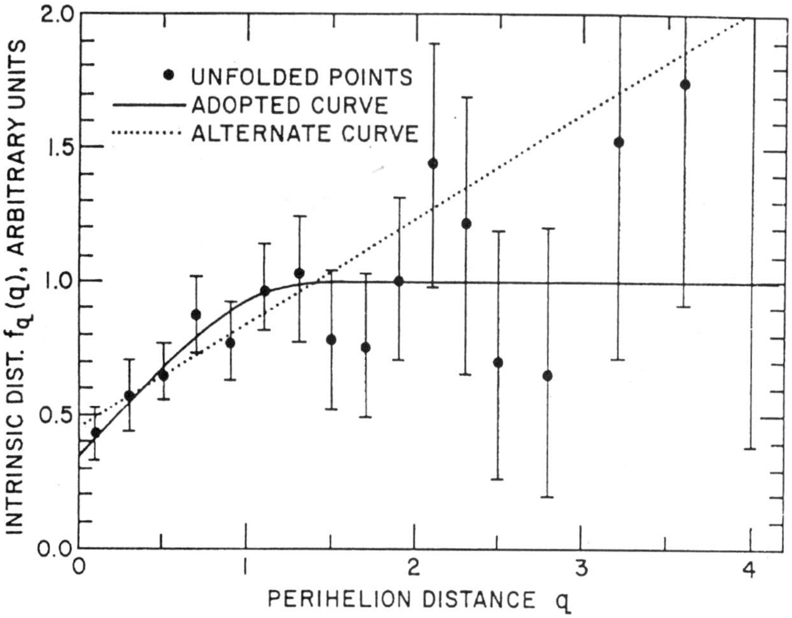

FIGURE 1. The perihelion distribution of the long-period comets, corrected for observational selection effects, by Everhart [12]. The distribution is not well determined for $q > 1$ AU. However, Fernández [15] and Weissman [75] showed that the number of comets likely continues to increase with increasing heliocentric distance.

THE OORT CLOUD

About two-fifths of observed LP comets passing through the planetary region (for which non-parabolic orbit solutions are possible) are on osculating (instantaneous) hyperbolic orbits. Early on, this led to the suggestion that the LP comets came from interstellar space. However, integration of the orbits backward in time to a point before the comets entered the planetary system, and correction to a barycentric rather than a heliocentric coordinate system, showed that virtually all of the orbits were originally elliptical. Thus, the LP comets are members of the solar system, though their semimajor axes in many cases are quite large, ~10^4 AU or more (corresponding to orbital periods ≥ 10^6 years).

A useful way to display the semimajor axis distribution of the LP comets is to use the inverse of the semimajor axis of the original orbit, $1/a_o$, prior to the comets having entered the planetary system. The inverse semimajor axis is proportional to the energy of the comet's orbit. Positive values of $1/a_o$ correspond to negative energies, or bound, elliptical orbits. Negative $1/a_o$ values correspond to positive energies, or unbound, hyperbolic orbits. The distribution of $1/a_o$ for 291 non-parabolic orbits from Marsden and Williams [42] is shown in Fig. 2.

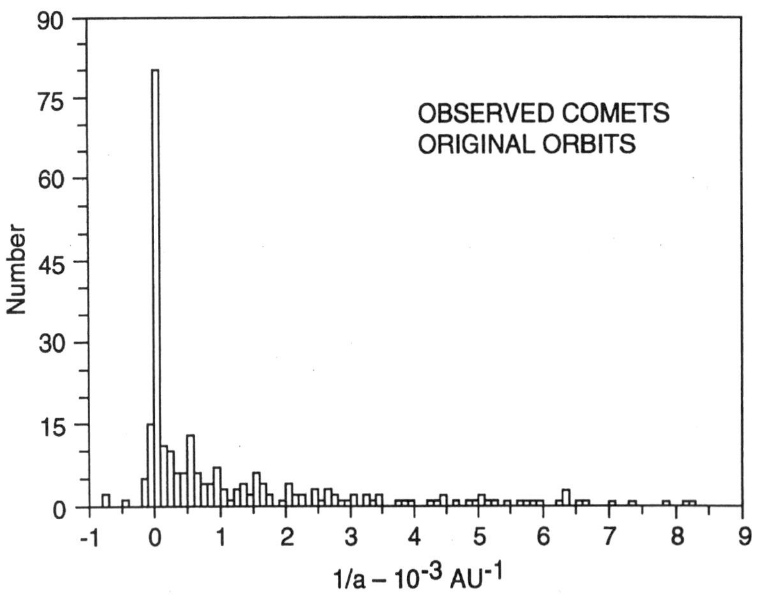

FIGURE 2. The distribution of inverse original semimajor axis, $1/a_o$, for 291 long-period comets. Data from Marsden and Williams [42]. The tall spike at the left represents long-period comets from the Oort cloud making their first perihelion passage through the planetary system.

The distribution is characterized by a large "spike" of comets at near-zero, but bound energies (with semimajor axes > 10^4 AU), and a low continuous distribution of bound orbits at increasing values of $1/a_o$. A modest number of comets have weakly hyperbolic orbits. These hyperbolic orbits are generally discounted as due to errors in the orbit determinations, and due to the effects of nongravitational forces from cometary jetting (see below), which tend to make the orbits appear more eccentric than they actually are.

The large spike of comets at near-zero energies does not appear in the distribution of osculating orbits of comets passing through the planetary region, or in the distribution of future orbits of comets integrated forward in time until they have passed well beyond the gravitational perturbations of the individual planets. Oort [49] recognized that the spike in the original energy distribution represents the source of the long-period comets, a vast spherical cloud of comets surrounding the planetary system at very large heliocentric distances, ~50,000–150,000 AU, but still gravitationally bound to the Sun. Most of the comets in the "spike" are making their first perihelion pass through the planetary region. Oort showed that on subsequent returns, planetary perturbations scatter the comets in orbital energy, allowing them to diffuse in $1/a_o$ and forming the low continuous distribution in Fig. 2.

Oort also showed that the orbits of comets in the cometary cloud were so weakly bound to the Sun that they were perturbed by random passing stars. On the order of 6–10 stars pass within 1 parsec of the Sun every million years [72], [14]. Stellar perturbations cause the perihelia of the very eccentric cometary orbits in the cloud to random walk back into the planetary region where they can be observed as "dynamically new" long-period comets. Once a comet enters the planetary region, its motion tends to be dominated by distant planetary perturbations, primarily by Jupiter, causing the orbits to random walk in orbital energy, $1/a_o$, and thus diffusing comets to larger or smaller semimajor axes.

The typical Jupiter energy perturbation is $\pm 630 \times 10^{-6}$ AU^{-1}, or more than six times the width of the spike in Fig. 2. Thus, comets rapidly diffuse away from the "Oort cloud" to either more tightly bound, shorter period orbits, or to unbound, hyperbolic orbits where they are ejected to interstellar space. About one-half of the "dynamically new" comets from the Oort cloud, that is, comets making their first perihelion passage through the planetary region, are ejected to hyperbolic orbits and do not return. The other half are scattered to smaller semimajor axes where they return some time in the future, to be scattered in orbital energy once again. A small fraction, ~5%, are returned to orbits with semimajor axes in the Oort cloud. The motion of the comets is thus a random walk in $1/a_o$, diffusing both to the left and to the right in Fig. 2, but being lost if they achieve negative values of $1/a_o$.

The few "hyperbolic" comets in Fig. 2 are most likely the result of small errors in their orbit determinations, or unmodeled nongravitational forces resulting from jetting of volatiles on the surfaces of the cometary nuclei (which make the orbits appear more eccentric than they actually are, as shown by Marsden et al. [39]). Since the Sun and solar system are moving with a velocity of ~16.5 km sec^{-1} with respect to the local stars, interstellar comets entering the planetary system would be expected to have comparable hyperbolic excess velocities. The largest hyperbolic excess velocity for any of the comets represented in Fig. 2 is 0.81 km sec^{-1} for comet Sato, C/1975 X1. That comet had a perihelion distance of 0.864 AU and the errors in its orbit could easily be due to unmodeled nongravitational forces, which were not included in its orbit solution. Kresák [31] pointed out other problems with the orbit solution, including the very short interval over which comet 1975 X1 was observed (less than 2 months), the lack of a central condensation in the coma, and the intrinsic faintness of the coma ($H_{10} = 11.2$). Similar problems exist with many of the other hyperbolic original orbit solutions. Thus, it is highly doubtful that any of the observed LP comets are on truly hyperbolic orbits.

Oort proposed that the cloud of comets had been created by material dynamically ejected from the planetary region by the growing proto-planets. Oort showed that as objects were driven to larger and larger orbits by close encounters with the planets, achieving aphelia of $\sim 2 \times 10^4$ AU or more, perturbations by random passing stars would become significant. The stellar perturbations have two effects. First, they cause the perihelion distances of the comets to random walk, with many perihelia increasing to distances greater than the semimajor axes of the giant planets, effectively detaching the cometary orbits from the planetary region. Second, stellar perturbations randomize the inclinations of the cometary orbits, which are initially close to the ecliptic plane, creating a spherical Oort cloud.

Once the comet orbits become detached from the planetary region, they continue to evolve under the influence of random stellar perturbations. In addition, giant molecular clouds (GMC's) in the galaxy, unknown at the time of Oort's work, serve as a second major perturber of the comet cloud. GMC encounters are much less frequent than stellar encounters, occurring perhaps once every 3×10^8 years, but can be quite significant because of the huge mass of interstellar material in the GMC's [27].

Repeated stellar and GMC perturbations cause the orbits of the comets to diffuse in velocity phase space, slowly increasing the energy and angular momentum of the Oort cloud comets and causing the cometary perihelia to diffuse away from the planetary region. However, a third important perturbation comes from the integrated mass of the galactic disk, known as the galactic tide, causing the perihelia and inclinations of the orbits to oscillate [6], [24]. The galactic tide can thus drive the perihelia of orbits out of, or back into the planetary region. Stellar and GMC perturbations also contribute to the random walk of the perihelia back into the planetary region, though the galactic tide appears to dominate the process.

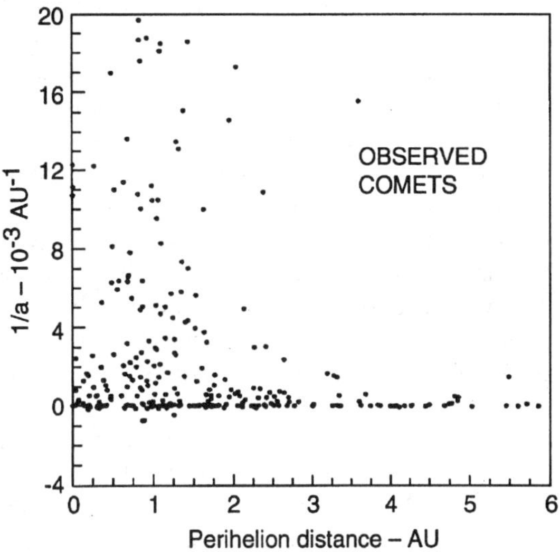

FIGURE 3. Scatter diagram in perihelion distance and inverse original semimajor axis for the observed long-period comets. The horizontal band of comets at near-zero $1/a_o$ are comets making their first perihelion passage from the Oort cloud. Comets diffuse up and down in the diagram as a result of planetary perturbations, primarily by Jupiter. Data from Marsden and Williams [42].

Although Oort suggested that the original source of the comets was the asteroid belt, Kuiper [32] pointed out that the icy nature of comets required that they had formed farther out from the Sun, among the giant planets, where it was cold enough for volatile ices to condense. Subsequent work by Safronov [55] showed that comets formed in the Jupiter-Saturn zone tended to be ejected to interstellar space by these massive giant planets, rather than placed on distant ellipses which could be captured into the Oort cloud. However, Safronov also showed that the smaller masses of Uranus and Neptune resulted in fewer ejections and more efficient capture into distant orbits. Thus, the Uranus-Neptune zone is generally recognized as the source of the Oort cloud comets, though some contribution from the Jupiter and Saturn zones is possible.

The dynamical evolution of LP comets from the Oort cloud is illustrated by the scatter diagram in Fig. 3, which plots each observed LP comet as a function of its perihelion distance and inverse original semimajor axis (omitting those comets for which only parabolic orbital solutions are available). The dynamically new, Oort cloud comets appear as a horizontal band of objects near zero $1/a_o$. As comets are perturbed by the giant planets they random walk vertically in the figure (planetary perturbations on the perihelion distance, q, are typically small unless a very close approach occurs). However, if a comet random walks to a negative value of $1/a_o$, it escapes the solar system and does not return.

An important observational selection effect is illustrated in Fig. 3. Although Oort cloud comets are quite bright and have been discovered with perihelion distances up to 6 AU and more, returning LP comets (ones with $1/a_o > 10^{-4}$ AU^{-1}) are rarely discovered unless their perihelion distances are less than 3 AU (the recently discovered comet Hale-Bopp, C/1995 O1, which passed perihelion on April 1, 1997 is a notable exception to this rule). Three AU is the distance at which water ice begins to sublimate at a sufficient rate to produce an easily visible coma. There is no dynamical mechanism which will selectively remove returning LP comets with perihelia > 3 AU, so the effect must be one of observability, as first noted by Marsden and Sekanina [38].

Several explanations have been proposed for the anomalous brightness of dynamically new LP comets from the Oort cloud. Whipple [86] proposed that the comets were covered with a “frosting” of highly volatile interstellar molecules, accreted during the comets’ long residence time in the Oort cloud. However, Stern [67] showed that microcratering by interstellar dust grains will likely erode away this frosting faster than it can form. In addition, galactic cosmic rays and solar protons [29] will sputter this volatile material away.

A more plausible explanation of the anomalous brightening of new comets involves the conversion of amorphous water ice in the cometary nuclei to crystalline ice, which occurs as the nucleus surface is warmed to temperatures between 110 and 150 K [52]. This occurs at about 5–8 AU as the comet is inbound to the Sun for the first time. If comets formed cold in the outer solar system, at temperatures < 60 K, then the water ice in the nuclei was initially in an amorphous state. The conversion from amorphous to crystalline ice is an exothermic reaction, so it is somewhat self-sustaining, though it eventually dies out because of the presence of other, nonvolatile material in the comet’s icy-conglomerate matrix. Typically, a layer several meters thick on the nucleus surface is converted to crystalline ice, and cometary materials may be blown off the surface. In addition, volatile gases such as CO in the near surface layers of the nucleus will be liberated during the conversion process, and will carry entrained dust off the nucleus surface, leading to a visible cometary coma. This greatly increases the brightness of the comet over that of its bare nucleus, and thus increases the chance of discovery by terrestrial observers.

Oort estimated a population for the cometary cloud of 1.9×10^{11} comets. However, using Everhart’s [12] estimates of the flux of LP comets corrected for observational selection effects, and more detailed modeling of the Oort cloud by Weissman [75] and Heisler [23], the population is now estimated to be ~10^{12} comets with $H_{10} < 11$ ([78]; $H_{10} < 11$ corresponds to nuclei with radii > 1.2 km; see below).

Oort proposed that there would be no comets in the comet cloud with orbital semimajor axes less than ~10^4 AU, because of the declining effect of stellar perturbations on orbits that close to the Sun. However, the recognition that the galactic tidal field also plays a role in perturbing the perihelia of distant comets caused a reappraisal of Oort's conclusion. Duncan et al. [10] modeled the dynamical evolution of comets ejected out of the proto-planetary region under the influence of planetary, galactic, and stellar perturbations. They showed that comets with semimajor axes as small as ~3×10^3 AU could have their perihelia raised out of the planetary region by the galactic tide. These comets would be captured into the Oort cloud and would continue to diffuse in velocity phase space, slowly increasing their orbital energies as well as randomizing their orbital inclinations. Duncan et al. showed that the inner Oort cloud retains some flattening for comets with semimajor axes less than ~6×10^3 AU, but that beyond that distance the orbital inclinations of the comets are essentially random.

Duncan et al. [10] estimated that the population of the inner Oort cloud was about five times that of the outer Oort cloud, or about 5×10^{12} comets, given the Oort cloud population estimates above. Taking Weissman's [77] estimate of an average cometary nucleus mass of 3.8×10^{16} g, the total mass in the inner and outer Oort clouds would be ~2.3×10^{29} g, or 38 Earth masses. A hypothetical illustration of the Oort cloud, using the orbital energy distribution found by Duncan et al., is shown in Fig. 4.

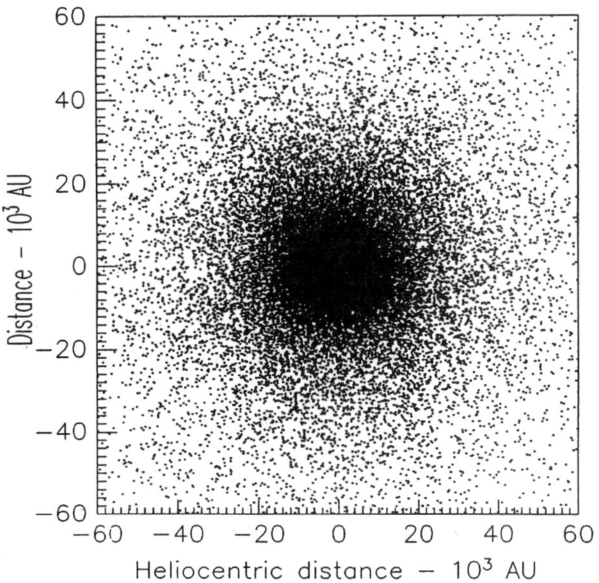

FIGURE 4. Hypothetical view of the Oort cloud of comets surrounding the solar system, based on the orbital energy distribution found by Duncan et al. [10].

PHYSICAL AND DYNAMICAL EVOLUTION OF LONG-PERIOD COMETS

Dynamical simulations by Weissman [70] and Fernández [16] showed that the average LP comet makes five perihelion passages through the planetary system before being removed by one of several physical and dynamical processes. About 65% of all LP comets are ejected to interstellar space or to semimajor axes beyond the Sun's sphere of influence ($a > 10^5$ AU). Another 8% of all LP comets are returned to orbits with aphelion distances in the Oort cloud, where stellar and galactic perturbations can once again raise their perihelion distances out of the planetary region. About 1% of LP comets are perturbed to perihelion distances within one solar radius of the Sun, where they will likely be tidally disrupted, as has been observed for the Kreutz group of comets. The mean lifetime of a LP comet is $\sim 6 \times 10^5$ years following its first perihelion passage [70].

Additionally, 27% of LP comets are lost to random disruption, i.e., splitting of nuclei. Random disruption is a poorly understood process. Although likely associated with heating of the nuclei during their perihelion passages, disruption events do not show a strong correlation with time of perihelion, passage through the ecliptic plane, or any other easily recognized element of the comets' orbits. Based on observed events, Weissman [73] estimated that about 10% of dynamically new LP comets from the Oort cloud randomly disrupted, while only 4% of returning LP comets broke up, and only 1% of short-period comets, per perihelion passage. These statistics suggest some sort of binding mechanism or gradation of cometary strengths, such that weak comets break up rather soon after first appearing, whereas stronger comets continue to return and can survive the many returns necessary to evolve to short-period orbits.

This scenario is complicated by the fact that most splitting events do not appear to involve disruption of the entire nucleus, but rather the breaking off of small fragments from a main nucleus [57]. Although the minor fragments typically fade and disappear in a few weeks to months, the main nucleus usually persists, and in the case of SP comets, may continue to make regular returns. One famous exception to this was comet Biela, an Earth-crossing SP comet which was first observed in 1772. In 1846 it was observed to have split into two comets. Both comets returned in 1852 but were never seen again, despite repeated searches. However, there were massive displays of the comets' associated meteor shower, the Andromedids, in later years.

This raises the question of the physical end-states of cometary nuclei, which is an important one for the hazard problem. If cometary nuclei disintegrate entirely into relatively small, meteoroidal particles, they will no longer pose a significant hazard to the Earth. However, it has been suggested that comets can evolve to dormant, asteroidal objects due to the build-up of nonvolatile lag deposits on their surfaces [5], [13], [80]. Levison and Duncan [33] estimate that SP comets are active for only ~10% of their dynamical lifetimes in the planetary region; the remainder of the time they are likely dormant, asteroidal objects. These "dead" or "extinct" comets will continue to orbit the Sun and pose a hazard to the Earth. Up to 50% of the near-Earth asteroids are estimated to be extinct cometary nuclei [84].

At present there is no clear estimate of the proportion of comets arriving at each of these two physical end-states. However, this is a less important concern for the LP comets because they are dynamically removed on a relatively rapid time scale, before they can physically evolve to dormant objects.

Nongravitational forces from jetting of volatiles can also perturb the orbits of the LP comets, in particular if they have perihelia < 0.5 AU. However, Weissman [70], [71] showed that nongravitational forces likely do not play a systematic role in the long-term dynamical evolution of the LP comet orbits. Weissman did propose that nongravitational forces could provide an explanation for the unusually small semimajor axes of the Kreutz group of Sun-grazing comets.

Nongravitational forces also make it difficult to predict accurately the return of LP comets and to estimate their precise paths through the planetary region. This uncertainty adds to the dilemma posed by the LP comets for the impact hazard problem, because of the difficulty in predicting the difference between a near-miss and a potential impact, and the difficulty in intercepting the comet with a destructive device.

TEMPORAL VARIATIONS

Current estimates of the flux of long-period comets crossing the Earth's orbit are based on the observed flux over the last 200 years, a relatively short time interval as compared with the orbital periods of the comets or the dynamical timescales for gravitationally perturbing the Oort cloud. Thus, it is worth considering how the cometary flux may vary with time, and whether the current flux is indeed typical, or possibly perturbed.

Hills [26] suggested that stars passing through the Oort cloud could initiate showers of comets into the planetary region, particularly if the Oort cloud had a dense inner core of comets, the inner Oort cloud. The effect of the star passage is to perturb so many comets in perihelia that the comets fill the "loss cone," the region in velocity phase space where the comet orbits have perihelia in the planetary region. In effect, the inner planetary region is flooded with comets. The perihelion distribution becomes uniform with heliocentric distance, and the number of comets passing within any perihelion distance is given by [26]

$$f = 2q/a\,(1 - q/2a) \approx 2q/a \quad \text{for } q \ll a, \tag{1}$$

where f is the fraction of the comet population with semimajor axis, a, that has perihelion distance < q, and q is the perihelion distance. Given an outer

Oort cloud population of 10^{12} comets [78] and a typical semimajor axis of 2.2×10^4 AU [40], Eq. (1) predicts a uniform perihelion distribution in the planetary region of 28 dynamically new comets per AU per year. This is 14 times the current estimated flux of dynamically new comets (see Section 2).

If the star passage severely perturbs the inner Oort cloud, assuming a population of 5×10^{12} comets and a typical semimajor axis of 3×10^3 AU, the density of perihelia predicted by Eq. (1) is 3.3×10^9 AU^{-1}. Because that semimajor axis corresponds to an orbital period of $\sim 1.6 \times 10^5$ years, the flux versus time is $\sim 2.0 \times 10^4$ AU^{-1} yr^{-1}, about 700 times the value obtained above for perturbing only the outer Oort cloud. However, Weissman [77] showed that even a close penetrating stellar passage is typically not sufficient to fill the loss cone for inner Oort cloud comets, and the peak increase in the LP comet flux at the Earth's orbit would be by a factor of ~300.

Hut et al. [28] and Fernández and Ip [18] modeled the dynamical evolution of cometary showers. The most intense showers would be caused by stars penetrating the Oort cloud to $\sim 3 \times 10^3$ AU, which would only be expected to occur about once every 5×10^8 years. More modest showers might occur from star passages at 10^4 AU, which would occur on average every 5×10^7 years The relative number of shower comets passing perihelion and the number of surviving comets versus time for a major shower is shown in Fig. 5 from Hut et al. [28]. The excess comets in the shower are dynamically removed due to planetary perturbations in about 2 to 3×10^6 years.

Weissman [77] estimated that cometary showers raised the total integrated flux at the Earth's orbit by about a factor of three over the history of the solar system, assuming that the solar system is currently not in a cometary shower. Fernández [17] suggested that the fact that the signature of the galactic tide is observable in the distribution of long-period comet orbits is evidence that the solar system is currently not experiencing a cometary shower (the signature is a deficit of dynamically new LP comets with aphelion directions towards galactic latitudes of 0° and ±90°, as pointed out by Delsemme and Patmiou [9]). Weissman [79] reached the same conclusion based on the $1/a_o$ distribution for the LP comets, which shows no evidence of a recent stellar perturbation of the inner Oort cloud.

Heisler [23] simulated the dynamical evolution of comets in the Oort cloud using a Monte Carlo model which included both random stellar perturbations and the galactic tide, modeling the flux of comets into the planetary region versus time. An example is shown in Fig. 6 which gives the relative number of dynamically new comets passing within 2 AU of the Sun versus time. The large spikes in the flux are the result of close stellar passages: the largest spike is for a star passing at only 7,200 AU from the Sun. The small scale variations in the steady-state flux of comets in Fig. 6 are most likely due to the statistics of the Monte Carlo model, and probably do not reflect real variations in the comet flux.

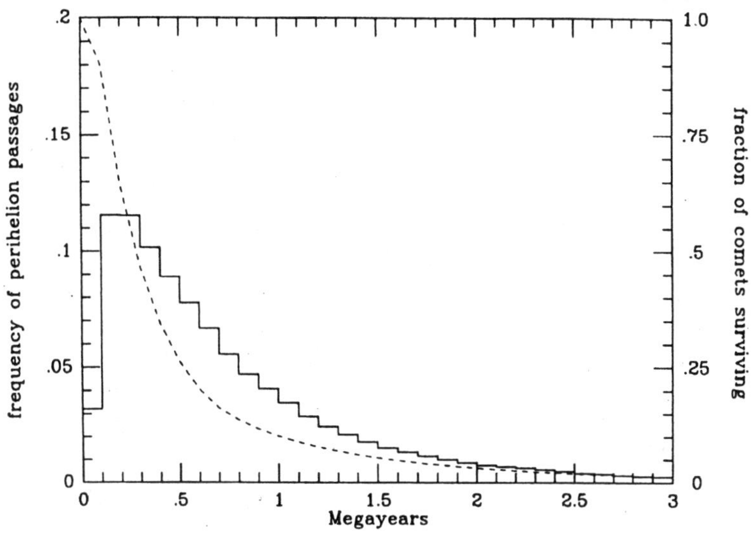

FIGURE 5. Dynamical evolution of the flux of comets through the planetary system during a major cometary shower, caused by a penetrating stellar passage through the inner Oort cloud. The solid histogram shows the relative number of comets passing perihelion versus time. The dashed curve gives the number of comets from the shower still dynamically evolving through the system. From Hut et al. [28].

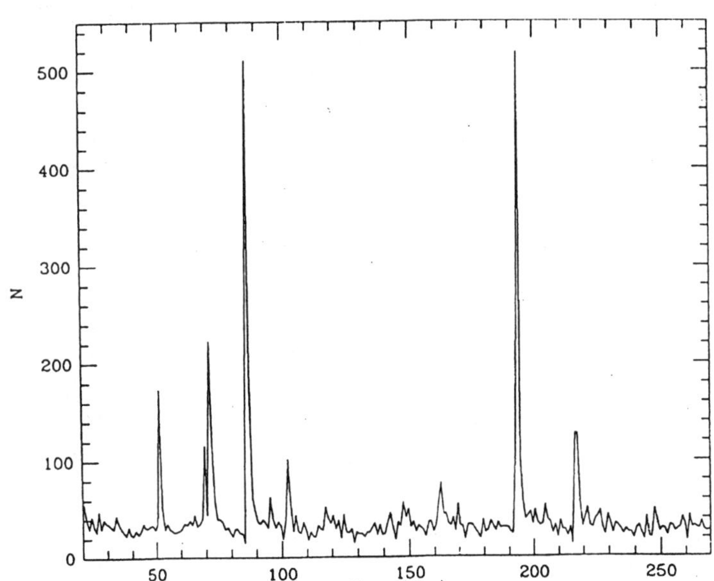

FIGURE 6. Hypothetical flux of dynamically new comets into orbits with perihelia < 2 AU as found by Heisler [23]. The large spikes are comet showers resulting from close stellar passages.

A second mechanism for varying the cometary flux from the Oort cloud was investigated by Matese et al. [43]. They estimated a factor of four variation in the cometary flux due to the solar system's harmonic motion above and below the galactic plane, which currently carries the planetary system ~75–85 parsecs out of the galactic plane. As the solar system moves into less dense regions of the galactic disk above and below the central plane, it experiences a smaller galactic tidal force, resulting in fewer comets being perturbed into the planetary region. This sinusoidal variation has a period of ~64–68 Myr [1].

Matese et al.'s dynamical model did not include stellar perturbations. It is possible that stellar perturbations act to mitigate the variation in the galactic tidal perturbations, though they actually may accentuate the variation due to the similarly varying space density of stars above and below the galactic plane. At present the solar system has just passed through the galactic plane in the last few million years, so the current steady-state flux is likely at a local maximum. This then suggests that the current steady-state LP comet flux is twice the long-term average flux.

An alleged periodicity in the times of biological extinction events on the Earth [54], [59] led researchers to suggest several hypotheses to produce periodic perturbations on the Oort cloud, in order to initiate periodic cometary showers into the planetary region. These include: (1) an unseen, dwarf companion star to the Sun passing through the Oort cloud [87], [8]; (2) the Sun's harmonic motion above and below the galactic plane [56], [53] ; and (3) a 10th planet in a precessing orbit beyond Neptune [88]. There are substantial dynamical problems with all of these hypotheses and many of their implications are clearly not borne out. If periodic extinctions are real, the physical mechanism for invoking them has not yet been identified. The reader is referred to Tremaine [68] and Weissman [76], [78] for more detailed discussions of these issues.

TERRESTRIAL IMPACT PROBABILITIES AND FLUXES

The potential hazard posed by the impact of comets on the Earth was first pointed out by Halley [21] in his classic paper on comets that included the first catalog of 24 cometary orbits. Writing about a comet which had passed near the Earth's orbit a short distance ahead of the Earth in its motion, Halley speculated,

> "But what might be the consequences of so near an appulse; or of a contact; or lastly, of a shock of the celestial bodies (which is by no mean impossible to come to pass), I leave to be discussed by the studious of physical matters."

It is a tribute to Halley's genius that he foresaw the attention today focussed on the threat of comet and asteroid impacts on the Earth. Halley also suggested that the comets in his catalog, for which he could only derive parabolic orbits, were likely members of the solar system and revolving in very long-period ellipses around the Sun. Of course, Halley's 1705 paper is best remembered for his recognition that the similar orbits for the comets of 1531, 1607, and 1682 were one and the same comet returning at periodic intervals, now known as Comet Halley.

The probability that any long-period comet will impact the Earth can be calculated as a function of the comet's orbital elements using equations derived by Öpik [50], [51]. The probability of impact, p, per perihelion passage is given by

$$p = s^2 U / p \sin i \, |U_x| \tag{2}$$

where s is the capture radius of the target planet, U is the hyperbolic encounter velocity of the comet relative to the planet (in units of the planet's circular orbital velocity about the Sun), i is the inclination of the comet's orbit to the plane of the planet's orbit, and U_x is the component of the hyperbolic encounter velocity in the radial (i.e., heliocentric) direction. The parameters U, U_x, and s are given by

$$U = [\, 3 - 1/a - 2\sqrt{a(1-e^2)} \cos i \,]^{1/2} \tag{3}$$

$$U_x = [\, 2 - 1/a - a(1 - e^2) \,]^{1/2} \tag{4}$$

$$s = (R/a_o)\,(1 + v_e^2/U^2)^{1/2}, \tag{5}$$

where a, e, and i, respectively, are the semimajor axis (in units of the target planet's semimajor axis, a_o), eccentricity, and inclination of the comet's orbit, and R, a_o, and v_e are the radius, semimajor axis, and escape velocity of the target planet. Equation 5 includes the effect of gravitational focussing by the target planet to "pull" comets onto collision paths. Note that in the case of a comet in a near-parabolic orbit, i.e., a long-period comet with eccentricity close to 1, the quantity $a(1 - e^2) = a(1 - e)(1 + e) \approx 2q$, where q is the comet's perihelion distance.

Öpik's equations assume that the target planet is in a circular orbit, a reasonable assumption for the Earth whose orbital eccentricity is only 0.0167. They also assume that the planet itself is spherical. In addition, the perihelion distance of the comet must be less than or equal to the planet's semimajor axis.

A weakness in Öpik's equations is that the impact probability goes to infinity as the inclination of the comet's orbit approaches 0° or 180°, or as the perihelion distance of the comet approaches the semimajor axis of the planet, i.e, as $q \rightarrow 1$ AU for the Earth. Additionally, Öpik's equations do not allow for the eccentricity of the orbit of the target planet. More accurate (though also more complex) methods for calculating impact probabilities were devised by Wetherill [83] and Kessler [30] involving the volumes swept out by the planet and the comet in their orbits. However, Öpik's equations perform satisfactorily so long as grazing orbits, $0.98 < q < 1$ AU, and orbits close to the ecliptic, $i < 1^{\circ}$ and $i > 179^{\circ}$ are excluded.

Long-period comet impact probabilities on the Earth have been calculated by a number of researchers. Weissman [74] found a mean impact probability of 2.2×10^{-9} per cometary perihelion passage, using Öpik's equations to average 2.5×10^{5} random cometary orbits, uniformly distributed in perihelion distance between 0.005 (the radius of the Sun) and 0.99 AU. Marsden and Steel [41] used Kessler's method to find a similar mean impact probability of 2.21×10^{-9}. Zimbelman [91] found values of 1.33×10^{-9} and 1.48×10^{-9} using analytic and numerical approximations to Öpik's equations. An examination of Zimbelman's calculations [65] showed that he erred and somehow lost a factor of $\pi/2$ (the mean value of sin i evaluated from 0 to 180°), which would then bring his estimates into general agreement with those above.

Shoemaker [61] found a higher impact probability of 3.3×10^{-9} by averaging encounter probabilities for the observed Earth-crossing LP comets. However, the observed LP comets are a biased sample, and the same factors which lead to an increased discovery probability also lead to an increase in the impact probability for LP comets. The observed sample is biased towards comets with perihelia near the Earth's orbit and inclinations close to the ecliptic plane, and the resulting mean impact probability is substantially higher than the true average. Similar results for 411 observed LP comets were found by Marsden and Steel [41], shown in Fig. 7.

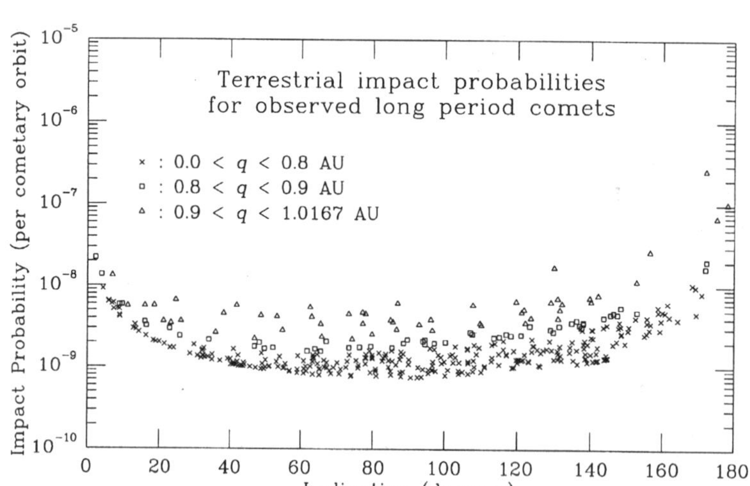

FIGURE 7. Impact probability on the Earth versus orbital inclination for 411 observed long-period comets. From Marsden and Steel [41].

The mean hyperbolic encounter velocity for the LP comets is 50.5 km sec^{-1}, and the mean impact velocity (considering the Earth's escape velocity of 11.2 km sec^{-1}) is 51.8 km sec^{-1} [74]. However, the most relevant parameter for the impact hazard is the energy of the impactor, proportional to its impact velocity squared. One can also weight the impact energy by the probability of an impact by a comet on that particular orbit. Doing so yields a representative impact velocity of 56.4 km sec^{-1} for the most probable LP comet impactor energy on the Earth. These values all assume a uniform perihelion distribution and randomly inclined cometary orbits.

As noted in Section 2, the perihelion distribution of the LP comets is likely not uniform interior to 1 AU. Weissman [75] showed that the perihelion distribution of the LP comets rises sharply throughout the planetary region because Jupiter and Saturn act as a dynamical barrier to the diffusion of LP comets into the inner planets region. If one assumes the perihelion distribution found by Everhart [12], shown in Fig. 1, then the mean impact probability rises to 2.4×10^{-9} and the most probable impact energy corresponds to a velocity of 56.7 km sec^{-1}. Using a perihelion distribution that varied as $q^{1/2}$ [15], Marsden and Steel [41] found a mean impact probability of 2.5×10^{-9}.

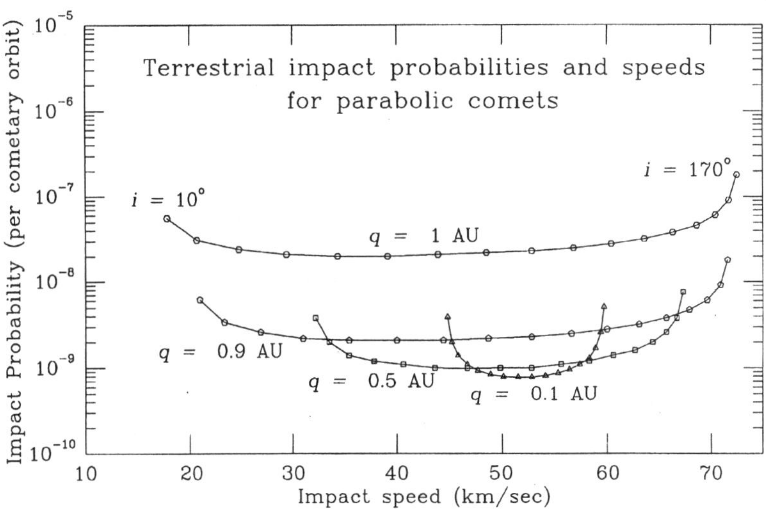

FIGURE 8. Impact probability on the Earth versus impact velocity for long-period comets in near-parabolic orbits, as a function of perihelion distance, q, and orbital inclination, i. From Marsden and Steel [41].

The variation of impact probability and impact velocity versus perihelion distance and orbital inclination is shown in Fig. 8 (from Marsden and Steel [41]). It can be seen that the impact probability increases as the comet's perihelion distance approaches 1 AU, and as the plane of the comet's orbit approaches the ecliptic. Impact velocity increases with increasing orbital inclination. Also, note that impact probabilities are generally slightly higher for retrograde orbits versus direct orbits. It is this correlation between impact probability and impact velocity which leads to the most probable impactor velocity being somewhat higher than the mean impact velocity.

A more difficult problem is that of estimating the actual impactor flux and crater production rate by LP comets on the Earth. This is because of the very poor present knowledge of the size and mass distributions of the LP comets. Cometary nuclei are difficult to observe directly since when they are close to the Sun they are enveloped in bright dust and gas comae, whereas when they are inactive they are usually too far from the Sun to be easily observed. Also, it is not always clear when nuclei are inactive as comae can be unresolved at the large distances involved. Unresolved comae can still contribute to both the visual and thermal radiation from the comets, making accurate determinations of nucleus dimensions and albedos quite difficult.

Nevertheless, several researchers have attempted to derive the size distribution of cometary nuclei. One example by Weissman [77] is shown in Fig. 9. Weissman used the observed distribution of LP comet absolute magnitudes, H_{10}, corrected for observational selection effects, as found by Everhart [12]. By comparing the H_{10} values for specific comets with observations of their nuclei at large heliocentric distances when they were presumably inactive, Weissman obtained the relationship

$$\log M = 20.0 - 0.4\,H_{10}, \tag{6}$$

where M is the nucleus mass in grams and a density of 0.6 g cm^{-3} was assumed. Using the magnitude distribution in Fig. 9, the average cometary nucleus brighter than $H_{10} = 11$ has a mass of 3.8×10^{16} g. Bailey and Stagg [2] derived a similar mass-magnitude relationship

$$\log M = 19.9 - 0.5\,H_{10}, \tag{7}$$

though with a steeper slope such that mass decreases more sharply with increasing absolute magnitude. The cometary size distribution has also been estimated by Shoemaker and Wolfe [62].

The size distribution in Fig. 9 is known as a broken power law distribution, where the number of comets in any logarithmically spaced size bin can be expressed as a power of the mass or radius of the cometary nuclei, and the size distribution is broken into two ranges with different characteristic exponents. A broken power law size distribution was recently suggested for

Kuiper belt comets (comets in a ring beyond the orbit of Neptune) by Weissman and Levison [82]. It has also been suggested by models of the accretion of planetesimals in the early solar system by Greenberg et al. [19] and Wetherill and Stewart [85].

Weissman [74], [77] used Shoemaker's [60] crater-energy scaling relationship to estimate crater sizes on the Earth for LP comets hitting at the most probable impact energy representative velocity of 56.4 km sec^{-1}. He showed that a comet with a mass of 1.9×10^{13} g was required to form a crater 10 km in diameter. Such a comet would have a radius of 0.20 km, assuming a bulk density of 0.6 g cm^{-3}. Using the crater scaling relationship of Melosh [44] one finds a similar mass, 2.1×10^{13} g, required for a LP comet to form a 10 km crater on the Earth. Using the mass-magnitude relationship in Eq.(6), Weissman [77] showed that the mass of 1.9×10^{13} g corresponded to an absolute magnitude of $H_{10} = 16.8$. By then using the size distribution shown in Fig. 9 and the estimate from Everhart's work of ~10 Earth-crossing LP comets per year brighter than $H_{10} = 11$, Weissman found that there were ~415 LP comets per year which could create 10 km craters if they impacted the Earth. Combining this with the mean impact probability from assuming Everhart's perihelion distribution in Fig. 1, one finds that a LP comet creates a 10 km diameter or larger crater on the Earth, on average, once every 1.0×10^6 years. In terms of the surface area of the Earth, this works out to a cratering rate of 0.20×10^{-14} km^{-2} yr^{-1}. For comparison, Shoemaker et al. [63] found a mean impact interval for Earth-crossing asteroids creating 10 km or larger craters of 1.1×10^5 years, or a rate of 1.8×10^{-14} km^{-2} yr^{-1}. Thus, LP comets constitute ~10% of the present cratering flux on the Earth.

As noted in the previous section, approximately two-thirds of all LP comets passing through the inner planets region arrive as part of cometary showers. Thus, the long-term integrated cratering rate from LP comets is ~3 times the value found above, equal to 0.59×10^{-14} km^{-2} yr^{-1}, or one impact every 3.3×10^5 years, on average. Since we are not currently in a cometary shower [79], [17], and the onset time for a cometary shower is on the order of 10^5 years, the appropriate value to use in evaluating the current impact hazard is that found above for the steady-state flux of one impact every 10^6 years.

Shoemaker et al. [63] find a somewhat higher impact frequency of a LP comet creating a 10 km diameter crater every 3.6×10^5 years, or a cratering rate of 0.55×10^{-14} km^{-2} yr^{-1}. Unfortunately, Shoemaker et al. [63] do not explain all the steps in their calculation so it is not possible to determine why their result is different from Weissman's above. However, Shoemaker et al. also noted that the relatively modest sized comets required to form 10 km craters would probably not survive atmospheric entry intact. Modeling of atmospheric entry for asteroidal and cometary objects by Chyba et al. [7] showed that at the sizes necessary to match the characteristics of the 1908 Tunguska event, cometary nuclei would explode at altitudes of 20–40 km, which would result in relatively little effect at ground level.

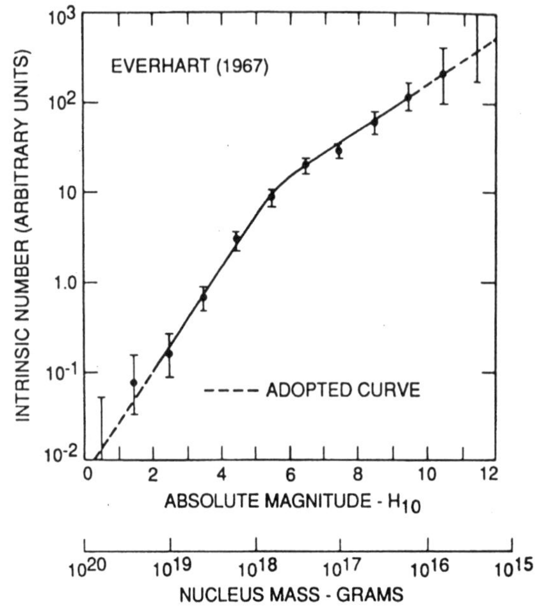

FIGURE 9. The mass distribution of the long-period comets as found by Weissman [77] using Everhart's [12] distribution of the number of comets as a function of their absolute magnitudes, H_{10}, corrected for observational selection effects. The mass scale assumes a bulk density for the nuclei of 0.6 g cm^{-3}.

P. Thomas (personal communication) has estimated that a cometary nucleus with a diameter of about 350 meters is required for the object to reach the Earth's surface before exploding. This is only slightly smaller than the size estimated above to form a 10 km crater. Thus, comets of this dimension may burst just above or at the surface of the Earth, creating either a modest or indistinct crater, as has been observed for many crater-like structures on Venus (see, for example, [25]). One possibility illustrated by the Venus data is that the comet may disrupt into several large fragments during atmospheric entry and form multiple, overlapping craters.

Olsson-Steel [48] found that LP comets accounted for 6% of impacts forming craters with diameters ≥ 10 km on the Earth, which suggests a mean impact interval of 2.2×10^6 years. Again, the details of Olsson-Steel's calculation are not completely explained in his paper so it is difficult to assess why his results are different from those above. Marsden [36] estimated that LP comets constitute only 2% of the Earth impactor flux by number, based on the statistics of observed close approaches by comets to the Earth. This implies an impact interval of ~5×10^6 years. Marsden did not consider observational selection effects which may have prevented the discovery of some close approaching comets, or the possibility that the flux includes small LP comets too faint to detect, but still sufficiently large enough to form 10 km craters. Both these factors contribute to the lower impact frequency found by Marsden. Also, note that Marsden stated his estimate in terms of the number of impactors, and did not normalize for the much higher impact velocities of the LP comets.

A potential weakness in all of these crater rate estimates (except Marsden's) involves the extrapolation of the cometary size distribution down to the relatively small, faint nuclei required to form 10 km craters on the Earth. Comets with that mass, $M \approx 2 \times 10^{13}$ g, are approximately two orders of magnitude smaller than the faintest LP comets currently observed (if we can believe the size distributions in Eqs. (6) or (7)). Some researchers have proposed that there are lower limits to the sizes of cometary nuclei (e.g., [58]), but it is difficult to identify whether the failure to observe such objects is real or only an observational selection effect.

One can also estimate the minimum mass LP comet that might cause a global climatic disturbance. Morrison et al. [45] find the most likely threshold energy for such an event is 3×10^5 megatons, or 1.3×10^{28} ergs. For a LP comet impacting at the most probable impact velocity of 56.4 km sec^{-1}, this corresponds to a mass of 8×10^{14} g, or a radius of ~0.7 km. Using the size distribution in Fig. 9, such a comet would have an absolute magnitude of 12.7. The Earth-crossing flux would be ~26 such comets per year and the mean impact interval would be ~1.6×10^7 years.

In summary, the current mean impact rate of LP comets sufficiently massive to create 10 km or larger craters on the Earth is estimated to be between 5×10^{-7} and 2.8×10^{-6} per year, with a most likely value of ~1.0×10^{-6} per year. The most probable impact energy corresponds to velocities of 56–58 km sec^{-1}.

DISCUSSION

The long-period comets pose a unique dilemma for the impact hazard problem. Because of their very long orbital periods and large distances from the Sun (most of the time) they cannot be surveyed and catalogued in the same manner as the near-Earth asteroids and short-period comets. Even with current astrometric capabilities, returns of comets with periods greater than 200 years cannot be predicted with the accuracy necessary to know whether a comet will pose a significant threat to the Earth. At it's mean orbital velocity of 29.78 km sec^{-1}, the Earth moves through its own diameter in only 7.1 minutes, so the difference between a very close approach and an impact event is truly very small.

A long-period comet approaching the inner planets region will move from Jupiter's orbit to 1 AU in a time interval of just under 1 year. In general, LP comets are relatively inactive at the distance of Jupiter's orbit, unless they are dynamically new comets making their first passage through the planetary system. The problem is one of detecting small, low albedo objects at 5 AU or more. A 1 km radius nucleus with an albedo of 0.04, viewed at opposition (0° phase), would have an R magnitude of $m_R = 23.7$ at 5 AU from the Sun. This is just within the capabilities of current CCD cameras on large aperture ground-based telescopes. An additional problem with regard to the LP comets is the fact that their orbits are randomly oriented on the sky, as opposed to the near-ecliptic inclinations of most near-Earth asteroids. Thus, the entire sky must be surveyed continuously in order to detect approaching LP comets at the earliest possible moment.

The problem of searching for LP comets is discussed in more detail in Marsden and Steel [41]. They estimate that LP comets on impact trajectories would pass opposition ~250–500 days prior to Earth encounter for preperihelion impacts (ignoring comets which reach opposition < 75 days before impact), and ~100–425 days for postperihelion impacts. Thus, possible warning times would range from a little over 3 months to approximately 16 months. Marsden and Steel also emphasize the need to search all ecliptic latitudes in order not to miss any LP comets.

Since no such deep searches for LP comets are currently underway, actual warning times found from current discoveries are typically much smaller. Only 12% of all LP comets were discovered more than 100 days prior to perihelion [41], though that number is improving as photographic and CCD discoveries play a larger and larger role. Some Earth-approaching LP comets have had remarkably short warning times. The recent bright, naked-eye comet Hyakutake, C/1996 B2, was found 8 weeks prior to closest approach, and made a close approach to the Earth of only 0.102 AU. Comet IRAS-Araki-Alcock, C/1983 H1, was discovered only 15 days prior to a close approach to within 0.031 AU from the Earth. This diffuse comet was discovered by the IRAS spacecraft but was not reported to the Central Bureau for Astronomical Telegrams until a week later, when it was independently discovered by two ground-based observers, only 8 days before closest approach! The nucleus of comet IRAS-Araki-Alcock was estimated to have a radius of ~4 km [20] so it would have likely resulted in a major global catastrophe if it had struck the Earth.

In contrast, the recent bright LP comet Hale-Bopp, C/1995 O1, which is on an Earth-crossing orbit, was found more than 20 months before perihelion at a heliocentric distance of 7.15 AU. Although the large estimated radius for this nucleus, 10–20 km, makes it a major potential hazard, its closest approach distance to the Earth was 1.32 AU, quite far away.

The long-period comets will continue to present a difficult problem for those concerned about the possibility of a major cometary impact on the Earth. Warning and response times will, in general, remain much shorter than for other potential impactors. Current technologies are insufficient to permit early detection, say at the edge of the planetary system, 30 AU, where the infall time is ~12 years. On the other hand, the mean interval of 10^6 years between significant impacts, and the 16×10^6 years mean interval between potentially catastrophic LP comet impacts, makes the likelihood of an impact quite low in the next century. During that time, routine improvements in several different technologies may make this problem trivial to solve.

ACKNOWLEDGMENTS

The author thanks Brian Marsden and Alan Harris for useful discussions during the course of this work. Brian Marsden also provided a very helpful review of an earlier draft of this paper. The author thanks John Remo for his encouragement and his patience.

REFERENCES

1. BAHCALL, J.N. & S. BAHCALL. 1985. The Sun's motion perpendicular to the galactic plane. Nature. **316:** 706–708.
2. BAILEY, M.A. & C.R. STAGG. 1988. Cratering constraints on the inner Oort cloud and implications for cometary origins. Mon. Not. Roy. Astron. Soc. **235:** 1–32.
3. BIERMANN, L., W.F. HUEBNER, & RH. LÜST. 1983. Aphelion clustering of "new" comets: Star tracks through the Oort cloud. Proc. Natl. Acad. Sci. USA. **80:** 5151–5155.
4. BOGART, R.S. & P.D. NOERDLINGER. 1982. On the distribution of orbits among long-period comets. Astron. J. **87:** 911–917.
5. BRIN, G.D. & D.A. MENDIS. 1979. Dust release and mantle development in comets. Astrophys. J. **229**:402–408.
6. BYL, J. 1983. Galactic perturbations on nearly parabolic cometary orbits. Moon & Planets. **29:** 121–137.
7. CHYBA, C.F., P.J. THOMAS, & K.J. ZAHNLE. 1993. The 1908 Tunguska explosion: atmospheric disruption of a stony asteroid. Nature. **361:** 41–44.
8. DAVIS, M., P. HUT, AND R.A. MULLER. 1984. Extinction of species by periodic comet showers. Nature. **308:** 715–717.
9. DELSEMME, A.H. & M. PATMIOU. 1986. Galactic tides affect the Oort cloud: An observational confirmation. *In* 20th ESLAB Symposium on the Exploration of Halley's Comet. B. Battrick, E. J. Rolfe, & R. Reinhard, Eds. ESA SP-250. **2:** 409–412.
10. DUNCAN, M., T. QUINN, & S. TREMAINE. 1987. The formation and extent of the solar system comet cloud. Astron. J. **94:** 1330–1338.
11. EVERHART, E. 1967a. Comet discoveries and observational selection. Astron. J. **72:** 716–726.
12. EVERHART, E. 1967b. Intrinsic distributions of cometary perihelia and magnitudes. Astron. J. **72:** 1002–1011.

13. FANALE, F. P., & J. R. SALVAIL. 1984. An idealized short-period comet model: surface insolation, H_2O flux, dust flux, and mantle evolution. Icarus **60:** 476–511.
14. FERNÁNDEZ, J.A. 1980. Evolution of comet orbits under the perturbing influence of the giant planets and stars. Icarus. **42:** 406–421.
15. FERNÁNDEZ, J.A. 1981. New and evolved comets in the solar system. Astron. Astrophys. **96:** 26–35.
16. FERNÁNDEZ, J.A. 1982. Dynamical aspects of the origin of comets. Astron. J. **87:** 1318–1332.
17. FERNÁNDEZ, J.A. 1994. Dynamics of comets: Recent developments and new challenges. *In* Asteroids, Comets, Meteors 1993, A. Milani, M. Di Martino, & A. Cellino, Eds.: 223–240. Kluwer. Dordrecht, The Netherlands.
18. FERNÁNDEZ, J.A. & Ip, W.-H. 1987. Time dependent injection of Oort cloud comets into Earth-crossing orbits. Icarus. **71:** 46–56.
19. GREENBERG, R., S.J. WEIDENSCHILLING, C.R. CHAPMAN, & D.R. DAVIS. 1984. From icy planetesimals to planets: Numerical simulations of collisional evolution. Icarus. **59:** 87–113.
20. HANNER, M., D.K. AITKEN, R. KNACKE, S. MCCORKLE, P.F. ROCHE, & A.T. TOKUNAGA. 1985. Infrared spectrophotometry of comet IRAS-Araki-Alcock (1983d): A bare nucleus revealed? Icarus. **62:** 97–109.
21. HALLEY, E. 1705. A Synopsis of the Astronomy of Comets. London. 24 pp.
22. HASEGAWA, I. 1980. Catalogue of ancient and naked-eye comets. Vistas in Astron. **24:** 59–102.
23. HEISLER, J. 1990. Monte Carlo simulations of the Oort cometary cloud. Icarus. **88:** 104–121.
24. HEISLER, J. & S. TREMAINE. 1986. The influence of the galactic tidal field on the Oort comet cloud. Icarus. **65:** 13–26.
25. HERRICK, R.R. & R.J. PHILLIPS. 1994. Implications of a global survey of venusian impact craters. Icarus. **111:** 387–416.
26. HILLS, J.G. 1981. Comet showers and the steady-state infall of comets from the Oort cloud. Astron. J. **86:** 1730–1740.
27. HUT, P. & S. TREMAINE. 1985. Have interstellar clouds disrupted the Oort comet cloud? Astron. J. **90:** 1548–1557.
28. HUT, P., W. ALVAREZ, W.P. ELDER, T. HANSEN, E.G. KAUFFMAN, G. KELLER, E.M. SHOEMAKER, & P.R. WEISSMAN. 1987. Comet showers as possible causes of stepwise mass extinctions. Nature **329:** 118–126.
29. JOHNSON, R.E., J.F. COOPER, L.J. LANZEROTTI, & G. STRAZZULA. 1987. Radiation formation of a nonvolatile comet crust. Astron. Astrophys. **187:** 889–892.
30. KESSLER, D.J. 1981. Derivation of the collision probability between orbiting objects: The lifetimes of Jupiter's outer moons. Icarus. **48:** 39–48.
31. KRESÁK, L. 1992. The peculiar orbit of comet 1976 I Sato. Astron. Astrophys. **259:** 692–695.
32. KUIPER, G. P. 1951. On the origin of the solar system. *In* Astrophysics, J.A. Hynek, Ed.: 357–424. McGraw Hill. New York.
33. LEVISON, H.F., & M.J. DUNCAN. 1994. The long-term dynamical behavior of short-period comets. Icarus. **108:** 18–36.
34. MARSDEN, B.G. 1967. The Sun-grazing comet group. Astron. J. **72**, 1170–1183.
35. MARSDEN, B.G. 1989. The Sun-grazing comet group. II. Astron. J. **98**, 2306–2321.
36. MARSDEN, B.G. 1993. To hit or not to hit. *In* Near-Earth Objects Interception Workshop, G.H. Canavan, J.C. Solem, & J.D.G. Rather, Eds.: 67–71. Los Alamos National Laboratory. Los Alamos, New Mexico.
37. MARSDEN, B.G. & E. Roemer. 1982. Basic information and references. *In* Comets, L.L. Wilkening, Ed.: 707–733. Univ. Arizona Press. Tucson, Arizona.

38. MARSDEN, B.G. & Z. SEKANINA. 1973. On the distribution of "original" orbits of comets of large perihelion distance. Astron. J. **78:** 1118–1124.
39. MARSDEN, B.G., Z. SEKANINA, & D.K. YEOMANS. 1973. Comets and nongravitational forces. V. Astron. J. **78:** 211–225.
40. MARSDEN, B.G., Z. SEKANINA, & E. EVERHART, E. 1978. New osculating orbits for 110 comets and the analysis of the original orbits of 200 comets. Astron J. **83:** 64–71.
41. MARSDEN, B.G. & D.I. STEEL. 1995. Warning times and impact probabilities for long-period comets. *In* Hazards Due to Comets and Asteroids, T. Gehrels, Ed.: 221–239. Univ. Arizona Press. Tucson, Arizona.
42. MARSDEN, B.G. & G.V. Williams. 1996. Catalogue of Cometary Orbits. 11th Edition, 110 pp. Smithsonian Astrophysical Observatory. Cambridge, Massachusetts.
43. MATESE, J.J., P.G. WHITMAN, K.A. INNANEN, & M.J. VALTONEN. 1995. Periodic modulation of the Oort cloud comet flux by the adiabatically changing galactic tide. Icarus. **116:** 255–268.
44. MELOSH, H. J. 1989. Impact Cratering: A Geologic Process: 284 pp. Oxford Univ. Press. New York, NY.
45. MORRISON, D., C.R. CHAPMAN, & P. SLOVIC. 1995. The impact hazard. *In* Hazards Due to Comets and Asteroids, T. Gehrels, Ed.: 59–91. Univ. Arizona Press. Tucson, Arizona.
46. MUMMA, M.J., P.R. WEISSMAN, & S.A. STERN. 1993. Comets and the origin of the solar system. *In* Planets and Protostars III, E.H. Levy & J.L. Lunine, Eds.: 1177–1252. Univ. Arizona Press. Tucson, Arizona.
47. NESLUŠAN, L. 1996. Perihelion point preferred direction of long-period comets and the north-south asymmetry of comet discoveries from the Earth's surface. Astron. & Astrophys. **306:** 981–990.
48. OLSSON-STEEL, D. 1987. Collisions in the solar system — IV. Cometary impacts upon the planets. Mon. Not. Roy. Astron. Soc. **227:** 501–524.
49. OORT, J.H. 1950. The structure of the cloud of comets surrounding the solar system and a hypothesis concerning its origin. Bull. Astron. Inst. Neth. **11:** 91–110.
50. ÖPIK, E.J. 1951. Collisional probabilities with the planets and the distribution of interplanetary matter. Proc. Roy. Irish. Acad. A. **54:** 165–199.
51. ÖPIK, E.J. 1976. Interplanetary Encounters: 155pp. Elsevier. Amsterdam, The Netherlands.
52. PRIALNIK, D. & A. BAR-NUN. 1987. On the evolution and activity of cometary nuclei. Astrophys. J. **313:** 893–905.
53. RAMPINO, M.R. & R.B. STOTHERS. 1984. Terrestrial mass extinctions, cometary impacts, and the Sun's motion perpendicular to the galactic plane. Nature. **308:** 709–712.
54. RAUP, D.M. & J.J. SEPKOSKI. 1984. Periodicity of extinctions in the geologic past. Proc. Natl. Acad. Sci. USA. **81:** 801–805.
55. SAFRONOV, V.S. 1969. Evolution of the Protoplanetary Cloud and Formation of the Earth and Planets, NASA TT-F-677, 1972. Nauka Press. Moscow, Russia.
56. SCHWARTZ, R.D. & P.B. JAMES. 1984. Periodic mass extinctions and the Sun's oscillation about the galactic plane. Nature. **308:** 712–713.
57. SEKANINA, Z. 1982. The problem of split comets in review. *In* Comets, L.L. Wilkening, Ed.: 251–287. Univ. Arizona Press. Tucson, Arizona.
58. SEKANINA, Z. & D.K. YEOMANS. 1984. Close encounters and collisions of comets with the Earth. Astron. J. **89:** 154–161.
59. SEPKOSKI, J.J. 1990. The taxonomic structure of periodic extinction. *In* Global Catastrophes in Earth History, V.L. Sharpton & P.D. Ward, Eds.: **247**: 33–44. Geol. Soc. Amer. SP.

60. SHOEMAKER, E.M. 1977. Astronomically observable crater forming projectiles. *In* Impact and Explosion Cratering, D.J. Roddy, R.O. Pepin, & R.B. Merrill, Eds.: 617–628. Pergamon. Elmsford, New York.
61. SHOEMAKER, E.M. 1984. Large body impacts through geologic time. *In* Patterns of Change in Earth Evolution, H.D. Holland & A.F. Trendall, Eds.: 15–40. Springer-Verlag. Berlin, Germany.
62. SHOEMAKER, E.M. & R.F. WOLFE. 1982. Cratering time scales for the Galilean satellites. *In* Satellites of Jupiter, D. Morrison, Ed.: 277–339. Univ. Arizona Press. Tucson, Arizona.
63. SHOEMAKER, E.M., R.F. Wolfe, & C.S. Shoemaker. 1990. Asteroid and comet flux in the neighborhood of Earth. *In* Global Catastrophes in Earth History, V.L. Sharpton & P.D. Ward, Eds. **247:** 155–170. Geol. Soc. Amer. SP .
64. SHOEMAKER, E.M., P.R. WEISSMAN, & C.S. SHOEMAKER. 1995. The flux of periodic comets near Earth. *In* Hazards Due to Comets and Asteroids, T. Gehrels, Ed.: 313–335. Univ. Arizona Press. Tucson, Arizona.
65. STEEL, D.I. 1993. Collisions in the solar system—V. Terrestrial impact probabilities for parabolic comets. Mon. Not. Roy. Astron. Soc. **264:** 813–817.
66. STEPHENSON, F.R., K.K.C. YAU, & H. HUNGER. 1985. Records of Halley's comet on Babylonian tablets. Nature. **314:** 587–592.
67. STERN, S.A. 1990. ISM induced erosion and gas dynamical drag in the Oort cloud. Icarus. **84:** 447–466.
68. TREMAINE, S. 1986. Is there evidence for a solar companion. *In* The Galaxy and the Solar System. R. Smoluchowski, J.N. Bahcall, & M.S. Matthews, Eds.: 409–416. Univ. Arizona Press. Tucson, Arizona.
69. TYROR, J.G. 1957. The distribution of the directions of perihelia of long-period comets. Mon. Not. Roy. Astron. Soc. **117:** 370–379.
70. WEISSMAN, P.R. 1979a. Physical and dynamical evolution of long-period comets. *In* Dynamics of the Solar System, R.L. Duncombe, Ed.: 277–282. D. Reidel. Dordrecht, The Netherlands.
71. WEISSMAN, P.R. 1979b. Nongravitational perturbations of long-period comets. Astron. J. **84:** 580–584.
72. WEISSMAN, P.R. 1980a. Stellar perturbations of the cometary cloud. Nature. **288:** 242–243.
73. WEISSMAN, P.R. 1980b. Physical loss of long-period comets. Astron. Astrophys. **85:** 191–196.
74. WEISSMAN, P.R. 1982. Terrestrial impact rates for long and short-period comets. *In* Geological Implications of Impacts of Large Asteroids and Comets on the Earth, L.T. Silver & P.H. Schultz, Eds.: 15–24. Geol. Soc. Amer. SP **190**.
75. WEISSMAN, P.R. 1985. Dynamical evolution of the Oort cloud. *In* Dynamics of Comets: Their Origin and Evolution, A. Carusi & G.D. Valsecchi, Eds.: 87–96. D. Reidel. Dordrecht, The Netherlands.
76. WEISSMAN, P.R. 1986. The Oort cloud and the galaxy: Dynamical interactions. *In* The Galaxy and the Solar System, R. Smoluchowski, J.N. Bahcall, & M.S. Matthews, Eds.: 204–237. Univ. Arizona Press, Tuscon, Arizona.
77. WEISSMAN, P.R. 1990. The cometary impactor flux at the Earth. *In* Global Catastrophes in Earth History, V.L. Sharpton & P D. Ward, Eds.: 171–180. Geol. Soc. Amer. SP **247**.
78. WEISSMAN, P.R. 1991. Dynamical history of the Oort cloud. *In* Comets in the Post-Halley Era, R.L. Newburn, Jr., M. Neugebauer, & J. Rahe, Eds.: **1:** 463–486. Kluwer. Dordrecht, The Netherlands.
79. WEISSMAN, P.R. 1993. No, we are not in a cometary shower. Bull. Amer. Astron. Soc. **25:** 1063 (abstract).

80. WEISSMAN, P.R., M.F. A'Hearn, L.A. McFadden, & H. Rickman. 1989. Evolution of comets into asteroids. *In* Asteroids II, R.P. Binzel, T. Gehrels, & M.S. Matthews, Eds.: 880–920. Univ. Arizona Press. Tucson, Arizona.
81. WEISSMAN, P.R. & H. Campins. 1993. Short-period comets. *In* Resources of Near-Earth Space, J. Lewis, M.S. Matthews, & M. Guerrieri, Eds.: 569–617. Univ. Arizona Press. Tucson, Arizona.
82. WEISSMAN, P.R. & H.F. Levison. 1996. The size distribution of cometary nuclei. Lun. Planet. Sci. Conf. **27:** 1409–1410 (abstract).
83. WETHERILL, G.W. 1967. Collisions in the asteroid belt. J. Geophys. Res. **72:** 2429–2444.
84. WETHERILL, G.W. 1991. End products of cometary evolution: Cometary origin of Earth-crossing bodies of asteroidal appearance. *In* Comets in the Post-Halley Era, R.L. Newburn, M. Neugebauer, & J. Rahe, Eds.: 537–556. Kluwer. Dordrecht, the Netherlands.
85. WETHERILL, G.W. & G.R. STEWART. 1993. Formation of planetary embryos: Effects of fragmentation, low relative velocity, and independent variation of eccentricity and inclination. Icarus. **106:** 190–209.
86. WHIPPLE, F.L. 1978. Cometary brightness variation and nucleus structure. Moon & Planets. **18:** 343–359.
87. WHITMIRE, D.P. & A.A. JACKSON. 1984. Are periodic mass extinctions driven by a distant solar companion? Nature. **308:** 713–715.
88. WHITMIRE, D.P. & J.J. MATESE. 1985. Periodic comet showers and planet X. Nature. **313:** 36–38.
89. YAU, K., D. YEOMANS & P. WEISSMAN. 1994. The past and future motion of comet Swift-Tuttle. Mon. Not. Roy. Astron. Soc. **266:** 305–316.
90. YEOMANS, D.K., K. YAU & P.R. WEISSMAN. 1996. The impending approach of periodic comet Tempel-Tuttle and the Leonid meteors. Icarus. **124**: 407–413.
91. ZIMBELMAN, J.R. 1984. Planetary impact probabilities for long-period comets. Icarus. **57:** 48–54.

Comets as Porous Aggregates of Interstellar Dust

J. MAYO GREENBERG[a] AND JOHN L. REMO[b]

[a]*Department of Astrophysics*
Leiden University
Leiden, The Netherlands
and
[b]*Quantametrics, Inc.*
1 Brackenwood Path
Head of Harbor
St. James, New York 11780

ABSTRACT: A comet model is derived based on the interstellar dust chemical composition in dense molecular and diffuse clouds resulting from their subsequent chemical interactions and UV photoprocessing. The collapsing presolar nebula leads to a porous aggregate model for comet nuclei, from which is derived certain physical properties that include thermal conductivity and tensile strength. The porous morphological structure is also shown to imply anomalous (expansion rather than contraction) behavior when subjected to strong shock waves, which is supported by recent shock experiments on (porous) carbonaceous chondrite material.

1. INTRODUCTION

It is generally regarded that comets originated from the cyclic evolution of dust in the interstellar medium, and that their material properties approximate that of the aggregates of interstellar dust in the primordial solar nebula. Therefore, the study of the material properties of comets is essential to understanding the relationship between stellar evolution, interstellar nebulae, and planetary formation. However, it is also important to try to characterize the unique character and composition of cometary nuclei material relative to all other solar system bodies for several additional reasons:

(1) Assist in interpreting observational spectra of active comet classification and evolution;
(2) Understand the nongravitational forces that could be activated by comet (thermo-chemical) activity, and the ensuing effects on the comet orbital dynamics;
(3) Provide a model to support the design of comet probes and material recovery missions, such as Stardust (NASA) and Rosetta (ESA) Roland Champollion types of probes;
(4) Model the dynamic interaction and fragmentation of comets within planetary atmospheres;

(5) Better interpret the cratering record and structure of specific craters in terms of comet impact on a variety of planetary surfaces;
(6) Assess the levels of environmental devastation that comets of various sizes can induce in the terrestrial biosphere; and
(7) Provide a basis for understanding the effect of mechanical, thermal, and radiation interactions for mitigating a possible comet (impact) threat to the Earth.

In the following section, a summary of the material properties needed for treating comets as a class of near-Earth objects (NEO's) is presented. In preparation for understanding comet nuclei, we shall describe in Section 3 the chemical and physical properties of interstellar dust as it appears both in diffuse clouds as well as in dense molecular clouds. A general model of comet nuclei as aggregates of dust in the proto solar nebula will be given in Section 4. The justification of this model will be related to observable comet properties as determined from both ground- and space-based observations. These observations will lead to a "complete" description of the chemical, physical and morphological structure of cometary material within the framework of the observations at hand. In Section 5 we list and derive some of the physical properties such as strength and thermal conductivity. Section 6 is intended to provide a basis for future studies of the effects of shocks on comet nuclei.

2. NEO MATERIALS AND ANALOGS

One of the advantages of asteroid modelling relative to comets is that asteroids may often be categorized, within broad limits, in terms of their mineral properties as measured by reflectance spectra in terms of well-known minerals, such as pyroxene, olivine, feldspar, and iron-nickel. Also, since asteroids are widely regarded as the source of many, if not most, classes of meteorites which are predominantly composed of the same minerals observed in the asteroid reflectance spectra, samples of material related to some types of asteroids can be carried out in a laboratory environment, allowing the determination of various physical and chemical properties and their dynamic interactions (see [44], also other references). Such an understanding of the dynamic interactions on NEOs in general is absolutely necessary before any NEO mitigation interactions could be contemplated [43].

Unfortunately, unlike the relationship between asteroids and their meteorite surrogates, there are not currently available any samples of comet material, other than some microscopic dust particles collected in the Earth's atmosphere, which are believed to be cometary in origin. The lack of any substantial and relatively unaltered comet surrogate material for laboratory studies severely hampers any effort to understand both their microscopic as well as their macroscopic material properties. On the other hand, we are fortunate in that recent laboratory experiments supported by observations allows us to better understand some of the unique characteristics of comet nuclei in terms of physically and chemically definable properties from which a comet nucleus model can be obtained. In this paper we will describe how a possible comet model can be derived by making use of the empirical and observational data combined with material properties on the microstructural level. The model developed will be a consequence of comet nuclei being formed from aggregates of interstellar dust which have undergone sequences of complex chemical reactions between gas and solid phases while subjected to UV irradiation, before entering the proto planetary phase. This leads to precometary grains with unique microstructural and chemical characteristics that may ultimately form a porous comet nucleus material. In particular, it will be assumed that comets were formed directly out of microscopic (submicron) particles of interstellar dust which accreted from the condensible atoms and molecules in the latest stages of contraction of these gas/dust clouds, which also ultimately formed the Sun and the planets in our own solar system . In such a case, it is anticipated that comet nuclei, in the absence of high temperature interactions, should preserve the relative proportions of such elements as O, C, N, Fe, Ni, Si, S, ..., as they were 4.55 billion or more years ago, and thereby assist in defining the solar system abundances. The sequence of interstellar processes leading to the dust properties in the proto solar nebula is shown in Fig. 1. The implications of a porous aggregate as the proto-comet building material has a profound effect upon the interpretation of comet observations, evolution, and interactions [24], [30], [52], [12], [50], [19], [40].

3. INTERSTELLAR DUST IN COMETS

When the interstellar cloud collapsed to form the solar system, small solid particles of interstellar dust, as a result of collisions and molecular attraction (such as London-van der Waals) forces, accreted at a rate proportional to the density of the cloud. The major question is then, how much of the dust in the collapsing cloud survived intact before aggregating to form comets? The answer to this question is related to the physical conditions prevailing during the collapse phase, such as: did accretion shocks evaporate the volatiles? Was a significant fraction of the dust which approached the sun evaporated and subsequently mixed with pristine dust to form the cometary nucleus? These are questions whose details remain to be worked out. However, all the present indications are that comet nuclei exhibit, in their comas, molecule species in proportions which more closely resemble those of interstellar dust than those which could have resulted from solar system chemistry of molecules evaporated from the dust; i.e., that comet nuclei are derived from chemically intact interstellar dust [2], [39]. In this section we shall outline what are believed to be the initial properties of interstellar dust in the various stages leading to solar system formation.

a. Cyclic Evolution of Dust and the Interstellar Medium

The interstellar medium is in a state of constant flux. New material is continually being added in the form of atoms, molecules, and solid particles by old stars and by super novae while new stars consume the material from the interstellar medium. What we see are regions of quite varied densities which are usually divided into diffuse and molecular regions primarily defined in terms of whether the hydrogen is in atomic or molecular form. The so-called "diffuse clouds" have atomic hydrogen densities of $n_0 \leq 10^2$–10^3 cm^{-3}, while for molecular clouds $n_0 \geq 10^3$ cm^{-3}, where n_0 is the number of hydrogens in all forms — atoms, ions, molecules. In the evolutionary picture of interstellar dust, small refractory particles of silicates are formed in, and blown out of, atmospheres of cool evolved (old) stars. These particles which emerge into the surrounding medium are instantly cooled to temperatures ~10 K. When the region in which they are suspended becomes sufficiently dense, the condensible atoms, O, C, N, accrete on the silicates while, at the same time, interacting in the gas to make new molecules of which H_2 and CO are the most abundant. The molecular mantles on the dust are subjected to energetic processing, mostly by ultraviolet radiation, which leads to the formation of both entirely new molecules as well as complex organic molecules. When the cloud of dust and gas collapses to form stars, some of the material is ejected and emerges into the low density regions where destructive processes lead to the elimination of the volatile components of the dust leaving the silicate cores surrounded by a mantle of complex organic refractory material. This organic matter is processed in the diffuse medium so that the dust in the cloud which *then* contracts to become a molecular cloud consists initially of silicate core-organic refractory mantle particles on which condensation from the gas can occur. The mean molecular cloud lifetimes are $\approx 3 \times 10^7$ years and the dust spends about equal times in and out of the molecular clouds. Since the mean turnover time of the interstellar medium to create stars is $\approx 5 \times 10^9$ years, the dust we observe has, on the average, undergone ~50 cycles before being destroyed by forming a star or solar system. In the processing of forming stars, sometimes a planetary system is also created and in the case of our Sun, we know that comets were also formed. In fact, there is evidence for the formation of comets around other stars such as has been suggested [27] for the case of the disk of β Pictoris. These comets are probably like the precursors of those in our solar system which ultimately became part of the Oort cloud population by dynamical ejection from the inner solar system. A summary of the cyclic processing of interstellar dust is shown in Fig. 1.

Let us now describe first the dust as observed in diffuse clouds and then provide some of the evidence for mantles and growth in molecular clouds.

Dust in Space
A brief history of a dust grain

Start

a) Small silicate particles condense in the oxygen rich atmospheres of cool stars and are ejected into space (9.7μm excess infrared emission is characteristic of the SiO stretch in silicates)

Molecular clouds

b) Silicate particles cool to T=10–15K and act as condensation nuclei for accretion of gas atoms + molecules as a mantle of frost

c) Complex chemical interactions between gas and solid leads to an H_2O and CO dominated grain mantle.

d) Ultraviolet radiation photoprocessing of mantle breaks simple molecules leading to complex molecules (laboratory analog results compared with infrared observations) } Lab. analog

Diffuse clouds

e) Star formation blows away clouds. Volatile icy mantles evaporate or are destroyed leaving complex organic refractory mantle only. Pieces of organic refractory break off. } ERA analog

Back to b,c,d,+e and repeat many times until grain consumed by star formation or becomes part of a comet.

Total mean lifetime 5×10^9 yr, =Turnover time for the interstellar medium into and out of stars

FIGURE 1. Sequential summary of the cyclic interstellar processes leading to the dust properties in the proto-solar nebula.

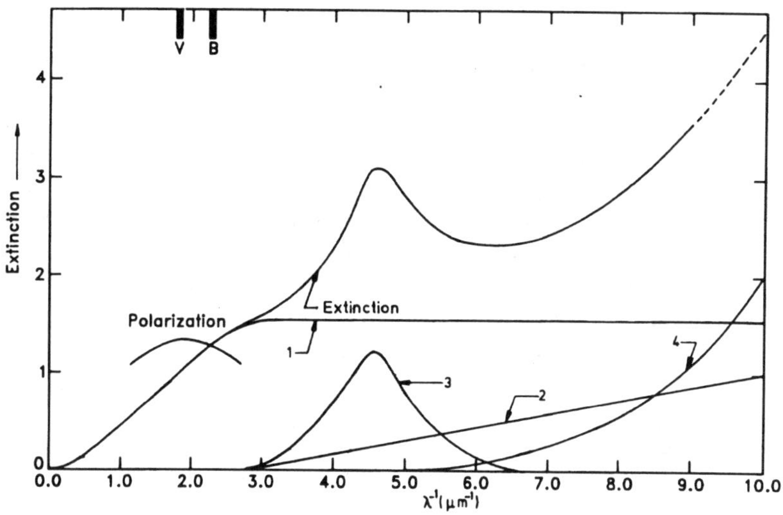

FIGURE 2. Average diffuse interstellar medium extinction curve.

b. Dust in Diffuse Clouds

A unified model for the dust in diffuse clouds has been developed which is applicable to the current epoch of the Milky Way [37]. The dust as it existed when the solar system was formed was subject to the same general physical interactions but could have been somewhat different because the abundance of the condensible atoms which make up the dust were possibly about 30% higher than today. But probably the difference was small because the elements had about the same relative proportions [29], [7]. A model of the dust in diffuse clouds which fits presently known constraints is given in Table 1, where the relative numbers and sizes of the dust particles in the present diffuse interstellar medium are listed. The core-mantle particles are represented for computational purposes by silicate core-organic refractory mantle finite cylinders to take account of their elongation. An explanation of the trimodal distribution of particles is based essentially on accounting for the shape of the extinction curve shown in Fig. 2.

It is beyond the scope of this article to explain in detail how the theory of scattering of light by small particles is applied to the explanation of the break-up of the distribution of the interstellar particles into three components. The details of this are to be found in Li and Greenberg [37] and has evolved over many years of investigation going back to the first such particle size decomposition [16]. The core-mantle particles consist of 0.07 radius silicate cores mantled by an organic refractory material. The hump particles are likely to be a form of carbonaceous material which may have been produced as a result of shattering of a portion of the mantles of the core mantle particles by the grain-grain collisions such as would be induced by supernova shocks. These "hump" particles are subjected to relatively intense ultraviolet radiation fields in the diffuse interstellar medium and because of photodissociation become highly carbonized-approaching hydrogenated amorphous carbon (HAC) or graphite-like material. The source of the far ultraviolet extinction is really still in question but is highly likely to be due to the absorption of light by medium to large polycyclic aromatic hydrocarbon molecules (PAH). In fact, such material has been used to account for a set of interstellar infrared emission features as well as being identified in meteorites [1]. It has also been found in mass spectrometer analyses of the organic refractory material produced by laboratory analogue experiments. [18]. The fit to the extinction curve by the particle sizes shown in Table 1 is shown in Fig. 3.

TABLE 1. Sizes, Numbers and Masses of Each Dust Component in the Diffuse Cloud Determined from the Finite Cylinder Model ($A_v/N_H = 5.3 \times 10^{-22}$ mag cm^2).

Mean Grain Size	Core-mantle $a_c = 0.070$ $\langle a_m \rangle = 0.0866$	Hump $\langle a_H \rangle = 26.7$ Å	PAH $\langle a_{PAH} \rangle = 8.6$ Å
n_d/n_H	3.89×10^{-13}	2.03×10^{-9}	3.11×10^{-7}
n_d/n_{cm}	1.00	5.22×10^3	7.99×10^5
m_d/m_{cm}	1.00	0.11	0.12

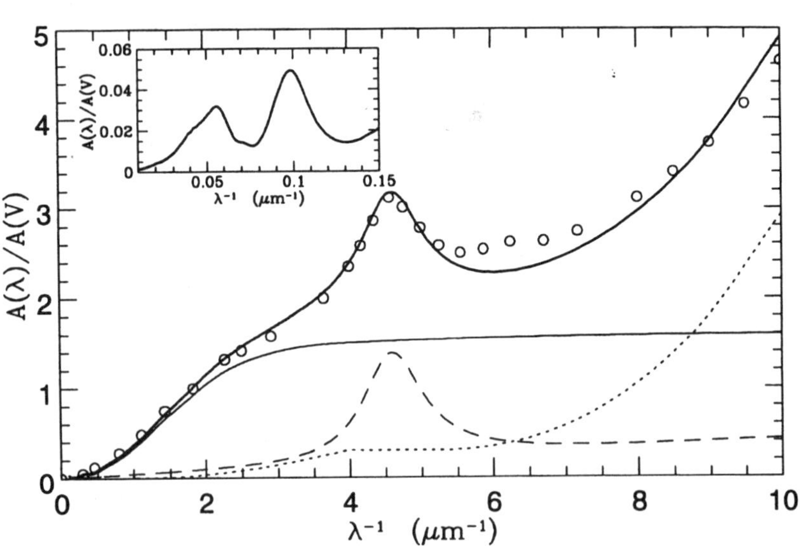

FIGURE 3. Fit to the average diffuse interstellar medium extinction curve by the dust distribution in Table 1.

TABLE 2. Molecules Observed in Interstellar Ice Mantles

Molecule	Feature(s), μm	Abundance[a]
H_2O	3.05, 6.0	100
CO(apolar)	4.673	10–40
CO(polar)	4.681	1–10
CH_3OH	3.54, 3.90, 9.75	[≤ 4]–10
CO_2	15.2	[≤ 0.4]–10
H_2CO	3.475	[≤ 1]–10
H_2	2.415	≥ 1
CH_4	7.68	~ 2
NH_3	2.96	≤ 10
O_3	9.63	≤ 2
OCN^-	4.615	?
OCS/XCS	4.91	?

[a]Abundances refer to observations of background sources if available, see [47].

c. Dust in Molecular Clouds

The dust in molecular clouds accretes atoms and molecules from the gas and this shows up in several ways. First the wavelength dependence of the extinction and polarization in the visible indicate particles whose sizes are larger than in diffuse clouds. Second, the relative contributions of the hump and the far ultraviolet extinction are reduced, suggesting that the particles responsible for these features have accreted on the large core-mantle particles. Third, one observes infrared absorption features corresponding to a variety of frozen molecules in various amounts. The presence of such frozen volatiles is accompanied by the well-known presence of an extremely wide variety of molecules in the gas phase. The overall interaction of the gas and the solid molecules is very complex but it should be noted that what is found as mantles in the solid is not a simple consequence of accretion from the gas [51]. Sometimes the molecules in the gas and in the mantles are in different proportions. For example, in the gas the ratio of water to carbon monoxide is $H_2O:CO < 0.1$ and often even < 0.01, while in the solid phase $H_2O:CO \gtrsim 5:1$ [10]. This is because H_2O is predominantly made *on* the dust, not in the gas [20]. Similarly, the CO_2:CO ratio is higher in the dust than in the gas simply because CO_2 is predominantly made *in* the dust (by ultraviolet photoprocessing of H_2O and CO) and not in the gas [18]. This implies that while *many* of the molecules in the gas also may be found on the dust, we may not assume that their relative proportions are the same in the solid phase. In Table 2 are shown the presently identified molecules in the dust with some representative relative proportions. The observations are limited by selection effects either to quiescent molecular clouds or to regions of star formation but in no case can it be said that we have direct information of what the dust looks like *just before* star formation. Nevertheless, it is from such data that one must attempt to extrapolate to the condition prevailing where comets form. It is to be noted that the water ice band as observed is generally characteristic of low temperature amorphous ice, the exceptions being near protostellar objects or in disks where there is evidence for heating [6]. Because of the way molecules are exchanged between gas and dust during the chemical evolution of a molecular cloud the overall relative proportions do not uniquely define how they are mixed within the dust mantles.

4. COMET NUCLEI AS AGGREGATES OF INTERSTELLAR DUST

a. Chemical Composition

We first present a generalized model of an aggregate of interstellar dust in which each particle is an average of the size distribution of the large dust grains which have accreted all condensible atoms, molecules, *and* small particles. The distribution of the chemical elements in various components will be obtained by assuming first that all Mg, Si, and Fe are contained in the silicates and that the organic mantle mass is about equal to the silicate mass. This has been inferred from the modelling of the silicate polarization feature at 10 and 20 μm [22]. The rest of the atoms are distributed by starting with the assumption that the H_2O is about 5 times as abundant as CO and CO_2 in the final stage of cloud collapse. This was the basis of the approach applied for the first time quantitatively in 1982 with the constraint that in this final state the dust contains *all* condensible elements in their solar system abundance proportions [24]. The solar system abundances have been rederived recently with the main difference being in the C abundance relative to O but we believe that this is still open to question so that we feel reasonably confident in the relative proportions as derived earlier and shown in Table 3.

TABLE 3. Mass Fraction Distribution of the Basic Chemical Components of an Aggregate of Proto Solar Interstellar Dust as Composing a Comet Nucleus.

Component	Mass fraction
Silicate	0.20
Small carbonaceous particles and large PAHs	0.06
Organic refactories	0.19
H_2O	0.40
CO	0.03
CO_2	0.02
Other (CH_3OH, H_2CO, CH_4, etc.)	0.10

FIGURE 4. "Bird's nest" model of a porous aggregate of submicron presolar nebula interstellar particles.

The mean particle dimensions for the fully accreted core-mantle particles are of the order of 0.25 μm thick and 0.5 μm long if the 2:1 ratio is preserved throughout. This will be approximated in some applications by a mean spherical *radius* of 0.15 μm. A "bird's nest" model of such particles is shown in Fig. 4, where the elongation was then assumed to be 3:1 rather than 2:1, as now appears to provide a better match to observations [37].

In Fig. 4, the "bird's nest" model of a porous aggregate of 100 submicron presolar nebula interstellar particles represents a 3 micron diameter piece of a comet nucleus. Each particle consists of a silicate core, an inner mantle of organic refractory molecules and an outer mantle of ices (predominantly water ice) in which are embedded very small carbonaceous particles and large molecules (polycyclic aromatic hydrocarbons). The porosity is 0.83 and the density is 0.28 g cm^{-3}. The scale of this figure conforms with the most recent interstellar dust model sizes [37] and the comet dust model of Greenberg and Hage [23]. The basic ideas incorporated in the "piece of comet" model shown in Fig. 4 are that the individual particles consist of silicate cores with organic refractory mantles and outer mantles of ices (predominantly H_2O but with substantial amounts of CO and CO_2) in which are embedded the *thousands* of carbonaceous and PAH particles/molecules of extremely small size. There are thus two basically different sorts of material: the refractories (silicates and organic refractory) and the volatiles (ices plus a *fraction* of the organics). The organic refractory component was confirmed in the comet Halley dust [34].

It is from the observations of the coma produced as the heated surface material comes off the comet that one must deduce the nucleus composition. Just as the interstellar dust which aggregates to form comets has to be *forward* extrapolated from the dust observed in diffuse and molecular clouds to the protososolar nebula cloud to *predict* the comet nucleus composition, so the comet coma molecules, ions and dust must be extrapolated *backwards* to *deduce* the comet nucleus composition. One should always bear these facts in mind when comparing comets with interstellar dust. But even with this proviso the observational evidence supports a very tight connection. The molecules listed in Table 4 have been observed in various comets. It is to be noted that they are not *uniformly* observed but in all cases the most abundant molecule is H_2O which generally constitutes about 80–90% of the total volatile component.

The fact there is not a one-to-one correspondence between dust volatile molecules as we know them and coma parent molecules is obvious. Nevertheless some molecular proportions strongly indicate that comets were made of unmodified interstellar dust which had not been evaporated and reconstituted in the solar nebula. The ratio of CO_2 to CH_4 and the ratio of HNC to HCN provide such evidence. The proportions observed are totally inconsistent with those which would have been provided by chemical interactions in the solar nebula [13] if a significant fraction of of the dust had evaporated [2], [32]. Since the evidence for such processing is counter indicative we are left with the ice as seen in interstellar space. From this we can infer a really important physical property of comet ice; namely, that the H_2O in comet nuclei must have been, *at least initially*, amorphous. It is only by preserving interstellar ice in its pristine form without evaporation that this can have occurred because water which recondensed anywhere in the solar nebula would automatically be crystalline [36].

TABLE 4. Molecular Abundances in Comets (from Bockelee-Morvan [2])

Molecule	T [K]	C/Hyakutake at 1 AU	Others at 1 AU	C/Hale-Bopp at 4 AU
H_2O	152	100.	100.	100.
CO	24	5.–30.	2.–20.	80.
CO_2	72	7.	3.–6.	30.$^{\text{at 4.6 AU}}$
CH_4	31	0.7	≤ 0.5–2.	
C_2H_2	54	0.3–0.9		
C_2H_6	44	0.4		
CH_3OH	99	2.	1.–7.	6.
H_2CO	64	0.2–1.	0.05–4.	0.1–0.2.
NH_3	78	0.5	0.4–0.9	
N_2	22		0.02	
HCN	95	0.15	0.1–0.2	0.6
HNC		0.01		
CH_3CN	93	0.01		
HC_3N	74		≤ 0.02	
H_2S	57	0.6	0.3	6.
OCS	57	0.3	≤ 0.5	
S_2		0.005	0.02–0.2	
SO_2	83		≤ 0.001	

b. Porosity and Density of the Comet Nucleus

One way to deduce the porosity of the comet nucleus is from the density. If one takes the proportions given in Table 4, one arrives at the density of fully packed interstellar components $\rho_0 = 1.54$ g cm^{-3} [23]. If all the volatiles *and* small carbonaceous particles/molecules are removed, the residual skeleton structure has a density $\rho' = 0.6$ g cm^{-3}. If comet dust had this density, it would correspond to a comet nucleus with porosity 0. However, comet dust with such a *high* density may be shown to be incabable of providing either the source of distributed CO in comet comae or the shape and strength of the silicate emission feature in the coma of comets. It is well known that a particle made up of material which absorbs at a certain wavelength can only reproduce the shape of the absorption feature (in absorption or emission, if heated) if its size is small compared with the wavelength [3]. It can also be demonstrated that a large mass particle acts like a *sum* of small particles if it is a fluffy aggregate of those particles — think of the limit of a cloud. Applying these two principles quantitatively and using the comet dust mass distribution observed for Comet Halley [38] it was demonstrated that the silicate emission feature could not have been observed in the comet coma unless the particles were fluffy aggregates of tenth micron particles [23]. Furthermore, the production of CO as a distributed molecule in the coma demands the higher temperatures of fluffy particles relative to those which are compact [26]. Fluffy particles of high porosity act like the tenth micron particles in the aggregates, and therefore, are substantially hotter than compact particles of the same mass. It turns out that in order to provide the proper conditions to satisfy both of these constraints the porosity of the comet dust is at least 0.975 and the inferred (by refilling in the volatiles) density of the comet nucleus is about 0.28 g cm^{-3}. The model of a piece of comet shown in Fig. 4 has this porosity. Another way to deduce comet densities has been from the effects of the nongravitational forces. Variously applied theories have led to desnities as low as 0.28 g cm^{-3} [46] and seldom higher than 0.6 g cm^{-3}. The latter density leads to a comet porosity value $P_C = 0.68$. We believe that the dust constraints leading to $P_C = 0.83$ provide a value more consistent with observation and this is, in fact, consistent with the comet density demanded by Rickman [46].

5. PHYSICAL PROPERTIES OF COMET NUCLEI

In the following, we discusss the kinds of considerations involved in deriving the thermal conductivity and strength of comet nuclei.

a. Thermal Conductivity

When a comet is heated slowly, as when it is subjected to solar radiation, the principal means of transferring heat is by the diffusion of evaporated gases — principally CO. In the context of extremely *rapid* injection of energy as characteristic of comet impacts on the atmosphere of a planet or its solid surface or being subjected to artificially induced shocks, the difusion time scales are too long to significantly affect thermal conduction. If volatiles are impulsively generated from external sources, it is most likely to lead to shattering or ablation, as will be discussed in Sections 6 and 7.

The thermal conductivity of highly porous objects is strongly influenced by the degree of porosity. However, for completeness, we shall derive the conductivity of the material components which consist of the tenth micron prestellar interstellar grains. As already noted these have a silicate core, an inner mantle of organic refractory material, and an outer mantle of ice (predominantly water ice). Table 5 gives a sample of some representative thermal conductivity coefficients. Although, it should be mentioned that conductivity values for the same substance often were not the same for different references.

TABLE 5. Thermal Conductivities of Representative Grain Material Components

Material	Estimated Thermal Conductivity κ (erg/cm/sec/°K)	Source
H_2O_{amorph}	7.1×10^{-3} T	[36]
H_2O_{cryst}	5.67×10^7 / T (T > 25)	[35]
Glass, Silica	$\sim 10^5$	Handbook Chem. & Phys.
Fused Quartz	10^7	[31]
Organic (paraffine, vaseline)	$\leq 10^4$	Handbook Chem. & Phys.

If we let the radii of the different components be a_1 (silicate), a_2 (organic refractory), a_3 (ice), it turns out that we can approximate the thermal conductivity of the core mantle structure by the same form as that derived for the dielectric function in the Maxwell-Garnet approximation [30]. The effective thermal conductivity of the silicate core-organic refractory mantle units in th nucleus is then given by

$$\kappa_{12} = \kappa_2\{1 + 3f_{12}^3(\kappa_1 - \kappa_2)/[\kappa_1 + 2\kappa_2 - f_{12}^3(\kappa_1 - \kappa_2)]\}, \tag{1}$$

where $f_{12} = a_1/a_2 < 1$, and if we use this silicate core-organic mantle refractory particle to define the "core" of radius a_2 for an ice mantle of radius a_3, the effective thermal conductivity of the double layered particle is

$$\kappa_{123} = \kappa_3\{1 + 3f_{23}^3(\kappa_{12} - \kappa_3)/[\kappa_{12} + 2\kappa_3 - f_{23}^3(\kappa_{12} - \kappa_3)]\} \tag{2}$$

where $f_{23} = a_2/a_3$.

As noted in Haruyama et al. [30], the above is strictly valid only when $f_{12}, f_{23} \ll 1$. When $\kappa_3 \ll \kappa_{12}$, as is true for amorphous ice mantles, Eq. (2) reduces to $\kappa_{123} = \kappa_3$ within a factor ~1 so that the thermal conductivity of a grain is dominated by its ice mantle if the ice is amorphous. In the final analysis, the thermal conductivity also depends on the thermal continuity of the aggregate, and this is given by the contact area per unit area of the particles. This we call the thermal conductivity reduction factor, R. Assuming no sintering to have occurred when the grains aggregated, we may calculate the contact area as the surface within which intermolecular forces act. In the next section on the tensile strength the ratio of the contact area to the total grain area for a very porous aggregate of spherical particles of radius a is given by (see Figs. 4 and 5).

$$R = \Delta A/A = 1/2\beta\ d/a, \tag{3}$$

where d = typical molecular force range, β = number of contact points per particle. For considerations of thermal conductivity, which is essentially unidirectional, the effective value of β is 1. For a mean particle radius a =

0.15 μm and a characteristic d = 2–3 × 10^{-8} cm the reduction factor is R = 0.0007–0.0010. Expressing the thermal conductivity in terms of the aggregate reduction factor brings

$$\kappa' = \kappa R. \tag{4}$$

b. Tensile Strength of Nuclei

The splitting of comets by planetary gravitational fields within the Roche limit, as exemplified by comet Shoemaker-Levy 9 (hereafter, SL-9) when it underwent multiply splitting as it passed near Jupiter, may not be an uncommon occurrence, because the comet nucleus is fragile; i.e., its tensile strength is small compared with that of solid materials. The tensile strength of a model comet can be quantified by regarding the nucleus, for simpliciity, as a homogeneous distribution of small spherical particles *all* covered with ice mantles as would be true for protostellar dust so that ice is the dominant material at the *interfaces* where binding forces act between molecules. The tensile strength of the aggregate material will then depend on the number of interacting molecules at the contact surfaces. Since this number is substantially less per unit volume than that in solid ice the tensile strength is reduced accordingly. The tensile strength of bulk material can be used to depict a mean molecular interaction energy, E, where

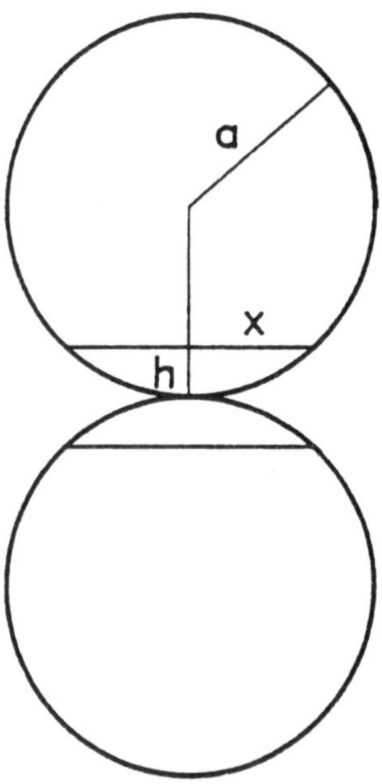

FIGURE 5. Two contacting spherical submicron particles of radius a. The height $h \ll a$ is of the order of the diameter of the molecules at the surface. Diagram is not to scale.

$$E = \alpha 10^{-2} \text{ eV molecule}^{-1} = 1.6 \times 10^{-14} \alpha \text{ erg molecule}^{-1} \quad (5)$$

for hydrogen or van der Waals bonding (depending on the material), where α is to be determined. We let the comet nucleus consist of a porous aggregate of spherical ice mantled particles of radius a. In the loosest possible aggregate the number of contact points per particle is 1. The number of contact points per particle are defined by β, such that $1 \leq \beta \leq 12$, and where $\beta = 12$ corresponds to a cubic lattice. Two spheres in contact are shown in Fig. 5. It is assumed that only those molecules around the point of contact and at a distance of the order of the molecular diameter $\sim 3 \times 10^{-8}$ cm will contribute to the tensile strength. Thus all those molecules within the sagitta defined by $h = 3 \times 10^{-8}$ cm are included in the contact area $\Delta A = \pi x^2 \simeq 2\pi a h$ where $h \ll a$ (in the final analysis it will turn out that $h/a \ll 10^{-3}$). Taking into account the number of contact points per particle, the total number of interacting molecules per particle is $\sim \beta \Delta A/h^2 \simeq 2\pi\beta\, a/h$.

If the nucleus has a porosity P, the number of particles per unit volume is

$$n = (1 - P)/[(4/3)\pi a^3] \quad (6)$$

so that the number, N, of interacting molecules per unit volume is

$$N = 3\beta(1 - P)/2a^2 h, \quad (7)$$

and the ideal tensile strength, T, of the aggregate is then

$$T = NE = 3\beta(1 - P)E/2a^2 h, \quad (8)$$

where E is the *average* molecular interaction energy at the particle interfaces. We then estimate the tensile strength of the aggregate to be

$$T = 6.1 \times 10^3 (1 - P)\beta(a/0.1\ \mu\text{m})^{-2} \text{dyne cm}^{-2}, \quad (9)$$

where $\alpha = 0.76 \times 10^{-2}$ eV.

It should be noted from (8) that the tensile strength is inversely proportional to the square of the particle radius.

The aggregated dust model shown in Fig. 4 as the basis for ascribing a value to the mean number of contact points per interstellar grain as $\beta = 5$. The porosity designed into the model is $P = 0.8$, which is consistent with the comet dust porosities derived from the infrared emission properties of comet Halley dust [23] and the comet nucleus density for this porosity is $\rho = 0.28$ g cm^{-3} which is consistent with that deduced by Rickman from non-gravitational forces [45]. The density assumed by Sekanina et al. [49] is even lower ($\sim$0.2 g cm^{-3}). The maximum mean interstellar dust grain radius before aggregation to form comets is a $\simeq$0.15 μm as derived from the complete accretion of all cosmically abundant elements on the typical (or average) large diffuse cloud dust grain of radius $a \simeq 0.1$ μm.

It also should be pointed out that the interstellar dust chemical, as well as morphological structure of this size and type, provided a quantitative basis for explaining the physical and radiative properties of comet Halley dust [23], [8] as well as having predicted many of the so-called "surprising" properties of the nucleus [17]. Using $\beta = 5$, $P = 0.8$, $a = 0.15$ μm in Eq. (4) gives,

$$T = 2.7 \times 10^3 \text{ dyne cm}^{-2}.$$

This is about 10^4 times smaller than the tensile strength of solid ice [30a] which one recalls is about 2×10^7 dyne cm^{-2}.

The disruption of comet SL-9 has been treated by many authors. It has been suggested by Scotti and Melosh [48] that the parent body of comet SL-9 was about 2 km in radius and split by tidal forces into many pieces at or around perijove. This would imply a very high tensile strength compared with other estimates [49] which are rather like that given in Eq. (5). On the other hand, Boss [5] has proposed a tidal fracture well *before* closest approach leading to the production of a rubble pile structure which subsequently breaks up without appreciable shear forces into many fragments. Based on the results of Weaver et al. [53], the effective radii of most of the fragments are at least 1–2 km which implies an initial radius of the order of 10 km which is 5 times larger than that used by Scotti and Melosh. Using the formalism of sequential tidal fragmentation, Greenberg et al. [21] showed that with a derived tensile strength of 0.0027 bar and a plausible bulk density of ~0.28 g cm^{-3}, the nucleus radius of Shoemaker-Levy 9 had to be at least 7–8 km. The detailed discussion of Sekanina [50] on the nuclear train of SL-9 condensations as well as the degree of penetration of the condensations in Jupiter's atmosphere definitely eliminates the strengthless "rubble pile" model of the nucleus as loose assemblages of hundred-meter sized cometesimals. His conclusion, based on an assumed nucleus density of 0.2 g cm^{-3} (a little lower than ours), is that the tensile strength was, on the average, about the value calculated above, and the parent nucleus was $\gg 2$ km in diameter. In fact, the comet could not have split into the observed number of major fragments unless it was initially at least 9–10 km [49].

6. SHOCK WAVE PROPAGATION THROUGH A POROUS MATERIAL

In the previous sections we have presented arguments that interpret comet nuclei in terms of a porous aggregate material derived from interstellar dust. The porosity and the relatively weak tensile strength (as compared to solid ice) are likely to give rise to anomalous effects when such nuclei interact with intense radiation or material at high velocity, that introduce shock waves to the comet material. In this section we outline how a shock wave propagating through porous material can induce either contraction or expansion in the medum through which it is propagating. The expansion is an anomalous and somewhat counter-intuitive effect, most likely to occur at high shock pressures [11]. Contraction or dilation will depend on the shock energy density and shape as well as on the material (composition and morphology) properties, which in turn will determine the internal energy partition [17], [18]. In the particular case of a strong shock wave traveling through a porous material containing a mixture of materials, as may be encountered with a comet nucleus, very complicated processes can ensue due to the heterogeneity of comet material and the typically large velocities encountered in a planetary environment. Large amounts of energy are (adiabatically) deposited within the porous comet structure, thereby initiating both endothermic and exothermic reactions, depending on the particular structure and composition of the porous material and its effect on the Hugoniot [4]. This trapped energy will give rise to material melting and vaporization, depending on the specific heats and degree of porosity. If all the energy cannot be absorbed by the local melting and vaporization, expansion will occur, giving rise to a local decrease in the density and an overall dilation of the shocked region. In such a situation, the shock wave is said to give rise to an anomalous Hugoniot [57].

7. ANALYTIC INTERPRETATION

An analytic interpretation of this effect has been developed by Zel'dovich and Raizer [56], [57], where the Hugoniot pressure, P_H, for the porous material lies above the Hugoniot pressure curve for a continuous material P_{CM}. This is because a higher operational pressure is required to compress a porous material. The higher the porosity, the steeper is the Hugoniot curve.

$$P_H = [(K - 1)P_{CM} - 2E_M/V]/[K - V_0/V] \quad (10)$$

where E_M is the elastic energy, and assuming that the Grüneisen coefficient is constant,

$$K = 2/\Gamma + 1, \quad (11)$$

and

$$\Gamma = \gamma - 1, \quad (12)$$

where γ is the ratio of the specific heats. V_0 is the initial volume, V is the final volume, and

$$\alpha = V_0/V \tag{13}$$

is compressibility.

In the limit $P_H \rightarrow \infty$, the compressibility for very strong shock waves is, from Eq. (10),

$$\alpha = (\gamma - 1)/(\gamma + 1). \tag{14}$$

Compressibility will occur when $\alpha < K$, and dilation will occur when $\alpha > K$. This result indicates the significance of the material properties and chemistry of the comet nuclei in determining whether it will undergo compression or dilation under a strong shock pressure wave. It is noted that for exceedingly large pressures the electronic properties of the material should also be taken into account through the Grüneisen coefficient. However, here our aim is to show that under certain circumstances, porous materials, such as comet nuclei materials can expand under intense shock pressure, leading to a net decrease in density. For this reason it is important to develop accurate comet nuclei models that take into account both the porosity and the exact chemical compositions which determine the specific heats.

8. RELATED EXPERIMENTAL WORK

In future work [42], we expect to explore the ramifications of the anomalous Hugoniot for porous materials in terms of comet and some types of meteorite evolution. In such models, the chemical composition, structural strength, and physical morphology will play major roles. Recent high intensity shock wave experiments on several types of meteorites have indicated that the low density carbonaceous chondrite and other related classes of stony meteorites expanded and then disintegrated into a fine dust with little evidence of melting. The (stronger) igneous-type of stony meteorites were fractured into small pieces, while the metallic meteorites were compressed under intense shock pressures [41]. Fractures, in general, are found to initiate from voids or pores where energy is trapped. Additional high intensity shock experiments are currently being planned.

9. CONCLUSIONS

Through a combination of data from available observations, laboratory experiments, and theoretical modelling, one can reach the conclusion that unmetamorphosed (pristine) comet nuclei are composed primarily of a porous aggregate material. The implications of such a model suggest a very low thermal conductivity and tensile strength. Such a comet nucleus material, upon being subjected to a strong shock wave, is likely to undergo — depending upon the chemical composition — an explosive expansion rather than a compression. It is therefore of great importance to understand the nature of comet nuclei in terms of the origin and evolution of the solar system and the interstellar medium, probable comet evolutionary paths, and their reaction to external forces both in order to assist in predicting orbit movements, tidal break-up, and interaction by external forces either for exploration or orbital adjustments.

REFERENCES

1. ALLAMANDOLA, L.J. 1996. *In* The Cosmic Dust Connection, J.M. Greenberg, Ed.: 81. Kluwer. Dordrecht, The Netherlands.
2. BOCKELEE-MORVAN, D. 1997. *In* IAU Symp. Molecules in Astrophysics: Probes and Processes, E. van Dishoeck, Ed. In press.
3. BOHREN, C.F. & D.R. HUFFMAN. 1983. Absorption and Scattering of Light by Small Particles. Wiley. New York, NY.
4. BOSLOUGH, M. 1990. J.Chem. Phys. **92:** 1939.
5. BOSS, A.P. 1994. Icarus. **107:** 422–426.
6. VAN DE BULT, C.E.P.M., J.M. GREENBERG, & D.C.B. WHITTET. 1984. Mon. Not. Roy. Astron. Soc. **214:** 289–305
7. CARDELLI, J.A., O.N. MEYER, M. JURA, & B.D. SAVAGE. 1996. Ap. J. **467:** 334.
8. CHAPMAN, C.R. 1990. Nature. **348:** 486–487.
9. DESERT, F.X., E. BOULANGER, & J.L. PUGET. 1990. Astron. & Astrophys. **237:** 215.
10. D'HENDECOURT, L., L.J. ALLAMANDOLA, & J.M. GREENBERG. 1985. Astron. & Astrophys. **152:** 130.
11. DREMIN, A.N. & O.N. BREUSOV. 1968. Russian Chem. Rev. **37:** 392.
12. ESPINASSE, S., J. KLINGER, C. RITZ, & B. SCHMITT. 1991. Icarus. **92:** 350.
13. FEGLEY, B. & R.G. PRINN. 1989. *In* Formation and Evolution of Planetary Systems, H.A. Weaver & L. Danly, Eds.: 171–205. Cambridge Univ. Press.
14. FURNISH, M.D. & J.L. REMO. 1997. Ice issues, porosity, and snow experiments for dynamic NEO and comet modelling. *In* Proceedings of the International Symposium on Near-Earth Objects, United Nations/Explorers Club, New York, April 24–26, 1995, in this volume.
15. GREENBERG, J.M. & C.X. MENDOZA-GOMEZ. 1992. Adv. Sp. Res. **12:** 169.
16. GREENBERG, J.M. 1973. *In* Molecules in the Galactic Environment, Vol. 93, M.A. Gordon & L.E. Snyder, Eds. J. Wiley, New York, NY.
17. GREENBERG, J.M. 1986. *In* The Galaxy and the Solar System, Vol. 103, R. Smoluchowski, J.N. Bahcall, & M.S. Mathews, Eds. University of Arizona Press. Tucson, Arizona.
18. GREENBERG, J.M., C.X. MENDOZA-GOMEZ, M.S. DEGROOT, & R. BREUKERS. 1993. *In* Dust & Chemistry in Astronomy, T.J. Millar & D.A. Williams, Eds.: 265. I.O.P. Publ. Ltd. Bristol, England.
19. GREENBERG, J.M. & J.L. REMO. 1994. A current working model of a Comet Nucleus and Implications for NEO Interactions. *In* Hazards due to Comets and Asteroids, T. Gehrels, Ed.: 635–647. University of Arizona Press. Tucson, Arizona.
20. GREENBERG, J.M. & O.M. SHALABIEA. 1993. *In* Asteroids, Comets, Meteors, A. Milani et al., Eds.: 327–342.

21. GREENBERG, J.M., H. MITZUTANI, & T. YAMAMOTO. 1995. Astron. and Astrophys. **295:** L35.
22. GREENBERG, J.M. & A. LI. 1996. Astron. and Astrophys. **309:** 258.
23. GREENBERG, J.M. & J.I. HAGE. 1990. Ap. J. **361:** 260.
24. GREENBERG, J.M. 1982. *In* Comets, L.L.Wilkening, Ed.: 131–161. University of Arizona Press. Tucson, Arizona.
25. GREENBERG, J.M. 1978. *In* Cosmic Dust, J.A.M. McDonnell, Ed.: Chap. 4, 187.
26. GREENBERG, J.M. & A. LI. 1997. *In* Solar System Ices, B.Schmitt, C. deBergh, M. Festou, Eds. Chap. 13. Academic Press, New York.
27. GREENBERG, J.M. & A. LI. 1996. *In* The role of Dust in the Formation of Stars, H.U.Kaufl & R.Siebenmorgen, Eds.: 161. Springer. Heidleberg, Germany.
28. GREENBERG, J.M. 1986. Nature. **321:** 385.
29. GREVESSE, N., A. NOELS, & A.J. SAURAL. 1996. *In* Cosmic Abundances, S. Holt & G. Sonneborn, Eds.: Vol. 99: 117. ASP Conference Series.
30. HARUYAMA, J., T. YAMAMOTO, H. MIZUTANI, & J.M. GREENBERG. 1993. J.Geophys. Res. (Planetary). **98E8:** 15079.
30a. HOBBS, P.V. 1974. Ice Physics. Claredon Press, Oxford.
31. HORAI, K. & J. SUSAKI. 1989. Phys. Earth. Planet. Inter. **55:** 292.
32. IRVINE, W.M., ET AL. 1996. Nature. **383:** 418.
33. JENNISKENS, P., G.A. BARATTA, A.KOUCHI, M.S. DEGROOT, J.M. GREENBERG, & G. STRAZZULLA. 1993. Astron. and Astrophys. **273:** 583.
34. KISSEL, J. & F.R. KRUEGER. 1987. Nature. **326:** 755–760.
35. KLINGER, J. 1975. J. Glaciol. **14:** 517.
36. KOUCHI, A., T. YAMAMOTO, J.KOZASA, R. KURODA, & J.M. GREENBERG. 1994. Astron. and Astrophys. **290:** 1009.
37. LI, A. & J.M. GREENBERG. 1997. Astron. and Astrophys., in press
38. MCDONNELL, J.A.M., P.L. LAMY, & G.S. PANKIEWICZ. 1991. *In* Comets in the Post-Halley Era, R.I. Newburn, M. Neugebauer, & J. Rahe, Eds.: 1043. Kluwer. Dordrecht, The Netherlands.
39. MUMMA, M.J., P.R. WEISSMAN, & S.A. STERN. 1993. *In* Protostars and Planets, III, E.H. Levy, J.I. Lunine, & M.S. Matthews, Eds.: 1175–1252. University of Arizona Press. Tucson, Arizona.
40. PRIALNIK, D. & A. BAR-NUN. 1990. Ap. J. **363:** 274.
41. REMO, J.L., A. FARNSWORTH, & W. TROTT. 1997. Laser Induced Shock Waves on Meteorite Material. In preparation.
42. REMO, J.L. 1997. High Pressure Shock Wave Interactions in Porous Media. In preparation.
43. REMO, J.L. 1996. Policy perspectives from the UN International Conference on near-Earth Objects. Space Policy. **12:** 13–17.
44. REMO, J.L. 1994. Classifiying and modeling NEO material properties and interactions. *In* Hazards Due to Comets and Asteroids, T. Gehrels, Ed.: 551–596. University of Arizona Press. Tucson, Arizona.
45. RICKMAN, H. 1986. *In* Comet Nucleus Sample Return Mission, O.Melitta, Ed.: 195. ESA SP-249,.Noordwijk, The Netherlands.
46. RICKMAN, H. 1991. *In* Comets in the Post-Halley Era, R.L.Newburn, M. Neugebauer, & J.Rahe, Eds.: 733. Kluwer. Dordrecht, The Netherlands.
47. SCHUTTE, W. 1996. *In* The Cosmic Dust Connection, J.M.Greenberg, Ed.: 1. Kluwer. Dordrecht, The Netherlands.
48. SCOTTI, J.V. & H.J. MELOSH. 1993. Nature. **365:** 733–735.
49. SEKANINA, Z., P. CHODES, & D. YEOMANS. 1994. Astron. and Astrophys. **289:** 607
50. SEKANINA, Z. 1996. *In* IAU Colloquium Comet Shoemaker-Levy, K.S. Noll, H.A. Weaver, & P.D. Feldman, Eds.: 55. Cambridge Univ. Press.

51. SHALABIEA, D.M. & J.M. GREENBERG. 1994. Astron. and Astrophys. **290:** 266.
52. TANCREDI, G., H. RICKMAN, & J.M. GREENBERG. 1994. Astron. and Astrophys. **286:** 659–682.
53. WEAVER, H.A., ET AL. 1994. Science. **263:** 787–791.
54. WHITTET, D.C.B. 1996. *In* The Cosmic Dust Connection, J.M. Greenberg, Ed.: 133. Kluwer. Dordrecht, The Netherlands. .
55. WHITTET, D.C.C.B. 1992. Dust in the Galactic Environment. I.O.P. Publ. Ltd. Bristol, England.
56. ZEL'DOVICH, Y.B. & YU.P. RAIZER. 1966. Physics of Shock Waves and High-Temperature Hydrodynamic Phenomena, Vol.1.
57. ZEL'DOVICH, Y.B. & YU.P. RAIZER. 1967. Physics of Shock Waves and High-Temperature Hydrodynamic Phenomena, Vol.2.

Radar Reconnaissance of Near-Earth Objects at the Dawn of the Next Millennium[a]

STEVEN J. OSTRO[b]

300-233, Jet Propulsion Laboratory
California Institute of Technology
Pasadena, Califoria 91109-8099

ABSTRACT: Radar, the most powerful ground-based technique for postdiscovery investigation of NEOs (near-Earth objects), can contribute a great deal to their exploration as well as to identification and mitigation of hazardous objects. My intentions in this article are to review the current state of NEO radar reconnaissance, examine the imminent prospects for this work as upgraded instrumentation becomes available, and propose construction of a next-generation radar telescope that, unlike any existing radar instrument, would be optimized for, and dedicated to, NEO radar.

Radar is uniquely capable of resolving NEOs spatially by measuring the distribution of echo power in time delay (range) and Doppler frequency (line-of-sight velocity) with very fine precision in each coordinate, as shown in Table 1. Delay-Doppler images provide otherwise unavailable information about the target's physical properties, including size, shape, rotation, multiplicity, near-surface bulk density and roughness, and internal density distribution [23].

TABLE 1. Radar Measurement Precision

	Range (meters)	Line-of-Sight Velocity (meters/sec)
Best radar resolution	~10	0.0001
Asteroid "size"	~1,000	0.01 to 1
Asteroid "location"	~10,000,000,000	10,000

The delay-Doppler projection is somewhat counter-intuitive and is subject to global aliasing, but given adequate orientational coverage, an imaging sequence can be inverted to yield a detailed, unambiguous model of the target [12]. Hudson has calibrated the accuracy of the "reconstruction inversion" with his laboratory laser radar [1], which serves as a scaled-down model of the Arecibo (13 cm, 2380 MHz) and Goldstone (3.5 cm, 8510

[a]This research was conducted at the Jet Propulsion Laboratory, California Institute of Technology under contract with the National Aeronautics and Space Administration.
[b]E-mail: ostro@echo.jpl.nasa.gov.

MHz) asteroid-imaging radars. (The inversion is much like the template-fitting process used to generate radar maps of Venus from ground-based or spacecraft delay-Doppler images.)

Additional remote-sensing advantages accrue from radar wavelengths' sensitivity to near surface bulk density and roughness. For comets, radar waves can penetrate optically opaque comas to examine the nucleus and can also disclose the presence of macroscopic coma particles.

The fractional precision of delay-Doppler positional measurements plus their orthogonality to optical astrometry makes them invaluable for refining orbits and prediction ephemerides [36]. A single radar detection secures the orbit well enough to prevent "loss" of the object, shrinking the object's instantaneous positional uncertainties by orders of magnitude with respect to an optical-only orbit and greatly improving the accuracy of long-term trajectory predictions [34], [35]. During the past two decades, observations of NEOs [21] have revealed range errors from ~100 km to ~100,000 km in pre-radar range predictions.

HIGHLIGHTS OF NEO RADAR RESULTS

Echoes from 42 NEAs (see list at http://echo.jpl.nasa.gov) have furnished new information about these objects' physical and dynamical properties. Reflectivity and polarization signatures reveal striking diversity in NEA characteristics. Asteroid 6178 (1986 DA) is significantly more reflective than any other radar-detected asteroid; it may be a 2 km, highly nonconvex (bifurcated?) piece of NiFe metal derived from the interior of a much larger object that melted, differentiated, cooled, and subsequently was disrupted in a catastrophic collision [22]. At the other extreme, the radar signature of 1986 JK suggests a surface bulk density within a factor of two of 0.9 g cm^{-3} [20]. Similarly, NEA circular polarization ratios, a measure of the severity of small-scale roughness, range from ~0.1 (e.g., 1685 Toro and 1989 JA) to 1.0 (2101 Adonis and 1992 QN); see [21], references therein, and [3].

Radar observations of six comets have provided useful constraints on nuclear dimensions, spin vectors, and surface morphologies. The echoes from comet IRAS-Araki-Alcock, which came within 0.03 AU of Earth in 1983 [9], [8], show a narrowband component from the nucleus as well as a much weaker broadband component attributed to large particles ejected mostly from the sunlit side of the nucleus. Models of the echoes suggest that the nucleus is very rough on scales larger than a meter, that its maximum overall dimension is within a factor of two of 10 km, and that its spin period is 2–3 days. The particles are probably several centimeters in size and account for a significant fraction of the particulate mass loss from the nucleus. Most of them appear to be distributed within ~1000 km of the nucleus, in the volume filled by particles ejected at several meters per second over a few days. Radar spectra of comet Halley [4] are broad and show no hint of a component as narrow as that expected from the nucleus, whose dimensions and spin vector were constrained by Giotto and Vega spacecraft images. The echo's bandwidth and radar cross section suggest that it arises predominantly from coma particles with radii > 2 cm. The nondetection of Halley's nucleus implies a surface with a very low bulk density. Echoes from the coma of comet 1996 B2 Hyakutake (Fig. 1) are about ten times stronger than those from the nucleus, whose size probably is a few kilometers [10].

Observations of 1620 Geographos [25], [28] yield ~400 images with ~100 m resolution (Fig. 2). The pole-on silhouette's extreme dimensions are in a ratio, 2.76 ± 0.21, that establishes Geographos as the most elongated solar system object imaged so far. The images show craters as well as indications of other sorts of large-scale topographic relief, including a prominent central indentation. Protuberances at the asteroid's ends may be related to the pattern of ejecta removal and deposition caused by the asteroid's gravity field.

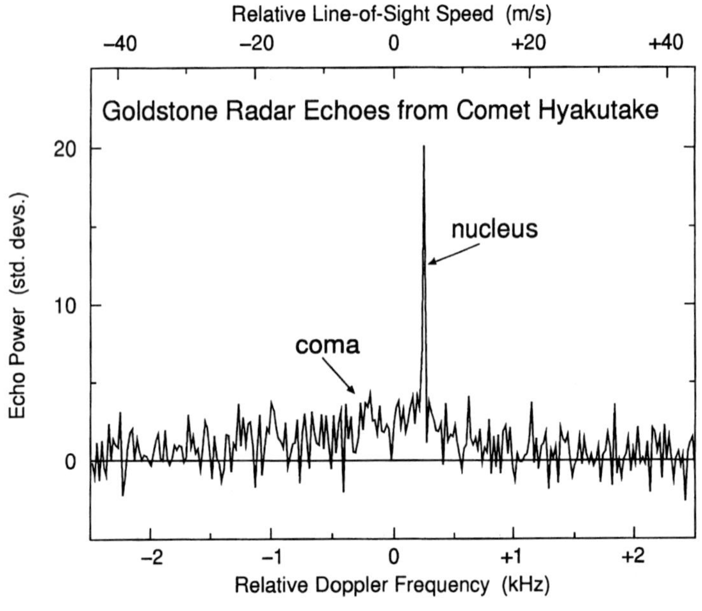

FIGURE 1. Echo power spectrum from comet 1996 B2 Hyakutake. See [10].

A sequence of delay-Doppler images of 4769 Castalia (1989 PB) (Fig. 3; [20]), obtained two weeks after its Aug. 1989 discovery, reveal it to consist of two kilometer-sized lobes in contact. Least-squares estimation of Castalia's three-dimensional shape (Fig. 4) from the radar images supports the hypothesis that Castalia is a contact-binary asteroid formed from a gentle collision of the two lobes and also constrains the object's surface morphology and pole direction [13].

Delay-Doppler imaging of 4179 Toutatis in Dec. 1992 [26] achieved resolutions as fine as 125 ns (19 m in range) and 8.3 mHz (0.15 mm/s in radial velocity), placing hundreds to thousands of pixels on the asteroid. The images reveal this asteroid to be in a highly unusual, nonprincipal-axis (NPA) spin state with several-day characteristic time scales. This data set provides physical and dynamical information that is unprecedented for an Earth crossing object. Hudson and Ostro [14] used a comprehensive physical model to invert the lower-resolution images (Fig. 5) to estimate the asteroid's shape and inertia tensor, initial conditions for the asteroid's spin and orientation, the radar scattering properties of the surface, and the delay-Doppler trajectory of the center of mass. Figure 6 presents four views of the low resolution model that show shallow craters, linear ridges and a deep topographic "neck" whose geologic origin is not known. It may have been sculpted by impacts into a single, coherent body, or Toutatis might actually consist of two separate objects that came together in a gentle collision. Toutatis is rotating in a long-axis mode (Fig. 7) characterized by periods of 5.41 days (rotation about the long axis) and 7.35 days (average for long-axis precession about the angular momentum vector). The asteroid's principal moments of inertia are in ratios within 3% of 3.19 and 3.01, and the inertia tensor is indistinguishable from that of a homogeneous body. Such information has yet to be determined for any other asteroid or comet, and probably is impossible to acquire in a fast spacecraft flyby. Higher-resolution images (e.g., Fig. 8) from the 1992 experiment are now being used to refine the Toutatis model.

PHYSICAL STUDIES OF NEOS USING RADAR-BASED MODELS

Accurate shape models of near-Earth asteroids open the door to a wide variety of theoretical investigations that previously have been impossible or have used simplistic models (spheres or ellipsoids). For example, with detailed models of real objects, it is possible to explore the evolution and stability of close orbits, with direct application to the design of robotic and piloted spacecraft missions, to studies of retention and redistribution of impact ejecta, and to questions about plausible origins and lifetimes of asteroidal satellites.

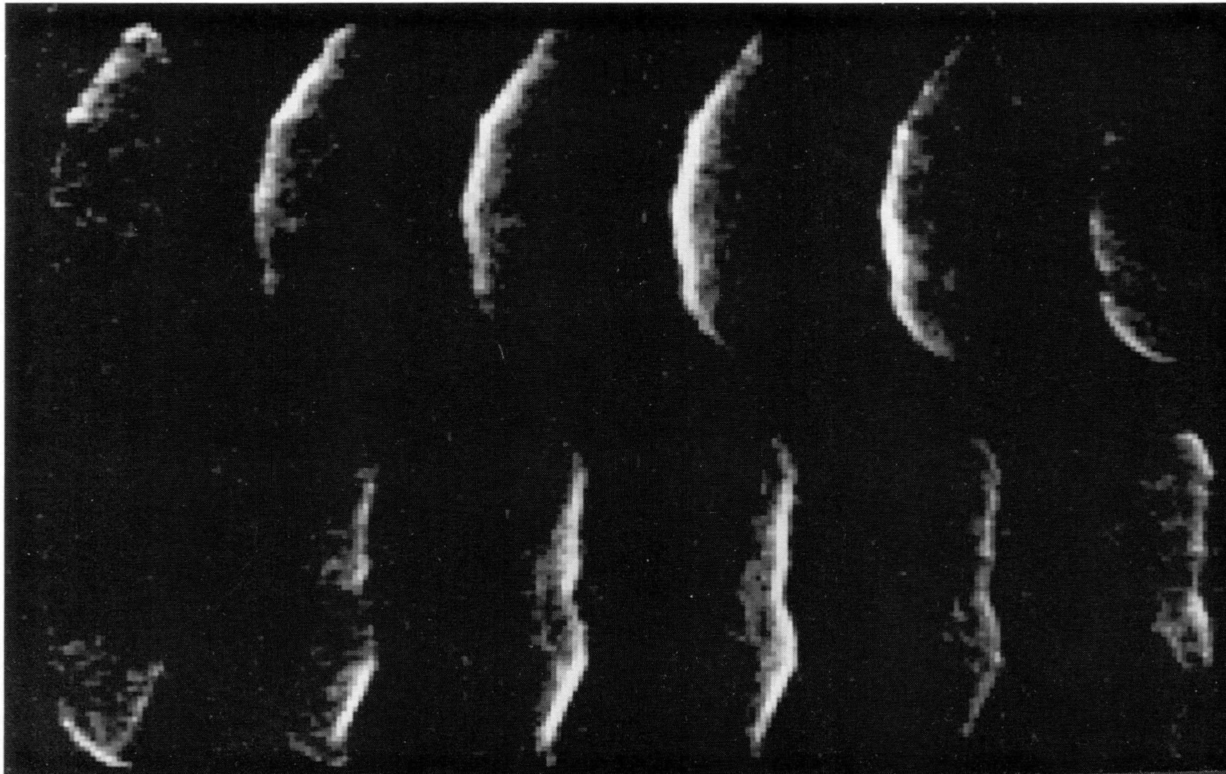

FIGURE 2. Delay-Doppler images of asteroid 1620 Geographos [28]. The resolution is 0.5 μs × 1.64 Hz (75 m × 87 m). These frames are multi-run sums of coregistered images within twelve independent, 30°-wide rotation-phase blocks. The view is along the pole and the asteroid has the same rotational orientation in each frame, but the direction to the radar is different in each frame. In the top row, the asteroid-to-radar vector points toward 12 o'clock in the far left image (frame 1), toward 11 o'clock in the next image (frame 2), etc.

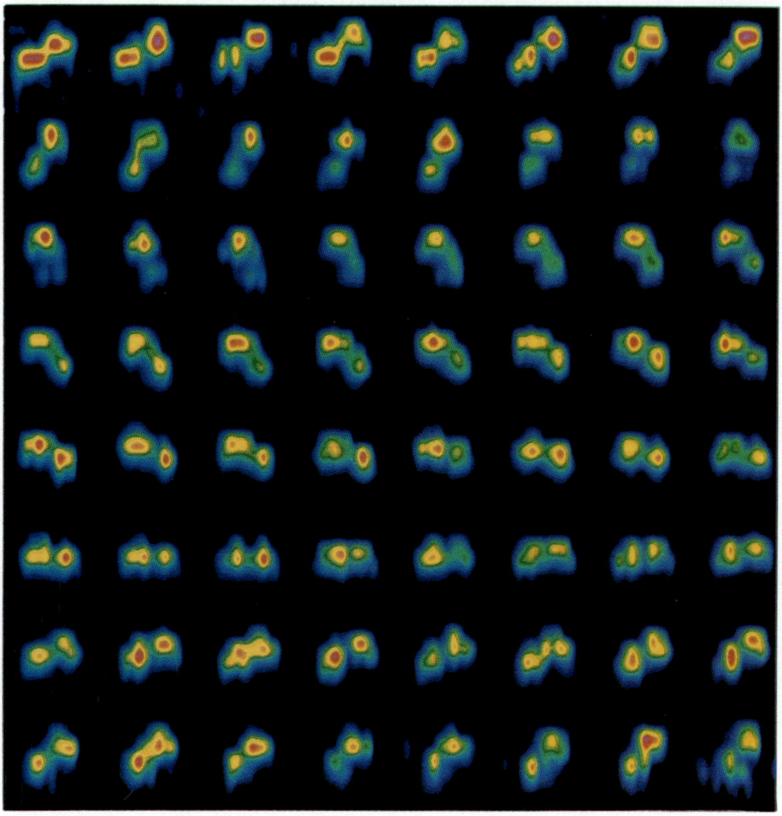

FIGURE 3. Radar images of asteroid 4769 Castalia (1989 PB). This 64-frame Arecibo "movie" is to be read like a book (left to right in the top row, etc.). The radar lies toward the top of the figure in the image plane, which probably is about 35° from the asteroid's equator. In each frame, echo power (i.e., the brightness seen by the radar) is plotted vs. time delay (increasing from top to bottom) and Doppler frequency (increasing from left to right). The radar illumination comes from the top of the figure, so parts of the asteroid facing toward the bottom are not seen in these images. The object, each of whose lobes is about a kilometer in diameter, is seen rotating through about 220° during the 2.5 h sequence. These images have been smoothed from the original resolution of 2 μs × 1 Hz (0.3 x 0.17 km). (Ostro et al. [20], copyright 1990 by the AAAS.)

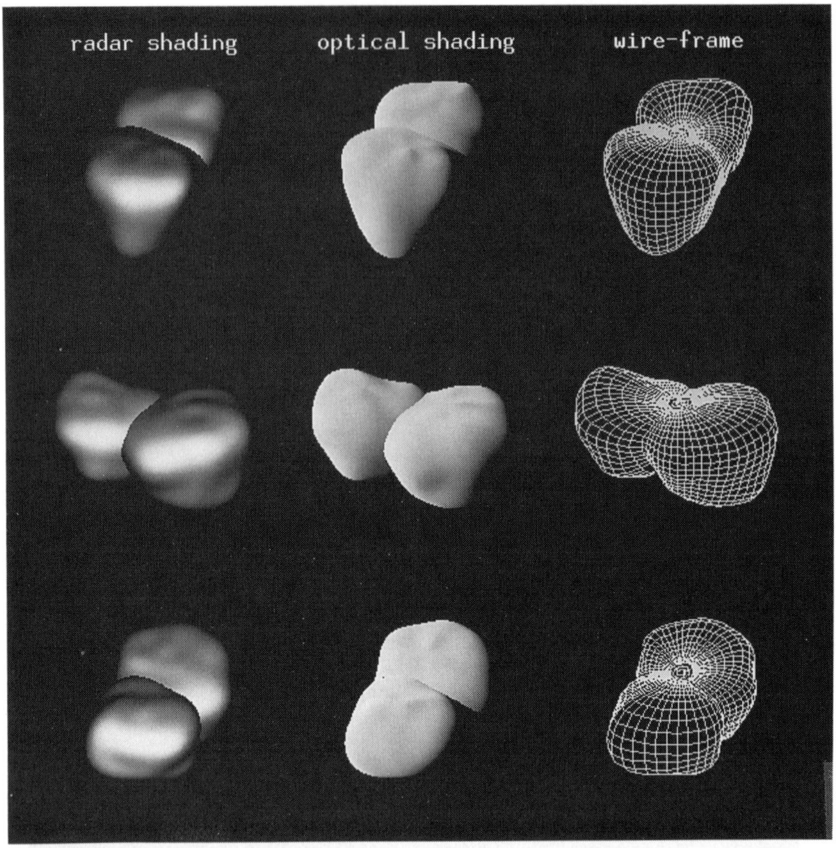

FIGURE 4. Three-dimensional computer model of asteroid 4769 Castalia (1989 PB), derived by Hudson and Ostro [13] from the radar images in Fig. 3. The object is about 1.8 km across at its widest. The effective resolution of the reconstruction is about 100 m. The left column uses the estimated radar backscattering law, the center column uses a realistic optical scattering law and a Sun-asteroid-viewer angle of 37°, and the right column shows a wire-grid representation. In each column the renderings are 90° of rotation phase apart.

Scheeres et al. [29], [30] have used the Castalia and Toutatis models to investigate close-orbit dynamics around these objects. For each body, both stable and unstable periodic orbits exist as closed orbits in an asteroid-fixed frame. The fate of material launched from the surface is a strong function of initial velocity and surface location. Return, escape, periodic, and realistic landing orbits about these objects have been rendered (e.g., Fig. 9) and animated by de Jong and Suzuki [6]. Similar studies [31] are exploring close orbits about 433 Eros [15].

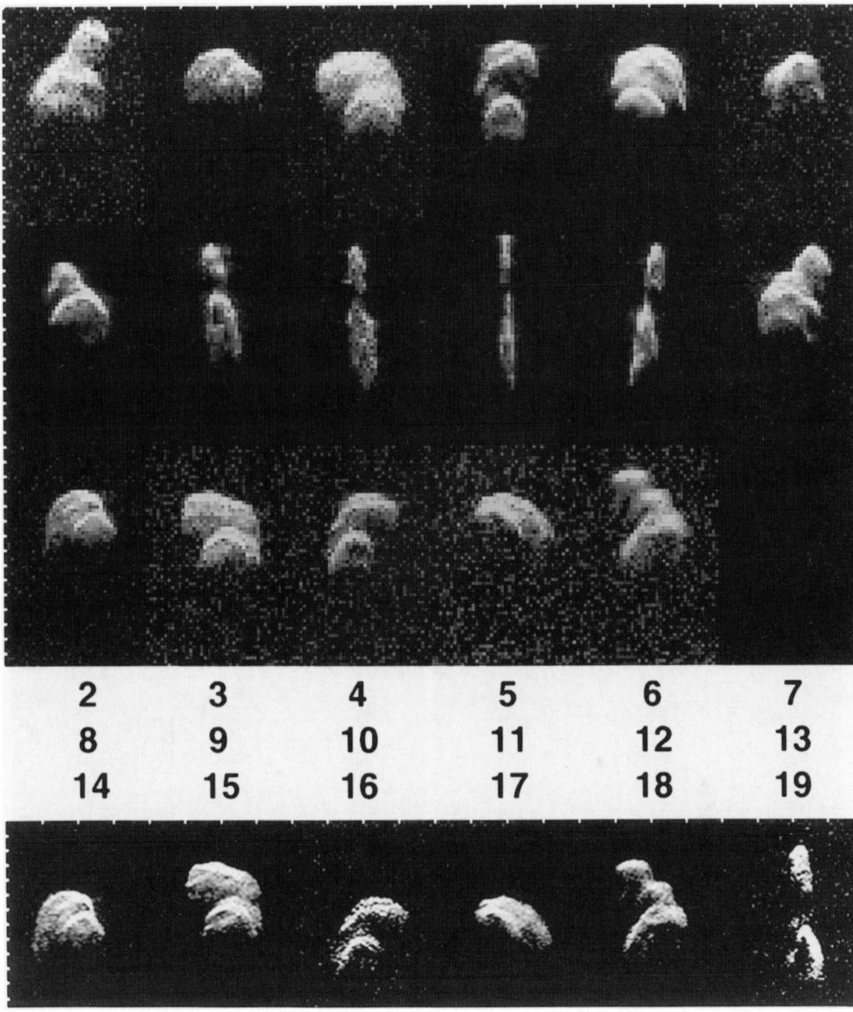

FIGURE 5. Radar images of asteroid 4179 Toutatis from 1992. Goldstone low-resolution images from Dec. 2–18 (top three rows) and Arecibo images from Dec. 14–19 (bottom row) are plotted with time delay increasing toward the bottom and Doppler frequency increasing toward the left. On the vertical sides, ticks are 2 μs apart. Two horizontal sides have ticks separated by 1 Hz for Goldstone and 0.28 Hz for Arecibo; those intervals correspond to a radial velocity difference of 18 mm/s. (Ostro et al. [26], copyright 1995 by the AAAS.)

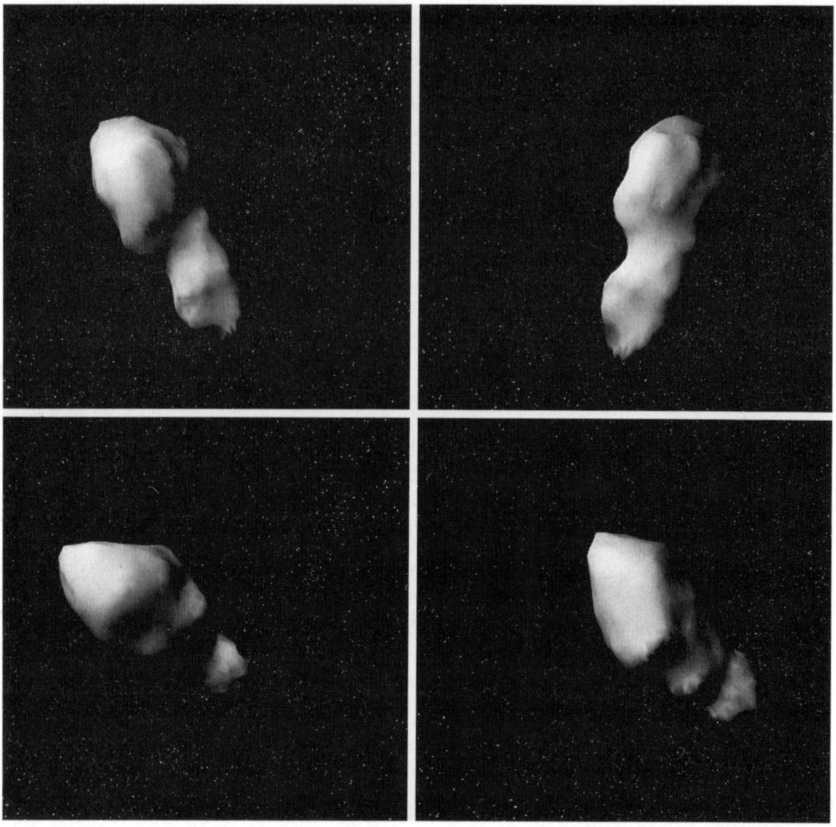

FIGURE 6. Computer model of asteroid 4179 Toutatis from radar images in Fig. 5. The object's dimensions along the principal axes are 1.92, 2.40, and 4.60 km. The resolution of the computer model is about 84 m. (Hudson and Ostro [14], copyright 1995 by the AAAS.)

Radar-derived models also allow realistic investigations of the effects of collisions in various energy regimes on the object's rotation state, surface topography, regolith, and internal structure. Simulations of impacts into Castalia using smooth particle hydrodynamics code have begun to suggest how surface and interior damage depends on impact energy, impact location, and the equation of state of the asteroidal material [2]. These computer investigations have clear ramifications for our understanding of asteroid collisional history, for exploitation of asteroid resources, and eventually for deflection/destruction of objects found to be on a collision course with Earth.

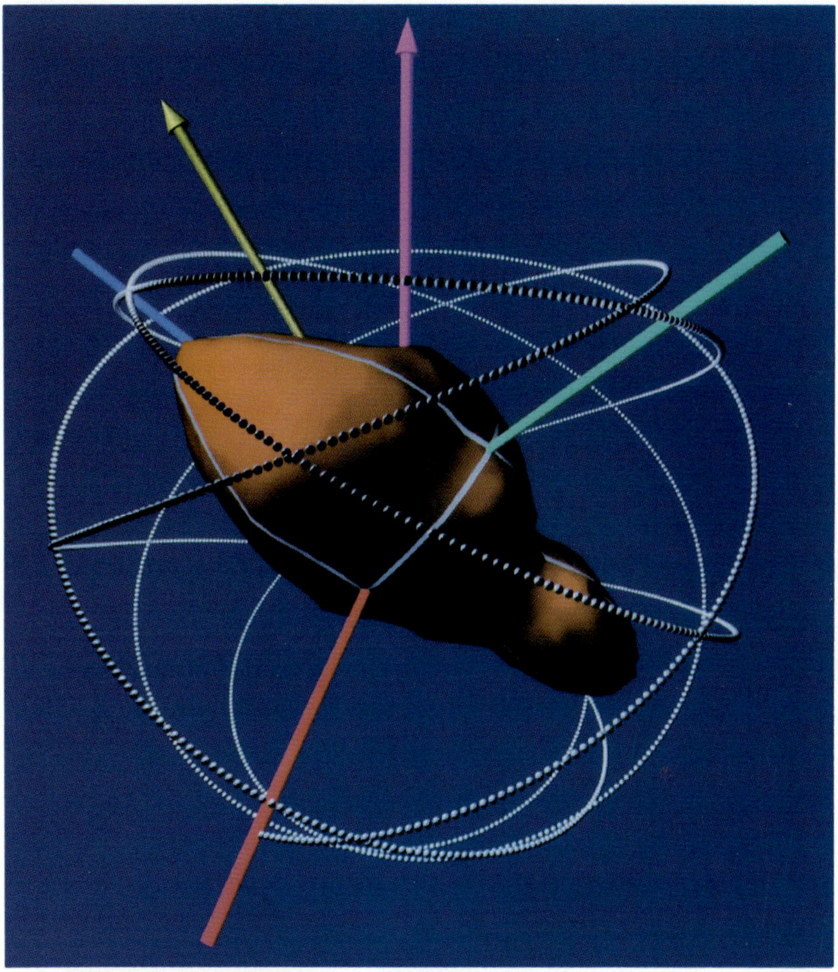

FIGURE 7. Spin state of Toutatis from the Hudson-Ostro [14] modeling of radar images. Here the white contours on the asteroid's surface are loosely analogous to an equator and 0° and 90° meridians. Red, green, and blue axes are the object's short, intermediate, and long principal axes of inertia. The violet arrow is the angular momentum vector, and the yellow arrow is the instantaneous spin vector. The small spheres trace the path in an inertial frame of a point on the red axis at 15-minute intervals for 20 days into the future.

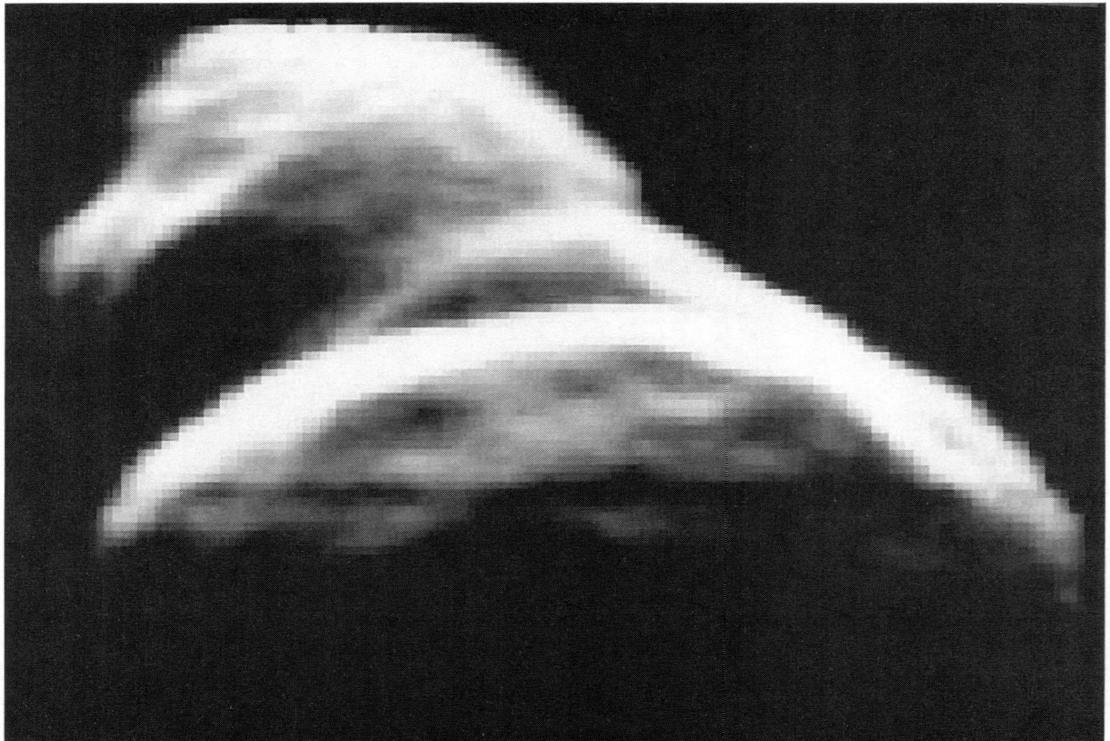

FIGURE 8. A high-resolution Goldstone image of Toutatis from Dec. 8, 1992, dilated according to km/Hz conversion factors predicted by the Hudson-Ostro [14] model for the asteroid's spin state, with look-to-look translational smearing removed according to the model-based estimates of the trajectory of the asteroid's center of mass with respect to the delay-Doppler prediction ephemerides. Geometry is as in Fig. 5. Perimeter ticks are 1 km apart. (Ostro et al. [26], copyright 1995 by the AAAS.)

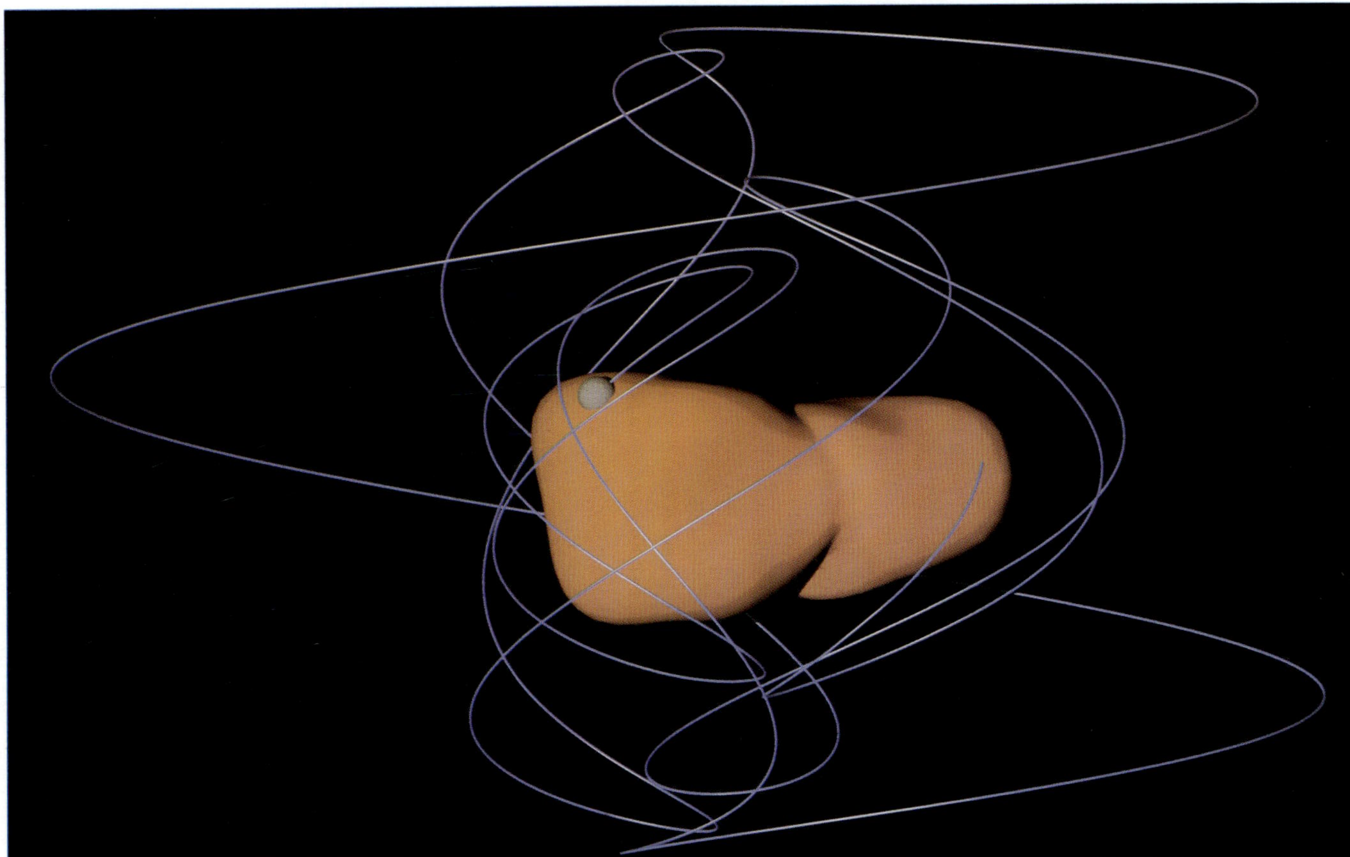

FIGURE 9. Return orbit about asteroid 4769 Castalia, calculated by Scheeres et al. [30] using the Hudson-Ostro [13] radar-based model. The particle was launched at 0.2 m/s and returns 1.4 days later at 0.2 m/s. This is one frame from a library of animations of close orbits around Castalia [6].

CAPABILITIES OF CURRENT RADAR TELESCOPES

The Arecibo Observatory (Fig. 10) has just undergone a major upgrade and is about to resume normal operations. By 1997 it will be more than 20 times as sensitive as Goldstone, see twice as far, and cover three times as much volume. The more fully steerable Goldstone instrument (Fig. 11), with a solid angle window twice the size of Arecibo's and tracking times at least several times longer than Arecibo's for any given target, will serve a complementary role, especially for newly discovered objects.

Table 2 lists the radar-astrometric range limits expected for Arecibo and Goldstone by 1997. Goldstone also can be used as the transmitter in a bistatic configuration with the 27 antenna Very Large Array (VLA) as an aperture-synthesis receiver [7]. The G-VLA system can measure angular positions of NEAs with uncertainties that approach 0.01 arcsec, or an order of magnitude finer than the best optical astrometry.

Table 3 lists statistics of NEA opportunities at Arecibo as a function of signal-to-noise ratio (SNR) achievable on a single date, before and after the upgrade and also after a "Spaceguard Survey" of the NEA population [16]. (Comet radar opportunities are very roughly 10% as frequent as NEA opportunities.) SNRs > 20 are quite adequate for detection and marginal resolution of the echoes. SNRs > 100 let one achieve enough resolution to be able to make simple statements about shape, and data sets with SNRs > 1000 can reveal geologic features. SNR increases as the square root of the integration time and can be expected to be of the same order as the number of useful imaging pixels and the number of parameters in a shape reconstruction. The highest-SNR radar investigations can be as informative as the cheapest NEO flybys under consideration. It is not exaggerating to claim that in principle, radar is to physical characterization of asteroids as the Hubble Space Telescope is to physical characterization of the rest of the Universe.

TABLE 2. Range Limits (AU) for Radar Astrometry of New NEOs, Corresponding to a Single Date SNR of 20 and Telescope Sensitivities Expected by 1997

Asteroid Diameter	Arecibo	Goldstone
10 m	0.043	0.016
100 m	0.10	0.041
1 km	0.24	0.10
10 km	0.59	0.24

TABLE 3. NEA Radar Opportunities with the Upgraded Arecibo Telescope

	Number of objects with a single-date SNR greater than			
	20	100	1000	5000
Annual average before upgrade (using actual statistics since 1980)	3	0.9	0.3	0.1
Annual average after upgrade (using current NEA pool)	10	6	2	0.6
Annual average after Spaceguard Survey				
(lower bounds) > 1 km	80	20	10	4
> 300 m	160	80	20	7

FIGURE 10a. The 305 m Arecibo telescope in Puerto Rico. Aerial view.

FIGURE 10b. The 305-m Arecibo telescope in Puerto Rico. Feed plafform, which is suspended ~170 m above the main reflector. The 26 m-diameter radome encloses the new Gregorian feed assembly and radar transmitting/receiving equipment.

RADAR AND THE NEO EXPLORATION STRATEGY

This volume is based on a conference that was cosponsored by the Explorers' Club. Exploration of new frontiers energizes the human spirit — it is an antidote to the human condition. It serves as a vehicle for international cooperation and is important to the future of human civilization. But exploration of space, which truly is the final frontier, is also horribly expensive. For economic reasons alone, those NEAs whose orbits make them the cheapest targets of robotic or piloted space missions are destined to be central to our future in space.

A cost-effective NEO exploration strategy can be visualized as a pyramid whose layers represent efforts that become more informative, more expensive, and less widely applied toward the top. Optical search programs are the pyramid's foundation. Above them are optical astrometric follow-up efforts and the infrastructure for orbit and ephemeris calculation, and then physical and chemical characterization by VIS/IR/UV techniques. Radar's attributes and limitations place it between other ground-based efforts and robotic space missions, which include numerous cheap flybys and then, higher up, rendezvous missions (orbiters, landers, sample returns). The pyramid is topped by missions with human crews.

FIGURE 11. The 70 m antenna (DSS 14) of the Goldstone Solar System Radar in California. Radar transmitting and front-end receiving equipment is in the nearest of the three cones at the secondary focus.

Radar can lower the risk and cost of space missions by making interplanetary navigation easier and by providing *a priori* characterization of the target. For example, the Goldstone imaging of Geographos in August 1994 [28] was given special impetus by the Clementine mission, which planned a 10.7 km/s flyby of Geographos on Aug. 31 at a miss distance of approximately 100 km. We planned to use radar astrometry to improve the preencounter ephemeris and to use imaging to optimize the postencounter physical model of the asteroid. Although a computer malfunction led to cancellation of the flyby, the Goldstone experiment provided practice for supporting future spacecraft reconnaissance of near-Earth asteroids. Before the observations, the asteroid's positional error ellipsoid had a typical overall dimension of ~11 km. Ranging on Aug. 28–29 and a preliminary shape model collapsed the ellipsoid's size along the line of sight to several hundred meters; its projection toward Clementine on its inbound leg would have been 11 × 2 km (D.K. Yeomans, personal communication). Goldstone-VLA plane-of-sky astrometry could have shrunk the error ellipsoid's longest dimension to about 1 km, about half of Geographos's shortest overall dimension.

The experience with Geographos and a similar one with 6489 Golevka (1991 JX) [26] suggest that with adequate radar reconnaissance, it would be possible for a spacecraft lacking onboard optical navigation to be guided into orbit around, or collision course with, an asteroid. Even if on-board "opnav" capability were available, radar would nonetheless shrink the navigation uncertainties significantly, perhaps reducing the mission's fuel requirements enough to allow a larger mass budget for science instrumentation. Indeed, the U.S. Department of Defense is planning a follow-up mission, Clementine II [11], that may fly by NEAs on dates that follow Arecibo/Goldstone radar opportunities.

TABLE 4. Effect of Delay/Doppler Measurements on Extrapolation of the Orbit of 6489 Golevka (1991 JX) from Discovery-Apparition Astrometry*

	Positional uncertainty		
Dataset	kilometers	Earth radii	lunar distances
Optical	8,000,000	1260	21
Optical + Radar	25,000	4	0.07

*Uncertainties in the asteroid's April 2019 position predicted from 54 optical measurements from May 9 to July 3, 1991, are in the top row. The bottom row gives uncertainties for a prediction that includes 24 delay-Doppler measurements from June 5–15, 1991. (D.K. Yeomans, personal communication).

RADAR AND THE IMPACT HAZARD

One of the far-reaching motivations for surveying and exploring asteroids and comets is the impact hazard [17]. Radar can reduce both kinds of post-discovery uncertainty about potential catastrophic NEO collisions: orbits and physical properties [24]. As demonstrated in Table 4, the uncertainty in a newly discovered NEO's position years or decades in the future can be hundreds of times smaller if radar astrometry is available. Golevka, the example in that table, was observed exhaustively with radar again in 1995 [27] and is known to make no dangerously close approaches to Earth for at least seven centuries (D.K. Yeomans, personal communication). But imagine a prediction that in 25 years a 500 m asteroid (a 10,000 megaton impactor, at the small end of the range of estimates for the transition from local devastation to global catastrophe) will miss Earth by 1 ± 30 lunar distances. That prediction would generate much more anxiety than, say, the report in spring 1996 that Eros might collide with Earth in a few million years (which garnered a full-page headline in the Apr. 25, 1996, *New York Times*), or the claim several years ago that comet Swift-Tuttle might hit the Earth on Aug. 14, 2126; see the cover story in the Nov. 23, 1992, *Newsweek*).

In fact, cislunar misses by objects 500 m or larger occur at average intervals of about 18 years. A Spaceguard Survey would discover ~6400 such objects (~70% of the ~9200 in the population). One of those objects passes one lunar distance from Earth once every 25 years on average, so four such events per century are likely to be predicted after a Spaceguard Survey. However, uncertainties in the predictions may not permit confident distinction of close-call trajectories from impact trajectories. Assessment of the danger would be dramatically more efficient and more reliable if radar astrometry were available to shrink the uncertainties [36]. The availability of radar measurements could easily be the difference between knowing that an object will pass "within several Earth-Moon distances of Earth" in a few decades and knowing that it will "hit the Earth."

Whereas the warning time for an asteroid impact in the post-Spaceguard era is likely to be at least a century and hence more than adequate for mitigation at a comfortable pace, the warning time for an impact by a long-period comet (LPC) would probably be less than one year. Moreover, LPC trajectory extrapolation will be hampered by obscuration of the nucleus and by uncertainties about nongravitational forces, both of which increase as a comet approaches the inner solar system. For example, Chodas [5] found that had comet Hyakutake (which missed Earth by 39 lunar distances) been on a collision course with Earth, the predicted close approach distance would have exceeded one Earth radius until just one month before impact.

Cislunar approaches by LPCs can be expected every few centuries. The uncertainty in trajectory extrapolation after discovery of these objects could be terrifying, and any number of panic scenarios are possible. A prediction of a comet or asteroid impact might provoke an economically or psychologically destructive societal response, even if the predicted collision were known to be insufficiently energetic to affect the global ecology. Such a prediction could lead to hasty and costly development of a mitigation system whose existence would in itself introduce a significant risk to civilization [33]. Society will surely want to reduce uncertainties associated with predictions of extremely close approaches of NEOs. Radar astrometry could provide insurance against false alarms. In the event of certain prediction of a collision, radar characterization of the threatening projectile would lend valuable support to the mitigation effort.

Fortunately, the odds against a globally catastrophic NEO collision are very large. The most likely outcome of a Spaceguard Survey will be that no object will turn out to be hazardous during at least the next century. Regardless of the outcome, the survey will identify tens of thousands of small worlds that, at least for a century, are not on collision course with Earth. The potential promise of these "friendly" NEOs is highly significant for a variety of reasons, most of which stem from their closeness to Earth and their accessibility (see [32, Chap. 11]). As noted in Table 3, radar echoes from thousands of NEOs discovered in a Spaceguard Survey would be detectable at high SNRs. However, it will be difficult to take advantage of most of the opportunities summarized in that table. Radar usage of Arecibo, historically less than 5% (vs. ~75% for radio astronomy and ~20% for ionospheric science), is unlikely to rise above 10%. The Goldstone 70 m antenna, DSS-14, which has the transmitter, is part of NASA's Deep Space Network, which exists for spacecraft telemetry; the antenna's total radar astronomy usage is unlikely to rise above ~5%. Neither Arecibo nor Goldstone were optimized for NEO radar.

Over the next decade, it will become increasingly clear that most of the NEO reconnaissance that is technically achievable with Arecibo and Goldstone is precluded by the limited accessibility of those instruments, and that a dedicated NEO radar instrument is desirable. In this light, it seems appropriate to begin to assess the potential design, cost, and value of such an instrument.

A DEDICATED NEO RADAR TELESCOPE FOR THE NEW MILLENNIUM

From preliminary studies carried out at JPL (G. Resch, personal communication), it appears that an ideal NEO radar telescope would probably consist of two large, fully steerable antennas (like the 100 m NRAO Greenbank Telescope now under construction in West Virginia; see http:// www.gb.nrao.edu/GBT/), one for transmitting and one for receiving, separated by a few tens of kilometers, operating at a wavelength between 0.9 and 3.5 cm (Ka and X band). A two-antenna (bistatic) configuration obviates the frequent transmit/receive alternation required in single antenna observations of NEOs — the round trip travel time (RTT) of the signal between the radar and the target in seconds is numerically nearly equal to the target's distance in mAU, e.g., 10 seconds for a target 0.01 AU away. Advantages of a two-antenna telescope for NEOs include:

1. no on/off switching of the transmitter, which ages klystron amplifier tubes;
2. no interruption of data acquisition or of coherent integration;
3. no interruption of orientational coverage;
4. no switching of antenna pointing between transmit and receive directions, which can differ by more than a beam width for a rapidly moving target;
5. doubled data integration time;
6. accessibility of arbitrarily close targets, including those within a few Earth-Moon distances (RTT < 10 s), using same configuration as for distant targets.

The capital cost of this system, as calibrated by the experience with the GBT, would be within 10% of $180M, comparable to the cost of a Discovery mission. (Three such systems suitably spaced in longitude and latitude to ensure constant visibility of any part of the sky could be built for less than the cost of one launch of the Space Shuttle.)

Since this instrument would be dedicated to NEOs, it would compile more NEO observation time in one year than Arecibo and Goldstone could in a decade. Given the current state of the art of antennas, receivers, and transmitters, it could be an order of magnitude more sensitive than the upgraded Arecibo telescope. The telescope's lifetime would probably be at least several decades, during which it would do flyby-level science (delay-Doppler imaging placing thousands of pixels on the target, over enough orientations to allow three-dimensional reconstruction of the shape in geological detail) of over a thousand of the optically discoverable NEAs, as well as several-hundred-pixel imaging (and several-hundred-parameter shape reconstruction) of a few hundred main belt asteroids, not to mention orbit-securing follow-up astrometry on many thousands of newly discovered NEOs. The scientific return per dollar of investment would be extraordinary.

This radar telescope could be built and operational in a decade. By then, the pool of radar-accessible NEAs would allow radar imaging at a level adequate to produce a useful 3-D model of a different asteroid almost every day. The radar images, movies, and models could be made available on the

World Wide Web in real time, allowing nearly global participation in the excitement of the radar flybys. Ten years into the next millennium, a first-time encounter with a newly discovered, nearby world could be part of the daily experience of the average person in most of the countries of the world.

REFERENCES

1. ANDREWS, A.K., R.S. HUDSON, & D. PSALTIS. 1995. Optical-radar imaging of scale models for studies in asteroid astronomy. Optics Letters. **20:** 2327–2329.
2. ASPHAUG, E., S.J. OSTRO, E.M. DE JONG, D.J. SCHEERES, S. SUZUKI, R.S. HUDSON, & W. BENZ. 1996. Disruptive impacts into small asteroids. BAAS. **28:** 1102.
3. BENNER, L.A.M., S.J. OSTRO, J.D. GIORGINI, R.F. JURGENS, D.L. MITCHELL, R. ROSE, K.D. ROSEMA, M.A. SLADE, R. WINKLER, D.K. YEOMANS, D.B. CAMPBELL, J.F. CHANDLER, AND I.I. SHAPIRO. 1996. Radar observations of asteroids 2062 Aten, 2101 Adonis, 3103 Eger, 4544 Xanthus, and 1992 QN. BAAS. **28:** 1105.
4. CAMPBELL, D.B., J.K. HARMON, & I.I. SHAPIRO. 1989. Radar observations of comet Halley. Astrophys. J. **338:** 1094–1105.
5. CHODAS, P.W. 1996. Warning time for long-period comets on Earth impact trajectories. Abstracts of Asteroids Comets Meteors 96. COSPAR Coll. **10:** p. 74.
6. DE JONG, E. & S. SUZUKI. 1996. Visualization of Earth-approaching asteroids, AVC-97-097. Catalog on-line at http://www.jpl.nasa.gov/archive/vidcat.html. (video tape).
7. DE PATER, I., P. PALMER, D.L. MITCHELL, S.J. OSTRO, D.K. YEOMANS, & L.E. SNYDER. 1994. Radar aperture synthesis observations of asteroids. Icarus **111:** 489–502.
8. GOLDSTEIN, R.M., R.F. JURGENS, & Z. SEKANINA. 1984. A radar study of comet IRAS-Araki-Alcock 1983d. Astron. J. **89:** 1745–1754.
9. HARMON, J.K., D.B. CAMPBELL, A.A. HINE, I.I. SHAPIRO, & B.G. MARSDEN. 1989. Radar observations of comet IRAS-Araki-Alcock 1983d. Astroph. J. **338:** 1071–1093.
10. HARMON, J.K., S.J. OSTRO, K.D. ROSEMA, R.F JURGENS, R. WINKLER, D.K. YEOMANS, D. CHOATE, L.A.M. BENNER, R. CORMIER, J.D. GIORGINI, D.L. MITCHELL, R. ROSE, M.A. SLADE, & M.L. THOMAS. 1996. Radar echoes from comet Hyakutake (C/1996 B2). BAAS. **28:** 1195.
11. HOPE, A., B. KAUFMAN, & D. BAKERIS. 1996. A Clementine II mission to the asteroids. *In* Abstracts of IAU Coll. 165, Dynamics and Astrometry of Natural and Artificial Solar System Bodies, pp. 55–56.
12. HUDSON, R.S. 1993. Three-dimensional reconstruction of asteroids from radar observations. Remote Sensing Reviews. **8:** 195–203.
13. HUDSON, R.S. & S.J. OSTRO. 1994. Shape of asteroid 4769 Castalia (1989 PB) from inversion of radar images. Science **263:** 940–943.
14. HUDSON, R.S. & S.J. OSTRO. 1995. Shape and non-principal axis spin state of asteroid 4179 Toutatis. Science. **270:** 84–86.
15. MITCHELL, D.L., R.S. HUDSON, S.J. OSTRO, & K.D. ROSEMA. 1996. Shape of asteroid 433 Eros from inversion of Goldstone radar Doppler spectra. Icarus. Submitted.
16. MORRISON, D., Ed. 1992. The Spaceguard Survey: Report of the NASA International Near-Earth Object Detection Workshop. Available at http://george.arc.nasa.gov/sst/spaceguard.html.
17. MORRISON, D., C.R. CHAPMAN, & P. SLOVIC. 1994. The impact hazard. *In* Hazards Due to Comets and Asteroids, T. Gehrels & M.S. Matthews, Eds.: 59–91. Univ. of Arizona Press. Tuscon, Arizona.

18. OSTRO, S.J., D.B. CAMPBELL, & I.I. SHAPIRO. 1983. Radar observations of asteroid 1685 Toro. Astron. J. **88:** 565–576.
19. OSTRO, S.J., D.K. YEOMANS, P.W. CHODAS, R.M. GOLDSTEIN, R.F. JURGENS, & T.W. THOMPSON. 1989. Radar observations of asteroid 1986 JK. Icarus. **78:** 382–394.
20. OSTRO, S.J., J.F. CHANDLER, A.A. HINE, I.I. SHAPIRO, K.D. ROSEMA, & D.K. YEOMANS. 1990. Radar images of asteroid 1989 PB. Science. **248:** 1523–1528.
21. OSTRO, S.J., D.B. CAMPBELL, J. F. CHANDLER, I.I. SHAPIRO, A.A. HINE, R. VELEZ, R.F. JURGENS, K.D. ROSEMA, R. WINKLER, & D.K. YEOMANS. 1991. Asteroid radar astrometry. Astron. J. **102:** 1490–1502.
22. OSTRO, S.J., D.B. CAMPBELL, J. F. CHANDLER, A.A. HINE, R.S. HUDSON, K.D. ROSEMA, & I.I. SHAPIRO. 1991. Asteroid 1986 DA: Radar evidence for a metallic composition. Science. **252:** 1399–1404.
23. OSTRO, S.J. 1993. Planetary radar astronomy. Rev. Modern Phys. **65:** 1235–1279.
24. OSTRO, S.J. 1994. The role of groundbased radar in near-Earth object hazard identification and mitigation. *In* Hazards Due to Comets and Asteroids, T. Gehrels & M.S. Matthews, Eds.: 259–282. Univ. of Arizona Press. Tuscon, Arizona.
25. OSTRO, S.J., K.D. ROSEMA, R.S. HUDSON, R.F. JURGENS, J.D. GIORGINI, R. WINKLER, D.K. YEOMANS, D. CHOATE, R. ROSE, M.A. SLADE, S.D. HOWARD, & D.L. MITCHELL. 1995. Extreme elongation of asteroid 1620 Geographos from radar images. Nature. **375:** 474–477.
26. OSTRO, S.J., R.S. HUDSON, R.F. JURGENS, K.D. ROSEMA, R. WINKLER, D. HOWARD, R. ROSE, M.A. SLADE, D.K. YEOMANS, J.D. GIORGINI, D.B. CAMPBELL, P. PERILLAT, J.F. CHANDLER, & I.I. SHAPIRO. 1995. Radar images of asteroid 4179 Toutatis. Science. **270:** 80–83.
27. OSTRO, S.J., D. CHOATE, R.A. CORMIER, C.R. FRANCK, R. FRYE, J. GIORGINI, D. HOWARD, R.P. JURGENS, R. LITTLEFAIR, D.L. MITCHELL, R. ROSE, K.D. ROSEMA, M.A. SLADE, D.R. STROBERT, R. WINKLER, D.K. YEOMANS, R.S. HUDSON, P. PALMER, L.E. SNYDER, A. ZAITSEV, S. IGNATOV, Y. KOYAMA, & A. NAKAMURA. 1995c. Asteroid 1991 JX: The 1995 Goldstone radar experimerlt. BAAS. **27:** 1063.
28. OSTRO, S.J., R.F. JURGENS, K.D. ROSEMA, R.S. HUDSON, J.D. GIORGINI, R. WINKLER, D.K. YEOMANS, D.J. SCHEERES, D. CHOATE, R. ROSE, M.A. SLADE, S.D. HOWARD, & D.L. MITCHELL. 1996. Radar images of Geographos. Icarus. **121:** 44–66.
29. SCHEERES, D.J., S.J. OSTRO, & R.S. HUDSON. 1995. Orbits around asteroid 4179 Toutatis. BAAS. **27:** 1075.
30. SCHEERES, D.J., S.J. OSTRO, R.S. HUDSON, & R.A. WERNER. 1996. Orbits close to asteroid 4769 Castalia. Icarus. **121:** 67–87.
31. SCHEERES, D.J., S.J. OSTRO, D.L. MITCHELL, & R.S. HUDSON. 1996. Ejecta distribution patterns using multiple 433 Eros models. BAAS. **28:** 1104.
32. SHOEMAKER, E.M., G. CANAVAN, J. DARRAH, A. HARRIS, D. MORRISON, M. MUMMA, D. RABINOWITZ, J. RAHE, C. CHAPMAN, B. MARSDEN, S. OSTRO, & D. YEOMANS. 1995. Report of the Near-Earth Objects Survey Working Group. NASA. Available at http://ccf.arc.nasa.gov/sst/ report/neo_rpt.html.
33. SAGAN, C. & S.J. OSTRO. 1994. Long-range consequences of interplanetary collision hazards. Issues in Science and Technology. **10:** (4) 67–72.
34. YEOMANS, D.K., S.J. OSTRO, & P.W. CHODAS. 1987. Radar astrometry of near-Earth asteroids. Astron. J. **94:** 189–200.
35. YEOMANS, D.K., P.W. CHODAS, M.S. KEESEY, S.J. OSTRO, J.F. CHANDLER, & I.I. SHAPIRO. 1992. Asteroid and comet orbits using radar data. Astron. J. **103:** 303–317.
36. YEOMANS, D.K. & CHODAS, P.W. 1994. Predicting close approaches of asteroids and comets to Earth. *In* Hazards Due to Comets and Asteroids, T. Gehrels, Ed.: 241–258. Univ. of Arizona Press. Tucson, Arizona.

Colliding Asteroids from Blind Directions

SYUZO ISOBE[a] AND MAKOTO YOSHIKAWA[b]

[a]*National Astronomical Observatory*
Mitaka, Tokyo 181, Japan
and
[b]*Communications Research Laboratory*
Hirai, Kashima, Ibaraki, 314, Japan

ABSTRACT: The Spaceguard Project proposes to set up a network of five ground-based telescopes, each with an aperture of 2–3 m. With ground-based telescopes, there is a blind celestial area surrounding the Sun (say, within 30 degrees), in which an asteroid approaching the Earth cannot be detected optically because of bright sky background. Our study shows that the number of close approaches of asteroids coming from this area to within a minimum distance of less than 0.01 AU of the Earth is more than 30% of the total.

Since 1989, when a program for detecting near-Earth asteroids (NEA) was started with the Spacewatch telescope of the University of Arizona, the number of newly-detected NEAs has risen to 30–40 per year. Most of the NEAs have relatively large orbital eccentricities.

One of us (M.Yoshikawa) has calculated the orbital motions of many asteroids and analyzed close encounters between asteroids and planets [1], [2]. Yoshikawa [2] calculated positions of 188 NEAs detected prior to 1993 up until the year 4600. For our present study, we analyzed close encounters between asteroids and the Earth using these results.

Figure 1 shows the motions of three typical asteroids relative to the Earth, the negative X-axis being in the direction of the Sun. Dashed and dot-dashed lines illustrate two asteroids that overtake the Earth, the relative velocities of the asteroid and the Earth being small in each case. For the case shown by a solid line, the orbital eccentricity is large and the orbit of the asteroid crosses that of the Earth quite steeply.

It is clear that ground-based observations cannot detect asteroids coming from the direction of the Sun and small solar elongations because of the bright sky background. Considering that the asteroids near the Sun are at low altitude when the sky is dark after sunset and before sunrise, we define as a working condition the blind area for asteroid detection as being within 30 degrees of the solar direction.

The number of close approaches of known asteroids to the Earth during the period 1994–4600 obviously depends on the distances one considers to be close. We find 6694 close approaches within 0.10 AU. The number decreases to 5689, 4702, 3799, 2918, 2146, 1469, 864, 378, and 94 approaches within 0.09, 0.08, 0.07, 0.06, 0.05, 0.04, 0.03, 0.02, and 0.01 AU, respectively. Extrapolation to 0.000043 AU (6400 km, the radius of the Earth) suggests a hit once every 1.3×10^6 years.

Some NEAs cross the blind zone at their closest approaches to the Earth and stay there for only a few days. In such cases there is a possibility of detecting the NEAs several days before their closest approaches. However, if an asteroid approaches the Earth directly from the blind zone, it stays in the zone for more than 10 days prior to the closest approach. Figure 2 shows two examples of asteroids leaving the zone. One can see that 1991 BA will leave the zone on the day of closest approach. This leads to the conclusion that there is a possibility of asteroid collisions without warning.

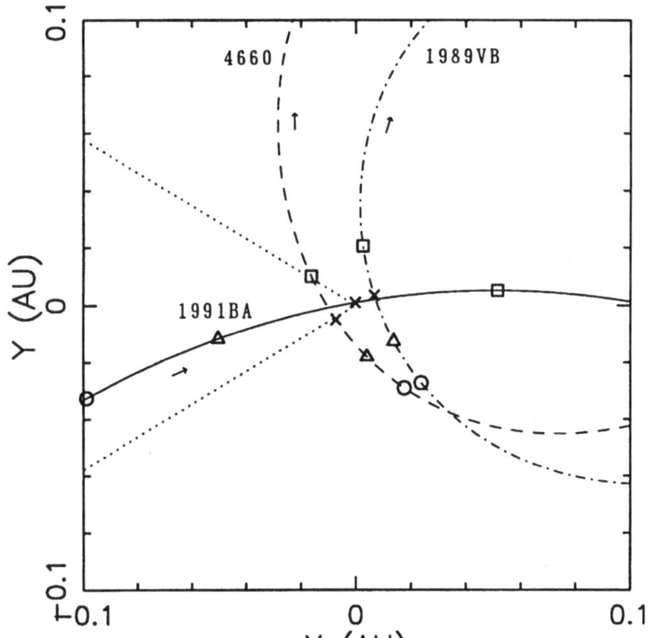

FIGURE 1. Relative motions of three asteroids near closest approach to the Earth. The Earth is at $(X, Y) = (0, 0)$. Positive values of X and Y are in the anti-solar direction and orbital direction of the Earth, respectively. Dotted lines are at 30 degrees of solar elongation. The solid line represents the close approach (at a distance of 0.0011 AU) on July 6, 2843, of the asteroid 1991 BA (semimajor axis, $a = 2.267$ AU, eccentricity $e = 0.683$, inclination $i = 2.116$ degrees to the J2000.0 ecliptic). Dashed and dot-dashed lines represent, respectively, approaches to the Earth of 0.0089 AU on January 13, 3269, for the asteroid (4660) ($a = 1.520$ AU, $e = 0.370$, $i = 1.171$ degrees) and 0.0093 AU on November 1, 4096, for the asteroid 1989 VB ($a = 1.856$ AU, $e = 0.460$, $i = 1.981$ degrees). Circles, triangles, crosses, and squares show points at –10, –5, 0, and +5 days with respect to closest approach.

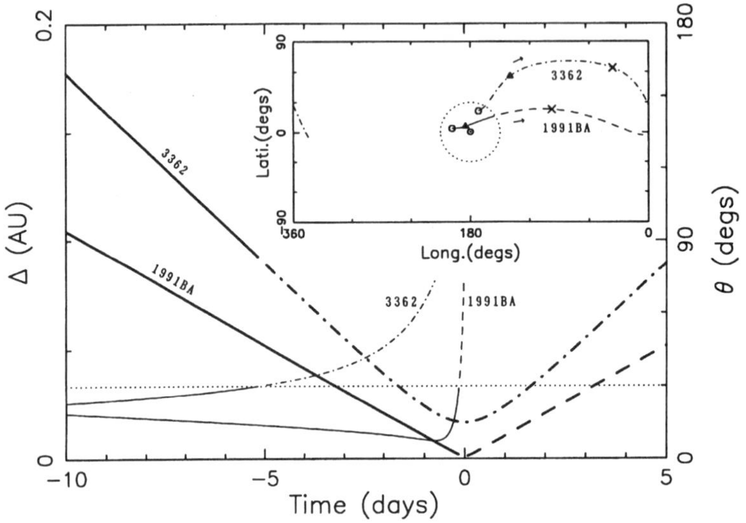

FIGURE 2. Change in distance with time from closest approach. The left ordinate (corresponding to thick lines) refers to linear geocentric distance. The right ordinate (corresponding to thin lines) refers to the angular elongation from the Sun, the horizontal dotted line representing an elongation of 30 degrees. The dashed and dot-dashed lines are for asteroids of 1991 BA and 3362, respectively, although they are shown as solid lines when they are inside the blind zone. The superposed figure depicts the motions of the two asteroids in a longitude-latitude plane, relative to the anti-Sun (the Sun being at 0 degree latitude and 180 degrees longitude). Circles, triangles and crosses are points indicating 10 days, 1 day and 0 day before closest approach, respectively.

The brightness of an asteroid depends on the radius and albedo, the distance from the Earth and the phase angle. It is very difficult to detect a small NEA more than 10 days prior to collision or closest approach because of its faintness. We define an asteroid leaving the blind zone a few days prior to its closest approach as a "dangerous" one.

The fraction of close approaches of asteroids leaving the blind zone each day prior to minimum approach, relative to the total number of estimated close approches of asteroids, is shown in Fig. 3. The fraction of "dangerous" close approaches is less than 3% when the minimum approach is 0.10 AU, but it is over 30% when the minimum approach is 0.01 AU. This means that a large fraction of asteroid collisions are "dangerous." It is clear that, even if missiles and weapons are kept on a standby condition, we have no possibility of deflecting a "dangerous" asteroid on a collision course.

The most effective way of avoiding such a collision is to find all the NEAs and determine their orbits. The Spaceguard project would search for asteroids larger than 1 km (a collision with which would certainly cause a global catastrophe). If the orbit of a "dangerous" asteroid is well determined many decades prior to an expected collision, we could avoid the collision by changing the asteroid's velocity by perhaps only 1 cm/s several tens of revolutions prior to the collision.

The Spaceguard project is not a perfect one, because only 90% of NEAs with a diameter larger than 1 km are expected to be detected in 30 years, and it would take much longer to reach 99% completeness. Most of the NEAs with diameter between 100 and 400 m that might cause a catastrophe on a continental scale would simply not be detected.

One way to find "dangerous" asteroids missed by Spaceguard is to deploy telescopes in space, where the sky brightness is very dark except the immediate vicinity of the Sun. There are discussions on building a lunar-based station by the middle of the 21st century under an international collaboration: if such a project becomes reality, a high priority should be to build and place telescopes there for detection of dangerous asteroids. To maximize the warning time, it would be desirable to search as much as a steradian of the sky each day.

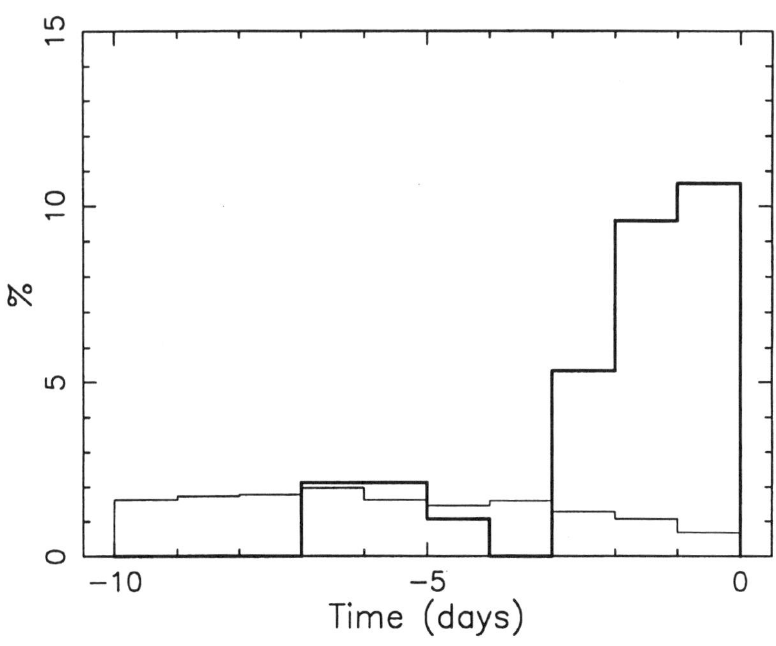

FIGURE 3. The fraction (out of the total number of close approaches) of close approaches of asteroids leaving the blind zone on each day (from 1 to 10 days) prior to their minimum approaches. The thin line refers to approaches within 0.10 AU, the thick line to approaches within 0.01 AU.

REFERENCES

1. YOSHIKAWA, M. 1994. Close Encounters of Asteroids. *In* Seventy-Five Years of Hirayama Asteroid Families, ASP Conference Series. **63:** 28–38.
2. YOSHIKAWA, M. & T. NAKAMURA. 1994. Near-Misses in Orbital Motion of Asteroids. Icarus. **108:** 298–308.

Correlation of Ground- and Space-based Bolides[a]

Z. CEPLECHA,[b] C. JACOBS,[c] AND C. ZAFFERY[c]

[b]*Academy of Sciences*
25165 Ondrejov Observatory
Czech Republic
and
[c]*Sandia National Laboratories*
Albuquerque, New Mexico 87185

ABSTRACT: Data on large meteoroid impacts into the Earth's atmosphere are available up to over ten meter sizes from global satellite observations by optical sensors, while data from ground-based photographic observations of meteoric fireballs (bolides) are available only up to meter sizes. However, the ground-based observations yield very precise data on the motion and ablation of these bodies, as well as their light curves. The space-based observations yield very precise light curves. Using data of the brightest Prairie Network bolides, ablation coefficients are determined for 48 bolides brighter than magnitude –10 and compared to relative time changes of their brightness. This way the known classification scheme for photographic bolides (based mostly on observed ablation coefficient) is related to maximum increase and decrease of their brightness. The same classification scheme is applied to 16 light curves of somewhat larger bodies observed globally by DOD satellites. Size range of these bodies is found to be from 2 to 15 m with median value of 3 to 4 m. Only separation of type I + II (38%) from type IIIA + IIIB (62%) is possible. Increase of relative strength of populations of cometary meteoroids (namely of the IIIB type) observed for photographic meteors up to a size of 5 m continues to larger sizes up to 15 m. Majority of meteoroids in the size range from 2 to 15 m are cometary bodies with the weakest known structure.

1. INTRODUCTION

Right now we experience a multitude of collisions with near-Earth objects (NEOs), with the small-size fraction of them. Meteoroids are near-Earth objects we can observe at collision, when they penetrate deep into the Earth's atmosphere. How many of them we can observe at their collisions depends on how large they are [2], [3], [5]. This is the only significant limit put on our rather precise observations and recordings of small size NEOs at their collision with the Earth. At the moment we have ground based photographic records of 14 objects larger than 50 cm at their entry [4]. Space-based records are available up to ten meter sizes; they marginally overlap the size range of the ground-based observations [13]. At the moment there are no

[a]This work has been supported by Contract of the Grant Agency of Czech Republic (205/94/1862) and by the Sandia National Laboratories.

common precise records of the same bolide from ground and from space. We are limited to statistical comparisons. Ground based observations of bolides by multistation photographic techniques yield very precise data with heights and distances often measured better than ±10 m. Our theoretical models can explain their motion and ablation in the atmosphere with about the same precision. Values of ablation coefficients determined from these observations are rather large values, which speaks about continuous fragmentation as the main process of ablation. We use the term "ablation" in the sense of all mass removed from a body, whether as gas, fluid, or solid fragments. These bodies we record from ground-based photographic stations (up to several meters in size) break in the atmosphere under very low dynamic pressures one or two orders of magnitude lower than is the strength of stony meteorites. We can ascribe the reason for this fragmentation to the structure of these bodies, which perhaps may be composed mostly of material highly shocked due to mutual collisions in the interplanetary space.

In this contribution we will study the precise ground-based photographic data on the motion of these bodies, and relate the dynamic data to their light curves. The results will be applied to the light curves recorded for larger meteoroids from space. Because we can sort the smaller meteoroids from ground-based observations according to their structure and composition, and we can get the ratio of cometary and asteroidal component, we hope to do the same for larger bodies, when only their precisely recorded light curves observed from space are available.

2. GROUND-BASED BOLIDES

First, I introduce the classification scheme of the largest bolides observed photographically from the ground. Classification in Table 1 is based on precise analysis of several hundred individual objects. The differences of atmospheric trajectories among bodies of individual populations are not trivial: if we compare the type I trajectory of a body with the same initial mass, the same velocity, and the same zenith angle, to type IIIB body, the difference expressed in air densities is 1:1000. This difference can also be well demonstrated comparing two very bright bolides photographed within the European Fireball Network (Fig. 1). The first one is the brightest object on photographic records, the Šumava Fireball of –21.5 absolute maximum magnitude terminating at 59 km height; the second one is the first meteorite fall on photographic records, the Příbram bolide of –19 absolute maximum magnitude, which came down to 13 km height: four meteorites, ordinary chondrites were recovered. Both of these bolides do not differ too much in the initial mass, in the initial velocity, or in the slope of the trajectory, but they terminated in wildly different heights.

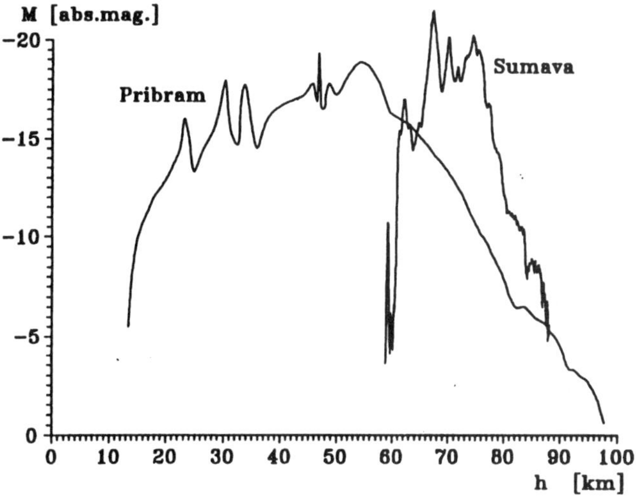

FIGURE 1. Different atmospheric penetration of meteoroids is well demonstrated by comparing the light curves of the Příbram bolide (type I with meteorite falls) with the light curve of Šumava bolide (type IIIB; the brightest meteor on photographic records so far). Absolute stellar magnitude, *M*, is plotted against height, *h*. Not much different in maximum brightness (an also in initial mass, velocity and other parameters) these objects differ more than 40 km in their therminal height.

We analyzed and classified 48 PN (Prairie Network) bolides brighter than absolute magnitude −10, which are actually all of such objects with reliable light curves available [9], [10], [6]. We classified all of these objects individually, after applying a continuous- and gross-fragmentation model to the observed dynamical data [8]. We then compared the classification of each bolide with its light curve. The differences found among the light curves were also very substantial, as demonstrated in Fig. 2, where there are plotted the light curves for selected characteristic cases belonging to individual dynamically resolved types. It is evident that the type of the light curve depends on the individual ablation coefficient.

TABLE 1. Populations of the Largest Observed Meteoroids: Fireballs (Bolides) Recorded Photographically (ρ, bulk density; σ, ablation coefficient)

Population	ρ g/cm^3	σ s^2/km^2	Characteristic orbit for 5–50 cm size *a*	*e*	*i*	Assumed composition
I	3.7	.014	2.4	.68	6°	stony
II	2.0	.042	2.3	.61	5°	carbonaceous
IIIA	0.75	.10	2.4	.82	4°	cometary
IIIAi	0.75	.10	$\approx\infty$	.99	random	cometary
IIIA[C3]	0.75	.10	2.7	.67	random	cometary
IIIB	0.27	.21	3.0	.70	13°	soft cometary

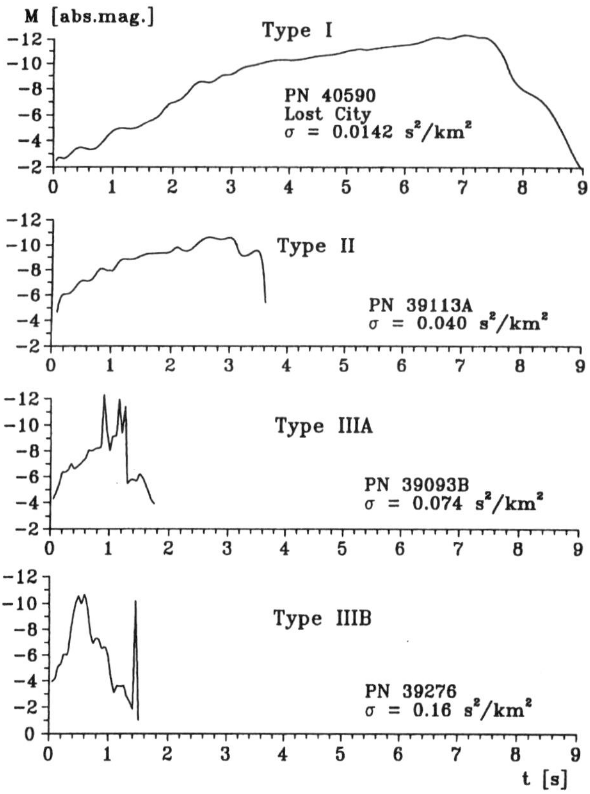

FIGURE 2. Light curves for different types of bolides differ significantly as demonstrated in this plot containing characteristic cases for types I, II, IIIA, and IIIB chosen among the brightest PN bolides. Ablation coefficients, σ, given in the plot were computed from dynamical data on these bolides applying the gross-fragmentation model to observed heights and distances as function of time.

After seeing such a high, but only qualitatively expressed difference, we tried to make some quantification for all 48 cases of PN fireballs brighter than magnitude –10. The best we found for description of differences among light curves of different types of bolides was based on changes of light with time, i.e., on the slopes of the light curves. We defined the relative time changes of light as $(dI/dt)/I$ in units of s^{-1}. We computed these values for each measured point of all 48 PN bolides (the measured points are mostly separated by time intervals of 0.05 s). The analysis proved that the extreme values of $(dI/dt)/I$ are the most sensitive to the different types of bolides. Thus we have two partly independent possibilities, the maximum relative increase and the maximum relative decrease of brightness as classification criteria.

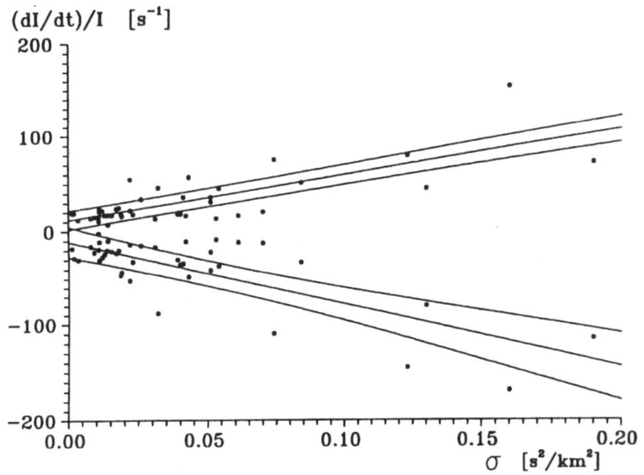

FIGURE 3. Maximum and minimum values of $(dI/dt)/I$ for individual PN bolides are plotted against the individual ablation coefficient determined dynamically. Average straight lines with their standard deviation limits are plotted for both of them.

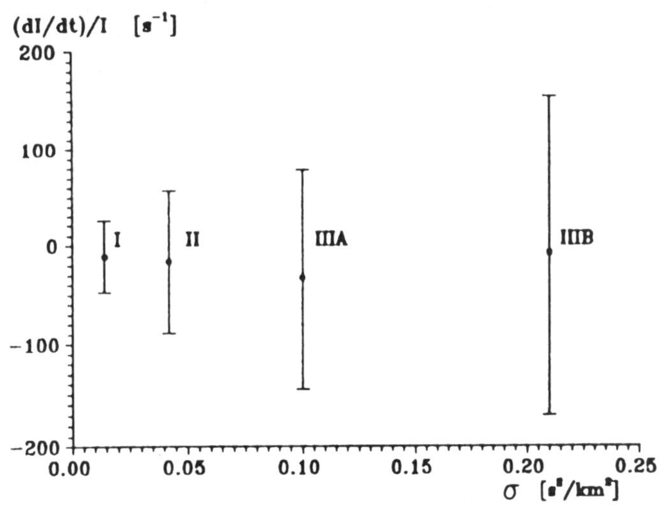

FIGURE 4. Average total ranges of $(dI/dt)/I$ values for different types of PN bolides are plotted against the average ablation coefficient, σ, belonging to the respective type.

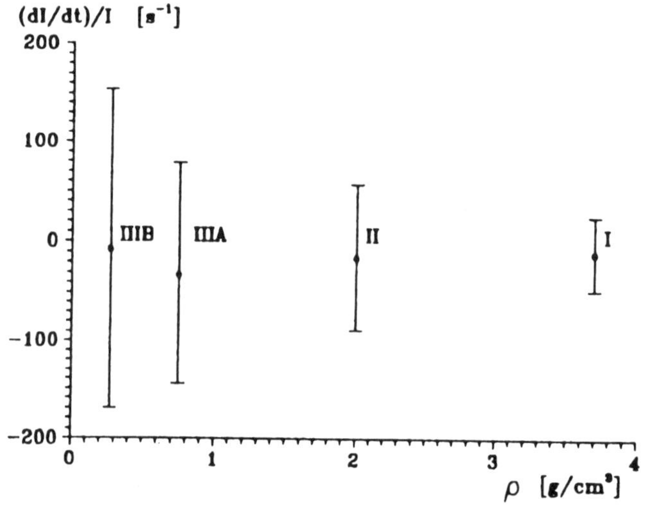

FIGURE 5. Average total ranges of $(dI/dt)/I$ values for different types of PN bolides are plotted against the average bulk density, ρ, belonging to the respective type.

We also studied the dependence of individual maximum changes of brightness on the ablation coefficient of individual meteoroids as given in Fig. 3, where the points represent maximum increase and maximum decrease plotted against the ablation coefficient, σ, with standard deviations included. If we limit ourselves to averages over the individual types of bolides as defined in Table 1, we see in Figs. 4 and 5 the differences and also the statistical overlappings of ranges of relative time changes of brightness for individual types.

TABLE 2. Classification from Maximum Increase of Brightness (Table contains percent of cases inside the given intervals)

Maximal$(dI/dt)/I$ (s^{-1})		Type (%)			
from	to	I	II	IIIA	IIIB
0	10	66	34	0	0
10	20	60	40	0	0
20	30	55	45	0	0
30	40	0	32	41	27
40	50	0	20	37	43
50	60	0	18	51	31
60	70	0	0	69	31
70	80	0	0	64	36
80	90	0	0	46	54
90	100	0	0	0	100
> 100		0	0	0	100

TABLE 3. Classification from Maximum Decrease of Brightness
(Table contains percent of cases inside the given intervals)

Minimal$(dI/dt)/I$ (s^{-1})		Type (%)			
from	to	I	II	IIIA	IIIB
0	–10	65	35	0	0
–10	–20	39	61	0	0
–20	–30	73	27	0	0
–30	–40	11	19	49	21
–40	–50	22	24	28	26
–50	–60	0	22	34	44
–60	–70	0	13	41	46
–70	–80	0	10	39	51
–80	–90	0	6	43	51
–90	–100	0	0	42	58
–100	–110	0	0	45	55
–110	–120	0	0	40	60
–120	–130	0	0	44	56
–130	–140	0	0	41	59
–140	–150	0	0	43	57
–150	–160	0	0	38	62
–160	–170	0	0	0	100
< –170		0	0	0	100

The statistical distribution of maximal and minimal increase as function of the type enabled to compute two classification tables (Tables 2 and 3). They contain percent of probability that the body with a given extreme value of $(dI/dt)/I$ belongs to a given type. The decrease of brightness separates the types somewhat better, and combination of both tables can test the reliability of this classification. The separation of individual types by using the light curves of bolides is mostly a statistical one, but values on extreme slopes can nicely and almost completely separate asteroidal from cometary bodies, except few intermediate intervals. The satellite observed bolides are on an average brighter than the PN bolides, but the difference is not too dramatic, and the proposed classification method can be applied to them also.

3. SPACE-BASED BOLIDES

We tried this method of classification on 16 bolides recorded by optical sensors on board of DOD satellites, most of them recorded within the last year. Peak power of emitted light, total radiated energy, and values of $(dI/dt)/I$ are given in Table 4. Even with limited statistics, we can apply the classification scheme of Tables 2 and 3. The proposed classification is in the last column of Table 4. If the probability of a single-type classification is above 2/3, it is given as unique; if unique classification is not possible, then, if sum of probabilities of classification of two most probable types is greater than 2/3, these two types are given as alternatives. Classification of 94166 is I or II, and this case is certainly type I, because it was accompanied by St.Robert meteorite fall [1].

TABLE 4. Data on Bolides Recorded on Board of DOD Satellites
$(P_R)_{max}$, peak power of radiation [w/sr]; E_R, total radiated energy [J]

No.	$(P_R)_{max}$	E_R	$\left(\frac{1}{I}\frac{dI}{dt}\right)_{max}$	$\left(\frac{1}{I}\frac{dI}{dt}\right)_{min}$	Type
88106	1.4×10^{12}	7.1×10^{12}	24	−19	I or II
90274	3.5×10^{11}	2.5×10^{12}	20	−10	I or II
91277	6.5×10^{10}	5.5×10^{11}	34	−90	IIIA or IIIB
93304	6.3×10^{9}	4.0×10^{10}	≈ 80	≈ −50	IIIA or IIIB
93333	1.0×10^{10}	2.6×10^{10}	140	−55	IIIB
94032	3.19×10^{12}	1.82×10^{13}	23	−25	I or (II)
94149	1.2×10^{11}	4.0×10^{11}	48	−30	IIIA or IIIB
94154	1.56×10^{11}	4.97×10^{11}	120	−55	IIIB
94166	4.00×10^{9}	1.3×10^{10}	12	−18	I or II
94227	2.71×10^{10}	1.78×10^{11}	110	−80	IIIB
94293	1.41×10^{10}	8.69×10^{10}	100	−50	IIIB
94300	8.42×10^{10}	3.22×10^{11}	180	−60	IIIB
94307	4.0×10^{11}	2.6×10^{12}	23	−15	I or II
94347	1.13×10^{10}	1.23×10^{10}	200	−100	IIIB
94350	3.0×10^{10}	5.0×10^{10}	380	−50	IIIB
95004	1.93×10^{10}	1.86×10^{11}	28	−18	I or II

4. RESULTS

We can compare these numbers of satellite observed bolides with numbers based on photographic observations from the ground [3], and on observations of 10 m size bodies and larger by Spacewatch telescope [12]. This comparison yields a guess on the size range of the bodies in Table 1, i.e., from 2 m to 15 m with median value of about 3 to 4 m [7]. This is in general agreement with the luminous-efficiency derived sizes by Nemtchinov et al. [11].

Even if we have only 16 cases, statistics of individual types as given in Table 4 may bring some hint on composition of bodies in this size range. We see that 38% of bolides in Table 4 belong to types I or II (partly of asteroidal origin), and 62% belong to IIIA or IIIB types, to the bodies of clearly cometary origin (44% have unique classification as IIIB bodies). We can compare these numbers with numbers derived from ground-based photographic observations of bolides (Table 5). Data in Table 5 take advantage of the detailed knowledge of the types of meteoroids in the 0.05–0.5 m size range. They can be extrapolated to the 0.5–5 m size range and also compared to the few statistics of 14 largest photographically observed meteoroids [4]. Calibration of sizes of photographic bolides from a recent analysis of the Lost City meteorite fall [5] resulted in correction of sizes given in Ceplecha [4] to smaller values by a factor of 2. Table 5 already contains the corrected sizes.

Satellite observations of bolides seem to indicate that the continuous increase in strength of the populations of cometary meteoroids (namely of the IIIB type) observed for photographic meteors up to a size of 5 m also continues to increase for larger sizes up to 15 m. Moreover, we can derive a median value of the size of the satellite observed meteoroids just comparing their relative numbers to the extrapolated relative numbers for photographic meteors. The resulting median size is half-way between 2.5 and 5 m, which is in a good agreement with the already mentioned median size derived from fluxes.

Majority of meteoroids in the size range from 2 to 15 m are cometary bodies with the weakest known structure.

TABLE 5. Data on Relative Importance of Populations of Bolides Derived from Photographic Ground-based Observations Compared to Data Derived from Satellite-observed Bolides (Table contains percent of meteoroids belonging to individual types for different sizes)

		Photographic								Satellite
		Observed			Extrapolated				Observed	Observed
Size	[m]	0.1	0.25	0.5	1	2.5	5	0.5 to 5	0.5 to 5	2 to 15
I and II	%	58	61	58	53	45	31	53	62	38
IIIA and IIIB	%	42	39	42	47	55	69	47	38	62

REFERENCES

1. BROWN, P., A.R. HILDEBRAND, D.W.E. GREEN, D. PAGÉ, C. JACOBS, D. REVELLE, E. TAGLIAFERRI, & J. WACKER. 1996. Meteoritics & Planetary Science. **31:** 502.
2. CEPLECHA, Z. 1988. Bull.Astron. Inst. Czechosl. **39:** 221
3. CEPLECHA, Z. 1992. Astron. Astrophys. **263:** 361
4. CEPLECHA, Z. 1994. Astron. Astrophys. **286:** 967
5. CEPLECHA, Z. 1996. Astron. Astrophys. **311:** 329
6. CEPLECHA, Z. & R.E. MCCROSKY. 1996. Prairie network fireballs: Data on height, distance, and brightness for each measured time-mark. ftp SUNKL.ASU.CAS.CZ, login: anonymous, directory: pub/PNSAO.DAT).

7. CEPLECHA, Z., R.E. SPALDING, C. JACOBS, E. TAGLIAFERRI. 1996. Sizes and masses of satellite observed meteoroids. *In* Space 96 Comet Day II, Albuquerque, New Mexico.
8. CEPLECHA, Z., P. SPURNY, J. BOROVIČKA, & J. KECLÍKOVÁ. 1993. Astron. Astrophys. **279:** 615.
9. MCCROSKY, R.E., C.-Y. SHAO, & A. POSEN. 1976. Center for Astrophys. Prepr. Ser. **665:** 1
10. MCCROSKY, R.E., C.-Y. SHAO, & A. POSEN. 1977. Center for Astrophys. Prepr. Ser. **721:** 1
11. NEMTCHINOV, I.V., O.P. POPOVA, V.V. SHUVALOV, & V.V. SVETSOV. 1995. Solar System Research. **28:** (2) 1.
12. RABINOWITZ, D.L., 1993, Astrophys. J. **407:** 412
13. TAGLIAFERRI, E., R. SPALDING, C. JACOBS, S.P. WORDEN, & A. ERLICH. 1994. *In* Hazards Due to Comets and Asteroids, T. Gehrels, Ed.: 199. University of Arizona Press. Tucson, Arizona.

Comet Shoemaker-Levy 9 Fragment Size Estimates: How Big Was the Parent Body?[a]

DAVID A. CRAWFORD

Computational Physics and Mechanics Department 9232
P. O. Box 5800, MS 0820
Sandia National Laboratories
Albuquerque, New Mexico 87185-0820

ABSTRACT: The impact of Comet Shoemaker-Levy 9 on Jupiter in July, 1994 was the largest, most energetic impact event on a planet ever witnessed. Because it broke up during a close encounter with Jupiter in 1992, it was bright enough to be discovered more than a year prior to impact, allowing the scientific community an unprecedented opportunity to assess the effects such an event would have. Many excellent observations were made from Earth-based telescopes, the Hubble Space Telescope (HST), and the Galileo spacecraft en route to Jupiter. In this paper, these observations are used in conjunction with computational simulations performed with the CTH shock-physics hydrocode to determine the sizes of the fifteen fragments that made discernible impact features on the planet. To do this, CTH was equipped with a radiative ablation model and a postprocessing radiative ray-trace capability that enabled light-flux predictions (often called the impact flash) for the viewing geometries of Galileo and ground-based observers. The five events recorded by Galileo were calibrated to give fragment size estimates. Compared against ground-based and HST observations, these estimates were extended using a least-squares analysis to assess the impacts of the remaining ten fragments. Some of the largest impacts (L, G, and K) were greater that 1 km in diameter, but the density of the fragments was low, about 0.25 g/cm^3. The volume of the combined fifteen fragments would make a sphere 1.8 km in diameter. Assuming a prebreakup density of 0.5 g/cm^3, the parent body of Shoemaker-Levy 9 had a probable diameter of 1.4 km. The total kinetic energy of all the impacts was equivalent to the explosive yield of 300 Gigatons of TNT.

INTRODUCTION

In early July, 1992, periodic comet Shoemaker-Levy 9 broke up during a close encounter with Jupiter. For a brief two-year period, about 20 large fragments and associated debris followed one last orbit about Jupiter before striking the planet at an estimated velocity of 60 km/s. The largest fragments entered the Jovian atmosphere during the week of July 16–22, 1994. Although the impact sites were located just beyond the limb of Jupiter and were not directly visible from Earth, the Galileo spacecraft was positioned for direct viewing of the impact sites. While impact phenomena were not spatially

[a]This work was supported by the United States Department of Energy under Contract DE-AC04-94AL85000. Sandia is a multiprogram laboratory operated by Sandia Corporation, a Lockheed Martin Company, for the United States Department of Energy. Additional funding came from NASA.

resolved by the spacecraft, its timing, spectral, and luminosity data are invaluable for comparison with analytical and numerical models. Fireballs and plumes generated by the impacts were visible in line-of-sight from Earth within a minute (Fig. 1) and the impact locations themselves rotated into view within 7–20 minutes [8]. The wealth of data provided by this fortuitous event gives us an opportunity to assess models of meteoroid entry into planetary atmospheres and, in the context of this paper, to estimate the size of the Shoemaker-Levy 9 parent body based on observations of the radiated light flux observed by the Galileo spacecraft and by Earth-based telescopes. The eventual goal of our modeling effort is to provide a mechanism to assess the specific hazard associated with a large atmospheric entry event on Earth [1].

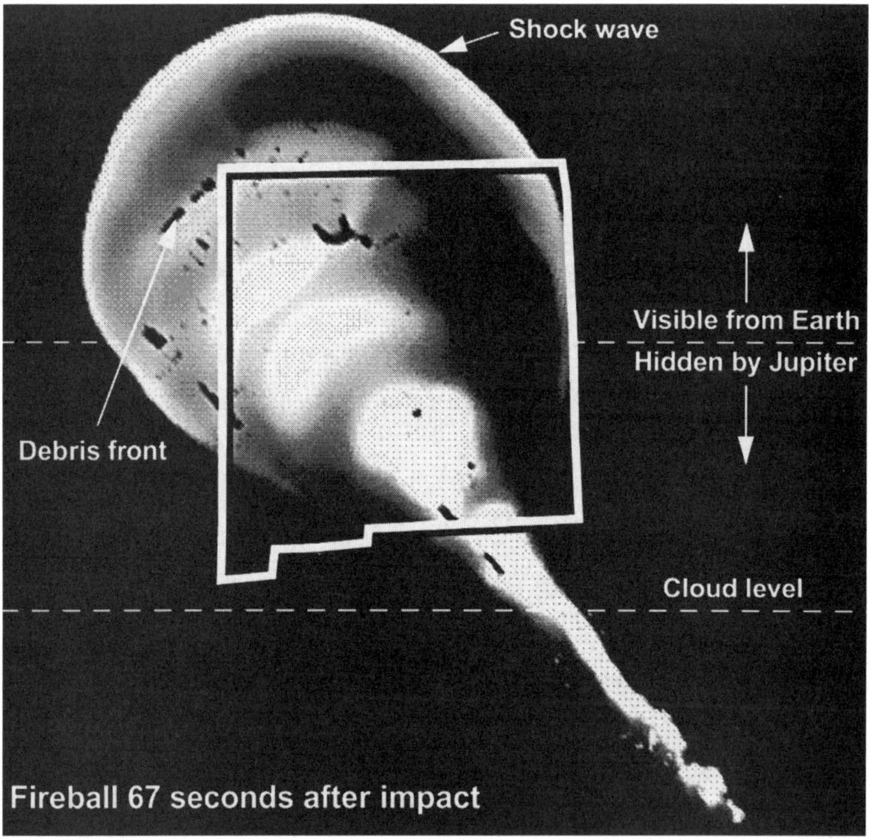

FIGURE 1. Simulated fireball from the impact of a 3 km diameter fragment into Jupiter at 60 km/s shown in cross-section, 67 seconds after impact, with the state of New Mexico (about 500 km tall) superimposed for scale. Temperature is represented by gray scale, with highest temperatures at this time step of about 2200 K.

ENTRY INTO THE ATMOSPHERE

Understanding the mechanisms of kinetic energy loss during meteoroid traversal of planetary atmospheres is crucial for understanding the development of fireballs and plumes that were observed during the impact of Comet Shoemaker-Levy 9 on Jupiter [7] and entry events in Earth's atmosphere [1]. Analytical models of the deceleration, ablation, deformation, and breakup of meteoroids during passage through planetary atmospheres have been proposed and refined by many researchers [11], [23], [10], [3], [19]. Here, a modified ablation model is proposed to describe comet entry at high altitudes. The model properly satisfies conservation of energy during the ablative process and can be reconciled with observations of terrestrial meteors. At lower altitudes, the dynamic pressure experienced by the fragment exceeds the compressive yield strength and the fragment deforms hydrodynamically according to the well-known "pancake model" of Zahnle [23]. The current comprehensive model, calibrated against numerical simulations, describes meteoroid entry over a wide range of velocities and spatial scales.

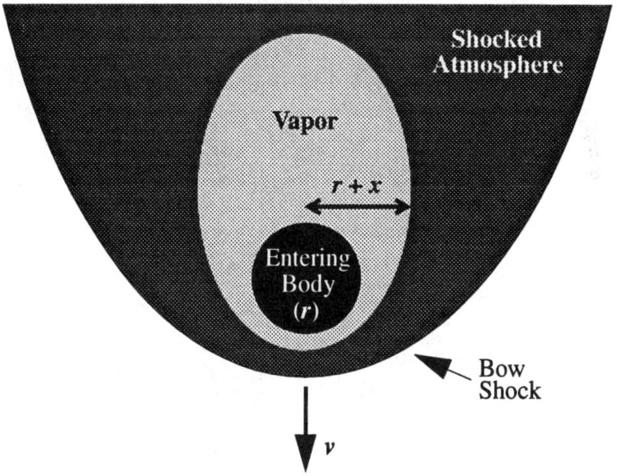

FIGURE 2. Schematic diagram of the analytical entry model that explicitly models the development and behavior of an ablative vapor layer. The increased interaction cross section, $\pi(r + x)^2$, due to the presence of the vapor, accommodates energy conservation during the ablative process while retaining the potentially large energy deposition rates of previous models. The thickness of the vapor layer, which changes dynamically as the impactor penetrates, is a function of impactor composition, the temperature at the bow shock, and the opacity of the vapor and atmospheric gases.

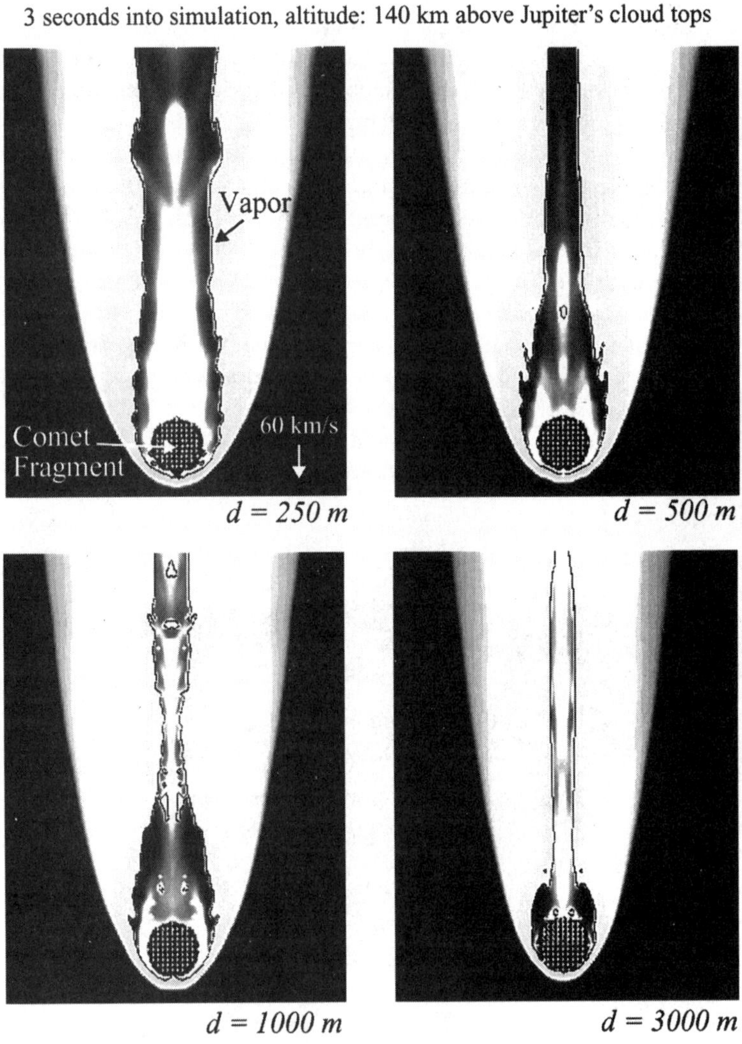

FIGURE 3. Comparison of four shock physics simulations (using the CTH hydrocode) showing the entry of SL-9 fragments varying from 250 m to 3 km in diameter (d = 250, 500, 1000, and 3000 m). In addition to the normal hydrodynamics that describe the deformation and eventual breakup of the fragments, the simulations include radiative ablation by computing the flux of electromagnetic energy impinging on the fragment surface and vaporizing a corresponding portion of the impactor material. There are three materials present in the calculations: Jovian atmosphere, water ice (an analog for pristine cometary material) and water vapor (formed by radiative ablation). In these 2D cross-sectional plots, gray scale represents temperature.

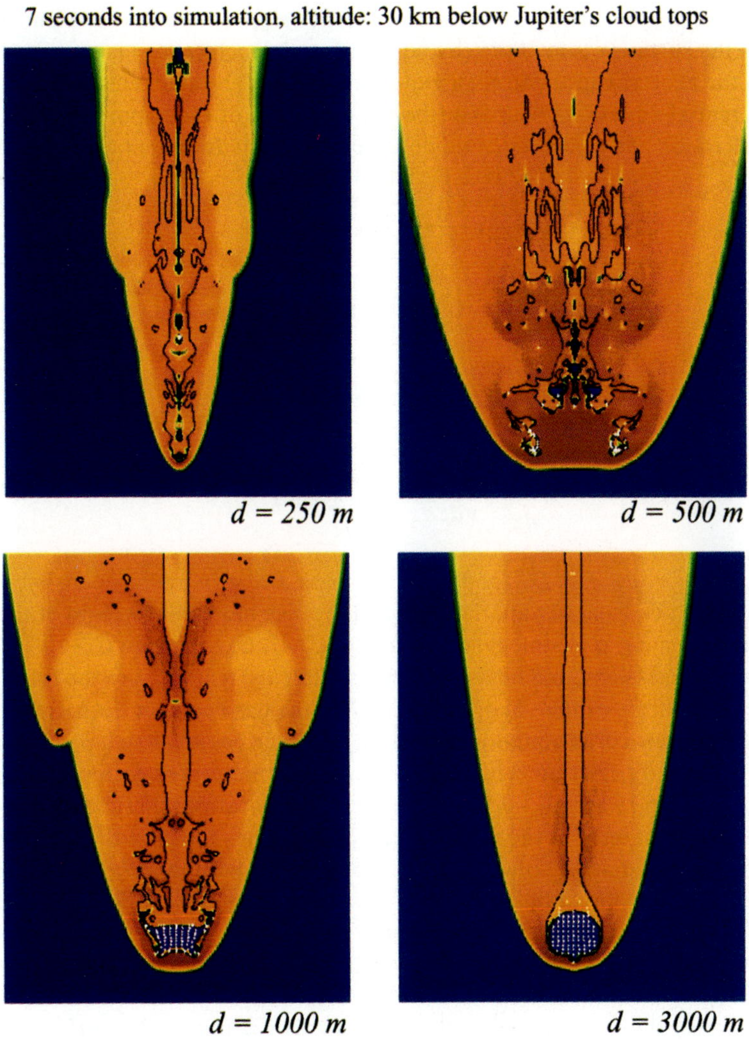

FIGURE 4. Comparison of four CTH simulations showing the entry of SL-9 fragments four seconds after the time shown in Fig. 3. At this altitude, just below Jupiter's cloud-tops, the aerodynamic pressure is causing severe deformation of most fragments. The 250 m fragment has disrupted, the 500 m fragment is in the process of disrupting, the 1 km fragment is severely deformed and the 3 km fragment is slightly deformed. Fragments smaller than about 200 m disrupt completely above the Jovian cloud tops, whereas the largest fragments may penetrate another 100–150 km before final disruption.

A comet fragment entering Jupiter's atmosphere at hypervelocity produces a parabolic bow shock (Fig. 2). Some of the irreversible heating at the shock is radiatively coupled to the fragment surface causing vaporization of the cometary materials. Because the energy required to remove a unit mass from the surface of the body (by stripping and deceleration) is much greater than the heat of vaporization, a vapor layer forms and thickens. The coupled differential equations that describe the evolution of the vapor layer and the deceleration and hydrodynamic deformation of the fragment are written as:

$$m\frac{dv}{dz} = \frac{C_d}{2}\pi(r+x)^2\rho v\sec\theta \tag{1}$$

$$\frac{dm}{dz} = C_m\pi(r+x)^2\rho\sec\theta \tag{2}$$

$$\frac{dx}{dz} = \frac{C_h R}{8Q_0 A}T_v\left(\frac{r}{r+x}\right)^2\sec\theta - \frac{1}{4}\frac{C_m RT_v}{Av^2}\sec\theta \tag{3}$$

$$r\frac{d^2r}{dz^2} + \frac{r}{v}\frac{dv}{dz}\frac{dr}{dz} = \frac{C_d\rho}{2\rho_f}(\sec\theta)^2 \tag{4}$$

where m, v, r, ρ_f, θ, z, Q_0, and A are the mass, velocity, radius, density, entry angle, altitude, heat of vaporization, and average molecular weight of the entering fragment; ρ is atmospheric density; C_d~1, C_m~0.5, and C_h~10^{-4}–10^{-2} are drag, mass-loss, and heat-transfer coefficients; x and T_v are the thickness and temperature of the vapor layer and R is the gas constant. Equation (1) is the well-known drag equation. Equation (2) describes mass stripped from the vapor layer by friction with the Jovian atmosphere. The stripped mass is equal to a fraction (C_m~0.5) of the encountered atmospheric mass at any time. The first term to the right of Eq. (3) describes the growth of the vapor layer due to radiative ablation. It is obtained by equating the vapor pressure with the kinematic "ram" pressure (ρv^2). The second term describes deflation of the vapor layer due to the mass loss of Eq. (2). Finally, Eq. (4) describes the mechanical flattening of the fragment due to hydrodynamic forces near the termination of its flight (after [23]).

Because the radiant energy that drives vaporization at the surface of the meteoroid must propagate through the vapor layer from the bow shock, growth of the vapor layer will slow as the vapor thickens. This leads to a scaling law for the heat transfer coefficient (C_h) which, as we will see, is inversely proportional to fragment size. Growth of the vapor layer is slowed by mass loss from small-wavelength Kelvin-Helmholtz instabilities that develop along the interface with the atmosphere [17]. The vapor reaches an equilibrium thickness (x_m) found by equating the two right-hand terms of Eq. (3):

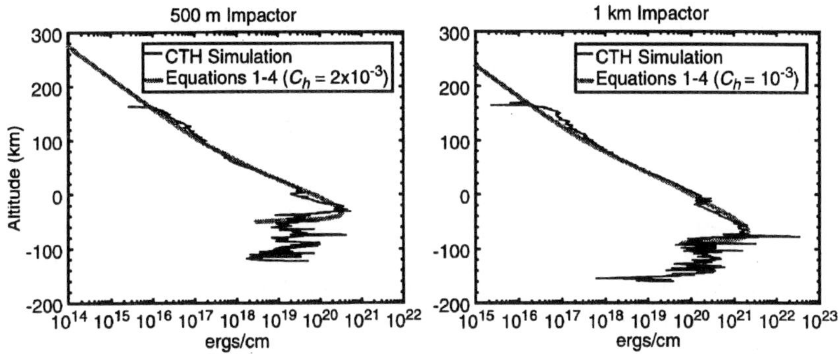

FIGURE 5. Energy deposition from 500 m and 1 km ice fragments entering Jupiter's atmosphere. The solid curves are from CTH simulations incorporating radiative ablation. Results of the semianalytical model (with C_h values indicated) are superimposed (gray curves).

$$x_m = \left(v\sqrt{\frac{C_h}{2C_m Q_0}} - 1\right)r . \tag{5}$$

For small meteoroids, x_m may be several times the radius of the entering body.

Equations (1)–(4) are numerically integrated to determine the deceleration, ablation, vaporization, and hydrodynamic deformation experienced by an entering body. The heat transfer coefficient (C_h) is calibrated against numerical multimaterial shock physics simulations (a specially-modified version of the CTH shock-physics code [26]) incorporating radiative ablation. The reduction in vapor volume with increasing fragment size apparent in these simulations (Figs. 3 and 4), is caused by a reduction in heat transfer efficiency as the vapor layer obstructs radiation from the shock front. The heat transfer coefficient is determined by fitting Eqs. (1)–(4) to the energy deposition per unit altitude derived from the CTH simulations (Fig. 5) and is found to be $C_h = 1/d$ (where d is fragment diameter in meters). This scaling law is consistent with fitted values of C_h for terrestrial meteors (C_h = 0.03–0.3; d < 10 m). In the following sections, we use the semianalytical model described by Eqs. (1)–(4) to provide driving conditions for further CTH simulations of Shoemaker-Levy 9 fireballs and plumes.

PLUME MORPHOLOGY: COMPARISON WITH OBSERVATIONS

Simulations performed prior to the impact of Comet Shoemaker-Levy 9 using the CTH shock-physics code compare well qualitatively with many of the observations [5], [6]. Nevertheless, good quantitative models of the event are required in order to extract useful information from observations of the events. Fortunately, models are strongly constrained by the excellent data collected, especially the direct light-flux observations made by the Galileo spacecraft [4], the seemingly contradictory observations that plumes observed by the Hubble Space Telescope (HST) all had approximately the same maximum altitude, yet the pattern of dark ejecta they left in the Jovian stratosphere varied considerably in albedo and lateral extent [8], [9] and the Earth-based telescope observations of the infrared impact light flashes [16]. By using the semianalytical meteoroid entry model as initial conditions for plume evolution models, we can investigate the Shoemaker-Levy 9 events in the context of matching light-flux and plume-height observations.

The technique we use to drive CTH simulations with the results of the semianalytical entry model is illustrated in Fig. 6. With the simulations driven in this way, we can efficiently run computational fireball simulations to investigate differences due to the size of the impactor without having to run time-consuming fully resolved entry calculations for each case. We have performed simulations for ten test cases with comet fragments ranging in size from 250 to 1250 m in diameter. In all cases, we assumed the comets were of density 0.95 g/cm^3 or 0.25 g/cm^3 with an equation-of-state (including melting, vaporization, dissociation, and ionization) appropriate for water ice. A two-dimensional representation, symmetric to the 45° entry angle, was used to simulate the first few minutes of fireball evolution in CTH. Tracer particles were added to the simulations to represent the Jovian cloud layers.

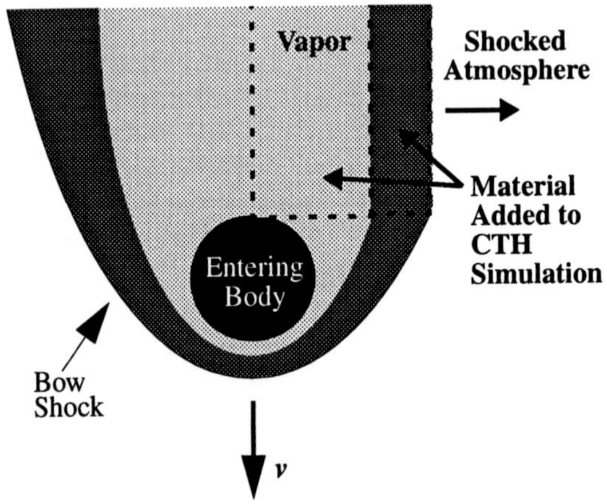

FIGURE 6. Schematic diagram showing the technique used to initialize CTH fireball simulations with the semianalytical model. Shocked atmosphere and vapor are added to the simulation with density, temperature, and momentum appropriate for the altitude of the entering fragment.

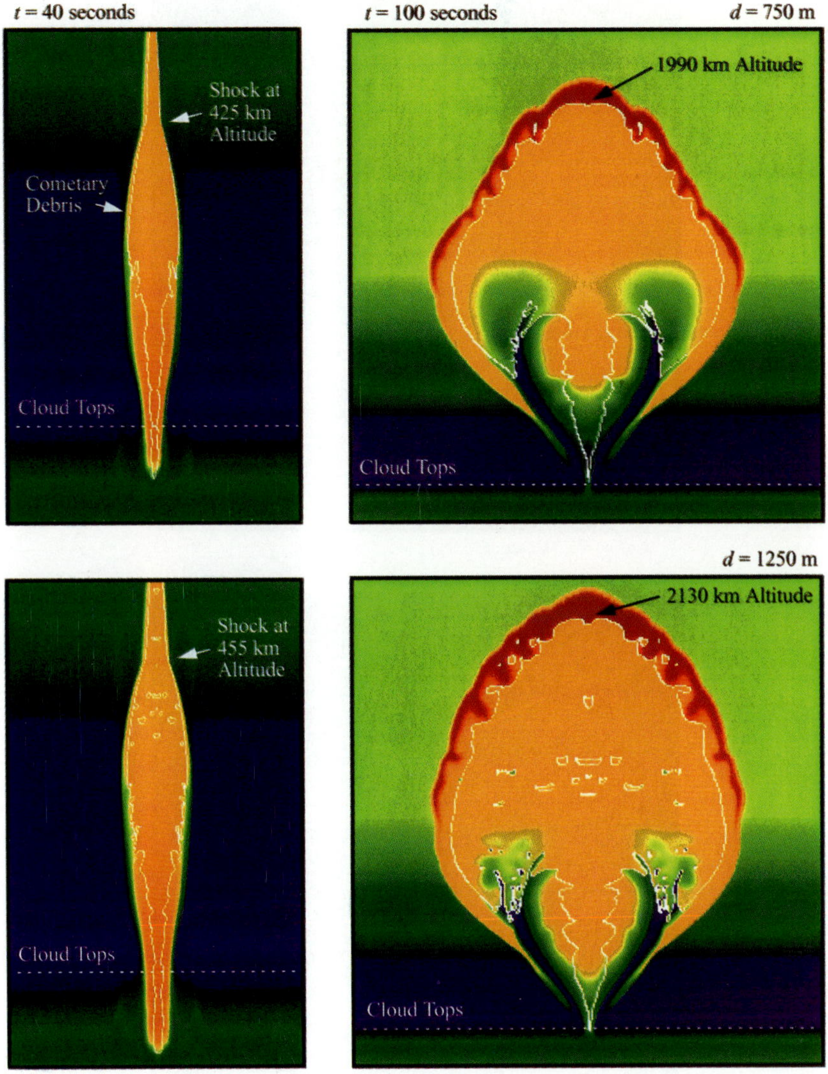

FIGURE 7. Plots of the ascending fireball from 750 m and 1250 m diameter (d) fragments seen 40 and 100 seconds after impact. A shock wave develops above the clouds and accelerates upwards along the debris column. After 100 seconds, debris from the comet has reached altitudes of 2000 km and is traveling upwards at 18–25 km/s. The plume has a distinctive mushroom shape which, when landed as ejecta on the planet, produces the crescent-shaped feature seen in telescope images. Color represents temperature. The debris front is indicated by the solid white curve.

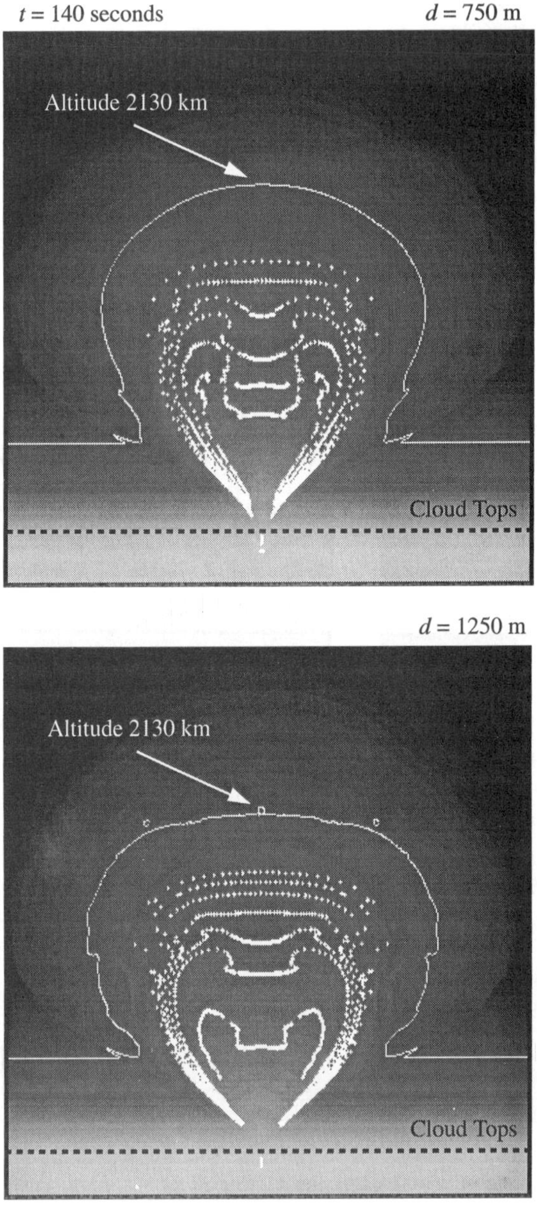

FIGURE 8. Plots of the fireball from 750 m and 1250 m fragments seen 140 seconds after impact. Gray scale represents log(density). The altitude of the 10^{-13} g/cm^3 isodensity contour (shown by the white curve) and the altitude of material from the Jovian cloud layers, represented by tracer particles (white dots) are invariant with respect to fragment size.

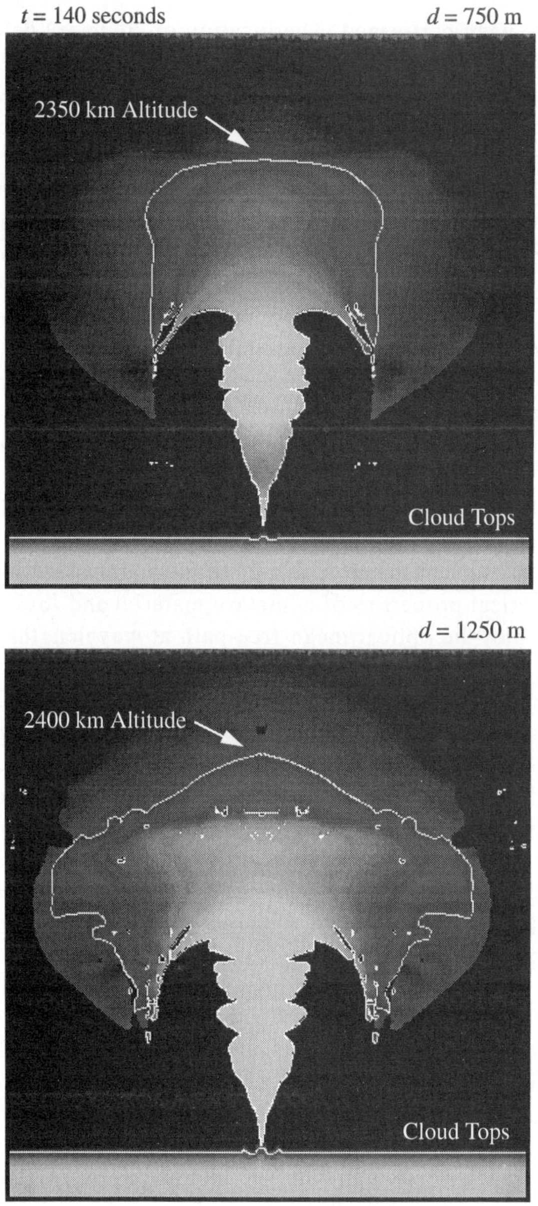

FIGURE 9. Cross-sectional plots showing plumes resulting from impact of 750 m and 1250 m fragments as they would appear in reflected sunlight. Gray scale represents –log(mean-free-path). The altitude of the 10 m mean-free-path contour (white curve) is nearly invariant with respect to fragment size. The albedo of cometary debris is unknown but probably very low, so the brightness of these plots is only a relative indicator.

Figures 7, 8, and 9 show results from two of the numerical fireball simulations that we have performed. The location of cometary debris within the developing plume is shown in the temperature plots of Fig. 7. In all cases, the debris front is traveling upwards (along the 45° trajectory) too quickly to make a plume only 3300 km high, the maximum altitude observed by HST. The plots in Fig. 8 show that material originating from the Jovian cloud layers (also shown in Fig. 8) is climbing at nearly the correct velocity to produce 3300 km plumes and their altitude seems to be relatively independent of impactor size. However, opaque materials from the cloud tops (NH_3, for example) is less than 1%, by mass, of the atmospheric gases within the clouds. By contrast, opaque material makes up about 30% of the mass within the debris front, hence cloud materials are unlikely to compete with cometary debris as a significant source of opacity for the plumes. The maximum altitude of a specific isodensity contour within the cometary debris ejected as part of the plume is an invariant function of fragment size and mass. Hence, it is probable that the tops of the HST plumes, depending on the opacity of condensed cometary materials, are represented by a specific isodensity contour.

To get a better idea of the morphology of the plumes seen by HST, we can estimate the optical properties of cometary material and Jovian atmospheric materials and plot the optical mean-free-path at wavelengths typical of the HST imagery. Figure 9 shows the inverse mean-free-path (at 890 nm wavelength) for plumes produced by 750-m and 1250-m fragments approximately two minutes after impact. The plots have a mean-free-path upper bound of 10 m (bright white) and are representative of what the plumes would look like if fully illuminated by sunlight. (The actual events occurred in darkness on Jupiter's far side but tops of the plumes climbed high enough to shine in reflected sunlight and be seen from Earth.)

The opacity model for cometary debris is based on the work of Nemtchinov et al. [14], modified to include opacity of 1 μm condensate particles (using simple geometric cross section) at temperatures below 3200 K. While condensation of cometary debris is important for matching the temperature data of the late-time plumes [2], it has only marginal effect on entry light curves, as will be shown later. The opacity model for Jovian air is from Chevalier and Sarazin [25]. Figure 9 demonstrates that the optical properties of the Shoemaker-Levy 9 impact plumes coupled with a dynamic impact process that produces invariant isodensity contours, generate constant maximum plume height even though cometary debris can be found on trajectories well above the apparent plume top.

The model plumes have a characteristic mushroom shape. If ballistically extrapolated forward in time until reimpact with the planet, these plumes are crescent-shaped and have a radial extent (dependent on impactor size and composition) that agree with the observed ejecta patterns. With confidence that our model can describe the morphology of the Shoemaker-Levy 9 plum-forming impacts, we can provide quantitative estimates of fragment sizes by comparing simulated radiative output with observations from the Galileo spacecraft and Earth-based telescopes.

RADIATIVE OUTPUT: COMPARISON WITH OBSERVATIONS

Figure 10 shows the scheme used to estimate light flux from a typical plume-forming simulation. To estimate light output at a particular wavelength, we integrate the Planck equation radially inwards until the optical limit is reached (at approximately one mean-free-path). During this process, absorption of light by intervening layers of material is taken into account. A realistic outcome to this process requires accurate opacity tables which, generally, depend on density and temperature for the materials involved. For cometary vapor, the opacity tables of Nemtchinov et al. [14] are used. The opacity of clean Jovian air is modeled with mean opacity data from Chevalier and Sarazin [25].

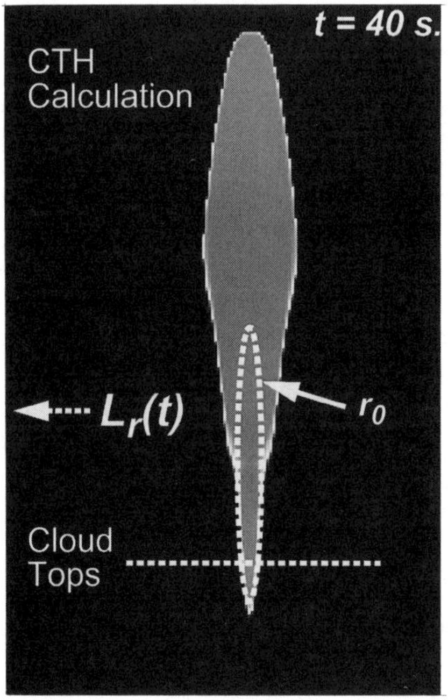

FIGURE 10. Scheme used to calculate the light flux, $L_r(t)$, from a CTH shock-physics calculation. The Planck equation with appropriate density- and temperature-dependent opacity tables is integrated radially inwards until the optical limit is reached (at r_0). Radiative output from below the cloud layers is excluded.

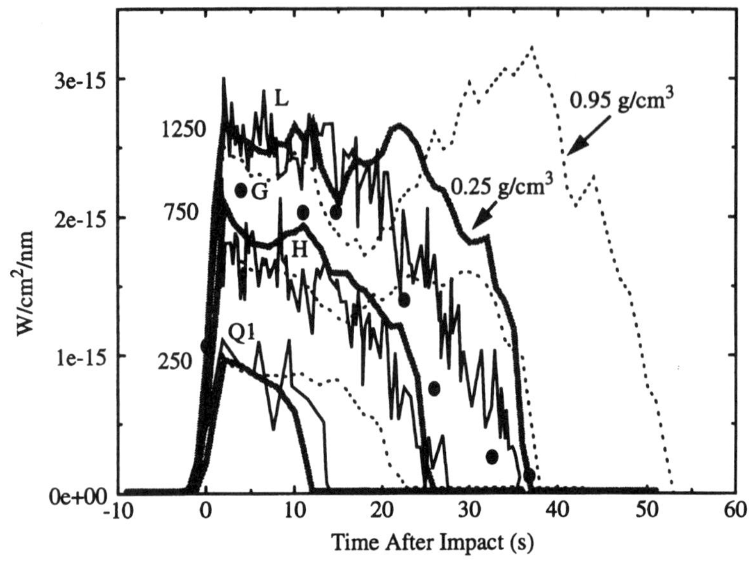

FIGURE 11. Comparison of CTH-simulated light curves (for 25O, 500, 750, 1000, and 1250 m fragments) against Galileo spacecraft observations (at 945 nm) of the impacts of fragments L, H, and Q1 (solid curves) and G (dots). Gray lines and dashed lines are simulations performed assuming fragment density of 0.25 and 0.95 g/cm^3, respectively.

Figure 11 shows light curves, scaled for distance to the Galileo spacecraft, for three of the five plume-forming events that we have simulated. Shown on the same figure are light curves for several events observed by Galileo's PPR sensor at a wavelength of 945 nm. By comparing with our simulations, we estimate that fragments L, G, H, and Q1 were 1250, 1000, 650, and 250 meters in diameter, respectively.

Surprisingly, fragment density plays a significant role in the character of the light curves. Near infrared light emitted from fireballs formed from relatively high density impactors is enhanced at late times (Fig. 11). Since the Galileo light curves did not exhibit significant enhancement at late times, this suggests that the fragments had low average density (0.25 g/cm^3) at time of impact.

The light flux shown in Fig. 11 exhibits $d^{1/2}$ scaling, where d is fragment diameter. This occurs because light is emitted from a column that has length solely dependent on the scale height (H) of the atmosphere and a radius proportional to the cometary debris radius ($r_d = r + x_m$). From Eq. (5):

$$r_d = r + x_m = vr\sqrt{\frac{C_h}{2C_mQ_0}} = v\sqrt{\frac{C_0 r}{2C_mQ_0}} \tag{6}$$

where $C_h = C_0/r$ (C_0 *is* independent of radius) and the area visible to Galileo ($\propto r_d H$) is proportional to $r^{1/2}$.

Many researchers have concluded that dust, mostly material from the impactors, was widely disseminated by the Shoemaker-Levy 9 impact sites [21] and may be important during the evolution of the fireballs [2]. Figure 12 shows light curve simulations assuming a simple model that neglects radiative energy loss and assumes that a single temperature adequately describes the materials that are emitting light. After the initial meteor entry rise and a brief plateau, the light flux begins to rise until it reaches unrealistic levels. In comparison, a more realistic model is produced assuming a two-temperature model where cometary vapor below 3200 K is allowed to condense. Because the condensates are probably very opaque, consisting of various metallic oxides, metals, and silicates, they are efficient emitters of radiation and cool rapidly. The loss of energy (now approximately accounted for) prevents the dramatic late-time rise (Fig. 12). The differences between the two models yields a small uncertainty (10–20%) of the fragment size estimates. Other sources of uncertainty include noise in the data and the enhanced signal due to light reflected off Jupiter's clouds (to add an additional uncertainty of 10–20%).

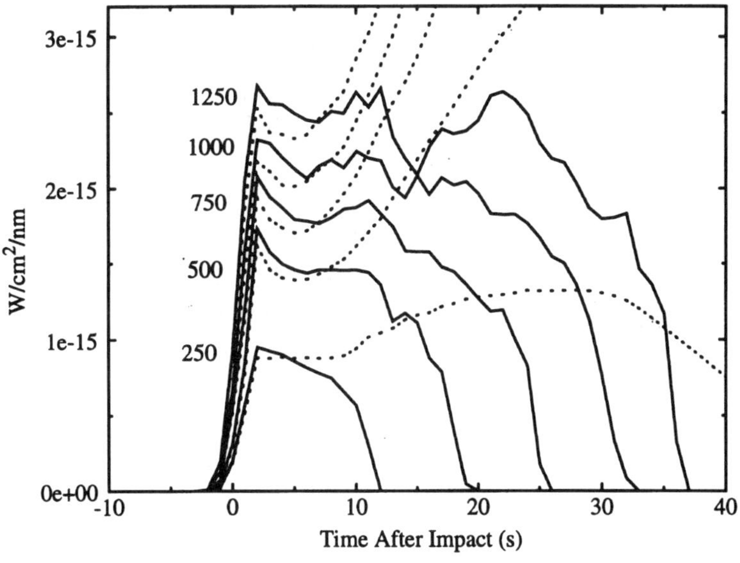

FIGURE 12. Comparison of CTH-simulated light curves using molecular opacity tables for cometary vapor from Nemtchinov et al. [14] (dashed curves) and modified opacity tables that accommodate condensation (and consequent radiative decoupling) for temperatures less than 3200 K (solid curves). Fragment diameters of 250, 500, 750, 1000, and 1250 meters are shown. Both sets of curves demonstrate that light flux is proportional to $r^{1/2}$. The discrepancy between the two opacity models in the magnitude of the initial rise yields a 20% uncertainty in fragment size.

TABLE 1. Shoemaker-Levy 9 Fragment and Parent Body Size Estimates

Fragment[a]	HST Class[b]	2.3 μm Peak (Jy)[c]	12 μm Peak (Jy)[c]	Diameter (m)[d]
A	2a	20	1,200	450
B	3			(50)
C	2a	25		380
D	3	4		200
E	2a	100	1,550	610
G	1	460		1,000•
H	2a	100	2,400	660•
K	1	460		1,000•
L	1	1,000	12,700	1,270•
N	3	0.04		45•
Q2	3			(50)
Q1	2b	90	1,700	600•
R	2b	70	990	530
S	2c	120		640
W	2c	60		490•
Parent	—	—	—	1,400[e]

[a]Fragments F, P2, T, U, and V produced no discernible impact features. Fragments J, M, and P1 faded from view before impact [9]. The letters I and O were not used.

[b]From Hammel [9]. Based on first view of the impact site with the Hubble Space Telescope. Class 1 = large dark feature (>10,000 km radius), Class 2 = medium dark feature (4000–8000 km radius), Class 3 = small dark feature (<3000 km radius).

[c]Observed light flux of the main infrared event seen from Earth at 2.3 and 12 μm [12], [20], [13], [16].

[d]Diameter is estimated from a least squares analysis incorporating Galileo light-flux observations (marked with • symbol and calibrated against CTH light-output calculations) and assuming r^3 dependence of the peak flux (for a given wavelength) observed from Earth-based telescopes [16]. Values in parentheses are estimates assuming diameter equivalence within HST class 3. Uncertainty (1 σ) of individual fragment diameter is 15% for fragments A, E, H, L, Q1, and R and 30% for fragments B, C, D, G, K, N, Q2, S, and W. Best fitting fragment density is about 0.25 g/cm^3.

[e]Assuming a density of 0.5 g/cm^3 before breakup in 1992. The total volume of the fragments would make a sphere 1760 m in diameter (at 0.25 g/cm^3). Uncertainty of these estimates is dominated by uncertainties of the largest fragments (L, K, and G), about 25%.

Table 1 shows our estimated sizes for the Shoemaker-Levy 9 fragments that produced discernible features on the planet. The fragment diameter is estimated from a least-squares analysis incorporating Galileo light-flux observations (calibrated against CTH light-output calculations and assuming r^3 dependence (mass scaling) of the peak flux for a given wavelength observed from Earth-based telescopes [12], [20], [13], [16]. For fragments that produced the smallest impact features on the planet, we assumed a constant diameter within the site classification derived from HST imagery [9].

For most cases where sufficient overlap between the Galileo and Earth-based observations exists, the r^3 scaling law holds fairly well. Normal variations due to density heterogeneity among the fragments is expected. There are a couple of notable exceptions, however. Fragments Q1 and W were relatively small by Galileo light flux estimates but of moderate size based on Earth-based observations. This could be explained if they were denser than the other fragments (e.g., fully dense silicate vs. fluffy, muddy water ice).

Other researchers have used the Galileo light flux observations to make estimates of fragment size. Nemtchinov et al. [15] estimated diameters of 800, 400, and 300 m for fragments K, W, and N, respectively, based on comparisons of the entry flash with radiative hydrodynamic simulations. These are comparable to the estimates of this study, especially when one considers that the K event light curve (measured with Galileo's solid-state imaging camera), has a distinctly different character than light curves of other comparable events (e.g., L and H) measured with a different instrument. Our estimate for fragment K diameter would agree almost exactly with the results of Nemtchinov et al. if *only* the entry light flash was modeled. A comparison with terrestrial observations, however, suggests that the K and G fragments are of comparable size and that the K light curve is curiously truncated (perhaps from high altitude clouds?).

Carlson et al. [2] have estimated the size of the G impactor as 200–400 meters based on a detailed semianalytical investigation using multi-wavelength time-dependent observations made by Galileo of this event. In their analysis, they assume that the debris front emitted radiant energy as a blackbody and was optically thick along its entire length. This is a good constraint for a lower bound of the size estimate for that reason. More realistic, density- and temperature-dependent opacity models (performed by Nemtchinov et al. [15] and this study) seem to produce less light for a given impactor diameter, requiring larger impactors.

For fragment G to be only 200–400 meters would require smaller fragments like H and K to explode above Jupiter's cloud tops according to most hydrodynamic deformation models. Normally, the rapid rise of the entry flash is truncated by arrival at the cloud tops (which occludes radiation from deeper) and the distinctive plateau of the light curve is produced. Explosions above the clouds would produce distinctly different light curves for these events (there would be no truncation or plateau), yet the fragment H and K light curves exhibited similar behavior as the larger events (Fig. 11).

CONCLUSION

The fragments of Shoemaker-Levy 9 that hit Jupiter with discernible effect have a total volume that would make a 1.8 ± 0.5 km diameter sphere at an average density of 0.25 g/cm^3. This is consistent with the total volume of dusty debris seen on the planet after the impacts, corresponding approximately to a 1 km diameter sphere [21], assuming a typical debris particle density of (1–2 g/cm^3). With a prebreakup density of 0.5 g/cm^3 [18], [24], the parent body had a diameter of 1.4 ± 0.4 km. With these parameters, the Shoemaker-Levy 9 family pummeled Jupiter with the energy equivalent of 300 Gigatons of TNT.

REFERENCES

1. BOSLOUGH, M.B. & D.A. CRAWFORD. 1997. Shoemaker-Levy 9 and plume-forming collisons on Earth. Proceedings of the International Symposium on Near-Earth Objects. United Nations/Explorers Club, New York. April 24–26, 1995. Annals of the New York Academy of Sciences. This volume.
2. CARLSON, B., ET AL. 1997. Temperature, Size, and Energy of the Shoemaker-Levy 9 G-Impact fireball. Icarus. In press.
3. CEPLECHA, Z., ET AL. 1993. Atmospheric fragmentation of meteoroids. Astron. and Astrophys. **279:** 615–626.
4. CHAPMAN, C.R. 1996. Galileo observations of the impact. *In* The Collision of Comet Shoemaker-Levy 9 and Jupiter, K.S. Noll, H.A. Weaver, & P.D. Feldman, Eds.: 121–132. Cambridge Univ. Press.
5. CRAWFORD, D.A., ET AL. 1994. The impact of Comet Shoemaker-Levy 9 on Jupiter. Shock Waves. **4:** 47–50.
6. CRAWFORD, D.A., ET AL. 1995. The impact of periodic Comet Shoemaker-Levy 9 on Jupiter. Int. J. Impact Engin. **17:** 253–262.
7. CRAWFORD, D.A. 1996. Models of fragment penetration and fireball evolution. *In* The Collision of Comet Shoemaker-Levy 9 and Jupiter, K.S. Noll, H.A. Weaver, & P. D. Feldman, Eds.: 133–156. Cambridge Univ. Press.
8. HAMMEL, H.B., ET AL. 1995. HST Imaging of atmospheric phenomena created by the impact of Comet Shoemaker-Levy 9. Science. **267:** 1288–1296.
9. HAMMEL, H.B., ET AL. 1996. HST imaging of Jupiter shortly after each impact: Plumes and fresh sites. *In* The Collision of Comet Shoemaker-Levy 9 and Jupiter, K.S. Noll, H.A. Weaver, & P.D. Feldman, Eds.: 111–120. Cambridge Univ. Press.
10. HILLS, J.G. & M.P. GODA. 1993. The fragmentation of small asteroids in the atmosphere. Astro. J. **105:** 1114–1144.
11. IVANOV, B.A. & O. YU. 1988. Simple hydrodynamic model of atmospheric breakup of hypervelocity projectiles. Lun. Planet. Sci. **19:** 535–536.
12. LAGAGE, P.O., ET AL. 1995. SL-9 fragments A, E, H, L, Q1 collision on to Jupiter: Mid-infrared light curves. Geophys. Res. Lett. **22:** 1773.
13. MCGREGOR, P., ET AL. 1996. CASPIR observations of the collision of Comet Shoemaker-Levy 9 with Jupiter. Icarus. **121:** 361–388.
14. NEMTCHINOV, I.V., ET AL. 1996. Spectral opacities and ablation rates for recognized meteoroid materials, Interim Report 001 on task 1.0 under the Sandia National Lab-

oratories Contract AS-6889, Institute for Dynamics of Geospheres, Russian Academy of Sciences, Moscow. 136 pp.

15. NEMTCHINOV, I.V., ET AL. 1997. Assesment of Comet Shoemaker-Levy fragment sizes using light curves measured by Galileo spacecraft instruments. Planet Space Sci. **45:** 311–326.
16. NICHOLSON, P.D. 1996. Earth-based observations of impact phenomena. *In* The Collision of Comet Shoemaker-Levy 9 and Jupiter, K.S. Noll, H.A. Weaver, & P.D. Feldman, Eds.: 81–109. Cambridge Univ. Press.
17. ROULSTON, M.S. & T.J. AHRENS. 1997. Impact Mechanics and frequency of SL9-type events on Jupiter. Icarus. **126:** 138–147..
18. SCOTTI, J.V. & H.J. MELOSH. 1993. Tidal breakup and dispersion of P/Shoemaker-Levy 9: Estimate of Progenitor Size. Nature. **365:** 7333.
19. SEKANINA, Z. 1993. Disintegration phenomena expected during the forthcoming collision of periodic comet Shoemaker-Levy 9 with Jupiter. Science. **262:** 382.
20. TAKEUCHI, S., ET AL. 1995. Near-IR imaging observations of the cometary impact into Jupiter: Time variation of radiation from impacts of fragments C, D and K. Geophys. Res. Lett. **22:** 1581
21. WEST, R.A. 1996. Particulate matter in Jupiter's atmosphere from the impacts of Comet P/Shoemaker-Levy 9. *In* The Collision of Comet Shoemaker-Levy 9 and Jupiter, K.S. Noll, H.A. Weaver, & P.D. Feldman, Eds.: 269–292. Cambridge Univ. Press.
22. ZAHNLE, K. & M.M. MacLow. 1994. The Collision of Jupiter and Comet Shoemaker-Levy 9. Icarus. **108:** 1–17.
23. ZAHNLE, K. 1992. Airburst origin of dark shadows on Venus. J. Geophys. Res. **97:** 10243–10255.
24. ASPHAUG, E. & W. BENZ. 1994. Density of Comet Shoemaker-Levy 9 deduced by modelling breakup of the parent "ruble pile." Nature. **370:** 120–124.
25. CHEVALIER, R.A. & C.L. SARAZIN. 1994. Explosions of infalling comets in Jupiter's atmosphere. Astrophys. J. **429:** 863–875.
26. MCGLAUN, J.M., S.L. THOMPSON, & M.G. ELRICK. 1990. CTH — A three-dimensional shock-wave physics code. Int. J. Impact Engin. **10:** 351.

Observations and Studies of NEOs and the SL-9 Impact at the Purple Mountain Observatory

HEQI ZHANG, JIAXIANG ZHANG, SICHAO WANG, QI WANG, AND JIEXING YANG

Purple Mountain Observatory
Academia Sinica
Nanjing 210008
Peoples Republic of China

Beginning in 1964, a program for the systematic search of asteroids and their further study was conducted by Y.C. Zhang, the former director of the PMO (Purple Mountain Observatory), and Jiaxiang Zhang. Photographic observations with a 40 cm double astrograph yielded the discovery of three comets and more than 100 asteroids that qualified for numbering. Two of those asteroids, (2077) Riangau and (2078) Nanking, are Mars-crossers.

In 1987, a program for the systematic search and study of NEOs was proposed by Sichao Wang. In the course of this search, Qi Wang, Yongliang Ge, and other PMO astronomers observed on several occasions with the 60 cm Schmidt telescope at the Xinglong station of the Beijing Observatory. Using a stereo comparator to examine a pair of plates obtained with this telescope, which has a field of 4.8 degrees, Ge and Wang discovered comet P/1988 V1 and the fast-moving object 1991 BH. The program became affiliated with the International Near-Earth Asteroid Survey (INAS), and the study of appropriate search strategy still continues, the main difficulty being that the observing time available on the telescope is very limited. Observations of some unusual NEOs, including (1566) Icarus and (4581) Asclepius, have also been made using a CCD attached to the 1 m reflector at the Yunnan Observatory.

In addition to the NEO searches, software has been developed for use in the dynamical modeling of bodies in the solar system. A predicted set of orbital elements and Jupiter-impact parameters for comet SL-9 produced on 1994 July 7 (from observations through July 2) was disseminated via the SL-9 e-mail exploder at the University of Maryland and was found to be in good agreement with the subsequent observations.

During 1994 July 13 through September 14, observations were made of Jupiter using the 60 cm f/16 reflector and an MTV 1881 EX CCD video camera. More than 800,000 high-quality images (unfiltered or with an R filter) were recorded using an M2 video recorder, and more than 900 images were sampled using a multimedia computer. The dark spots induced by the nuclei G, H, K, L, Q1, R, S, and W were monitored. The size, low albedo and characteristic appearance of a central region with a ring-like structure, suggest that the impact events were very strong. A spot could have been produced by a high-altitude cloud in the stratosphere, the cloud consisting of dark particles and other material that condensed from the impact plume. The dark feature to the south of a spot was perhaps caused by splash-back from the impact plume.

Yuehua Ma and his colleagues have studied the processes of the comet's ablation and dissociation in the jovian atmosphere. The cometary fragments were variously considered to have masses of 10^{12}, 10^{13}, and 10^{14} kg and densities of 3.0, 1.0, 0.8, 0.5 and 0.2 g/cm^3. The results show that these fragments exploded below the ammonia cloud layer, near the 1 bar level, thus explaining the observed presence of ammonia and absence of water. It can also be estimated that, as was observed, these explosions would lift atmospheric material by up to 8000 km above the top cloud layer.

In other work, Sichao Wang, Pinxin Xu, and Jianming Wang have studied the shock effects of high-velocity impacts among solid bodies in the solar system, and J.H. Lu has computed the collision frequencies between 41 Aten-Apollo-Amor objects and the Earth.

The impact of comet SL-9 on Jupiter attracted a lot of attention and greatly aroused public interest in NEOs in China. The PMO has recently proposed a plan to construct a new telescope of aperture 1 m or so and 2048 × 2048 CCD equipment at a station about 400 km north of Nanjing. The site has good weather and low light-pollution, and the telescope will perhaps be completed in 2–3 years. It is to be hoped that the establishment of such an instrument at the indispensable longitude of 8 hours east of Greenwich will, before long, make an important contribution to the international NEO survey.

NEO Orbits and Nonlinear Dynamics: A Brief Overview and Interpretations

JOHN L. REMO

Quantametrics, Inc.
1 Brackenwood Path, Head of Harbor
St. James, New York 11780

ABSTRACT: The development of nonlinear dynamics is outlined with emphasis on the concepts leading to the development of a numerical technique known as "fuzzy boundary theory" or FBT. Origins of this methodology are a direct outgrowth of concepts derived from the Hamiltonian formulation, KAM theory, and "Mather regions." The significance of FBT as a computational methodology for NEO studies includes the weak capture and fast resonance shifting applied to NEO orbits and spacecraft rendezvous as described by a paper by Belbruno in this volume. Further implications of the FBT methodology suggest that some NEO capture cross-sections with planetary bodies may have to be recomputed.

INTRODUCTION

The following paper in this volume, "Fast Resonance Shifting as a Mechanism of Dynamic Instability Illustrated by Comets and CHE Trajectories," by E. Belbruno, describes an innovative technique for the computation of several new dynamic mechanisms involving three- and four-body problems, including ballistic capture in general and fast resonance shifting at both the Moon and Jupiter. The CHE dynamics referred to in the title describes a weak capture (C) mechanism, followed by a finite sequence of hops (H), and ending with an ejection or escape (E) within a finite period of time, hence CHE. To some readers without a strong background in the mathematics associated with nonlinear equations, this work may appear to be esoteric. However, the potential significance of this and related work [23] on nonlinear orbits in celestial mechanics and chaotic evolution in the solar system [22], in terms of NEO studies in general, and resonances in comet orbits in particular, is difficult to underestimate. Therefore, in order to assist some of those who wish to understand better the Belbruno paper and speculate on some of its implications for NEO studies, a brief sketch of some of the mathematical progress in nonlinear analysis during the twentieth century that has led up to the current stage of development of the so called "fuzzy boundary theory" (FBT) computational method is presented. A limited set of references is also provided to lead the reader to some key papers in this and related areas of nonlinear analysis. It is hoped that this brief overview will encourage interest and foster insight into the significance of the FBT computational approach in orbital mechanics and astronautics.

Low Energy Orbital Trajectories

On October 2, 1991, an operational Japanese spacecraft called Hiten, which was never designed to be placed in lunar orbit, reached the lunar surface from an Earth orbit using only a very small amount of residual propellant. Unlike traditional trajectories to the Moon, which take about three days and use significant amounts of fuel at lunar arrival to become captured, this trajectory takes 80 days, and requires no fuel to be captured. In carrying out this maneuver, Hiten was the first spacecraft to execute a lunar boundary transfer using a trajectory computed from "fuzzy boundary theory" [4]. Although the time of flight for such fuzzy boundary transfers is approximately three months, the propellant required is minimal, being about 25% of that required for the (traditional) Hohmann transfer to achieve a low circular lunar orbit.

Following along the same lines of analysis that provided the trajectory for the Hiten lunar maneuver, a recent study [6] has suggested a low energy mechanism to achieve a comet rendezvous by means of a resonance transition. Orbital trajectories from the Earth that utilize such a mechanism are computed to (ideally) save up to 86% of the energy that would be used in the traditional trajectory design. In addition to its pragmatic significance for planning mission to NEOs, what is so fascinating about this mechanism is that it is thought to be capable of occurring in regions about Jupiter called fuzzy boundaries which are thought to possess chaotic properties. While there may be no generally accepted definition of chaos, a basis for such chaotic motion nonetheless lies within classical particle dynamics which is a highly developed branch of physics that has provided the (initial) basis for understanding several types of nonlinear motion.

CONCEPTS IN NONLINEAR DYNAMICS

Historical Perspectives on Dynamical Systems

During the eighteenth and nineteenth centuries, Newtonian mechanics, using a law of motion described by acceleration (a second-order differential) provided an orderly and useful framework to solve most of the celestial mechanical problems of interest for that time. In order to specify the initial state of a Newtonian system, the velocities and positions of each particle must be specified. However, in the mid-nineteenth century, Hamilton recast the formulation of dynamical systems by introducing the so-called Hamiltonian function, H, which represents the total energy of the system expressed in terms of the position and momentum, which is a first-order differential. This first-order aspect of the Hamiltonian, which represents a universal formalism for modeling dynamical systems in physics, implies a determinism for classical systems, as well as a link to quantum mechanics through the Bohr correspondence principle.

Chaotic Motion

By the early 1900's, Poincaré understood that the classical Newtonian three-body problem gave rise to a complicated set of dynamics that was very sensitive to dependence on initial conditions, which today is referred to as "chaos theory." The origin of chaotic motion can be traced back to classical (Hamiltonian) mechanics which is the foundation of (modern) classical physics. In particular, it was nonintegrable Hamiltonian mechanics and the associated nonlinear problems which posed both the dilemma and ultimately the insight into the occurrence of randomness and unpredictability in apparently completely deterministic systems. It was, perhaps, the embarrassment of riches provided by the brilliant insights of quantum mechanics and relativistic physics during the first half of the twentieth century that relegated the nonintegrable Hamiltonian mechanics from the mainstream of physics research. Also, the advent of the computer provided the tools which were hitherto lacking to earlier researchers, such as Poincaré. Nonetheless, with the development of computational methodology combined with deep intuitive insights, the early 1960's gave rise to the formulation of the KAM theorem, named after A.N. Kolmogorov, V.I. Arnold, and J. Moser, that provided the conditions for randomness and unpredictability for nearly nonintegrable Hamiltonian systems.

Within the framework of current thinking, almost synonymous with certain classes of nonlinear problems is the so-called chaotic behavior. Chaos is not just simply disorder, but rather an order without periodicity. An interesting and revealing aspect of chaotic behavior is that it can appear random when the generating algorithms are finite, as described by the so-called logistic equations.

Chaotic motion is important for astrophysical (orbital) problems in particular, simply because very often within generally chaotic domains, patterns of ordered motion can be interspersed with chaotic activity at smaller scales. Because of the scale characteristics, the key element is to achieve sufficiently high resolving power in the numerical computation in order to describe precisely the quantitative behavior that can reveal certain types of chaotic activity. Such precision is required because instead of the much more familiar spatial or temporal periodicity, a type of scale invariance manifests itself. This scale invariance, discovered by Feigenbaum [10], [11] for one-dimensional mappings, provided for the possibility of analyzing renormalization group considerations within chaotic transitions.

Ergodicity, Randomness, and Determinism

Insights into stochastic mechanics have also been derived from related developments in nonlinear analysis, such as the relationship between nonlinear dynamics and modern ergodic theory. For example, if time averages along a trajectory on an energy surface are equal to the ensemble averages over the entire energy surface, a system is said to be ergodic on its energy surface. In the case of classical systems, randomness is closely related to ergodicity [12], [2]. When characterizing attractors in dissipative systems, similarities to ergodic behavior are encountered.

Perhaps one of the most familiar examples of a system's inherent randomness is the work of E.N. Lorenz [17] on thermal convection, which demonstrated that completely deterministic systems of three ordinary differential equations underwent irregular fluctuations. Such bounded, nonperiodic solutions which are unstable can introduce turbulence, and hence the apellation "chaos," which connotes the apparent random motion of some mappings. For dissipative systems the term "strange attractor," was introduced [20]. One test that can be used to distinguish chaos from true randomness is through invocation of algorithmic complexity; if one has a sequence of length n, and the length of the shortest algorithm that reproduces that sequence is also approximately n, then the algebraic complexity is maximal. Simply put, a random sequence of zeros and ones can only be reproduced by copying the entire sequence; periodicity is of no assistance.

HAMILTONIAN MOTION

The Hamiltonian formulation seeks to describe motion in terms of first-order equations of motion. The usefulness of the Hamilitonian viewpoint lies in providing a framework for the theoretical extensions into many physical models, foremost among which is celestial mechanics. We note that the Hamiltonian equations hold for both special and general (in a modified form) relativity. Furthermore, within classical mechanics it forms the basis for further development, such as the familiar Hamilton-Jacobi method and, of even greater extension, the basis for perturbation methods. This latter aspect of Hamiltonian theory will provide a starting point for the analytical discussions to follow in this brief outline.

Hamiltonian Systems

As already mentioned, the Hamiltonian formulation basically seeks to describe motion in terms of first-order equations of motion. The number of initial conditions and independent variables determining the motion must be $2N$ first-order equations expressed in terms of $2N$ independent variables in one-dimensional phase space, and $6N$ particles in three-dimensional phase space.

That is, the double set of independent variables are usually selected equally from N generalized coordinates, $q_{i,}$ and the from the N generalized conjugate momenta, $p_{i,}$ which are known as the canonical variables. The Hamiltonian function $H = H(p_i; q_i)$ can be described by the canonical equations for N degrees of freedom, $I = 1, 2, ..., N$, such that

$$dq_i/dt = \partial H / \partial p_i \tag{1}$$

and

$$dp_i/dt = -\partial H / \partial q_i. \tag{2}$$

If successive canonical transformations exist which map (change) the set (p_i, q_i) into a new set of canonical variables (J_i, Q_i) such that the Hamiltonian function depends only on the J_i, but not on the Q_i, then $H = H(J_i)$ alone, and the Q_i become periodic variables. The corresponding canonical equations now become

$$dQ_i/dt = \partial H/\partial J_i = \Pi_i(J_i) \tag{3}$$

and

$$dJ_i/dt = -\partial H / \partial Q_i = 0, \tag{4}$$

which clearly shows the cyclic nature of the Q_i, by identifying Q_i (Mod 2π). That is, the Q_i define angular variables on an N-dimensional torus. Integration of the above yields

$$Q_i(t) = \Pi_i t + Q_i(0), \tag{5}$$

$$J_i(t) = J_i(0). \tag{6}$$

The motion of the above integrable system can be interpreted as a system of coupled oscillators. In order to maintain motion within a finite region of the phase space, the Q_i must appear as arguments of periodic functions, or else they will increase. For $N = 1$, the motion is rotational about a circle of radius $\sqrt{2}\, J(0)$ with a constant angular velocity Π. That is, the bounded motion of a system with one degree of freedom is periodic. Similar initial states will predictably lead to similar final states. Therefore, an important attribute of the Hamiltonian causality is that it implies that similar causes lead to similar effects.

For $N = 2$, the radii are determined by $J_1(0)$ and $J_2(0)$, and the angular velocities by Π_1 and Π_2; such motion is confined to a two-dimensional toroidal surface whose radii are defined by the two J's. In addition to the periodic motion there also will exists the possibility that quasi-periodic or conditionally-periodic motion can occur. If the ratio of Π_1/Π_2 is an irrational number, the motion will no longer be periodic. In such a case, the trajectory winds up the torus densely and endlessly. Generally, the motion of an integrable Hamiltonian system with N degrees of freedom is periodic and confined to the N-torus as shown in Fig. 1. Hamiltonian systems for which $N = 1$ are all integrable, while the vast majority of systems with $N \geq 2$ become nonintegrable.

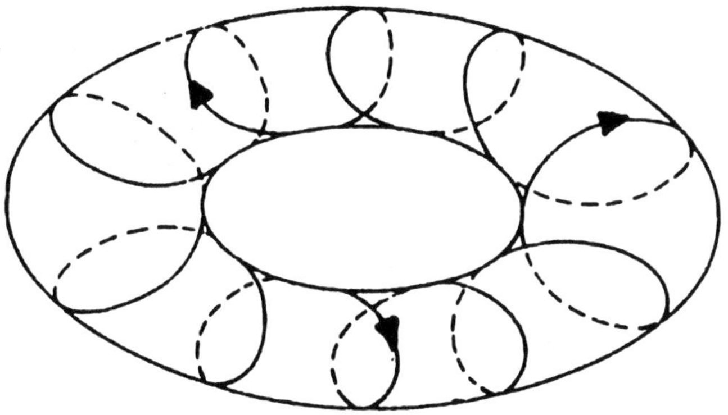

FIGURE 1. Description of an integrable system with two degrees of freedom on a torus, an a closed orbit of a trajectory. The KAM tori are concentric versions of the single torus.

Simply restating the above, an integral of motion which makes it possible to reduce the order of a set of equations, is called the first integral. To integrate a set of differential equations of the order $2N$, that same number of integrals are generally required, except in the case of the Hamiltonian equations of motion, where N integrals are sufficient. This also can be expressed in terms of the Liouville theorem, which states that, "any region of phase space must remain constant under any (integrable) Hamiltonian formalism." The phase space region can change its shape, but not its phase space volume. Therefore, for any conservative dynamical system, such as planetary motion or pendula that does not have an attracting point, the phase space must remain constant. The relationship between the phase space and the integrals of motion are expressed through the Liouville-Arnold theorem [3].

THE KAM THEOREM, ARNOLD DIFFUSION, AND MATHER REGIONS

Another outcome of the Hamiltonian formulation, which started out as a formulation for regular motion, is now the implication of the existence of irregular and chaotic trajectories. As already mentioned, Poincaré realized that nonintegrable, classical, three-body systems could lead to chaotic trajectories (1892). It is important to understand that chaotic behavior is due neither to a large number of degrees of freedom nor to any initial numerical imprecision. Chaotic behavior arises from a nonlinearity in the Hamiltonian equations with initially close trajectories that separate exponentially fast into a bounded region of phase space. Since initial conditions can only be measured with a finite accuracy, and the errors propagate at an exponential rate, the long range behavior of these systems cannot be predicted. That is, initial numerical conditions are usually characterized by real values which may be irrational numbers (with an infinite sequence of digits).

Returning to the Hamiltonian formalism, we first look at the effects of perturbations in establishing regions of nonintegrability.

Perturbed Hamiltonian

If the number of degrees of the Hamiltonian system is larger than two, integrability is a singular property. Integrable Hamiltonians are so uncommon that it is generally impossible to approximate nonintegrable Hamiltonians with a series of integrable ones. The interesting situation from the point of view of orbital mechanics, is when a system is made only slightly nonintegrable. For such cases, the variation in the Hamiltonian can often be expressed in terms of a perturbation parameter, α, such that

$$H = H_0 + \alpha V, \tag{7}$$

where the unperturbed Hamiltonian, H_0, is integrable. How can the motion of the system be described for a (weak) perturbation? The answer lies within the KAM theorem which was originally stated by Kolmogorov [15] and rigorously proven by Arnold [1] and Moser [18], which required an analysis of perturbative solutions to the classical many-body problem.

KAM Theorem

Essentially, the KAM theorem states that provided the perturbation, αV, which is responsible for the nonintegrability of the Hamiltonian is small, and that the frequencies, Π_i, of the nondegenerate unperturbed integrable Hamiltonian, H_0, satisfy the nonresonance condition,

$$\det \left[\partial^2 \mathrm{H}_0 / \partial J_i \, \partial J_i\right] \neq 0, \tag{8}$$

then the dynamic motion is still confined to an N-torus except for a negligible set of measure zero of initial conditions which may lead to a wandering motion on the energy surface. This wandering motion is chaotic, implying a great sensitivity to initial conditions. However, it is often a very gradual function of time. The N-tori, in this case, are known as KAM surfaces. When observed as plane sections they are often called KAM curves. These surfaces and curves may be slightly distorted (perturbed) as compared to the case where $\alpha = 0$ and no perturbation is present. When the two conditions given above hold, the motion of the unperturbed system is similar to that of the perturbed system. That is, for a sufficiently small conservative Hamiltonian perturbation, most of the nonresonant invariant tori will not vanish, but will undergo a slight deformation, such that in the perturbed system phase space there are also invariant tori, filled by phase curves, which are conditionally periodic.

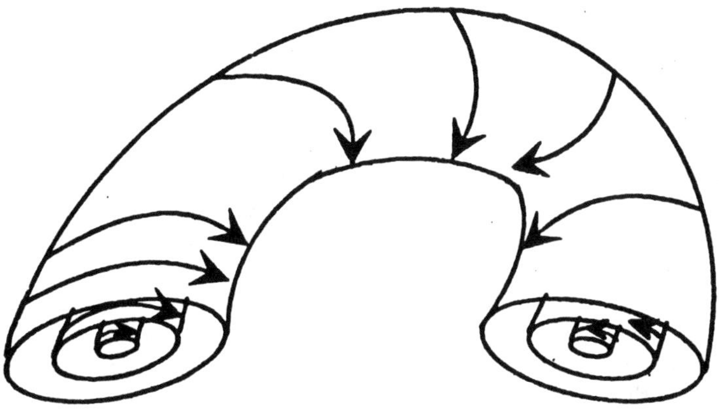

FIGURE 2. Depiction of a set of KAM invariant tori on the surface of which lie (elliptic) integrable solutions. The nonintegrable solutions, irregular paths, which are hyperbolic in nature lie in between the invariant tori in so-called resonant zones, which are also sometimes referred to as stochastic zones.

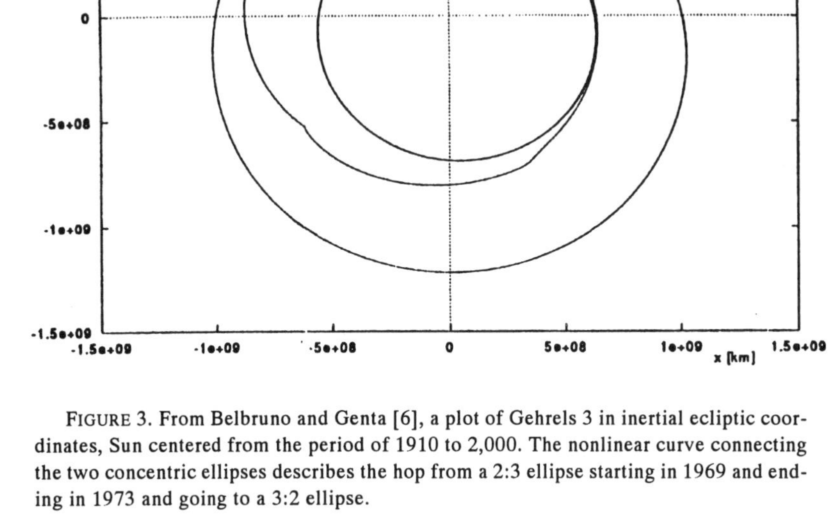

FIGURE 3. From Belbruno and Genta [6], a plot of Gehrels 3 in inertial ecliptic coordinates, Sun centered from the period of 1910 to 2,000. The nonlinear curve connecting the two concentric ellipses describes the hop from a 2:3 ellipse starting in 1969 and ending in 1973 and going to a 3:2 ellipse.

Arnold Diffusion

It is noted that for $N \leq 2$, the chaotic motion which is not necessarily confined to the N-torus is trapped between neighboring tori. In this sense the motion is stable. For systems with N degrees of freedom, $N > 2$, a qualitatively new phenomenon appears. This motion is no longer trapped between tori by virtue of the (increased) dimensions, and the particle can slowly deviate away from nearby tori. This instability is known as Arnold Diffusion [3], and follows from the fact that constant energy boundary surfaces must be of dimension $2N - 2$. If $N < 2N - 2$. The N-dimensional KAM torus cannot provide boundaries separating the energy surface into regions impenetrable for wandering trajectories. When $N > 2$, these trajectories may wander along the whole energy surface and give rise to a new randomness mechanism referred to as Arnold diffusion [1], which is not generally detectable under the KAM conditions. J. Wisdom [23], utilizing a very long integration method, numerically indicated that slow chaotic instabilities are likely to exist in the solar system, which can be interpreted in terms of Arnold diffusion. In order to illustrate some of the key concepts associated with the KAM torus, a very simple example of a KAM torus, the well-studied mathematical pendulum as applied to two (weakly) coupled oscillators that are nearly integrable, is given in Appendix A.

For alpha sufficiently small, the separation between the tori is correspondingly small. As alpha increases, the N-tori start to disappear, and the gaps widen. For $N = 2$, Mather's [17a] work implies, in general, that for alpha sufficiently large, no more N-tori remain, and the resulting motion is chaotic. When the N-tori disappear, other geometries appear as ghosts of these tori, which are called Aubrey-Mather sets [17a], [14].

Mather Regions

The KAM results were further extended in an important manner through the results of J. Mather [17a], [17b]. Essentially, KAM theory treats motions and related orbits that are very close to being well behaved and stable. Since KAM theory is basically a perturbation analysis, by its very nature the perturbation constant must be very small ($\alpha \ll 1$). Strong departures from the original operator through the perturbation parameter will invalidate the use of the original eigenfunctions used to generate the set of perturbed eigenfunctions. Mather's work goes much further, in that it can analyze unstable motions which are far from being well behaved. The perturbation can be relatively strong, and entirely new eigenfunctions (solutions) can be generated. In addition to the mathematical significance of Mathers results, the practical importance of this work for planetary orbit, escape, and capture is that the dynamics are applicable to those regions in phase space (Mather regions) associated with three- and four-body problems. Mather essentially proved that for chaotic regions in lower (two) dimensions for any conservative Hamiltonian system, there exists (remains) only elliptical orbits which are unstable. In terms of NEO (near-Earth object) issues, KAM and Mather regions are important for describing both the orbital motions of comets [7] as well as for planning fuel conserving ballistic (flyby, rendezvous, and interception) trajectories to comets and other NEOs. Belbruno's direct relationship to Mather's work computationally demonstrates that it can be generalized to three degrees of freedom. Furthermore, Belbruno's work implies that the (planetoid) resonant transitions can occur as rapidly as around 30 days as opposed to an order of thousands of years. Application to other areas of the astronomical and space sciences are likely to follow. It is for these reasons that that the FBT methodology and application to comet orbits and other related topics were discussed at the U.N. NEO conference, and the reason for the inclusion of this background paper in the conference proceedings.

FUZZY BOUNDARY THEORY

The ballistic capture of the Hiten spacecraft by the Moon was an example of the application of weak capture together with "Mather regions" in phase space trajectory for ballistic capture through fast resonance shifting. The actual methodology for the application of the Mather regions low energy transfers is called "fuzzy boundary theory" or FBT. Simply stated, the fuzzy boundary is a region in the phase space where stable motion does not exist [4], [5a], and it is conjectured to be chaotic. While an analysis of the FBT method is beyond the scope of this communication (one is referred to [4] and [7]), the FBT approach is to find regions in the phase space where the KAM dynamics is not occurring, thereby implying, under certain conditions, that such regions be candidates for the Mather dynamics. This can be achieved when the dynamics becomes sufficiently unstable and there is evidence of "special resonance motion."

The FBT method differs from perturbation approaches starting with Hamiltonian dynamics which are explicitly solvable, and then introduce a perturbation and evaluate the effects by analyzing the ratio of relative orbital frequencies, f. When f is rational, a single closed curve is traced on the torus; if f is irrational, the particle motion moves densely on the torus such as shown in Fig. 2.

As already mentioned, Arnold diffusion can permit a trajectory to travel from one resonance state to another, but for planetary problems it takes an astronomically long period of time. The advantage of using FBT for space craft trajectories, is that it takes advantage of the highly unstable orbits which are far from well behaved in the classical (Hamiltonian) sense. On of the most important attributes of FBT is that it predicts the existence of the so-called "hop" [4], [5] which occurs on a much faster time scale than Arnold diffusion. The hop is computed to occur in a time period of 1–10 years for comets orbiting the Sun, versus about 10^{10} years for Arnold diffusion [23]. It is believed that resonance effects of high order, a well as double resonance effects could play a significant role in comet orbit evolution. This issue is currently being addressed by recent work by Belbruno and Marsden [7] which interprets resonance hopping in comets for the motions of ten short-period comets strongly perturbed by Jupiter. In this work the "fuzzy boundary" concept is utilized to suggest that objects associated with the 3:2 and 2:3 resonances are those that can become involved in weak or quasi-weak capture as temporary satellites of Jupiter. An interesting case is the comet Lexell, which passed the Earth in 1770 and had a revolutionary period close to half that of Jupiter. What is unusual about comet Lexell is that it had a revolution period much longer than Jupiter's before 1767 and after 1779. Analysis confirmed these effects of the 2:1 mean-motion resonance and also showed that the comet's orbital trajectory was unpredictable after 1779. Such motion can be characterized as chaotic.

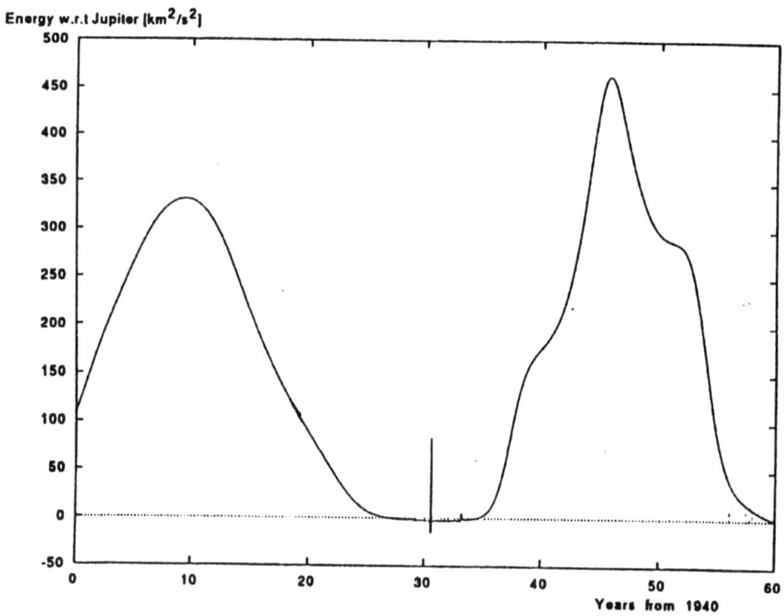

FIGURE 4. From Belbruno and Genta [6], plots the Kepler energy for Gehrels 3 with respect to Jupiter. The negative values indicate that a hop is occurring.

An example of "hop" is the motion of comet Gehrels 3 from a 2:3 ellipse initiating in 1969 and ending in 1973 at a 3:2 resonance is shown in Fig. 3 [6]. Figure 4 gives a plot of the Kepler energy of Gehrels 3 with respect to Jupiter; note that the negative values indicate that a hop is occurring. The hops are pursued where the bounded new resonance orbit becomes highly eccentric and falls toward the Sun, where in doing so it can cross the orbit of an inner planet. If the orbit is unbounded, the comet, which is in an *m*:*n* resonance will now "hop" to escape into a hyperbolic orbit at a very high velocity. An example of this ejection behavior is given by Belbruno in the following paper in this proceedings.

SIGNIFICANCE OF NONLINEAR ORBITAL ANALYSIS AND FBT FOR NEO STUDIES

We have briefly discussed some of the nonlinear methodologies utilized for the analysis of planetary orbits, from the original concepts of Poincaré, the Hamiltonian formulation, KAM theory, along with Mather regions and their utilization for the FBT method. It should be emphasized that instability does not necessarily mean a breakdown of the system, but instead can become a pathway to intrinsic stochasticity which is a fundamental property of nonintegrable Hamiltonian systems. Essentially, stability can be interpreted to imply (analytical) determinism, while instability implies randomness.

Chaotic Evolution and Catastrophic Events

The observational universe consisting, in part, of galaxies, stellar systems, planets, and planetesimals, serves as the laboratory of celestial mechanics. Often the motions of these objects can be described in terms of regular motion, the simplest case being Keplerian orbital motion. However, computations carried out over a time scale of nearly 100×10^6 y by Sussman and Wisdom [22] suggest that the evolution of the solar system as a whole is chaotic, over a time scale of exponential divergence on the order of 4×10^6 y. Calculations on the Jupiter system suggests the existence of both chaotic motion as well as quasi-periodic motion, depending on the variations to the system. Belbruno's results suggest a fast, new (chaotic) resonance shifting instability that occurs in hitherto unexplored regions of phase space.

The above results indicate that it is possible a deterministic chaos also coexists with regular motions in local regions, and can initiate departures from regularity which result in an anomalous event. When these anomalous or nonlinear events are perceived to occur in an otherwise regular (mechanical) universe and introduce a secondary effect, such as a planetary impact, the observer can claim that a catastrophe has occurred. Mathematically speaking, one can say that a local irregularity has occurred. This, of course, gives scant comfort to those species (mammalian or otherwise) that may happen to undergo extinction as a result of an irregularity or "hop." It may be that an orbital "hop" was responsible for a comet orbit to intersect the Earth's orbit, from which the ensuing impact was the agent for the Cretaceous-Tertiary extinctions. The key, in terms of hazard mitigation from NEOs, is to use the tools of nonlinear analysis combined with traditional computational and observational methods to presage the possibility of an impending NEO impact, as well as to understand the dynamics of NEO orbits in general.

CONCLUSION

While the FBT and related methods of computing rapid resonance shifts will likely make an impact on several areas of science and technology, two areas relevant to this proceedings are briefly discussed below; space travel and orbital collision probability and warning time.

Space Travel

With regard to utilizing the FBT approach to design spacecraft missions, one must understand the risk/benefit play-off. While the FBT application to spacecraft trajectories exploits the gravitational potential of planets to a greater extent than does the Hohmann transfer, and thereby conserves a substantial amount of fuel, there is at least one penalty. That penalty is most likely a substantially increased mission time, and its associated liabilities, particularly with respect to exposure of the crew to cosmic rays and solar flares [16]. For example, one study on the reliability analysis of propulsion systems indicates that the reliability during three cumulative hours of active thrusting is probably less important than 9,120 hours of standby (dormancy) between thrust phases [21]. Therefore, if the FBT method is to used to plan either a planetary orbital launch, flyby, rendezvous, or interception mission, careful consideration must be given to the risk/benefit of an extended trajectory time in terms of the required (mass of) support systems and failure probability. For manned missions or certain payloads or mission objectives, this may become problematical. On the other hand, for long term automated or robotic missions, an FBT computed trajectory may very well provide the answer to low cost missions to the solar system and beyond, including (planetary) orbital launches as well as deep space missions.

The Hiten experience has clearly demonstrated that missions can be salvaged through innovative trajectory analysis. Perhaps, sometime in the future a mission rescue can be effected through the utilization of a FBT-like computation. This spacecraft computer could have the capability to reprogram its trajectory to take advantage of three- or four-body dynamics to save itself if its fuel ran perilously low. This would be a "digital lifeboat©." However, for each mission, there must be a careful analysis carried out on the basis of additional risks incurred versus the benefit, which is usually a cost saving.

Predicting NEO Orbits

In terms of predicting NEO orbits, it should be emphasized that classical mechanics is not necessarily fully deterministic. One example of this is Lexell, a comet orbit that may have undergone an FBT type of dynamics. Ballistic capture of Hiten by the Moon and the resonance shifting of Lexell, for example, also imply that fast resonance shifting can enhance our understanding of the frequency of cometary impacts with the Earth. However, the computational results must be carefully interpreted in terms of the known cratering record on the Earth and other planets in the solar system. The problem that naturally arises is the difficulty in predicting and observing such orbital resonance "hops" into an orbit of Jupiter, which are thought to occur on such a brief time scale. Also, this introduces a major uncertainty in the prediction, i.e., warning time, for NEOs, since a chaotic aspect to NEO impact probabilities may exist. Hopefully, this will not be a major problem in predicting a close NEO encounter, because such hop resonances are thought to be rare. If such a hop would occur, it would, for bounded (elliptical) orbits, still constrain the comet to be in an orbit dominated by the Sun and Jupiter. In particular, Belbruno and Marsden [7], whose collaboration among many others, was initiated at the U.N. Conference on Near-Earth Objects, suggest that objects associated with the 3:2 and 2:3 resonances are those that can become weakly captured as temporary satellites of Jupiter. Nonetheless, in matters involving the possibility of planetary catastrophe, one must err on the side of caution, since the unbounded orbits, according to FBT can escape the Sun–Jupiter system at extremely high velocities. It is therefore strongly suggested that, based on the correctness of the FBT "hop" and related computational methods as applied to the comet Lexell orbit and related planetesimals, NEO collision parameters, especially those of comets, must, to be on the safe side, be reworked in the light of the implications of FBT.

One further comment regards the applicability of fast resonance shifting to other areas of the physical sciences, such as those that could be encountered in (laser) Fabry–Perot cavity resonances with externally coupled cavities, where resonance shifts can be forced over relatively short time scales for a variety of small perturbations. In these cases, one would not need to rely on (astronomically) long time scales, and thereby further test the methodology. Successful demonstrations of extended generality would not only further establish the validity of the methodology, but it also would provide a powerful analytic tool for the physical sciences in general.

APPENDIX A

The Mathematical Pendulum

As a simple example of the KAM torus which can nonetheless illustrate some of the key ideas relating to nonlinear orbits, consider the well-studied mathematical pendulum as described by the following nonlinear differential equation,

$$d^2\phi/dt^2 + \lambda^2 \sin\phi = 0, \qquad \text{(A.1)}$$

where the angle ϕ and the angular velocity $d\phi/dt$ are mapped onto a two-dimensional phase plane for the region $-\pi \leq \phi < \pi$, as shown in Fig. 1A. The connecting points $(-\pi, 0)$ and $(\pi, 0)$ provides the separatrix which divides the phase plane into three regions of motion: oscillation (A) and rotation in opposite directions ($+/-B$). At the coordinate (0, 0), the stable equilibrium point is surrounded by concentric ellipses, while the unstable equilibrium point, $(\pi, 0)$, is a hyperbolic or saddle point. $(\pi, 0)$ and $(-\pi, 0)$ correspond to equivalent points in the "stable" and "unstable" directions. The intersection of stable and unstable directions can be generalized into intersections of stable and unstable manifolds.

If we now consider two (weakly) coupled oscillators that are nearly integrable, they appear to resemble a combination of the single mathematical pendulum of Fig. 1A, which is now shown as a section of phase space, J_i, in Fig. 2A. Again, in the so-called combined region, there are apparently equivalent regions of oscillation and positive and negative rotation, which can be thought of as resonance zones.

As the coupling between the pendula increase, the conditions of the KAM theorem weaken and the resonance zones begin to overlap and the original separatrices become stochastic layers of finite width, as is shown in Fig. 3A. In addition to the growth of the (finite) stochastic layers, the hitherto simple closed oscillation curves shown in Figs. 1A and 2A, are now split along a series of elliptic and hyperbolic points on a relatively smaller scale which overlap one another, and appear to possess stochastic characteristics [8], [9].

As departure from the KAM condition continues, the KAM curves evolve from the qualitative representations in Fig. 2A to Fig. 3A. If a transverse intersection of stable and unstable manifolds results in the destruction of the same KAM curve, it is called a homoclinic point, and if the stable and unstable manifolds come from different curves, the transverse intersection is considered a heteroclinic point. The intersection of stable and unstable manifolds is sketched in Fig. 4A. The heteroclinic and homoclinic points establish the regions of chaotic motion and the properties of associated dissipative systems. An important property of a homoclinic point is that because of the nature of the stable and unstable manifolds, the existence of a single homoclinic point implies the existence of an infinite number of homoclinic points.

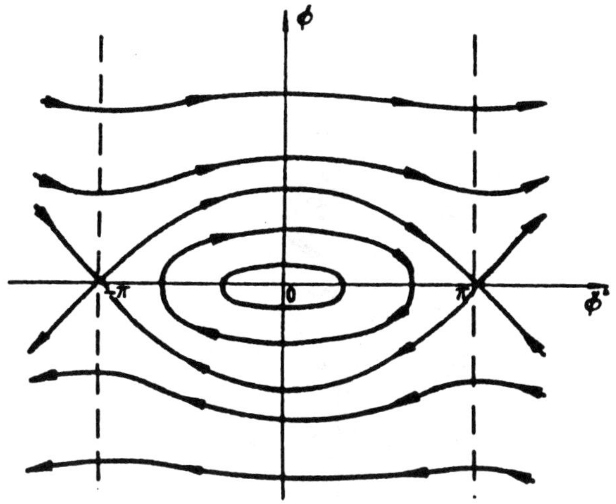

FIGURE 1A. Phase plane of the simple mathematical pendulum described in the text can be interpreted as planar intersections of the KAM torus in phase space.

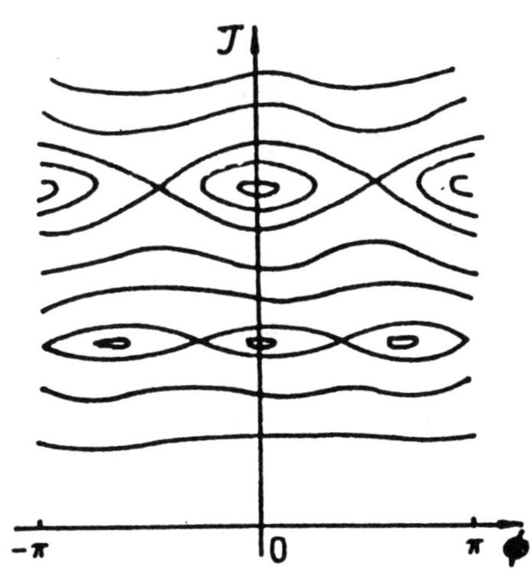

FIGURE 2A. Resonance zones of coupled oscillators.

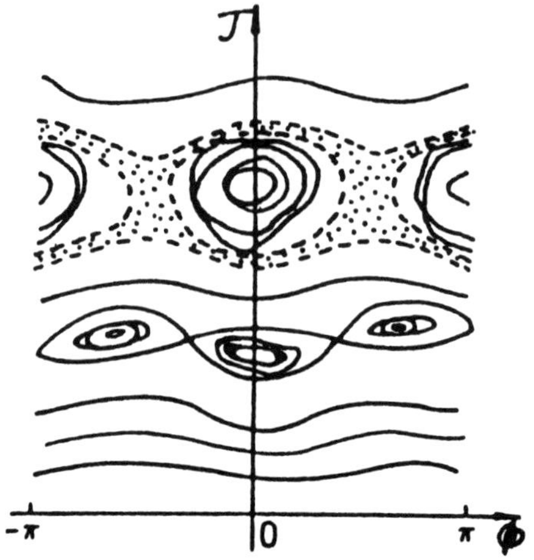

FIGURE 3A. Stochastic layers over the interval $-\pi$ to $+\pi$.

FIGURE 4A. Schematic interpretation of the intersections of stable and unstable manifolds.

ACKNOWLEDGMENTS

I thank E. Belbruno, B.G. Marsden, P. Hammerling, and an anonymous reviewer for their helpful comments.

REFERENCES

1. ARNOLD, V.I. 1963. Proof of a theorem of A.N. Kolmogorov on the invariance of periodic motions under small perturbations of the Hamiltonian. Russ. Math Surv. **18:** 9.
2. ARNOLD, V.I. & A. AVEZ. 1968. Ergodic Problems of Classical Mechanics. Addison-Wesley. Reading, MA.
3. ARNOLD, V.I. 1978. Mathematical Methods in Classical Mechanics. Springer. New York, NY.
4. BELBRUNO, E.A. 1990. The nonlinear dynamics of ballsitic capture and escape in the Earth-Moon system. Proceedings AIAA Astrodynamics Conf., Aug., no. AIAA 90-2891.
5. BELBRUNO, E. & J. MILLER. 1993. Sun-perturbed Earth-to-Moon transfers with ballistic capture. J. Guid., Control and Dynamics. **16:** (4) 770–775.

5a. BELBRUNO, E. 1994. Ballistic lunar capture transfers using the fuzzy boundary and solar perturbations: A survey. J. Brit. Inter. Soc. **47:** 73–80.

6. BELBRUNO, E. & G. GENTA. 1996. Low energy comet rendezvous using resonance transitions. First IAA Symposium on Realistic Near-Term Advanced Scientific Space Missions, International Academy of Astronautics, Torino, June 25–27.
7. BELBRUNO, E. & B.G. MARSDEN. 1996. Resonance hopping in comets. Preprint .
8. CHIRIKOV, B.V. 1971. Research concerning the theory of nonlinear resonance and stochasticity. Nuclear Phys. Inst. of Siberian section of Acad. Sci, USSR, no. **267**, CERN Trans. 71-40 by A.T. Sanders, Geneva.
9. CHIRIKOV, B.V. 1979. A universal instability of many oscillator systems. Phys. Repts. **52:** 265.
10. FEIGENBAUM, M.J. 1978. Quantitative universality for a class of nonlinear transformations. J. Stat. Phys. **19:** 25.
11. FEIGENBAUM, M.J. 1979. The universal metric properties of nonlinear transformations, J. Stat. Phys. **21:** 669.
12. FORD, J. 1983. How random is a coin toss? Phys. Today. April, 40.
13. HENON, M. & C. HEILES. The applicability of the third integral of the motion : Some numerical experiments. Astron. J. **69:** 73.
14. KATOK, A.B. 1982. Lyapunov exponents, entropy, and periodic points for diffeomorphisms. Publ. Math. IHES. **51:** 137.
15. KOLMOGOROV, A.N., 1954. Preservation of conditionally periodic movements with small change in the Hamiltonian function. Akad. Nayk. SSSR Doklady. **98:** 527. Translation 1979: *In* Stochastic Behavior in Classical and Quantum Hamiltonian Systems, G. Casati & J. Ford, Eds.: 51. Springer Verlag. New York, NY.
16. LAZARETH, O., M. DIVADEENAM, E. SCHMIDT, H. LUDEWIG, & J.R. POWELL. 1991. Prediction of the radiation dose to human beings during a voyage to Mars. AIAA no. 91-3409, AIAA/NASA/OAI Conference on Advanced SEI Technologies, September 4–6.
17. LORENZ, E.N., 1963. Deterministic nonperiodic flow. J. Atmos. Sci. **20:** 130.

17a. MATHER, J.N. 1982. A criterion for the non-existence of invariant circles. Publ. Math. IHES. **63:** 153–204.

17b. MATHER, J.N. 1993. Variational construction of connecting orbits. Ann. Inst. Fourier, Grenoble. **43:** (5) 1349–1386.
18. MOSER, J. 1962. On invariant curves of area-preserving mappings of an annulus. Nachr. Akad. Wiss. Gottingen Math Phys. Kl. **2:** 1.
19. POINCARÉ, H. 1892. Les Methodes Nouvelles de la Mechanique Celeste. Gauthier-Villars. Paris.
20. RUELLE, D. & K. TAKENS. 1971. On the nature of turbulence. Commun. Math. Phys. **20:** 167.
21. SFORZA, P.M., M.L. SHOOMAN, & D.G. PELACCIO. 1993. A Safety and reliability analysis for space nuclear thermal propulsion systems. Acta Astronaut. **30:** 67–83.
22. SUSSMAN, G.J. & J. WISDOM. 1992. Chaotic evolution of the solar system. Science. **257:** 56–62, (1992)
23. WISDOM, J. 1980. Astron. J. **85:** 1122.

Fast Resonance Shifting as a Mechanism of Dynamic Instability Illustrated by Comets and CHE Trajectories

EDWARD BELBRUNO[a]

The Geometry Center
University of Minnesota
1300 South 2nd Street
Minneapolis, Minnesota 55455

ABSTRACT: An interesting process is described where a particle can very rapidly jump between different resonance orbits about the Earth, where the resonances are with respect to the Moon. This was discovered numerically in 1988, and until recently there was no way to prove its validity. Recent results with Brian Marsden and the author have proven that comets are performing this dynamics about the Sun in resonances with Jupiter. The fast resonance shift, termed the hop, occurs in a region of sensitive dynamics where *weak capture* occurs. The hop provides an interesting mechanism for instability of motion in the n-body problem, $n > 2$. Although most of this paper is survey in nature, some new results are presented on special capture-hop-escape trajectories, CHEs, which may have relevance to the *Birkhoff conjecture*. The significance of the weak capture region about the Moon was demonstrated in 1991 when a Japanese spacecraft arrived there on a new low-energy transfer.

1. INTRODUCTION

The three-body problem under the Newtonian gravitational force law has been a source of providing insights into dynamics of motion in conservative systems. Poincaré in the late 19th century obtained deep insights into dynamics by the use of this problem [25]–[27]. From his studies has evolved the field of dynamical systems which includes the area of chaos theory. Poincaré noticed that a very sensitive type of motion was possible in the three-body problem whose complexity seemed infinite. It was not until the 1960's through the work of Arnold and Moser that the nature of this motion was understood, and is known as the Kolmogorov-Arnold-Moser Theorem [2], [23], [24]. This theorem gives deep insight into the way the planets move around the Sun, and shows that although they move in a predictable stable fashion, there can be an element of chaos to this motion. However, it is very slight and in general will not effect the stability. If it did, it would be very slow acting and take perhaps billions of years to cause an appreciable effect. This slow chaotic instability is called *Arnold diffusion* and was proven to exist in 1964 [4]. Until that time, that was the only mechanism known

[a]Also with the Department of Mathematics, Princeton University, Princeton, New Jersey.

for elliptic type motion of say, a planet or comet about the Sun to become unstable.

In 1988 another chaotic instability was discovered by means of computer investigations in the three-body problem for the elliptic motion of a particle of negligible mass about the Earth whose motion is perturbed by the Moon [ll]. Unlike Arnold diffusion, it is extremely fast, taking only days. It is called *the hop*. It takes place between resonance ellipses about the Earth with the Moon. A particle moving one of these ellipses when passing near the Moon in a special region can dramatically shift from one ellipse to another. Until 1995 there was no way to prove that the hop was a real motion, and not just a computer artifact. This changed in April 1995 when it was brought to the author's attention by Brian Marsden in New York at the International NEO Conference that the comet Oterma was shown to rapidly pass between two different resonances with Jupiter about the Sun by passing near Jupiter [20]. Since then, it has been verified that Oterma as well as nine other comets are performing the hop [5]. Thus, the hop is a verifiable process, and causes dramatic effects in the way some comets move about the Sun. It is natural to ask if other objects in the solar system are performing this dynamics. For example, many of the newly-discovered Kuiper belt objects seem to be in resonance with Neptune [19] and could therefore rapidly shift to a different resonance ellipse by passing near to Neptune.

When an object transitions from one resonance ellipse to another it does so in a special region where sensitive or chaotic dynamics occurs about the secondary planet, and is referred to as the *fuzzy boundary* (fb), or *weak stability boundary* [9], [11], [12]. Mathematically speaking, it is conjectured to be a so-called *hyperbolic tangle* [5], [8]. A tangle is an infinitely complex structure where the motion is unstable and first hypothesized in general to exist by Poincaré [25]. In the lunar case, the the fb surrounds the Moon and extends out to a maximal distance. An object moving there is in near escape with the Moon and has no central body, i.e., the Earth or Moon. It is captured temporarily and is very weakly gravitationally bound. This is called *weak capture*. The idea of weak capture originally defined in 1986 [12] was motivated by the work of Mather on the nonexistence of certain types of motions [22]. In Section 2, the fb and weak capture are described in more detail. The comets studied in [5] perform their hops, or resonance transitions during weak capture in the fb of Jupiter. This is discussed in Section 3.

An interesting dynamics is described in Section 4 which was recently generated. It occurs in the three-body problem between the Earth-Moon-particle. Here a particle is captured from infinity by the Moon and falls into a sequence of 8 sets of hops about the Earth. It then escapes the Earth-Moon system. This entire process of capture-hopping-escape is only 3000 days. A process like this is important because it provides a mechanism for a particle to become captured into the Earth-Moon system and then aquire enough energy to escape in a very short period of time. The mechanism of Arnold diffusion would yield estimates of many millions of years for this to occur. The process of capture and escape through transient resonant motion provides a tool to understand the global dynamics. If the Sun is included as a gravitational perturbation, then a capture-escape trajectory with respect to the Earth-Moon system can also be found, and the duration is reduced to 450 days with only 3 hops. This is described in more detail in Section 4. In Section 5 the connection of this process is discussed in its relevance to deep theoretical issues in the n-body problem for $n > 2$ such as the *Birkhoff conjecture* [13], [14], [15].

The reason an object would suddenly transition from one resonance to another resonance after weak capture is not understood. A physical argument is given in [5]. A mathematical proof of the existence of the hop may involve the deep results of Mather on dynamics which are far from integrable [21], [22]. This is discussed in Section 6.

The significance of the fb and weak lunar capture was underscored in 1990 when a low-energy transfer from the Earth to the Moon was found for the Japanese spacecraft Hiten by Belbruno, assisted by J. Miller, using a method developed in 1986 [10], [12]. This transfer was used by Hiten in 1991 which enabled it to arrive at the Moon in October of that year [9], [10], [11]. This is described in Section 7. A low-energy transfer to comets has recently been discovered which is done while a comet is performing a hop [6]. This is also discussed in Section 7.

Most of this paper is of a survey nature and some new results are given in Sections 4 and 5.

For those readers not familiar with some of the concepts mentioned here, see [28] in this Proceedings.

2. WEAK CAPTURE AND THE HOP

Throughout this paper, we will model the motion of a small body, $\wp$, of negligible mass m_0 by the restricted three-, four-, and twelve-body problems. In these problems, $\wp$ moves in the Newtonian gravitational fields of two, three, or eleven other bodies, respectively, which are moving in prescribed elliptical orbits about their centers of mass. In the case of the three- and four-body problems, the main perturbing bodies are the Earth, E, Moon, M, and Sun, S, of masses m_i, respectively, $i = 1, 2, 3$. The three-body problem that will be considered is, E, M, $\wp$. The four-body problem will be E, M, S, $\wp$. In these systems, $\wp$ can be viewed as any small body of negligible mass with respect to E, S, M. For example, an asteroid. The twelve-body problem considered models the solar system. The additional eight bodies of masses m_i, $i = 4, 5, \ldots, 11$, are Mercury, Venus, Mars, Jupiter, Saturn, Neptune, Uranus,

Pluto. In this system, $\wp$ represents a comet. In a coordinate system centered at the Sun the positions of the planets are prescribed by the planetary ephemeris DE403 and are given by the functions $\mathbf{x}_i(t) \in \Re^3$, where t is time, $i = 1, 2, \ldots, 11$. Let $\mathbf{x} = \mathbf{x}(t) \in \Re^3$ denote the position of $\wp$. The differential equations which describe the motion of $\wp$ are given by

$$\ddot{\mathbf{x}} = \sum_{i=1}^{n} Gm_i(\mathbf{x} - \mathbf{x}_i)|\mathbf{x} - \mathbf{x}_i|^{-3}, \tag{1}$$

where $\cdot \equiv d/dt$, and where $n = 2, 3, 11$ for the three-, four-, and twelve-body problems, respectively. In the latter case of the full solar system modeling, the gravitational effect of Jupiter and the Sun are dominant, and the other planets are negligible for the class of comets considered.

Weak capture of $\wp$ occurs at one of the smaller bodies of the main perturbing planetary bodies. For example, in the E, M system, it occurs at M. Weak capture is a temporary capture of $\wp$ at M where the motion of $\wp$ satsifies two conditions:

(a) $H_M = -\lambda^2$, λ near 0,
(b) $\wp$ is near stability transition.

H_M is the Kepler energy with respect to M,

$$H_M = \frac{1}{2}|\mathbf{v}|^2 - \frac{Gm_2}{|\mathbf{x}|},$$

where $\mathbf{x}$, $\mathbf{v}$ are the position, velocity, respectively, of $\wp$ with respect to M, and G is the universal gravitational constant. The condition of $\wp$ being near stability transition, ST, means that it is near a region where escape from M occurs within one cycle of motion about M. Without this condition, Condition (a) specifies that $\wp$ is near escape, but not to the degree. Condition (b) is a refinement of this. The conditon of ST is now more precisely defined. This is done very briefly here since the details can be found in the literature [5], [11], [12].

Fix a line l which extends from the center of M. Let $\wp$ be the distance along l. Fix a velocity vector $\mathbf{v}$ which is based along l whose magnitude $v = v(p)$ is adjusted such that the eccentricity $e = e_0$ is fixed, $0 < e_0 < 1$. At each point along l numerically integrate System (1) with $\mathbf{v}$ perpendicular to l. Two things generally happen. Let Ξ be the unique plane through l, perpendicular to $\mathbf{v}$. When System (1) is numerically integrated, either $\wp$ will make a complete cycle about M without first having moved about E, or $\wp$ will move about E before returning to Ξ. In the former case, we say that the motion of $\wp$ is *stable*, and in the latter, *unstable*. Intuitively, it is seen that for $\rho \sim 0$ the motion should be stable, and for ρ sufficiently large, unstable. This is observed on the computer, and, moreover, it is found by interation that a well-defined value $\rho = \rho^*$ exists such that the motion is stable for $\rho < \rho^*$, and unstable for $\rho > \rho^*$.

This critical value of ρ^* defines the ST region. In the case where the motion of E, M are defined where they perform mutually uniform circular motion about their common center of mass, ρ^* depends on four parameters. When they perfrom mutually elliptic motion, then there are five parameters. The region of ST as defined here is referred to as the *fuzzy boundary*. It is remarked that when $\wp$ starts its motion *near* ST, it is possible that $H_M = +\delta^2$. If this happens, we say that $\wp$ is in *quasi-weak capture*. It is noted that being near ST implies that $\wp$ has no central body. The region near ST where both weak and quasi-weak capture occur is defined here to be a *Mather region*.

When $\wp$ starts in weak or quasi-weak capture, it is found that it stays near this state for a short period of time, and then abruptly escapes where, $H_M > 0$ and $\wp$ is no longer near ST. For example, see Fig. 1.

As it turns out, weak capture gives rise to interesting dynamics and applications. Because weak capture, WC, is a state which can be reached from trajectories starting outside of it, it is then a place where capture can occur with minimal energy. This property makes it suitable for the capture of spacecraft from the Earth with substantially less energy than the classical methods. This is discussed in Section 7.

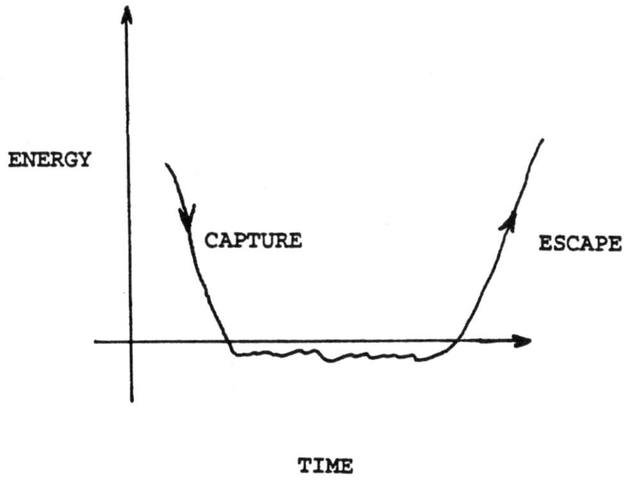

FIGURE 1. Weak capture.

An interesting dynamics called the *hop* discussed in Sections 3 and 4. Let $\wp$ be in an m:n resonant elliptical orbit with respect to the Earth, in resonance with respect to the Moon, where $m \neq n$. That is, the period of $\wp$ equals n/m times the period of the Moon, where n, m are positive integers. Then $\wp$ hops from the m:n ellipse to an i:j resonance ellipse, $i \neq j$, if it transitions through WC or QWC at the Moon. This was first discovered in 1988 [11]. If a hop occurs by transitioning from QWC, we refer to it as a *quasi-hop*. Although the above definitions have been made with respect to the Moon, they carry over to the smaller of any two primaries. For example, Jupiter where the larger primary is the Sun. The hop was found to be associated with a class of short-period comets in 1995 [5].

It is remarked that the resonances referred to in this paper are not exact and are near resonances. That is, if $\varphi = \varphi(t) = (\mathbf{x}(t), \dot{\mathbf{x}}(t))$ is a resonant orbit of (1) with resonance period $T_R = (n/m)T_L$, T_L = one lunar period, then we require that $\varphi(T_R)$ lie in a neighborhood of $\varphi(0)$

3. HOPPING COMETS

A set of ten short-period comets are shown to be performing the hop in a recent study completed in 1996 [5]. Some of the results are discussed here.

The first comet demonstrated to be performing a fast resonance shift was 39P/Oterma. Originally discovered in 1943 by Oterma herself, it was noticed in 1943 that after its first encounter with Jupiter in 1937 it was in a near 3:2 resonance with Jupiter, while prior to 1937 it was in a 2:3 resonance. Again in 1961, 39P/Oterma reencountered Jupiter, and after this encounter it shifted back to a 2:3 resonance which is discussed by Marsden [20]. An analysis of these resonance transitions in 1995 showed that hops were occuring.

Some other selected comets from [5] shown to be performing hops or quasi-hops are 82P/Gehrels 3, 14P/Wolf, D/1770 Ll(Lexell), 74P/Smirnova-Chernykh. The duration of the hops vary from a few months to 10 years. As discussed in Section 6, this is vastly faster than the classical method of shifting resonances, Arnold diffusion, where the duration of such shifts is often on the scale of millions or even billions of years.

Many of the comets in [5] are actually performing quasi-hops which tend to be associated with higher order resonances. For example, the comet 74P/Smirnova-Chernykh is in this category, where a shift 6:13 → 7:5 occurs. The results indicate that the more stable shifts are the hops where the resonances 2:3, 3:2 generally appear. For example, 39P/Oterma, 82P/Gehrels 3 are in this category.

Figures 2A, 2B, 2C show typical behavior during a hop. These figures are for the comet 74P/Smirnova-Chernykh which performed a quasi-hop starting on June 1, 1955 and lasting until April 13, 1956. Figure 3A shows the

6:13 ellipse outside of Jupiter's orbit which has a nonlinear transition to the smaller 7:5 ellipse. It is integrated from 1900–2000. The variation of the energy with respect to the Sun, H_S, has the characteristic level values for the different resonances with an abrupt jump between them. The plot in Fig. 2B shows the period with respect to the Sun which has the same behavior as H_S. Figure 2C shows how the energy with respect to Jupiter, H_J, moves very near to zero typical for QWC during the quasi-hop.

As is discussed in [5], the hop seems to be a robust process. This implies that it is not a dynamics which occurs with a zero probability, but rather is not too difficult to achieve. Mathematically, it seems to occupy a set of positive measure in the phase space. A sketch of a proof explaining how the hop can occur is also given. Although it is not rigerous, it indicates a physical mechanism.

Two final results are mentioned from [5]. One relates to the comet 14P/Wolf, and the other to D/1770 Ll(Lexell). 14P/Wolf is performing a surprising double resonance. That is, it is oscillating between 3:2 and 4:3. This implies that perhaps other interesting types of resonance oscillations may be possible. D/1770 Ll(Lexell) shows that hopping comets can be Earth crossers. In 1767 this comet did a quasi-hop from a 4:3 to a 2:1 and passed within 2.3 million km from the Earth in 1770. Moreover, in 1779 it did a close flyby of Jupiter and possibly gained enough energy to be on an ellipse of an apoapsis of 92 AU which is within the Kuiper belt distance. Therefore, this comet makes a link between comets moving out within the Kuiper belt and those which can become Earth crossers via the hop.

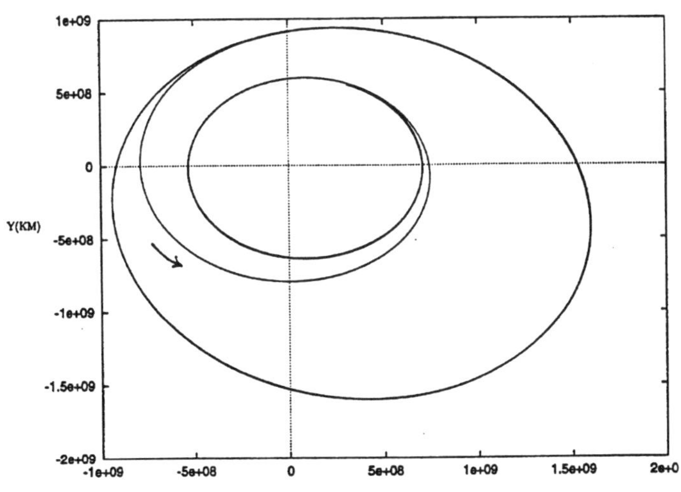

FIGURE 2A. Inertial plot of 74P/Smirnova-Chernykh, ecliptic projection, x-axis points to the vernal equinox of 2000 (X,Y-Ecliptic2000). Integrated from 1900 to 2000. S centered.

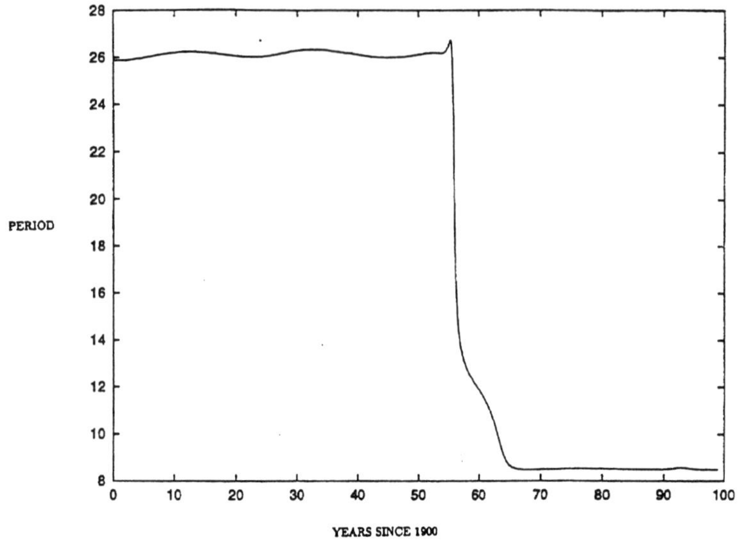

FIGURE 2B. Period of 74P in years.

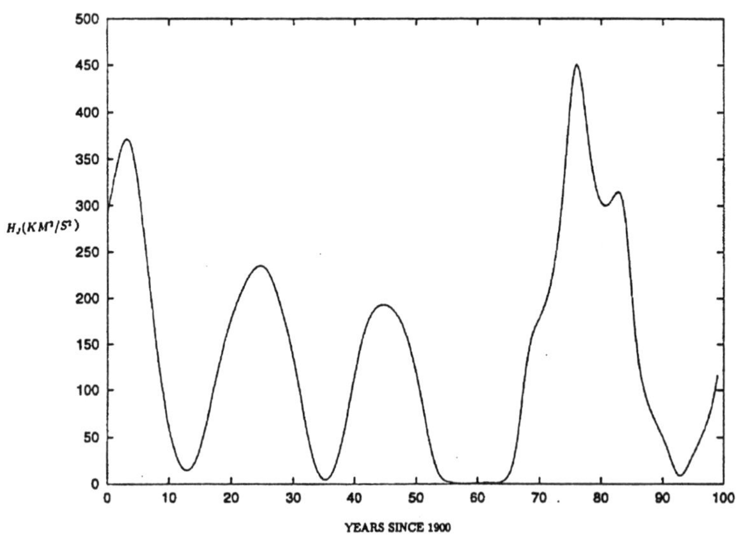

FIGURE 2C. H_J, (KM^2/S^2).

These results are obtained by numerical integration of System (1) for the solar system modeling with $n = 11$. They are started from observational data, and monitored for roundoff error by checking the positions of the comets with other data and by ensuring that forward and backward integrations agree. Because of the sensitivities of the hop dynamics, continuous numerical integrations over two hundred years are not done, unless by shadowing methods. A numerical integrator of order 10 is used together with the planetary ephemeris DE403.

It was shown in [5] that the hop persists when only S, J are modeled and in the idealized *elliptic restricted three-body problem*, E^3, between Jupiter, J, and S, $\wp$, where J and S move on uniform ellipses about their common center of mass. In fact, it also persists in the case J, S are moving uniformly on circles, which is called the *circullar restricted three-body problem*, C^3. For nomenclature, only S, J are modeled using DE403, we refer to this as the *pseudo-elliptic restricted three-body problem*, PE^3. As far as numerical integration and modeling is concerned for the time spans considered here, the difference between E^3 and PE^3 is negligible.

4. CAPTURE AND EJECTION THROUGH HOPS IN THE THREE- AND FOUR-BODY PROBLEMS: CHE TRAJECTORIES

Fast resonance shifting was intitially reported in 1990 [11], using System (1) for $n = 2, 3$, i.e., the three- and four-body problems, between E, M, $\wp$ and E, M, S, $\wp$, respectively, using the DE403 ephemeris. Extended numerical integrations in forward and backward time in these two problems starting from weak lunar capture suggests the following general behavior, which we refer to as *Capture-Hop-Escape*(*CHE*) *dynamics*,

> $\wp$ is initially hyperbolic and becomes captured into the E, M-system. It then performs a finite sequence of hops, and is ejected from the E, M-system in a finite period of time.

CHE dynamics is conjectured to occur, in general, in the E,M-system starting from WC initial conditions at M.

As in the previous section, DE403 can be replaced by uniform elliptic motion for E, M, S. The differences between DE403 and uniform elliptic motion for the time spans considered here are numerically negligible as far as numerical integration is concerned, especially since $\wp$ does not pass too close to E, M, or S. Following the nomenclature in the previous section for $n = 2$, we have the E^3 and PE^3 problems bewteen E, M, $\wp$ for DE403 and uniform elliptic motion, respectively. All our results in this section use DE403.

Capture into the E, M-system at $t = t_0$ means that,

$$H_E = \frac{1}{2}|\mathbf{v}|^2 - \frac{Gm_1}{|\mathbf{x}|} < 0 \ ,$$

where **x** and **v** are the position and velocity, respectively, of $\wp$ with respect to E. This is not necessarily weak capture. Ejection occurs when $H_E > 0$ at a time $t_1 \neq t_0$, where $t_1 < t_0$, or $t_1 > t_0$.

In the ejected states for $n = 2$, it is found that $|\mathbf{x}| \to \infty$ as $t \to \pm\infty$. This property is not easy to achieve.

For $n = 3$, although $\wp$ eventually escapes E in forward and backward time, $|\mathbf{x}| \not\to \infty$. This is because it has in general not escaped S. To escape S would be very difficult. Once escaped the E, M-system, its energy is only slightly higher than the Kepler energy of E about S. Thus it is in a near E orbit about S. To escape S, $\wp$ must repetitively return to near E and fly by it so as to pump up the energy with respect to S to decrease the periapsis so the particle can eventually fly by S where it is possible to increase the velocity of $\wp$ enough to escape S as well. This process could take many millions of years.

The following nomenclature is used. If $\wp$ escapes the E, M-system in forward and backward time from hopping, but it doesn't necessarily escape infinitely far from E in forward and backward time, as in the case of $n = 3$, then the resulting trajectory is called a *CHE trajectory*, or a *CHE*, for brevity. If it is captured from infinitely far from E, and escapes infinitely far from E as in the case of $n = 2$, then it is referred to as a *bi-infinite CHE*. It is remarked that by measure theoretic considerations, most initial conditions that lead to capture, will also lead to ejection, and conversely, which goes back to Schwarzschild [30], [31].

Two numerical examples are given illustrating a CHE for $n = 3$, and a bi-infinite CHE for $n = 2$. The figures for the first case are Figs. 3A–3F. The figures for the second case are Figs. 4A–4G, 5A–5G, 6A–6D.

A. CHE, $n = 3$

The sequence of events are,

$$\text{capture} \to 7{:}3 \to 2{:}1 \to \text{escape.}$$

Figure 3A shows the complete CHE. The coordinate system is centered at E, and S is near the negative x-axis when the integration is begun at the initial point labeled by **o**. The end point is labeled **n**. This is an inertial plot, where E, M, S are all in motion. Only x,y-plots are given for brevity, although there is significant out-of-plane motion. However, this has no bearing on our discussion. The start date at **o** is March 27, 1987, and the end date is May 22, 1988, which is 422 days later. The trajectory starts hyperbolically at **o** and after about 125 days it first moves close to L_1 and then L_2 to begin a sequence of hops. The distance of $\wp$ to L_2 is shown in Fig. 3D where the x-axis is days from the initial date. Figure 3B shows a blow up of what is going on after capture. The 7 : 3 motion is broken up into two elliptic bands indicated. Then the particle moves onto the 2 : 1 where it eventually escapes. This resonant motion is verified by comparing Figs. 3C, 3E, 3F. In Fig. 3C, the particle is seen moving near L_1 and L_2. It then moves onto the 7:3 which precesses in the fixed-coordinate system. In this system, M is located on the x-axis at $-1 + \mu$ units, where 1 unit equals the mean E, M distance of approximately 386,000 km and $\mu \cong .0123$. The lobes indicate the portion of the 7 : 3 motion the particle is on, and they are consecutively numbered. After completing the seventh circuit about E, it moves onto a 2 : 1 which is labeled. Figure 3E shows H_E is initially positive, then abruptly dropping indicating capture. From about 130 days to about 215 days the energy varies so as to be near the 7 : 3, and then jumps to the 2 : 1 energy value before escaping. Figure 3F shows H_M. The 7 : 3, 2 : 1 break up into characteristic oscillatory peaks and valleys.

B. Bi-infinite CHE, $n = 2$

The sequence of events are,

$$\textit{capture} \rightarrow 1{:}2(11) \rightarrow 1{:}2 \rightarrow 7{:}3 \rightarrow 1{:}2(9) \rightarrow 7{:}3 \rightarrow 2{:}1(3) \rightarrow 4{:}5 \rightarrow 2{:}1(13) \rightarrow \text{escape}.$$

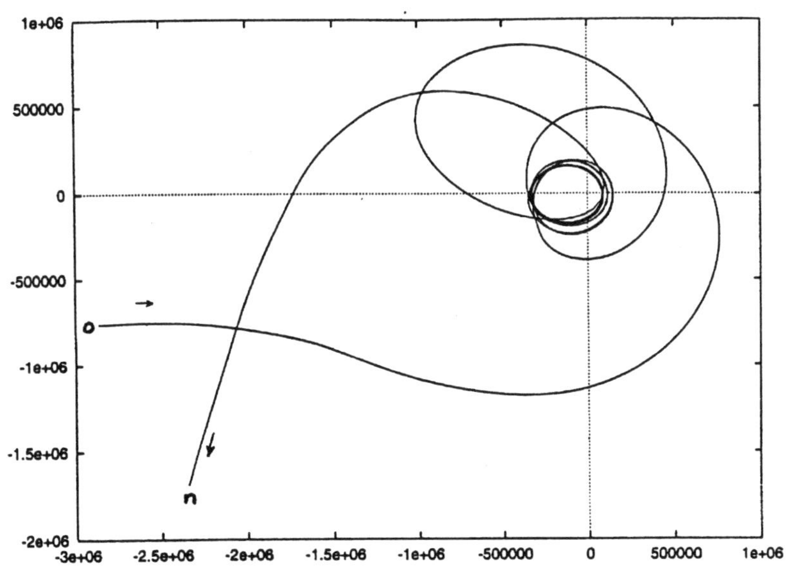

FIGURE 3A. Inertial plot, X,Y-Ecliptic2000. Units, KM. E centered.

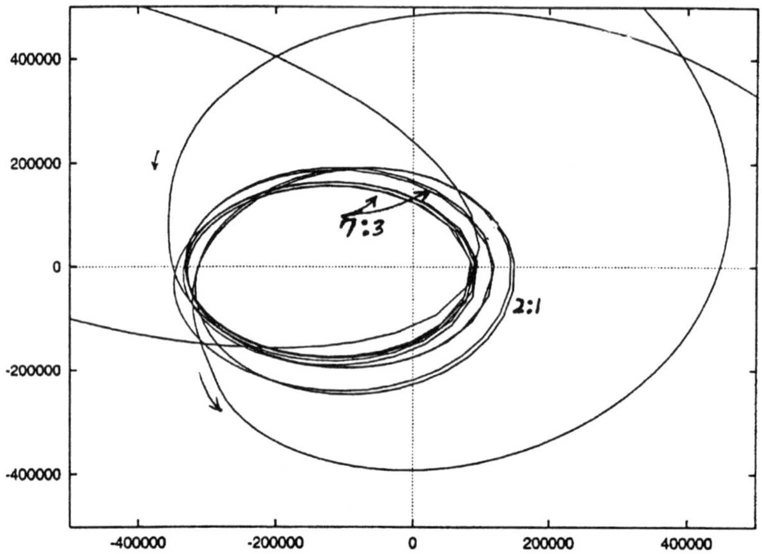

FIGURE 3B. Inertial plot, X,Y-Ecliptic2000. Units, KM. E centered.

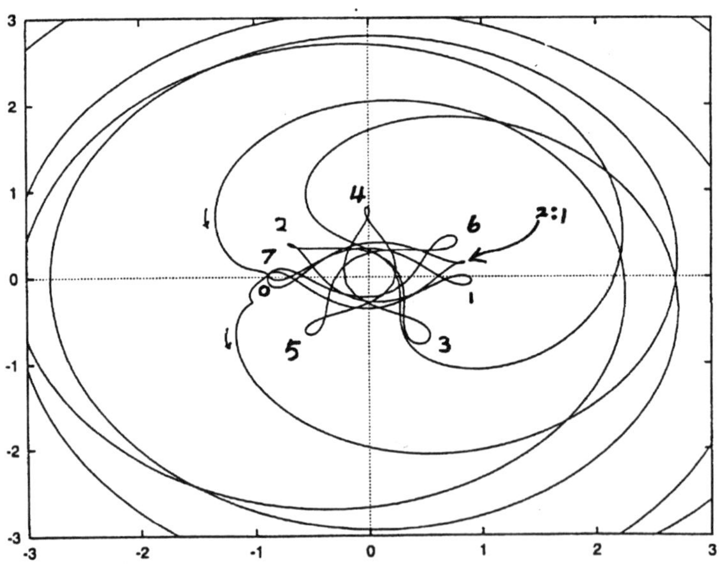

FIGURE 3C. Fixed coordinate system centered at cm of E, M. Dimensionless units. One unit is E, M distance. M at $X \cong -.9877$.

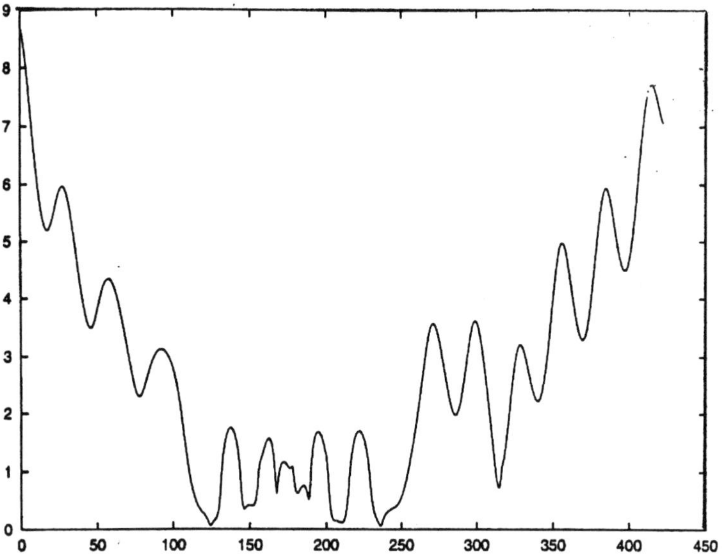

FIGURE 3D. Distance to L_2 in dimensionless units. X-axis is days.

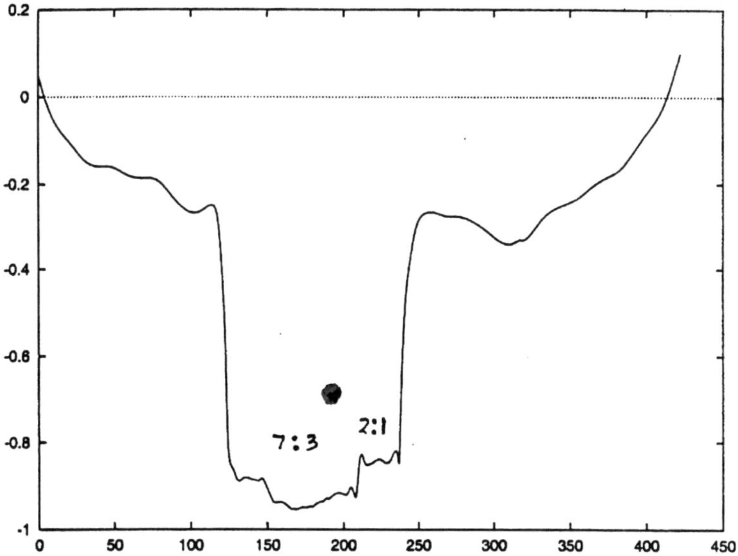

FIGURE 3E. H_E, (KM^2/S^2).

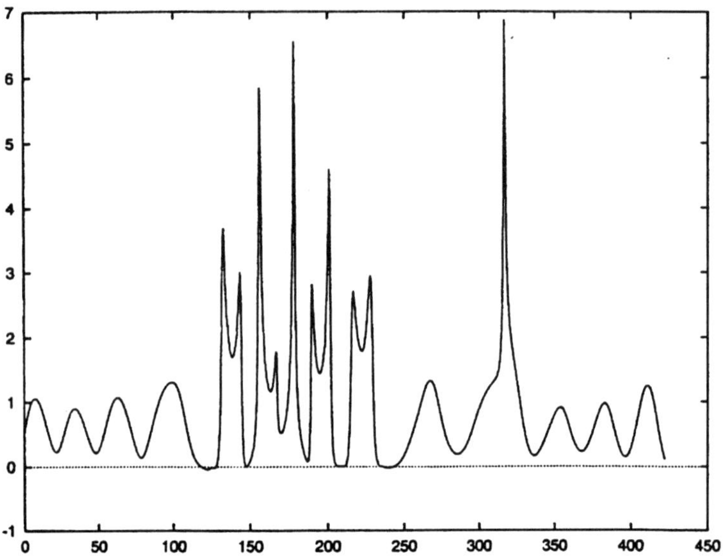

FIGURE 3F. H_M, same units as E.

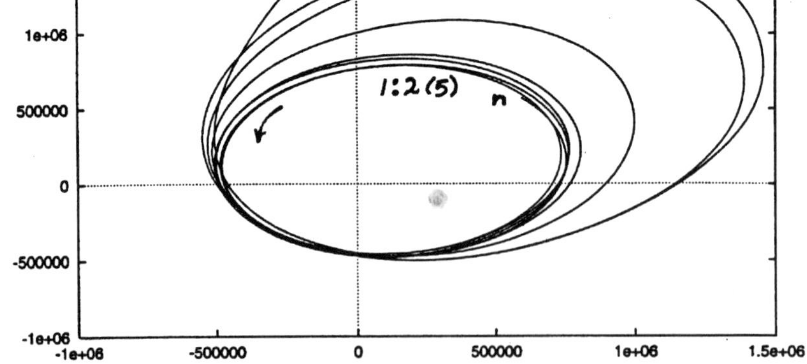

FIGURE 4A. 4A–4G: X,Y-Ecliptic2000, KM. E centered. Plots consecutive.

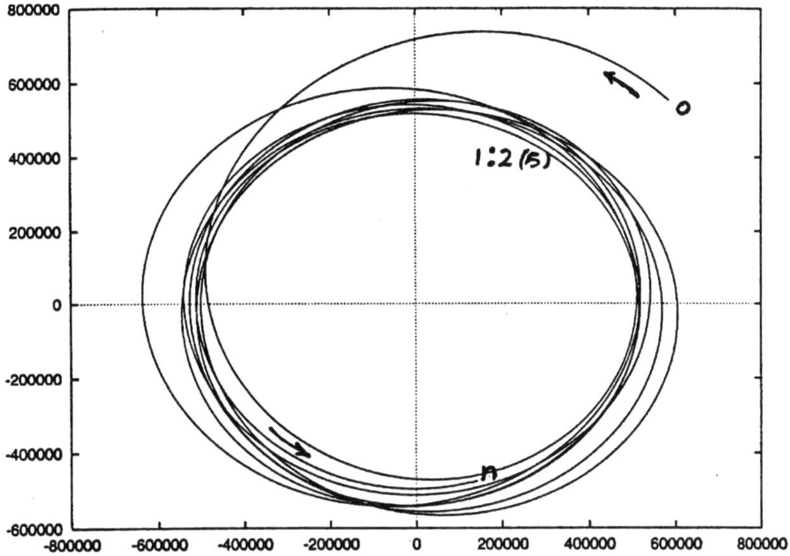

FIGURE 4B

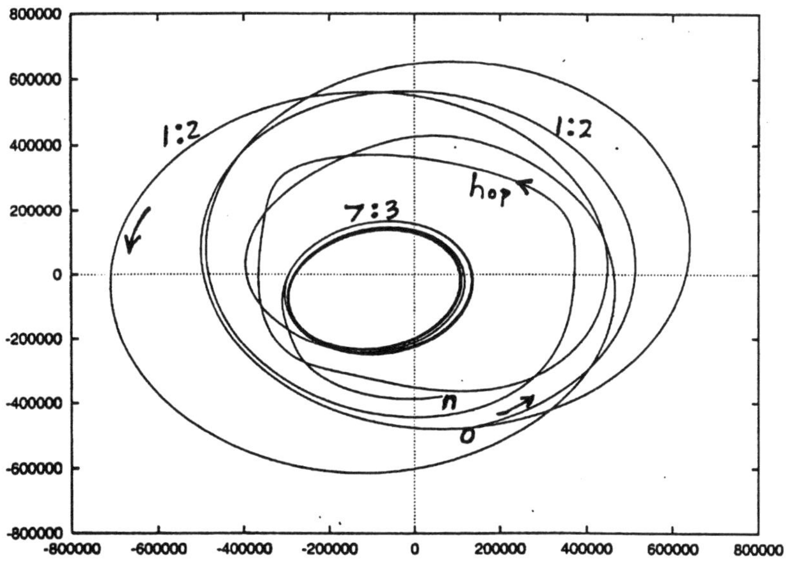

FIGURE 4C

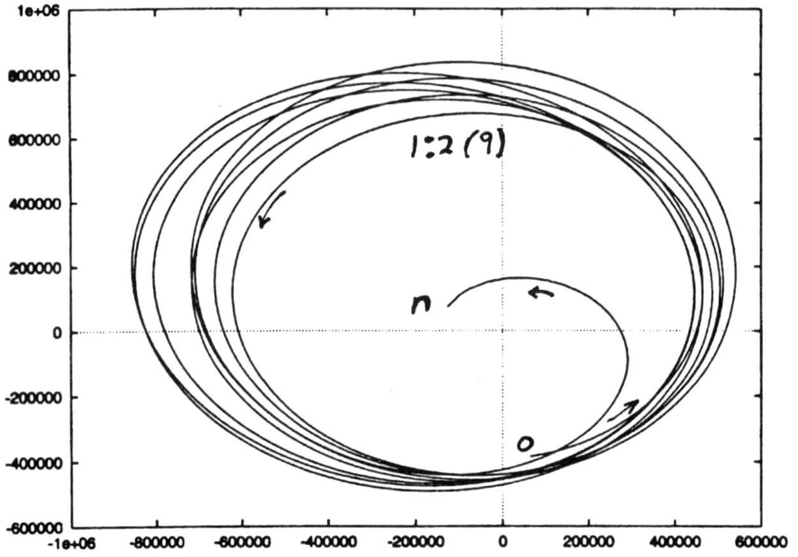

FIGURE 4D

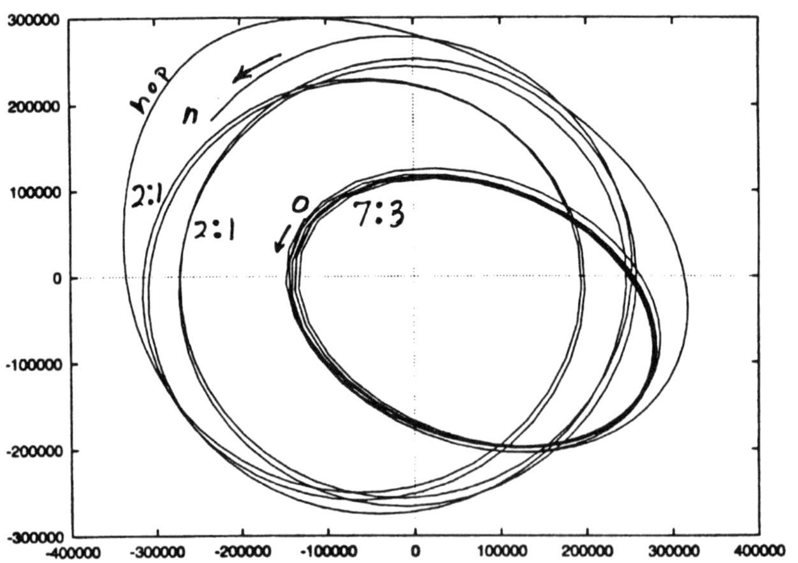

FIGURE 4E

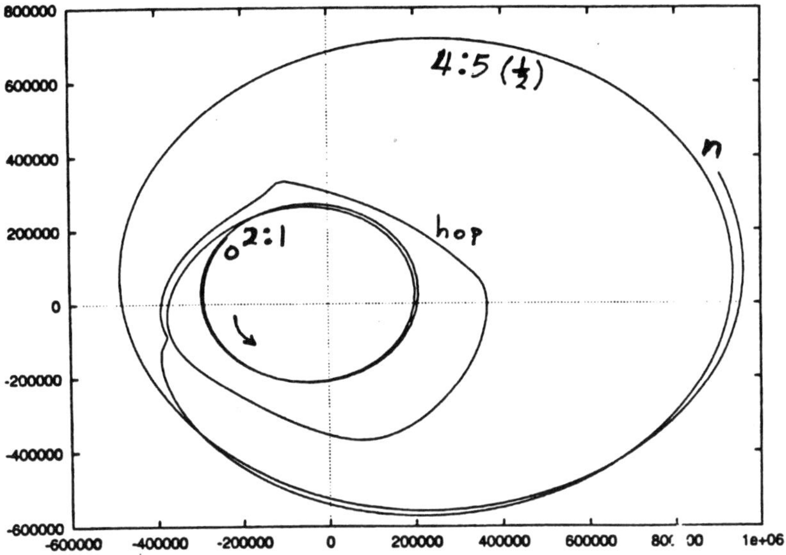

FIGURE 4F

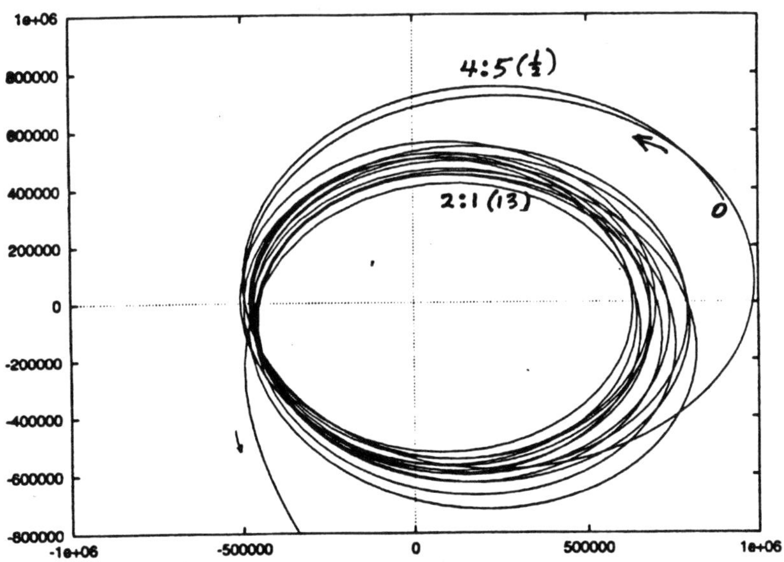

FIGURE 4G

The ability of a particle to escape the E, M-system without the the gravitational perturbation of S is more difficult, and a longer time is needed. Upon escaping, $\wp$ can continue to move infinitely far from E. An example is documented here in which $\wp$ moves infinitely far from E in forwards and backward time. While captured in the E, M-system, it performs a complex sequence of hops listed in the above sequence. To fully explain this in detail here would be too lengthy. Most of the fixed coordinate system plots are not included, nor most of the plots for H_M. The initial investigations on a portion of this bi-infinite CHE were initially done in 1990 [12] for the 2:1 → 4:5 shown in Fig. 4F. In 1995, the portion of the sequence from Fig. 4E to Fig. 4G was animated [7], showing the sequence 7:3 → 2:1(3) → 4:5 → 2:1(13) → *escape*.

The entire sequence is shown in Figs. 4A–4G. The same conventions for labeling are used here as in Case A. The start date for this example is April 25, 1984, and it ends on August 6, 1992. The particle starts hyperbolically as shown in Fig. 5A and ends hyperbolically shown in Fig. 5G, where approximately 3000 days have past.

This is briefly described. The figures shown have the respective dynamics; Figure 4A, *capture* → 1:2(5); Figure 4B, 1:2(5); Figure 4C, 1:2 → 1:2 → 7:3; Figure 4D, 1:2(9); Figure 4E, 7:3 → 2:1(2); Figure 4F, 2:1 → 4:5(1.5); Figure 4G, 4:5(1.5) → 2:1(13) → *escape*.

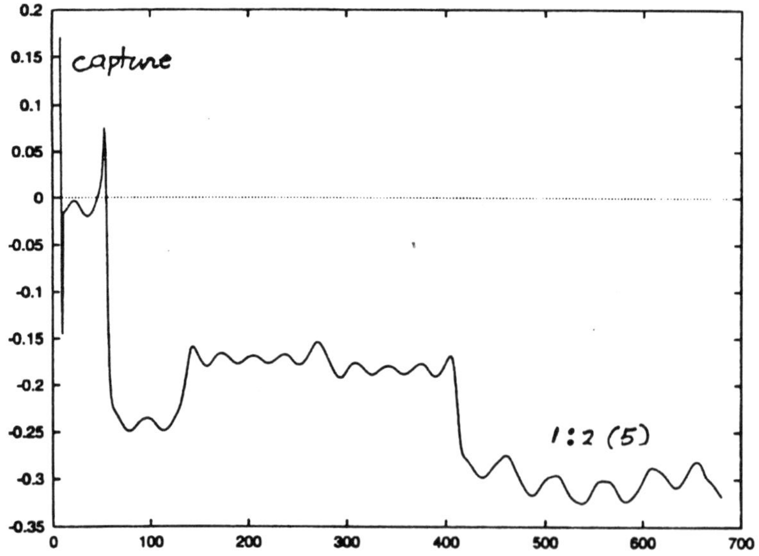

FIGURE 5A. A–G: H_E. Same units as 3E. X-axis, days.

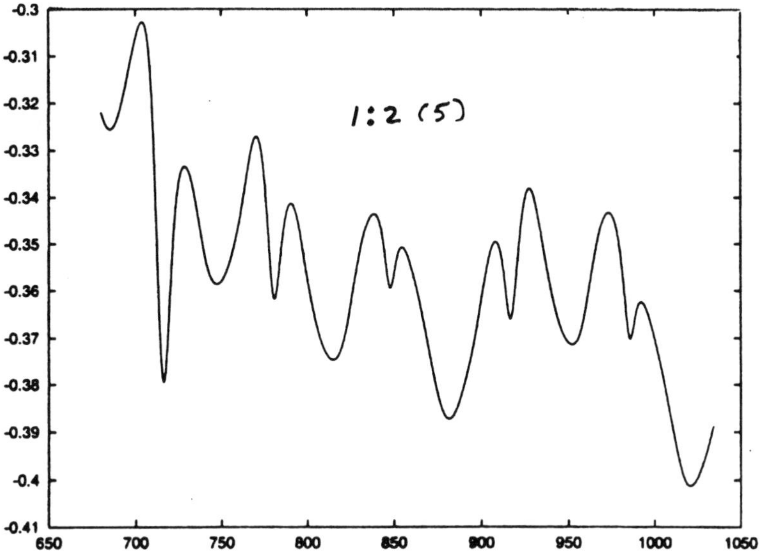

FIGURE 5B

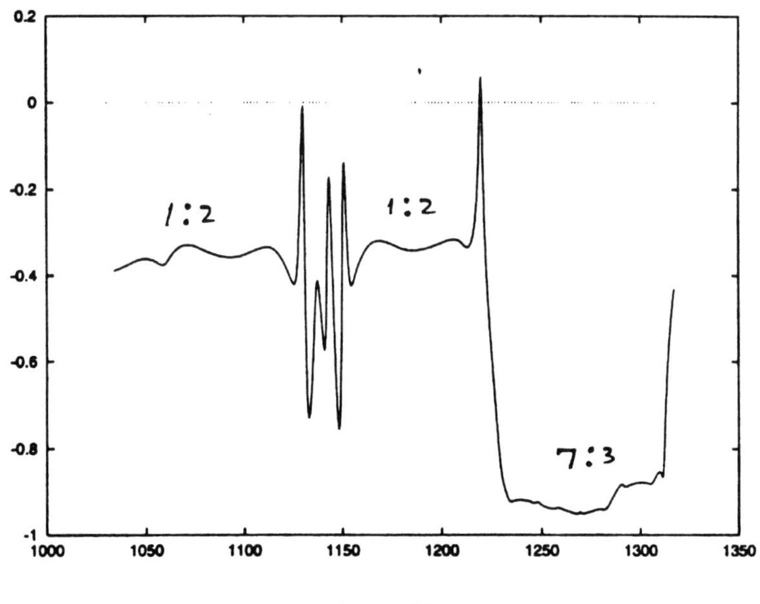

FIGURE 5C

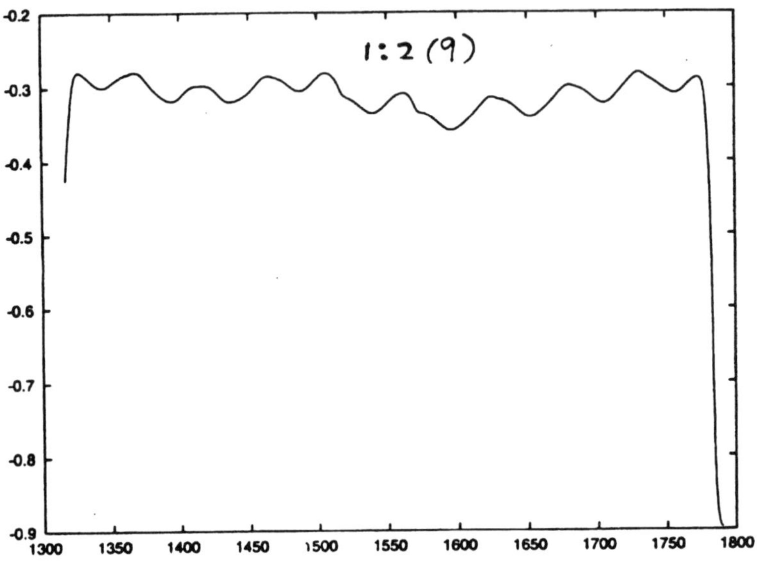

FIGURE 5D

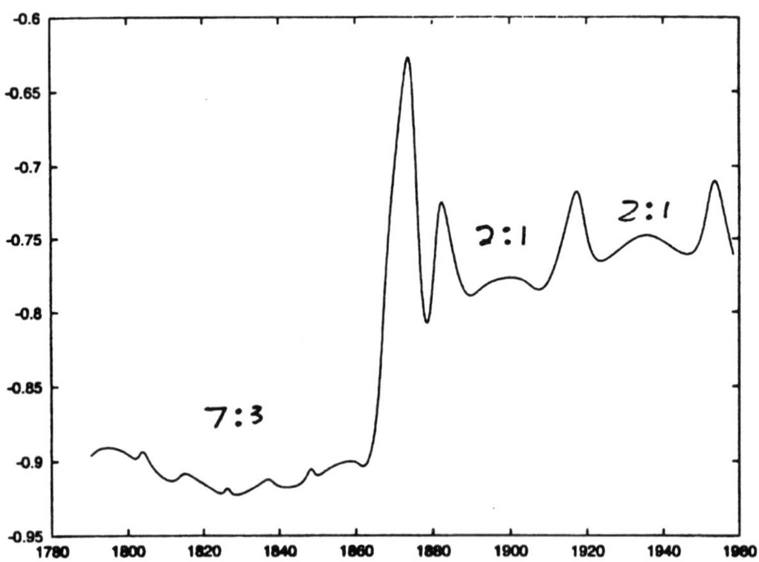

FIGURE 5E

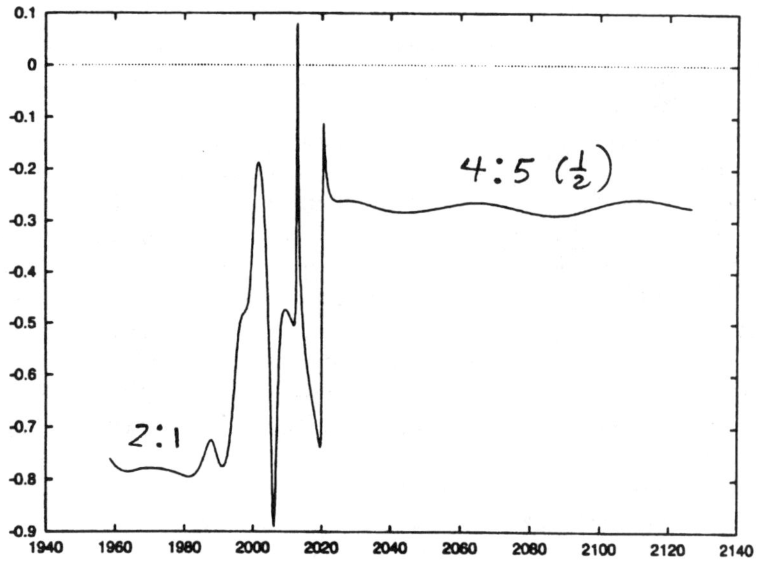

FIGURE 5F

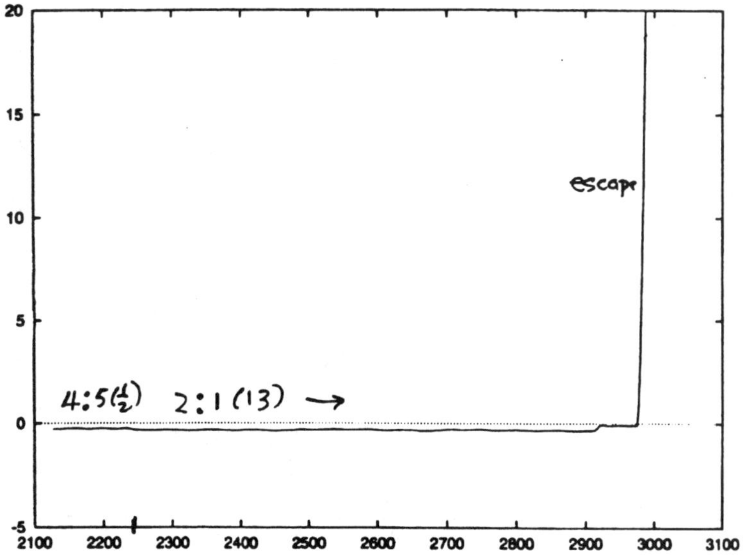

FIGURE 5G

The plots of H_E; are shown in Figs. 5A–5G. They are labeled to indicate the resonances, in the same respective order as Figs. 4A–4G for comparison. Figure 4A shows the capture. Comparing this to Fig. 5A shows H_E abruptly go from positive to negative. Figure 5C gives a correlation with the dynamics in Fig. 4C. The escape is shown in Fig. 4G. Note how H_E is near zero in Fig. 5G. Corresponding to Fig. 5G is Fig. 6B showing H_M.

In order to escape the E, M-system with no near collisions, a very subtle energy pumping mechanism is required to gradually increase H_E from negative to positive, so that $\wp$ not only escapes the E, M-system, but will move infinitely far away. This process is seen clearly in Fig. 5G where the energy of H_E is made to gradually increase in a stepwise fashion as $\wp$ makes 13 revolutions about E on 1:2 ellipses. A tiny amount of energy increase is obtained as $\wp$ flys by M. The brief interactions with M are shown in Fig. 6B. It is remarked that as $\wp$ moves from capture to escape, H_E stays relatively near to zero, in negative values. This is required so that the particle can have the necessary energy to be able to hop. It is seen that many cycles of $\wp$ are required about E in order to gradually pump down the energy, as is seen in Figs. 4A, 5A. The energy can't be pumped down too much or too little, otherwise $\wp$ will not be able to start hopping. This dynamics is seen to be very delicately balanced, however, as is commented on later, numerically appears to be robust. This dynamics is nontrivial.

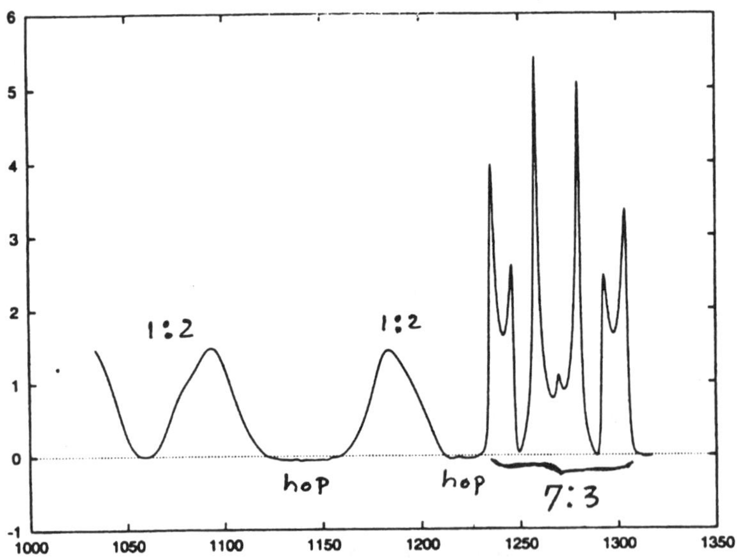

FIGURE 6A. H_M for 4C.

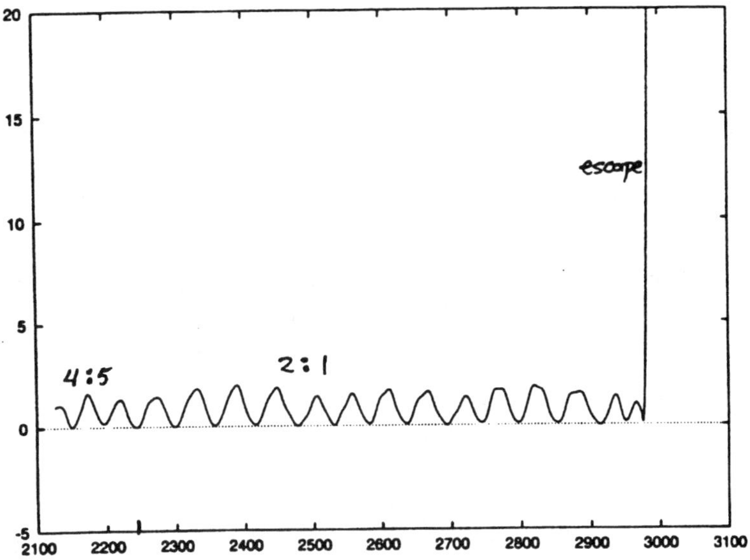

FIGURE 6B. H_M for 4G.

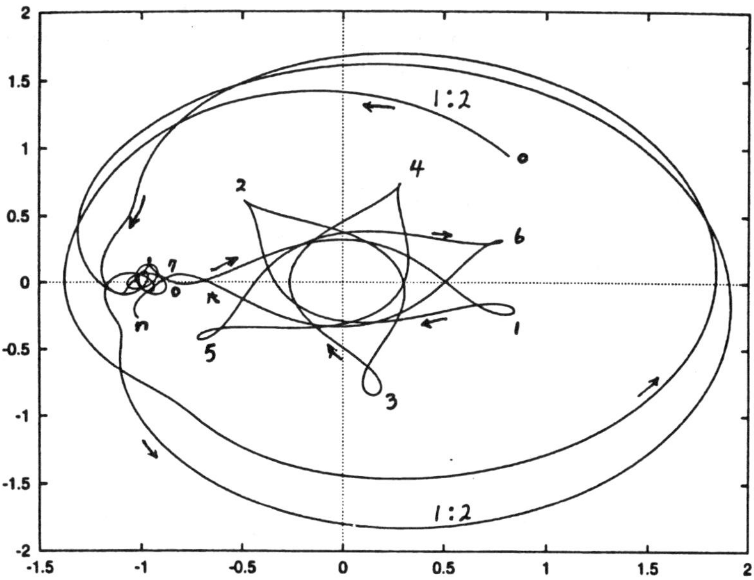

FIGURE 6C. Fixed plot for 4C. Same units as 3C.

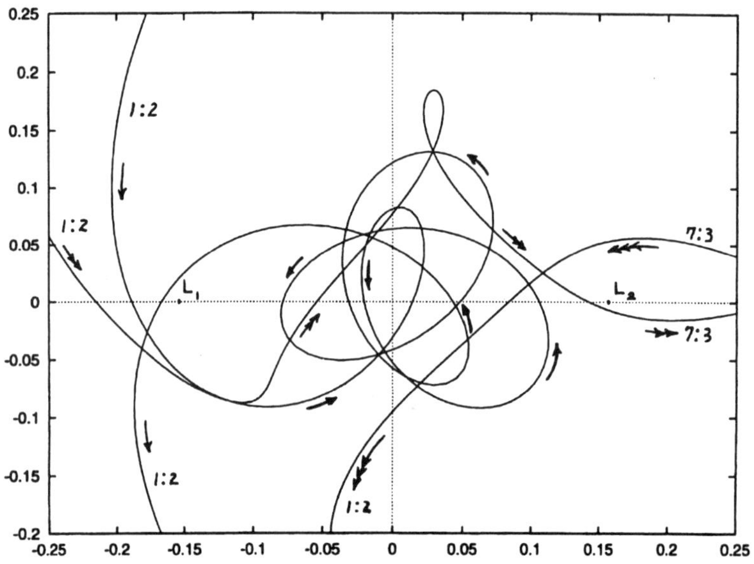

FIGURE 6D. Magnification of lunar neighborhood for C. Typical motion near a tangle.

We conclude with a discussion of Fig. 4C which is done in slightly more detail. The plot of H_M in Fig. 6A shows a clear division of the different resonances. The corresponding plot H_E is shown in Fig. 5C. The plot of Fig. 4C in a fixed coordinate system is shown in Fig. 6C. As with the discussion of Fig. 3C in Case A, the 7:3 in this example is indicated. During the hops from 1:2 to 1:2 and then to 7:3, note the complicated dynamics around M in the region of ST. This is magnified in Fig. 6D. This is an M-centered fixed-coordinate system. The two collinear Lagrange points, L_i, $i = 1, 2$ on the x-axis are approximately d units from M,

$$d = (\mu/3)^{1/3}, \quad d \cong .16.$$

The particle is seen approaching near to L_1 on the upper left after completing over one complete circuit about M on a 1:2 ellipse after leaving **o**. It then performs a complicated motion about M. It is performing a hop while on this motion. The trajectory is marked using a single arrow. Upon exiting, $\wp$ passes near to L_1. It exits on a 1:2 ellipse also. Another circuit about M brings it back to near L_1 where it is now on the center left, where a double arrow is used to track it. It moves nonlinearly about M in another hop and exits near L_2 where it passes onto a 7:3. Returning near to L_2 after approximately three lunar periods, it passes near to L_2 again, being marked with three arrows, where it hops back to a 1:2. This motion about M is typical to motion near a so-called *hyperbolic tangle* which is explained in Section 6. The fact $\wp$ exits enters and exits hops in resonances is suggested by the work of Mather described in the Section 6 also.

5. CHE TRAJECTORIES AND THE BIRKHOFF QUASI-ERGODIC HYPOTHESIS

Understanding the ultimate fate of N bodies in the general N-body problem, $N > 2$, under the inverse square Newtonian gravitational force law for an arbitrary set of initial conditions has been an outstanding problem in dynamics going back to Lagrange in the late 18th century [31]. This is an enormously difficult problem, and qualitative results were started by Poincaré using the planar restricted three-body problem, C^2, where $\wp$ moves in the same plane as two primaries P_1, P_2 where the mass of P_1 is much larger than the mass of P_2. His results were obtained for periodic motions of $\wp$ about P_1 [25]. One can imagine that P_1 = S, P_2 = J, and $\wp$ is an asteroid. Of course, the motions in the C^2 or even C^3 problems don't have to be periodic. If $\wp$ is given an initial position and velocity, in say the C^3 problem, it can move in many different ways. Some of these motions are bounded, some are unbounded, some are oscillatory, recurrent, near collision, etc. Of fundamental interest are those solutions which become unbounded in phase space. Thought about in terms of position space, this is a very difficult problem because the route to becoming unbounded can be extremely complicated, and very slow. $\wp$ could wander all over the physical space and eventually make its way to infinity, or it may stay bounded. Viewed in position space, Birkhoff conjectured that in the general 3-body problem, the set of solutions tending to infinity were, in fact, dense and he indicated the possibility of this being true for $N > 3$ as well [13].

This is known as the *Birkhoff quasi-ergodic hypothesis* or the *Birkhoff conjecture* [14], [15]. We state it now in terms of the general phase space. As N-bodies are moving, the Hamiltonian energy H is conserved [31]. This yields a $6N - 1$ dimensional surface $\Sigma = \{H(p, q) = c\}$, $c \in \Re$, where $q \in \Re^{3N}$, $p \in \Re^{3N}$, are the position and momentum variables. Let $\varphi(x, t)$ be the flow associated with the Hamiltonian H on Σ obtained from the Hamiltonian differential equations,

$$\dot{q} = \frac{\partial H}{\partial p}, \quad \dot{p} = \frac{-\partial H}{\partial q},$$

where $x \in \Sigma$, $\varphi(t) = (q(t), p(t))$.

> The Birkhoff conjecture asserts that the set of $x \in \Sigma$ such that $\varphi(x, t)$ becomes unbounded as $t \to \pm\infty$ is dense in Σ.

This is very difficult to answer since for a given set of initial conditions, $\varphi(t)$ may become trapped in an open region of the phase space thus disproving the conjecture. This is indeed the case for the C^2 problem which is described below, which is due to the fact that this problem has only two degrees of freedom. The mechanism of Arnold diffusion *may* allow $\varphi(t)$ to become unbounded, for a large set of initial conditions in the C^3or E^3 problem, however, very slowly as is described below. In fact, this mechanism is so slow, that numerical integrations used to verify it are prohibitively long.

In order to shed light on the Birkhoff conjecture, a mechanism is needed which will allow $\varphi(t)$ to rapidly move between large regions of the phase space which may move near elliptic-type motion.

The bi-infinite CHE for $n = 2$, we refer to as Case B, described above for the PE^3 problem, and also the E^3 problem, provides such a mechanism, which is numerically verifiable. This trajectory becomes unbounded as $t \to \pm\infty$. Numerical investigations show that Case B is robust under small perturbations. That is, under small perturbations another bi-infinite CHE is obtained where, in general, the sequence of hops are different. This yields an open tube T_B in phase space about Case B all points of which become unbounded in forward and backward time.

> Thus, $T_B \subset \Sigma$ provides an open region on Σ where the Birkhoff conjecture is true.

Although T_B is only a small portion of Σ, Case B trajectories may be promising to investigate further.

Also, it is important to note that the existence of Case B is numerically established, and an analytic proof is not yet available. It would seem to be difficult to prove the existence of such orbits.

As is discussed in Section 6, it is very likely that bi-infinite CHEs exist in regions of the phase space devoid of KAM tori, and the regions where hopping is occuring may be related to more general regions studied by Mather. A theory to describe a way to move through large regions of phase space is given in [21].

A DISCUSSION OF ARNOLD DIFFUSION

The question of what happens to $\wp$ in the C^2 problem if $\wp$ starts its motion with elliptic initial conditions about S was ultimately solved by the celebrated KAM theorem, where KAM means Kolmogorov-Arnold-Moser [2], [23]. Basically, $\wp$ will move in a stable elliptic way, not varying much from the initial elliptic state, provided it does not move too close to resonance with P_2. It just keeps cycling around and around P_1, and its orbital elements will almost be constant. That is, P_2 doesn't change things too much. Even when $\wp$ moves in near resonance with P_2 the motion is also stable and the orbital elements change very little. However, under closer examination, $\wp$'s elliptic orbit will have a very small chaotic character to it. Geometrically what is going on is that $\wp$ is moving on or near so-called two-dimensional KAM tori in the four-dimensional phase space of position and velocity supporting quasi-periodic motion of two frequencies. When the motion is not too close to resonance, it moves on one of these tori. Otherwise, it moves in between them. In this case, it can't go very far because the KAM tori are dense and block the motion of $\wp$. They block the motion because as $\wp$ moves, the energy H is a constant c. In suitable coordinates H depends on four variables — two positon and two velocity. Thus, $\wp$ moves on a so-called energy surface $H = c$ of three dimensions. Since the tori are two-dimensional, concentric tori trap trajectories in between them. The C^3 problem is not so clear.

In this case, $\wp$ doesn't have to move on the P_1, P_2 orbital plane and moves in in three dimensions with three position and three velocity coordinates. Thus, H depends on six variables and $H = c$ yields a five-dimensional surface. As $\wp$ moves about P_1 in an ellipse whose elements don't change very much and which is sufficiently far from resonance with P_2, it is moving on a torus of three dimensions. However, since these tori are two dimensions less than the dimensions of the of the energy surface, motion off one of these tori is not blocked by neighboring tori. Thus, near resonant motion can drift away from its initial elliptic state.

This drifting is referred to as *Arnold diffnsion*, and first analytically proven to exist for a simpler problem in 1964 [4]. It is very slow because as $\wp$ moves away from KAM tori it starts very close to them, and this keeps it in their neighborhood for long periods of time. It moves from torus to torus by passing near a complex set of hyperbolic invariant manifolds [2]. These set of tori are referred to as a *transition chain*. $\wp$ is actually moving in a region near the tori called a *resonance zone* where the three frequencies associated to the motion are not rationally independent. As an example, one can imagine that an asteroid moving about S under the perturbation of J, $\wp$ will slowly become unstable due to motion in the resonance zone. These rates are estimated to be tens of millions of years.

Estimates by Nekoroshev [29] and others for the C^3 problem indicate that to move significant distances in physical space from the initial elliptic motion, that time spans on the order of 10^{10} years are required. This is because the rate of velocity of $\wp$ is on the order of

$$e^{-1/\mu^{\kappa}},$$

where $0 < \kappa < 1$, μ is the mass ratio of the two primaries, $\mu = .001$ in the S, J-system. Typically, $\kappa = ½$. As seen in Section 3, the time to span significant distances through the hop is on the order of 10^1 years. This is a decrease by nine orders of magnitude. This factor is more relevant in the S, J-system where the mass ratio is fairly small, which is required for Arnold diffusion theory. In the E, M-system, $\mu = .0123$.

6. IMPLICATIONS AND REASONS FOR FAST RESONANCE SHIFTING

The speed of the hop has been seen to be approximately *nine* orders of magnitude faster than Arnold diffusion. Observationally, this gives rise to the abrupt changes in the elliptical orbits of certain comets [5]. It also shows how Case B in the previous section can move relatively quickly through capture and escape. The hop and CHEs are mechanisms which provide examples of how change can occur rapidly in the solar system from elliptic orbits.

When the hop occurs, two things of note are taking place. First, the particle $\wp$ is able to quickly move from one resonance orbit to another which can be quite far apart. For example, the comet 74P/Smirnova-Chernykh which goes from a 6:13 → 7:5 shown in Fig. 2A. The nonlinear transition through the hop moves through a large region of the phase space, and only over approximately one year. Second, the particle is going from one resonance to another.

The former case can be understood in a straightforward way by first showing that the two resonance orbits that the particle is transitioning between are unstable. This is done by computing the variational matrix, $\mathcal{M}$, associated with each orbit, and show it has purely real eigenvalues indicating unstable hyperbolic behavior. For example, if the orbit is given by $\varphi = \varphi(\varphi_0, t) \in \Re^6$, where $\varphi(\varphi_0, 0) = \varphi_0$, then

$$\mathcal{M} = \left.\frac{\partial\varphi}{\partial\varphi_0}\right|_{t = T_R},$$

where T_R is the period of φ. There are six eigenvalues, λ_i, $i = 1, 2, \ldots, 6$ occuring in complex conjugate pairs. Two of these are real, one being the inverse of the other. This is carried out, for example, in [11] for the case shown in Fig. 4F. Associated to these two eigenvalues are a stable and unstable manifold each of two dimensions. Locally, they are are tangent to the space spanned by the two eigenvectors associated to the unstable eigenvalues. The hop indicates that these manifolds intersect. For example, referring to Fig. 2A, the unstable manifold of the 6:13 should intersect the stable manifold of the 7:5. Recent numerical investigations show this is indeed the case. It is necessary to numerically extend the stable and unstable manifolds until their intersection is observed.

The intersection of these stable and unstable manifolds implies that they intersect infinitely often in a very complex structure called a hyperbolic tangle [2], [23]. The motion near such a tangle should be very complicated. This manifests itself in the fact that the hop dynamics should be very nonlinear. For example, see Fig. 6D. There are times where the hop may not appear to be too nonlinear. This is because it is occuring in a short period of time, and is fairly direct. However, a closer examination shows a complicated behavior.

It is noted that the hops occur near the smaller of the two primaries in all the cases considered. This is evident in the fixed-coordinate plots included here. Moreover, the hops seem to exit and enter near the Lagrange points L_1, L_2. More precisely, in all cases considered, the hops occur near the apoapsis or perapsis locations of all the resonance orbits with respect to the larger primary, which lie near these Lagrange points. This is discussed in much more detail in [5], where an argument is given to provide a physical mechanism of why the hop is occuring. It turns out that near these Largange points in phase space are retrograde unstable periodic orbits called *Lyapunov orbits*, which also have two-dimensional stable and unstable manifolds [8], [18]. Numerical investigations indicate the possiblility that these manifolds may intersect, and form a hyperbolic tangle that surrounds the smaller of the two primaries, say M or J. The physical locations of L_1, L_2 with respect to M are shown in Fig. 6D. This tangle is conjectured to be the ST region, or fb. The tangle formed from the resonant period orbits lies within this tangle.

It is conjectured that the ST region is devoid of KAM tori. Thus, when a particle enters it, the particle should not be slowed down by approaching a KAM torus. It should be able to move through this region relatively quickly [5].

The second property of note is the fact that the particle moves near resonance states after escaping from ST. A probable reason why this happens is related to the work of Mather in a lower dimensional setting corresponding to two degrees of freedom [22]. Translated to the C^2 problem on a two-dimensional energy surface Σ, defined in Section 5, his results prove that in the absence of two-dimensional KAM tori, the phase space consists of unstable resonant orbits, with associated stable and unstable manifolds which in general intersect, therefore forming hyperbolic tangles. These unstable resonant orbits are referred to as *Aubry-Mather sets*, or *Birkhoff periodic orbits* [17], [22]. Translating this to a higher dimensional setting, such as the E^3, or PE^3 problems would help explain the resonance shifting. There are partial results towards a higher dimensional version of Mather's work [17], [22].

7. APPLICATIONS

In this brief section we present a few applications of this work, one of which has been used to enable a Japanese spacecraft to reach the Moon and go into orbit in 1991 [9], [10], [11], [16]. This illustrates use of weak capture. The second application shows how resonance transitions can be used, and the third show the applicability of CHEs.

In January 1990, Japan sent two spacecraft linked together into Earth orbit. The larger, called *Hiten*, stayed in Earth orbit while the smaller one went to the Moon in early February of that year. This smaller spacecraft lost communication with the Earth as it arrived at the Moon, so the mission was not complete. In order to achieve the goal of putting a Japanese spacecraft in orbit around the Moon, thereby making Japan the third country in history to do this, it was decided to try to get Hiten to the Moon. However, Hiten had too little fuel to do this. To go to the Moon by the standard method, using a so-called *Hohmann transfer* [9], [16], it takes about three days, and upon arrival the excess velocity is about 1 km/s. This requires a significant amount of fuel to minimally achieve lunar orbit. Hiten had only a small fraction of this. Thus, it was impossible by standard means. I was asked to look into this problem in mid-1990, and it was found that lunar capture could be achieved with essentially no fuel if Hiten went to weak capture conditions at the desired altitude of 100 km, where the osculating eccentricity turns out to be .94. Starting at a weak capture state, and integrating in backward time using the four-body problem modeling of E, M, S, $\wp$, where $\wp$ = Hiten, it was found that the trajectory could be made to not only come back to the Earth, but satisfy the correct initial conditions for Hiten with little energy. I was assisted by J. Miller who helped with some of the numerical calculations [10], [16]. The resulting trajectory for Hiten carried it out to one million km from the Earth, where it fell back to weak capture. The route took about 120 days. A negligible amount of fuel was needed to allow Hiten to go onto this route, sometimes referred to as a BCT (Ballistic Capture Transfer), or WSB Transfer, where WSB stands for *weak stability boundary*.

It was used beginning in April 1991, and Hiten successfully reached the Moon on October 2, 1991 with most of its fuel to spare. BCT's are now a new way to reach the Moon and plan to be used again by both the US and Japan in 1998–99. The US mission is being planned by the US Air Force Academy as a student project in 1999, called Blue Moon.

A recent result with Genta shows that transferring to a comet while it is performing a hop at Jupiter can reduce the rendezvous energy by 86% as compared to rendezvousing without the influence of Jupiter [6]. The comet this was demonstrated for was 82P/Gehrels 3. The reason for this is that while the comet is hopping, it is weakly captured at Jupiter. Gravity losses at Jupiter provide the energy savings.

The last application to mention here concerns CHEs. In particular, the type with Case A for $n = 3$. Referring to Fig. 3A, it is seen that $\wp$ is able to escape the E, M-system by performing the resonance shifting from a 2:1. This means that if $\wp$ were a spacecraft, no fuel would be required to escape the E, M-system from the 2:1 orbit. Recent work shows that $\wp$ could transfer to the 2:1 and match its initial conditions with no energy using a BCT. Such an ejection process from a BCT to a 2:1 to escape has promise to leave the E,M-system with significant savings of energy.

ACKNOWLEDGMENTS

I thank the organizer of the International NEO Conference, John Remo for inviting me to attend and to give a lecture at the United Nations. During this conference, I had the opportunity to meet Brian Marsden, who introduced me to the welcome fact that comets may be demonstrating the hop. This paper was written while I had a visiting position at the Department of Mathematics at Princeton University, and I thank John Mather for informative discussions. I am also grateful to Michel Herman who invited me to speak in his seminar at Columbia University in Fall of 1996. I also thank Stuart Levy of the Geometry Center for a great deal of computer assistance. Finally, my thanks to Richard McGehee and The Geometry Center staff for their support.

REFERENCES

1. ALEXEYEV, V.M. & YU.S. OSIPOV. 1982. Accuracy of Kepler approximation for fly-by orbits near an attracting centre. Ergod. Th. and Dynam. Sys. **2:** 263–300.
2. ARNOLD, V.I. 1989. Mathematical Methods of Classical Mechanics. GTM Series, no. 60. Springer-Verlag. New York, NY.
3. ARNOLD, V.I. & A. AVEZ. 1968. Ergodic Problems of Classical Mechanics. W.A. Benjamin, Inc. New York, NY.
4. ARNOLD, V.I. 1964. Instability of dynamical systems with several degrees of freedom. Soviet Mathematics. **5:** (3) 581–585.
5. BELBRUNO, E. & B. MARSDEN. 1996. Resonance hopping in comets. Astron. J. **113:** (4) 1433–1444.
6. BELBRUNO, E. & G. GENTA. 1996. Low energy comet rendezvous using resonance transitions. *In* Proceedings of First IAA Symposium Missions to the Outer Solar System and Beyond. Liberia Editrice Universitaria. Levrotto & Bella Torino.
7. BELBRUNO, E. 1995. Visualization of Hop Dynamics in the Three-Body Problem", Geometry Center Tape no. S-106, April 1995.
8. BELBRUNO, E. 1994. The dynamical mechanism of ballistic lunar capture transfers in the four-body problem from the perspective of invariant manifolds and Hill's regions. Centre de Recerca Matematica, Institute d'Estudis Catalans, Report no. 270.
9. BELBRUNO, E. 1994. Ballistic lunar capture transfers using the fuzzy boundary and solar perturbations: A survey. J. Brit. Inter. Soc. **47:** 73–80.
10. BELBRUNO, E. & J. MILLER. 1993. Sun-perturbed Earth-to-Moon transfers with ballistic capture. J. Guid., Control and Dyn. **16:** 73–80.
11. BELBRUNO, E. 1990. Examples of the nonlinear dynamics of ballistic capture and escape in the Earth-Moon system. *In* Proc. AIAA Conf. on Astrodynamics, AIAA Paper no. 90-2896.
12. BELBRUNO, E. 1987. Lunar capture orbits, a method of constructing Earth-Moon trajectories and the lunar GAS mission. *In* Proceedings 19th AIAA/DGLR/JSASS Inter. Elec. Propl. Conf., AIAA Paper no. 87-1054, Col. Springs. 87-1054,
13. BIRKHOFF, G.D. 1927. Dynamical systems. Amer. Math. Soc. Coll. Publ. **9:** 290–291.
14. HERMAN, M. 1996. Hamiltonian dynamics and the ergodic hypothesis. Notes from Samnel Eilenberg Lectures at Columbia University.

15. HERMAN, M. 1992. Stabilitie topologique des systemes dynamiques conservatifs, a paraitre dans Atas do 18° Col. Bras. Mat.
16. FRANK, A. 1994. Gravity's rim: Riding chaos to the Moon. Discover: September, 74–79.
17. KATOK, A. & D. BERSTEIN. 1987. Birkhoff periodic orbits for small perturbations of completely integrable Hamiltonian systems with convex Hamiltonians. Invent. Math. **88:** 225–241.
18. LLIBRE, J., R. MARTINEZ, & C. SIMO. 1985. Transversality of the invariant manifolds associated to the Lyapunov family of periodic orbits near L2 in the restricted three-body problem. J. Differential Equations. **58:** 104–156.
19. MARSDEN, B.G. 1997. Kuiper Belt: Securing adequate orbital data. *In* Asteroids, Comets, Meteors 1996, A.C. Levasseur-Regourd, Ed. Elsevier Science Ltd. Oxford.
20. MARSDEN, B.G. 1961. The orbit and ephemeris of periodic comet Oterma. Astron. J. **66:** 246–248.
21. MATHER, J.N. 1993. Variational construction of connecting orbits. Ann. Inst. Fourier, Grenoble. **43:** (5) 1349–1386.
22. MATHER, J.N. 1982. A Criterion for the non-existence of invariant circles. Publ. Math. IHES. **63:** 153–204.
23. MOSER, J.1973. Stable and Random Motions in Dynamical Systems. Ann. Math. Studies. **77:** Princeton Univ. Press. Princeton, NJ.
24. MOSER, J. 1962. On invariant curves of area-preserving mappings of the annulus. Nachr. Akad. Wiss. Gottingen, math.-phys. **1:** 1–10.
25. POINCARÉ, H. 1899. Les Methods Nouvelles de la Mechanique Celeste, Vol. 3, Paris.
26. POINCARÉ, H.1893. Les Methods Nouvelles de la Mechanique Celeste, Vol. 2, Paris.
27. POINCARÉ, H. 1892. Les Methods Nouvelles de la Mechanique Celeste, Vol. 1, Paris.
28. REMO, J. 1997. NEO orbits and nonlinear dynamics: A brief overview. *In* Proceedings of the Inter. Near Earth Objects Conference. Annals of the New York Academy of Sciences. This volume.
29. NEKOROSHEV, N.N. 1971. The behavior of Hamiltonian systems that are close to integrable ones. Functional Analysis and Appl. **5: (4)** 1971.
30. SCHWARTZSCHILD, K. 1896. Uber die stabilitat der bewegung eines durch Jupiter gefangenen kometen. Astr. Nachr. **141:** 1–8.
31. SIEGEL, C.L.& J.K. MOSER. 1971. Lectures on Celestial Mechanics. Springer-Verlag, Grundlehren Series: Vol. 187.
32. TABACMAN, E. 1995. Variational computation of homoclinic orbits for twist maps. Phys. D. **85:** 548–562.

Gaspra and Ida: Implications of Spacecraft Reconnaissance for NEO Issues

CLARK R. CHAPMAN

Southwest Research Institute
1050 Walnut St., Suite 426
Boulder, Colorado 80302 USA

ABSTRACT: The first spacecraft encounters with asteroids were flybys of Gaspra and Ida by Galileo. They emphasize the diversity of asteroid properties and permit us to consider the probable physical nature of near-Earth objects (NEOs).

INTRODUCTION

On October 29, 1991, the Galileo spacecraft flew past the 18 × 9 km main-belt asteroid 951 Gaspra, and sent a stream of images and other data back to Earth. It was the first time that an asteroid had ever been seen "up close." For most of the nearly two centuries since the first asteroid was discovered, these small, distant objects had been nothing more than points-of-light even in our largest telescopes. During the decade before Galileo's encounter with Gaspra, ground-based radar had begun to give us delay-doppler maps of some small Earth-approaching asteroids, but they revealed little more than the objects' gross shapes. Even so, Gaspra is a considerably larger asteroid than most near-Earth objects (NEOs), so it was a striking revelation to suddenly see images showing not only its extremely angular shape, but an array of craters and other geological features on its surface.

Nearly two years later, on August 28, 1993, Galileo obtained even higher resolution images of the still-larger main-belt asteroid 243 Ida. In the years to come, we may expect analogous images to be obtained for some asteroids from the recently refurbished Arecibo radar observatory. We expect even more detailed data from 433 Eros (imaging, geochemical, and much more), which will be orbited by the Near-Earth Asteroid Rendezvous (NEAR) mission in calendar year 1999. (NEAR will make a Gaspra-like flyby reconnaissance of 253 Mathilde in late June 1997.) Additional asteroid missions are planned by several countries. However, until these new missions are developed, launched, and return data, our knowledge of the physical nature of asteroids — a population already known from astronomical data to be extraordinarily diverse — will be based largely on Galileo's observations of Gaspra and Ida.

The purpose of this short paper is not to repeat the summaries and detailed research papers describing these two asteroids that are already published in the widely available literature. Most of the detailed research results may be found in special issues of *Icarus* dealing with each asteroid: for Gaspra see *Icarus*, Vol. 107 (1), 1994; for Ida see *Icarus*, Vol. 120 (1), 1996. The images are all available over the Internet and in other ways from NASA's Planetary Data System.

PHYSICAL TRAITS AND NEO ISSUES

In applying our knowledge about Gaspra and Ida to the NEO population, we must consider several different types of issues concerning NEOs. While many NEOs may be comets, the majority are believed to be fragments from the main asteroid belt. Hence the scientific investigation of NEOs, especially by spacecraft, can consider relevant attributes of Gaspra and Ida to be possibly pertinent to studies of some NEOs. In addition, NEOs are of particular interest as potential way stations for astronauts in a Mars exploration program, as sources of raw materials for deep space operations, and as potentially threatening bodies (if one were to be found on a possible Earth-impact trajectory, we would want to know about the physical nature of such objects in order to better characterize the threatening body and plan effective mitigation).

Evaluation of the cosmic impact hazard depends, of course, on our perceptions of the nature of the threatening projectiles. The question of *how hazardous* asteroids and comets might be has evolved in recent decades due to new perceptions about *how many* projectiles (and resulting impacts) there are as well as about *how serious* the environmental consequences of impacts are. Because, to first order, the environmental consequences of an impact depend on the kinetic energy of the impactor rather than on the specifics of its physical nature, much of the discussion of comets and asteroids has, appropriately, centered on their numbers: how many are there of various masses and what are their orbits?

Nevertheless, a closer examination of the hazard issue reveals significant reasons for considering the physical natures of the threatening bodies. Most obviously, any attempt to deflect or destroy such a body may depend critically on its physical nature. For example, depending on such properties as strength and cohesiveness, one might not dare to give the object a single impulse (for purposes of deflection) if it might, instead, break the body up into a potentially more dangerous rain of fragments.

But even evaluation of the hazard itself is affected by the nature of the bodies. Small bodies have a very wide variety of material compositions (ranging from "snow" to iron-nickel alloy); furthermore, they probably vary widely in cohesion (ranging from solid objects to rubble piles bound together only by gravity). There are other possibly relevant physical parameters, as well, including shape (ranging from spherical to markedly asymmetric) and spin (fast or slow, principal axis rotation or tumbling). These can all have second-order (or occasionally first-order) effects on the consequences of an impact.

As I discuss below, Galileo's studies of Gaspra and Ida tell us, in some detail, about these particular bodies. In a more philosophical sense, Galileo's explorations — particularly the unexpected discovery of Ida's satellite Dactyl — also remind us of how little we really know about asteroids and comets. Since Gaspra and Ida are both S-type asteroids, we can readily imagine that other types of asteroids (and comets!) might be very different.

Let me amplify on the presence of Dactyl. It is one concrete example that asteroids and comets are not necessarily solitary bodies. There is, in fact, much observational and theoretical work suggesting that small asteroids may commonly be multicomponent bodies held together only by gravity (i.e., rubble piles). The main element of the theoretical arguments (supported by experimental results scaled up to the asteroid case as well as hydrocode modelling) is that it takes far more collisional energy to disperse fragments from an impact than it takes simply to fracture and break up the body. Indeed, some asteroids are known to be immersed in "dust" bands (discovered by the IRAS satellite) and some comets have been observed by ground-based radar to be immersed in debris trails composed of macroscopic objects (numerous particles that are large compared with the multi-cm wavelength of the radar). As I explain elsewhere [5], ground-based astronomers are nearly blind to the presence of objects in the meter-to-kilometer size range, so we have little knowledge of how common they are. So-called dust bands could just be the small-particle component of an invisible population — dominating the mass — of much bigger objects, just as for Saturn's rings. The upper portion of this size range includes objects large enough to penetrate the Earth's atmosphere and produce severe local or regional damage.

Even if dangerous accompanying objects of these sizes are uncommon, the break up of Comet Shoemaker-Levy 9 reminds us that bodies which are, in fact, very weak (or are loose, gravitationally-bound aggregates) can be tidally disrupted by close passages to larger objects, like Jupiter, the Earth, or the Sun. A spray of large objects — whether generated by natural processes or by attempts to deflect a threatening solid object — could be very damaging by distributing effects over a broader region of the Earth.

Studies of an asteroid also help us define the distinctions between observations of the surficial layer (optical surface, or somewhat deeper layers when investigated at long wavelengths) and the properties of the whole body. While surface properties are most readily characterized (and may be relevant, for example, for attempts to interact with the asteroid in *in situ* exploration, resource acquisition, or certain kinds of mitigation measures), the bulk of the body will dominate the consequences of an impact on the Earth as well as some kinds of mining or threat-mitigation measures. Some direct measurements (of the gravity field, for example) and indirect inferences from the geology of the surface tell us about the interiors of a small body. Galileo's observations of Ida and Gaspra provide clues about these issues. Even if the data fail to resolve them, they highlight the kinds of evidence that may be important to obtain in future exploration of asteroids and comets.

GASPRA AND IDA

It is important to emphasize that virtually all of our rigorous knowledge about the natures of Gaspra and Ida pertain to their optical surfaces. Studies of Dactyl's motion have provided a crude estimate for the bulk density of Ida, but no instrument probed these objects more deeply than their superficial thermal skin depths. Nevertheless, photogeologists have honed their skills on the Earth (where there is the opportunity for ground truth reality checks), the Moon, and other planets. By interpretation of morphological surface features, some of which "dig" into the interior, combined with spatial correlations with compositionally-relevant remote-sensing data (e.g., infrared absorption bands indicative of mineralogy), geologists can make significant inferences about regolith depth, interior configuration, and other matters that are important to future exploration, mining, and hazard mitigation of asteroids.

Such inferences can be more powerful if data are obtained that are more comprehensive than were possible from the brief, serendipitous flybys of Gaspra and Ida. Nevertheless, it is interesting to consider some of the inferences made from the Galileo data.

Gaspra

Distant images of Gaspra show it, from some perspectives, to resemble a peanutlike object, as though composed of two objects stuck together with the contacts subsequently smoothed out. Despite Gaspra's generally lumpy appearance from various perspectives, the last several images (at highest resolution) portray a very angular, misshapen body, whose shape is prominently formed by planar elements (called "facets") connected by ridges. It remains possible that the narrow "end" of Gaspra is, indeed, a second component of the body, but the impression is stronger (at least to many members of the Galileo Imaging Team, cf. [9]) that Gaspra is a monolithic object whose present shape has been formed by exogenic spallation or "chipping away" from its original form.

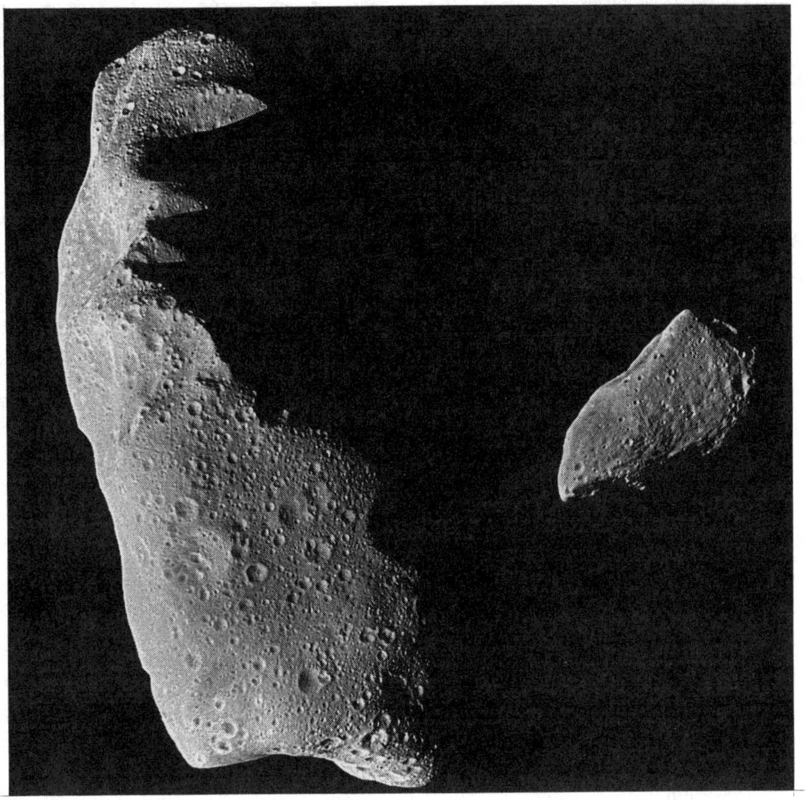

FIGURE 1. Galileo images of Ida and Gaspra are shown with the objects in approximately correct relative sizes.

Following receipt of images from Ida, which show Ida to look very different from Gaspra in some important respects, I have become increasingly convinced that Gaspra is probably a strong, metallic object. (See [4], [3], for brief earlier discussions of this idea.) Several independent pieces of evidence lend credence to this idea. First, it has long been known that Gaspra is unusual among S-type asteroids in having a reflectance spectrum that is dominated by olivine rather than pyroxene. This means that it probably is not chondritic in composition but, rather, is of geochemically differentiated composition. A portion of the interior of a large, differentiated precursor body, near the core-mantle interface, would plausibly exhibit a reflection spectrum like Gaspra's. So it is easy to draw an analogy between Gaspra and olivine-rich stony-iron meteorites (pallasites).

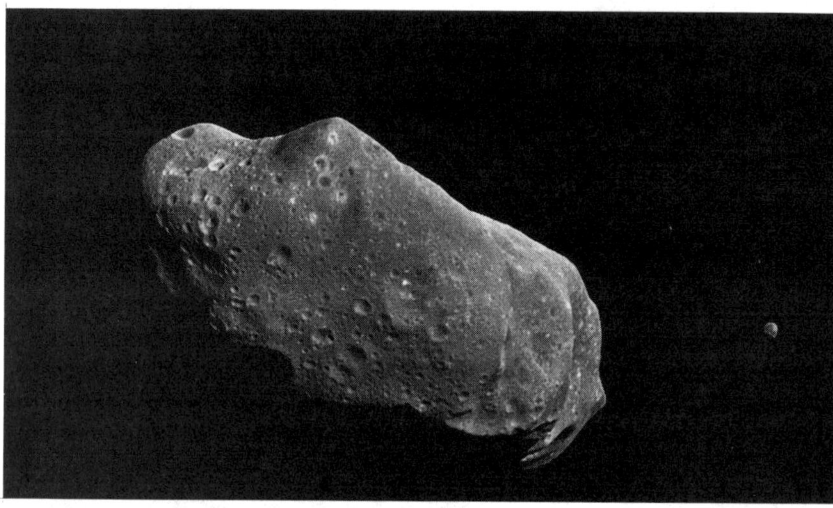

FIGURE 2. This portrait of Ida and its small moon, Dactyl, was taken by the solid state imaging system of the Galileo spacecraft during its encounter on 28 August 1993.

Another feature of Gaspra is its very low crater density compared with Ida. Since Gaspra and Ida are expected to be in roughly the same collisional environment, due to the appreciable eccentricities of asteroid orbits, one would ordinarily expect the crater populations to be similar. One possibility is that Gaspra's surface is, by chance, rather young; that is, it may be that a large impact event obliterated much of Gaspra's preexisting impact record comparatively recently, and Gaspra's surface has not had time to accumulate a saturation density of subsequent craters. An alternative explanation is that Gaspra's surface is very strong (e.g., metallic) and thus a given projectile forms craters that are much smaller on its surface than on the surface of a rocky body.

Taken together, this combination of attributes of Gaspra — its monolithic shape, its low crater density, and its apparently differentiated composition, plus the fact that Gaspra differs from Ida — all suggest to me that Gaspra is a strong, essentially metallic object. Since there are alternative explanations for all of these attributes, my conclusion may be wrong. However, the possibility that an object as large as Gaspra might have this kind of physical make-up is sufficient to consider Gaspra as an example of one end of the spectrum of kinds of bodies that might potentially impact the Earth. Unlike bodies of weaker strength, monolithic metallic bodies of any size can penetrate the Earth's atmosphere and cause damage. Gaspra represents a starting point for modelling how mitigation measures might deal with objects like Gaspra, or with smaller versions of Gaspra.

Ida

Ida is a larger body than Gaspra and, in some ways, even more misshapen, with maximum and minimum dimensions of 56 × 15 km. Ida is not so angular and faceted as Gaspra, however, and Thomas et al. [10] conclude that it is made predominantly of two different pieces. The general configuration of Ida can be described as two lobes separated by a "waist." The surface topography is notably different on the two portions. For instance, although both portions are saturated with small and moderate sized craters, one portion appears to have been heavily bombarded even by the rarer largest projectiles whereas the other portion lacks very large craters; a younger surface age for that portion could explain why it has not been bombarded so heavily.

It is uncertain whether Ida's gross shape and dichotomous geology reflects formation of Ida from two different objects or, instead, reflects inhomogeneities inherited from a precursor. Ida *appears* to be a coarse "rubble pile." I emphasize "appears" because, in fact, Ida's internal make-up remains conjectural. Ida has a deeper regolith than Gaspra (estimated at 50–100 m by Sullivan et al. [8]), and emplacement mechanisms have been modelled and described [6]. How the regolith grades into a coarse megaregolith or rubble-pile internal structure is difficult to assess from images of the exterior. The features discussed by Thomas and others in the special Ida issue of *Icarus* are suggestive of a multicomponent internal structure, but do not prove it conclusively.

Indeed, Asphaug et al. [1] propose a possible mechanism for the formation of grooves, observed on one part of Ida, that would preclude Ida's internal structure from being a classical rubble pile. Asphaug et al. note that the cracked or grooved terrain is roughly on the opposite side of Ida from a peculiar large feature called Vienna Regio. If Vienna Regio were caused by a large impact, then fractures in the opposing Pola Regio could result, according to hydrocode simulations, but only if elastic stresses could be propagated through the interior of the body, which seems unlikely for a rubble-pile.

Ida's low density implies that it has appreciable void space, since none of the possible mineralogical assemblages inferred from its reflectance spectrum have densities so low as Ida. But, within error bars, the void space could be small enough that a rubble pile structure would not necessarily be implied.

Unlike Gaspra, Ida appears to be composed of minerals like the ordinary chondrite meteorites, which have physical properties similar to ordinary terrestrial rocks. Thus Ida's strength is probably like rock — whether solid rock or, as preferred, broken-up rubblized rock, overlain with a rocky soil (regolith) up to 100 m deep.

CONCLUSIONS

Following the Galileo encounters, we have far better ideas about the physical properties of asteroids than we had before — certainly better than "spherical cow" models from the epoch when we had astronomical observations only. Asteroids are clearly very irregular in shape, perhaps reflecting inhomogeneities in their interiors. Although it is not conclusive, these first two asteroids observed close up probably have a very wide range of bulk physical properties — monolithic fragment with the properties of steel (Gaspra) or a multicomponent, gravitationally bound assemblage of rocky materials, overlain with up to 100 meters of soil (Ida).

Because of its presumed strength, a small version of Gaspra would not be expected to break up by tidal disruption or stresses during atmospheric penetration. Such bodies would be very damaging in the vicinity of their impact point, but would not be as efficient at widely distributed damage as a body that might break up before or during entry and cause widely dispersed damage. Obviously, if the object were larger than a couple of kilometers in diameter, then worldwide effects would be expected from the great volume of ejecta alone [2], [7].

Ida, on the other hand, reminds us of the widespread evidence that many asteroids are probably rocky rubble piles. Because Ida is larger than any known Earth-approaching object, it is important to recognize that smaller rocky rubble piles may be different from Ida. A variety of studies, including radar observations, have indicated that small rubble piles may be common among Earth-approaching objects. In particular, it is expected that bodies a few km in diameter and smaller will tend to lack surface layers of soil or regolith. Such objects stand some chance of breaking up prior to entry and small ones (like the object that caused Tunguska) may explode entirely in the lower atmosphere, unlike a metallic object.

While Galileo studied main-belt asteroids, appreciably larger than the vast majority of potential Earth impactors, it has nevertheless provided a reality check to our conceptions of asteroids that were previously based on unresolved telescopic studies with a healthy dose of theoretical extrapolation. Forthcoming missions, some directed at actual near-Earth objects (both asteroids and comets, some missions in planning stages with others like the Near-Earth Asteroid Rendezvous mission already launched) should greatly augment our understanding of various small bodies. Because of the diversity of small bodies, however, it will always be essential to perform detailed physical observations of any *particular* body slated for exploration, mining, or deflection, prior to undertaking such an interactive mission.

ACKNOWLEDGMENTS

I thank David Morrison and Alan Harris for discussions, the Galileo Project and Galileo Imaging Team for accomplishing these encounters and interpretations, and John Remo for encouraging me to complete this manuscript.

REFERENCES

1. ASPHAUG E., J.M. MOORE, D. MORRISON, W. BENZ, M.C. NOLAN, & R.J. SULLIVAN. 1996. Mechanical and geological effects of impact cratering on Ida. Icarus. **120:** 158–184.
2. CHAPMAN, C.R. & D. MORRISON. 1994. Impacts on the Earth by asteroids and comets: assessing the hazard. Nature. **367:** 33–40.
3. CHAPMAN, C.R., E.V. RYAN, W.J. MERLINE, G. NEUKUM, R. WAGNER, P.C. THOMAS, J. VEVERKA, & R.J. SULLIVAN. 1996. Cratering on Ida. Icarus. **120:** 77–86.
4. CHAPMAN, C.R. 1996. S-type asteroids, ordinary chondrites, and space weathering: the evidence from *Galileo*'s fly-bys of Gaspra and Ida. Meteoritics & Plan. Sci. **31:** 699–726.
5. CHAPMAN, C.R. 1997. Asteroids: relationships with other small bodies. *Presented at* Asteroids, Comets, Meteors 1996. Plan. Spa. Sci. Submitted.
6. GEISSLER, P., J-M. PETIT, D.D. DURDA, R. GREENBERG, W. BOTTKE, & M. NOLAN. 1996. Erosion and ejecta reaccretion on 243 Ida and its moon. Icarus. **120:** 140–157.
7. D. MORRISON, C.R. CHAPMAN, & P. SLOVIC. 1994. The impact hazard. *In* Hazards Due to Comets & Asteroids, T. Gehrels, Ed.: 59–91. Univ. of Arizona Press. Tucson, AZ.
8. SULLIVAN, R., R. GREELEY, R. PAPPALARDO, E. ASPHAUG, J.M. MOORE, J. MORRISON D., M.J.S. BELTON, M. CARR, C.R. CHAPMAN, P. GEISSLER, R. GREENBERG, J. GRANAHAN, J.R. HEAD III, R. KIRK, A. MCEWEN, P. LEE, P.C. THOMAS, & J. VEVERKA. 1996. Geology of 243 Ida. Icarus. **120:** 119–139.
9. THOMAS, P.C., J. VEVERKA, D. SIMONELLI, P. HELFENSTEIN, B. CARCICH, M.J.S. BELTON, M.E. DAVIES, & C. CHAPMAN. 1994. The shape of Gaspra. Icarus **107:** 23–36.
10. THOMAS, P.C., M.J.S. BELTON, B. CARCICH, C.R. CHAPMAN, M.E. DAVIES, R. SULLIVAN, & J. VEVERKA. 1996. The shape of Ida. Icarus. **120:** 20–32.

Shoemaker-Levy 9 and Plume-forming Collisions on Earth[a]

MARK B.E. BOSLOUGH AND DAVID A. CRAWFORD

Department of Computational Physics and Mechanics
Sandia National Laboratories
Albuquerque, New Mexico 87185-0820

ABSTRACT: Computational models for the July, 1994 collision of comet Shoemaker-Levy 9 with Jupiter have provided a framework for interpreting the observational data. Imaging, photometry, and spectroscopy data from ground-based, Hubble Space Telescope, and Galileo spacecraft instruments are consistent with phenomena that were dominated by the generation of incandescent fireballs that were ballistically ejected to high altitudes, where they formed plumes that subsequently collapsed over large areas of Jupiter's atmosphere. Applications of similar computational models to collisions into Earth's atmosphere show that a very similar sequence of events should take place for NEO impacts with energies as low as 3 megatons, recurring on 100 year timescales or less. This result suggests that the 1908 Tunguska event was a plume-forming atmospheric explosion, and that some of the phenomena associated with it might be related to the ejection and collapse of a high plume. Hazards associated with plume growth and collapse should be included in the evaluation of the impact threat to Earth, and opportunities should be sought for observational validation of atmospheric impact models by exploiting data already being collected from the natural flux of multikiloton to megaton sized objects that constantly enter Earth's atmosphere on annual to decadal timescales.

INTRODUCTION

When pieces of the broken comet Shoemaker-Levy 9 slammed one-by-one into Jupiter in 1994, the overall physical consequences were out of proportion with even the most enthusiastic predictions. The fallout included a transformation of public perceptions about the impact threat to Earth, because the reality of planetary collisions no longer could be dismissed. Commentators described the event as a wake-up call, and it led some policy makers — for the first time — to take seriously the hazards from comets and asteroids. Some members of the scientific community also changed their opinions because the visible consequences of the collision were so much greater than anticipated. Proponents of detection and mitigation pursuits are now able to point to the impact on Jupiter when they make the case for an

[a]This work was funded by NASA and by the Sandia National Laboratories LDRD program. Sandia is a multiprogram laboratory operated by Sandia Corporation, a Lockheed Martin Company, for the United States Department of Energy under Contract DE-AC04-94AL85000.

increase in effort directed at discovering near-Earth objects, or for developing a means to protect our planet from them.

By contrast, the main scientific lesson learned from the devastation on Jupiter was not that large, fast objects collide with planets. That fact had been established decades earlier by analyzing the cratered planetary surfaces, and its importance had been brought home by more recent research into the geologic record of Earth's craters, impact catastrophes, and mass extinction of life. For scientists, the principal revelation from the crash of SL-9 was that the direct aftermath — especially the complex sequence of phenomena that followed — far exceeded our capability to predict. Fortunately, no lives depended on any decisions that were based on the preimpact predictions of the Jovian catastrophe. If the demise of this comet is to be taken as a warning, it should be that even the most sophisticated modeling and theoretical studies, by themselves, do not yield a framework of knowledge that is sufficiently reliable for making major policy decisions. Conversely, the data from observed impacts such as these, by themselves, cannot be fully interpreted and understood without the framework that can be provided by computational models. Models and simulations of phenomena that take place over many orders of magnitude in time, distance, and energy scales need to be checked against actual data. In the case of collisions into planetary atmospheres, we now have been able to take advantage of a validation "experiment" in the form of a rare impact on another planet.

Even though SL-9 provides the only real benchmark data for large-scale hypervelocity collisions with atmospheres, our understanding of such events will continue to improve along with our modeling capabilities. It is humbling to note that — even when extensive data are available — there can still be a wide variation in interpretations and quantitative assessments. Postimpact estimates of mass (or energy) have varied by several orders of magnitude among modelers who are considering the same set of astronomical imagery and lightcurve data. As the models begin to converge, the best estimates still vary over a wide range. This observation naturally leads one to question the quantitative accuracy of estimates of the consequences for our own planet of an impact of a given size. An obvious response would be that the observable phenomena that resulted from the impact on Jupiter are not the same type of phenomena that would kill people and cause property damage on Earth. For example, a Tunguska-scale impact on Earth might be a plume-forming, lightcurve-generating event, but the threat to surface-dwellers would be from the blast wave, which is easier to relate to the yield. Nevertheless, the impact of SL-9 underscores the fact that the most important effects resulting from a complex sequence triggered by a single event are not always those that are anticipated.

For this reason, we have begun to take a closer look at the Tunguska event by applying what we have learned from SL-9. The Tunguska explosion has also been attributed to an impactor with an estimated mass and yield that varies by several orders of magnitude. The most commonly quoted magnitude of the explosion (about 12 megatons) is still not universally accepted, and many researchers have argued for a larger or smaller value. By applying the simulation methods we used for SL-9, we are beginning to develop more insight into the energy deposition into the atmosphere at Tunguska. We have found that Tunguska is very much a terrestrial analogue of the SL-9 impact. It was entirely an atmospheric event, in which the Earth's surface played very little role other than acting as a "witness plate," covered with trees that would be knocked down and charred for later examination. Because atmospheric entry physics is similar on Earth, the Tunguska event probably generated a plume very similar to those that were seen on Jupiter. This paper reviews the impact of SL-9, interprets it in the context of the computational models, and presents Tunguska as a terrestrial example of the same process.

SHOEMAKER-LEVY 9

The multiple impacts on Jupiter in July 1994 provided a remarkable first opportunity to observe the immediate results of a hypervelocity collision between a comet and a planet. Astronomers and modelers have analyzed in detail the collision phenomena and aftermath, leading to advances in our understanding of comets, of Jupiter, and of the collisional processes that shaped the solar system. We are now beginning to take advantage of this improved understanding to develop better models for the assessment of the impact threat to Earth.

The SL-9 impacts provided a free "natural experiment" that we could exploit to validate computational simulations (involving velocities, masses, and kinetic energies many orders of magnitude higher than had ever before been witnessed), using the shock physics codes (CTH and PCTH) that had been developed at Sandia National Laboratories [27]. Traditionally, experiments or scaled tests are used to provide data to check the validity of a complex computational simulation against the real world. The natural astronomical event allowed us to use observational data for validating our simulations. As a spin-off, we were able to provide predictions to help guide astronomical observations of the event. These predictions led to our primary preimpact recommendation to observational astronomers: that the most useful Earth-based measurements to be made within a few minutes after impact would be time-resolved photometry at the Jovian limb [3]. More recently, we have used our models and simulations to assist astronomers and planetary scientists in interpreting the observational data.

Prior to impact, the computational effort was focused primarily on making predictions. In the period between the recognition in mid-1993 that SL-9 would strike Jupiter, and the availability of new astrometric data after Jupiter emerged from solar conjunction at the end of that year, the estimated point of impact was far on Jupiter's back side (the history of the preimpact orbital calculations is reviewed by Chodas and Yeomans[11]). There was little hope for a direct view from Earth, and many astronomers were resigned to the very limited possibilities of observing reflections from Jupiter's inner satellites, or of seeing residual atmospheric disturbances that would rotate into view several hours after impact. After the better orbital data revealed that the impact point for some of the brightest fragments would be less than six degrees over the Jovian limb, various modeling groups focused more attention on the fireball/plume phenomena that had the greatest potential for being observable. By the time the comet arrived at Jupiter, there was general agreement among the impact modelers that (for sufficiently large impactors), debris ejected by the collisions would quickly rise above the Jovian limb into line of sight for Earth-based telescopes [51], [42], [1], [2], [3], [41]. A consensus was never reached, however, on how bright the fireballs would be, how high the plumes would rise, nor on how large the fragments would need to be to generate visible plumes. The uncertainty in plume visibility as a function of fragment size, coupled with the uncertainty in the sizes of the actual fragments [48], resulted in a spectrum of recommendations from modelers. In most cases, the impact models focused on fragment entry and early plume growth.

Astronomers indeed observed the fireballs and plumes predicted by the models, and in doing so transformed them from theoretical constructions to observational facts. The actual event, however, produced a much richer array of consequences than anyone had anticipated. Because of a massive international effort, an overwhelming amount of high-quality observational data was collected during the impact week. Some of the new phenomena were quickly explained and are fully consistent with the models; interpretation of other observations continues to require further analysis and synthesis of the data. Computational models continue to provide guidance and contribute to our understanding of this event.

Moreover, the simulations of the SL-9 impact — coupled with interpretations of the observational data — will ultimately enhance our ability to predict the consequences of a comet or asteroid impact on Earth, leading to improved risk assessments. This section briefly summarizes the most useful observations of the collisions, provides a description of our simulations, and outlines our current "big picture" interpretation. The framework presented here is an elaboration on and update of the scenario presented by Boslough et al. [4]. It is consistent with the consensus that was reached at the 1995 IAU meeting in Baltimore (summarized by Chapman [10]), and incorporates elements of models of Svettsov [74], Crawford [12], Fortov et al. [61], Mac Low [23], Zahnle [52], and others. This scheme attempts to conform with as much of the observational data as possible, and incorporates the most recently published observational interpretations. As with any working model, the details of the interpretations may evolve as the research efforts continue.

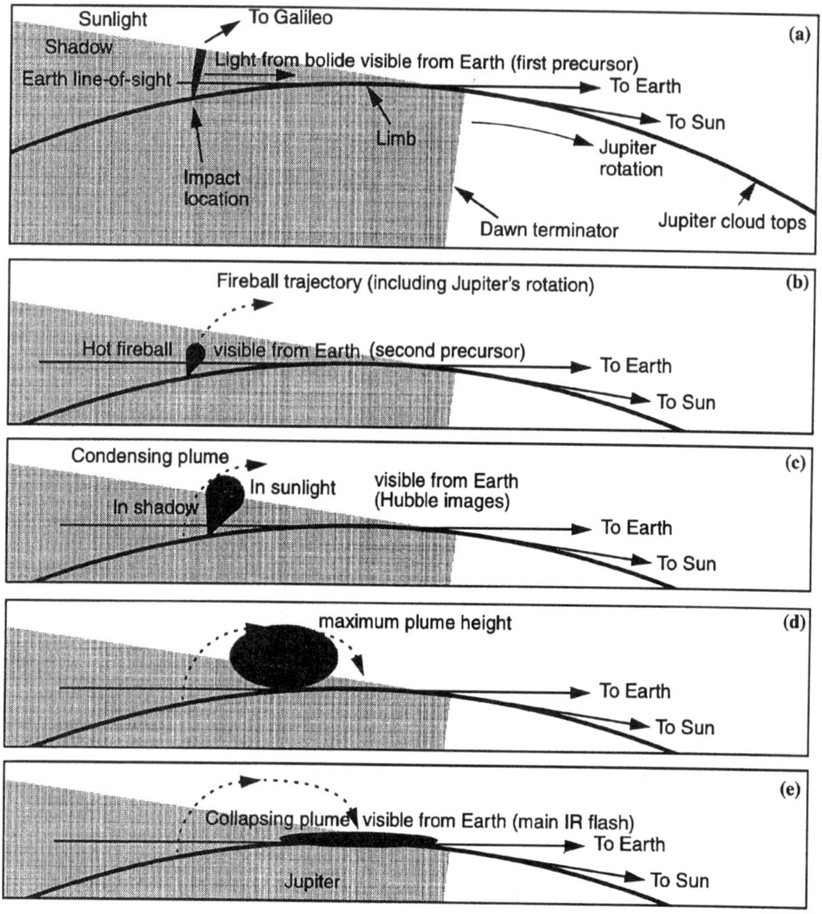

FIGURE 1. Geometry of impacts, just beyond Jupiter's limb (horizon) with respect to Earth: (a) Dust and main fragment enter Jupiter's atmosphere, generating entry flash; (b) incandescent fireball rises into view from Earth; (c) condensing plume ballistically rises into sunlight; (d) plume reaches maximum altitude of about 3300 km above clouds; (e) Plume collapses and rotates into view (modified from [3]).

SL-9 OBSERVATIONS

It turned out, by good fortune, that the geometry and timing of the series of impacts was nearly ideal for making useful observations from Earth-based instruments and from the Galileo spacecraft (Fig. 1). With the impact location only a few degrees beyond Jupiter's limb, as predicted by dynamicists, the hot debris ejected by each collision had to rise only a few hundred kilometers to become visible. It could then be seen in profile, making it possible to observe its shape and size. The vantage point from Earth was close to perpendicular to the trajectory of the fragments, so that the effect of impact obliquity could be seen. Because the impact point was beyond Jupiter's limb, the arrival time of debris above the line-of-sight altitude could be measured for some of the fragments. Combining this information with the time of impact extracted from direct measurements from Galileo (and in some cases from Earth), the early fireball trajectory could be determined. The position of Jupiter (near quadrature) meant that the luminous debris was in shadow when it first rose into view, making it possible to estimate its brightness. This configuration also meant that additional trajectory information could, in principle, have come from the time of arrival of the fireball into sunlight. As the cold, condensed plume rose into the sunlight it scattered enough light to be imaged so that its projected shape could be determined, and out-of-plane morphological information could be extracted from the shadow-line on it. Furthermore, each impact site was on the side of Jupiter (near local dawn) that immediately rotated into view from Earth. This gave the fireball a velocity component toward the limb, and made it possible to observe the pattern of debris and wave phenomena immediately after impact, and for an entire Jovian half-rotation (from local Earth-rise to Earth-set) afterward. Finally, there wasn't just a single collision. A whole string of about twenty sizable fragments impacted at a rate of several per day for nearly a week, allowing observatories around the world to collect data with many types of instrumentation, thereby providing multiple chances to improve the observations and to employ different modes of data collection. These multiple opportunities were particularly fortunate for the Hubble Space Telescope (HST), which is in low-Earth orbit and blocked from viewing Jupiter about half the time. Likewise, the Galileo spacecraft was able to take advantage of several different data collection strategies. This nearly best-case impact situation — like a well-designed series of experiments — has allowed many direct comparisons to be made between simulations and observations. It would not have been possible if the vantage point had been different, or if there had only been a single event.

The collision also took place at a time that was quite advantageous in terms of the instruments that could be used to make observations of the impact sites; the recently-refurbished Hubble Space Telescope (HST) and the approaching Galileo spacecraft were both available to complement the Earth-based observations in a way that would not have been possible only a year earlier. HST happened to be in position to capture direct images of the plumes from four impacts, and Galileo had a direct line of sight for all. HST images have provided information about the size, shape, evolution, and optical properties of the fireballs and plumes, and about the structure and evolution of the impact sites after plume collapse [19], [63]. HST and Earth-based spectroscopy yielded evidence for water, ammonia, iron, silicon, magnesium, sodium, calcium, lithium, potassium, diatomic sulfur other sulfur compounds, carbon monoxide, and hydrogen cyanide at the impact sites [36], [55], [76], [24], [26], [39]. Galileo's Ultraviolet Spectrometer (UVS), Photopolarimeter Radiometer (PPR), Near-Infrared Mapping Spectrometer (NIMS), and Solid State Imaging camera (SSI) instruments provided precise timing for the entry of many of the fragments into the atmosphere, as well as information about the expansion rates, cooling histories, and upward velocities of the incandescent, ballistic fireballs [26], [6], [9], [20]. The event timing as determined by Galileo measurements has provided a strong basis for interpreting the multiple flashes observed from Earth in a way that is consistent with the computational models (Fig. 2).

PREIMPACT SL-9 SIMULATIONS

Crawford et al. [14] used the CTH Eulerian shock-physics code to simulate two- and three-dimensional representations of the impact events. The 2D computations were of the impactor penetration, simulating the entry, deformation and breakup of the comet fragments. The calculations were performed in a "reverse ballistic" sense using a Jovian atmosphere moving upward at 60 km/sec impinging upon an initially stationary fragment. The Eulerian mesh extended 100 km radially and 1000 km above and below the comet. The fragment was maintained in a high resolution portion of the mesh (equivalent to 25 computational zones across the projectile radius and extending 10 km vertically and 5 km radially) by Galilean transformations of the entire mesh every 0.1 seconds of simulation time. Zone size gradually increased away from the high resolution portion of the mesh to preserve all the materials of the calculation yet maintain computational efficiency.

The comet fragments in the simulations were composed of water ice with initial density and temperature of 0.95 g/cm^3 and 100 K, respectively, using tabularized version of the ANEOS equation-of-state which allows melting and vaporization [46]. The atmospheric stratigraphy in the calculations matched Voyager data for Jupiter at high altitudes (Orton, unpublished data) and extended adiabatically to lower altitudes. The atmosphere consisted of 89% hydrogen and 11% helium at all altitudes and was modeled with a tabular equation-of-state allowing dissociation and ionization [21]. It was scaled vertically by a factor of 1.41 (to account for the approximate 45° entry angle) and inserted into the lower portion of the computational mesh. The atmosphere propagated into the upper portion of the mesh as the comet deformed and broke up in the higher-pressure regions of the lower atmosphere.

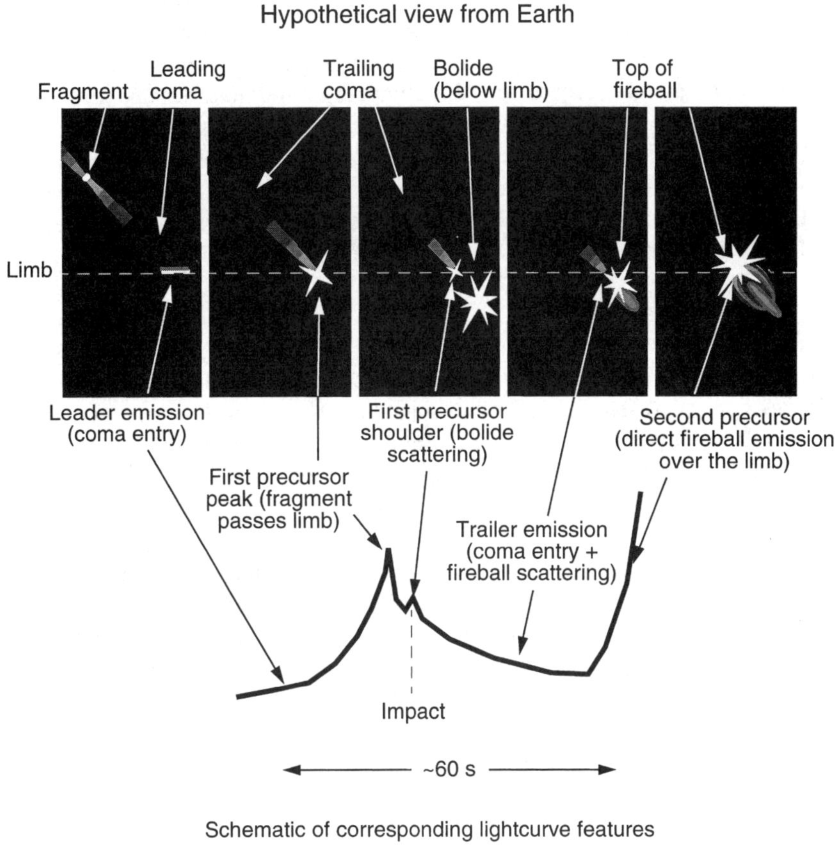

FIGURE 2. Cartoon of the first minute of impact phenomena as observed from Earth, relating events at the limb of Jupiter to an idealized high-resolution lightcurve. Only light radiated or scattered from above the limb was detected on Earth.

Before the impacts, we focused our attention on modeling large fragments with diameters between 1 and 3 km. This is the size range bracketed by the most optimistic preimpact estimates (summarized by Weaver et al. [48]), and would be more likely to yield observable phenomena than smaller fragments. During entry into the low density outermost reaches of the atmosphere, the projectile forms a clean bow shock. Atmospheric temperatures at the leading edge of the projectile reach values as great as 35,000 K. During deformation, the projectile thins and the leading edge flattens. Acceleration instabilities develop in a manner described by Swegle and Robinson [43]. Eventually, projectile thinning meets with the growing instabilities and breakup occurs. During penetration, the projectile continuously gives up kinetic energy to heating and deflection of the Jovian atmosphere (a relatively small amount goes toward internal heating of the cometary constituents).

Crawford et al. [14] determined the total energy deposited by hypothetical 1, 2, and 3 km diameter cometary fragments during their penetration of the Jovian atmosphere. An important result for fragments between 1 and 3 km in diameter was that most of their kinetic energy and mass was deposited beneath Jupiter's outermost visible cloud layer, which is about 10–20 km above the reference altitude at 1 bar. For the 3 km diameter fragments most of the impactor's mass is deposited at depth, and less than 1% is entrained in the upwardly growing fireball. Crawford et al. [14] also investigated the influence of fragment body shape on the energy deposition profile, and tested for sensitivity to numerics by performing the same calculations with different resolutions.

For the 3D calculations, the results from the two-dimensional entry, deformation and breakup studies were inclined at 45° and mapped into a three-dimensional representation. Three-dimensional, bilaterally symmetric simulations most accurately rendered fireball evolution beginning about 10–15 seconds after first contact of the fragment with Jupiter's atmosphere. Density, temperature, fluid velocity, and pressure of the cometary debris and shocked Jovian atmospheric constituents were preserved in a spatially average sense while total energy was conserved. Some calculations were allowed to evolve for up to 120 seconds. Generally, the simulation results indicated that early-time fireball growth is predominantly directed outward along the incoming bolide trajectory but is redirected, at later time, towards growth dominated by the vertical gradient of the Jovian atmosphere.

In an attempt to attain adequate resolution for these large fireball simulations, the calculations were performed on the 1840-processor Intel Paragon massively parallel supercomputer at Sandia National Laboratories. The simulation of the fireball formed by the impact of a 3 km comet fragment consisted of 8 million 5 km cubical zones. Lower energy events, formed by the impact of 1 or 2 km fragments, were modeled with more finely resolved, but less spatially extensive, simulations. More recent 2D simulations indicate that these preimpact 3D simulations were somewhat under-resolved, despite having been carried out on the world's most powerful computer. The qualitative aspects of the simulated fireballs and plumes remain valid, but the lack of sufficient resolution limited their quantitative accuracy, and under-predicted the ejection velocity of material during early fireball growth.

In the 3 km simulation, the fragment deposited (as internal energy of H_2, He and H_2O vapor) more than 95% of its kinetic energy (6 million megatons) during its penetration of the Jovian atmosphere (a comparatively small amount remained as kinetic energy of cometary water vapor). The resulting fireball and surrounding shock wave is shown in cross-section, 67 seconds after impact, in Crawford's Fig. 1 [58]. The spherical shock wave is advancing upward at a velocity of 25 km/sec. It has reached a diameter of 700 km and an altitude of 900 km above the clouds. For reference, the Jovian cloud tops are located at an altitude of 10–20 km and the projection of the limb of

Jupiter (as seen from Earth) varies as a function of impact location. The limb projections for impacts occurring 4 and 6 degrees beyond the limb were at 200 and 400 km altitude, respectively. The fireball itself is a rapidly rising cloud of cometary debris and Jovian atmosphere at high temperature. About a minute after the impact of a 3 km cometary fragment, the fireball is still at 1700 K, and the shock wave temperature is 2300 K. An optically-thick fireball would have had an apparent bolometric magnitude (as viewed from Earth) of about 2 at this time. The observed fireballs were significantly dimmer, implying they were not optically thick and/or that the fragments were smaller than 3 km in diameter.

The term "fireball" refers to the mass of hot gases consisting of a mixture of Jovian atmosphere and cometary material that is ballistically shot upward by the impact. In the first moments after impact it is very hot, incandescent, and radiating in the visible and near infrared. The fireball is preceded by the "entry flash" or "bolide" phase, during which time the comet fragment (and associated debris) deposits its energy in Jupiter's atmosphere. The entry flash as seen from Earth may be dominated by the deposition of energy at very high altitude (above the limb) by small particles in the coma surrounding the main fragment, and for that reason its use as a timing fiducial is not as straightforward as we had hoped [2]. The main fragment mass does not begin to deposit a large amount of energy until it has passed beyond the limb, after which it begins to heat an entry column of atmosphere as it loses its kinetic energy. It is that column of gas that explosively expands and becomes the fireball. The word "plume" describes the debris cloud after it has expanded, cooled adiabatically, and begun to condense. Clearly, there are no precise temporal demarcations separating bolide, fireball, and plume phases, but various well-defined "events" associated with these phases, and related to the viewing geometry, can be extracted from the best time-resolved lightcurves (Fig. 2).

The differences between the fireballs described here, and other closely-related phenomena should be understood. Analogies to the fireball associated with the detonation of a nuclear device are limited. The development of a nuclear fireball is dominated by interior radiative transport at temperatures of tens of millions of degrees. Some fraction of this energy forms a shock wave in the atmosphere, which separates from the fireball but can still be luminous if strong enough. The shock wave generated by the impact fireball is similar to the outer, mechanically-driven nuclear blast wave. However, the temperature of a shock wave can be much higher at a given propagation distance from an impact that is a million times more energetic than a megaton-scale nuclear device. The fireball itself is a ballistically-rising mixture of shocked atmosphere and vaporized cometary material. A nuclear fireball that is small compared to the scale height of the atmosphere will be driven upwards by buoyant forces because it is less dense than the surrounding atmosphere. This upwelling gives rise to the toroidal "mushroom cloud" usually associated with atmospheric explosions. A large impact fireball can be much greater than the scale height of the Jovian atmosphere. Because the atmospheric pressure is much greater at the bottom than at the top, it is contained at depth and relatively uncontained at altitude. It, therefore, accelerates upwards as if shot from a gun. Simulations and observations indicate that later, when most of the fireball mass is unconstrained, it begins a more homologous upward expansion. Even though its density is much greater than the surrounding atmosphere at the top, its inertia will carry it on a ballistic trajectory which rises as much as several thousand kilometers above the clouds.

POSTIMPACT SL-9 SIMULATIONS

Subsequent to the collision, Crawford [12] has improved the computational models by developing an analytical entry model that allows the numerical simulations of the fireballs and plumes to be initialized without having to perform a computationally-intensive hypervelocity entry simulation every time. This paper showed that previously-ignored radiative ablation makes a significant contribution to the rate at which energy and mass are deposited at high altitude, which controls the formation and early evolution of the fireball. By performing a large set of two dimensional simulations, Crawford [12] called into question our earlier interpretation of a fireball that transforms directly to a plume as it grows, cools, and condenses. Instead, the early, incandescent "entry fireball" grows from and expanding entry column, fades, and becomes transparent before it reaches an altitude of 6,000 to 10,000 km. Imbedded within the fireball, however, is material excavated from deeper within Jupiter's atmosphere, which condenses to form the visible outer surface of the plume that rises up to about 3,000 km.

One key result of this paper was an explanation for why all the observed visible plumes rose to about the same height. Both the energy deposited per unit distance along the path, and the mass into which it is deposited, are proportional to a projectile's interaction cross-section. The energy density within the entry column is therefore independent of fragment size down to the depth at which its velocity begins to appreciably drop off. Since the expansion of the hot vapor in the column converts internal energy of each packet of mass to gravitational potential energy, the maximum height each packet reaches is independent of fragment size, as long as it comes from above the level at which the fragment begins decelerating. If this level is beneath the cloud layer that forms the visible envelope of the plume, the plumes will all appear to rise to the same height. The threshold diameter above which fragments will penetrate the ammonia clouds before slowing appreciably yields the lower limit for the size of the fragments that generated the high plumes. According to Crawford [12], this lower bound is about 200 m.

The most recent computational models [58] have provided another means for estimating the size of individual comet fragments, by determining light emission from the impacts as a function of time and viewing geometry. By varying parameters and comparing the synthetic lightcurves to those recorded by instruments aboard the Galileo Spacecraft, the sizes of five fragments were estimated. These values were correlated with HST images and Earth-based photometry to arrive at approximate sizes for the other largest fragments, which in turn allowed an overall progenitor diameter to be determined assuming a density for SL-9, based only on the impact flash. For densities of 0.25 and 0.50 g/cm^3, the best estimates are 1.8 km and 1.4 km, respectively.

SL-9 INTERPRETATIONS

Figures 1–4 are schematic portrayals of the sequence of events associated with an idealized SL-9 impact. These representations are inferred from the composite of Earth-based photometry data, Galileo lightcurves, and HST imagery. Figure 1 depicts a side view, looking from a direction perpendicular to the line between Jupiter and Earth. Figure 2 is a cartoon depicting the view from Earth during the first minute or so, with the features of a generic high-resolution lightcurve identified with various events at the limb. The lightcurve is modeled after the time-resolved data collected at 2.3 μm from the L impact by Hamilton et al. [64] at Calar Alto Observatory, Spain. Figure 3 is a plan view of the impact site from a stationary (nonrotating) vantage point, with snapshots of a map projection of the evolving impact sites at various time steps after impact. As the point of impact rotates from west to east, it moves from left to right in the stationary field of view of the illustration. Jovian north is up; the approximately vertical lines represent the minimum line-of-sight altitudes to the Earth and Sun. The figure is not intended to depict the exact geometry, nor is it supposed to represent a particular impact, but is a composite of features observed from various events. Figure 4 shows a simplified side-view of the fireball/plume evolution. In reality, the ejecta cloud is not a discrete packet, but a continuum with widely varying temperatures, densities, and pressures. As described earlier, the impacts were not "clean experiments," but probably involved closely-spaced multiple impactors embedded within a dusty, light-scattering cloud of smaller particles (coma), which also had a hypervelocity, light-producing interaction with Jupiter's atmosphere.

In the Fig. 3 inset are some idealized examples of Galileo and Earth-based lightcurves. The Galileo PPR curve is based on the measurements of several impacts at 945 nm [26]. Since Galileo had a direct view of the impact, the sharp rise in the PPR curve defines the time of impact for the events for which PPR data were collected (UVS and SSI instruments on Galileo determined timing for some of the other impacts; see Chodas and Yeomans [11] for precise timing information). The upper Earth-based lightcurve is a schematic adaptation of data at 3.5 μm collected at the Palomar Observatory by Nicholson et al. [35] for the R impact; qualitatively similar curves were obtained by Graham et al. [18] at 2.3 μm with the Keck Telescope, and for the K impact at 2.35 μm at the Okayama Astrophysical Observatory by Takeuchi et al. [45]. The lower curve is based on 10 μm data collected at the European Southern Observatory by Livengood et al. [22].

The following is a generic description of the sequence of events, with letters corresponding to those in Figs. 3 and 4. The various phases are defined primarily for conceptual purposes; there are not distinct boundaries between them, and in many cases they overlap. Approximate times are given relative to fragment impact time (defined as its arrival at the 100 mbar level by Chodas and Yeomans [11]. These descriptions relate the phenomena associated with each phase to corresponding observational data that were collected from at least one of the impacts, acknowledging the fact that a different suite of instruments was used for each impact, and that some impacts were too weak to be observed. In reality, every impact has its own unique set of data.

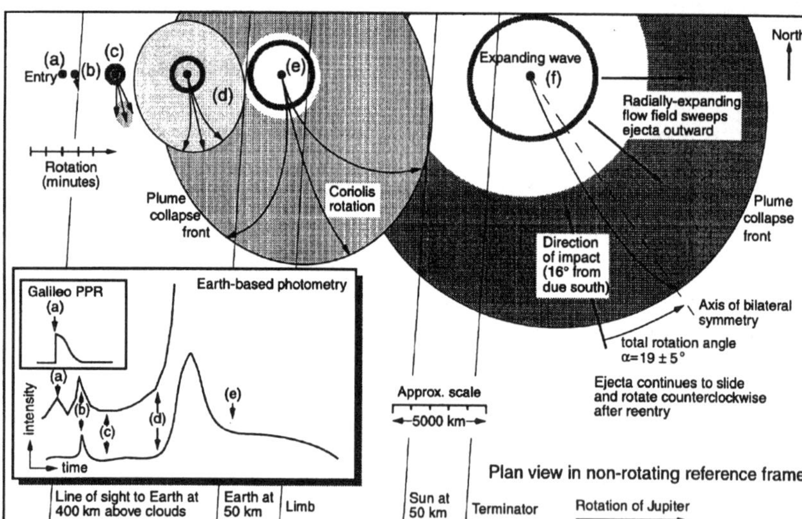

FIGURE 3. Plan view (map projection) of idealized impact site from a stationary (non-rotating) vantage point, with snapshots of a planar projection of its evolution, interpreted using the conceptual framework provided by computational simulations. This figure schematically represents features that were seen after several of the larger impacts. See text for detailed explanation (from [4]).

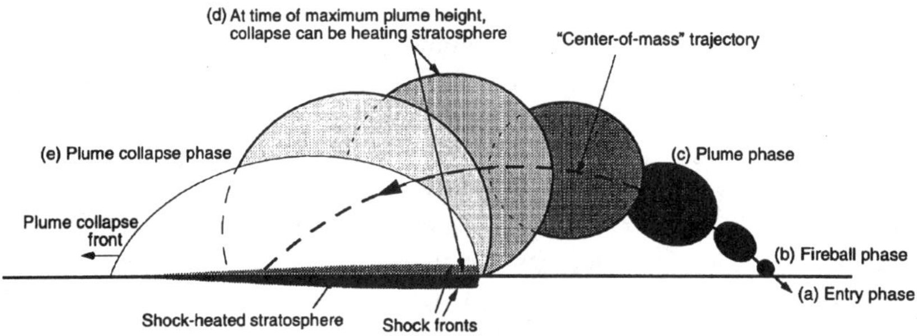

FIGURE 4. Side view of highly idealized fireball/plume evolution which leads to a "hypervelocity splat" when the plume collapses, heating Jupiter's upper atmosphere over a very large area. This schematic shows the ejecta cloud as a discrete packet, rather than the continuum it really is (modified from [4]).

A. Entry Phase (Preimpact to 10 or 15 Seconds)

A fragment (or tightly-spaced cluster) enters Jupiter's atmosphere, depositing energy and leaving a debris column that consists of a mixture of Jovian atmosphere and cometary vapor at high temperatures and pressures. Thermal radiation from this column is dominated by the very high temperatures at the front of the bow shock before it drops beneath the uppermost cloud layer, and is seen directly in the ultraviolet, visible, and infrared wavelengths by Galileo's detectors, which recorded color temperatures of around 8,000 K during this first part of the entry phase for the G impact [20]. The temperature of the column at a given point along the trajectory of the fragment begins to drop as soon as the energy is deposited into the bow shock, and it immediately starts its expansion. As the entry phase continues beneath the clouds, where kinetic energy of the fragment is converted to internal energy, the upper part of the column begins its expansion into an incipient fireball. The emission during the entry and early fireball phase, as viewed from Earth, was complicated by viewing geometry and scattering effects. But since time-resolved data were collected with a direct view by Galileo's instruments, they can be compared directly to theoretical lightcurves to provide estimates of fragment sizes. This aspect of the data is thoroughly discussed by Crawford [58].

In some cases the onset of the emission seen from Earth was not sudden, as would have been expected for a single fragment, but exhibited a gradual increase in intensity (the "leader" [35]). The best explanation for the leader is the hypervelocity collision of the leading part of the coma into Jupiter's upper atmosphere. Because the small particles that make up the coma are so far ahead of the main fragment, they begin to impact Jupiter many seconds — or even minutes — in advance. McGregor et al. [70] determined that the time-resolved infrared emission at 2.34 μm, from the vicinity of the limb just above the impact point of G and K fragments, began to rise above background levels at about 1.5 and 3.5 minutes before fragment impact, respectively. This implies that localized meteor storms were becoming bright enough to be seen from Earth when these two fragments were still at respective distances greater than 5,000 and 12,000 km from their point of impact (Fig. 2). The 2.3 μm lightcurve [64] shows a leader beginning about 30 seconds before the peak, when the L fragment was still nearly 2,000 km above the cloud tops. These distances are much less than the total length of the comae, which had been stretched out to lengths tens of thousands of kilometers in their approximate direction of motion as the fragments neared Jupiter due to differential acceleration by the planet's gravitational gradient [48].

The shape of the first precursor probably has more to do with the density profile of the stretched-out coma than with anything else. The Earth-based data [64] were collected from the L impact with sufficiently small sampling intervals (1.2 seconds) to resolve the structure of the coma on a spatial scale of less than 100 km as it collided with the rarefied upper reaches of Jupiter's atmosphere at 60 km/sec. At any instant in time, the emission from the entering coma as seen from Earth is dominated by meteors within a scale height of the line-of-sight altitude above the limb (about 350 km above the clouds for the L impact) so the time-resolved emission is proportional to the number of meteors passing through that altitude, assuming a uniform size distribution. The main peak in the L lightcurve has a half-height width of about 12 seconds, which is consistent with a debris cloud having a density concentration within about 700 km along the direction of motion. This peak appears 8 seconds before the impact time as determined by Galileo instruments [26], precisely when the main fragment was passing the limb at about 350 km altitude. A second, smaller peak in the first precursor of the L lightcurve appears within one sample interval of the Galileo impact time, when the bolide is at its brightest just before reaching the cloud tops. This short, bright entry flash would not be visible from Earth unless there were still scattering material above the limb at the same instant. The fact that the first peak had not yet disappeared by this time implies that the densest inner part of the trailing portion of the coma was still crossing the limb at the time of impact, providing a scattering medium in Earth's line of sight.

The "trailer" of the first precursor probably contains contributions from both the trailing portion of the coma generating high altitude line-of-sight meteors, and the same portion of the coma scattering light emitted from the fireball rising from beneath. There are many complications due to multiple sources and dependence of scattering and emission on the details of the unknown density profile of the coma in the immediate vicinity of the main fragment. Thus, the structure and intensity of the first precursor in Earth-based lightcurves cannot be used to determine fragment properties. Moreover, only the L lightcurve [64] is sufficiently well resolved to distinguish the limb-crossing coma peak from the bolide flash peak. In many of the lightcurves, these two events are unresolved, and together form the "first precursor." In other lightcurves, no first precursor was observed at all. A reasonable explanation is that in addition to depending on wavelength, sampling interval, and sensitivity, its presence depends on coma density and location of impact point beyond the limb. The latter is important because it determines the altitude and density of the atmosphere into which meteors form in line of sight from Earth, and dictates the distance between the bolide and the scattering medium.

B. Fireball Phase (5 or 10 Seconds to 3 or 4 Minutes)

After energy is deposited in the wake of the fragment, the column explosively expands upward and outward along the atmospheric density gradient, cooling isentropically as it rises. The expansion begins instantaneously, before the entry phase is complete. This cooling is seen by Galileo instruments as a decrease in radiative intensity, and a shift toward longer wavelengths in thermal emission. In the infrared, the NIMS data exhibit a decrease in the intensity of methane absorption bands with time for the G fireball, consistent with a rapidly rising light source within Jupiter's atmosphere. Carlson et al. [6] used these time-dependent absorption band changes to determine that the vertical velocity component of the main near-infrared light source from the G fireball had reached 2–3 km/sec within 20–30 seconds of impact.

Computational models show that the fireball growth is not a simple isotropic expansion, but that it starts near the tropopause and is guided up the high-temperature, low-density, radially expanding entry column [12], with an appearance reminiscent of the inflation of a long narrow balloon. In less than one minute or so, the incandescent fireball rises to a few hundred kilometers and becomes visible from Earth, appearing as a second precursor in some photometry data. The exact timing depends on both the point of impact and on fragment size although the most recent models suggest that, over the range of sizes of the fireball-producing SL-9 fragments. the latter dependence is weak [58]. The entry fireball is preceded by several seconds by a shock wave. Earth-based detection of this shock would have provided further validation for the computational models, but it was either too weak to have been seen as an independent precursor, or else it was simply unresolved. In the absence of a nearby light-scattering medium, the arrival time of hot material above the Jovian limb would have been seen as a sudden increase in the lightcurve. In many cases the second precursor was not sharp, suggesting that the sudden fireball arrival was preceded by a more gradual increase in light scattered over the limb. The most likely source of light is from the rising fireball itself, before it reaches the altitude where it can be seen directly. The most probable light scattering material that would be above the limb at this time (and close to the light source) would be the trailing part of the coma. Some of the lightcurves exhibit sharp second precursor onset — such as is the 2.3 μm L data [64]. At this time, the rising fireball approaching the limb is colliding head-on with the descending coma that had been trailing the L fragment by about 2,300 km. After this early fireball rises above the limb, it continues to grow and cool, leading to a trailing off of the second precursor.

Information about the Earth-based field of view during the entry and early fireball phases for various impact locations can be extracted from Fig. 5. The line-of-site altitude from Earth at a spot directly above the impact point is plotted as a function of time for various fragments. As the rotation of Jupiter brings the impact point closer to the limb, the line-of-sight altitude moves lower, so the curves (provided by Paul Chodas, personal communication) are decreasing functions of time. We originally used this diagram to argue for the visibility of fireballs by plotting ballistic trajectories extrapolated from our 3D shock physics simulations [2], and showing that for impactors greater than 1 km in diameter, fireballs would rise above the line-of-sight altitude within a few minutes. More recent, higher-resolution 2D simulations by Crawford [12] has revealed that our original 3 km fireball trajectory was about right, but that our 1 km trajectory was too low due to insufficient numerical resolution. Furthermore, these simulations demonstrated that plumes from impactors greater than about 200 m in diameter should all rise to about the same height, so the ballistic trajectories would all be about the same. Their time of arrival above the limb would therefore be insensitive to fragment size for a given impact geometry, in contradiction to our expectations [2]. When the calculated trajectories are plotted on this diagram, their intersection with the line-of-sight curves should define the times of arrival of the plume-forming surface above the limb. If that same surface evolved from the leading edge of the incandescent fireball, then the beginning of the second precursor (in the absence of scattering) should appear at this time. However, the second precursor start times extracted from a compendium of measurements [11] do not show any clear correlation. This inconstancy may be the result of a combination of light scattering and limited temporal resolution.

The Hamilton et al. [64] lightcurve at 2.3 μm for the L impact provides the best published example of an emergent second precursor with good time resolution. This precursor starts about 36 seconds after impact, which is about 14 seconds before the visible plume-forming surface reaches line-of-sight. Our early work suggested that it was the expanding debris front that defined both the leading edge of the incandescent fireball at early times, and the top of the plume at late times [4]. This L lightcurve, however, strongly supports the Crawford [12] view that the top of the incandescent "entry" fireball precedes the visible plume to that altitude, and later becomes transparent, with the visible plume embedded inside.

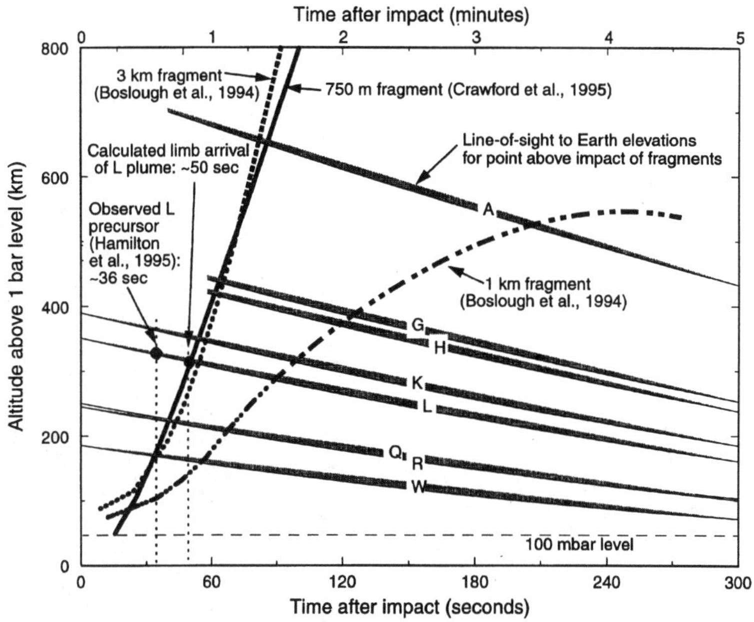

FIGURE 5. Calculated trajectories of fireballs from 3D simulations for 1 and 3 km diameter ice fragments [3], and for a 750 m diameter fragment [12]. Approximate line-of-sight elevations to the Earth are based on the impact location for several fragments (identified by letter).

C. Plume Phase (3 or 4 Minutes to 10 or 15 Minutes)

The debris continues to rise ballistically. It expands and cools, and begins to condense. Crawford [12] favored the idea that the outer surface of the visible plumes was mapped from a horizontal surface near the cloud tops of the preimpact atmosphere. In this view, the visible surface of the plume is a condensation surface derived from the Jovian clouds. In the most recent models [58], it appears that the surface translated by the explosion from a region near the tropopause is accompanied by a steep density and temperature gradient. The opacity that gives definition to the plumes may be due to a threshold density of particles of condensation. This would imply that the visible plumes were rather diffuse, and supports the view that they were actually nested within a transparent plume that rose to much greater altitudes. The outer, transparent layers would be derived from the debris in the upper part of the entry column which defined the hot, incandescent entry fireball. When the visible plume reaches an altitude greater than one or two thousand kilometers (depending on the point of impact), it enters sunlight. We had anticipated that this would be detectable from Earth, and would provide another point on the ballistic trajectory [3], but the precise sunlight arrival time was not obvious from the lightcurves. The plume's emergence from Jupiter's shadow, however, enabled it to be imaged for several impacts by the HST [19]. A sequence of HST images of the growth and collapse of the G plume over a period of about 20 minutes is shown in Fig. 6, along with the results of a 3D simulation of fireball/plume evolution after the impact of a 3 km diameter fragment.

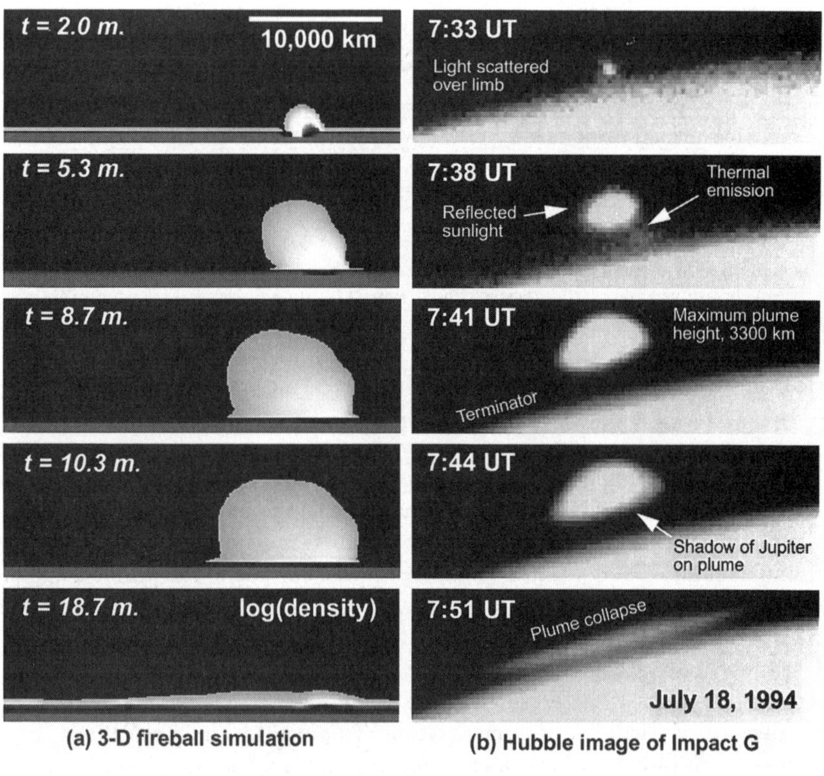

FIGURE 6. (a) Simulation of 3D fireball/plume evolution after the impact of a 3 km diameter fragment. Shading indicates log (density) with a cutoff at 10–12 g/cm^3; times are in minutes after impact (from [15]). (b) Sequence of G plume images collected by Hubble Space Telescope (from [19]).

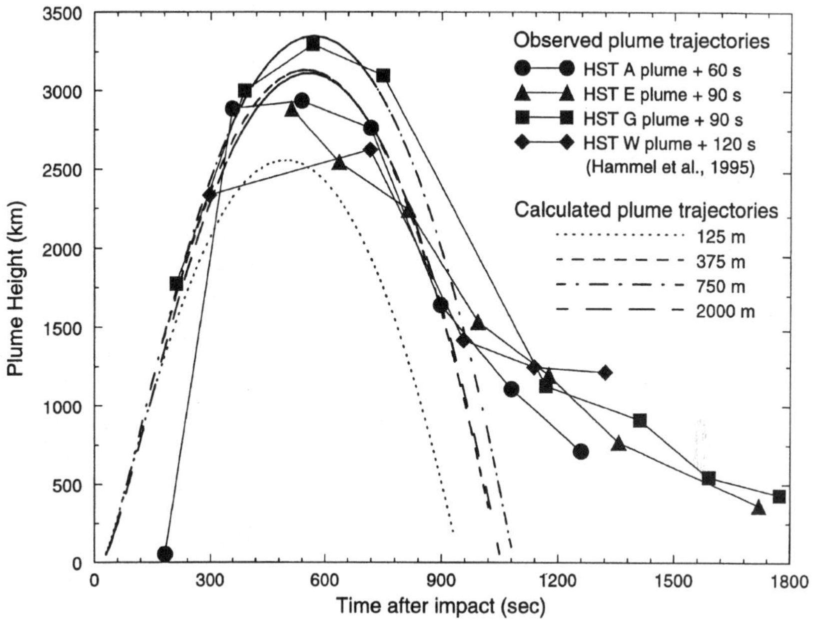

FIGURE 7. Plume heights determined from the Hubble Space Telescope as a function of time [64] plotted along with plume heights determined from a ballistic extrapolation of material ejected by impactorts of various sizes.

D. Maximum Height (about 9 Minutes)

At six minutes after impact, the plume is still growing to its maximum height, according to HST images. At this time, the lower part of the plume begins to collapse and heat Jupiter's stratosphere to a sufficiently high temperature over a large enough area that its emission begins to appear as the "main event" in both Earth-based [18], [34] in and Galileo NIMS [6] lightcurves. Because the plume is still expanding as it rises, the collapse phase can actually begin before the top of the plume reaches its maximum height (Fig. 4). Figure 7 plots the altitude of the highest tracer particle derived from the 1 bar level, as a function of time, for impactors of four different sizes simulated by Crawford [12]. This figure demonstrates how insensitive the maximum plume height is to fragment diameter once a threshold about 125 m is reached. Even though the plumes were ejected along the entry track at an angle of 45 degrees, they also expanded laterally away from the track, and upward, so the that tops of the visible plumes were not comprised of the material that was ejected at highest velocity with a horizontal component. Since the top of the plume is defined by a slowly rotating surface, its position is not described by a single Lagrangian point on a ballistic trajectory; however, the difference is minuscule. The plume heights determined from HST images [19] are also plotted in Fig. 7, with the time scales shifted to optimize their alignment with the calculated trajectories. The differences between the HST plume heights and the calculated ballistic plume height are probably attributable to a combination of variable sensitivities and wavelengths, as well as exposure times that were long compared to the timescales of plume motion.

E. Plume Collapse (Splashback) and Transition Phase (about 5 to 15 Minutes)

As the still-expanding debris cloud begins to fall back, it compresses and heats a large area of the Jovian stratosphere. The heated region grows rapidly. The peak of the main event in the Earth-based lightcurves is probably determined by a combination of competing effects, including increasing area, radiative and decompressional cooling, and viewing geometry. However, Nicholson [35] and others have pointed out that the data of McGregor et al. [70] and Lagage et al. [69] demonstrate that the main event began about six minutes after impact in all cases, regardless of viewing geometry. The fact that this six-minute interval before the main event is an intrinsic property of the plume collapse was later confirmed by the Galileo NIMS data [6]. The most reasonable explanation for a six minute interval has been provided by Zahnle [52], who suggested that there is a threshold vertical velocity component that the re-entering plume must reach before the shock temperature is high enough to synthesize the organic compounds that serve as the required opacity source for emission. This threshold velocity is about 4–5 km/sec, which is reached after about six minutes.

Another transition takes place near the peak of the main event, and was observed as an emergent CO emission band 12 minutes after the K impact in the spectral data at 2.3 μm by Meadows and Crisp [79]. Excitation of this band requires a temperature of at least 2000 K, which implies that it must be coming from the shocked layer where the plume is collapsing onto the top of Jupiter's atmosphere. Nicholson [35] attributes its sudden appearance to the time at which the vertical component of the plume re-entry velocity approaches 9 km/sec. That explanation, however, may not completely account for the suddenness of its onset, which may be alternatively explained by a morphological transition associated with the plume collapse process (Fig. 8). When the plume is growing, it consists of nested shells of Jovian atmosphere derived from various initial depths. As it begins to collapse, the outer shells collide with Jupiter's atmosphere first, and a zone of inverted stratigraphy begins to grow laterally uprange from the point of impact. These collapsed layers maintain their horizontal velocity and slide outward before losing their kinetic energy by dissipative processes. Since the collapse zone is shocked, it radiates at a much higher temperature than the surrounding plume, but because the outermost visible shell of the plume has some degree of opacity, this source is obscured. As the plume continues to collapse, a time is reached after which the outer shell folds back upon itself and remains upright, and the collapse front begins to recede from the outer limit of the collapse zone, exposing it to view from outside and above. This is the time at which observers could expect to see a rapid increase in radiation from the collapse zone, and might explain the rapid onset of the hot CO emission band.

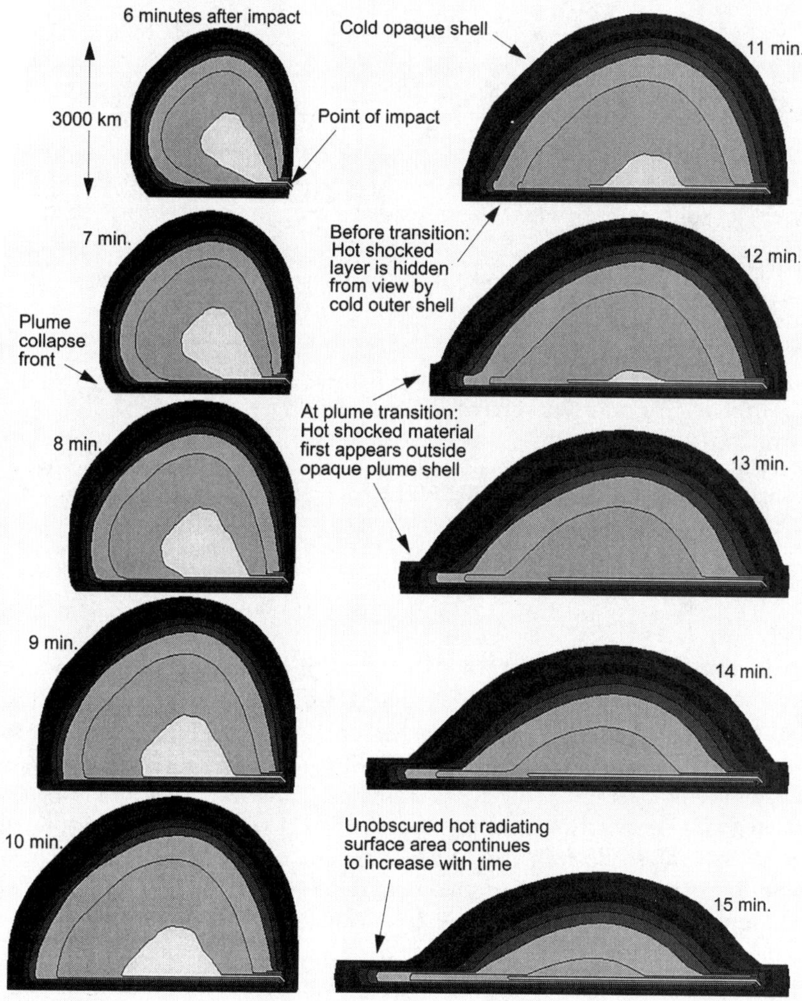

FIGURE 8. Rotated ballistic extrapolation of the ejected shells of material from the Crawford [12] 2D fireball simulations of a 1 km fragment, showing that a morphological transition takes place between 10 and 11 minutes after impact, when the leading edge of the collapse front begins folding back on itself, exposing high temperature shocked material to view from above. The collapse zone is vertically exaggerated in this figure to show how the atmospheric layers that re-enter before the transition are inverted.

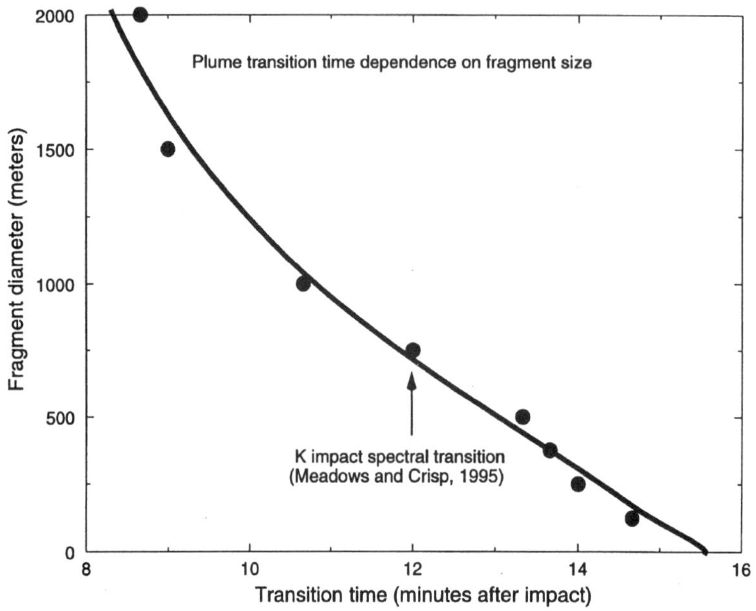

FIGURE 9. Transition time plotted versus fragment size, based on ballistic extrapolations of rotated 2D simulations. This dependence is primarily due to the variation of plume shape with impactor size, but the actual numerical relationship between diameter and transition time may depend on 3D effects and on precise position of the outermost opaque layer within the plume.

It is notable that this morphological transition time has a strong dependence on plume shape, which in turn depends on fragment sizes. The larger fragments produce wider plumes than the smaller ones [12], which is apparent in the HST images (the G plume is significantly wider than the W plume, for example [64]). Figure 9 was generated by ballistically extrapolating the ejected shells of material from the Crawford [12] series of 2D simulations, in a rotated reference frame to account for the oblique impact angle. The plume transition times were plotted versus fragment diameter. It would be premature to use this curve with high confidence to determine fragment diameters from observed transition times, because so many assumptions (e.g., origin depth of outer opaque layer) and approximations (e.g., rotated 2D simulations) were involved. However, there is a strong dependence of transition time on diameter, which suggests that a refined curve based on 3D simulations will ultimately be useful in providing an independent determination of fragment sizes. It is encouraging that the observed K transition time of 12 minutes yields a fragment diameter of 750 m, within the uncertainty of the estimate based on synthetic lightcurves [58].

After the primary peak in the main event appear subsequent shoulders and smaller peaks. In the infrared lightcurves of Nicholson et al. [34] and Graham et al. [18], these arise about 10 minutes after the main peak, and are almost certainly due to a much smaller, second compression of the atmosphere after the plume rebounds from the first collapse with a 10 minute period [16], [23]. This is followed by a third, and perhaps even a fourth bounce seen in the R impact lightcurves of Graham et al. [18].

F. Postcollapse "Splat" Phase (about 15 to 45 Minutes)

As the expanding debris cloud collapses it rotates counterclockwise in Jupiter's reference frame, due to the Coriolis Effect. This causes the ejecta footprint's symmetry axis to rotate out of alignment with the incoming fragment trajectory, which approached from 16° east of due south in Jovecentric coordinates. Since the highest parts of the ballistic plumes had as long as 18 minutes of flight time (Fig. 7), they rotated with respect to Jupiter by about 7.5° before re-entering. However, the HST image of the G impact reveals a total rotation of 23° [19] which would have required another 30 minutes of flight. However, one would not expect a fluid plume, collapsing onto a fluid atmosphere, to immediately decelerate, or "stick." Because it lands with a high horizontal velocity component, it maintains its momentum and "slides," continuing to rotate counterclockwise with respect to Jupiter. The total rotation angle of 23° actually implies a slide time of much greater than 30 minutes, because the mean sliding velocity was less than the re-entry velocity owing to its ultimate deceleration to a stop. This postcollapse rotation indicates that the ejecta blanket also expanded radially by the same mechanism. The outwardly-directed velocity component sets up a radially-expanding flow field that sweeps condensed matter outward, which may contribute to the crescent-shaped postcollapse debris pattern. This pattern remains to be fully explained, however, and is likely related to the structure of the plume itself [58].

A linear, radially expanding wave is made visible by an unknown mechanism, possibly condensation in the rarefaction part of the wave. These waves were observed in HST images of several impact sites as rings expanding at a constant velocity of 450–500 m/sec [19]. They have been modeled as inertia-gravity waves by Ingersoll and Kanamori [66] who predicted an enrichment in the water content of the troposphere by as much as ten times the solar abundance in order to explain the high wave velocities. However, the Galileo probe subsequently showed that Jupiter's atmosphere is too dry to support this model, hence this phenomenon remains unexplained. An alternative model of Zahnle [52] involves impact debris which is caught up in breaking nonlinear stratospheric waves. As the wave expands, the debris surfs along in front for thousands of kilometers, but slows down to a velocity below what was observed.

G. Aftereffects (Minutes to Hours)

The computational models indicate that there is also an upwelling phase. Careful examination of the 3D simulations of Crawford et al. [14] reveals that for massive, deeply-penetrating impactors, a bubble (or several bubbles) of hot Jovian atmosphere mixed with cometary vapor rises buoyantly from the depth of maximum energy deposition. This is between 200 and 300 km beneath the 1 bar level for 2–3 km diameter fragments. At 82 seconds after impact for a 1 km impactor there are three or four instabilities developing between about 50 and 200 km below the 1 bar level. At this time, they are 20–30 km in diameter, and have risen by about that distance from their starting point. These bubbles are analogous to buoyant nuclear explosion fireballs. Extrapolation of their upward motion suggests that they will begin arriving at the ammonia cloud layer within minutes, after having adiabatically expanded to many times their size. The resulting massive displacement of atmosphere is a likely source for the expanding wave, perhaps enhanced by a release of the latent heat of condensation of the entrained water vapor as suggested by Zahnle [52]. The upwelling might also manifest itself as thermal brightening or appearance of new spectroscopic signatures at the impact sites. It may be possible to extract information about the penetration depth (and therefore fragment mass) from the timing, temperature, and composition of any buoyantly-upwelling material.

Another late-time phenomenon was the giant ring that was observed at 3–4 μm by McGregor et al. [70], following two or three of the impacts. These rings continued to expand for up to 2 hours after impact, and were seen propagating as far as 18,000 km away from the impact site. McGregor et al. [70] suggested a couple of possibilities for the giant rings; that they were either a sliding phenomenon or second wave. One other possibility is that these rings are related to the collapse of the outer, invisible shells of the plume which were ejected to much higher altitudes than the visible plume, with correspondingly longer flight times and re-entry velocities. This cause might explain the sizes, locations, and velocities of the rings, but a plausible mechanism for emission is still lacking.

SL-9 Interpretation Summary

The description presented above is designed to provide a conceptual framework for relating the observations to the physical phenomena that were observed, and is not intended to be exhaustive or exclusive. The focus here has been on phenomena that are closely related to the shock physics modeling effort, and that can be identified with features or predictions of those models. The primary intent of this framework is to serve as a descriptive summary of the sequence of events, and to serve as a guide for data interpretation and prioritization of future directions for computational work. Other phenomena associated with the collision, such as the stimulated auroral processes [57],

[72], have less obvious connections to the shock physics models which have not yet been sorted out. More rigorous reviews of the physics of comet entry, fireball growth, and plume collapse, along with a more comprehensive discussion of the related observations, can be found in Noll et al. [36].

PLUME-FORMING COLLISIONS ON EARTH

Plume-forming collisions with Earth had already been investigated theoretically prior to the discovery of SL-9, but the nature of these plumes was somewhat different from those observed on Jupiter. The plumes reviewed by Melosh [28] are dominated by vapor released from an impact with the ground. At the point of impact, strong shock waves are driven into the target and back into the impactor, generating a volume of material at high temperature and pressure that vaporizes upon expansion. As the resulting cloud of vapor is ejected, it sweeps aside the ambient atmosphere and forms a strong blast wave that propagates outward and upward. For small impacts, when the vapor cloud reaches pressure equilibrium with the surrounding atmosphere, its radius is less that of the atmospheric scale height, and it rises buoyantly like a nuclear fireball. Above a threshold energy of about 150 megatons (corresponding to a 3 km diameter crater, and recurring once every several thousand years) the radius of the vapor cloud exceeds the scale height, and the expanding fireball blows out the top of the atmosphere and goes ballistic. One result of such a plume might be the formation of tektite strewn fields as entrained melt quenches to glass and returns to Earth thousands of kilometers away. At even larger scales, such as the Cretaceous/Tertiary extinction event, so much material can be ejected by this mechanism that it can be entrained in the stratosphere and distributed globally. The density of atmospheric debris can be so high that it reduces light levels at Earth's surface, resulting in global cooling [37]. Another serious consequence of the solids ejected with the plume is that when material re-enters the atmosphere it reaches high enough temperatures over a large enough area that wildfires can be ignited by radiative coupling to the surface [29], [50]. Large-scale plumes may also blow off massive quantities of atmosphere, resulting in a net reduction of air, and long-term atmospheric erosion [30].

The SL-9 plumes are fundamentally different, because Jupiter has no surface from which solid material can be shocked to form a dense vapor cloud. Instead, the kinetic energy of the projectile is deposited in an atmospheric wake, which then explodes downward, outward, and upward to form the plume. Shoemaker-Levy 9 is therefore a better analogue to terrestrial impactors with kinetic energies of less than 50 megatons or so, that deposit their energy into the atmosphere and do not reach the ground. Figure 10 makes a direct comparison between the Crawford et al. [14] 3D fireball simulation and an eyewitness artist's depiction of the February 12, 1947 Sikhote-Alin fireball in Siberia, suggesting that the physics of atmospheric entry and plume generation is similar over many orders of magnitude in the scale of impactor kinetic energy and physical size. The Sikhote-Alin impact energy was 10–20 kilotons [33], whereas the simulated fireball is from a 6 million megaton impact event, nearly a billion times more energetic. The simulation image shows the explosion 50 seconds after the impact of a 3 km diameter fragment on Jupiter. The illustration of the Siberian event, commemorated on a tenth-anniversary Soviet postage stamp, was recorded on canvas by artist P.I. Medvedev. According to most popular accounts, the artist was an eyewitness who recorded his observation immediately after the fall.[1] If the feature depicted by Medvedev represents the debris in the entry column and rising incipient plume, then its resemblance to the Jovian event implies that

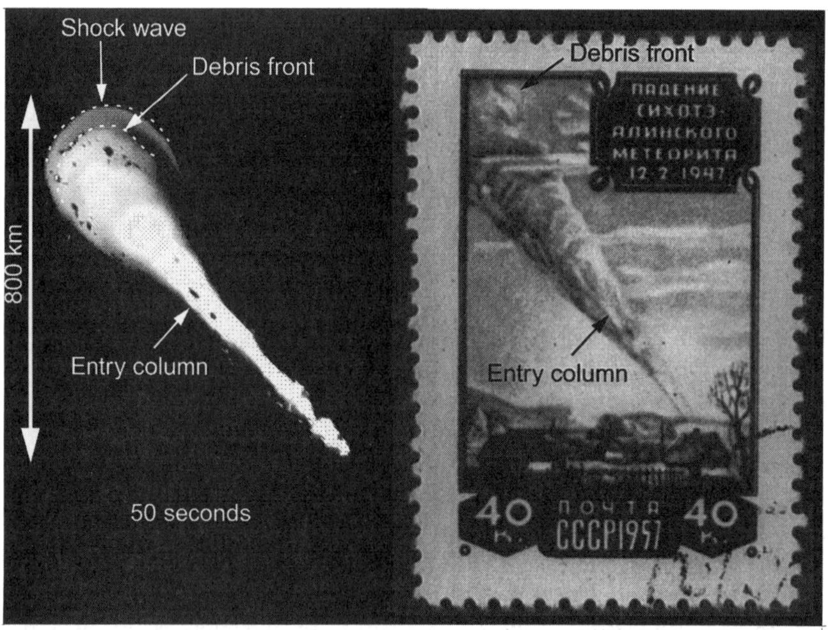

FIGURE 10. Comparison of (a) 3D simulation of impact of 3 km diameter fragment on Jupiter, 50 seconds after entry (from [14]) with (b) artist's depiction of 1947 Sikhote-Alin impact.

[1]In part because of publicity associated with this year's fiftieth anniversary of Sikhote-Alin, some doubts have surfaced about this account of the origin of the Medvedev painting, which has been called the "*Mona Lisa* of meteoritical science " [71]. There is now evidence for at least three versions of the painting, and some speculation that the original version was painted with the plume oriented in the wrong direction. Further investigations, however, have demonstrated that the plume was oriented properly (from upper left to lower right) in the original painting, which is actively being sought (O.R. Norton, personal communication).

atmospheric impact explosions behave similarly over many orders of magnitude. The following subsections summarize work in progress to assess the implications of plume-forming collisions on Earth as a hazard to satellites, a collective phenomenon from bolide swarms, a model for the Tunguska explosion, and a mechanism for enhancing transport of material to the upper atmosphere.

Plume-related Satellite Hazards

The simulations and observations of the impact of SL-9 raise some issues that relate to the impact threat to Earth, demonstrating that ballistic impact fireballs and plumes are ejected to very high altitudes, and that explosive expansion of shocked atmosphere along the entry column is highly directional and poorly modeled by point explosions. These observations lead to the suggestion that satellites in low-Earth orbit (LEO) may be vulnerable to ejection of material into their environment by an impact into the atmosphere. Because of the high orbital velocities of these satellites (about 7 km/sec), even a very low-density plume ejected into their path would be catastrophic. For a vapor plume, a satellite/plume interaction would be similar to an atmospheric reentry. At best, an interaction with a very low-density plume would cause a change in the attitude and orbit of the satellite. A worse outcome would result from a higher-density plume, which could cause premature reentry or otherwise destroy the satellite. For a plume containing particles of condensation, like those generated by SL-9, the interaction would involve numerous hypervelocity impacts similar to those occasionally experienced with space debris and micrometeorites. This would most likely end the life of the satellite.

To test this idea, we have performed preliminary 2D simulations of the plume generated by a 34 m diameter stone (density = 3 g/cm^3) impacting at 20 km/sec with vertical incidence. The kinetic energy of the impactor is equivalent to an explosive yield of 3 megatons of TNT, and the expected frequency of such an event is about once per century [31]. The simulation was performed using the same methods as our SL-9 simulations, inserting an appropriate impact energy deposition curve into a gravitationally-stabilized, stratified Earth atmosphere, using the 1976 U.S. Standard Atmosphere (NOAA, NASA, USAF, Washington D.C., 1976) density profile. The energy deposition curve was calculated using the model of Crawford [14] which implies near-negligible radiative ablation for objects greater than about 10 m in diameter (as required by momentum conservation) and hydrodynamic deformation governed by long-wavelength Kelvin-Helmholtz instability. This model has been calibrated with results of entry simulations using CTH. The energy was inserted as 100 discrete energy sources with appropriate magnitudes along the axis of symmetry from 120 km to the surface. The individual "point charges" were set off sequentially, beginning at the top, and initiating downward with the velocity of the projectile, accounting for its deceleration as it descends into the atmosphere. This method of energy insertion effectively simulates the entry column as a line charge of varying energy density.

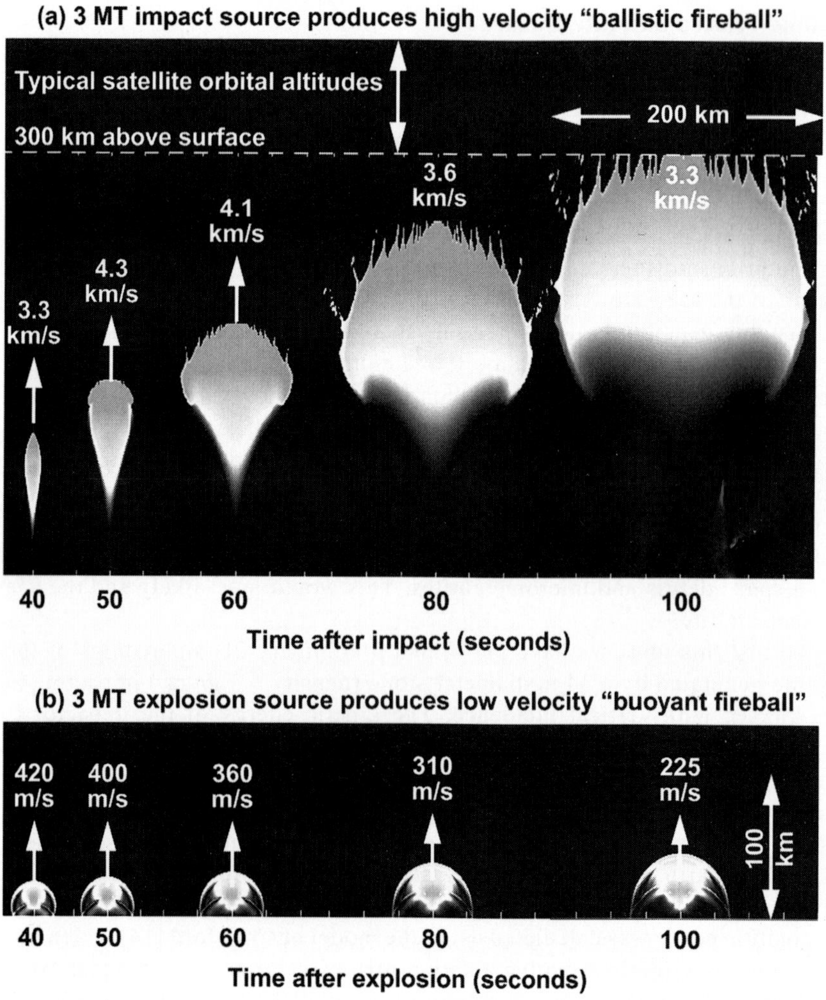

FIGURE 11. Comparison on the same scale of evolution of (a) ballistic fireball generated by 3 megaton impact to (b) buoyant fireball generated by a 3 megaton explosion at same altitude of maximum energy deposition (7 km). Shading indicates velocity magnitude of the air as indicated in the text.

A time sequence showing the results of the simulation is given in Fig. 11. The shading indicates material velocity magnitudes, ranging from 100 m/sec to 3 km/sec. Immediately after entry, the meteoroid has deposited its energy in a long column, with a sharp peak in the energy deposition curve at an altitude of 7 km. The column begins to expand explosively the instant that energy is deposited in the column, so by the time the object reaches the bottom, the top of the column has had time to expand and accelerate upwards. When the main fireball begins to develop lower down, it expands most easily up along the low-density, high-sound-speed column that is already moving upward. In this way, much of its mass is accelerated and launched into space as a "ballistic fireball." At about 100 seconds after atmospheric entry, the top of the plume (as defined by the 120 km density contour) has reached an altitude of nearly 300 km, and is still moving upward at a velocity of about 3 km/sec (the "fingers" protruding from the top of the plume are an artifact of instabilities resulting from the relatively low resolution used for this simulation). Ballistic extrapolation indicates a peak plume height at about 800 km, putting it across the path of many satellites in LEO. The plume will have a "hang time" greater than 10 minutes; during this interval it will obstruct any satellite passing overhead. Furthermore, it will continue to expand radially during this time to cover a region thousands of kilometers in diameter. Observations and simulations of the plume collapse on Jupiter imply that the atmospheric density contours in the region of collapse will remain elevated for a much longer period of time, posing a further risk to satellites entering the area.

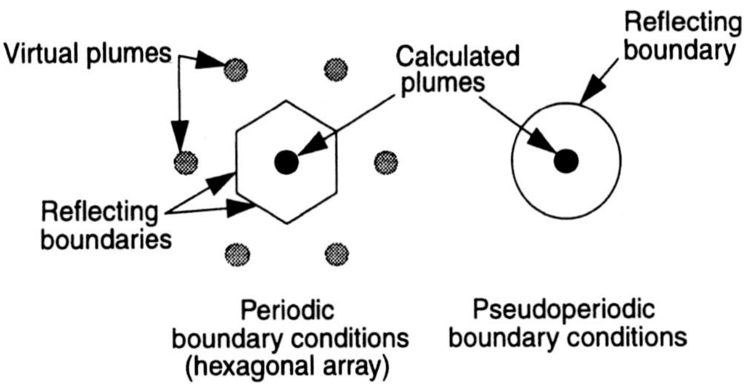

FIGURE 12. Diagrammatic explanation of 2D pseudoperiodic computational scheme (plan view).

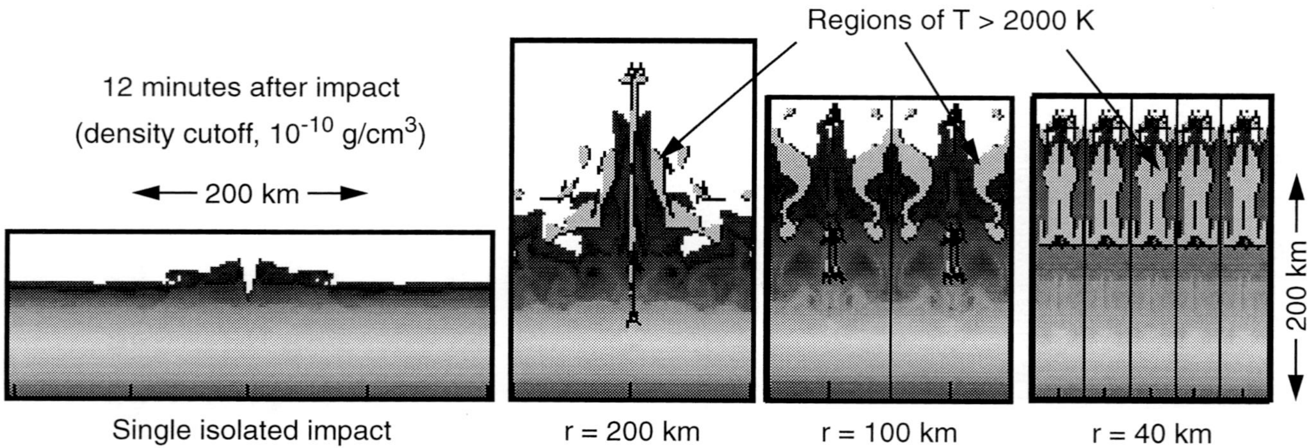

FIGURE 13. Plumes from clusters of 3 MT impacts, showing density bands of collapsed plume at 12 minutes after impact, with regions where temperature exceeds 2000 K; *r* is the interaction distance between neighboring impacts.

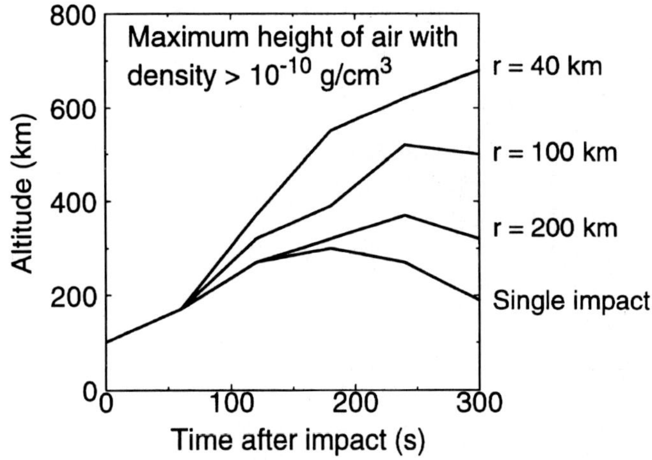

FIGURE 14. Plot of maximum height of 10–10 g/cm^3 isodensity surface as a function of time, for various plume separation distances resulting from bolide swarm.

Plume Clusters

We have also simulated the interaction of atmospheric impact plumes generated by an array of simultaneous impact events on Earth. Wasson [47] has suggested that atmospheric deposition of energy from a disrupted comet or asteroid may have led to the formation of layered (Muong-Nong) tektites by a process of radiative coupling, so there is possible evidence in the geologic record for such atmospheric plume phenomena. In addition, NEO threat pulverization strategies such as those proposed by Wood et al. [49], could potentially produce broad clusters of closely spaced impactors if there was insufficient time to disperse the fragments. As a first step toward modeling an array of plumes, we approximated a swarm of equally-spaced impacts by performing an axisymmetric simulation in a cylinder with reflecting boundary conditions. The diameter of the cylinder is roughly the distance between plumes in a hexagonal array with periodic boundary conditions, so these boundary conditions can be considered "pseudoperiodic" (Fig. 12). In this highly simplified geometry, the lateral extent of the cluster is infinite, but for a 1000 km cluster the effects of the edge will not reach the center of the collective plume until after it has collapsed.

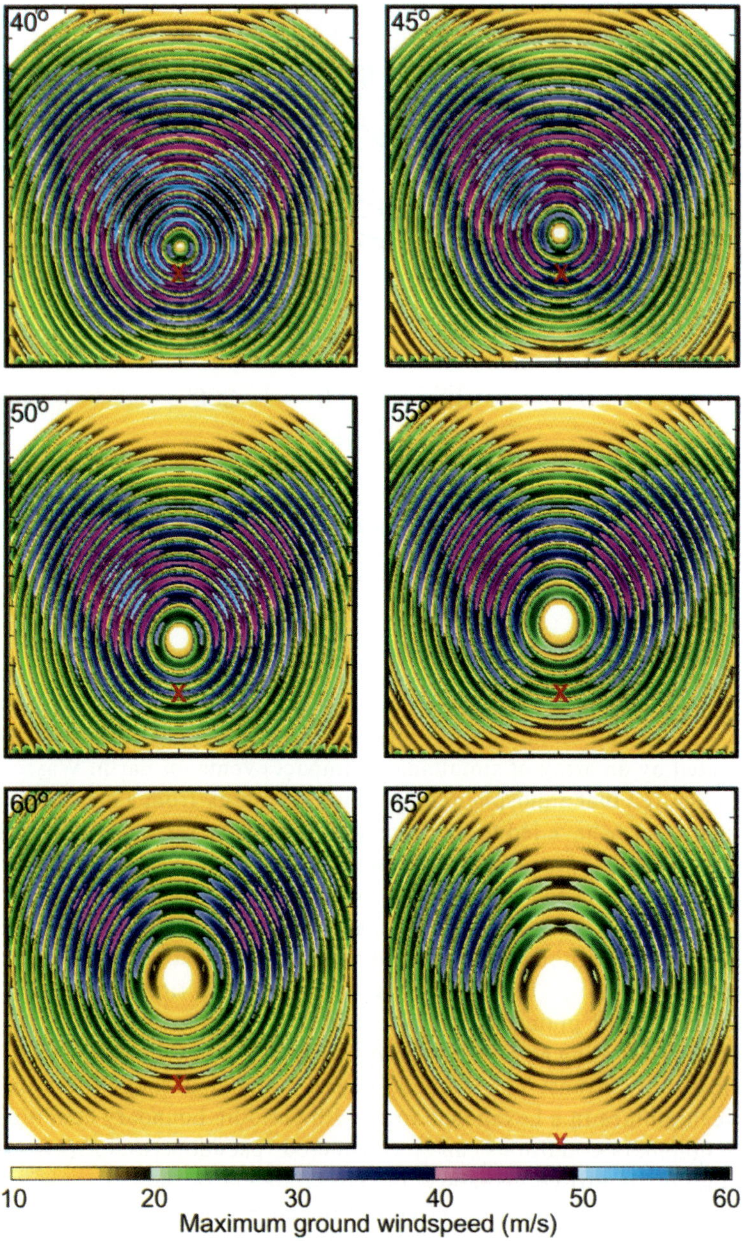

FIGURE 15. Plan view of peak surface windspeeds determined from 3D simulations of 12 megaton plume-forming collisions. Concentric zones show the magnitude of radially-directed winds at 5-second intervals. Various oblique entry angles (given from zenith) were simulated.

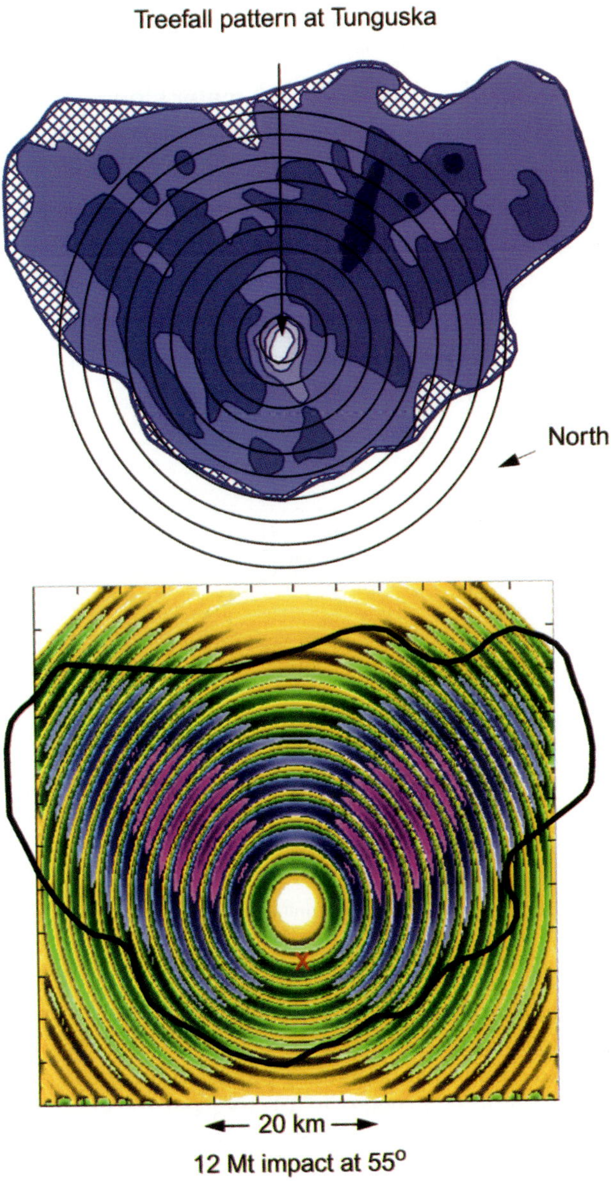

FIGURE 16. Comparison of butterfly-shaped treefall pattern at Tunguska [17] to map of peak surface windspeeds calculated from the collision of a 12 megaton object entering the atmosphere at an oblique angle, 55° from the zenith.

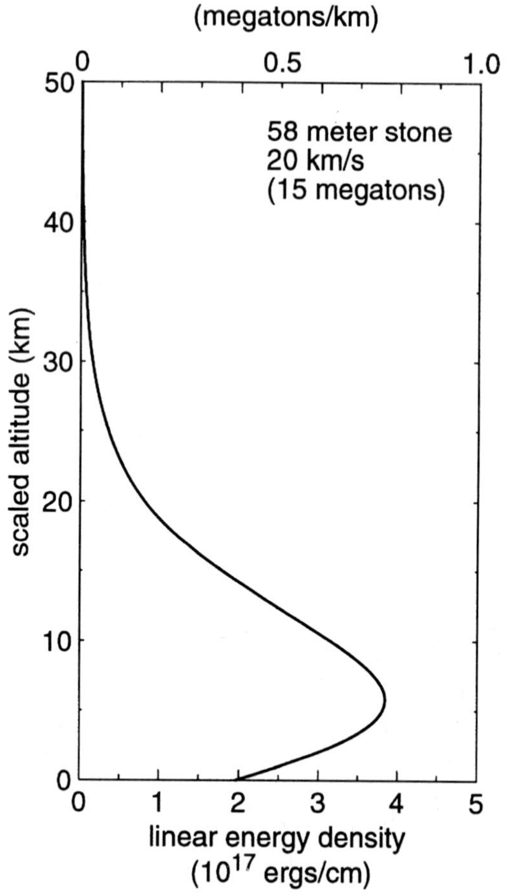

FIGURE 17. Energy deposition curve from a 15 megaton impact at a 45° entry angle, used to generate the plume depicted in Figs. 18–19.

We simulated a plume from the impact of a 20 km/sec, 34 meter-diameter asteroid (kinetic energy of 3 megatons) at vertical incidence, and compared its evolution and collapse to interacting plumes from arrays of impacts with near-neighbor separation distances of 40, 100, and 200 km. As expected, the closer-packed arrays lead to denser plumes that reach higher altitudes and generate higher temperatures within denser air upon collapse. The time steps at 12 minutes after impact of each simulation are shown in Fig. 13, where the shading is proportional to log(density) with a cutoff density of 10–10 g/ cm^3. The regions where the temperature is greater that 2,000 K are also shown. Figure 17 provides maximum altitude as a function of time of air with density greater than 10^{-10} g/cm^3. Because more mass is ejected per unit area for the tighter cluster, the density of the plume is higher and much more atmospheric mass is heated upon its collapse. The energy transport in these simulations neglects radiation, so the actual late-time atmospheric temperatures may be lower. To fully assess the effects of radiative energy coupling to the ground will require a simulation that includes detailed opacity calculations.

Tunguska

The 1908 Tunguska explosion was almost certainly a plume-forming impact. Estimates for the magnitude of the event vary by a wide margin, but by virtually all accounts the yield was significantly more than 3 megatons, implying that the scale of the resulting plume was at least comparable to that shown in Figure 11. The similarity between the terrestrial and Jovian impact plume simulations suggests that phenomena associated with the Tunguska event should be re-examined in the context of these models, and that estimates for its energy should be revisited. Our model incorporates most of the physics that was first applied to the Tunguska event by Chyba et al. [56], but in addition couples the momentum and energy of the impacting body to the atmosphere along the entire entry path, leading to the extended "line explosion" that results in plume formation as well as enhanced momentum and energy coupling to the ground.

For reasons of practicality, most of the published models for Tunguska have assumed that the explosion was essentially a point source, and the most widely quoted yield estimates are based on this approximation. These calculations made use of seismic records, barograms, and measurements of fallen trees, and are calibrated with nuclear airburst data (see Sekanina [73] for a review of the relevant data). In all cases, however, the point source approximation coupled with other simplifications may have led to serious overestimates of the yield of the Tunguska explosion. One of the most solidly accepted values for the energy of Tunguska is that of Ben-Menahem [54], who analyzed old records from four seismic stations and compared them to modern seismograms recorded during Soviet and Chinese atmospheric nuclear tests. He concluded that the ground motion resulted from a vertical point impulse of 7×10^{18} dyn sec, which would be generated by a point explosion of about 12.5 megatons at an altitude of 8.5 km. However, our simulations of a 3 megaton event reveal that within the first minute of impact, the upward-directed momentum associated with the atmospheric plume has reached 7×10^{18} dyn sec, requiring that a reaction impulse sufficient to generate the observed seismic records has been coupled to the ground. This implies that, from a seismic standpoint, a 3 megaton plume-forming event gives the appearance of being a much larger point source explosion. Likewise, the strong momentum anisotropy associated with plume formation was not taken into consideration for estimates based on atmospheric waves [65], [54].

There are many discrepancies among the various estimates based on treefall [78], [77], [80], [54], [73], which in all cases make comparisons between the damage at Tunguska and the criteria for blast damage to forests established by the nuclear weapons effects literature, and range from "a few megatons" up to 50 megatons. This is the most reasonable approach, but it is important to recognize the limitations associated with using the weapons tables, which apply to living coniferous forests [62] and implicitly assume flat topography. There is cause for skepticism of these yield estimates, because it is clear from the photographs of Krinov [68], that slopes of 15–20 degrees and greater are typical of the terrain around the impact site. A 15 degree upslope results in a secondary Mach stem (in addition to the one that forms from the flat-ground reflection) that increases the dynamic pressure by a factor of 3 for weak shock, so the topography at Tunguska would naturally lead to concentrations of blast wave energy far beyond the distance that would be calculated assuming flat terrain. One must assume that the most distant fallen trees were in such areas, so a yield estimate based on the extent of treefall would be too high. Krinov [68] indeed described a treefall pattern that was dependent on local topography, and in all his photographs showing the forest devastation, there are many trees that remain standing.

Another reason to be skeptical of treefall-based yield estimates is that none of them take into account the pre-impact condition of the Siberian forest at the site of the explosion. One result of a 1961 expedition to the site [60], was that "the region of the forest flattened in 1908 was not one of homogeneous primeval intact taiga," and that according to an expert from the Forestry and Lumber Institute of the Siberian Branch of the Academy of Sciences, "... the region of meteorite impact in 1908 was basically a fire-devastated area ... a partly flattened dead and rotting forest was standing in this area ... ." Because of this observation, Florenskiy [60] issued the apparently unheeded advice that "... an estimate of the force of the shock wave that is based on the number of flattened trees must necessarily take into consideration the condition of the forest at that time." Fortunately, he provided some of the data required for that task, in the form of dynamometry measurements of the force required to fell trees of various sizes and conditions. A key graph in his paper shows that the felling moment of trees from a dead timber stand of 1908 are about one third that for living trees of the same diameter. Even after accounting for lower wind resistance of dead timber, one would expect the forest at Tunguska to be blown down by a significantly smaller explosion than would be required to destroy the same area of a healthy forest. Dynamic pressure scales as the square of wind speed, so the peak wind speeds for flattening the forest could be reduced by as much a 40% from the values previously assumed.

To provide some quantitative estimates of wind speed at ground level, we performed a series of 3D simulations, with 500 meter resolution, for various entry angles of a 12 megaton impact. The results (Fig. 15) show the characteristic butterfly pattern that has been reproduced experimentally by Zotkin and Tsikulin [78], and analytically by Korobeinicov et al. [67]. The peak wind speeds immediately behind the shock front are plotted at 5-second intervals. For comparison, the peak wind speed is 40 m/sec at a distance of 23 km from a 12 megaton point explosion at its optimum height of burst. According to Glasstone [62], this is the threshold for severe treefall (defined as 90% blow down of a conifer forest) by aerodynamic drag pressure from a 12 megaton explosion. Similarly, Glasstone's radius to moderate damage (30% blow down) for a 12 megaton air burst is 29 km, where the maximum wind speed is about 30 m/sec. If the requisite wind speeds can be reduced by 40%, as suggested by the dynamometer data Florenskiy [60], then the required point source energy (at optimum height) to achieve the same radius for severe damage is about 3.5 megatons [62].

Since the maximum wind speed contours are dependent on entry angle, the calculated butterfly patterns can be used to estimate the angle by comparing them to the treefall data of Fast [17]. These data are presented in Figure 16, where the contours represent the average scatter in the deviation of the angle of uprooted trees from the radial direction, a parameter that presumably depends on aerodynamic pressure. The best agreement is for a zenith angle of about 55°, which is very close to the 60° angle favored by Zotkin and Tsikulin [79]. The contour map of Fast [17] shows a high degree of bilateral symmetry, but there are also significant deviations that are probably related both to topography and pre-impact inhomogeneities in the taiga. A useful next step would be to perform high resolution 3D simulations of impacts, using actual topography data from Tunguska.

One of the most unusual phenomena associated with Tunguska was the bright sky over an extensive region of Europe and Asia for several consecutive nights after the explosion. Many of the early papers, especially those by Soviet researchers [59], argue that the Eurasian bright nights were evidence for a cometary tail that was directed southwestward from the point of the impact, giving rise to the observed geographic distribution which extended southward about two thousand kilometers, and westward through Europe to the Atlantic Ocean. More recently, investigators have attempted to explain the bright sky in terms of noctilucent clouds made up of water ice and dust generated by the impact and transported westward by high-altitude horizontal winds. Noctilucent clouds reside in the mesosphere, at altitudes ranging from 75 to 90 km, and are most visible at high latitudes during midsummer when they are in direct sunlight all night long on account of low solar declinations. The Tunguska explosion took place at 60° north, and only 9 days after the summer solstice. Turco et al. [75] suggest that the clouds were condensed primarily from the vapor ablated at very high altitude from the entering meteor. However, this model requires an extremely low density meteor of less than 0.01 gm/cm^3 in order to deposit sufficient mass (about 3×10^9 g) above 50 km. Turco et al. [75] point out that deposition above this altitude would be necessary for the cloud to have been transported westward rapidly enough by easterly winds to have been visible 6000 km away, in England, within 12 hours. Chyba et al. [56] were able to find a mechanism capable of transporting as much as 3×10^{12} g of water vapor aloft, reasoning that atmospheric humidity could be lifted by a buoyant fireball that rose to about 40 km from the 15 megaton terminal explosion.

We agree that the light nights were caused by noctilucent clouds made up of water and dust generated by the impact, but the recognition that Tunguska was a plume-forming event simplifies the explanation significantly. The easiest argument is made simply by analogy to the aftermath of the impact on Jupiter. At time (e) in Figs. 1 and 3, the geometry of the collapsed plume from SL-9 can be seen to be on the nighttime side of the terminator, yet parts of it are at high enough altitude to be illuminated by the Sun. The HST image taken at 7:51 UT on July 18, 1994 of the collapsed plume from impact G is shown in Fig. 6(b). which reveals a sunlit cloud extending above and beyond Jupiter's terminator. This feature is, by definition, a noctilucent cloud.

We chose to model the formation of the plume at Tunguska by depositing the energy of a 58 m diameter stony asteroid (3 g/cm^3) striking the atmosphere at 20 km/sec at a 45° entry angle, using the model of Crawford [12]. The energy deposition curve is shown in Fig. 17, and represents a total energy of 15 megatons. These were 2D axisymmetric simulations, with the atmosphere scaled by a factor of 1.414 to accommodate the actual density gradient encountered by the oblique impact. The resulting plume, 100 seconds after impact, is shown in Fig. 18, with tracer particles showing the ballistic trajectories of air packets derived from altitudes of 350 m, 1.4 km, 7.1 km, and 90 km. Material within a cylinder nearly 1 km in diameter has been accelerated to sufficient speed to escape the mesosphere, and it is this mixture of vaporized meteor and tropospheric air that we would expect to condense in the upper atmosphere to form the noctilucent clouds observed across Europe. We can approximate the mass of entrained water by estimating the volume of ejected air and assuming a humid troposphere as was done by Chyba et al. [56]. Our result is that about 10^{11} g of water collapses with the plume on top of the mesosphere, along with virtually the entire 3×10^{11} g mass of the vaporized meteor. Ballistic extrapolation of the plume reveals that, in the absence of significant atmospheric dissipation, the ejected material extended more than 2000 km from the impact site within 15 minutes (Fig. 19). If the bolide arrived from the southeast, as most evidence indicates, then the plume was ejected back in that direction, and may have reentered the atmosphere as far away as Lake Baikal. The lateral extent of this collapsed plume is fully consistent with that of the noctilucent cloud, extending as far south as 45°N latitude.

All the modeling on which the above reinterpretations are based is preliminary, and it is too early to make strong quantitative conclusions. Many of the simulations were 2D, and the 3D simulations were probably insufficiently resolved. Radiative transport was not included, and significant work is required to develop a complete entry model that accounts for all the physics. Nevertheless, the formation of atmospheric plumes by deposition of an energy profile that resembles Fig.17 is a highly robust result. Plume formation does not appear to be strongly dependent upon the details of the assumptions, leaving little doubt that Tunguska-class impacts are plume-forming events.

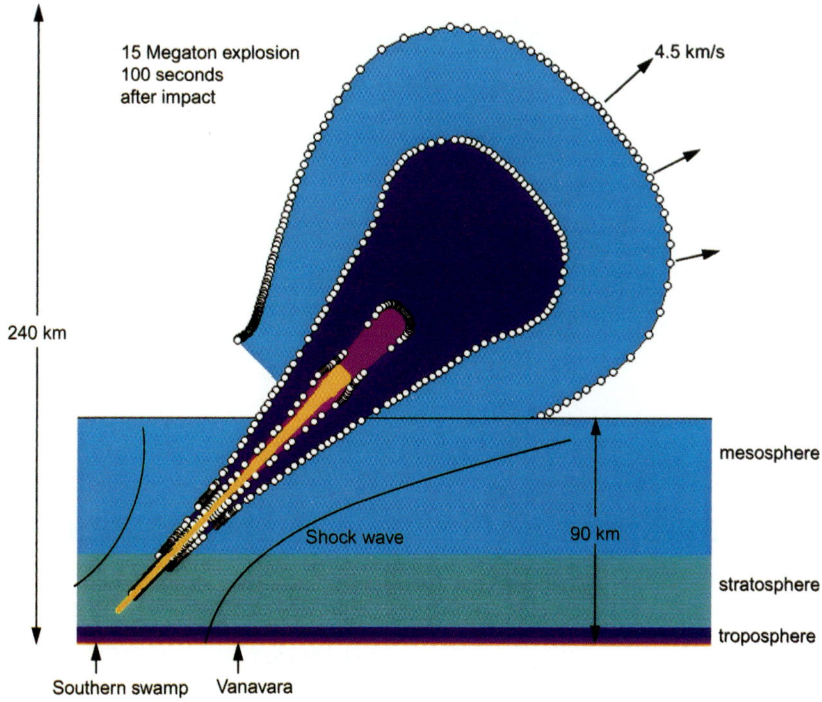

FIGURE 18. Anatomy of a 15 megaton plume, 100 seconds after impact. Material within all but the outermost shell has been ejected from within the troposphere, and contains the mass of the impactor, as well as water from the humid lower atmosphere.

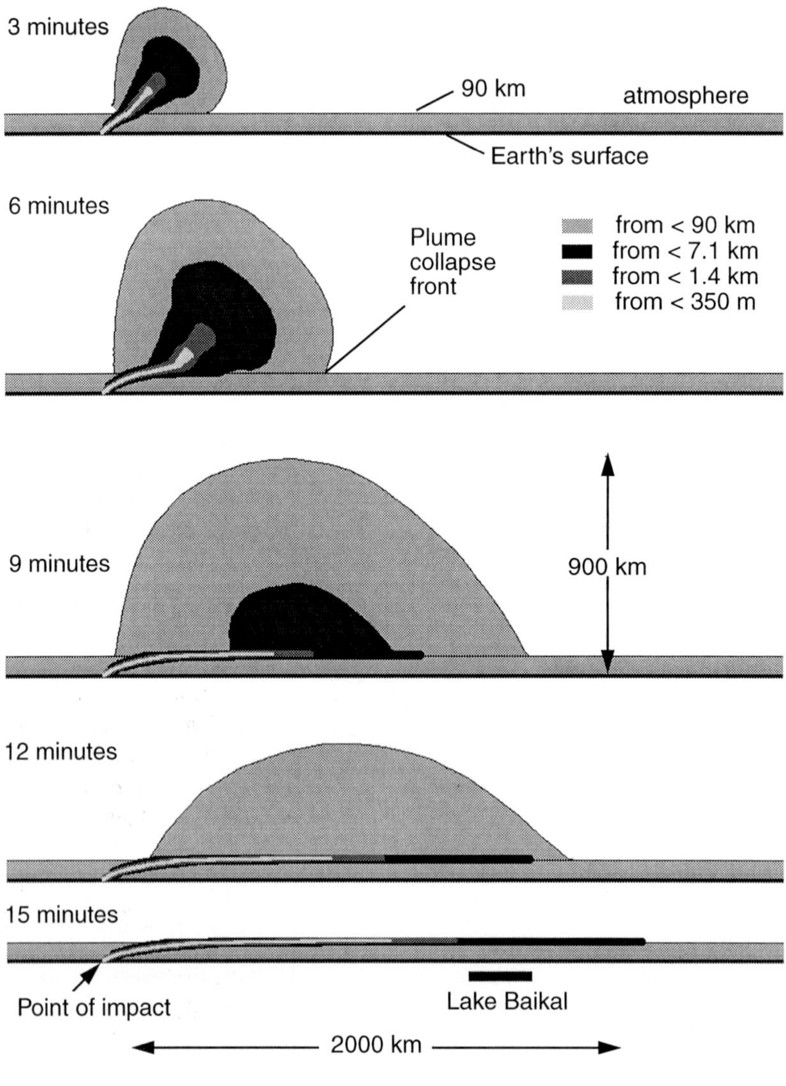

FIGURE 19. Growth and collapse of the 15 megaton plume over a period of 15 minutes, showing that tropospheric water and meteoritic debris can be distributed over distances of thousands of kilometers at mesospheric altitudes.

Observational Validation

The impact of SL-9 underscores the importance of observational validation of impact modeling. It is probably not realistic to expect another opportunity to watch a comet collide with Jupiter, an event that only happens once in a period ranging from 200 to 10,000 years [41]. A more reasonable plan would be to attempt to gather as much information as possible about the smaller atmospheric impacts that are continuously taking place on Earth. If the physics of atmospheric entry and plume formation is indeed similar at scales that are different by more than 9 orders of magnitude, then careful, quantitative measurements of impacts that take place at intervals of a year or less would help give entry and impact models the degree of validity necessary to allow their use as a basis for hazard assessment for the smallest Earth-threatening impacts. To accomplish this task, we recommend a coordinated campaign based on a combination of satellite sensors and Earth-based observations. Infrared and visible-light sensors on satellites operated by the U.S. Department of Defense have already detected over a hundred events [44], including the February 1, 1994 bolide over the South Pacific with an estimated energy of 40–70 kilotons [32]. Microbarograph records are particularly useful in characterizing the explosive yield of large impact events, and have recorded an explosion as large as 1.1 megatons for the August 3, 1963 bolide (possibly in the plume-forming class) over the ocean south of Africa [38]. Other sources of data that can be used to validate the simulations are photographic and video images (e.g., [8], [5]), infrasound [38], and seismic data.

There is one major drawback in all the above methods as they have been used to date: the events they observed were not predicted in advance. Therefore, the instrumentation was operated in "open shutter" mode, or set to trigger off of the event, or relied on serendipity. By contrast, the impact of SL-9 was predicted a year in advance, and observing plans were carefully assembled well in advance of the event, resulting in the collection of vast amounts of high-quality data. We suggest a similar strategy would be most useful for validation of Earth impact models. A ground-based search system capable of providing short notice of an impact in the 100 kiloton range would mean that the approaching object could be characterized before impact. Moreover, arrays of sensors, cameras, and satellite observing plans to be quickly put in place so that data from the event could be captured and used to provide quantitative validation.

The enhancement in modeling sophistication brought about by further increases in high performance computing capabilities will allow impact models to continue to be advanced and refined, which should reduce the uncertainties and improve the risk assessments. However, until the next large impact takes place, the data from SL-9 will be the only true reality check for models of collisions with planetary atmospheres.

CONCLUSIONS

The demise of comet Shoemaker-Levy 9 was the first time a plume-forming atmospheric impact was caught in the act, and it generated a wealth of information leading to a new understanding about this important class of planetary phenomena. Much of the observational data can be explained, and are fully consistent with computationally-based models, but there are still many associated phenomena that are not well understood. Even as the models continue to be improved, coming into accord with many disparate sets of observations, it is unlikely that we will ever confidently know the mass (and equivalent explosive energy) of the comet by much better than an order of magnitude. An equivalent statement would be: Even after the experience of directly observing a series of collisions on another planet, our estimates of the observable consequences of an impact of known size could be significantly in error. This great uncertainty should be kept in mind when assessing the potential consequences of impacts on Earth, when comparing the relative threat from different impact classes, and when performing cost-benefit analysis in support of policy decisions.

One previously-neglected aspect of the Earth-impact threat that has been highlighted by Shoemaker-Levy 9 is the atmospheric plume phenomenon. Simulations based on lessons learned from SL-9 suggest that relatively small impacts into Earth's atmosphere form plumes which transport water and dust into the upper atmosphere, with possible climatic effects. Many of the phenomena associated with the 1908 Tunguska explosion are undoubtedly related to the formation and collapse of an atmosphere plume. There is a high probability that another plume-forming impact will take place on Earth within our own lifetimes. The next such event should be anticipated as potential hazard, but also as an opportunity to gather more information on the physics of atmospheric entry, ballistic fireball growth, and plume collapse. Computational modeling with careful benchmarks against observed impacts will allow the hazard to be estimated for the smaller, high-probability impacts.

ACKNOWLEDGMENTS

We are grateful to the astronomical community for generously sharing SL-9 impact data prior to its publication. We thank Giuseppe Longo and Zdenek Sekanina for providing information and helping interpret data from the Tunguska event.

REFERENCES

1. AHRENS, T.J., T. TAKATA, J.D. O'KEEFE, & , G.S. ORTON. 1994. Radiative signatures from impact of Comet Shoemaker-Levy 9 on Jupiter. Geophys. Res. Lett. **21:** (14) 1551–1553.
2. BOSLOUGH, M.B., D.A. CRAWFORD, A.C. ROBINSON, & T.G. TRUCANO. 1994. Mass and penetration depth of Shoemaker-Levy 9 fragments from time-resolved photometry. Geophys. Res. Lett. **21:** (14) 1555–1558.
3. BOSLOUGH, M.B., D.A. CRAWFORD, A.C. ROBINSON, & T.G. TRUCANO. 1994. Watching for fireballs on Jupiter. Eos. **75:** 305–310.
4. BOSLOUGH, M.B., D.A. CRAWFORD, T.G. TRUCANO, & A.C. ROBINSON. 1995. Numerical modeling of Shoemaker-Levy 9 impacts as a framework for interpreting observations. Geophys. Res. Lett. **22:** (13) 1821–1824.
5. BROWN, P., Z. CEPLECHA, R.L. HAWKES, G. WETHERILL, M. BEECH, & K. MOSSMAN. 1994. The orbit and atmospheric trajectory of the Peekskill meteorite from video records. Nature. **367:** 624–626.
6. CARLSON, R.W., P.R. WEISSMAN, M. SEGURA, & THE NIMS SCIENCE TEAM. 1995. Galileo infrared observations of the Shoemaker-Levy 9 G impact fireball: A preliminary report. Geophys. Res. Lett. **22:** (12) 1557–1560.
7. CARLSON, R.W., P. DROSSART, TH. ENCRENAZ, P.R. WEISSMAN, J. HUI, & M. SEGURA. 1997. Temperature, size and energy of the Shoemaker-Levy 9 G-impact fireball. Icarus. In press.
8. CEPLECHA, Z. 1994. Earth-grazing daylight fireball of August 10, 1972. Astron. Astrophys. **283:** 287–288.
9. CHAPMAN, C.R., W.J. MERLINE, K. KLAASEN, & THE GALILEO IMAGING TEAM. 1995. Preliminary results of Galileo direct imaging of S-L 9 impacts. Geophys. Res. Lett. **22:** 1561–1564.
10. CHAPMAN, C.R. 1996. Galileo observations of the impacts. *In* The Collision of Comet Shoemaker-Levy 9 and Jupiter, K.S. Noll, H.A. Weaver, P.D. Feldman, Eds.: 121–132. Cambridge University Press.
11. CHODAS, P.K. & D.K. YEOMANS. 1996. The orbital motion and impact circumstances of comet Shoemaker-Levy 9. *In* The Collision of Comet Shoemaker-Levy 9 and Jupiter, K.S. Noll, H.A. Weaver, & P.D. Feldman, Eds.: 1–30. Cambridge University Press.
12. CRAWFORD, D.A. 1996. Models of fragment penetration and fireball evolution. *In* The Collision of Comet Shoemaker-Levy 9 and Jupiter, K.S. Noll, H.A. Weaver, & P.D. Feldman, Eds.: 133–156. Cambridge University Press.
13. CRAWFORD, D.A., M.B. BOSLOUGH, T.G. TRUCANO, & A.C. ROBINSON. 1994. The impact of comet Shoemaker-Levy 9 on Jupiter. Shock Waves. **4:** (1) 47–50.
14. CRAWFORD, D.A., M.B. BOSLOUGH, A.C. ROBINSON, & T.G. TRUCANO. 1995. Dependence of Shoemaker-Levy 9 impact fireball evolution on fragment size and mass. Lunar and Planetary Science XXVI: 291–292.
15. CRAWFORD, D.A., M.B. BOSLOUGH, T.G. TRUCANO, & A.C. ROBINSON. 1995. The impact of periodic comet Shoemaker-Levy 9 on Jupiter. Int. J. Impact. Engng. **17:** 253–262.
16. DEMING, D. & J. HARRINGTON. 1995. Unpublished presentation. IAU Colloquium 156, Baltimore, MD.
17. FAST, V.G. 1967. Statisticheskij analiz parametrov Tungusskogo vyvala. *In* Problema Tungusskogo Meteorita, Part 2: 40–61. Izdatelstvo Tomskogo Universiteta, Tomsk.
18. GRAHAM, J.R., I. DE PATER, J.G. JERNIGAN, M.C. LIU, & M.E. BROWN. 1995. The fragment R collision: W. M. Keck telescope observations of SL-9. Science. **267:** 1320–1323.

19. HAMMEL, H.B., R.F. BEEBE, A.P. INGERSOLL, ET AL. 1995. HST Imaging of atmospheric phenomena created by the impact of Comet Shoemaker-Levy 9. Science. **267:** 1288–1296.
20. HORD, C.W., W.R. PRYOR, A.I.F. STEWART, ET AL. 1995. Direct observations of the comet Shoemaker-Levy 9 fragment G impact by Galileo UVS. Geophys. Res. Lett. **22:** (13) 1565–1568.
21. KERLEY, G.I. 1991. User's Manual for PANDA II: A Computer Code for Calculating Equations of State: SAND88-2291. Sandia National Laboratories.
22. LIVENGOOD, T.A., G. BJORAKER, T. KOSTIUK, ET AL. 1994. Bull. Am. Astr. Soc. **26:** supplement.
23. MAC LOW, M.-M. 1996. Entry and fireball models vs. observations: What have we learned? *In* The Collision of Comet Shoemaker-Levy 9 and Jupiter, K.S. Noll, H.A. Weaver, & P.D. Feldman, Eds.: 157–182. Cambridge University Press.
24. MAILLARD, J.-P., P. DROSSART, B. BEZARD, ET AL. 1995. Methane and carbon monoxide infrared emissions observed at the Canada-France-Hawaii telescope during the collision of Comet SL-9 with Jupiter. Geophys. Res. Lett. **22:** (12) 1573–1576.
25. MARTEN, A., D. GAUTIER, M.J. GRIFFEN, ET AL. 1995. The collision of Comet Shoemaker-Levy 9 with Jupiter: Detection and evolution of HCN in the stratosphere of the planet. Geophys. Res. Lett. **22:** (12) 1589–1592.
26. MARTIN, T.Z., G.S. ORTON, L.D. TRAVIS, L.K. TAMPPARI, & I. CLAYPOOL. 1995. Observation of Shoemaker-Levy impacts by the Galileo photopolarimeter radiometer. Science. **268:** 1875–1878.
27. MCGLAUN, J.M., S.L. THOMPSON, & M.G. ELRICK. 1990. CTH: A three-dimensional shock-wave physics code. Int. J. Impact. Engineering. **10:** 351.
28. MELOSH, H.J. 1989. Impact Cratering a Geologic Process. Oxford University press. New York, NY. 245 pp.
29. MELOSH, H.J., N.M. SCHNEIDER, K. ZAHNLE, & D. LATHAM. 1990. Ignition of global wildfires at the Cretaceous/Tertiary boundary. Nature. **343:** 251–254.
30. MELOSH, H.J. 1991. Atmospheric impact processes. Adv. Space Res. **11:**(6) 87–93.
31. MORRISON, D., C.R. CHAPMAN, & P. SLOVIC. 1995. The impact hazard. *In* Hazards Due to Comets and Asteroids, T. Gehrels, Ed.: 59–91. 1995. Univ. Arizona Press, Tucson, Arizonia.
32. NEMCHINOV, I.V. 1994. Assessment of the 1 February 1994 Event. Interim Report 006 to Sandia National Laboratories.
33. NEMCHINOV, I.V. 1995. Analysis of the 12 February 1947 Event. Interim Report 01A to Sandia National Laboratories.
34. NICHOLSON, P.D., ET AL. Palomar Observations of the R Impact of Comet Shoemaker-Levy 9: I. Light Curves," Geophys. Res. Lett., 22(12), 1589-1592, 1995.
35. NICHOLSON, P.D. 1996. Earth-based observations of impact phenomena. *In* The Collision of Comet Shoemaker-Levy 9 and Jupiter, K.S. Noll, H.A. Weaver, & P.D. Feldman, Eds.a; 81–109. Cambridge University Press.
36. NOLL, K.S., H.A. WEAVER, P.D. FELDMAN, EDS. 1996. The Collision of Comet Shoemaker-Levy 9 and Jupiter. Cambridge University Press.
37. O'KEEFE, J.D. & T.J. AHRENS. 1982. The interaction of the Cretaceous/Tertiary Extinction Bolide with the atmosphere, ocean, and solid Earth. *In* Geological Implications of Impacts of Large Asteroids and Comets on the Earth, L.T. Silver & P.H. Schultz, Eds.: pp. 103–120. Geological Soc. of America Special Paper 190, Boulder: Geological Soc. of America.
38. REVELLE, D.O. 1997. Historical detection of atmospheric impacts by large bolides using acoustic-gravity waves. *In* Near-Earth Objects: The United Nations Conference on Near-Earth Objects, New York, April 24–26, 1995. Annals of the New York Academy of Sciences. This *volume*, **822:** 284–302.

39. ROOS-SEROTE, M., A. BARUCCI, J. CROVISIER, P. DROSSART, M. FULCHIGNONI, J. LECACHEUX, & F. ROQUES. 1995. Metallic emission lines during the impacts L and Q1 of Comet P/Shoemaker Levy 9 in Jupiter. Geophys. Res. Lett. **22:** (12) 1621–1624.
40. SHOEMAKER, E.M. 1995. Comet Shoemaker-Levy 9 at Jupiter. Geophys. Res. Lett., **22:** (12) 1555–1556.
41. SHOEMAKER, E.M., P.J. HASSIG, & D.J. RODDY. 1995. Numerical simulations of the Shoemaker-Levy 9 impact plumes and clouds: A progress report. Geophys. Res. Lett. **22:** (13) 1825–1828.
42. STELLINGWERF, R.F., N.M. HOFFMAN, & C.A. WINGATE. 1994. Bull. Am. Astr. Soc. **26:** 880.
43. SWEGLE, J.W. & A.C. ROBINSON. 1989. Acceleration instability in elastic-plastic solid. I. Numerical simulations of plate acceleration. J. Appl. Phys. **66:** 2838.
44. TAGLIAFERRI, E., R. SPALDING, C. JACOBS, P. WORDEN, & A. ERLICH. 1995. Detection of meteoroid impacts by optical sensors in Earth orbit. *In* Hazards Due to Comets and Asteroids, T. Gehrels, Ed.: 199–220. Univ. Arizona Press. Tucson, Arizona.
45. TAKEUCHI, S., H. HASEGAWA, J. WATANABE, ET AL. 1995. Near-IR imaging observations of the cometary impact into Jupiter. Geophys. Res. Lett. **22:** (12) 1581–1584.
46. THOMPSON, S.L. 1989. ANEOS Analytic Equations of State for Shock Physics Codes Input Manual: SAND89-2951. Sandia National Laboratories.
47. WASSON, J.T. 1995. *In* Lunar Planet Sci. XXVI: 1469–1470.
48. WEAVER, H.A., ET AL. 1995. The Hubble Space Telescope (HST) observing campaign on Comet Shoemaker-Levy 9. Science. **267:** 1282–1288.
49. WOOD, L., R. HYDE, M. ISHIKAWA, & E. TELLER. 1995. Cosmic Bombardment V: Threat Object-Dispersing Approaches to Active Planetary Defense, Planetary Defense Conference, LLNL.
50. ZAHNLE, K. 1990. Atmospheric chemistry by large impacts. *In* Global Catastrophes in Earth History, V.L. Krypton & P.D. Ward, Eds.: 271–288, Geological Soc. of America special Paper 247, Boulder: Geological Soc. of America.
51. ZAHNLE, K., & M.-M. Mac Low. 1994. The Collision of Jupiter and Comet Shoemaker-Levy-9. Icarus. **108:** 1–17.
52. ZAHNLE, K. 1996. Dynamics and chemistry of SL9 plumes. *In* The Collision of Comet Shoemaker-Levy 9 and Jupiter, K.S. Noll, H.A. Weaver, & P.D. Feldman, Eds. 183–212. Cambridge University Press.
53. ADUSHKIN, V.V. & I.V. NEMCHNOV. 1994. Consequences of impacts of cosmic bodies on the surface of the Earth. *In* Hazards Due to Comets and Asteroids, T. Gehrels, Ed.: 721–778. Univ. Arizona Press. Tucson, Arizona.
54. BEN-MENAHEM, A. 1975. Source parameters of the Siberian explosion of June 30, 1908, from analysis and synthesis of signals at four stations. Phys. Earth Planet. Inter. **11:** 1–35.
55. BJORAKER, G.L., S.R. STOLOVY, T.L. HERTER, G.E. GULL, & B.E. PIRGER. 1996. Detection of water after the collision of Fragment-G and Fragment-K of comet Shoemaker-Levy 9 with Jupiter. Icarus. **121:** 411–421.
56. CHYBA, C.F., P.J. THOMAS, & K.J. ZAHNLE. 1993. The 1908 Tunguska explosion: Atmospheric disruption of a stony asteriod. Nature. **361:** 40–44.
57. CLARKE, J.T., R. PRANG, G.E. BALLESTER, ET AL. 1995. HST far-ultraviolet imaging of Jupiter during the impacts of comet Shoemaker-Levy 9. Science. **267:** 1302–1307.
58. CRAWFORD, D.A. 1997. Comet Shoemaker-Levy 9 fragment size estimates: How big was the parent body? *In* Near-Earth Objects: The United Nations Conference on Near-Earth Objects, New York, April 24–26, 1995. Annals of the New York Academy of Sciences. This *volume*, **822:** 155–173.

59. FESENKOV, V.G. 1962. On the cometary nature of the Tunguska meteorite. Sov. Astron. English Translation. **5:** 441–451.
60. FLORENSKIY, K.P. 1963. Preliminary results from the 1961 combined Tunguska meteorite expedition. Meteoritica. **XXIII:** 3–37.
61. FORTOV, V.E., YU.N. GNEDIN, M.F. IVANOV. A.V. IVLEV, & B.A. KLUMOV. 1966. Collision of comet Shoemaker-Levy 9 with Jupiter: What did we see? Physics Uspekhi. **39:** 363–392.
62. GLASSTONE, S., ED. 1977. The Effects of Nuclear Weapons. U.S. Department of Defense/Department of Energy, Washington, D.C. 653 pp.
63. HAMMEL, H.B. 1996. HST imaging of Jupiter shortly after each impact: Plumes and fresh sites. *In* The Collision of Comet Shoemaker-Levy 9 and Jupiter, K.S. Noll, H.A. Weaver, & P.D. Feldman, Eds.: 111–120. Cambridge University Press.
64. HAMILTON, D.P., T.M. HERBST, A. RICHICHI, H. BÖHNHARDT, & J.L. ORTIZ. 1995. Calar Alto Observations of Shoemaker-Levy 9: Characteristics of the H and L Impacts. Geophys. Res. Lett. **22:** 2417.
65. HUNT, J.N., R. PALMER, & W. PENNEY. 1960. Atmospheric waves caused by large explosions. Philos. J.R. Soc. London. Sec. A. **252:** 275–315.
66. INGERSOLL, A. & H. KANAMORI. 1995. Waves from the impacts of Shoemaker-Levy 9 with Jupiter. Nature. **374:** 706–708.
67. KOROBEINICOV, V.P., P.I. CHUSHKIN, & L.V. SHURSHALOV. 1976. Mathematical model and computation of the Tunguska meteorite explosion. Acta Astronautica. **3:** 615–622.
68. KRINOV, E.L. 1963. The Tunguska and Sikhote-Alin meteorites. In the Moon, Meteorites, and Comets, B.M. Middlehurst & G,P, Kuiper, Eds.: 208–234. University of Chicago. Chicago, IL.
69. LAGAGE, P.O., P. GALDEMARD, E. PANTIN, ET AL. 1995. Collision of Shoemaker-Levy 9 fragments A, E, H, L, and Q1 with Jupiter: Mid-infrared light curves. Geophys. Res. Lett. **22:** 1773–1776.
70. MCGREGOR, P.J., P.D. NICHOLSON, & M.G. ALLEN. 1995. CASPIR observations of comet Shoemaker-Levy-9 with Jupiter. **121:** 361–388.
71. NORTON, O.R. 1997. Something is wrong at Sikhote-Alin, a mystery. Meteorite! **3:** (2) 34–35.
72. PRANGÉ, R., I.M. ENGLE, J.T. CLARKE, ET AL. 1995. Auroral signature of comet Shoemaker-Levy 9 in the Jovian atmosphere. Science. **267:** 1317–1320.
73. SEKANINA, Z. 1983. The Tunguska event: No cometary signature in evidence. The Astronomical Journal. **88:** 1382–1414.
74. SVETTSOV, V.V. 1995. A comet fall into the Jovian atmosphere. Solar System Research. **29:** 287–294.
75. TURCO, R.P., O.B. TOON, C. PARK, R.C. WHITTEN, J.B. POLLACK, & P. NOERDLINGER. 1982. An analysis of the physical, chemical, optical, and historical impacts of the 1908 Tunguska meteor fall. Icarus. **50:** 1–52.
76. YELLE, R.V. & M.A. MCGRATH. 1996. Ultraviolet spectroscopy of the SL9 impact sites. 1. The 175–230 nm region. Icarus. **119:** 90–11.
77. ZOLOTOV, A.B. 1967. Estimation of parameters of the Tungus meteorite based on new data. Sov. Phys. Dokl., English Translation. **11:** 108–111.
78. ZOTKIN, I.T. & M.A. TSIKULIN. 1966. Simulation of the explosion of the Tungus meteorite. Sov. Phys. Dokl., English Translation. **11:** 183–186.
79. Meadows, V. & D. Crisp. 1995. Impact plume composition from near-infrared spectroscopy. *In* Proceedings of the European SL-9/Jupiter workshop, R. West & H. Bonhardt, Eds.: 239.
80. Reed, J.W. 1987. Air-pressure waves from Mount St. Helens eruptions. J. Geophysical. Res. **92:** 11979–11992.

Atmospheric Screening of Comet and Asteroid Impacts

H. J. MELOSH

University of Arizona
Lunar and Planetary Laboratory
Space Sciences Building # 92
Tuscon, Arizona 85721

Small comets and asteroids that strike a planet possessing an atmosphere do not invariably strike the surface. The atmosphere resists penetration by high velocity projectiles in two ways. First, the projectile must push aside the atmospheric gases to reach the surface. This requires that the mass of gas swept aside by the projectile is roughly comparable to its own mass. For a stony object striking the Earth, this means that rigid meteoroids up to a few meters in diameter will be stopped in the atmosphere. The second important effect of the atmosphere is to crush fast-incoming meteoroids. The pressure in front of a projectile traveling with velocity v is ρv^2, where ρ is the density of the ambient atmosphere. The pressure in the wake behind it is near zero. This differential pressure tends to crush the meteoroid and disperse its fragments. Stresses on typical incoming objects may reach tens of kilobars in the lower atmosphere, large enough to shatter all but the strongest iron meteoroids. When an object is crushed, its fragments experience much more drag than they would experience as an intact body, and thus have more difficulty reaching the surface. Stony objects less than about a hundred meters in diameter do not reach the surface, instead depositing their energy in an airburst similar to that of the 1908 Tunguska, Siberia explosion. This process is well documented by crater size distributions and strewn fields produced by fragmented meteorites on both the Earth and on Venus, although the 100 times thicker atmosphere of Venus is proportionately more effective in screening approaching asteroids and comets. The size of the impact-generated strewn fields is typically a kilometer or two on Earth, but averages about 20 km on Venus.

Historical Detection of Atmospheric Impacts by Large Bolides Using Acoustic-Gravity Waves[a]

DOUGLAS O. REVELLE

Earth and Environmental Sciences Division
Los Alamos National Laboratory
Los Alamos, New Mexico 87545

ABSTRACT: During the period from about 1960 to the early 1980's a number of large bolides (meteor-fireballs) entered the atmosphere which were sufficiently large to generate blast waves during their drag interaction with the air. For example, the remnant of the blast wave from a single kiloton class event was subsequently detected by up to six ground arrays of microbarographs which were operated by the U.S. Air Force during this presatellite period. Data have also been obtained from other sources during this period as well and are also discussed in this summary of the historical data. The Air Force data have been analyzed in terms of their observable properties in order to infer the influx rate of NEOs (near-Earth objects) in the energy range from 0.2 to 1100 kt. The determined influx is in reasonable agreement with that determined by other methods currently available such as Rabinowitz [21], Ceplecha[4], [5] and by Chapman and Morrison [8] despite the fact that due to sampling deficiencies only a portion of the "true" flux of large bodies has been obtained by this method, i.e., only sources at relatively low elevations have been detected. Thus the weak, fragile cometary bodies which do not penetrate the atmosphere as deeply are less likely to have been sampled by this type of detection system. Future work using the proposed CTBT (Comprehensive Test Ban Treaty) global scale infrasonic network will be likely to improve upon this early estimate of the global influx of NEOs considerably.

I. INTRODUCTION

A. The Meteoroid-Atmosphere Interaction Spectrum

We now know that in the size range of interest of the NEO (near-Earth object) population, i.e., from about 1 m to several km across, the bolide population is comprised of at least ordinary chondritic materials, carbonaceous chondrites, regular cometary materials, "soft" cometary materials, and finally nickel-iron materials [4], [31]. ReVelle [30] has recently analyzed the entire spectrum from small to large bodies using a unified energetics approach that can be used to infer the observed behavior of entering bodies in a self consistent way. The observed bolide phenomena fit within this framework in such a way that if only four independent parameters are specified the consequences of the drag interaction with the atmosphere can be immediately anticipated. From this energetics viewpoint, only six discrete interaction

[a]This work was supported in part by the Department of Energy, Office of Nonproliferation and National Security.

regimes are predicted to occur. This unified pattern of the meteoroid-atmosphere interaction spectrum, along with its predictors are given in Table 1. Only regimes (iv) through (vi) are of direct interest to the NEO population under consideration with regard to blast wave generation and its subsequent decay to acoustic-gravity waves that can be detected by ground based sensors. The specific body size that corresponds with regimes (iv) through (vi) depends on the meteoroid group properties themselves (bulk density, ablation coefficient, material strength, etc.) as well as on entry parameters directly, i.e., initial velocity, entry angle, atmospheric scale height, surface pressure, etc.

B. The Blast Wave Interaction Remnants at Great Ranges

During the continuum flow interaction and the impact and explosion cratering regimes of bolides with the atmosphere, i.e., (iv) to (vi) above, a line source explosion is generated which subsequently decays first from a strong to a weak shock front and eventually to a nearly linear acoustic-gravity wave. In addition, for some bodies a strong, point source terminal explosive fragmentation event can occur near the end of the visible trajectory as well. Multiple explosions are also possible (for example, [38]). The recent impact of Shoemaker-Levy 9 with Jupiter make the need for realistic solutions of the large body impact regime with a planetary atmosphere far more urgent than previously considered [9].

It has long been known that the fundamental period of observed explosive waveforms (independent of nonlinear propagation effects or of geometric or material dispersion phenomena) is inversely related to the energy deposited into the fluid, i.e., higher fundamental frequencies correspond to smaller energy releases in the atmosphere. Using this consideration and the predictions of spherical and cylindrical blast wave theory, we can anticipate that the signals of interest should range from frequencies of a few Hz down to periods approaching 1 minute or more depending on the magnitude and altitude of the energy release by the entering bolide.

II. DETECTION OF AIRWAVES FROM LARGE BOLIDES

A. Measurement Systems

Conventional arrays of microbarographs with Daniels-type ported-pipe noise filters, were operated at numerous stations globally by NOAA, the U.S. Air Force (previously operated by the U.S. Army) with the prescribed system bandpass generally in one of two separate frequency intervals depending on the expected signal amplitudes (or ranges) and associated frequencies of interest.

TABLE 1. The Meteoroid-Atmosphere Interaction Spectrum

Interaction type	Local Knudsen number	α*	Changes in dimensionless ablation effiency
(i) Thermoshpere micrometeoroid regime	$\gg 1$	$\gg 1$	$\ll 1$
(ii) Free molecule flow regime	$\gg 1$	$\gg 1$	$\gg 1$
(iii) Transition flow regime	0(1)	> 1	$\gg 1$
(iv) Continuum flow regime	$\ll 1$	0(1)	> 1
(v) Impact cratering regime	$\ll 1$	<1	0(1)
(vi) Explosive cratering regime	$\ll 1$	$\ll 1$	$\ll 1$

* Surface pressure/modified ballistic entry parameter.

Associated Phenomena for each regime:
(i) Negligible mass loss and light production.
(ii) Extensive mass loss and concomitant light production.
(iii) Moderate to extensive mass loss; diffuse shock wave formation and light curve flaring.
(iv) Mass loss, highly size and velocity dependent; strong blast wave formation and propagation-hypersonic booms.
(v) Low velocity and negligible mass loss with direct Earth or oceanic impact. Bolide kinetic energy depletion height intersects the Earth's surface. Blast wave interaction with the ocean/land interface; Electrophonic noise through interaction with the geomagnetic field.
(vi) Climatic change effects, explosive cratering, tektites. Bolide kinetic energy > Atmospheric potential energy ($3.8 \cdot 10^{23}$ J). Negligible mass loss with large atmospheric changes expected. Blast wave propagation, Surface fires; Tsunamis, Electrophonic noise.

The basic sensor for all these measurements is a relatively high frequency acoustic capacitance microphone whose low frequency response is adjusted using a large backing volume which is connected through a leak valve to the background air pressure. Operating below the normal auditory hearing limits of humans, these microbarographs responded to the so-called atmospheric wave "zoo," i.e., a very large number of natural and man-made events such as aurora (including auroral electrojet surges), volcanic eruptions, meteor-fireballs (bolides), earthquakes, mountain associated waves, severe weather (tornadoes, hurricanes, thunderstorm convective processes, etc.), atmospheric boundary layer processes such as nocturnal low-level jets and wind shear events like gust fronts, etc., microbaroms (with the same source as microseisms in the Earth) the Polar Tropospheric jet stream, sub- and supersonic aircraft, commercial mining and other man-made chemical and nuclear explosions, etc. [10].

The M4 Signal Monitor operated by the U.S. Air Force had the following frequency bandpass characteristics [3]:

(i) High Frequency Filter N10: 3 db down at 8.2 Hz and 25 sec, respectively.

(ii) Low Frequency Filter N9: 3 db down at 44 and 440 sec, respectively, but a significant frequency response extended as low as 15 min in period.

In addition, there was also a separate very low frequency and close range sensor incorporated into the overall M4 system as well, the Millibarograph N2 system which operated within nearly a constant gain (+\– 1 db) over the range from 10^3 to 10 sec and could detect pressures as great as 10 mb (M4 Signal Monitor Technical Manual, March 7, 1972: TI 141-1MA-1). The CORAL N4 system (correlator–analyzer) provided continuous cross-correlation of all the relevant signals of interest. The entire system was routinely calibrated in both frequency and amplitude to ensure reliable operation in both detection and location capability in as near to continuous operation as possible.

Such infrasound stations are still in operation today at only a few stations such as Los Alamos National Laboratory [18], [41], [42], at Uppsala, Sweden [15], in Japan, in France, in the Netherlands, etc., but the probability is quite good that within 5 years' time from 60 to 75 stations will be operating worldwide as one integral part of the four proposed technologies of the CTBT (Comprehensive Test Ban Treaty) monitoring program. These four element arrays are currently proposed to have wide passband operation with a nominal response from 50 seconds period to about 10 Hz.

TABLE 2. Summary of Airwave Signal Velocity as a Function of Season, Based on Guiding Ozonospheric Winds from 30–50 km in the Northern Hemisphere

Season	Date	Propagation type	Propagation direction (degrees)	Signal velocity (m/sec)
Winter	10/16 to 03/14	Upwind	195–345	280
Winter	10/16 to 03/14	Downwind	40–140	315
Winter	10/16 to 03/14	Crosswind	345–40 140–195	306
Spring	03/15 to 05/15	Fall/Spring type	—	300
Summer	05/16 to 08/14	Upwind	15–165	280
Summer	05/16 to 08/14	Downwind	220–320	315
Summer	05/16 to 08/14	Crosswind	320–15 and 165–220	306
Fall	08/15 to 10/15	Fall/Spring type	—	300

Using beamforming and cross-correlation techniques, the following parameters are continuously available for signals with sufficiently large signal-to-noise ratios at each element of the microbarograph array (usually conditions of low wind noise):

(a) Amplitude: Signal-to-noise ratio;
(b) Period: For Lamb, compressional and gravity waves;
(c) Signal duration;
(d) Elevation arrival angle: Through the characteristic velocity (horizontal trace velocity);
(e) Azimuth arrival angle: At maximum cross-correlation;
(f) Power spectral density.

Signals consist of generally lower frequency Lamb waves traveling at signal velocities (travel distance/travel time) that average about 0.34 km/sec [20], [19] followed by acoustic arrivals at average signal speeds from 0.29 km/sec for refractive returns from the ozonosphere (about 50 km aloft) to about 0.24 km/sec for refractive returns from the lower ionosphere (about 100 km aloft). These waves are indicative of a very large part of the acoustic-gravity wave spectrum, which can also be subdivided at angular frequencies much larger than the Brunt-Väisällä (buoyancy frequency) as a class of waves known as infrasound [10]. For such waves, the restoring force to the wave motion is not buoyancy, as it is at low frequencies large compared to the inertial frequency, but instead is dominated by the elasticity of the medium and modified by stratification effects as compared to simple, high frequency, longitudinal and irrotational, linear acoustic waves.

These signal velocity values are indicative of data taken at Los Alamos, but are generally confirmed by the U.S. Air Force results provided in Table 2.

Specific techniques have been developed to facilitate analysis of both the Lamb wave and of the separate acoustic arrivals as will be discussed.

B. Historical Database

In Tables 3–7, the historic database of signals detected by such global scale microbarographs has been summarized. In Table 3, some of the event energies have also been confirmed by other techniques as well (for example, by seismic detection by Earth impact and for air-coupled Rayleigh waves, etc.). Unfortunately only the U.S. Air Force data have a truly global sampling efficiency as will be seen in Table 8, so that the other detections can not readily be used in estimating the global influx of NEOs. Previously these events were tabulated and interpreted in ReVelle and Wetherill [33], [34].

The classic event that basically started researchers working on this class of problems was the unique event of June 1908 in the Tunguska region of Siberia in the former U.S.S.R. and reported by F.J.W. Whipple and by numerous other later authors.

TABLE 3. Summary of Basic Meteoroid Airwave Events Taken by the U.S. Air Force during the Period from 1960–1974

Date	Source location	Origin time (GMT)	Total range (mi)	Source* energy (kt)
1:† 11/02/60	9N, 43E	0022	2488	10
2: 09/26/62	30N, 35E	1545	688	20
3: 09/27/62	32N, 60E	1529	518	30
4: 08/03/63	51S, 24E	1645	7038	1100
			8590	1100
5: 11/30/64	18N, 123W	0310	3243	10
6: 01/03/65	21N, 68E	2151	2008	0.2
7: 04/01/65	49N, 117W	0548	1552	0.24–2.4‡
			2173	0.24–2.4‡
8: 06/12/66	51N, 164E	0905	1800	8
			2750	8
			3000	8
			4150	8
9: 01/08/71	30N, 40E	1826	8632	6
10: 04/14/72	13S, 78E	1613	2300	14
			2700	14
			3400	14
			4850	14
			8000	14
			8550	14

*1 kt TNT = $4.185 \cdot 10^{12}$ Joules; Assuming E_s = 2*Yield.

†Originally estimated by the Air Force to be 1 kt.

‡The average E_s of about 1.3 kt is from E.M. Shoemaker [36] using a multimodal analysis from Pfeffer and Zarichny (1963). E_s = 26 kt was used for the NEO influx calculations, but since Eq. (3) predicted 44.6 kt, our influx is likely to be too low. Also, in Table 6A, E_s = for Revelstoke was about 69 kt. Bayer and Jordan [1] located its ground impact using infrasonic and seismic waves from multiple stations in the U.S. and Canada.

Other groups have also detected meteor-fireballs on occasion by either the recognition of its large amplitude, low frequency signature, or by noting an exceptionally large characteristic velocity (horizontal trace velocity across the array), or by aircraft reports or other visual confirmation of exceptionally bright, fast objects in the night sky. On occasion an exceptionally bright bolide also occurs during daytime, such as the famous August 10, 1972 skip fireball that was witnessed by Jacchia and whose orbital change was calculated initially by Ceplecha and later refined based on refined mass estimates [5]. This event was also recorded on two U.S. infrared satellite systems [22]. Infrasonic signals were also recorded from this object which was estimated to be about 5 meters across, but to date no infrasonic data have been available from the U.S. Air Force on this unique event, even though the author was informed that acoustic signals from this event were detected [44]. A similar, but much larger bolide with higher velocity was also observed by both visible and infrared military satellite sensors on February 1, 1994 [38], but only a marginal detection directly confirming its associated infrasonic signals have been reported [43], probably due to a lack of such sensors operating currently and to the relatively high frequency passband of most currently operational systems (> about 0.1 Hz). A very definite detection of either the impact of part of the original body or of the blast wave itself was made by hydroacoustic sensors in the Pacific however [39].

Other data recorded by NOAA (National Oceanographic and Atmospheric Administration) at the Wave Propagation Laboratory by A. Bedard, V. Goerke, G. Greene, J. Young and others as well as signals detected by USGS (U.S. Geologic Survey) by Shoemaker and by researchers at the University of Alaska (C. Wilson) are given in Table 5.

Analyses of these events as summarized by ReVelle [24], [25] and interpreted for line source explosion geometry are given in Table 6A.

TABLE 4. Summary of Detailed Airwave Data Taken by the U.S. Air Force during the Period from 1960–1974 (Maximum values for all elements of all the arrays detecting the events)

Date	Maximum peak compression amplitude (microbar)	Lamb period (sec)	Maximum compression period (sec)	Total duration (min)
1: 11/02/60	2.4	ND*	11.1	18
2: 09/26/62	0.6	13.9	13.9	10+
3: 09/27/62	4.2	19.4	17.1	14+
4: 08/03/63	1.9	ND*	48.0	10+
5: 11/30/64	2.1+	13.3	10.0	6.0
6: 01/03/65	1.1	3.5	3.5	5.0+
7: 04/01/65	11.4	36.0	15.0	26
8: 06/12/66	2.1	15.0	10.0	13
9: 01/08/71	0.6	14.6	10.0	< 25
10: 04/14/72	4.3	34.5	13.5	27

*ND denotes no detection of the Lamb wave.

TABLE 5. Summary of Meteoroid Airwave Data Taken by NOAA Wave Propagation Laboratory (Boulder), by the U.S. Geological Survey (Flagstaff) and by the University of Alaska

Date*	Maximum amplitude (μbar)	Period at max. ampl. (sec)	Total duration (min)	Total range (km)
1: 04/01/65	4.0	16.0	20	1550
2a: 12/14/68 Flagstaff	0.25	4.0	12	134
2b: 12/14/68 Boulder	1.1	2.5	10	720
3: 09/17/66	1.25	54	34	2270
4: 12/19/69	2.3	12	> 12	327

*1: Revelstoke Meteorite [1].

2a: Holbrook: Observed from Boulder, Colorado (V.H. Goerke, 1972, personal communication).

2b: Holbrook: Observed from Flagstaff, Arizona (E.M. Shoemaker, 1972, personal communication).

3: Kincardine Fireball/Meteorite [45], [46].

4: College, Alaska Fireball (Johnson and Wilson, 1972, personal communication).

We can also compare these estimates by ReVelle [25] with those deduced using the semiempirical period at maximum amplitude relations given in Section III.C, Eq. (3). From this relation which has been extensively tested and confirmed by alternative detection methods, the doubled source energy for each of these events is given in Table 6B.

It would be very beneficial to further evaluate and refine these data by using both realistic multimodal analyses [19], i.e., including zonal and meridional steady state winds and by using the pressure amplitude wind normalization factor developed by Mutschlecner and Whitaker [18]. This refinement of the above source energy estimates could impact upon the NEO flux estimate and associated error limits discussed in Section III.E.

Still other data have been recorded using relatively high frequency sensors similar to those in use today at Los Alamos (Globe Universal Sciences Model 100 C.). For these sensors the –3 db bandpass is from about 10 to 0.1 Hz. These data were recorded at Boulder by N.O.A.A. researchers, by the University of Michigan in Sioux Falls, S. Dakota and at the National Research Council Canada at the now defunct Springhill Meteor Observatory. The first event was confirmed by visual and by aircraft reports on the night in question, the second was confirmed by association with the photographed event, PN42556 of the Smithsonian's U.S. Prairie Network, and the latter event was detected in association with a meteor patrol radar during the Geminid meteor shower in 1975. These events and their analyzed properties are detailed in Table 7.

TABLE 6A. Analysis of Table 5: Meteoroid Airwave Data (from [25]).

Event	R_o(m)	Diameter (m)	Mass (kg)	Kinetic energy (kt)
1:	518.6	13.7	$4 \cdot 10^6$	69.1
2a: *	177.4	5.01	$2.0 \cdot 10^4$ to $2.0 \cdot 10^5$	0.3–3.1
2b: *	54.1	1.53	$5.6 \cdot 10^2$ to $5.6 \cdot 10^3$	.01–0.1
3: *	1761	49.7	$1.9 \cdot 10^7$ to $1.9 \cdot 10^8$	285.7 to 2857.1
4: *	452	12.8	$3.3 \cdot 10^5$ to $3.3 \cdot 10^6$	4.76 to 47.6

*For these events the mass, diameter, and energy estimates were made using V = 11.2 km/sec. For V = 20 km/sec, the energy estimates would all be lowered by a factor of 4.

TABLE 6B. Comparison of Source Energy Estimates Using the Line-source Approach in Revelle [25] with the Semi-empirical Period at Maximum Amplitude Method in IIIC.

1: 55.3 kt — About 20% < the average value in Table 6A.
2a: 0.54 kt — About 3.15 × < the average value in Table 6A.
2b: 0.11 kt — About 2 × > the average value in Table 6A.
3: 3.22 MT — About 2 × > the average value in Table 6A. (Alternatively using Eq. (4) this becomes 7.30 MT or 4.65 × > the average value in Table 6A).
4: 21.2 kt — About 20% < the average value in Table 6A.

III. ANALYSIS OF AIRWAVE DATA FROM LARGE BOLIDES

A. Ray Theory Approaches

This approach allows the altitude of the source and the azimuth arrival angle to deduced. As noted earlier depending on the altitude of the ducted refractive return, differing signal velocities are expected on the basis of previous U.S. Air Force and Los Alamos experiences. Also in the case of PN42556 (U.S. Prairie Network Fireball Number), Kraemer and Bartman [12] were able to ray trace from the ground array back to the source with very great precision. After traveling about 120 km, the original back ray tracing missed the known three-dimensional trajectory by only 410 meters.

TABLE 7. Other Higher Frequency Airwave Meteor Data

Event	Maximum amplitude (μbar)	Period at max. ampl. (sec)	Total range (km)	Yield* (kt)
1:	9	1	250	$2.63 \cdot 10^{-3}$ $1.1 \cdot 10^{10}$ J
2:	6.6	0.5–1	240	$1.01 \cdot 10^{-3}$ $4.23 \cdot 10^{9}$ J
3:	1.4–2.3	0.21–.22	130	$1.55 \cdot 10^{-5}$ $6.51 \cdot 10^{7}$ J

1: Bedard and Greene [2]. An independent estimate of the source yield was not available, but the event was certainly quite energetic since visual magnitudes from –5 to brighter than the full moon were estimated by ground and airplane observers on April 22, 1975. This event was detected by two arrays, 40 km apart (at Boulder and Fraser, Colorado). One array received signals from both the ozonospheric and the lower ionospheric sound channels while the other array only detected the ozonospheric return.

2: McIntosh, Watson, and ReVelle [17]. On December 14, 1974 the same Geminid meteor was also detected by a patrol radar at the Springhill Meteor Observatory with V_∞ = 35 km/sec. The above infrasonic source energy (using Eq. (3) with E_s = 2*Yield) is 3.56 times greater at a period of 5.0 Hz than that computed by McIntosh et al. for this event.

3: Kraemer and Bartman [12]. In this case the infrasonic estimate of the doubled source yield is about 3.0 times greater than that deduced for the 320 gram (photometric mass) carbonaceous chondrite (Ceplecha and McCrosky fireball Group II) traveling at a measured initial velocity of 16.5 km/sec. Due to the unusual geometry of this event relative to the ground, the initial signals recorded on May 24, 1975 were from the upper part of the meteor trail (PN42556).

*Based on the period at maximum amplitude relation given in III.C, in Eq. (3).

For truly line source signatures the entry angle of the trajectory is very critical, since the steep angled entries, which are typical of smaller bodies produce corresponding ray paths which are less likely to reach ground level [25]. For corresponding shallower entries, ray paths are more nearly vertical and consequently are less likely to suffer from significant refraction effects. Such arguments are more useful at close range to the entry which seems most applicable to only one of the ten U.S. Air Force events (0.2 kt event).

Numerous acoustical data were reported from the first meteorite which was photographed and subsequently recovered on the ground, the Pribram meteorite in the Czech Republic (formerly Czechoslovakia) in 1958. These data were catalogued by Ceplecha and illustrated the well-known anomalous propagation of sound effects first noted dramatically at Queen Victoria's funeral in Great Britain. In such cases semi-spherical zones of audibility and of silence are found surrounding the acoustical source region. These zones are readily predictable on the basis of ray acoustics if the corresponding atmospheric properties, primarily the temperature and wind structure, are sufficiently well known.

B. Lamb Wave Approach

Pierce and Kinney [19] and co-workers have developed a prediction of source yield, Y, as a function of the arriving Lamb wave period, T, at corresponding amplitude, Δp, which can be written for distances, r, small compared to the Earth's radius in the limiting form:

$$\log(Y) = 1.5 \cdot \log(T) - 3.37 + \log(\Delta p) + 0.5 \cdot \log(r), \quad (1)$$

with Δp in μbars, T in seconds, r in km, and Y in kt TNT.

Lamb (or edge) waves, which are guided acoustic waves at high frequencies and guided gravity waves at low frequencies, take time to develop in association with a given source. These waves are vertically evanescent and are only of significant amplitude near the Earth's surface. Thus, for closer range and progressive higher frequency, these waves are less likely to be observable [32]. From the analysis of the meteor data, ReVelle and Delinger [32] were able to deduce an empirical fit to the arriving Lamb waves in the form:

$$\log(Y) = 2.0 \cdot \log(T) - 3.18 + \log(\Delta p) + 0.5 \cdot \log(r); \quad r^2 = 0.58. \quad (2)$$

C. Period at Maximum Amplitude Approach

This approach is semiempirical and relates the period at maximum amplitude to the source energy at sufficiently large range from the source. In a private communication between the author and Dr. G. Leies [13] the following relations were disclosed (deduced from the data from low altitude nuclear explosions listed in S. Glasstone, *Effects of Nuclear Weapons*, 1968), connecting the yield, Y, in kt for each explosion and the observed period, P, at the observed maximum amplitude of the arriving acoustic signal:

$$\log(Y) = 3.34 \cdot \log(P) - 2.58 : \quad Y \leq 100 \text{ kt}, \quad (3)$$

$$\log(Y) = 4.14 \cdot \log(P) - 3.61 : \quad Y > 40 \text{ kt}. \quad (4)$$

In the Air Force analysis of the meteor-fireball data, an equation similar to (3) was used for the analysis for events < 40 kt and (4) was used for events exceeding 40 kt. The yield in these equations is the source energy release, E_s, divided by 2, corresponding to the large fraction of radiation emitted during a low altitude nuclear explosion event. This is justified since the original microbarograph data recordings used in the empirical least-squares curve-fit all originated from low altitude nuclear explosion events.

Motivated by the Air Force meteoroid airwave data, ReVelle and Delinger [32] have also interconnected the Lamb wave approach and the period at maximum amplitude approach and derived an expression for the Lamb wave period as a function of the period at maximum amplitude of the arriving explosive source waveform.

D. Waveguide Normal Mode Analysis

Again, Posey and Pierce [20] and Pierce and Kinney [19] and other earlier co-workers have developed waveguide modal analyses that allow an estimation of the normal modes that are excited by a point source explosion in the atmosphere. Such a scheme was used by Shoemaker [36] to estimate the energy release of the Revelstoke meteorite [1]. Such analyses are typically of greatest utility far from the source region in contrast to the ray approaches which are more applicable at relatively close range. As discussed in ReVelle [28], Golitsyn and co-workers and Korobeinikov, Chuskin and Shurshalov have considered the theoretical modeling of bolides from the line source viewpoint and in the latter case with the addition of a terminal point source-like explosion at the end of the visible trajectory. In the latter case the emphasis was on the modeling of the Tunguska fireball of 1908.

For completeness sake, we should also note that numerous recordings have also been made of air-coupled Rayleigh waves from large bolides at sufficiently low altitudes [28]. Such waves are traveling in the Earth at the speed of sound in air and are readily detected by conventional seismic sensors.

TABLE 8. Summary of the U.S. Air Force, Global Airwave Detection Probability, P_D (Percent Coverage of Earth), versus Yield and Season

Season	Yield (kt)	Percent coverage of Earth (%)
Winter*	1	11
(10/16–03/14)	3	18
	10	27
	30	48
	100	71
	300	87
Summer†	1	6
(05/16–08/14)	3	36
	10	43
	30	63
	100	86
	300	94

*Winter — Least-squares curve-fit at maximum correlation:

$$P_D(\%) = 11.74 \cdot Y^{0.375};\ r^2 = 0.9928.$$

†Summer — Least-squares curve-fit at maximum correlation:

$$P_D(\%) = 11.35 + 15.19 \cdot \ln(Y);\ r^2 = 0.9870.$$

E. NEO Influx Rate Estimation

Using the percentage area coverage of the Earth information in Table 8 and the 10 events listed in Table 3, all provided by the U.S. Air Force, ReVelle [28] was able to predict the influx rate of NEOs as a function of source energy and also the cumulative mass influx per year over the Earth, assuming that these signals emanated from low altitude near-point source type explosions.

We have recently refined two of the E_s values in Table 3 (Events 1 and 7) and again have determined the influx rate of NEOs in the curve-fitted form (cumulative number of bodies with the source energy ≥ E_s (in kt) per year over the entire Earth; see Fig. 1):

$$N(E_s) = 7.17 \cdot E_s^{-0.731};\ r^2 = 0.9235, \tag{5a}$$

or in the CGS equivalent, cumulative flux forms, with E_s in ergs:

$$N_f(E_s) = 9.78 \cdot 10^{-12} \cdot E_s^{-0.731}, \tag{5b}$$

$$\log(N_f(E_s)) = -11.01 - 0.731 \cdot \log(E_s), \tag{5c}$$

with both of the latter results in units of number of bolides per square cm per second.

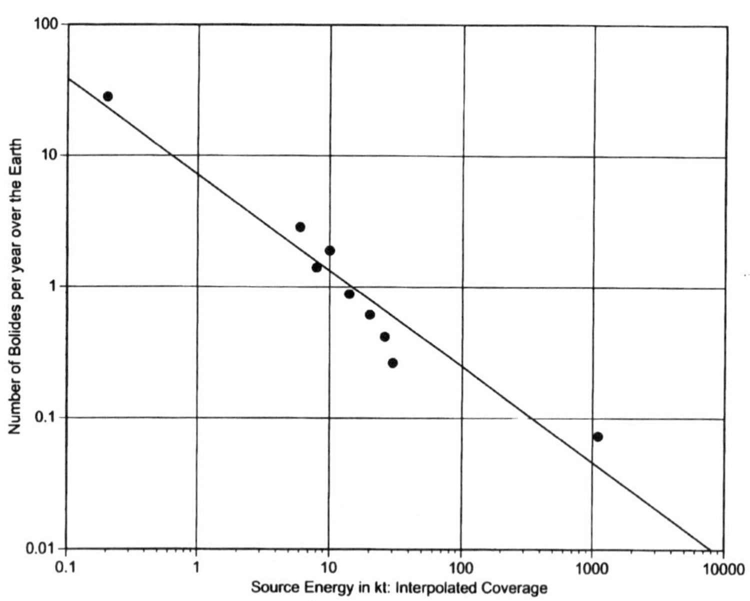

FIGURE 1. Predicted cumulative influx rate (number per year over the Earth) for source energies ≥ E_s (in kt).

Equation (5a), was obtained by computing the highest correlation, least-squares curve-fit of E_s versus the cumulative number of events per year on Earth whose source energies equalled or exceeded each of the individual event source energies (assuming constant coefficients over the entire source energy range). We have also tried various other extremes situations, such as excluding either the 1100 kt or the 0.2 kt events or excluding both during the curve-fitting process. The resulting equations were generally not greatly different from the above result which used all 10 of the events, with the exception of the case of excluding both extreme events from the data prior to curve-fitting, which in retrospect was clearly not a very realistic possibility. Clearly, a more refined analysis should now be determined using all the possible waveform analysis techniques discussed in this article along with a more refined discussion of all the possible errors.

Thus, by multiplying the cumulative number of events for each bolide by the interpolated value of the inverse probability of detection for each event in the appropriate season from Table 8 (the corresponding percent coverage of the Earth's surface at the deduced yield or $E_s/2$) and by dividing by the total time that the Air Force global network was in operation, i.e., 13.67 years, we were able to obtain the number of events expected at a specified E_s per year of observing over the entire Earth. Also, by multiplying each individual E_s by these same factors, we were able to determine the total integrated amount of bolide energy released per year over the entire Earth to be about 105.5 kt. This can be reduced to a cumulative mass value by assuming a mean entry velocity of about 17 km/sec (for the deeply penetrating part of the NEOs, i.e., the noncometary part of the flux) to a value of about $3.1 \cdot 10^6$ kg/year over the mass range from about 10^3 to 10^8 kg. Wetherill and ReVelle [33], [34] had earlier determined a preliminary value with this same data to be about 10^6 kg/year using an average velocity of 20 km/sec (for an integrated energy release over the Earth per year of 102.9 kt). The analysis of these data reveal that we should expect at least one 15 kt, deeply penetrating event to occur every year, on the average, over the entire Earth which is in excellent agreement with the estimates available from the analysis of DOD satellite data [39]. In addition, using 24 MT as the equivalent energy release of the Tunguska event (1908), as adopted in ReVelle [28], we can estimate using (5a) that events of this energy can reoccur once on a timescale of about every 222 years over the entire globe.

ReVelle [28] has studied the possible error limits for using differing source altitudes above the ground and of the overall yield-source energy relationship and concluded that an equation very similar to (5a) is as reliable as can be deduced given the observable data and its associated error bar uncertainties (due to yield uncertainties, statistical counting errors, etc.). Allowing for these various uncertainties in analyzing the data, we can conclude that the corresponding uncertainty in the mass influx is about a factor of 2 larger or smaller than the value given above over the stated mass range. Another factor not considered previously however is that if we include the other infrasonic detections that were also made during the period from 1960 to the middle of the 1970's, besides just the Air Force data, the influx rate will become even higher. For example, the Kincardine event in 1966 which most likely dropped meteorites into Lake Huron, was clearly a very significant multi-megaton event which apparently wasn't detected by the Air Force system for unknown reasons, but which was detected by NOAA in Boulder.

The associated yields were determined using a semiempirical method connected with the period at maximum amplitude of the arriving signals for each of the ten U.S. Air Force events given in Table 3 (a self-consistent global scale data set) and using other independently derived yields for some of the events. For example, in the case of the Revelstoke Meteorite, both atmospheric and seismic signals were reliably recorded [1] as noted earlier. An equation similar to (5a) was developed earlier by E.M. Shoemaker [36] using a somewhat similar data set, but with a different yield-period scaling law. As a result a great overestimation (by about 10 times) of the NEO influx was predicted by Shoemaker and colleagues in the late 1960's, just prior to the first U.S. manned landing on the Moon [37].

These results can be compared with those presented more recently by Rabinowitz [21], by Ceplecha [4], [6], by Chapman and Morrison [8] and also those recorded long ago during the operation of the U.S. Prairie Network [16]. Our data are in reasonable agreement with the values deduced by Rabinowitz and co-workers using the Spacewatch Telescope. For example, our least-squares result given in (5a) above, predicts that about one 15 kt event will occur over the Earth in a year's time as noted above. Ceplecha [4] has combined several meteor and fireball detection techniques in order to determine the cumulative influx rate. His cumulative mass/year over the entire Earth is about $1.7 \cdot 10^8$ kg, over the entire range from the micrometeoroids to the ten's of meters size range detected by satellite observing systems. This is also in reasonable agreement with our deduced value (about 55 times higher), if we consider that we have probably not completely sampled all the types of compositional materials arriving, and also, because we have only sampled over a subset of the above, immense size range.

Ceplecha [7] has also recently refined the global mass influx on the basis of a new analysis of the Lost City meteorite (a bronzite chondrite) and its luminous efficiency [27] in the presence of rotation. In the mass range which is relevant to the Air Force infrasound data set, this revision results in a predicted global mass influx which is smaller by about a factor of two compared to the earlier predictions. Also, in Ceplecha [5], it was confirmed that for bodies larger than about 10 meters across, the most probable arriving composition is of the Group IIIA/IIIB type (regular and weak cometary bodies). Such bodies are known in numerous studies not to penetrate the atmosphere very effectively and as such are far less likely to have been detected by the infrasonic global arrays.

Also, Chapman and Morrison [8] adopted Shoemaker's cumulative flux curve which is based in part on crater counting statistics for the Earth and the Moon and in part on the infrasonic data as well. Quite good agreement is found despite the fact that the acoustic-gravity wave estimate is likely to be deficient because of under-sampling of the cometary part of the flux whose atmospheric end height typically is much larger than that corresponding to either chondritic materials of reasonable strength or of nickel-iron materials [26], [27], [28], [29], [30].

Finally, we note that the form of our latest result is quite similar to that obtained by direct photographic techniques described in McCrosky [16] who determined the cumulative CGS influx (in number of bodies per square cm per second over the energy range from 10^{15} to 10^{19} ergs) to be:

$$\log(N_f(E_s)) = -10.5 - 0.67 \cdot \log(E_s). \tag{6}$$

Even though there are similarities, there are also serious discrepancies between the predictions made using (5c) and using (6), with the cumulative flux predicted being much larger in the latter case for a specified E_s. These differences are reduced somewhat by using E_s values in (6) that are about 10 times smaller than those proposed by McCrosky, with all other factors being held constant. This overestimation of the meteoroid source energy (mass) by a factor of about 10 proposed earlier by ReVelle and by ReVelle and Wetherill was caused by McCrosky's use of a luminous efficiency factor which was about 10 times too small during the integration of the meteoroid's initial photometric mass. The major factor, however, distinguishing predictions made using (6) from those made using (5c), is that McCrosky's result potentially includes all bolide compositional types. In the fireball range, the structurally weak, cometary type IIIA/IIIB bodies outnumber the type I chondrites by about a factor of three [40]. As shown by Ceplecha [5], this trend continues even into the very largest size range that presently can be observed as well.

IV. SUMMARY AND CONCLUSIONS

A. Airwaves from Large Bolides

During the period from the early 1960's to the early 1980's a large body of acoustic-gravity wave signatures from large bodies entering through the atmosphere have been obtained by a number of research groups. The largest self-consistent dataset by far was taken by the U.S. Air Force from a global scale network operating for about 14 years. From this data we have been able to deduce the global influx rate of NEOs with the limitation that higher altitude cometary sources have probably not been detected by any of these networks. Despite this limitation, the influx rate determined by this method is in reasonable agreement with results obtained more recently by Rabinowitz [21], by Ceplecha [4], [6], and by Chapman and Morrison [8].

B. Future Systems

In the near future the CTBT global scale effort for monitoring of explosive events will allow a great increase in the number of bolide events that can be observed annually. Since 60 to 75, 4 element arrays of broadband infrasound stations have been proposed as part of the International Monitoring System, the prospect for retrieving a revised estimate for the global influx rate of NEOs, at least of the deep penetrating part of the influx, i.e., chondrites and nickel-iron type materials, should be greatly increased.

REFERENCES

1. BAYER, K.C. & J.N. JORDAN. 1967. Seismic and acoustic waves from a meteor (abstract). J. Acoust. Soc. Amer. **41:** 1580.
2. BEDARD, A.J. & G.E. GREENE. 1981. G, Case study using arrays of infrasonic microphones to detect and locate meteors and meteorites. J. Acoust. Soc. Amer. **69:** 1277–1279.
3. BAKER, H.M. 1979. Personal communication.
4. CEPLECHA, Z. 1922. Influx of interplanetary bodies onto Earth. Astron. Astrophys. **263:** 361–366.
5. CEPLECHA, Z. 1994a. Earth-grazing daylight fireball of August 10, 1972. Astron. Astrophys. **283:** 287–288.
6. CEPLECHA, Z. 1994b. Impacts of meteoroids larger than 1 m into the Earth's atmosphere, Astron. Astrophys. **286:** 967–970.
7. CEPLECHA, Z. 1995. Personal communication.
8. CHAPMAN, C.R. & D. MORRISON. 1994. Impacts on the Earth by asteroids and comets: assessing the hazard. Nature. **367:** 33–40.
9. CRAWFORD, D.A., M.B. BOSLOUGH, T.G. TRUCANO, & A.C. ROBINSON. 1994. The impact of Comet Shoemaker-Levy 9 with Jupiter. Shock Waves. **4:** 47–50.
10. GEORGES, T.M. & J.M. YOUNG. 1972. Passive sensing of natural acoustic-gravity waves at the Earth's Surface. *In* Remote Sensing of the Troposphere, V.E. Derr, Ed.: 21-1-21-23. U.S. Government Printing Office.
11. GORDON, E., C.D. BARTKY, & J.W. WAGER. 1975. Dynamics of a large meteor, AIAA Paper 75-14, AIAA 13th Aerospace Sci. Meet., Pasadena, Ca., January 20–22: 1–6.
12. KRAEMER, D.R. & F.L. BARTMAN. 1981. Infrasound from accurately measured meteor trails. *In* Proceed. Intern. Sympos. Acoustic Remote Sensing of the Atmos. and the Oceans: V.31 to V.49. The Univ. of Calgary Press. Calgary, Alberta, Canada.
13. LEIES, G.1978. Personal communication.
14. LI, F., C.B. BARTKY, R.W. WAGER, & E. GORDON. 1975. Determination of the size of a large meteor. AIAA 13th Aerospace Sc. Meet., Jan. 20–22, Pasadena, Ca.: AIAA Paper 75-200.
15. LISKA, L. 1981. Studies of the stratospheric and mesospheric wind system and temperature distribution from observations of infrasonic waves generated during regular

concorde flights. *In* Proceed. Intern. Sympos. Acoustic Remote Sens. Atmos. and the Oceans: V.16 –V.30. The Univ. of Calgary Press. Calgary, Alberta, Canada.

16. McCrosky, R.E. 1968. Distribution of Large Meteoritic Bodies, Smithsonain Astrophysical Observatory, Special Report 280.
17. McIntosh, B.A., M.D. Watson, & D.O. ReVelle. 1976. Infrasound from a radar-observed meteor. Canad. J. Physics. **54:** 655–662.
18. Mutschlecner, J.P. & R.W. Whitaker. 1990. The correction of infrasound signals for upper atmospheric winds. *In* Fourth International Symposium on Long-Range Sound Propagation, NASA Conf. Publication 3101: 143–153. Compiled by W.L. Wilshire, NASA Langley Research Center, Hampton, Va.
19. Pierce, A.D. & W. A. Kinney. 1976. Computational techniques for the study of infrasound propagation in the atmosphere. AFGL Report TR-76-0056, 186 pp.
20. Posey, J.W. & A.D. Pierce. 1971. Estimation of nuclear explosion energies from microbarograph records. Nature. **232:** 253.
21. Rabinowitz, D.L. 1992. The flux of small asteroids near the Earth. *In* Asteroids, Comets, Meteors, 1991: 481–485. Flagstaff, Arizona, Lunar and Planet. Inst.
22. Rawcliffe, R.D., C.D. Bartky, F. Li, E. Gordon, & D. Carta. 1974. Meteor of August 10, 1972. Nature. **247:** 449–450.
23. Rawcliffe, R.W. 1979. Satellite Observations of meteors (abstract). Astrophys. J. **228:** 338.
24. ReVelle, D.O. 1975. Studies of sounds from meteors. Sky and Telescope. **49:** (2) 87–91.
25. ReVelle, D.O. 1976. On meteor-generated infrasound. J. Geophys. Res. **81:** 1217–1237.
26. ReVelle, D.O. 1979. A quasi-simple ablation model for large meteorite entry: Theory versus observations. J. Atmos. Terr. Phys. **41:** 453–473.
27. ReVelle, D.O. 1980a. A predictive macroscopic integral radiation efficiency model. J. Geophys. Res. **85:** 1803–1808.
28. ReVelle, D.O. 1980b. Interaction of Large Bodies with the Earth's Atmosphere, in Solid Particles in the Solar System, Intern. Astron. Sympos. No. 90, I. Halliday & B.A. McIntosh, Eds.: 185–198.
29. ReVelle, D.O. 1987. Theoretical fireball end heights. *In* Handbook for MAP, Vol. **25**, First GLOBMET Symposium (Dushanbe, U.S.S.R.), R.G. Roper, Ed.: 255–257. SCOSTEP Secretariot, Urbana, Illinois.
30. ReVelle, D.O. 1993. The meteoroid-atmosphere interaction spectrum. *In* Meteoroids and Their Parent Bodies, Internat. Astron. Sympos., Bratislava, Slovakia, J. Stohl & I.P. Williams, Eds.: 343–346.
31. ReVelle, D.O. & Z. Ceplecha. 1994. Analysis of identified iron meteoroids: Possible relation with M-type Earth-crossing asteroids? Astron. Astrophys. **292:** 330–336.
32. ReVelle, D.O. & W.G. Delinger. 1981. Passive acoustic remote sensing of infrasound of natural origin. Proceed. Internat. Sympos. Acoustic Remote Sens. of the Atmosphere and Oceans: V.6 –V.15. Univ. of Calgary Press, Calgary, Alberta, Canada.
33. ReVelle, D.O. & G.W. Wetherill. 1978a. Terrestrial microbarograph "airwave" recordings: The global influx rate of large meteoroids. Department of Terrestrial Magnetism, Carnegie Institution of Washington Yearbook, 490–492.
34. ReVelle, D.O. & G.W. Wetherill. 1978b. Influx of large bodies using terrestrial airwave data (abstract). J. Meteoritical Soc. **13:** 611.
35. ReVelle, D.O. & G.W. Wetherill. 1980. Atmospheric waves from large meteors (abstract), J. Acoust. Soc. Amer. Suppl. 1. **68:** S4.
36. Shoemaker, E.M. 1971. Personal communication.

37. SHOEMAKER, E.M. & C. J. LOWERY. 1967. Airwaves from large meteoroids (abstract). J. Meteorit. Soc. **3:** 123.
38. SPALDING, R., E. TAGLIEFERRI, & D.O. REVELLE. 1994. Preliminary Analysis of the February 1, 1994, Marshall Islands Fireball (abstract). 106th Meeting of Astron. Soc. Pacific, Preprint Vol. *In* Completing the Inventory of the Solar System, T. Bowell, Ed. (50th Anniversary of the Founding of the Lowell Observatory), Flagstaff, Arizona.
39. SPALDING, R.E. 1995. Personal Communication.
40. WETHERILL, G.W. & D.O. REVELLE. 1981. Which fireballs are meteorites? A study of the large meteor deceleration data. Icarus. **48:** 308–328.
41. WHITAKER, R.W., J.P. MUTSCHLECNER, M. DAVIDSON, & S.D. NOEL. 1990. Infrasonic observations of large-scale HE events. *In* Fourth International Symposium on Long-Range Sound Propagation, NASA Conf. Public. 3101: 133–141. Complied by W.L. Willshire, NASA Langley Res. Center, Hampton, Va.
42. WHITAKER, R.W., S.D. NOEL, & W.R. MEADOWS. 1994. Infrasonic observations and modeling of the minor uncle high explosive event. *In* Sixth International Symposium on Long-Range Sound Propagation, National Research Council Canada. 480–497. Compiled by D.I. Havelock and M.R. Stinson, Ottawa, Ontario, Canada.
43. WHITAKER, R.W. 1994. Personal communication.
44. WHIPPLE, F.L. 1975. Personal communication.
45. CHAMBERLIN, V.D. 1968. Meteorites of Michigan. Bull. No. 5 of the Geological Survey of Michigan. Lansing, Michigan.
46. GEORKE, V.H. 1966. Intrasonic observations of a fireball. Sky and Telescope. **32:** 313.

Analysis of Satellite Observations of Large Meteoroid Impacts

I.V. NEMTCHINOV,[a] C. JACOBS,[b] AND E. TAGLIAFERRI[c]

[a]*Russian Academy of Sciences*
Institute for Dynamics of Geospheres
38 Leninsky Prospect (Bldg. 6)
Moscow 117979, Russia

[b]*7740 Geneva Drive*
Gurnee, Illinois 60031
and
[c]*ET Space Systems*
5990 Worth Way
Camarillo, California 93012

ABSTRACT: Observations from high altitude (> 20,000 km) space-based infrared and optical sensors which have detected over 200 bright flashes in the atmosphere since 1972 are interpreted in terms of geographical distribution, lightcurves, aerodynamic interaction of the fragments, and heat transfer and luminosity coefficients. The luminosity efficiency for meteorite explosive disintegration is estimated to be about 2 to 3 times smaller than for energetically equivalent nuclear explosions. Explanations for this difference are offered.

INTRODUCTION

Space-based infrared and optical sensors operated by the United States Department of Defense have detected over 200 bright flashes in the atmosphere since 1972. These intense light impulses were caused by impacts of large meteoroids. The energy is released in the atmosphere upon explosive disintegration of the meteoroids due to the action of aerodynamic forces Usually meteoroids deposited their energy at high altitudes above the Earth's surface, mainly at the altitudes of 30 to 45 km, but some of them penetrated deep into the dense layers of the atmosphere, to the altitudes as low as 20 km.

A relatively small number of satellites at high altitude orbits (20,000 km or higher) provide coverage of most of the Earth's surface. It is possible to have essentially continuous, day and night, all weather detection of meteoroids over the entire surface of the Earth.

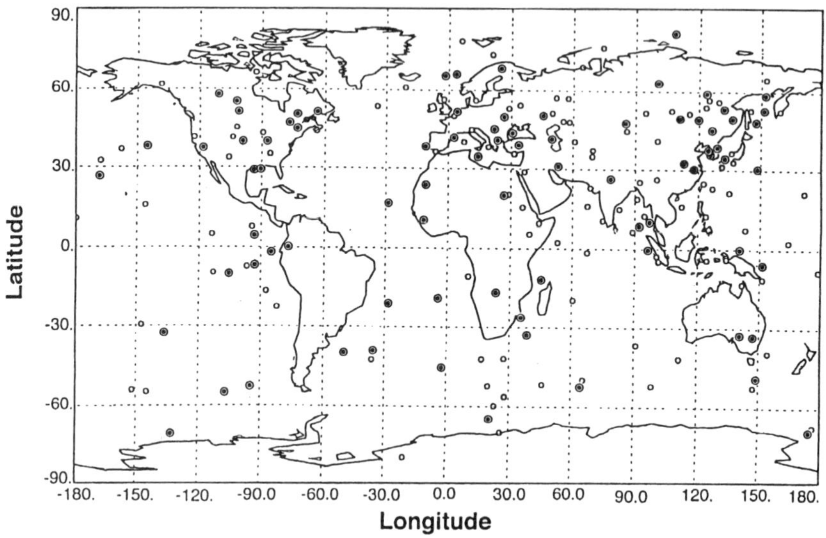

FIGURE 1. Solid circles represent night-time detections; open circles are daytime detections. Number of detections have averaged about 30 per year over the last three years.

OBSERVATIONS

In Fig. 1, geographic distribution of meteoroid fireballs detected by infrared sensors is presented. As it would be expected they are rather randomly distributed worldwide. Average number per year is about 30.

In addition to infrared radiation sensors, visible radiation wavelength sensors have recorded lightcurves for a subset of these events. In Fig. 2 three lightcurves are presented, for the events April 15, 1988; October 1, 1990; October 4, 1991 [5], [15].

Assuming source temperature of 6,000 K peak radiated power for these events have been estimated. It is 1.4×10^{12}, 3.5×10^{11}, and 6.5×10^{10} watts per steradian; total radiated energies for the period of observation were 7.1×10^{12}, 2.5×10^{12} and 5.5×10^{11} Joules. The amount of the emitted radiation is equivalent to the energy of 1.7 kilotons TNT, 0.57 kilotons TNT, and 0.14 kilotons TNT, respectively.

In Fig. 3 the lightcurve for 1 February 1994 event is given [14]. In this case, the first disintegration occurred at the altitude of 35 km and the major one at the altitude of 21 km; two clouds of debris were detected. For this deeply penetrating meteoroid the energy of the light flash is as high as 4.5 kt TNT. If we assume (as in the case of nuclear detonations) that about 30% of energy of the airblast is converted into the energy of the radiation, then the energy of the airblast which caused this flash is about 15 kt TNT, almost equal to the nuclear devices exploded above Hiroshima and Nagasaki. But there is no foundation for using this value of the luminous efficiency.

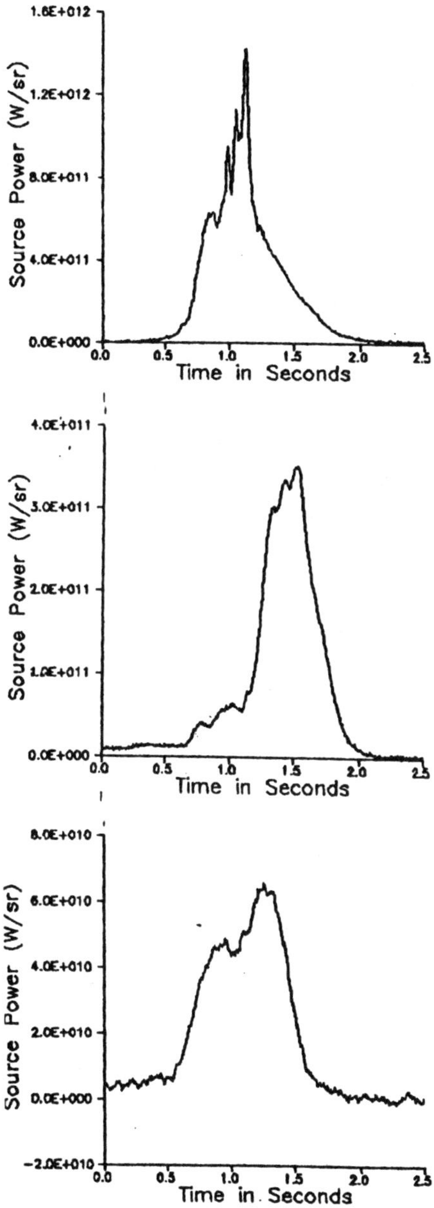

15 April 1988
$P_m = 1.4 \times 10^{12}$ W/sr
$h_m = 43$ km
$E_s = 1.70$ kt TNT
$E_k = 8–9$ kt TNT

1 October 1990
$P_m = 3.5 \times 10^{11}$ W/sr
$h_m = 30$ km
$E_s = 0.57$ kt TNT
$E_k = 5–8$ kt TNT

4 October 1991
$P_m = 6.5 \times 10^{10}$ W/sr
$h_m = 33$ km
$E_s = 0.14$ kt TNT
$E_k = 1–2$ kt TNT

FIGURE 2. Lightcurves for the subset of the events; visible sensors of Sandia National Laboritories.

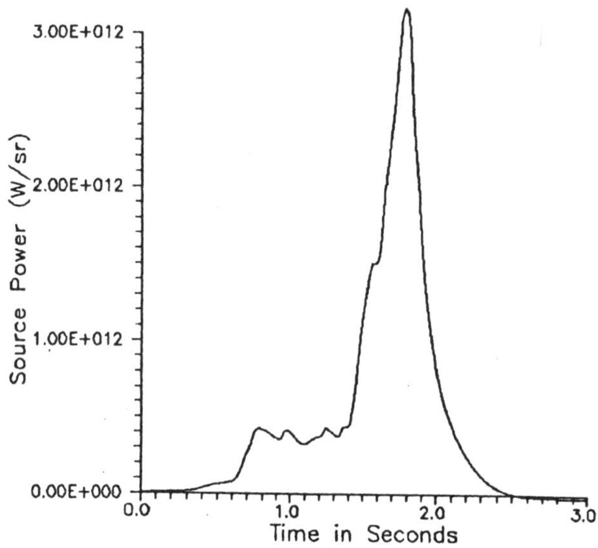

FIGURE 3. Lightcurve for the 1 February 1994 event.

TABLE 1. Main Characteristics of Meteoroids

Energy of light impulse (E_s), initial kinetic energy (E_k), mass (M), velocity (V), height of breakup (H_b), and of peak luminosity (H_m) for several events

	April 15, 1988	October 1, 1990	October 4, 1991	February 1, 1994
E_s, kt TNT	1.70	0.57	0.14	4.39
E_k, kt TNT	8–9	5–8	1.2–2	40–70
M, tons	25–45	100–300	25–75	1000–2500
V, km/sec	40–50	15–20	15–20	15–20
Luminous efficiency, %	19–21	7–11	7–11	7–11
H_b, km	60–65, 55–60	35	35–45	37, 24
H_m, km	43	30	33	34, 21
	1*	2*	1*	3*

1* Sunlit debris (errors 20–25%);
2* H_m is the height of the peak emission determined by a companion locator;
3* Two clouds of debris.

The luminous efficiency estimated by us is 7 to 11% (see Table 1). It is 2 to 3 times smaller than for nuclear detonations and the preatmospheric kinetic energy for the 1 February 1994 event in reality is as high as 40 to 70 kt TNT, several times higher than Hiroshima bomb. And fortunately it exploded above the Pacific Ocean, near Marshall Islands, not above some town, and at a rather high altitude (21 km). Kinetic energy of other 3 events mentioned is from 1–2 to 8–9 kt TNT, also a rather high energy. The height of breakup (h_b) is several kilometers higher than the height of peak emission determined by a companion locator or height of sunlit clouds of debris (h_m). We have also estimated the initial mass of the meteoroids (M), and initial velocity (V) (see Table 1).

INTERPRETATIONS

What is the reason for this big difference between luminous efficiency for nuclear detonations and for the meteoroids explosive disintegration?

The shape of the fireball for the meteoroid impact is quite different from the quasi-spherical shape of the nuclear detonation fireball. It is an elongated quasi-cylindrical luminous plasma column. It is more like a pencil or a slightly diverging cone, not a sphere. In Fig. 4 the flow pattern for a cylindrical intensely ablating body moving at the altitude of 40 km with the velocity of 20 km/sec is given. One can see the unstable boundary dividing vaporized material and the air. We do not know exact shape of the intruding bodies. But for sufficiently large bodies, especially moving in the dense layers of the atmosphere, this shape is relatively unimportant, as the plasma is optically thick, and we do not see the body itself. The radiation is emitted mainly from the shock wave and only the average radius, or the cross-section area are the important parameters.

The characteristic size of the meteoroid's fireball is different from that of a nuclear detonation with similar yield. For a 6 kt nuclear explosion at the altitude of 30 km the maximum fireball radius is about 0.3 km. For a 2 m in radius and 3 m in length stony body moving with the velocity of 20 km/sec the kinetic energy is also about 6 kt TNT, but the radius of the luminous wake is only about 10 m and the length is about 20 m. So the size of the luminous volume is much less than that in the case of nuclear detonation. The duration of the light impulse is about 1–2 sec. A cosmic body with the velocity of 20 km/sec covers the distance of 10 to 20 km, so the energy release is gradual. Thus while the fireball caused by the explosive disintegration of a meteoroid exhibits many features of the nuclear detonations with the same yield, there are specific differences.

On the other hand, for assessment of these large meteoroid characteristics we cannot use empirical values of luminosity efficiency obtained for small meteoroids or artificial meteors. The empirical values (less than 1–2%) are unsuitable, as the masses of the meteoroids detected by ground-based networks are usually by two or three orders of magnitude smaller. We shall not dwell on this problem, as data from ground-based networks is discussed by Ceplecha and Jacobs (in this volume [3]). Determination of the meteor luminous efficiency is a main step in the assessment of meteoroid's characteristics from lightcurves.

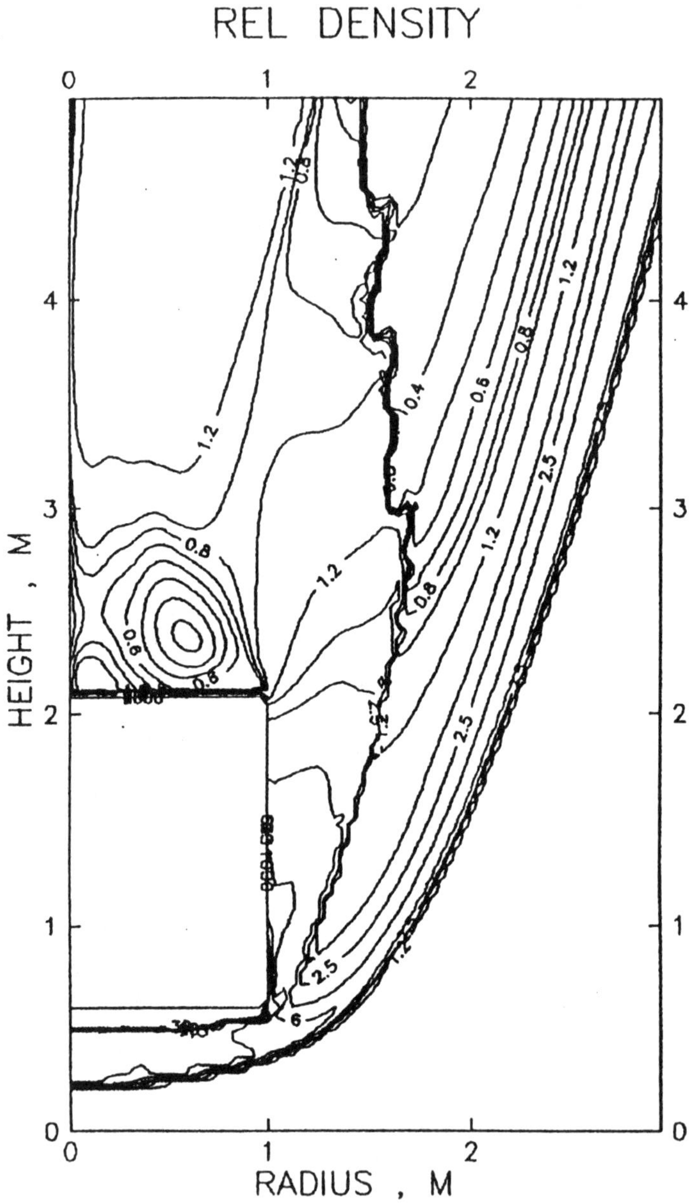

FIGURE 4. Flow pattern for an ablating meteoriod moving at the altitude of 40 km with the velocity of 20 km/sec. The exact shape of the body is relatively unimportant, since for sufficiently large bodies moving in the dense atmosphere, air and vapor plasma are optically thick. Radiation is emitted mainly from the shock wave.

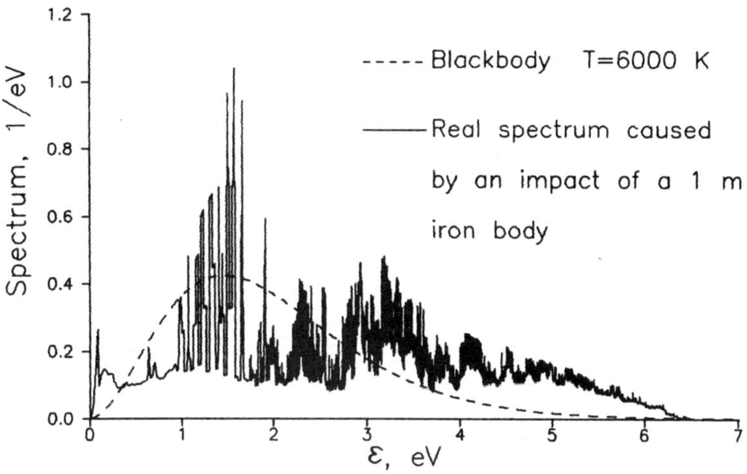

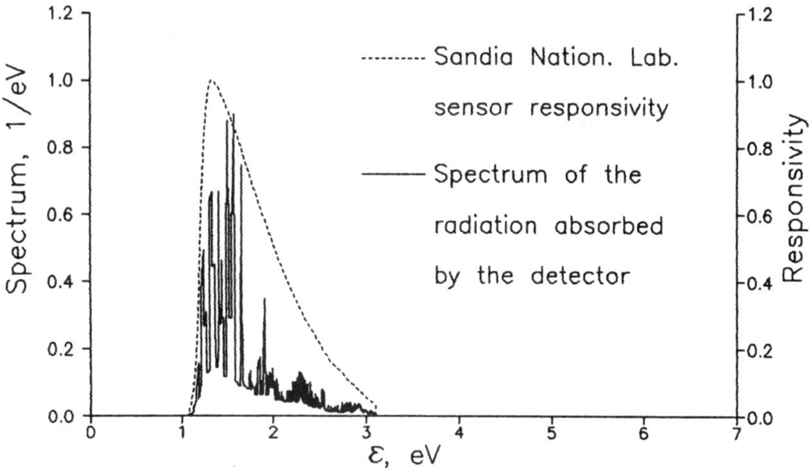

FIGURE 5. Responsivity of the Sandia National Laboratories visible light sensor. Spectrum of the radiation emitted by an iron meteoroid with 1 m radius (altitude 40 km, velocity 20 km/sec).

Several techniques for the assessment of the meteoroid characteristics have been developed [9], [10]. Let us briefly describe these techniques.

Radiation-hydrodynamic 1D, 2D, and even 3D numerical simulations of the flight in the atmosphere of meteoroids with different sizes, velocities, heights of flight have been conducted. They were based on the detailed tables of spectral opacities for hot air and ablated material of meteoroids (iron, H-chondrites, cometary material), which have been calculated for a wide range of temperatures, densities, and wavelengths. Equations of radiative transfer have been solved for as much as 10,000 wavelengths or photon energies both in the air and in the vapor of a meteoroid. An amount of meteoroid energy released in the atmosphere, radiation emitted at this altitude, and spectrum of radiation have been obtained. This is important because the visible sensors are spectrally selective. Typical responsivity as a function of wavelength is presented in Fig. 5 (lower panel, dotted curve) and a typical spectrum of radiation emitted by of a 1 m radius iron meteoroid moving at an altitude of 40 km with a velocity of 20 km/sec is also given. In Fig. 5 (upper panel) where the real spectrum obtained in our simulations is compared with that of the 6,000 K black body (dashed curve). One can see that the real spectrum substantially differs from that of a 6,000 K blackbody.

The results of a large number of numerical simulations are the tables of heat transfer coefficients and luminous efficiencies for different velocities, sizes of body, altitudes of flight, composition. As an example they are given in Table 2 for an iron body.

TABLE 2. Heat Transfer Coefficient and Luminous Efficiency Strongly Depend on the Height of Flight, the Size, and the Velocity of a Meteoroid

Iron Meteoroid						
	Heat transfer coefficient			Luminous efficiency		
V_b, km/s	$R_b = 0, 1$ m	$R_b = 1$ m	$R_b = 10$ m	$R_b = 0, 1$ m	$R_b = 1$ m	$R_b = 10$ m
$h = 40$ km						
50	0.13	0.059	0.029	0.23	0.40	0.50
30	0.19	0.10	0.063	0.19	0.33	0.46
20	0.24	0.13	0.11	0.16	0.24	0.35
15	0.30	0.13	0.12	0.12	0.21	0.28
10	0.22	0.15	0.093	0.081	0.067	0.14
$h = 30$ km						
50	0.055	0.028	0.015	0.28	0.42	0.42
30	0.082	0.044	0.024	0.21	0.35	0.40
20	0.11	0.067	0.041	0.18	0.29	0.38
15	0.14	0.079	0.050	0.15	0.24	0.33
10	0.13	0.077	0.053	0.056	0.09	0.17
$h = 20$ km						
50	0.023	0.011	0.0037	0.33	0.32	0.29
30	0.033	0.014	0.0057	0.27	0.30	0.28
20	0.049	0.020	0.0085	0.22	0.29	0.27
15	0.065	0.026	0.013	0.17	0.23	0.26
10	0.078	0.039	0.024	0.054	0.10	0.16

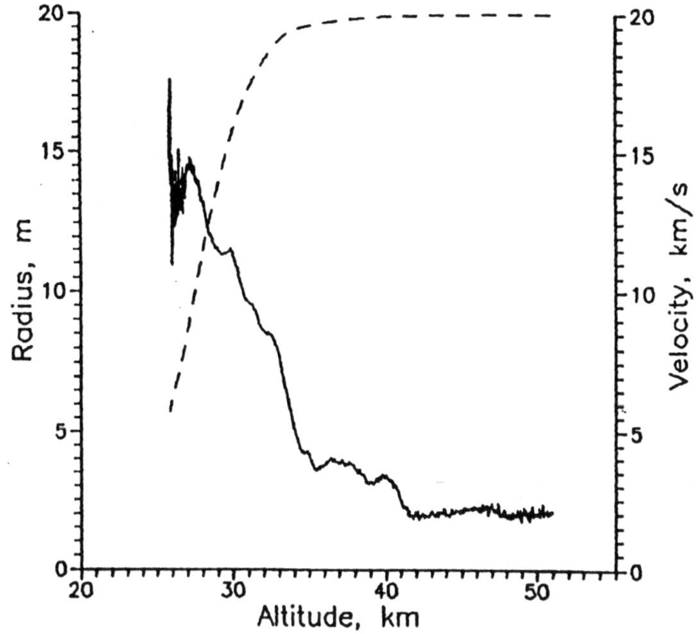

FIGURE 6. 1 October 1990 event.

Intensity of light is proportional to the cross-sectional area of the bolide and luminous efficiency. But value of luminous efficiency depends on the velocity of meteoroid and height of flight. In some cases the situation is simplified as the initial velocity has been determined by the observations. Companion burst-locating sensors can detect altitude of flashes for large events. For some fraction of events infrared sensors can provide the needed location of debris clouds.

Comparing the observed intensity of light with theoretical values for different sizes of the body we can estimate the instantaneous size of the body and can follow the increase in size of the cloud of fragments after the meteoroid's breakup and can even estimate rotation of the body which causes the variations of cross-section. Of course, in these estimates equations of motion of the meteoroid, taking into account deceleration and ablation and disintegration, are used.

The results of simulations for the 1 October 1990 event are given in Fig. 6. Here the velocity is the dashed curve, and effective radius of the body is the solid curve. At high altitudes this radius does not change in time. The meteoroid which caused 1 October 1990 event has an average radius of 2 to

3 m, initial mass 100 to 300 t, kinetic energy of 5 to 8 kt, and velocity 15 to 20 km/sec. At an altitude of 42 km, the radius begins to grow and the stagnation pressure at the blunt nose of the meteoroid at the moment of breakup gives us the apparent strength of the body. In this case, the meteoroid was a chondritic or stony body, its strength is 15 Mdyn/cm^2. At an altitude of 35 km, the radius increases 2 times, and then increases several times at altitudes of about 30 to 25 km. After the breakup, the meteoroid was heavily fragmented due to aerodynamic loads, but a single-body model of the cloud of fragments and vapor has still been used.

Sometimes we should analyze individual motion of large fragments. In Fig. 7 the results of simulations of the flow pattern for two fragments moving one near another are presented [1]. Aerodynamic interaction between the fragments leads to dispersion of the cloud of fragments with a lateral velocity which is about the velocity along the trajectory multiplied by the square root of the ratio of the air density to the density of the meteoroid [11], [8]. But if the number of fragments is large enough, the cloud of fragments and vapor expands like a single body until its radius increases 4 to 7 times — see Fig. 8 (15 April 1988 event). This leads to the increase in the cross-section by one, or even two, orders of magnitude and creates a very intense flash.

In Fig. 8 dashed and dotted curves are the results of numerical simulations using two models of cloud expansion [6], [4]. There is a rather good correlation between empirical and theoretical results, but to use theory one should use data on the strength of the body, which depends not only on the type of body, but also on its size, structure, shape, etc. Our analysis of the lightcurves gives us important information on the strength of the impacting meteorites. It also shows that effective cross-section is not only limited at some altitudes, but even begins to decrease with time.

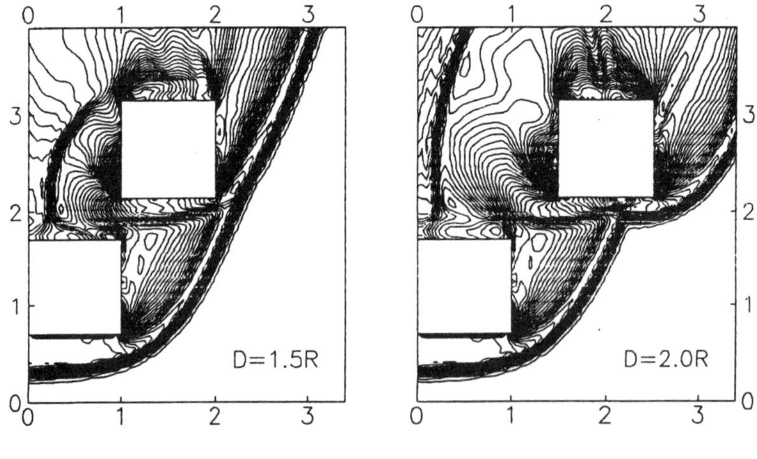

FIGURE 7. Aerodynamic interaction of the fragments. Interaction leads to dispersion of fragments with the lateral velocity $U \sim V \sqrt{\rho_a/\rho_b}$, where V is the velocity along the trajectory, ρ_a is the density of the atmosphere, and ρ_b is the density of the body.

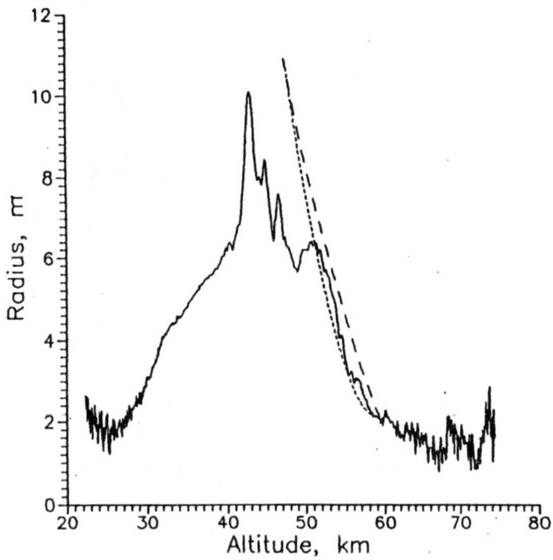

FIGURE 8. 15 April 1988 event.

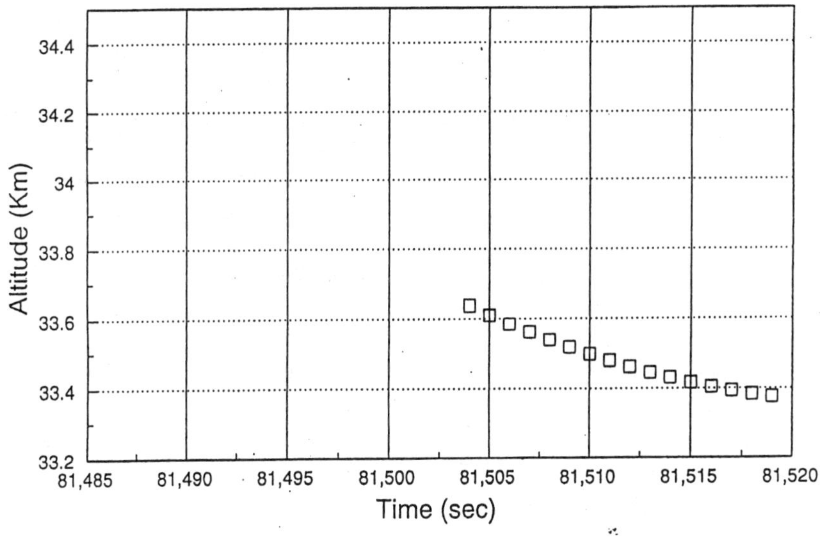

FIGURE 9. 94032: Altitude vs. time, high altitude object.

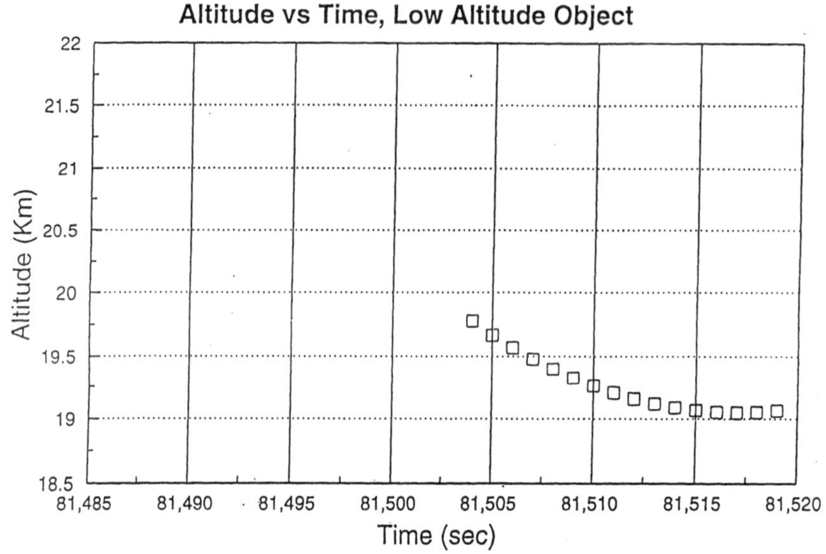

FIGURE 10. 94032: Altitude vs. time, low altitude object.

The meteoroid which caused the 15 April 1988 event has an initial kinetic energy of about 8 to 9 kt TNT and a rather high velocity of 40 to 50 km/sec. It broke up at 60–55 km; peak intensity was reached at the altitude of 43 km. It was a chondritic or stony body (strength 16 to 20 Mdyn/cm^2).

After explosive disintegration, a cloud of fragments expands and decelerates. Material that dispersed into the atmosphere as ashlike or sandlike particles may present a hazard to aircraft for some period of time. In Figs. 9 and 10 time altitude history of debris clouds for the event of 1 February, 1994 is presented. Two patches of debris has been detected, i.e., one at the altitude of 34 km and another at the altitude of 21 km. We see that in 15 sec the high altitude debris cloud falls to about 33 km due to gravity. The low altitude object stabilizes at an altitude of 19 km. The initial velocity (24 km/sec), and angle of trajectory inclination (45°), and event orbit parameters have been determined [7].

In Fig. 11, once more the lightcurve for 1 February 1994 event is given but in a logarithmic scale. A calculated radius exhibits two distinct maxima, due to the two-stage disintegration.

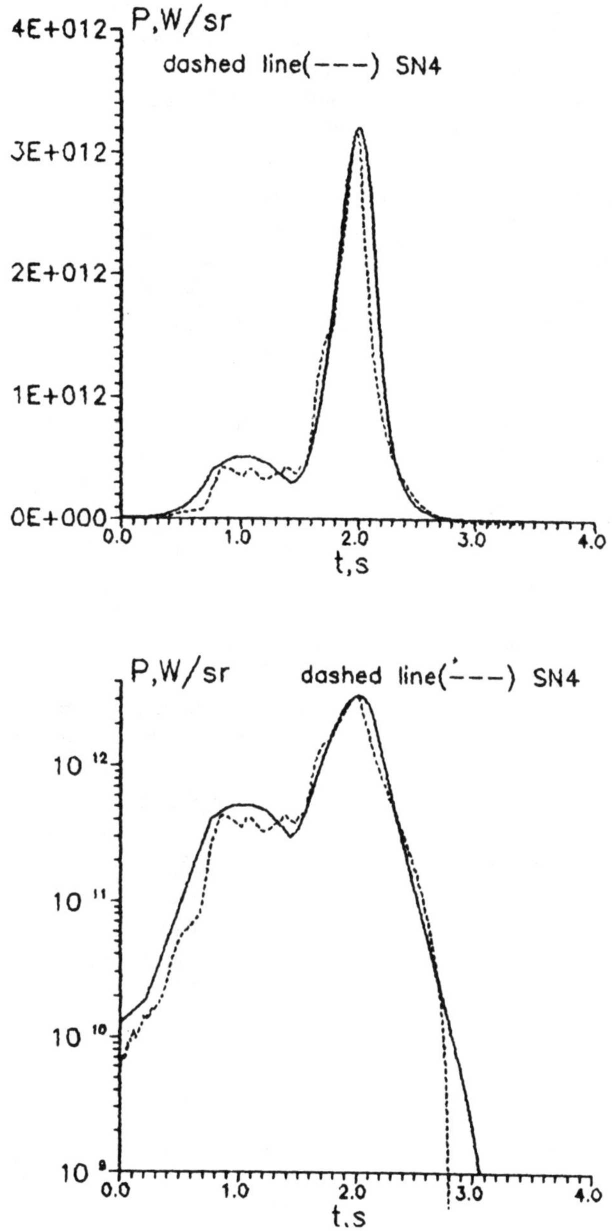

FIGURE 11. Comparison of the observed and modeling lightcurves for the 1 February 1994 event.

Assuming a two-stage disintegration, we have reproduced the lightcurve during the entire flight. If we do not use data on the initial velocity and trajectory inclination, we came to the conclusion that the 1 February 1994 event was caused by an iron meteoroid with the energy of about 40 to 70 kt TNT; initial mass was 1,000–2,500 tons, velocity 15 to 20 km/sec, angle of trajectory 40 to 50°, and strength 100–150 Mdyn/cm^2. Mass of the first fragment was about 500 tons, and the body which experienced major breakup has mass of about 1,400 tons. Using data on the velocity and trajectory, inclination we obtained that the initial kinetic energy was about 40 kt TNT, mass of about 520 tons; the strength of the second fragment is about 100 Mdyn/cm^2, its mass and energy are 430 t and 32 kt TNT, respectively. The first fragment had a very low apparent strength of about 5 Mdyn/cm^2, and probably the meteoroid was a binary object.

Increase in the initial velocity, in comparison to the assumption in our first simulations, leads to substantial decrease in mass but not very large decrease of the initial energy. This gives additional foundation to the estimates of kinetic energy for those events for which we do not know initial velocity. On the other hand, it clearly demonstrates that velocity/trajectory tracking substantially increase the precision of the assessment of meteoroid's characteristics.

Continuation of satellite-based observations and their analysis will give us statistically significant information on the size-frequency distribution of impacts and probability of the Tunguska-class airblasts. It can also elucidate the scientific problem of meteoroid origin and their relation to the near-Earth asteroid belt discovered by the ground-based Spacewatch system [13], [12], [2].

The precision of determination of meteoroid characteristics may be increased not only by usage of more sophisticated codes, which are now being developed, but by larger amounts of observational data, e.g., on the angle of trajectory inclination and the velocity of the meteoroid body along the trajectory. Spectral instruments being installed in upgraded satellite systems may also give very important information on the chemical composition of the impacting cosmic bodies.

REFERENCES

1. ARTEM'EVA, N.A. & V.V. SHUVALOV. 1996. Interaction of shock waves during the passage of a disrupted meteoroid through the atmosphere. Shock Waves. **5:** (6) 359–367.
2. CARUSI, A., T. GEHRELS, E.F. HELIN, B.G. MARSDEN, K.S. RUSSEL, C.S. SHOEMAKER, E.M. SHOEMAKER, & D.I. STEEL. 1994. Near-Earth objects: Present search programs. *In* Hazards Due to Comets and Asteroids, T.Gehrels, Ed.: 199–220. Univ. of Arizona Press. Tucson, Arizona.

3. CEPLECHA, Z. & C. JACOBS. 1995. Correlation of ground- and space-based bolides. Proc. UNited Nations International Conference on Near-Earth Objects, J. Remo, Ed.
4. CHYBA, C.F., P.J. THOMAS, & K.J. ZAHNLE. 1993. The 1908 Tunguska explosion: atmospheric disruption of a stony asteroid. Nature. **361:** 40–44.
5. JACOBS, C. & R. SPALDING. 1993. Fireball observation by satellite-based Earth-Monitoring optocal sensors. Hazards due to Comets and Asteroids, January 4–9, 1993. Abstract book, p. 45. Tucson, Arizona.
6. HILLS, J.G. & M.P. GODA. 1993. The fragmentation of small asteroids in the atmosphere. Astron. J. **105:** 1114–1144.
7. MCCORD, T.B., J. MORIS, D. PERSING, E. TAGLIEFERRI, R. SPALDING, L.A. GRADY, & R. SCHMIDT. Detection of a meteoroid entry into the Earth's atmosphere on February 1, 1994. JGR-Planets. **E2:** 3245–3249.
8. MELOSH, H.J. 1989. Impact cratering: A Geologic Process. Oxford Univ. Press. New York.
9. NEMTCHINOV, I.V., O.P. POPOVA, V.V. SHUVALOV, & V.V. SVETSOV. 1994. Radiation emitted during the flight of asteroids and comets through the atmosphere. Planet. Space. Sci. **42:** (.6) 491–506.
10. NEMTCHINOV, I.V., O.P. POPOVA, V.V. SVETSOV, & V.V. SHUVALOV. 1995. On photometric masses and radiation sizes of large meteoroids. Solar System Research. **29:** (2) 155–173.
11. PASSEY, Q.R. & H.J. MELOSH. 1980. Effects of atmospheric breakup on crater field formation. Icarus. **42:** 211–233.
12. RABINOWITZ, D.L., T. GEHRELS, J.V. SCOTTI, R.S. MCMILLAN, M.L. PERRY, W. WISHIEWSKI, S.M. LARSON, E.S. HOWELL, & B.E.A. MUELLER. 1993. Evidence for a near-Earth asteroid belt. Nature. **363:** 704–706.
13. SCOTTI, J.V., D. RABINOWITZ, & B. MARSDEN. 1991. Near miss of the Earth by a small asteroid. Nature. **354:** (6351) 287–289.
14. SPALDING, R. 1994. Personal communication.
15. TAGLIEFERRI, E., R. SPALDING, C. JACOBS, S.P. WORDEN, & A. ERLLCH. 1994. Detection of meteoroid impacts by optical sensors in Earth orbit. Hazards due to Comets and Asteroids, T.Gehrels, Ed.: 199–220. Univ. of Arizona Press. Tucson, Arizona.

Impact Record in the Solar System

GERHARD NEUKUM

DLR, Institute for Planetary Exploration
Rudower Chaussee 5
Berlin 12489 Germany

The terrestrial planets, the Earth's Moon, Mars, Venus, Mercury, and the asteroids Gaspra and Ida show striking similarities in their crater size distribution characteristics. The investigation of the crater populations on Gaspra and Ida yield *information* on the crater size distribution in the source region of the bodies which are largely responsible for cratering the inner solar system planets. Comparison of these data especially with the lunar impact record, which is our most reliable data base, confirms the complex shape of the crater size distribution curve (standard distribution), and lends strong support to the idea of a common population of bodies impacting the inner planets and largely stemming from the asteroid belt.

The magnitude and time dependence of impact rate can be determined by making use of the radiometric age data of lunar rocks from the Apollo landing sites in combination with the respective crater frequency measurements. Additional information can be obtained from the impact record of large craters ($D \geq 10$ km) on Earth over the past 500 Myr which are still preserved in their ground structures on the old shields. From those data, it is evident that there was an early high impact rate on the Moon and the Earth (orders of magnitude above the present rate), and which decayed exponentially until about 3 billion years ago and has remained nearly constant at the present rate since then.

For the other terrestrial planets the cratering record has been determined from high resolution imagery taken by planetary missions. From these investigations it appears that a cratering rate time-dependence similar to the one in the Earth-Moon system is valid, i.e., that the other terrestrial planets and that even the moons of the giant planets also experienced a heavy bombardment with a subsequent steep decay early on and leveling off at later times.

Current or past impact rates can be calculated from such information for any size of crater or projectile for the Earth-Moon system. The impact hazard for projectiles, e.g., capable of forming 1 km craters on solid surfaces of the Earth (if the atmosphere were absent) is assessed for the present and results in a production rate of one crater every 1600 yr for the surface area of the whole Earth, or every 6000 yr for the area of the continents; for 100 km craters, e.g., the corresponding number is one event every 27 Myr for the whole Earth.

Target Earth: Evidence for Large-scale Impact Events[a]

RICHARD A. F. GRIEVE

Geological Survey of Canada
Ottawa, Ontario
Canada, K1A 0Y3

ABSTRACT: Unlike the Moon, the Earth has retained only a small sample of its population of impact structures. Currently, over 150 impact structures are known and there are 15 instances of impact known from the stratigraphic record, some of which have been correlated with known impact structures. The terrestrial record is biased toward younger and larger structures on the stable cratonic areas of the crust, because of the effects of constant surface renewal on the Earth. The high level of endogenic geologic activity also affects the morphology and morphometry of terrestrial impact structures; although, the same general morphologic forms that occur on the other terrestrial planets can be observed. A terrestrial cratering rate of $5.6 \pm 2.8 \times 10^{-15}$ km^{-1} a^{-1} for structures $\geq$ 20 km in diameter can be derived, which is equivalent to that estimated from astronomical observations. Although there are claims to the contrary, the overall uncertainties in the ages of structures in the impact record preclude the determination of any periodicity in the record. Small terrestrial impact structures are the result of the impact of iron or stony iron bodies, with weaker stony and icy bodies being crushed on atmospheric passage. At larger structures (>1 km), trace element geochemistry suggests that ~50% of the impact flux is from chondritic bodies, but this may be a function of the signal:noise ratio of the meteoritic tracer elements. Evidence for impact in the stratigraphic record is both chemical and physical. Although currently small in number, there are indications that more evidence will be forthcoming with time. Such searches for evidence of impact have been stimulated by the chemical and physical evidence of the involvement of impact at the K/T boundary. There will, however, be problems in differentiating geochemically the signal of even relatively large impact events from the background cosmic flux of every day meteoritic debris. Even with these biases and difficulties, the terrestrial impact record is the dominant source of ground truth information on the details of the impact flux and its known and potential effects on the evolution of the Earth and its biosphere. For although the record is poorly known, what evidence there is represents an integration over considerable geologic time. On the timescales of 10^5–10^6 a, it is clear that impact represents a major threat to human civilization. Given the stochastic nature of impact, the timing of such an event is unknown.

INTRODUCTION

Planetary exploration has revealed the importance of hypervelocity impact by interplanetary bodies as a geologic agent. This is particularly evident on planetary bodies that have retained portions of their earliest crust, such

[a]Contribution from the Geological Survey of Canada 1996391.

as the Earth's Moon. The highly active geologic environment of the Earth, however, serves to remove, modify, and mask the terrestrial impact record, making it harder to read than on the other terrestrial planets. Nevertheless, this disadvantage is more than offset by the availability of ground-truth information on impact structures in the terrestrial environment. Therefore, in parallel with planetary studies, there has been an effort to study the terrestrial impact record, in large part, to provide ground-truth constraints on impact processes. During these developments, however, few in the larger (geo)science community regarded impact as a process affecting terrestrial evolution.

This tendency to discount impact processes as a factor in the Earth's geologic history has been severely challenged by the interpretation that elevated siderophile element abundances in Cretaceous-Tertiary (K/T) boundary sediments are due to a major impact event and that impact was the causal agent for the contemporaneous mass extinction event [3], [31], [119]. Interestingly, the occurrence of a geochemical anomaly in K/T boundary sediments had been described previously [10]. Similarly, the hypothesis that a major impact event could cause a mass extinction event was not new (e.g., [17], [81], [130]). What was new was the establishment of a linkage between the geochemical evidence at the K/T boundary and the impact process. This was possible through the level of knowledge, existing at the time, concerning impact processes and the known association of siderophile geochemical anomalies with some impact craters. Thus, the K/T impact hypothesis was readily accepted by the impact and planetary geoscience community.

The acceptance of the K/T impact hypothesis by the more general terrestrial geoscience community was less complete and considerable controversy and debate was generated. Aspects and progress in this debate are recorded in three compendia of work related to the development of the hypothesis that a major impact event was responsible for the K/T boundary, edited by Silver and Schultz [118], Sharpton and Ward [113], and Ryder et al. [105]. Today, there are few workers who would deny that there is abundant diagnostic evidence that a major impact event occurred at the K/T boundary. It is fair to say, however, there is less consensus on the role of impact as a killing mechanism for the associated mass extinction event, with some workers still having difficulty in accepting impact-related processes as the cause (e.g., [88]).

The study of impact craters is a relatively recent endeavor in the geosciences. Although the first recognition of a terrestrial impact crater (Meteor Crater, Arizona) was in the early 1900's [7], over 60% of the currently known terrestrial impact structures were discovered in the last 20 years. Apart from the known occurrence of some tektite and microtektite strewn fields, the recognition of impact-related materials in the terrestrial stratigraphic record is an even more recent endeavor, beginning largely subsequent to the K/T boundary discoveries. This review provides some of the general characteristics of the currently known terrestrial impact record, in terms of both impact structures and evidence for impact in the stratigraphic record. It also comments upon how this relates to questions of the cratering rate, nature of impacting bodies and on some of the suggestions linking major impact events and other terrestrial geologic phenomena.

TERRESTRIAL IMPACT STRUCTURES

The character of the known impact record is the result of the combination of impact and terrestrial geologic processes. Due to the active geologic environment and processes of surface renewal, the known terrestrial impact record is only a small sample of a larger population. If it were not for endogenic geologic processes causing surface renewal, the Earth would appear as scarred as the surface of the moon and it would not be an intellectual task to consider impact as a process affecting the terrestrial environment. A recent listing of known terrestrial impact structures can be found in Grieve et al. [44]. Since then, eight additional structures have been discovered or have had an impact origin confirmed. They are: Calvin (USA), Goyder (Australia), Karikkoselkä (Finland), Mjølnir (Norway), Morokweng (S. Africa), Mt. Toodina (Australia), Neugrund (Estonia), and Suavjari (Russia). By the time this contribution appears, there will undoubtedly be additional discoveries. Details of the locations and age and size estimates of these impact structures and others discussed specifically in this text are given in Table 1.

Approximately 30% of the known impact structures are buried by postimpact sediments and were noted initially as geophysical anomalies, which are associated with impact structures (e.g., [48], [95]). They were subsequently drilled for commercial and/or scientific purposes and an impact origin confirmed. With four exceptions, all known terrestrial impact structures are on land or partially on land. The exceptions are: Montagnais on the outer edge of the continental shelf off Nova Scotia, Canada [56], Mjølnir, in the western Barents Sea, north of Norway [22], Neugrund in the Baltic Sea, just off the coast of Estonia [126b], [126a], and Tvären also in the Baltic Sea, just south of Stockholm, Sweden [75]. No impact structures are known from the true ocean basins, reflecting, in part, the relatively young age of the ocean floors but, more importantly, our lack of detailed knowledge of the ocean floors.

TABLE 1. Basic Parameters of Terrestrial Impact Structures Mentioned in Text

Crater Name	Location	Latitude	Longitude	Age (Ma)	Diameter (km)
Barringer	USA	N 35°2′	W 111°1′	0.049 ± 0.003	1.19[a]
Bosumtwi	Ghana	N 6°30′	W 1°25′	1.03 ± 0.02	10.5
Brent	Canada	N 46°5′	W 78°29′	450 ± 30	3.8
Calvin	USA	N 41°50′	W 85°57′	450 ± 10	8.5
Charlevoix	Canada	N 47°32′	W 70°18′	357 ± 15	54
Chicxulub	Mexico	N 21°20′	W 89°30′	64.98 ± 0.05	180[b]
Chesapeake Bay	USA	N 37°17′	W 76°1′	35.5 ± 0.6	90
Clearwater (East)	Canada	N 56°5′	W 74°7′	290 ± 20	26
Clearwater (West)	Canada	N 56°13′	W 74°30′	290 ± 20	36
Eagle Butte	Canada	N 49°42′	W 110°30′	< 65	10
Gardnos	Norway	N 60°39′	E 9°0′	500 ± 10	5
Gosses Bluff	Australia	S 23°49′	E 132°9′	142.5 ± 0.8	22
Goyder	Australia	S 13°9′	E 135°2′	> 65	3
Gusev	Russia	N 48°21′	E 40°14′	65 ± 2	3.5
Iso-Naakkima	Finland	N 62°11′	E 27°9′	> 1000	3
Kalkkop	S. Africa	S 32°43′	E 24°34′	< 1.8	0.64
Kamensk	Russia	N 48°20′	E 40°15′	49 ± 18	25
Karrikkoselkä	Finland	N 63°13′	E 25°15′	< 1800	1.5
Lockne	Sweden	N 63°0′	E 14°48′	455 ± 5	7
Lumparn	Finland	N 60°12′	E 20°6′	~ 1000	9
Mjølnir	Norway	N 73°48′	E 29°40′	143 ± 20	40
Montagnais	Canada	N 42°53′	W 64°13′	50.50 ± 0.76	45
Morokweng	S. Africa	S 26°28′	E 23°32′	1400 ± 200	70
Mt. Toondina	Australia	S 27°57′	E 135°22′	< 110	8
Neugrund	Estonia	N 59°20′	E 23°40′	—	7
New Quebec	Canada	N 61°17′	W 101°58′	1.4 ± 0.1	3.2
Ries	Germany	N 48°53′	E 10°37′	15 ± 1	24
Rio Cuarto	Argentina	S 30°52′	W 64°14′	< 0.1	4.5[c]
Steinheim	Germany	N 48°2′	E 10°4′	15 ± 1	3.8
Suavjavi	Russia	N 63°7′	E 33°23′	< 2400	16
Suvasvesi	Finland	N 62°42′	E 28°0′	< 1000	4
Tvären	Sweden	N 58°46′	E 17°25′	> 455	2
Wanapitei	Canada	N 46°45′	W 80°45′	37 ± 2	7.5
Wolfe Creek	Australia	S 19°10′	E 127°48′	< 0.3	0.88

[a]Also known as Meteor Crater.
[b]The diameter of Chicxulub is the subject of current debate (see text).
[c]Length of largest oblique crater in this field of at least 10 craters.

FIGURE 1. Examples of terrestrial simple impact structures (Table 1). (a) Oblique aerial view of Barringer or Meteor Crater, Arizona, USA. (b) Oblique aerial view of New Quebec, Quebec, Canada. (c) Oblique aerial view of Wolfe Creek, W. Australia, Australia. (d) Vertical aerial view of Brent, Ontario, Canada. Brent is by far the oldest of these examples (Table 1). It has no rim and is filled by postimpact sediments. North is to the left.

(i) Morphology

Relatively uneroded terrestrial impact structures display the basic morphologic progression from simple to complex craterforms with increasing crater diameter, observed on the other terrestrial planets. Simple impact structures have the form of a bowl-shaped depression with a structurally upraised rim (Figs. 1 and 2). The rim area is overlain by ejecta deposits and the exposed crater floor represents the top of a subsurface lens of allochthonous breccia (Fig. 2). The canonical example of a near-pristine simple impact structure is Barringer or Meteor Crater, USA (Fig. 1). Simple impact structures occur up to diameters of ~2–4 km on Earth. Above 4 km, terrestrial impact structures generally have a complex form. When relatively uneroded, they are characterized by structurally complex and faulted rim areas, a flat annular trough, and topographically high central structures, which contain rocks uplifted from deeper levels (Figs. 3 and 4). Various lines of observational and experimental evidence indicate that complex structures are a highly modified craterform, with respect to simple structures (e.g., [111], [83], and references therein).

Terrestrial complex impact structures also show the secondary forms observed on other planetary bodies, such as central peak craters, peak-ring craters, and ring basins. Care must be exercised, however, when comparing morphologic elements between individual terrestrial impact structures and, in particular, when comparing terrestrial and planetary craters [94].

Original morphologic elements can be enhanced, modified, or removed by various erosional processes on Earth. Morphometric relations are also affected by terrestrial processes to varying degrees. Some of the basic relations, such as depth/diameter (e.g., Figs. 2 and 4), for relatively pristine terrestrial impact structures are given in Grieve and Pesonen [39].

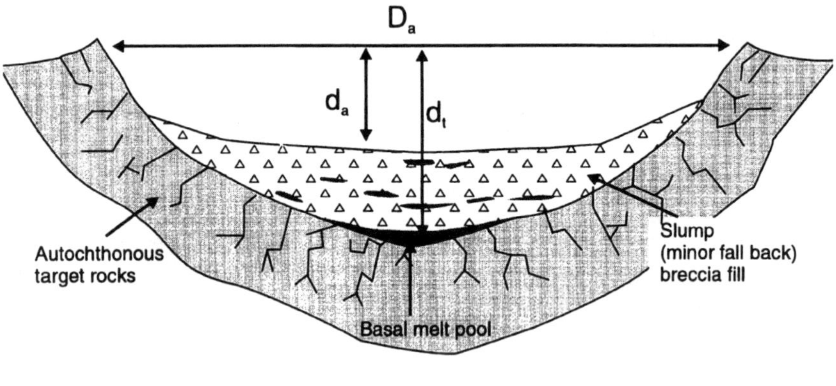

FIGURE 2. Schematic cross-section of a simple impact structure. The rim diameter (D_a), apparent (d_a) and real (d_t) depths are indicated. The interior of the structure is filled by a lens of allochthonous (slump) breccia (see text for more details).

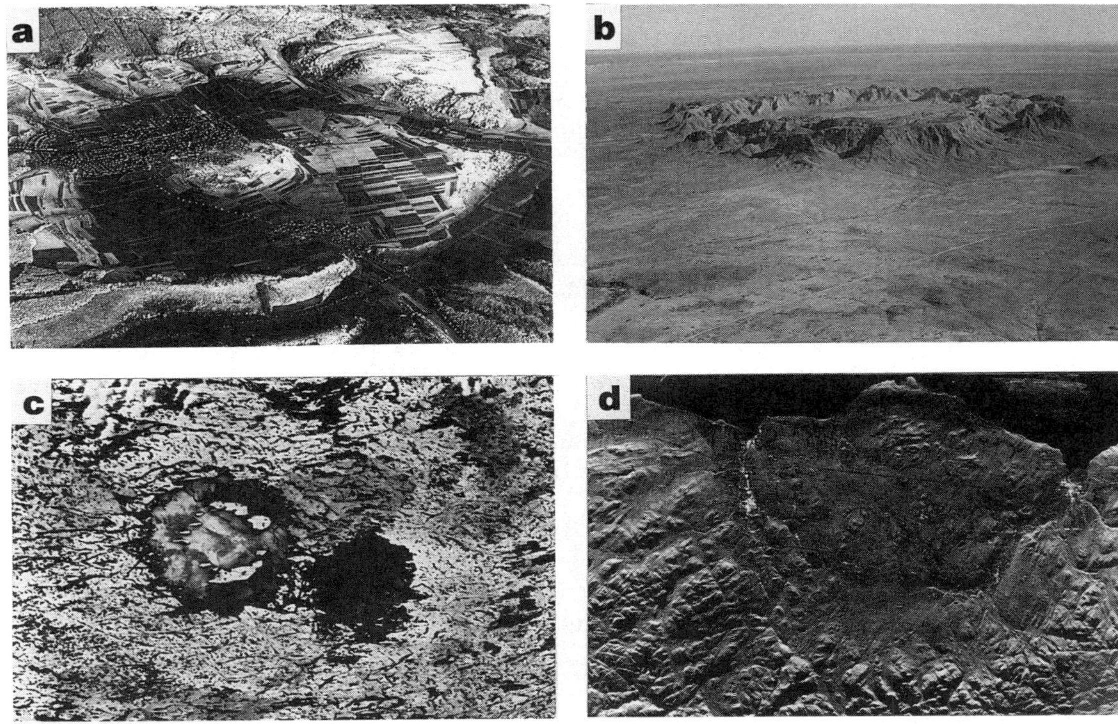

FIGURE 3. Examples of terrestrial complex impact structures (Table 1). (a) Oblique aerial view of Steinheim, Baden Wurttemberg, Germany. (b) Oblique aerial view of Gosses Bluff, N. Territory, Australia. Annular ring of hills is 5 km in diameter and is an erosional remnant of the central uplift. The original diameter of the rim is estimated to be 22 km (Table 1). (c) LANDSAT image of E. and W. Clearwater, Quebec, Canada. These structures represent the impact of a binary body. (d) Airborne radar (C-band) image of Charlevoix, Quebec, Canada. The eastern portion of the structure is truncated by the St. Lawrence River. This structure is eroded to below the original floor. North is to the left.

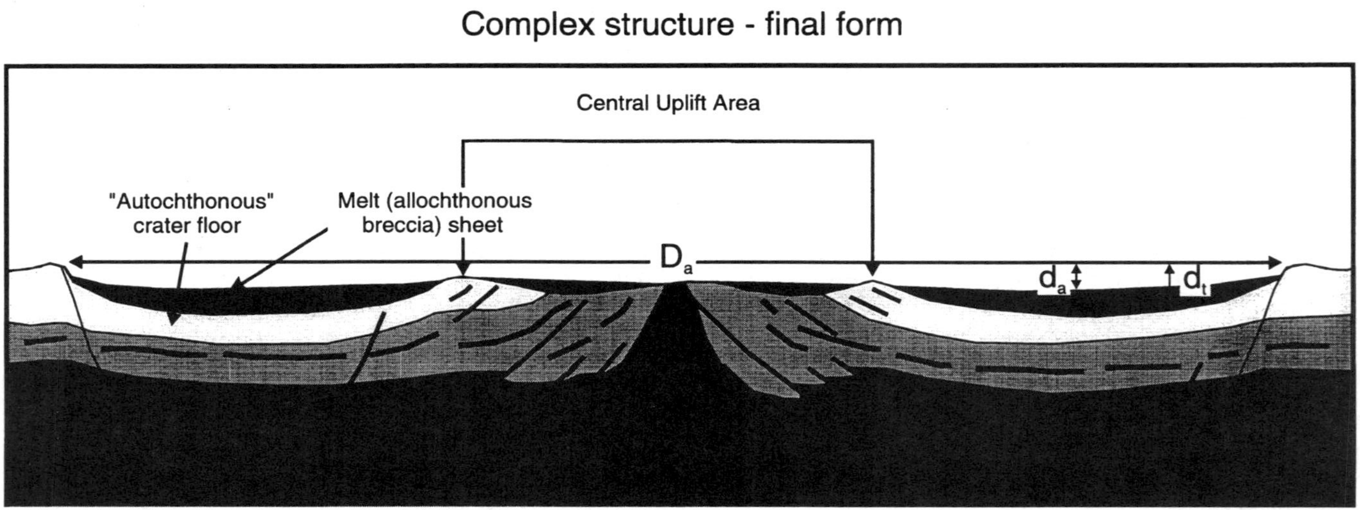

FIGURE 4. Schematic cross-section of a complex impact structure. The rim diameter (D_a), apparent (d_a) and real (d_t) depths, the amount of structural uplift (SU) and central uplift areas are indicated. Note the smaller depth:diameter ratio compared to simple structures (Fig. 2); although, the actual depth may be greater. The interior of the structure is filled with various allochthonous breccias and/or impact melt rocks.

Any listing of the diameters of terrestrial impact structures is a mix of interpretations from topographical, geological, and geophysical data. Individual diameter estimates can be different and controversial. For example, reasoned interpretations of the rim diameter of the buried Chicxulub structure, Mexico, which is the putative crater for the K/T boundary deposits, range from ~180 km [51], [97] to ~300 km [114]. As more and different types of data become available for individual impact structures, estimates of their original diameter are revised. This is particularly evident when considering larger and older impact structures. For example, the Sudbury structure was initially listed as 55 km in diameter by Dietz [21]. Recent geological and geophysical interpretations indicate considerable postimpact deformation (e.g., [85], [52], [13]) and, when combined with a greater understanding of cratering mechanics, have resulted in a much larger estimate of the original diameter of up to > 250 km [18]. Quantitative interpretations based on data compilations of rim diameters of terrestrial impact structures should be regarded with some caution. Such compilations are subject to revision.

(ii) Spatial Distribution

The locations of known terrestrial impact structures are shown in Fig. 5. These structures have evidence of impact origin through the documented occurrence of meteoritic material and/or shock metamorphic features. The details of the character of shock metamorphic effects can be found in papers in French and Short [30] and Roddy et al. [102] and in Stöffler [121], [122], Stöffler and Langenhorst [123] and Grieve et al. [45]. Shock metamorphic effects are diagnostic of impact structures and are the result of the highly transient compression, followed by adiabatic release, of the target rocks, as the energy of the impacting body is transferred to the target by a radially propagating shock wave. Readers interested in the physics of the impact process are referred to Melosh [83] and references therein.

There are concentrations of known impact structures in N. America, Australia, and Europe, particularly Fennoscandia, through to the eastern part of the former USSR (Fig. 5). This can be attributed to the fact that these are generally stable cratonic areas, where the level of endogenic geologic activity has been relatively low for a considerable period of geologic time. Thus, they are areas best suited to preserve evidence of near-surface geologic features, such as impact structures. These are also areas which have been subjected to more intense scrutiny for the occurrence of impact structures. Programs to identify impact structures have considerable influence on the local rate of discovery. For example, accelerated awareness of impact structures and their characteristics in Fennoscandia has led to the recent confirmation of an impact origin for Karikkoselkä, Finland [72], Gardnos, Norway [23], Lockne, Sweden [74], Iso-Naakkima, Finland [24], Lumparn, Finland [127], Neugrund, Estonia [126], Suavjarvi, Karelia, Russia [80], Suvasvesi, Finland [93] and Tvären, Sweden [89].

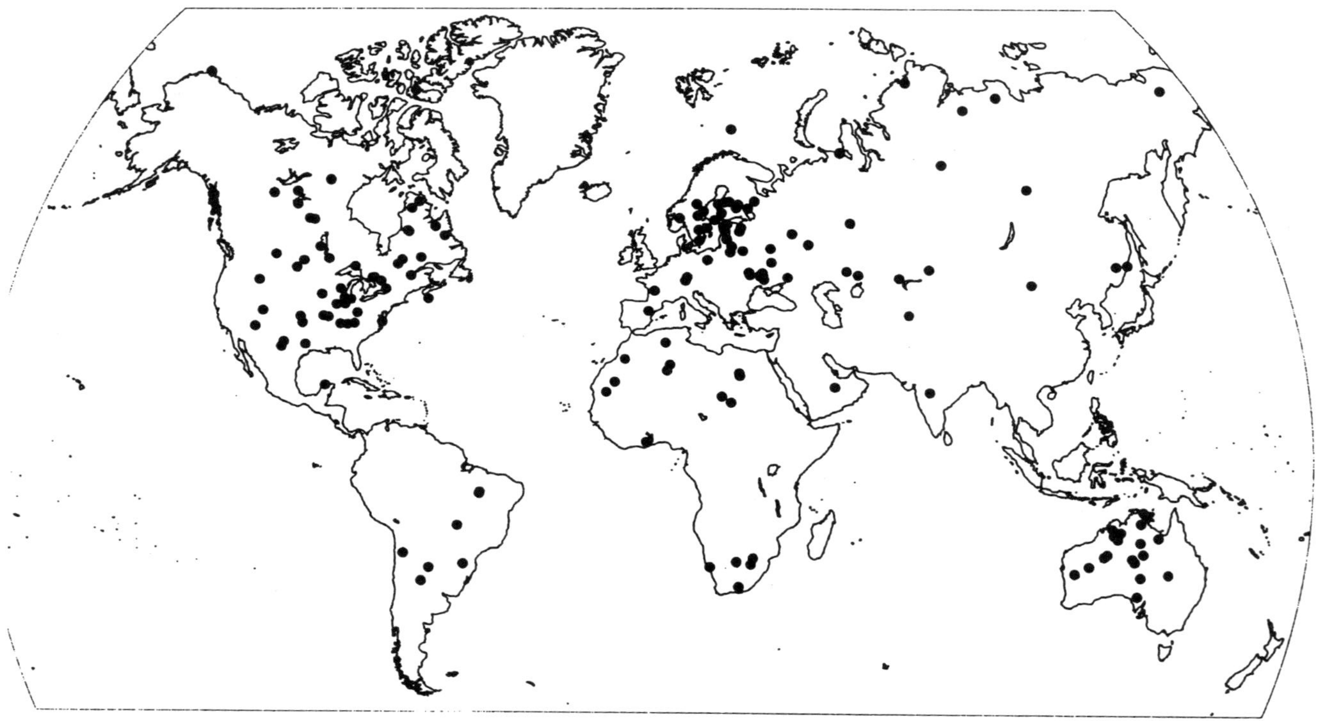

FIGURE 5. Spatial distribution of known terrestrial impact structures. Note apparent concentrations in N. America, Scandinavia, and Australia. See text for details.

(iii) Temporal Distribution

Approximately 40% of the known terrestrial impact structures have been dated isotopically, generally from the analysis of impact melt rocks. Most of the materials (90%) involved in an impact event, however, are subjected to insufficient shock pressures and postshock temperatures to significantly disturb isotopic dating systems [19].

The bulk of isotopic dates for terrestrial impact structures are K-Ar or, more recently, ^{40}Ar-^{39}Ar dates. Fine-grained, often clast-rich, impact melt rocks are not particularly easy to date isotopically by Argon isotope systematics, because of inherited Ar from the clasts. In some cases, single K-Ar dates can be in considerable error with respect to the age of the impact event (cf. [15], [78]).

In only a few cases are impact melt rocks of sufficient grain size or have sufficient compositional variation to permit such dating techniques as Rb-Sr mineral isochrons [19]. Precise U-Pb dates have been obtained from the analysis of the discordia of shocked zircons [65] or from new zircons crystallized from impact melt rocks [54]. The remainder of known terrestrial impact structures have biostratigraphic or stratigraphic dates. In some cases, the biostratigraphic dates, on such units as crater-filling sediments, are as accurate and precise as isotopic dates, e.g., at Lockne, Sweden [35]. Most are, however, minimum age estimates. In some cases, stratigraphic dates are maximum estimates of age, the age being listed only as less than the age of the target rocks; for example, Eagle Butte, Canada, is formed in Cretaceous rocks and listed as < 65 Ma. In some such cases, the degree of erosion can be used to provide an additional crude constraint on the age.

As with the spatial distribution, the temporal distribution of known terrestrial craters is not random. It is highly biased towards younger age estimates, with over 60% being younger than 200 Ma (Fig. 6). As terrestrial impact craters are, at least initially, surface features in a highly active geologic environment, they can be removed or buried from observation relatively rapidly. The rate at which this occurs varies with the geologic history of the area but it is a general tenet that older impact structures have a higher probability of being removed from the record by endogenic geologic processes.

(iv) Size-Frequency Distribution

There is also a bias in the size-frequency distribution of terrestrial impact structures. At diameters below ~20 km, the cumulative size-frequency distribution falls off (Fig. 7). This is generally taken to indicate the effects of erosion and, to a lesser extent, burial of smaller craters. The geological effects of impact at simple structures are visible to a depth of 1/3 the final rim diameter (Fig. 2; [39]). Thus, it is easy to appreciate why there is a deficit of small simple craters in the size-frequency distribution (Fig. 7), with the geologic evidence for the largest terrestrial simple impact structures being removed by < 1.5 km of erosion. There is also an atmospheric shielding effect for relatively small impacting bodies. Model calculations indicate that aerodynamic stresses are large enough to lead to atmospheric crushing, dispersal and velocity reduction of icy impacting bodies $\gtrsim$ 500 m in diameter. Somewhat stronger stony and, considerably stronger, iron bodies are less susceptible to atmospheric effects, although iron bodies < 200 m in diameter are still subjected to aerodynamic crushing and dispersal [82].

At larger complex structures, the depth:diameter relationship is relatively shallower (Fig. 4), but the absolute depths are often greater. In addition, the uplift of originally deeper lithologies in the center of complex structures provides an additional geologic manifestation of the impact event. The amount of stratigraphic uplift (SU in Fig. 4) undergone by the deepest lithologies exposed in the central structures of complex impact structures is ~1/10 the final rim diameter [41]. Thus, even when the topography and interior impact-related lithologies at a complex impact structures have been essentially removed by erosion, it will still be recognizable as a roughly circular geologic anomaly. The shape of the cumulative size-frequency distribution in Fig. 7 appears to be an inherent property of the terrestrial record, as it has persisted as more impact structures have been added to the known sample over the years.

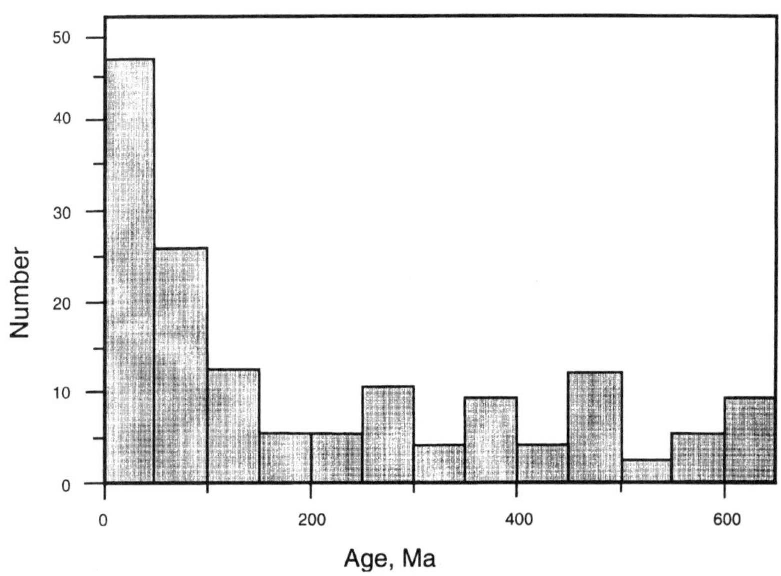

FIGURE 6. Histogram, binned by 50 Ma intervals, of ages of known terrestrial impact structures with ages ≤ 650 Ma. Note apparent increase in numbers in more recent geologic time. See text for details.

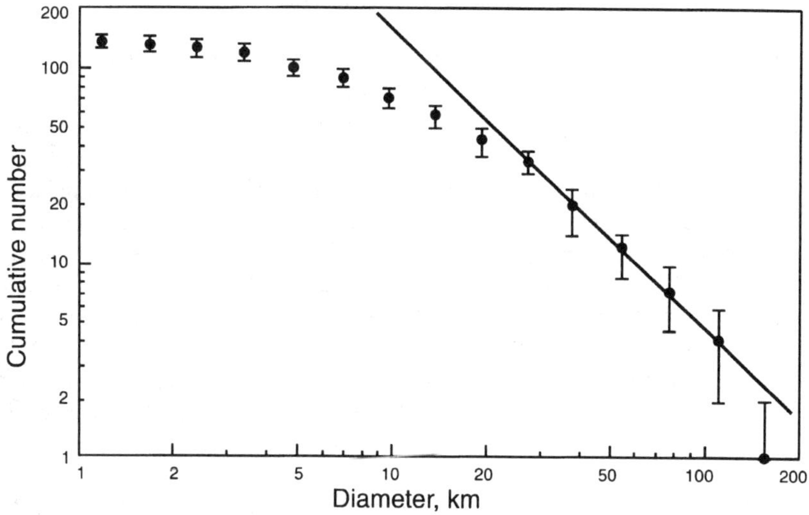

FIGURE 7. Cumulative size-frequency distribution of known Phanerozoic-aged terrestrial impact structures. Note deficit of structures at diameters ≳20 km. See text for details.

In principle, it is possible to use the observed size-frequency distribution of impact structures to derive the size-frequency distribution of impacting bodies. The size-frequency distribution, however, is not well-constrained because of small-number statistics (Fig. 7). In addition, the conversion from impact structure diameters to impacting body sizes must be accomplished through energy-scaling relations. Early energy scaling relations were based commonly on the results of underground nuclear explosions and generally considered the kinetic energy of impact as a single term. More recent experimental data and dimensional analyses indicate that there is dependence on impact velocity. Schmidt and Housen [106] provide a summary of recent relations for crater dimension scaling. There is another complication in that scaling equations refer generally to transient cavity dimensions; that is, the cavity formed directly by ejection and displacement of material due to the cratering flow field (e.g., [83]). It takes no account of cavity modification processes, which result in the final crater form and can be considerable (~50–60% enlargement in diameter) in the case of complex structures. To relate transient cavity dimensions to final crater dimensions, an additional empirical scaling relation is required. The most generally used modification scaling relation can be found in Croft [14].

CRATERING RATE

Due to the inherent biases in the known terrestrial impact record, care must be exercised in quantitative or statistical analyses of the record, or subsamples of the record. If the biases in the known impact record are accounted for, the terrestrial impact rate is estimated at $5.6 \pm 2.8 \times 10^{-15}$ km^{-2} a^{-1} for $D \geq 20$ km [42]. This rate is based on the occurrence of relatively young (≤ 120 Ma), large ($D \geq 20$ km) impact structures on the relatively well-searched cratonic areas of N. America and Fennoscandia through to European Russia. This rate estimate is similar to that based on astronomical observations of bodies with Earth-crossing orbits of $5.9 \pm 3.5 \times 10^{-15}$ km^{-2} a^{-1} for $D \geq 20$ km, corresponding approximately to impacting bodies $\gtrsim 1$ km in diameter [117], and that estimated from Proterozoic-aged impact structures in Australia of $3.8 \pm 1.9 \times 10^{-15}$ km^{-2} a^{-1} [115]. All these rate estimates have uncertainties of the order of ± 50%, reflecting concerns over completeness of search and small number statistics.

With the size-frequency distribution and the cratering rate, it is possible to estimate the expected number of craters above any given diameter in any time period. For example, the expected number of terrestrial impact structures with $D \geq 40$ km on the total land surface of the Earth ($\sim 15 \times 10^7$ km^2) in 100 Ma is ~20. Currently, only 5 are known in the last 100 Ma. The effect of geologic activity over time on the terrestrial impact record is evident in the fact that even fewer structures (3) are known with $D \geq 40$ km in the earlier period of ≥ 100 Ma – ≤ 200 Ma.

These estimates of the terrestrial cratering rate are higher than the average postmare cratering rate on the moon, when it is corrected to terrestrial conditions [87], [115]. With all the observational and model uncertainties associated with converting astronomical estimates of impacting body sizes to impact structure diameters and scaling between lunar and terrestrial impact conditions, it is probably unjustified to categorically state that the current cratering rate, based on terrestrial and astronomical data, is higher than that integrated over the last ~3.2 Ga, based on lunar cratering data. In addition, it is difficult to imagine a mechanism to increase the cratering rate with geologic time, beyond the increasing involvement of cometary impacts [116].

It would be heartening to believe that all that is required to improve the cratering rate statistics is the discovery and dating of a number of newly discovered terrestrial impact structures. Unfortunately, this is not the case. The discovery of additional large, young impact structures throughout the world will not help, unless they are in an area where there is also a fair degree of confidence that the majority of the large impact structures have been found. Conversely, the discovery of additional large, young impact structures within areas with relatively complete searches, such as those already used to derive cratering rate estimates, i.e., the N. American and Fennoscandia-

European Russian cratonic areas, would acerbate the problem of an apparently higher rate in relatively recent geologic time.

Considerable interest in the potential for impact to disrupt the biological balance on Earth followed the initial reports of evidence for the involvement of large-scale impact at the K/T boundary. Raup and Sepkoski [103] reported evidence for a periodicity to the marine extinction record over the past 250 Ma and this was followed by a number of works claiming a similar periodicity in the terrestrial cratering record (e.g., [2], [16], [99]). They suggested that this was the result of periodic cometary showers, due to a variety of astronomical mechanisms. These periodic cometary showers were linked statistically to terrestrial mass extinction events, and to a variety of other geological and geophysical phenomena [101]. The claims were based on time-series analyses of subsets of the terrestrial cratering record. There were a variety of selection criteria but they conformed generally to large (≥ 5 km), young (≤ 250 Ma) and well-dated (± 20 Ma) impact structures.

Grieve et al. [43] have argued against these conclusions, noting the biases in the terrestrial impact record and that, if the uncertainties in crater age estimates are taken into account, periodicities in the cratering record are questionable to weak. They also argued that, to have confidence in the reality of any period, the uncertainties attached to individual age estimates had to be < 10% of the period in question, which, in this case, was ~30 Ma. This relatively small degree of uncertainty in age is not a general property of age estimates in the terrestrial cratering record (e.g., Table 1). Heisler and Tremaine [47] reached a similar conclusion based on different statistical arguments and Baksi [6] detected no periodicity, if the selected impact structures were restricted to those with age estimates of sufficient accuracy and precision. Weissman [134] also found no evidence for periodic cometary showers and challenged the proposed astrophysical mechanisms for producing periodic cometary showers. He did not, however, rule out random cometary showers.

Despite these arguments, periodic cometary showers, as defined by time-series analysis of the terrestrial cratering record, are still featured and suggested as a causative agent for various geologic phenomena on Earth (e.g., [125], [124], [100]). The original listing used to define periodic cometary showers by most workers contained 103 known terrestrial impact structures. Grieve and Pesonen [40] using a more current listing of terrestrial impact structures repeated earlier time-series analyses. They concluded that, despite continuing assertions, there is no compelling evidence for periodic impacts in the terrestrial cratering record. Similarly, there are currently insufficient high quality data to comment upon whether or not the cratering record contains clusters of impacts of similar age that could reflect large-scale events in the asteroidal belt and relatively rapid resonant dynamics. There are, however, a number of cases (e.g., East and West Clearwater, Canada, and Kamensk and Gusev, Russia), where the impact structures were apparently formed by binary impacting bodies [84].

IMPACT BODY COMPOSITIONS

All small (≤ 1 km) terrestrial impact structures are the result of the impact of iron or stony-iron impacting bodies. This is due to the, previously noted, selective effect of the terrestrial atmosphere, which effectively destroys the weaker stony and icy bodies, as crater-forming bodies, by atmospheric crushing, followed by dispersal and velocity retardation [82]. These small impact structures, which often occur as crater fields due to atmospheric breakup, are Recent in age and it is often possible to recover pieces or, more commonly, fragments of the impacting body that spalled off prior to impact. Larger impacting bodies survive atmospheric passage and do so with impact velocities undiminished by atmospheric effects. In these cases, shock pressures and temperatures at the point of impact, which are a function of impact velocity, are sufficient to melt and/or vaporize the impacting body [1]. Estimates of the terrestrial r.m.s. impact velocity for asteroidal and long-period cometary bodies are ~18 km s^{-1} and ~58 km s^{-1}, respectively [116]. Thus, at sizes where crater-forming, stony impacting bodies are not crushed and dispersed by the atmosphere, their impact velocities are sufficient to guarantee that they do not survive as a physical entity on impact. There appears to be one exception. This is the < 100,000 year old Rio Cuarto cratering field, Argentina, where 3 pieces of a chondritic meteorite have been recovered [112]. The largest of the ten known impact structures of Rio Cuarto is 4.5 km in its maximum dimension and it is believed that the preservation of a piece of the impacting body is due to the very oblique angle of the impact event, which resulted in projectile decapitation and highly elongated impact structures [110].

Although the impacting body is destroyed physically in larger impact events, some of the vaporized impacting body material can be mixed into the impact lithologies; particularly, in impact melt rocks. Allochthonous impact lithologies, such as breccias and impact melt rocks, are physical and chemical mixtures of preexisting target lithologies. Least-squares mixing models of impact melt compositions for major and trace elements, in general, give statistically and geologically reasonable results in terms of a mixture of target rock compositions (e.g., [37], [104]). There are a number of cases, however, where siderophile elements, and sometimes chromium, in impact melt rocks are in excess of any plausible mixture of the target rocks. These small geochemical anomalies (at the ng/g level for most siderophiles) represent a chemical admixture of volatilized/melted projectile material, with overall projectile contamination generally at the level of a percent or less. Detailed geochemical studies have, in some cases, determined the type of impacting body relative to known meteorite compositional groups through interelemental ratios (e.g., Fig. 8; [91], [27]). A number of factors, however, can complicate the issue: relatively low elemental abundances, postimpact alteration, and corrections for indigenous components (e.g., [70]). In some cases, therefore, projectile identifications are relatively tentative and open to interpretation. It is generally assumed that a cometary impact would have a primitive (chondritic) signature. At this time, however, there is no practical mechanism for differentiating a cometary versus an asteroidal impact [5].

Although experimental studies indicate elemental fractionation during impact vaporization is possible [138], well-documented cases at terrestrial impact structures are generally lacking. This may be due to a small signal-to-noise ratio, with complex target compositions and postimpact alteration playing a role in complicating the geochemical evidence. Chemical fractionation in materials relatively divorced from the bulk of the impact lithologies and the impact site has been suggested for some variations in interelement abundances [86]. For example, variations in the Ru/Ir ratio in K/T boundary clays, from an average of 1.67 ± 0.38 in Europe to 0.76 ± 0.26 in N. America, have been attributed to fractionation of Ru from Ir, during condensation of ejecta vapor, and correlate with distance from the Chicxulub impact structure, Mexico [28]. In general, however, the possibility of fractionation of the geochemical signature of the impacting body in impact lithologies is neither well-documented nor understood (e.g., [107]).

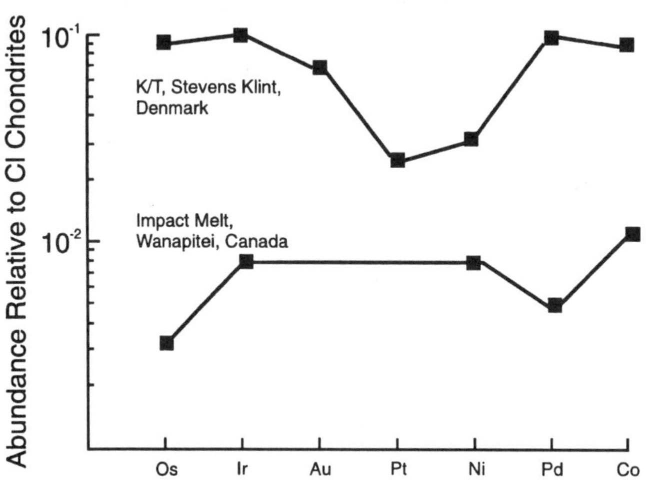

FIGURE 8. Chondrite-normalized pattern of siderophile abundances from impact melt rocks at Wanapitei and K/T boundary sediments from Denmark. Note relatively flat meteoritic patterns and the differences in relative abundance.

TABLE 2. Estimates of Impacting Body Types at Large Terrestrial Impact Structures[a]

Crater	Country	Diameter (km)	Body Type
New Quebec	Canada	3.4	Chondrite
Brent	Canada	3.8	Chondrite
Gow	Canada	4	Iron?
Rio Cuarto	Argentina	4.5[b]	Chondrite
Ilyinets	Ukraine	4.5	Stone?
Sääksjärvi	Finland	5	Chondrite
Wanapitei	Canada	7.5	Chondrite
La Moinerie	Canada	8	[c]
Mien	Sweden	9	Stone?
Bosumtwi	Ghana	10.5	Iron
Ternovka	Ukraine	12	Chondrite
Nicholson Lake	Canada	12.5	Achondrite
Zhamanshin	Kazakhstan	13.5	Chondrite?
Dellen	Sweden	15	Stone?
Obolon	Ukraine	15	Iron
Lappajärvi	Finland	17	Chondrite
El'gygytgyn	Russia	18	Achondrite
Clearwater East	Canada	22	Chondrite
Rochechouart	France	23	Chondrite? or Iron?
Ries	Germany	24	Achondrite? or Chondrite?
Boltysh	Ukraine	25	Chondrite
Mistastin	Canada	28	Iron? or Achondrite?
Clearwater West	Canada	32	[c]
Kara	Russia	65	Chondrite
Manicouagan	Canada	100	[c]
Popigai	Russia	100	Chondrite
Chicxulub	Mexico	180	Chondrite

[a]All smaller impact structures appear to have been formed by iron or stony iron bodies, due to atmospheric effects (see text). With the exception of Rio Cuarto (Argentina) all listed estimates of impacting body types are based on geochemical anomalies at the impact structure.

[b]Longest dimension of largest crater.

[c]Search for meteoritic signature unsuccessful, due either to a sampling problem or to an achondritic impacting body.

The estimates of impacting body types, taken from the literature without qualification, are dominated by bodies with chondritic compositions (Table 2). At face value, this is misleading. The reliability of impacting body identification at many of the impact structures in the former USSR is generally based on relatively few elements and interelemental ratios. Also, the geochemical signature of chondritic impacting bodies is the most easily identifiable, through their relatively high abundances of both siderophile elements and Cr. Identification of nonchondritic impacting bodies, e.g., Mistastin, Canada (Table 2), can be more of a negative result that the body was not a chondrite, rather than the firm indication that it was, as suggested, an iron body. Similar arguments apply to the identification of basaltic achondrites, which as differentiated bodies have compositions most similar to terrestrial rocks and tend to represent a negative result in terms of meteoritic contamination in impact melt rocks.

Recent interest has been revived in using Re/Os isotope systematics to detect a meteoritic component in impact melt rocks. They have been used to identify a meteoritic component in K/T boundary sediments [77]. New, more sensitive, analytical techniques have extended the utility of Re/Os isotopes; particularly, where the target rocks are already relatively enriched in siderophile or platinum group elements [61]. It is possible, that the increased sensitivity of Re-Os isotopic techniques may help detect additional cases of faint meteoritic signatures in impact lithologies. For example, the technique has been used at Kalkkop, S. Africa, where an admixture of 0.05% meteoritic material was detected in one sample of suevitic breccia [62]. Unfortunately, Re-Os systematics are, in themselves, not an effective discriminator between meteorite classes.

IMPACT IN THE STRATIGRAPHIC RECORD

Known occurrences of impact-related materials in the stratigraphic record are limited (Fig. 9). Prior to the discovery and interpretation that geochemical anomalies in K/T boundary sediments (Fig.8) were due to impact and to the subsequent realization that this was a global phenomenon, the known record of impact in the stratigraphic column was limited to the occurrence of the Australasian, Ivory Coast and N. American and moldavite tektite and microtektite strewn fields. The Ivory Coast microtektites are associated with the Bosumtwi structure, Ghana (e.g., [108]). The association of moldavite tektites with the Ries structure, Germany has also been recognized for some time (e.g., [25]). Most recently, the Chesapeake Bay structure, USA appears to be the best candidate for the source of the N. American strewn field [63]. The source of the Australasian field is currently unknown, although there have been a number of suggestions (e.g., [109], [133], [9]).

TABLE 3. Impact Debris in the Stratigraphic Column

Material/Horizon	Age, Ma	Evidence	Source[a]
Australasian microtektites	0.7	Microtektites, shocked minerals, small Ir anomaly	[58]
Tikal tektites, Guatemala	0.8	Tektites	[50]
Ivory Coast microtektites	1.1	Microtektites, small Ir anomaly, low $^{187}Os/^{188}Os$ ratio, association with Bosumtwi structure	[59]
N. Pacific microtektites	2.2	Microtektites, siderophile anomaly	[92]
Eltanin impact, S. Pacific	~2.5	Meteoritic debris, Ir, and Au anomaly	[68]
S. Ural tektite, Russia	6.2	Single tektite	[20]
Moldavites, Czechoslovakia	15	Tektites, association with Ries structure	[26]
Urengonite tektites, Russia	24	Tektites	[79]
N. American microtektites	35	Tektites, microtektites, shocked minerals	[34]
Clinopyroxene spherules	35	Microtektites (microkrystiles), small Ir anomaly	[33]
Late Eocene, Italy	35.7	Shocked quartz, small Ir anomaly	[71]
Cretaceous-Tertiary, global	65	Tektites, shocked minerals, siderophile anomaly, association with Chicxulub structure	[105]
Manson ejecta, USA	73.8	Shocked minerals	[53]
Mjølnir ejecta, Barents Sea	~143	Shocked minerals, siderophileanomaly	[22]
Chondritic spherules, Italy	~170	Meteoritic spherules, Ir anomaly	[57]
Qidong microtektites, China	~363	Microtektites, small Ir anomaly	[132]
Microtektites, Belgium	~365	Microtektites	[11], [12]
Alamo breccia, USA	~370	Shocked quartz	[73]
Acraman ejecta, Australia	588	Shocked minerals, siderophile anomaly	[131]
Early Archean, S. Africa and Australia	~3400	Spherule beds, siderophile anomaly	[76]

[a]Source is not necessarily original source, but rather a recent publication referring to the materials in question.

As a result of the interest generated by the K/T discoveries and possible connection between impact events and other short-term events in the geologic record, there have been searches for siderophile element (mostly Ir) anomalies at other major stratigraphic boundaries. This work has had mixed success, with positive (e.g., [137]), negative (e.g., [90]) and conflicting results (e.g., [55], [90]). In some cases, weak anomalies have been reported but, at this time, there is no compelling reason, or confirmatory evidence, to ascribe them to impact-related processes. None of these anomalies are global in scale and, in some cases, cannot be duplicated in replicate analyses. The most comprehensive summary of the search for Ir at stratigraphic boundaries can be found in Orth et al. [90], which cites Ir abundances at various stratigraphic boundaries in the range 0.002–0.2 ng/g. This can be compared with > 10 ng/g Ir at K/T boundary sites and in some melt rocks at terrestrial impact structures, with a very clear meteoritic signature [91].

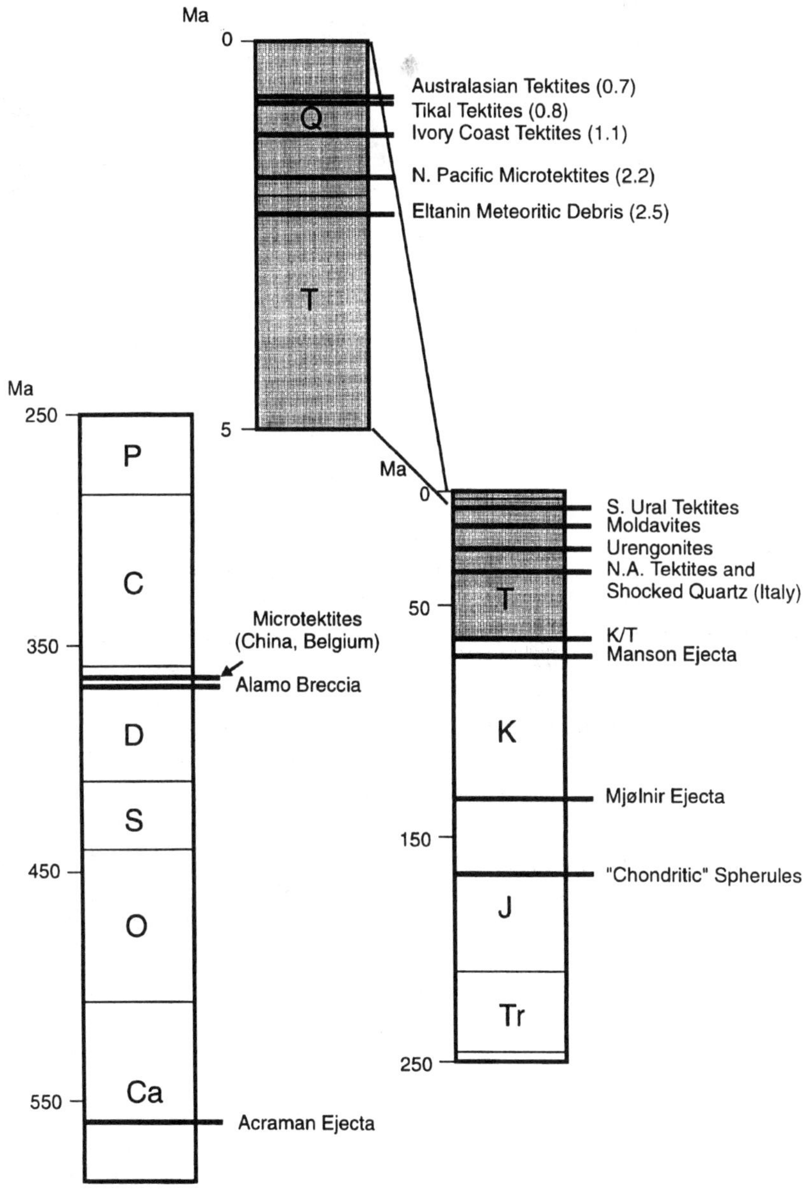

FIGURE 9. Location in time of known occurrences of impact-related materials in the stratigraphic column. See Table 3 and text for more details.

By far the majority of known impact events recorded in the stratigraphic column were recognized initially through the occurrence of physical, not geochemical, evidence of impact. Unlike the K/T, however, none are global in scale. The majority are the previously mentioned tektite or microtektite occurrences (Fig. 9; Table 3), some of which cover considerable areas.

For example, the Australasian strewn field occurs over an area of 50×10^6 km^2 [58]. Ejecta from Acraman, Australia (Table 3) has been recognized up to ~500 km from the structure, as a distinct clastic layer in shales of the Bunyeroo Formation of the Adelaide Geosyncline [135]. There is an associated siderophile anomaly, although there has been postimpact, low temperature remobilization [131]. In the case of the underwater Mjølnir structure in the Barents sea confirmation of an impact origin for the structure comes, in fact, from shocked materials and an ~1 ng/g Ir anomaly in ejecta encountered in a drill hole 30 km north-northeast of the structure [22].

What was an initial Ir and Au anomaly in Late Pliocene sediments from the southeast Pacific [67] is now known to be associated with meteoritic debris. This debris occurs over at least 600 km of the ocean floor and is the result of the impact of an ~0.5–2 km diameter achondritic impacting body, named Eltanin, which appears to have failed to make a crater on the ocean floor [68]. The deep water target and lack of a crater resulted in minimal projectile dilution by terrestrial materials and a clear geochemical anomaly.

The other occurrences of impact related materials in the stratigraphic column (Fig. 9; Table 3) represent relatively isolated discoveries and none, at present, are related to a known impact structure. For example, the Urengonite tektites, Siberia, were not recovered in place [79] and the Tikal tektites, Guatemala, were discovered in a museum archeological collection, during a search for K/T and Chicxulub-related materials in central America [50]. The spherules and irregular particles found in hard ground in northern Italy (Table 3) have an associated 100–500 ng/g Ir anomaly. They are believed to be debris from a large chondritic body, but they also represent an isolated occurrence [57]. The interpretation that early Archean spherule beds, with associated siderophile anomalies in S. Africa and Australia (Table 3), owe their origin to impact has been challenged [60] and are best considered an open question at this time. Not listed in Table 3 are reports of shocked quartz at Triassic-Jurassic boundary [4], [8]. These associations are not sufficiently documented or replicated at this time.

The relatively sparse evidence for impact-related material and geochemical anomalies in the stratigraphic record, however, is not as discouraging as it appears. Approximately 30% of the recognized impact materials occur in the most recent 2 Ma of the sedimentary record (Fig. 9; Table 3). The best documented events are recorded in material recovered from the sedimentary record of the present ocean basins. This is the segment of the stratigraphic record that has the highest signal-to-noise ratio, with respect to the possible detection of anomalous impact-related horizons. Detection of similar events in lithified materials from the paleoocean or paleocontinental environment will be more challenging. It is encouraging, however, that several of the (older) entries in Table 3 have been discovered recently and it suggests that others will be forthcoming.

Given the estimated terrestrial cratering rate, there are obviously many more impact events to be detected. Not all, however, will automatically result in an obvious signal in the stratigraphic record. A certain portion of the impacting bodies will be from differentiated parent bodies, such as basaltic achondrites. Such bodies would not result in an easily detectable geochemical anomaly, except for circumstances such as the case of the Late Pliocene debris in the S. Pacific (Table 3). The majority of terrestrial impact events will, in fact, occur in the world's oceans. While many of these will result in high-energy wave deposits from tsunamis running up paleocoastal areas, such as occurred along the K/T shoreline in the southern USA and Mexico (e.g., [120]), it is not likely that impact-generated deposits can be differentiated from seismic-generated tsunami deposits, unless other evidence of impact is present. In addition, the physical evidence of oceanic impacts may not be particularly obvious. Relatively inert materials from the target rocks, such as shocked quartz, will be absent. Mafic minerals from oceanic crust respond differently to shock pressures, relative to tectosilicates. Due to their greater compressibility, they retain more postshock waste heat and tend to undergo thermal decomposition at lower shock pressures [121], [122], [29]. They are also susceptible to subsequent alteration in the postimpact environment. The physical signature in the stratigraphic record of a major impact event into oceanic crust will be different from that of the K/T event and it will be more difficult to recognize.

The impact signal of the K/T event is recognizable globally, because large impact events have the capacity to produce atmospheric blow-out [83]. In essence, a hole is blown out in the atmosphere above the impact site. This permits some impact-related materials to be dispersed globally by entrainment in the impact fireball, which rises above the atmosphere, and, thus, they have the capacity to encircle the globe in relatively short-time periods. They eventually return to the surface and leave a signature in the stratigraphic record. Model calculations indicate that it does not require a Chicxulub-sized event to result in atmospheric blow-out. Relatively small impact events, resulting in impact structures in the ~20 km size range can produce atmospheric blow-out [83].

Kyte and Wasson [66] analysed a continuous 9 m section of piston core LL44 GPC3 for Ir from the Pacific that contained ~30 Ma of sedimentary history. They detected only one significant Ir anomaly and it was associated with the K/T boundary event (Fig. 10). Since then, an additional 15.3 m of core has been analysed and the K/T boundary remains unique in terms of a

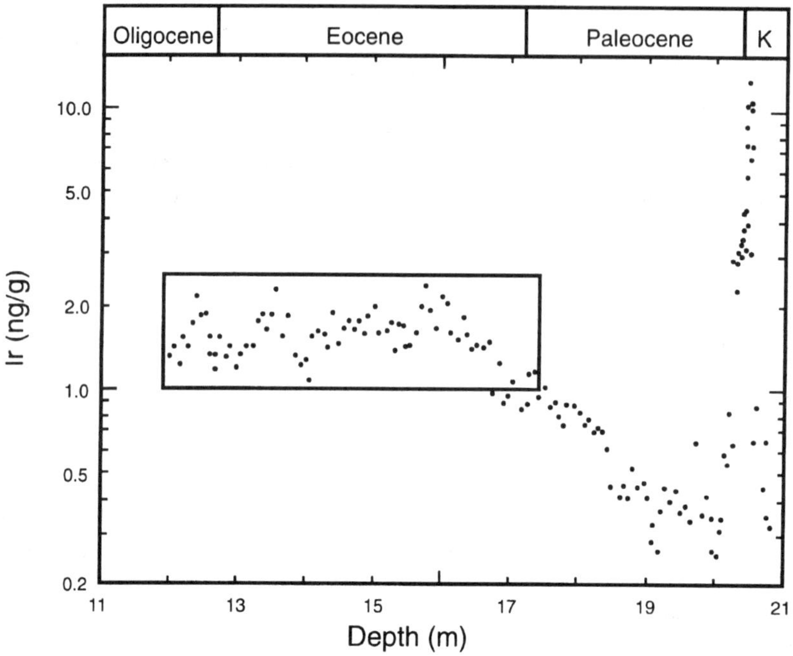

FIGURE 10. Ir abundances in deep sea core from the central Pacific. Note the single spike at the K/T boundary and the fluctuation of between ~1.0 > 2.0 ng/g (boxed area) during the Eocene. See text for the significance of this variation for the detection of large impact events, based solely on Ir abundances. Data from Kyte et al. [69].

clear spike in Ir content (Fig. 10; [69]). How does this reconcile with the relatively high frequency of impact events, approximately 2–4 events resulting in structures with D $\gtrsim$ 20 km per million years, based on the estimated cratering rate? These events have the potential to result in atmospheric blowout and, thus, global dissemination of impact-related materials.

The lack of other identifiable Ir peaks in this oceanic piston core may be related to the signal-to-noise ratio. The maximum Ir values associated with the K/T boundary in the Pacific deep-sea core are in order of 12 ng/g [66]. Taking these peak values as the result of the formation of the Chicxulub structure and assuming similar impacting conditions (projectile type, impact velocity, etc.), it is possible to estimate the expected peak Ir values associated with smaller impact events. Assuming ~180 km for the diameter of Chicxulub, dimensional scaling laws for impact crater formation (e.g., [106]) indicate that the mass of an impacting body resulting in an ~100 km diameter would be approximately an order of magnitude less than that of the body that produced Chicxulub. Assuming an equivalent sedimentary environment, as the central Pacific, this would result in an Ir peak of ~1–2 ng/g [66]. These Ir values are comparable to what appears to be the background influx of chondritic material in the form of cosmic dust and micrometeorites during the Eocene-Oligocene (Fig. 10). Thus, the signal of an impact event resulting in even a 100 km diameter impact structure may be difficult to detect against the background of Ir values resulting from the daily accretion of fine-grained cosmic material. Indeed two major impact structures, Chesapeake Bay and Popigai, are known to have formed just prior to the Eocene-Oligocene boundary at ~35 Ma. At least, Popigai is believed to have been formed by a chondritic body. Yet, there is no clear geochemical signal in the record from the central Pacific (Fig. 10). Given, therefore, the likelihood of even relatively large impact events to produce a relatively weak geochemical signal, depending on the sedimentary environment, it would seem that it is necessary to conduct a detailed physical examination for impact-related materials of any suggested impact horizons in the stratigraphic column, in concert with geochemical studies.

CONCLUDING REMARKS

This review has attempted to give the flavor of the terrestrial impact record. The record contains biases due to terrestrial processes, which are particularly evident in the poor sample of smaller and older impact structures. The known record of impact differs from astronomical observations in that it is not a current snapshot but represents an integration over considerable geologic time. Estimates of the terrestrial cratering rate can be derived, which are consistent with astronomical estimates. Given the nature of the current knowledge-base of terrestrial impact structures, however, only the most general statements can be made concerning the character of the impact flux. Although the discovery rate of new impact structures continues at a reasonable pace, new discoveries do not necessarily improve constraints on the character of the impact flux. The addition of a newly-discovered impact structure of uncertain age and geochemical character is, in itself, of no assistance. Systematic studies involving isotopic measurements for age estimates and trace element analyses for impacting body compositions are required, if there are to be major advances in constrainting the impact flux by studies of the terrestrial impact record.

The discovery of geochemical and physical anomalies related to impact in the stratigraphic record is a recent development in the geosciences. It is clear that many more anomalies remain to be discovered but it is likely that their discovery will initially have an element of controversy, similar to the past debate surrounding the K/T boundary. The study of distal ejecta from large impact events has not been a traditional concern of the community studying terrestrial impact structures. The cumulative knowledge gained from the study of these distal impact lithologies discovered in the stratigraphic record will undoubtedly provide new insights and constraints for cratering processes. Such impact-related horizons, which are deposited essentially instantaneously, have the potential to act as marker horizons in the stratigraphic record, at scales from local to global. As more are discovered, impact will increasingly be recognized as simply another natural process occurring on Earth that must be considered as a potential forcing function for change to the environment at scales from local to global.

This review has not considered the specifics of whether or not, or how, major impact events may be recorded in the stratigraphic record as extinction events. While this is, by far, the best working hypothesis for the K/T event, it is not self-evident, at this time, which one of the several proposed impact-related killing mechanisms is the most viable or whether it was a combination of mechanisms. It is even possible that the causal relationship between a major impact and the K/T mass extinction event was unique due to the presence of sulphates and carbonates in the target rocks of the Chicxulub structure [49] and the subsequent effects their volatilized products had on the climate and biosphere (e.g., [96]). Countering this argument are recent calculations that indicate sufficient sulfur is contained with the impacting bodies, themselves, to have deleterious effects to the environment [64].

While it is possible that the K/T impact event was unique in resulting in a mass extinction, it was clearly not a singular event. Large impact events have occurred throughout geologic history on the Earth and some of the effects of the K/T event on the terrestrial environment are generic to large impact events, e.g., dust loading of the atmosphere [3], [129], shock heating of the atmosphere resulting in nitric acid rain [98] and global wildfires [136]. It is evident that large impact events have the potential to result in rapid, short-term changes to the terrestrial environment. We have reached a point in understanding impact processes on Earth that the possibility that any rapid excursion in the biologic, climatic, or geologic record is the result of large-scale impact cannot be ignored. Establishing impact, however, as the best working hypothesis to the largely conservative geoscience community is another matter. The geochemical evidence of the K/T event is heavily reinforced by physical evidence in the form of shocked minerals in K/T boundary sediments and the occurrence of the Chicxulub impact structure.

This contribution also has not dealt with the hazard potential of impact events for human civilization. The geologic record of impact on the Earth indicates that 2–4 impact events occur every million years. These have the potential to produce the equivalent of a nuclear winter . Details of the hazard potential of terrestrial impacts can be found in papers in Gehrels [32] and this volume. Without some mitigatory action, our current civilization, if it persists long enough, will, undoubtedly, be severely affected and possibly destroyed by a terrestrial impact. Impacts on Earth, however, are not entirely negative in their consequences. For example, a major impact on the proto-Earth by a Mars-sized object is the best current working hypothesis for the origin of the moon [46]. Without the moon, there would be no ocean tides and the biological evolution of the Earth would likely be considerably different from what has occurred. Similarly, without the K/T impact, it is likely that the rise of mammals and, ultimately, the human species would not have been a foregone conclusion. In addition, as localized, high energy events producing distinctive structures, terrestrial impact structures also have considerable economic potential. As well as some major hydrocarbon-producing structures [36], two world-class mining camps, the Sudbury and Witwatersrand Basins in Canada and S. Africa, respectively, owe their existence to the Sudbury and Vredefort impact structures [38], [128].

ACKNOWLEDGMENTS

I would like to thank M. Ford and J. Smith for their assistance in manuscript preparation on short notice. I would also like to thank A. Therriault for reviewing an earlier verison of the manuscript.

REFERENCES

1. AHRENS, T.J. & J.D. O'KEEFE. 1977. Equations of state and impact-induced shock-wave attenuation on the moon. *In* Impact and Explosion Cratering, D.J. Roddy, R.O. Pepin, & R.B. Merrill, Eds.: pp. 639–656. Pergamon Press. Elmsford, New York.
2. ALVAREZ, W. & R. MULLER. 1984. Evidence from crater ages for periodic impact on the Earth. Nature. **308:** 718–720.
3. ALVAREZ, L.W., W. ALVAREZ, F. ASARO, & H.V. MICHEL. 1980. Extraterrestrial cause for the Cretaceous-Tertiary extinction. Science. **208:** 1095–1108.
4. BADJUKOV, D.D., H. LOBITZER, & M.A. NAZAROV. 1987. Quartz grains with planar features in the Triassic-Jurassic boundary sediments from Northern Limestone Alps (abstract). Lunar and Planetary Science. **XVIII:** 38–39.
5. BAILEY, M.E. 1991. Comet craters versus asteroid craters. Advances in Space Research. **11:** 43–62.
6. BAKSI, A.K. 1990. Search for periodicity in global events in the geologic record: Quo vadimus? Geology. **18:** 983–986.
7. BARRINGER, D.M. 1906. Coon Mountain and its crater. Proceedings Academy of Natural Sciences of Philadelphia. **57:** 861–886.
8. BICE, D.M., C.R. NEWTON, S. MCCAULEY, P.W. REINERS, & C.A. MCROBERTS. 1992. Shocked quartz at the Triassic-Jurassic Boundary in Italy. Science. **255:** 443–446.
9. BLUM, J.D., D.A. PAPANASTASSIOU, C. KOEBERL, & G.J. WASSERBURG. 1992. Neodymium and strontium isotopic study of Australasian tektites: New constraints on the provenance and age of target materials. Geochimica et Cosmochimica Acta. **56:** 483–492.

10. CHRISTENSEN, L., S. FREGERSLEV, A. SIMONSEN, & J. THIEDE.1973. Sedimentology and depositional environment of the lower Danian fish clay from Stevns Klint, Denmark. Bulletin of the Geological Society of Denmark. **22:** 193–216.
11. CLAEYS, P., J.CASIER, & S.V. MARGOLIS. 1992. Microtektites and mass extinctions: Evidence for a late Devonian asteroid impact. Science. **257:** 1102–1104.
12. CLAEYS, P., F.T. KYTE, F.T., & J.-G. CASIER. 1994. Frasnian-Famennian boundary: Mass extinctions, anoxic oceans, microtektite layers but not much Iridium (abstract). New Developments Regarding the KT Event and Other Catastrophes in Earth History. Lunar and Planetary Institute Contribution. **825:** 22–24.
13. COWAN, E.J. & W.M. SCHWERDTNER. 1994. Fold origin of the Sudbury Basin. *In* P.C. Lightfoot & A.J. Naldrett, Eds. Ontario Geological Survey Special Volume 5, P.C. Lightfoot & A.J. Naldrett, Eds.: 45–55. Proceedings of the Sudbury-Noril'sk Symposium. Ontario Ministry of Northern Development and Mines, Toronto.
14. CROFT, S.K. 1985. The scaling of complex craters. Proceedings of the 15th Lunar and Planetary Science Conference. 828–842.
15. CURRIE, K.L. 1971. Geology of the resurgent cryptoexplosion crater at Mistastin Lake, Labrador. Canada Geological Survey Bulletin. **207:** 62 pp.
16. DAVIS, M., P. HUT, & R.A. MÜLLER. 1984. Extinction of species by periodic comet showers. Nature. **308:** 715–717.
17. DE LAUBENFELS, M.W. 1956. Dinosaur extinction: One more hypothesis. J. Paleontology. **30:** 207–218.
18. DEUTSCH, A. & R.A.F. GRIEVE. 1994. The Sudbury Structure: Constraints on its genesis from Lithoprobe results. Geophysical Research Letters. **21:** 963–966.
19. DEUTSCH, A. & U. SCHÄRER. 1994. Dating terrestrial impact events. Meteoritics. **29:** 301–322.
20. DEUTSCH, A., M. OSTERMANN, & V.L. MASAITIS. 1996. Neodymium–strontium isotope systematics of impact-related glassy objects (Urengoites, South Ural glass, Zhamanshinites, Irghizites) (abstract). Meteoritics & Planetary Science. **31:** Supplement A37.
21. DIETZ, R.S. 1964. Sudbury structure as an astrobleme. J. Geology. **72:** 412–434.
22. DYPVIK, H., S.T. GUDLAUGSSON, F. TSIKALAS, M.J. ATTREP, R.E.J. FERRELL, D.H. KRINSLEY, A. MORK, J.I. FALEIDE, & J. NAGY. 1996. Mjølnir structure: An impact crater in the Barents sea. Geology. **24:** 779–782.
23. DONS, J.A. & J. NATERSTAD. 1992. The Gardnos impact structure, Norway (abstract). Meteoritics. **27:** 215.
24. ELO, S., T. KUIVASAARI, M. LEHTINEN, O.SARAPÄÄ, & A. UUTELA. 1993. Iso-Naakkima, a circular structure filled with Neoproterozoic sediments, Pieksämäki, southeastern Finland. Bulletin Geological Society Finland **65:** 3–30.
25. ENGELHARDT, W. VON & F. HÖRZ. 1965. Ries glasses and moldavite (in German). Geochimica et Cosmochimica Acta. **29:** 609–620.
26. ENGELHARDT, W. VON., E. LUFT, J. ARNDT, H. SCHOCK, & WEISKIRCHNER, W. 1987. Origin of moldavites. Geochimica et Cosmochimica Acta. **51:** 1425–1443.
27. EVANS, N.J., D.C.GREGOIRE, R.A.F. GRIEVE, W.D. GOODFELLOW, & J. VEIZER. 1993a. Use of platinum-group elements for impactor identification: Terrestrial impact craters and Cretaceous-Tertiary boundary. Geochimica et Cosmochimica Acta. **57:** 3737–3748.
28. EVANS, N.J., D.C. GREGOIRE, W.D. GOODFELLOW, B.I. MCINNES, N. MILES, & J. VEIZER. 1993b. Ru/Ir ratios at the Cretaceous-Tertiary boundary: Implications for PGE source and fractionation within the ejecta cloud. Geochimica et Cosmochimica Acta. **57:** 3149–3158.
29. FEL'DMAN, V.I. 1994. The conditions of shock metamorphism. Geological Society of America Special Paper 293: 121–132.

30. French, B.M. & N.M. Short, Eds. 1968. Shock Metamorphism of Natural Materials. Mono Book Corp. Baltimore. 644 pp.

31. Ganapathy, R., 1980. A major meteorite impact on the Earth 65 million years ago: Evidence from the Cretaceous-Tertiary boundary clay. Science. **209:** 921–923.

32. Gehrels, T. Ed. 1994. Hazards Due to Comets and Asteroids, University of Arizona Press, Tucson, 1300 pp.

33. Glass, B.P. & C.A. Burns. 1987. Late Eocene crystal-bearing spherules: Two layers or one? Meteoritics. **22:** 265–279.

34. Glass, B.P., C.A. Burns, J.R. Crosbie, & D.L. DuBois. 1985. Late Eocene North American microtektites and clinopyroxene-bearing spherules. Proceedings Lunar and Planetary Science Conference 16th. J. Geophysical Research, Supplement. **90:** D175–D196.

35. Grahn, Y. & J. Nõlvak. 1993. Chitinozoan dating of Ordovician impact events in Sweden and Estonia. A preliminary note. Geologiska Föreningens I Stockholm Förhandlingar. **115:** 263–264.

36. Grieve, R.A.F. 1997. Terrestrial impact structures: Basic characteristics and economic significance, with emphasis on hydrocarbon production. Oklahoma Geological Survey Special Paper. In press.

37. Grieve, R.A.F. & R.J. Floran. 1978. Manicouagan impact melt, Quebec 2. Chemical interrelations with basement and formational processes. J. Geophysical Research. **83:** 2761–2771.

38. Grieve, R.A.F. & V.L. Masaitis. 1994. The economic potential of terrestrial impact craters. International Geology Review. **36:** 105–151

39. Grieve, R.A.F. & L.J. Pesonen. 1992. The terrestrial impact cratering record. Tectonophysics. **216:** 1–30.

40. Grieve, R.A.F. & L.J. Pesonen. 1996. Terrestrial impact craters: Their spatial and temporal distribution and impacting bodies. Earth, Moon and Planets. **72:** 357–376.

41. Grieve, R.A.F. & M. Pilkington. 1996. The signature of terrestrial impacts. AGSO J. Australian Geology and Geophysics, **16:** 399–420.

42. Grieve, R.A.F. & E.M. Shoemaker. 1994. The record of past impacts on Earth. *In* Hazards Due to Asteroids and Comets, T. Gehrels, Ed.: 417–462. University of Arizona Press. Tucson, Arizona.

43. Grieve, R.A.F., V.L. Sharpton, J.D. Rupert, & A.K. Goodacre. 1988. Detecting a periodic signal in the terrestrial cratering record. Proceedings Lunar and Planetary Science Conference 18th, 375–382.

44. Grieve, R.A.F., J. Rupert, J. Smith, & A. Therriault. 1995. The record of terrestrial impact cratering. GSA TODAY. **5:** 189, 194–196.

45. Grieve, R.A.F., F. Langenhorst, & D. Stöffler. 1996. Shock metamorphism of quartz in nature and in experiment: II Significance in geoscience. Meteoritics. **31:** 6–35.

46. Hartmann, W.K., R.J. Phillips, & G.J. Taylor, Eds. 1986. Origin of the Moon. Lunar Planetary Institute, Houston. 781 pp.

47. Heisler, J. & S. Tremaine. 1989. How dating uncertainties affect the detection of periodicity in extinctions and craters. Icarus. **77:** 213–219.

48. Henkel, H. 1992. Geophysical aspects of meteorite impact craters in eroded shield environment, with special emphasis on electric resistivity. Tectonophysics. **216:** 63–90.

49. Hildebrand, A.R., G.T. Penfield, D.A. Kring, M. Pilkington, A.Z. Camargo, S.B. Jacabsen, & W.V. Boynton. 1991. Chicxulub crater: A possible Cretaceous/Tertiary boundary impact crater on the Yucatan Peninsula, Mexico. Geology. **19:** 867–871.

50. HILDEBRAND, A.R., H. MOHOLY-NAGY, C. KOEBERL, L. MAY, F. SENFTLE, A.N. THORPE, & P.E. SMITH. 1994. Tektites found in the ruins of the Maya city of Tikal, Guatemala (abstract). Lunar and Planetary Science. **XXV:** 549–550.
51. HILDEBRAND, A.R., M. PILKINGTON, M. CONNORS, C. ORTIZ-ALEMAN, & R.E. CHAVEZ. 1995. Size and structure of the Chicxulub crater revealed by horizontal gravity gradients and cenotes. Nature. **376:** 415–417.
52. HIRT, A.M., W. LOWRIE, W.S. CLENDENEN, & R. KLIGFIELD. 1993. Correlation of strain and the anisotropy of magnetic susceptibility in the Onaping Formation: Evidence for a near-circular origin of the Sudbury Basin. Tectonophysics. **225:** 231–254.
53. IZETT, G.A., W.A. COBBAN, J.D. OBRADOVICH, & M.J. KUNK. 1993. The Manson impact structure: 40Ar/39Ar age and its distal impact ejecta in the Pierre Shale in southeastern South Dakota. Science. **262:** 729–732.
54. HODYCH, J.P. & G.R. DUNNING. 1992. Did the Manicouagan impact trigger end-of-Triassic mass extinction? Geology. **20:** 51–54.
55. HSU, K.J., H. OBERHANSLI, J.Y. GAO, S. SUN, H. CHEN, & U. KRAHENBUL. 1985. “Strangelove ocean” before the Cambrian explosion. Nature. **316:** 809–811.
56. JANSA, L.F., G. PE-PIPER, P.B. ROBERTSON, & , O. FREIDENREICH. 1989. Montagnais: A submarine impact structure on the Scotian Shelf, eastern Canada. Geological Society of America Bulletin. **101:** 450–463.
57. JÉHANNO, C., D. BOCLET, P. BONTÉ, A. CASTELLARIN, & R. ROCCHIA. 1988. Identification of two populations of extra-terrestrial particles in a Jurassic hardground of the southern Alps. Proceedings of the 18th Lunar and Planetary Science Conference: 623–630.
58. KOEBERL, C. 1993. Extraterrestrial component associated with Australasian microtektites in a core from OPD Site 758B. Earth and Planetary Science Letters. **119:** 453–458.
59. KOEBERL, C. & S.B. SHIREY. 1993. Detection of a meteoritic component in Ivory Coast tektites with rhenium-osmium isotopes. Science. **261:** 595–598.
60. KOEBERL, C., W.U. REIMOLD, & R.H. BOER. 1993. Geochemistry and mineralogy of Archean spherule beds, Barberton Mountain Land, S. Africa: Evidence for origin by impact doubtful. Earth and Planetary Science Letters. **119:** 441–452.
61. KOEBERL, C., S.B. SHIREY, & W.U. REIMOLD. 1994a. Re-Os isotope systematics as a diagnostic tool for the study of impact craters (abstract). New Developments Regarding the K/T Event and Other Catastrophes in Earth History. Lunar and Planetary Institute Contribution. **825:** 61–63.
62. KOEBERL, C., W.U. REIMOLD, S.B. SHIREY, & F.G.L. ROUX. 1994b. Kalkkop crater, Cape Province, South Africa: Confirmation of impact origin using osmium isotope systematics. Geochimica et Cosmochimica Acta. **58:** 1229–1234.
63. KOEBERL, C., C.W. POAG, W.U. REIMOLD, & D. BRANDT. 1996. Impact origin of Chesapeake Bay structure and the source for the North American tektites. Science. **271:** 1263–1266.
64. KRING, D.A., H.J. MELOSH, & D.M. HUNTEN. 1996. Impact-induced perturbations of atmospheric sulfur. Earth and Planetary Science Letters. **140:** 201–212.
65. KROGH, T.E., S.L. KAMO, & B.F. BOHOR. 1993. Fingerprinting the K/T impact site and determining the time of impact by U-Pb dating of single shocked zircons from distal ejecta. Earth and Planetary Science Letters. **119:** 425–429.
66. KYTE, F.T. & J.T. WASSON. 1986. Accretion rate of extraterrestrial matter: Iridium deposited 33 to 67 million years ago. Science. **232:** 1225–1229.
67. KYTE, F.T., Z. ZHOU, & J.T. WASSON. 1981. High noble metal concentrations in late Pliocene sediment. Nature. **292:** 417–420.
68. KYTE, F.T., Z. ZHOU, & J.T. WASSON. 1988. New evidence on the size and possible effects of a late Pliocene oceanic asteroid impact. Science. **241:** 63–65.

69. KYTE, F.T., M. LEINEN, R.G. HEATH, & L. ZHOU. 1993. Cenozoic sedimentation history of the central Pacific: Inferences from the elemental geochemistry of core LL44–GPC3. Geochimica et Cosmochimica Acta. **57:** 1719–1740.
70. LAMBERT, P. 1982. Anomalies within the system: Rochechouart target rock meteorite. Geological Society of America Special Paper 190, 57–68.
71. LANGENHORST, F. 1996. Characteristics of shocked quartz in late Eocene impact ejecta from Massignano (Ancona, Italy): Clues to shock conditions and source crater. Geology. **24:** 487–490.
72. LEHTINEN, M., L.J. PESONEN, R. PURANEN, & A. DEUTSCH. 1996. Karikkoselkä — a new impact structure in Finland (abstract). Lunar and Planetary Science. **XXVII:** 739–740.
73. LEROUX, H., J.E. WARME, & , J.-C. DOUKHAN. 1995. Shocked quartz in the Alamo breccia, southern Nevada: Evidence for a Devonian impact event. Geology. **23:** 1003–1006.
74. LINDSTRÖM, M. & E.F.F. STURKELL. 1992. Geology of the early Palaeozoic Lockne impact structure, central Sweden. Tectonophysics. **216:** 169–185.
75. LINDSTRÖM, M., T. FLODEN, Y. GRAHN, & B. KATHOL. 1994. Post-impact deposits in Tvären, a marine Middle Ordovician crater south of Stockholm, Sweden. Geological Magazine. **131:** 91–103.
76. LOWE, D.R., G.R. BYERLY, F. ASARO, & F.T. KYTE. 1989. Geological and geochemical record of 3400-million-year-old terrestrial meteorite impacts. Science. **245:** 959–962.
77. LUCK, J.M. & K.K. TUREKIAN. 1983. Osmium-187/Osmium-186 in manganese nodules and the Cretaceous-Tertiary boundary. Science. **222:** 613–615.
78. MAK, E.K., D. YORK, R.A.F.GRIEVE, & M.R. DENCE. 1976. The age of the Mistastin Lake crater, Labrador, Canada. Earth and Planetary Science Letters. **31:** 345–357.
79. MASAITIS, V.L., M.A. IVANOV, V.A. EZERSKY, V.S. KOZLOV, & N.B. RESHETNYAK. 1989. The West Siberian dispersion area of tektite-like glasses (in Russian). Doklady Akademii Nauk SSSR. **304:** 1419–1423.
80. MASHCHAK, M.S. & M.V. NAUMOV. 1996. The Suavjarvi structure: An early Proterozoic impact site on the Fennoscandian shield (abstract). Lunar and Planetary Science. **XXVII:** 825–826.
81. MCLAREN, D.J. 1970. Presidential address: Time, life, and boundaries. J. Paleontology. **44:** 801–815.
82. MELOSH, H.J., 1981. Atmospheric breakup of terrestrial impactors. *In* Multi-Ring Basins, P.H. Schultz & P.B. Merrill, Eds.: 29–35. Pergamon Press. Elmsford, New York.
83. MELOSH, H.J. 1989. Impact Cratering: A Geologic Process. Oxford, New York. 245 pp.
84. MELOSH, H.J. & J. STANSBERRY. 1991. Doublet craters and the tidal disruption of binary asteroids (abstract). Meteoritics. **26:** 371–372.
85. MILKEREIT, B., A. GREEN, & THE SUDBURY WORKING GROUP. 1992. Deep geometry of the Sudbury structure from seismic reflection profiling. Geology. **20:** 807–811.
86. MITTLEFEHLDT, D.W., T.H. SEE, & F. HÖRZ. 1992. Dissemination and fractionation of projectile materials in the impact melts from Wabar Crater, Saudi Arabia. Meteoritics. **27:** 361–370.
87. NEUKUM, G. & B.A. IVANOV. 1994. Crater size distributions and impact probabilities on Earth from lunar, terrestrial, planet, and asteroid cratering data. *In* Hazards Due to Comets and Asteroids, T. Gehrels Ed.: 359–416. University of Arizona Press. Tucson, Arizona.
88. OFFICER, C.B. & J. PAGE. 1996. The Great Dinosaur Extinction Controversy. Helix. New York. 209 pp.

89. ORMÖ, J. & R. TÖRNBERG. 1994. The Tvären Bay impact crater-pre-impact stratigraphy and shock metamorphosis (abstract). European Science Foundation, Second International Conference. Impact Cratering and Evolution of Planet Earth. The Identification and Characterization of Impacts, Lockne, Sweden.
90. ORTH, C.J., M.J. ATTREP, & L.R. QUINTANA. 1990. Iridium abundances across bio-event horizons in the fossil record. Geological Society of America Special Paper 247, 45–60.
91. PALME, H. 1982. Identification of projectiles of large terrestrial impact craters and some implications for the interpretation of Ir-rich Cretaceous/Tertiary boundary layers. Geological Society of America Special Paper 190, 223–233.
92. PENG, H., 1994. An extraterrestrial event at the Tertiary-Quaternary boundary (abstract). New Developments Regarding the KT Event and Other Catastrophes in Earth History, Lunar and Planetary Institute Contribution 825, 88–89.
93. PESONEN, L.J., M. LEHTINEN, A. DEUTSCH, S. ELO, & H. LUKKARINEN. 1996. New geophysical and petrographic results of the Suvasvesi N impact structure, Finland (abstract). Lunar and Planetary Science. **XXVII:** 1021–1022.
94. PIKE, R.J. 1985. Some morphologic systematics of complex impact structures. Meteoritics. **20:** 49–68.
95. PILKINGTON, M. & R.A.F. GRIEVE. 1992. The geophysical signature of terrestrial impact craters. Reviews Geophysics. **30:** 161–181.
96. POPE, K.O., K.H. BAINES, A.C. OCAMPO, & B.A. IVANOV. 1994. Impact winter and the Cretaceous/Tertiary extinctions: Results of the Chicxulub asteroid impact model. Earth and Planetary Science Letters. **128:** 719–728.
97. POPE, K.O., A.C. OCAMPO, G.L. KINSLAND, & R. SMITH. 1996. Surface expression of the Chicxulub crater. Geology. **24:** 527–530.
98. PRINN, R.G. & B. FEGLEY. 1987. Bolide impacts, acid rain, and biospheric traumas at the Cretaceous-Tertiary boundary. Earth and Planetary Science Letters. **83:** 1–15.
99. RAMPINO, M.R. & B.M. HAGGERTY. 1994. Extra-terrestrial impacts and mass extinctions of life: *In* Hazards Due to Comets and Asteroids, T. Gehrels, Ed.: 827–857. University of Arizona Press. Tucson, Arizona.
100. RAMPINO, M.R. & B.M. HAGGERTY. 1996. The "Shiva hypothesis": Impacts, mass extinctions and the galaxy. Earth, Moon, and Planets. **72:** 441–460.
101. RAMPINO, M.R. & R.B. STOTHERS. 1984. Terrestrial mass extinction, cometary impacts and the Sun's motion perpendicular to the galactic plane. Nature. **308:** 709–712.
102. RODDY, D.J., R.O. PEPIN, & R.B. MERRILL, Eds. 1977. Impact and Explosion Cratering. Pergamon Press. Elmsford, New York. 1301 pp.
103. RAUP, D.M. & J.J. SEPKOSKI. 1984. Periodicity of extinctions in the geologic past. Proceedings National Academy of Science. **81:** 801–805.
104. REIMOLD, W.U. 1982. The Lappajärvi meteorite crater, Finland: Petrography, Rb-Sr, major and trace element geochemistry of the impact melt and basement rocks. Geochimica et Cosmochimica Acta. **46:** 1203–1225.
105. RYDER, G., D. FASTOVSKY, & D. GARDNER, Eds. 1996. The Cretaceous-Tertiary and Other Catastrophes in Earth History. Geological Society of America Special Paper 307. 569 pp.
106. SCHMIDT, R.M. & K.R. HOUSEN. 1987. Some recent advances in the scaling of impact and explosion cratering. International J. Impact Engineering. **5:** 543–560.
107. SCHMIDT, G., H. PALME, & K.L. KRATZ. 1995. The fractionation of highly siderophile elements (HSE) in impact melts and the determination of the meteoritic component (abstract). Meteoritics. **30:** 573–574.

108. SCHNETZLER, C.C., J.A. PHILPOTTS, & H.H. THOMAS. 1967. Rare-earth and barium abundances in Ivory Coast tektites and rocks from the Bosumtwi crater area, Ghana. Geochimica et Cosmochimica Acta. **31:** 1987–1993.

109. SCHNETZLER, C.C., L.S. WALKER, & J.G. MARSH. 1988. Source of the Australasian tektite strewn field: A possible off-shore impact site. Geophysical Research Letters. **15:** 357–360.

110. SCHULTZ, P.H. & R.E. LIANZA. 1992. Recent grazing impacts on the Earth recorded in the Rio Cuarto crater field, Argentina. Nature. **355:** 234–237.

111. SCHULTZ, P.H., & R.B. MERRILL, Eds. 1981. Multi-Ring Basins. Pergamon Press. New York. 295 pp.

112. SCHULTZ, P.H., C. KOEBERL, T. BUNCH, J. GRANT, & W. COLLINS. 1994. Ground truth for oblique impact processes: New insight from the Rio Cuarto, Argentina, crater field. Geology. **22:** 889–892.

113. SHARPTON, V.L. & P.D. WARD, Eds. 1990. Global Catastrophes in Earth History. Geological Society of America Special Paper 247. 631 pp.

114. SHARPTON, V.L., K. BURKE, A. CAMARGO-ZANOGUERA, S.A. HALL, D.S. LEE, L.E. MARIN, G. SUAREZ-REYNOSO, J.M. QUEZADA-MUNETON, P.D. SPUDIS, & J. URRUTIA-FUCUGAUCHI. 1993. Chicxulub multiring impact basin: Size and other characteristics derived from gravity analysis. Science. **261:** 1564–1567.

115. SHOEMAKER, E.M. & C.S. SHOEMAKER. 1996. The Proterozoic impact record of Australia. AGSO J. Australian Geology and Geophysics. **16:** 379–398.

116. SHOEMAKER, E.M., R.F. WOLFE, & C.S. SHOEMAKER. 1990. Asteroid and comet flux in the neighborhood of Earth. Geological Society of America Special Paper 247, 155–170.

117. SHOEMAKER, E.M., P.R. WEISMANN, & C.S. SHOEMAKER. 1994. The flux of periodic comets near Earth. *In* Hazaards Due to Comets and Asteroids, T. Gehrels, Ed.: 313–335. University of Arizona Press. Tucson, Arizona.

118. SILVER, L.T. & P.H. SCHULTZ, Eds. 1982. Geological Implications of Impacts of Large Asteroids and Comets on the Earth. Geological Society of America Special Paper 190. 528pp.

119. SMIT, J. & J. HERTOGEN. 1980. An extraterrestrial event at the Cretaceous-Tertiary boundary. Nature. **285:** 198–200.

120. SMIT, J., A. MONTANARI, N.H.M. SWINBURNE, W. ALVAREZ, A.R. HILDEBRAND, S.V. MARGOLIS, P. CLAEYS, W. LOWRIE, & F. ASARO. 1992. Tektite-bearing, deep-water clastic unit at the Cretaceous-Tertiary boundary in northeastern Mexico. Geology. **20:** 99–103.

121. STÖFFLER, D., 1972. Deformation and transformation of rock-forming minerals by natural and experimental shock processes. I. Behavior of minerals under shock compression. Förtschritte der Mineralogie. **49:** 50–113.

122. STÖFFLER, D. 1974. Deformation and transformation of rock-forming minerals by natural and experimental shock processes. II. Physical properties of shocked minerals. Förtschritte der Mineralogie. **51:** 256–289.

123. STÖFFLER, D. & F. LANGENHORST. 1994. Shock metamorphism of quartz in nature and experiment: 1. Basic observation and theory. Meteoritics. **29:** 155–181.

124. STOTHERS, R.B. 1993. Impact cratering at geologic stage boundaries. Geophysical Research Letters. **20:** 887–890.

125. STOTHERS, R.B. & M.R. RAMPINO. 1990. Periodicity in flood basalts, mass extinctions and impacts; a statistical view and a model. Geological Society of America Special Paper: 247. 9–18.

126. a. SUUROJA, K. 1996. Personal communication.

126. b. SUUROJA, K. & T. SAADRE. 1995. Gneiss breccia erratic boulders from northwestern Estonia as witnesses of an unknown impact structure (in Estonian). Bulletin of the Geological Survey of Estonia. **5:** (1) 26–28.
127. SVENSSON, N.B. 1994. Lumparn — an impact crater on Åland, SW Finland and some effects of shock wave in the surrounding bed rock (abstract). European Science Foundation, Second International Conference. Impact Cratering and Evolution of Planet Earth. The Identification and Characterization of Impacts.
128. THERRIAULT, A.M., R.A.F. GRIEVE, & W.U. REIMOLD. 1997. Original size of the Vredefort structure: Implications for the geological evolution of the Witwatersrand Basin. Meteoritics and Planetary Science. In press.
129. TOON, O.B., J.B. POLLACK, T.P. ACKERMAN, R.P. TURCO, C.P. MCKAY, & M.S. LIU. 1982. Evolution of an impact-generated dust cloud and its effects on the atmosphere. The Geological Society of America Special Paper 190. 187–200.
130. UREY, H. 1973. Comet collisions and geological periods. Nature. **242:** 32–33.
131. WALLACE, M.A., V.A. GOSTIN, & R.R. KEAYS. 1990. Acraman impact ejecta and host shales: Evidence for low-temperature mobilization of iridium and other platinoids. Geology. **18:** 132–135.
132. WANG, K., H.H.J. GELDSETZER, & B.D.E. CHALTERTON. 1994. A late Devonian extraterrestrial impact and extinction in eastern Gondwana: Geochemical, sedimentological, and faunal evidence. Geological Society of America Special Paper 293, 111–120.
133. WASSON, J.T. 1991. Layered tektites: A multiple impact origin for the Australasian tektites. Earth and Planetary Science Letters. **102:** 95–109.
134. WEISSMAN, P.R. 1990. The cometary impactor flux at the Earth. Geological Society of America Special Paper 247, 171–180.
135. WILLIAMS, G.E. 1994. Acraman: A major impact structure from the Neoproterozoic of Australia. Geological Society of America Special Paper 293, 209–224.
136. WOLBACH, W.S., I. GILMOUR, & E. ANDERS. 1990. Major wildfires at the Cretaceous/Tertiary boundary. Geological Society of America Special Paper 247, 391–400.
137. XU, D., Y. SUN, Q. ZHANG, Z. YAN, J. HE, & Z. CHAI. 1989. Astrogeological Events in China. Van Nostrand Reinhold. New York. 264 pp.
138. YAKOVLEV, O.I. & A.T. BASILEVSKY. 1994. Experimental studies of geochemical aspects of impact cratering. Geological Society of America Special Paper 293, 121–132.

The Cretaceous-Tertiary Impact Crater and the Cosmic Projectile That Produced It[a]

VIRGIL L. SHARPTON[a] AND LUIS E. MARÍN[b]

[a]*Lunar and Planetary Institute*
3600 Bay Area Boulevard
Houston, Texas 77058-1113
and
[b]*Universidad Nacional Autónoma de México*
Ciudad Universitaria
Mexico City, CP 04510, Mexico

ABSTRACT: Evidence gathered to date from topographic data, geophysical data, well logs, and drill-core samples indicates that the buried Chicxulub basin, the source crater for the ~65 Ma Cretaceous-Tertiary (K/T) boundary deposits, is ~300 km in diameter. A prominent topographic ridge and a ring of gravity anomalies mark the position of the basin rim at ~150 km from the center. Wells in this region recovered thick sequences of impact-generated breccias at 200–300 m below present sea level. Inside the rim, which has been severely modified by erosion following impact, the subsurface basin continues to deepen until near the center it is ~1 km deep. The best planetary analog for this crater appears to be the 270 km-diameter Mead basin on Venus. Seismic reflection data indicate that the central zone of downward displacement and excavation (the transient crater is ~130 km in diameter, consistent with previous studies of gravity anomaly data). Our analysis of projectile characteristics utilizes this information, coupled with conventional scaling relationships, and geochemical constraints on the mass of extraterrestrial material deposited within the K/T boundary layer. Results indicate that the Chicxulub crater would most likely be formed by a long-period comet composed primarily of nonsilicate materials (ice, hydrocarbons, etc.) and subordinate amounts ($\leq 50\%$) primitive chondritic material. This collision would have released the energy equivalent to between 4×10^8 and 4×10^9 megatons of TNT. Studies of terrestrial impact rates suggest that such an event would have a mean production rate of $\sim 1.25 \times 10^{-9}$ y^{-1}. This rate is considerably lower than that of the major mass extinctions over the last 250 million years ($\sim 5 \times 10^{-7}$ y^{-1}). Consequently, while there is substantial circumstantial evidence establishing the cause-effect link between the Chicxulub basin forming event and the K/T biological extinctions, the results of our analysis do not support models of impact as a common or singular causative agent of mass extinctions on Earth.

INTRODUCTION

Since the advent of plate tectonics, no issue in geology has stirred the Earth science community and intrigued the general public as much as the hy-

[a]Parts of this research were made possible through research grants provided by the National Aeronautics and Space Administration's Planetary Geology and Geophysics Program (to VLS), and the National Science Foundation's Continental Dynamics Program. The Lunar and Planetary Institute is operated by Universities Space Research Association under Contract NASW-4574 with NASA. This is Lunar and Planetary Institute Contribution Number 901.

pothesis of an impact origin for the K/T extinction event, 65 million years ago. Mass extinctions of a broad spectrum of lifeforms [44], a worldwide clay layer containing geochemical [2] mineralogical (e.g., [5]) traces of impact and isotopic anomalies (e.g., [32]), and tsunami deposits (e.g., [8]) all point to a major impact event at that time. The debate has continued since the pioneering study of Alvarez et al. [2] resulting in an impressive list of publications that have served, on whole, to support and modestly-refine this hypothesis. The discovery of Chicxulub and its link to the K/T boundary, has encouraged a major shift away from addressing the now banal question of "did it?" to the more fertile one of "how did it?"

Even at its smallest size estimate of ~170 km in diameter, the Chicxulub basin records an event that was almost incomprehensibly energetic, releasing roughly the equivalent of 60 million megatons of exploding TNT. Yet, theory and observations indicate that the vast majority of this energy is deposited in and very close to the impact crater. How such a point source could induce a global biological crisis has been a perplexing problem since the early studies by Alvarez et al. [2] who proposed that it could have been global darkening and cooling initiated by dust injected into the upper stratosphere that led to the extinctions. This is clearly a viable contributor, after all, there is the K/T boundary clay layer located around the world to testify to the amount of ash and dust that settled out from the Chicxulub-forming blast. Yet concerns about the atmospheric residence time of such dust [59], [41] (cf. [11]) have fueled the search for alternative ways impact could induce extinction. Early on, it was recognized that an oceanic impact would inject a considerable quantity of water vapor into the atmosphere perhaps leading to greenhouse warming [13] or ozone depletion [60]. Prinn and Fegley [43] explored the production of NO_x from shock-heating the terrestrial atmosphere and found this to be a viable mechanism of global change under certain circumstances. Soot produced in global wildfires would contribute to global darkening and cooling (e.g., [15]).

With recognition of Chicxulub as the K/T source crater, attention has turned to understanding the potentially devastating effects of vaporizing the Mesozoic platform sequence (e.g., [9]) which constitutes the upper 2 to 3 km of the target assemblage [52]. A major impact into such materials is expected to shock-vaporize and otherwise release substantial quantities of CO_2 (from carbonates) and SO_X (from evaporites) into the atmosphere. Carbon dioxide could contribute to global warming and sulfur compounds could contribute to light-blocking smog and acid rain. Pope et al. [42] have modeled the effects of sulfate vaporization and their calculations indicate that most is released as SO_2, which slowly converts to H_2SO_4 aerosol and consequently prolongs the global cooling and acidification effects. A determination of whether carbonate- and/or sulfate-vaporization contributed significantly to the K/T environmental crisis depends of course on the extent to which these materials were vaporized and injected into the atmosphere.

The first step toward understanding the killing potential and modus operandi of the Chicxulub basin-forming event is an evaluation of the energy available to vaporize volatile-rich sediments. In this paper, we present geological and geophysical constraints on the size and shape of the Chicxulub impact basin. We then use these characteristics to estimate the energy released upon impact and derive some information about the object that collided with Earth 65 million years ago and irrevocably changed the course of biological evolution. To introduce the reader to some of the concepts needed to evaluate this discussion, however, we first provide background information on the formation and appearance of planetary craters.

THE MORPHOLOGY AND STRUCTURE OF IMPACT-GENERATED LANDFORMS

Exploration of the inner solar system during the latter half of the twentieth century has shown that large-body impact played a major role in shaping ancient planetary landscapes such as those preserved on Mercury and the Moon. It is here, on the ancient surfaces of these small airless planets, that the best examples of large impact basins are observed (e.g., [57]). The largest basins exhibit multiple concentric rings and can exceed 2000 km in diameter. In contrast, Earth's active surface processes quickly destroy its impact record; only ~140 impact craters have been identified so far and many are small, bowl-shaped, or simple, craters. Larger complex structures, greater than about 4 km in diameter on Earth, are rarer, older, and more modified; the truly large impact basins, those greater than ~50 km in diameter, are extensively modified and have retained few, if any clues to their original surface configuration. But such erosional and tectonic activity, combined with extensive drilling and geophysical surveys, also affords the only information on crater substructure from which to understand the third dimension of large impact craters, anywhere in the solar system.

Complex Craters and Basins

The smallest pristine craters on any planet are simple bowl-shaped depressions with raised rims; these features are produced by the downward displacement and excavation flowfield initiated by a system of impact-driven compression and rarefaction waves (see [34] for a complete treatment of the physics of impact). But larger magnitude impacts create transient conditions within the target which initiate large scale, gravity-driven modification of this bowl-shaped craterform during the final stages of the impact event. Two processes take place immediately and almost simultaneously: (i) the floor of the initial or "transient crater" rebounds upward producing a central structure that, when exposed, takes the form of a central peak composed of the originally deep target rocks comprising the crater subfloor; and (ii) the crater margins collapse inward along either a series of normal faults (producing terraces) or along a single scarplike fault. This late-stage modification results in a final crater that is broader and shallower than the original crater of excavation and downward displacement (i.e., the transient crater). This "complex crater" is characterized (ideally) by the following topographic and structural elements: (i) a central peak, (ii) broad flat floor deposits composed of highly-shocked rock debris and melted rock deposits, and (iii) a terraced rim zone. The modified basin is filled with a mixture of melt rock and highly shocked rock debris, both originating near the impact site, as well as a considerable amount of weakly-shocked or unshocked debris, much of which reentered the basin during slumping and enlargement. At consecutively larger impact events, the central peak is replaced by clusters of peaks, and finally by a "peak ring." The peak ring, like the central peak, is composed of uplifted massifs of deep target rocks from below the floor of the original transient crater. Impact specialists typically refer to large craters manifesting peak rings as "peak-ring basins" or "double-ring basins."

The Multiring Basin Controversy

At the very largest diameters, impact basins are characterized by two or more concentric rings in addition to — and outside of — the central peak ring. The canonical multiring basin is the Orientale basin on the Moon (Fig. 1). Multiringed basins have been studied intensely over the course of 30 years of planetary investigations (e.g., [3], [47], [17], [36], [38], [34], [57]); yet even now they are not well understood. The surficial expression of basin rings, particularly the distal ones, can be quite subtle, especially when the basin has been modified by eons of subsequent geological activity as most have. There is even debate over whether or not multiring basins exist at all on some terrestrial planets (see discussion in [34, Chap. IX]) and one faction maintains that a planetary asthenosphere is necessary for multiring basin formation [36], [34]. This has led to a generally hesitant approach to interpreting even the largest impact features on Earth as the modified remains of multiring basins.

The recent Magellan mission provided the first high resolution unencumbered views of the surface of Venus. Recognition that multiring basins occur on this planet [45], [1] so different from the Moon and Mercury, yet similar to Earth in terms of its surface gravity, atmospheric conditions, and target strength, has strengthened the proposition that multiring basin morphology is an ultimate consequence of increased impact energy and not related to some planetary property such as the presence of an asthenosphere. The diameter at which the transition from peak ring basins to multiring basins takes place, appears to follow the same inverse dependence on planetary gravity as the transition from simple to complex craters, and the transition from central peak to peak ring basins. All impact basins greater than ~140 km on Venus exhibit the attributes of multiring basins; this and the basic similarities between Venus and Earth suggest that impact features of this size on Earth should correspond to multiring basins as well. There are three terrestrial craters identified to date that exceed this size: the Sudbury structure in Ontario, Vredefort structure in South Africa, and the Chicxulub structure.

CHICXULUB: THE K/T BOUNDARY SOURCE CRATER

The collision that produced the Chicxulub basin occurred 65 million years ago into the Yucatán platform, which at that time was an active carbonate depositional environment. Because the region was covered by a shallow sea, the basin was shielded somewhat from erosion, although, the high-standing rim and ejecta facies along the flanks of the basin most likely have been severely modified by wave action [52]. The basin is now buried by up to a kilometer of Cenozoic sediments and geopotential studies gave the first hints of a large structure underneath the present day carbonate surface of modest relief. Test wells were drilled by Petróleos Mexicanos (Pemex) beginning in the early 1950s and three, near the center, intercepted unusual occurrences of breccias and silicate crystalline rocks [52], initially interpreted to be of volcanic origin. An impact origin, however, was proposed as early as 1981 [37]. Subsequent analyses of samples recovered by these early drilling efforts reveal diagnostic evidence of shock metamorphism [18], [50] and clearly demonstrate that the crystalline rock sequence was produced in a large meteorite impact event, not by volcanic processes.

There is a compelling link between the Chicxulub basin and the Cretaceous-Tertiary (K/T) boundary layer across which is extensive paleontological evidence of an unusually abrupt and massive global extinction event. The crystallization age of melt rocks recovered from the center of the structure is ~65 Ma and, within limitations of the ^{40}Ar-^{39}Ar dating method, are coincident with the age of the K/T boundary [22], [58], [50]. Magnetic properties of the core samples indicate that these melt rocks cooled during an interval of reversed geomagnetic polarity, again consistent with a K/T connection (magnetochron 29R; [50], [61]). The compositions of melt rocks within the Chicxulub structure correspond precisely to the composition of the impact glass spherules found at K/T boundary sites in Haiti and northeastern Mexico [23], [54], [55], [27], [56], in terms of major elements, trace elements, and isotopic properties [18], [50], [27], [58], [4], [26]. Fragments of unmelted silicate basement in breccia samples from Chicxulub reveal a medium-to-high grade metamorphic granitic protolith similar in composi-

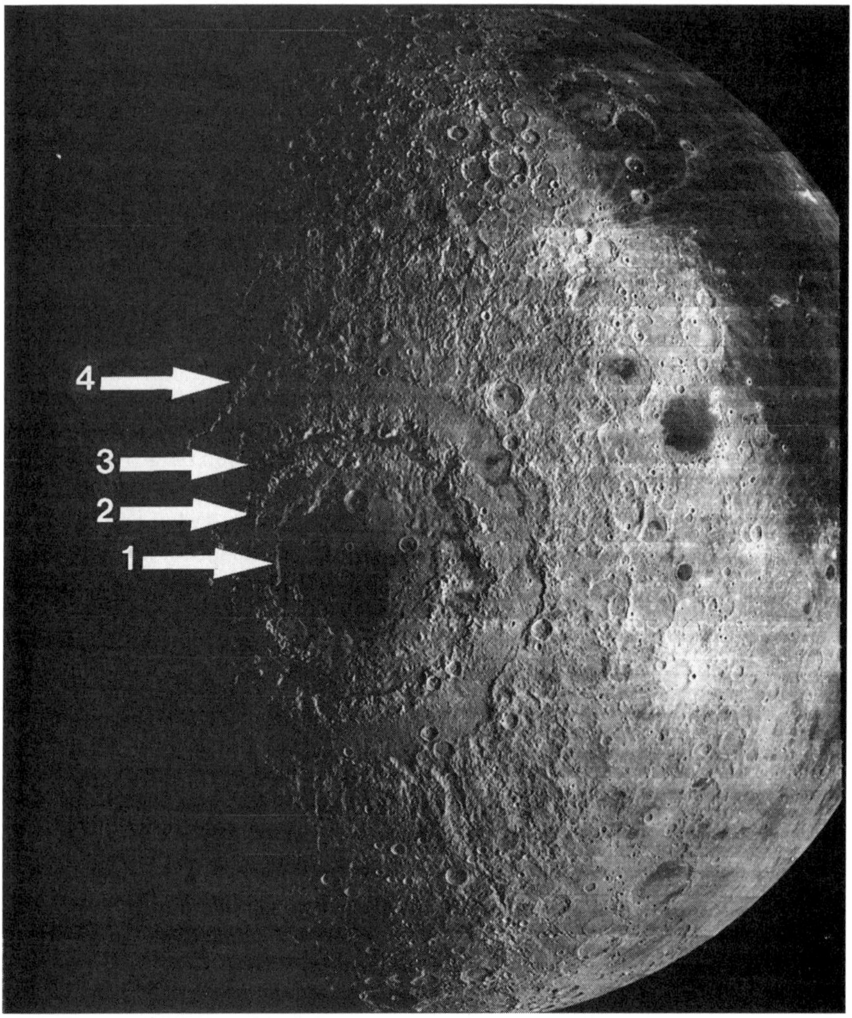

FIGURE 1. The 930 km Orientale multiring basin on the western limb of the Moon as seen from the Lunar Orbiter IV spacecraft. North is toward the top of the page. Numbered arrows indicate the four main basin rings (e.g., [57, Chap. 4]): 1 is the central peak ring; 2 and 3 are the Inner and Outer Rook Mountain rings, respectively; and 4 is the Cordilleran Scarp.

tion to shocked lithic fragments found at the K/T boundary worldwide [6], [5], [49], [50], [23]. U-Pb determinations on zircons indicate that the age of the Chicxulub basement matches precisely the 545 Ma age of unmelted shocked debris from the K/T boundary in Haiti and in the Western Interior of North America [28], [29], [7].

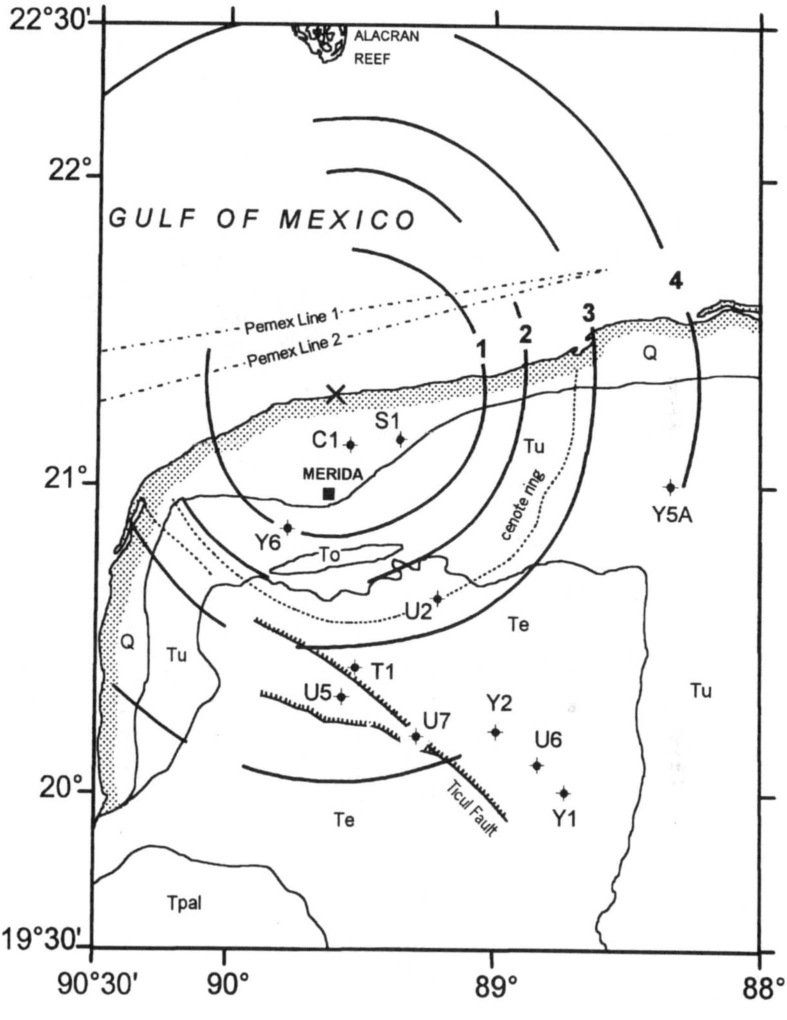

FIGURE 2. Surface geology, ring locations from gravity data, and wells in the vicinity of the Chicxulub multiring impact basin. The three wells that penetrated impact melt rocks and breccias beneath the carbonate cover rocks are C1 (Chicxulub 1), S1 (Sacapuc 1), and Y6 (Yucatan 6). Other Pemex well sites shown are Yucatan 1 (Y1), Yucatan 2 (Y2), Yucatan 5A (Y5A), and Ticul 1 (T1). The Yucatan 4 (Y4) well site is located off the map, ~65 km east of Y5A. Important sites of recent shallow drill-coring, led by UNAM scientists, are denoted by U2 through U7. Carbonate units at the surface are Q (Quaternary; < 1.6 Ma), Tu (Upper Tertiary; ~1.6 to 38 Ma), To (Oligocene; ~24 to ~38 Ma),Te (Eocene; ~38 to ~55 Ma), and Tpal (Paleocene; ~55 to ~66.0 Ma). Hatchured lines represent the Ticul fault system. Dashed lines indicate trends of ringlike zones of cenotes or sinkholes. Existing Pemex offshore seismic lines are shown by broken line patterns. Pemex seismic lines actually extend to the west of this map region: D92PR001 (northernmost line) extends to 21.55°N; 90.52°W and D92PR002 extends to 21.31°N; 90.75°W.

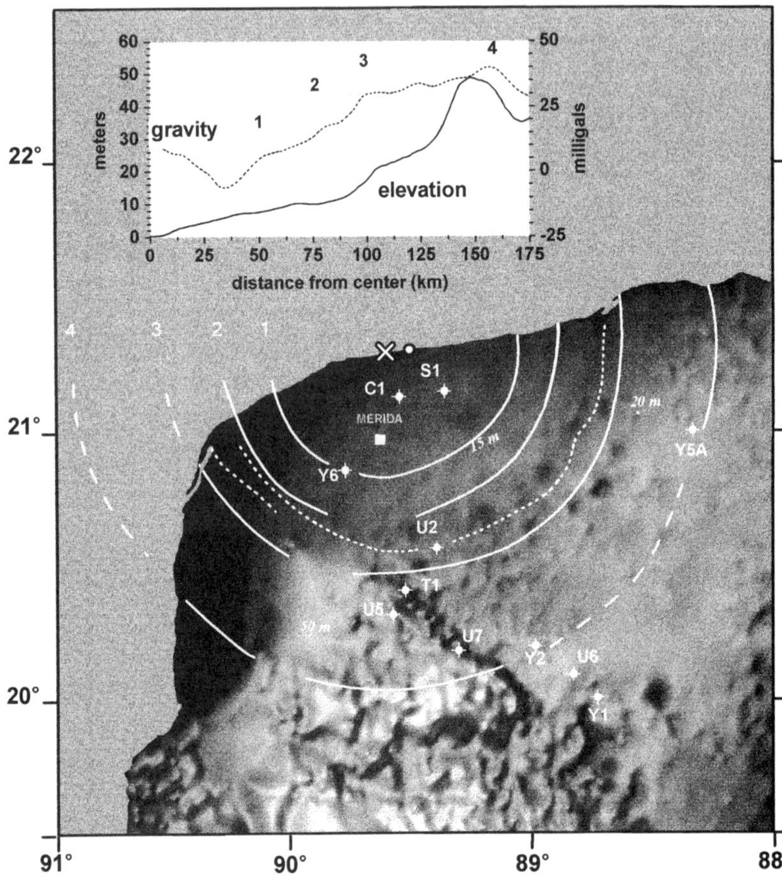

FIGURE 3. Shaded topographic relief map of Northern Yucatan, Mexico, showing locations of exploratory wells and traces of concentric gravity anomalies after Sharpton et al. [52]. Topography is color contoured at 5 meter intervals between 0 and 20 m, 10 m intervals between 20 and 50 m, and 25 m intervals above 50 m. Northwest-trending break in elevation near wells T1 and U7 represents the Ticul fault system. Dotted lines indicate the cenote ring. The center of the concentric gravity anomalies (×) is offset from the center of cenote ring (white dot). Inset shows an eastward profile of gravity anomaly data (90° azimuth) and the average elevation profile (from 25 equally spaced radial profiles between 90° and 220° azimuth), both measured from ×.

CONSTRAINTS ON CHICXULUB'S SIZE AND SHAPE

Surface Expression of the Buried Chicxulub Basin: Clues to the Basin Rim Diameter

The present-day surface of Northern Yucatan consists of a Paleocene-Quaternary carbonate rock sequence as shown in Fig. 2. Outcrop patterns reveal a modest northward-trending epirogenic rise located slightly west of the peninsula's center although there is no topographic expression of this structural feature. No outcrops of impact deposits such as breccias or melt rocks have been discovered in the region, suggesting that the impact crater is completely buried by younger sediments.

High-resolution topographic data assembled from ground-based gravity surveys (Fig. 3) show an arcuate zone of local and discontinuous depressions, 5–8 m deep, located approximately 85 km from the buried crater center that correlates with the ring of cenotes (water-filled sinkholes) that have been taken by some to indicate the rim of the buried crater [18], [40]. Perhaps more importantly, however, these topographic data also reveal a broad concentric ridge that defines the topographic rim of the arcuate basin extending over the northwest corner of the Yucatan Peninsula. East of the northeast-dipping Ticul normal fault (Fig. 2), the crest of this broad ridge is ~25 m above sea level and local relief exceeds 10 m. Across the fault, the ridge exceeds 170 m in elevation and averages ~40 m in local relief (Fig. 3).

The ridge corresponds closely to the position of the outermost gravity ring of Sharpton et al. [51], [52] as shown in Fig. 3. These authors concluded that this gravity ring, which encompasses the edge of the crater's broad gravity low and associated discontinuous local highs, approximates the position of the buried crater's topographic rim. Using the center of concentric gravity anomalies as the buried crater center, they estimated the mean distance to the outer ring to be 139 ± 11 km. Between azimuths of 90° to 220°, the mean distance of the topographic ridge from this same basin center is 146 ± 7 km.

This ridge, and the geophysical evidence that it reflects subsurface structure, strengthens the case for this being the location of the buried impact basin rim. When the full extent of the topographic ridge is considered, the best estimate for the diameter of the Chicxulub crater is ~300 km. Long-term subsidence of the impact-weakened crust could account for the subtle but sustained topographic expression of this buried impact basin. Concentrically oriented depressions south of the Ticul fault system, located at a distance of ~200 km from the crater's center, hint of an even larger ring, perhaps reflecting an additional basin ring outside the rim. Such "exterior" rings are commonly associated with multiring basins.

Stratigraphy from Wells and Core Samples: Additional Evidence of a 300 km Basin

Pemex Drilling

Well data from Pemex exploratory drilling indicate that the Chicxulub basin extends ~135 to 150 km from center to the vicinity of Yucatan-2, Yucatan-5A, and Yucatan-1 (Fig. 2), where the K/T boundary is nearest to the present surface (Fig. 4). This distance corresponds to the location of both the fourth (outermost) ring of gravity anomalies of Sharpton et al. [51], [52] and the topographic ridge discussed above. Filling the basin are Tertiary limestones and marls ranging in thickness from slightly over 1 km near the center to less than 300 m at the basin rim. The total thickness of platform rocks at the time of the impact is constrained by depth to basement outside the structure (removing the thickness of the Tertiary section) and is ~2 km in Yucatan-4 as well as in several other wells located east and south of the structure between 200 and 300 km from the center, such as Quintana Roo 1, Tower Hill 1 and 2, and the Basil Jones 1 wells [31].

The central deep portion of the Chicxulub basin, showing high-amplitude magnetic and gravity anomalies and deeply buried occurrences of suevite breccias and melt rocks, show no indications of the Mesozioc carbonate sequence; however, this central basin is flanked by a broad annular shelf where the complete (pre-Tertiary) target sequence is intact but depressed several hundred meters. In wells located between ~90 km and ~160 km of the center, that is, Ticul-1, Yucatan-1, Yucatan-5A, and Yucatan-2, the silicate basement appears to be 750 to 1,000 m deeper than in more distant wells (Fig. 4). The Upper Cretaceous-Lower Cretaceous contact is also several hundred meters deeper in wells inside the basin indicating that the whole pre-Tertiary crustal column was affected by this structural event. As there are no indications in this area of an Upper Cretaceous tectonic event, it seems most likely that these stratigraphic offsets primarily reflect the structural displacements induced during the collapse (modification) stage of this impact event [34], [12]. On the basis of these observations, the annular shelf appears to extend to ~150 km from the basin center.

Covering this annular shelf and the outlying areas is a breccia sequence consisting of centimeter-scale clasts of dolomite, limestone, and anhydrite in a clayey-micritic matrix. This unit was completely overlooked by Lopez-Ramos [31], who compiled his lithostratigraphic data for the Chicxulub region primarily from analyses of drill cuttings. Marshall [33] was the first to document this "clastic anhydrite-dolomite lithofacies," and Sharpton et al. [52] first discovered evidence that this unit is the continuous ejecta deposit surrounding and filling the Chicxulub impact basin. They found trace quantities (< 1% volume) of shocked but unmelted silicate basement clasts and melt-rock clasts with small xenocrysts of shocked quartz and feldspar throughout the anhydrite-dolomite breccia shown in Fig. 3. Furthermore, shocked feldspar crystals within larger melt clasts have compositions similar to those found within the Chicxulub suevite breccias and the K/T boundary layer [52]. Thus the anhydrite-dolomite breccia unit is analogous to the well-studied Bunte Breccia at the Ries Crater [19], [20] which represents a mixture of "primary" ejecta (the rock debris ejected from the transient crater) and local materials incorporated into the deposit as the ejecta was emplaced. It is likely that the proportion of silicate basement apparent in these breccia samples has been diminished by the effects of chemical alteration and secondary mineralization working on the highly vulnerable shocked and partial-

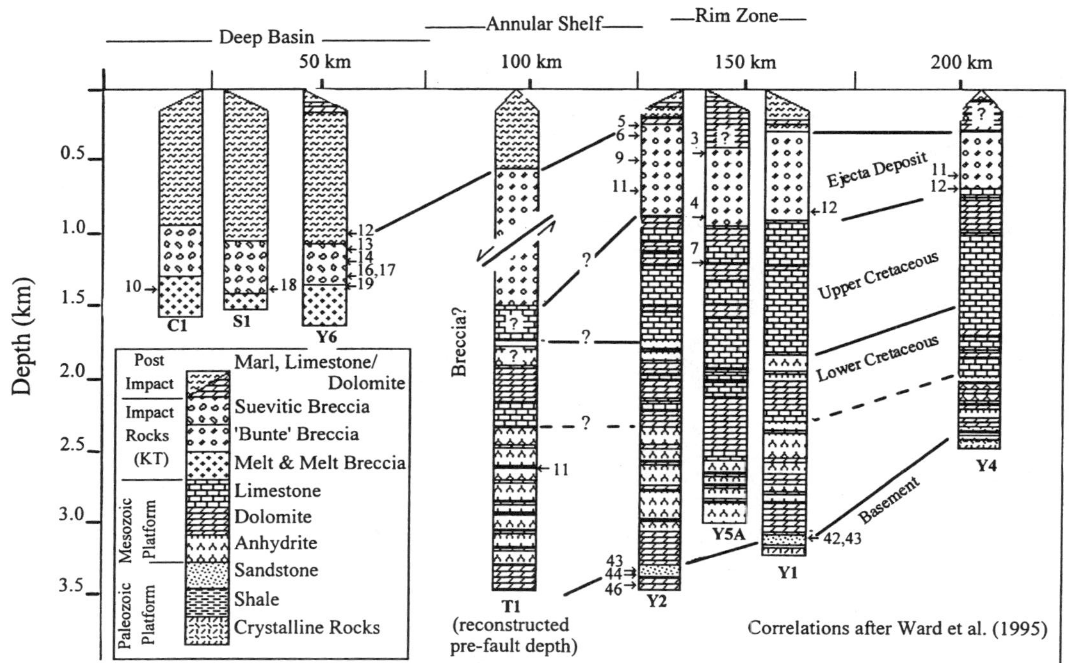

FIGURE 4. Stratigraphic correlations across the Chicxulub basin based on well logs and sample analysis. Wells are arranged in distance from the basin center. The samples we have studied are identified by arrows. Other lithological assignments are based on visual inspection of cores, updating previous published and unpublished reports. The stratigraphic thickness of the breccia unit in T1 reflects the prefault conditions we infer from analysis of well-log data and drilling reports. True (present) depths can be calculated by removing 200 m from the depths of all units below the fault shown. Major stratigraphic correlations are based on our own work and that of Ward et al. [63].

ly melted basement clasts. At the Yucatan-2 well site, this breccia unit is > 600 m thick, and its top is located ~250 m below sea level.

Because there is no evidence of Upper Cretaceous bedded evaporites, anhydrite clasts may provide a lower constraint on the proportion of primary ejecta. At Yucatan-2, anhydrite clasts, averaging ~0.5 mm to ~3 cm in length, account for 10 to 20% of the total volume in samples analyzed by Sharpton et al. [52]. If, as the well logs suggest, bedded anhydrite constitutes no more than about one-third of the total pre-Tertiary platform sequence, then the breccia at Yucatan-2 is at least 30 to 60% primary ejecta. Based on analyses of ejecta deposits at the Ries crater, such an enrichment is expected only very near the final basin rim of large impact structures [20].

Recent Shallow Drilling

In March, 1994, UNAM commenced a shallow drilling program at Chicxulub leading to the recovery of over 2800 m of core from seven sites along the inner flank of the topographic basin. Of particular interest to issues regarding the size and shape of the Chicxulub impact structure are the cores recovered from UNAM-2, -5, -6, and -7, the locations of which are shown in Fig. 2. UNAM-2, located ~85 km from the basin center, was drilled into the Cenote Ring, thought in previous models to mark the approximate location of the buried crater rim [40]. This well penetrated 570 m and bottomed in carbonate rocks of Oligocene age (< 36 Ma). Consequently, the crater is still deeply buried at this point, and the high-standing rim must be located some distance outside of the ~85 km radius.

The remaining three wells, all located beyond the Cenote Ring recovered core samples of impact deposits. Figure 5 shows the basic stratigraphy of the impact units recovered from these wells and their distance from the center. The UNAM-5 well (112 km from center) intercepted the uppermost portion of the impact sequence at a depth of ~330 m. This sequence consists of an ~30 m thick altered meltrock unit containing infrequent cobble-sized clasts of highly shocked and partially melted basement rocks. This unit sits conformably on crater suevite, a polymict, clastic-matrix breccia containing clasts of impact-generated glasses (often altered). This rock type is rich in highly shocked clasts representing the deeper target strata. Two basic types of suevite have been described from other terrestrial craters: (i) fallout suevite is a type of ejecta that characteristically contains sculpted glass clasts indicative of ballistic transport through the atmosphere, and (ii) crater suevite, a kind of allogenic breccia, deposited within the impact basin in which the glasses characteristically shows no evidence of atmospheric sculpting. The suevite unit at UNAM-5 contains abundant cm-scale clasts of partially altered to unaltered impact meltrock and larger highly shocked basement clasts. The absence of aerodynamically shaped glass clasts in this suevite deposit suggests that it is allogenic crater fill material and not ejecta deposited beyond the topographic rim. Total depth was reached at the UNAM-5 well while still in the suevite.

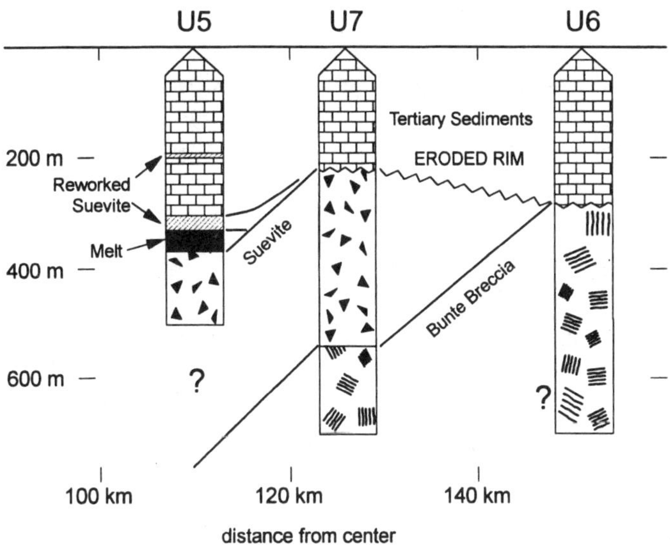

FIGURE 5. Stratigraphy of shallow wells penetrating the southern flank of the Chicxulub impact basin. For exact well locations see Fig. 2.

The UNAM-7 well (125 km from center) penetrated a thick sequence of suevite, but no impact melt rock was recovered. Instead, an erosional unconformity separates the truncated upper part of the suevite deposit from the overlying postcrater sediments. UNAM-6 (151 km from center) is located on the fourth gravity ring and the topographic ridge shown in Fig. 3. Core samples from this well revealed even more severe indications of erosion: the suevite breccia unit is completely absent (Fig. 5) and the uppermost impact unit is an anhydrite-dolomite breccia (bunte breccia) similar to that observed at Pemex wells of similar distances from the crater center [52].

Prior to impact, the region was covered by a shallow sea probably no deeper than ~50 m and so the rim and ejecta units were subjected to subaerial erosion and reworking. The preservation of suevite breccia and melt rocks at UNAM-5 and UNAM-7 suggests that their absence further out from crater center is due to their removal by erosion. Consequently, the UNAM-5 sequence was probably deposited within the final topographic basin formed by the impact event. The relative depths of the bunte breccia upper contacts at UNAM-6 and UNAM-7 reveal at least 250 m of stratigraphic relief across these two wells (Fig. 5); similarly the suevite upper contacts at UNAM-7 and UNAM-5 indicate an additional 150 m of relief. Therefore, immediately after the impact event, there was over 400 m of total stratigraphic relief from UNAM-5 to UNAM-6, with the top of the UNAM-5 impact sequence

~150 m below the K/T sea level (approximated by the unconformity) and the pre-erosional top of the UNAM-6 sequence ~350 m above sea level (ignoring radial thinning of the units). This paleobathymetry accounts for the preservation of an apparently complete sequence of impact units at UNAM-5 but not in wells located at greater distances. Consequently, the stratigraphic sequence revealed by the UNAM shallow drilling program supports the previous geological and geophysical observations that the Chicxulub impact basin was ~300 km in diameter.

Geophysical Constraints on the Size of the Transient Crater

Any attempt at reconstructing the dimensions of the "transient crater," i.e., the ephemeral bowl-shaped feature produced by excavation and downward displacement of the target upon impact, faces substantial challenges. A transient crater may never form completely in such large impact events as Chicxulub and even if it does, it is destroyed during uplift and inward slumping of the crater flanks associated with the late modification stage of the impact process. The present situation at Chicxulub is further complicated by postimpact erosion and redeposition prior to complete coverage of Tertiary carbonates [52]. Place on top of all that the preconceptions and pet theories that various theorists have for how multiring basins are formed and you have a problem of biblical proportions. Nonetheless, Sharpton et al. [52] proposed, that because brecciation, wholesale deformation and kilometer-scale uplift are confined primarily to the transient crater, there might be a recognizable signal in the gravity data, expressed as steep, inward-facing gradients.

Constraints from Gravity Data

The gridded Bouguer gravity anomaly map of the Yucatan region is shown in Fig. 6. Gravity anomalies in the vicinity of the Chicxulub basin (Plate 1C) range from –16.4 mgal (10^{-5} m s^{-2}) to +53.6 mgal. The Chicxulub basin is expressed as a broad, nearly circular region in which gravity values are 20 to 30 mgal lower than regional values. A distinct 15- to 20-mgal high occupies the geometric center at 21.3°N latitude and 89.6°W longitude. Figure 2 shows gravity profiles through the center of the structure for each 10° of azimuth and clearly reveals multiple rings expressed as local maxima in the gravity anomaly data. Like topographic rings at large multiring basins on other planets, these rings are spaced systematically such that distances between successive rings are approximately multiples of $\sqrt{2}$ [38], [57], [51]. Major topographic rings characteristic of multiring impact basins include (1) the innermost or central peak ring, thought to represent the topographic expression of the central uplift [16], [34, pp. 163–168], (2) one or more intermediate rings marking the margins of annular shelves stepping down toward the center, and (3) the topographic basin rim [57, p. 7, Fig. 1.2].

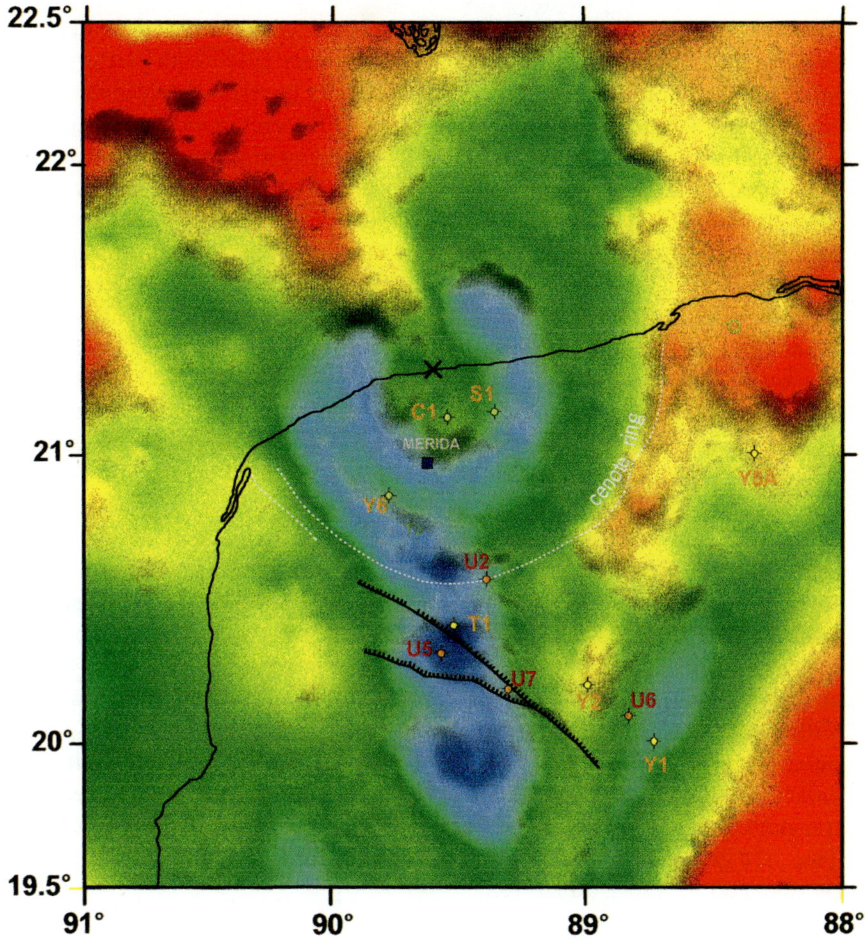

FIGURE 6. Relief-shaded Bouguer gravity anomaly data for the Yucatan region showing location of the cenote ring, Ticul Fault, basin center and exploratory wells. Colors are ordered spectrally with deep blue representing lowest gravity values and red representing the highest. Data were shaded using a low illumination angle from the north to emphasize broad low-amplitude anomalies. All data have been Bouguer corrected using a model slab density $\rho = 2200$ kg m^{-3}. Gravity anomalies in this region range from –16.4 mgal to 53.6 mgal.

Gravity values increase abruptly and significantly between ~70 and ~105 km from the center (Fig. 6); this region exhibits the steepest gravity gradients associated with the Chicxulub structure. Its distal edge defines a prominent gravity ring (Ring 3), which Hildebrand et al. [18] and Pilkington et al. [40] interpret as the basin rim. Such steep inward-facing gradients define a sharp boundary to the major mass deficiency in the basin center. But because the crater topography has been completely filled with platform sediments, there should only be a very small density contrast between the crater fill and the surrounding upper crustal rocks of the crater walls. Therefore, it seems unlikely that the steepest gradients observed in the structure would be associated with the rim zone, a region that, in other terrestrial impact structures, is characterized by shallow inward slumping (e.g., [39]). Furthermore, for other well-preserved impact craters, there is a sharp inward steepening in gradient near the edge of the central zone of excavation and deep deformation as determined by independent techniques [52]. Consequently, Rings 2 and 3 at Chicxulub probably correspond to the Inner and Outer Rook Mountain Rings at Orientale (Rings 2 and 3, Fig. 1) and that the steep gradients observed between these two rings could mark the outward extent of excavation and deep deformation associated with the collapsed transient crater. Thus, the transient crater boundary would be located between 70 and 105 km from the crater center.

This measurement should not be taken as a precise estimate of the "apparent" or "true" transient crater diameter (the diameter of the crater as measured from the preimpact surface) and as used by modellers to define scaling relationships (e.g., [34, Chap. VII]). During their collapse, the transient crater walls are severely deformed and the zone of intense deformation is broadened somewhat. This leads to an enlargement of the gradient ring associated with the boundary between the transient crater and the more distal slumped outer basin floor. It seems unlikely that this deformation would result in an enlargement of more than ~25% (equivalent to the difference between the transient crater's "true diameter" and its rim crest diameter; see [34, p.112]). Thus, the gravity data yield reasonable bounds for Chicxulub's transient crater diameter of 130 ± 30 km [52].

Constraints from Seismic Profiles

Reflection seismic data can provide a more direct and less ambiguous indication of substructure. Camargo and Suárez [10] presented two nearly east-west multichannel seismic lines within the offshore portion of the Chicxulub basin gathered by Pemex (Fig. 2). These two seismic profiles, D92RP001 and D92RP002 (hereafter referred to as Lines 1 and 2), have lengths of 192 and 224 km, respectively. The latter profile lies closer to the coastline and is practically parallel to it. A schematic interpretation of Line 1 is shown in Fig. 7. Neither line extended to gravity ring 4 of Sharpton et al. [51]. However, the inner rings were transected by both lines to the west of center.

Figure 7 clearly reveals the extent of the ~1 km deep basin filled by layered Cenozoic sediments. In this profile located ~40 km north of the basin center, the deep basin is ~115 km in width where the profile was taken, yielding a central basin diameter of ~130 km. The western edge of the basin floor is located just inboard of gravity ring 2. No laterally continuous coherent reflecting horizons are apparent beneath the postcrater sediments in this region and there is substantial relief on reflecting horizons that indicate the boundary of this zone is a region of substantial deformation (some seismic reflector packages are located at a depth of ~5 km, apparent in this line on the eastern side of the basin around shot point 6000). Reflections corresponding to Mesozoic platform rocks involved are apparent on the western side of the basin extending westward from shot point 1500 and to a depth of ~1.8 seconds.

At other terrestrial impact basins where high-quality reflection profiles have been acquired, the outward extent of the central zone of intense deformation and excavation (i.e., the transient crater) is constrained by the extent of coherent returns in the target strata (e.g., [39]). For Chicxulub, coherent returns from Mesozoic sediments become faint and disappear between gravity rings 2 and 3. This corresponds to a transient crater diameter of ~130 km, in excellent agreement with the 110–160 km transient crater estimate derived from gravity analysis [51], [52].

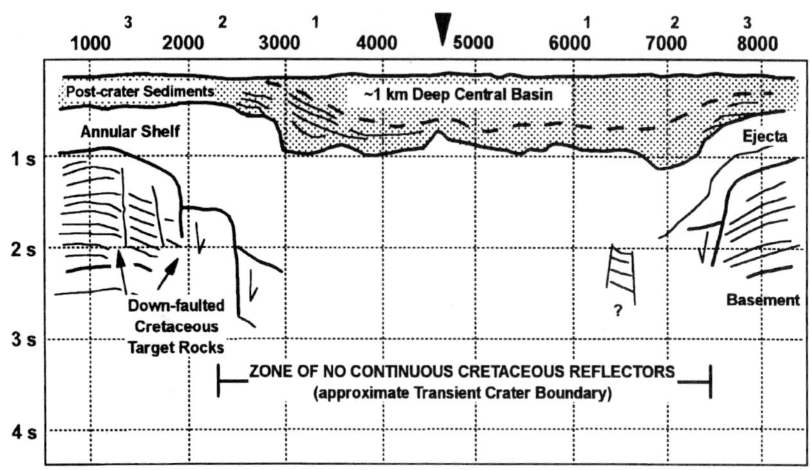

FIGURE 7. Schematic interpretation of Pemex seismic reflection line 1 across the northern portion of the Chicxulub impact basin (see Fig. 2). Intersections with the gravity rings of Sharpton et al. [51] are shown above the shot point numbers. Stippled zone represents postcrater sediments, which may include redeposited ejecta from the modified basin rim and flanks. Horizontal axis shows shot point locations; 1000 shot point intervals is equivalent to 25 km along track. Vertical axis is 2-way travel time.

Reconstructing the Original Appearance of the Chicxulub Impact Basin

The difficulties in reconstructing the shape of a completely buried impact basin are immense. Because the impact took place in a shallow sea, the situation at Chicxulub is further complicated by postimpact erosion and redeposition that destroyed the highstanding rim and smoothed the basin topography, a fact confirmed by the UNAM 5-7 drill core samples. Nonetheless, from our analysis of available geological and geophysical evidence we strongly suspect that Chicxulub is a multiring basin ~300 km in diameter. A model of how this basin may have evolved is presented in Fig. 8. The basic appearance of Chicxulub was probably similar to the ~270 km-diameter Mead multiring basin on Venus. In fact when the Mead multiring basin is resized and superimposed on the Chicxulub region, the correlation between expected characteristics and those gleaned from drilling and geophysics is surprisingly good (Fig. 9). The central basin would correspond to the deep inner basin associated with the transient crater, similar to the deep inner basin of Mead. The central basin is flanked by a broader, shallower shelf consisting of downfaulted Mesozoic rocks overlain by thick deposits of ejecta.

ESTIMATING THE ENERGY RELEASED DURING CRATER FORMATION

The most often employed way to constrain the energy released upon impact of a comet or asteroid begins with an estimate of the diameter of the final basin rim crest (D_{fb}) and employs a scaling factor to derive the transient crater diameter (D_{tc}). For multiring basins, however, the situation is somewhat more complicated because these impact features *may* have disproportionately small D_{tc}-to-D_{fb} ratios [36]. While this does not necessarily represent the mainstream view (or even an opinion with reliable empirical backing) it nonetheless represents the uncertainty surrounding multiring basin models and has to be addressed. Fortunately, for Chicxulub it is possible to derive estimates for both D_{fb} and D_{tc}.

Once D_{tc} is constrained, semiempirical or theoretical scaling laws can be employed to back out such projectile characteristics as impact velocity (v), projectile diameter (D_p), and mass (m), given certain target conditions such as planetary gravity (g) and rock density (ρ). Currently, the most frequently relied upon formulation is the Schmidt-Holsapple point-source relation [46] derived from a Pi-group scaling analysis:

$$D_{tc} = 1.44\left(\frac{v^2}{gD_p}\right)^{0.22}\left(\frac{m}{\rho}\right)^{1/3} \tag{1}$$

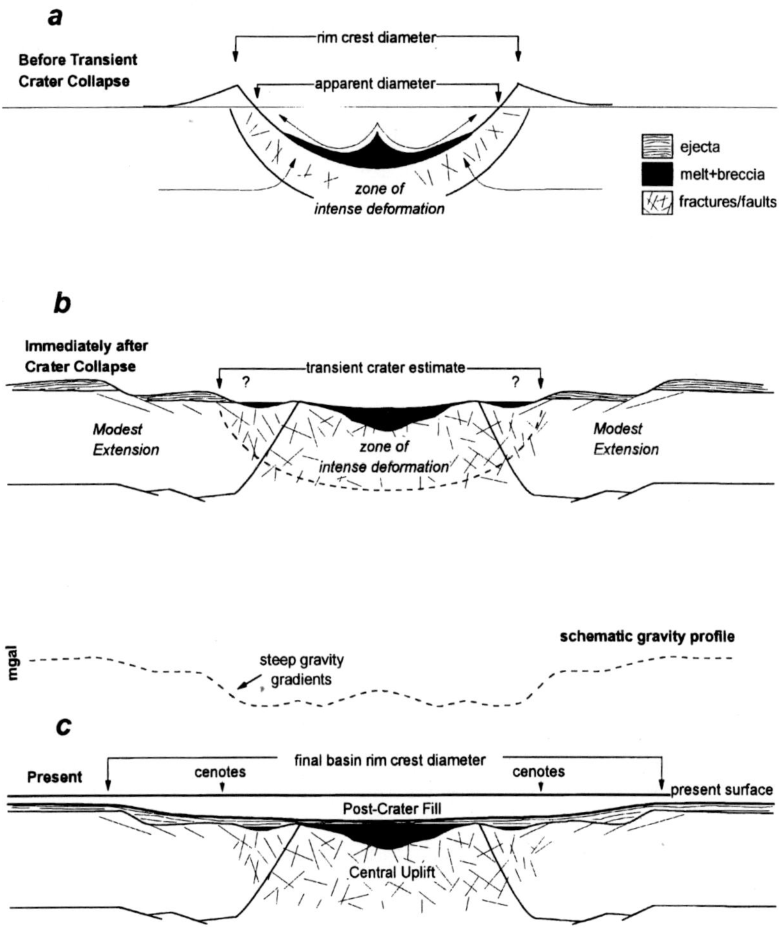

FIGURE 8. Schematic crater cross sections showing stages in development and evolution of the Chicxulub basin. (a) The transient crater, generated by excavation and downward displacement of target rocks during impact, is considerably smaller, but deeper, than the final basin. (b) The final impact basin resulting from uplift of the transient crater floor and widening of the basin flanks. The topographic model for this sketch was taken from the 270-km-diameter Mead basin on Venus. The zone of intense deformation, corresponding to the collapsed transient crater, is somewhat broader than the "apparent" transient crater in (a). (c) The ejecta deposits comprising the high-standing rim were quickly eroded and redeposited within and around the crater. This smoothed the original basin topography and accounts for the observation that all distal wells (from ~135 km to over 210 km from center) reach ejecta between ~250 and ~300 m depth). All wells closer in, including Ticul-1, 97 km from center, intercept breccia at considerably deeper levels, indicating that the topographic basin is between 250 km and 300 km wide. Figures 8b and 8c are based on reconstructions of Pemex and UNAM well data. Schematic gravity profile illustrates the gravity response to the deformation associated with the present basin (Figure 8c): broad, shallow anomalies along the flanks of the filled basin, steep gradients associated with the edge of the intensely deformed zone (transient crater), and additional concentric features associated with uplifted deep crust and thick deposits of melt and breccia.

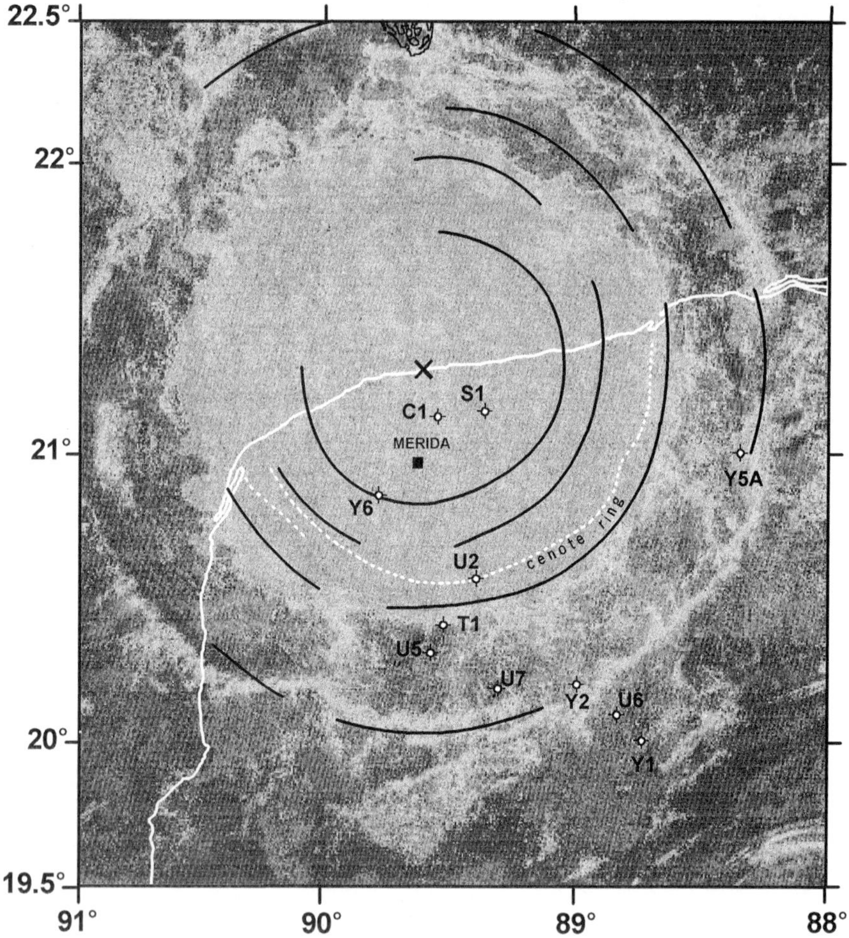

FIGURE 9. Overlay of Chicxulub location information, surface features, and geophysical trends onto the Magellan radar image of the Mead impact basin on Venus. The Mead image has been resized from ~270 km to ~300 km. The interpretation presented here is compatible with this general configuration, including information gather from well data and geophysics. Consequently, Mead may be a useful analog for understanding the original, unmodified appearance of the Chicxulub basin.

where impact angle (θ) is assumed to be vertical ($\theta = 90°$). For oblique angle impacts, Gault and Wedekind [14] found that the cratering efficiency follows a $(\sin\theta)^{1/3}$ dependence; in other words, lowering the impact angle results in the formation of increasingly smaller craters given that all other conditions remain the same. This yields a value for $\theta = 30°$ that is ~80% of that calculated from Eq. (1), and is thus within the margins of other uncertainties faced by this approach. For lower impact angles, some conspicuous asymmetries in the geophysical anomalies might be expected, and while there are some elongated features present in the geopotential data, these most likely represent the signature of the preexisting structural fabric upon which the impact signature was superimposed. In any event, Eq. (1) would provide the minimum projectile size and velocity needed to account for the K/T crater.

For the Chicxulub impact event, there is the added benefit of having a measurable projectile signature in the worldwide K/T boundary layer [2]. The average iridium fluence at the K/T boundary is ~50 ng cm^{-2} [30], yielding a globally integrated Ir mass of ~2.5×10^8 g. Assuming typical CI chondritic proportions (~500 ppb Ir), this yields ~5×10^{13} kg of chondritic material in the K/T projectile. This value represents a minimum because some of the projectile is partitioned into the melt rocks within the crater itself [48] and, particularly for such a large event, some may have escaped Earth during the ejection process [62].

Melosh [35] recently stated that the transient crater we measured requires a projectile that is inappropriately large, given the amount of iridium at the K/T boundary. Using the Schmidt-Holsapple point-source scaling relation [46] and an impact velocity of 20 km s^{-1} and he calculates a transient crater diameter of 72 km, considerably smaller than our estimate of ~130. Under these assumed conditions, he is correct. But such a low impact velocity is characteristic only of near-Earth asteroids whose rms velocities are ~20 km s^{-1} and there is no reason to favor an asteroid over a comet.

The projectile signature in the K/T boundary materials points to a primitive, undifferentiated object, suggestive of a comet. Furthermore, Shoemaker et al. [53] recently assessed the rate of terrestrial crater production over the last 100 million years and concluded that it is considerably more likely that craters the size of Chicxulub were produced by comet impacts rather than asteroid impacts. Comets are dramatically more energetic than asteroids. Long-period comets have average impact velocities of ~57 km s^{-1}; short-period comets impact at ~30 km s^{-1} (e.g., [53], [64]). Furthermore, a comet's chondritic mass is augmented by other nonsilicate components. The iridium at the K/T boundary only constrains the amount of rocky (i.e., chondritic) material contributing to the comet's mass. Recent flyby missions to Halley indicate that this short-period comet contains a surprising amount of refractory organic compounds, in addition to the rock-forming silicate materials, that contribute to its low albedo. The rocky components may not exceed 25% of the comet's mass [24], [21]. Long-period comets could be considerably more ice-rich.

Figure 10 shows the terrestrial transient crater sizes that are predicted from collisions with all possible Earth crossing objects. Given the estimated global abundance of Ir at the K/T boundary, long-period comets 10–21 km

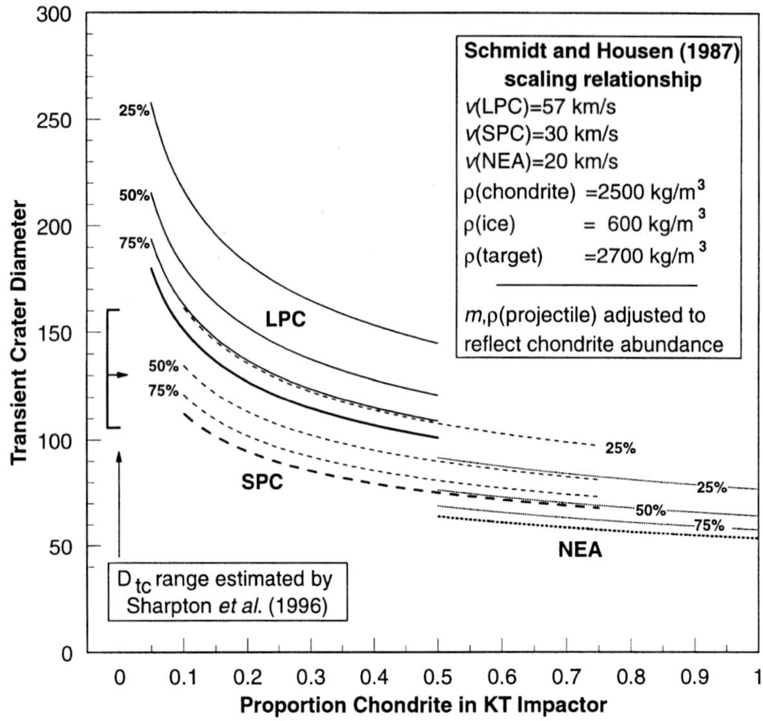

FIGURE 10. Calculated transient crater diameters as functions of chondrite/total projectile mass (horizontal axis), impact velocity (using rms velocities for long-period comets, short-period comets, and near-Earth asteroids), and the proportion of each impactor's chondritic mass deposited in the K/T boundary. Solid lines represent velocities typical of long-period comets (LPC), dashed lines represent impact velocities associated with short-period comets (SPC), and dotted lines indicate typical near-Earth asteroid (NEA) velocities. Bold lines show curves assuming all the projectiles chondritic material was deposited in the K/T boundary; percentages indicate proportion of total chondrite mass contained in the boundary layer. Bracket indicates transient crater range derived from the gravity analysis of Sharpton et al. [52] and horizontal arrow shows estimate derived from seismic profiles.

in diameter and containing at least 50% nonchondritic (volatile) material would deliver the proportion of Ir observed at the K/T boundary. Upon colliding with Earth, such objects would generate a scaled transient crater ~105–160 km in diameter according to the revised Schmidt-Holsapple scaling relationship. In all cases, the chondritic proportion of the projectile would be equivalent to a sphere 7.6 km in diameter. These collisions would have released the energy equivalent to between 4×10^8 and 4×10^9 megatons of TNT.

If our evaluation of the size of the Chicxulub crater is accurate (and scaling relationships hold for this size event), it is improbable that the impactor was a short-period comet, unless either a substantial proportion of the projectile's iridium was not partitioned into the distal ejecta at the K/T boundary or the comet was traveling considerably faster than the rms velocity of short-period comets. Finally, it would seem that a near-Earth asteroid could not have generated the K/T crater traveling at near the rms velocity for such objects.

TOWARD UNDERSTANDING IMPACT-INDUCED EXTINCTIONS: QUESTIONS THAT REMAIN AND HOW TO SOLVE THEM

What does it take to push the biosphere into crisis? Studies of terrestrial impact rates suggest that such a 300 km impact basin would, on average, be produced at a rate of $\sim 1.25 \times 10^{-9}$ y^{-1}, or about one every 800 million years [53]. This suggests then that the 1000-million-year-long lineage of complex life has experienced only one, maybe two, biosphere-threatening events. This rate is considerably lower than that of the major mass extinctions over the last 250 million years (5×10^{-7} y^{-1}). Consequently, while there is substantial circumstantial evidence establishing the cause-effect link between the Chicxulub basin-forming event and the K/T biological extinctions, the results of our analysis do not support models of impact as a common or singular causative agent of mass extinctions on Earth. To date there is only one example, the K/T boundary, where worldwide evidence of impact is coupled with a sudden, massive, and global biological crisis.

It is also essential to understand Chicxulub's target characteristics, in order to someday understand how the K/T extinctions are linked to impact and to gauge whether or not this particular impact event was uncharacteristically deadly for its size. Understanding the total thickness of the platform sequence at Chicxulub as well as the relative proportions and purity of the carbonate and evaporite units are critical to deriving more precise models of global change initiated by their impact-induced vaporization. To their credit, Pope et al. [42] consider anhydrite volumes corresponding to layer thicknesses ranging between 500 and 1500 m in their 2D hydrocode calculations. However, it is unclear if the proportion of anhydrite available for involvement in the impact even falls between this broad range. Pemex drilled 5 wells into the Mesozoic sequence of Northern Yucatan within 210 km from Chicxulub's center [52]. The integrated thickness of anhydrite in each well ranges from <400 to >1800 m. The wells on the flanks of the structure (Yucatan-1 and Yucatan-4) indicate between 300 and 800 m of evaporite in the platform section; however, the evaporites seem to thicken toward the center of the structure, specifically in the vicinity of Ticul-1. The fact that Ticul-1 is situated within a strong linear gravity low (Fig. 6) could indicate that the thick anhydrite deposits are associated with this feature and not typical of the regional stratigraphy. Alternatively, the perceived thickness of these units at Ticul-1 could be a reflection of structural thickening from outwardly directed compression associated with the impact event [52]. Furthermore, the purity of the anhydrite layers is not well constrained. In any event, it seems unlikely that there is as much anhydrite as previously estimated by Lopez-Ramos [31] who, relying upon reports based on analysis of well cuttings, mistook the anhydrite-dolomite breccia in the Upper Cretaceous section (Fig. 4) for interbedded anhydrite and dolomite layers. Sharpton et al. [52] have shown that this unit is debris ejected from the Chicxulub basin and deposited onto the preexisting Upper Cretaceous rocks around the impact site. Drilling, particularly along the western side of the crater where no lithological constraints currently exist, will contribute greatly to better estimates of how much of each kind of platform rock was involved in the impact.

ACKNOWLEDGMENTS

Sue Sharpton assisted in manuscript preparation, and Scott Lee, Steve Hokanson, and Debra Rueb assisted in preparation of graphics.

REFERENCES

1. ALEXOPOULOS, J.S. & W.B. MCKINNON. 1994. Large impact craters and basins on Venus, with implications for ring mechanics on the terrestrial planets. *In* Large Meteorite Impacts and Planetary Evolution: Boulder, Colorado, B.O. Dressler, R.A.F. Grieve, & V.L. Sharpton, Eds.: 29–50. Geological Society of America Special Paper 293.
2. ALVAREZ, L.W., W. ALVAREZ, F. ASARO, & H.V. MICHEL. 1980. Extraterrestrial cause for the Cretaceous-Tertiary extinction. Science. **208:** 1095–1108.
3. BALDWIN, R.B. 1963. The Measure of the Moon. University of Chicago Press. Chicago, Illinois. 488 pages.
4. BLUM, J.D., C.P. CHAMBERLAIN, M.P. HINGSTON, C. KOEBERL, L.E. MARÍN, B.C. SCHURAYTZ, & V.L. SHARPTON. 1993. Isotopic comparison of K/T boundary impact glass with melt rock from the Chicxulub and Manson impact structures. Nature. **364:** 325–327.
5. BOHOR, B.F. 1990. Shocked quartz and more; Impact signatures in Cretaceous/Tertiary boundary clays. *In* Global catastrophes in Earth history; An interdisciplinary conference on impacts, volcanism, and mass mortality, V.L. Sharpton & P.D. Ward, Eds.: 335–342. Geological Society of America Special Paper 247.
6. BOHOR, B.F., , E.E. FOORD, P.J. MODRESKI, & D.M. TRIPLEHORN. 1984. Mineralogical evidence for an impact event at the Cretaceous/Tertiary boundary. Science. **224:** 867–869.

7. BOHOR, B.F., W.J. BETTERTON, & T.E. KROGH. 1993. Impact-shocked zircons: discovery of shock-induced textures reflecting increasing degrees of shock metamorphism. Earth and Planetary Science Letters. **119:** 419–424.
8. BOURGEOIS, J., T.A. HANSEN, P.L. WIBERG, & E.G. KAUFFMAN. 1988. A tsunami deposit at the Cretaceous-Tertiary boundary in Texas. Science. **241:** 567–570.
9. BRETT, R. 1992, The Cretaceous-Tertiary extinctions; A lethal mechanism involving anhydrite target rocks: Geochimica et Cosmochimica Acta. **56:** 3603–3606.
10. CAMARGO, Z.A. & R.G. SUÁREZ. 1994, Evidencia sismica del crater de impacto de Chicxulub [Seismic evidence of the Chicxulub Impact Crater]: Boletín de la Asociacion Mexicana de Geofisicos de Exploracion. **34:** 1–28 (in Spanish).
11. COVEY, C., S.J. GHAN, & J.J. WALTON. 1990. Global environmental effects of impact-generated aerosols: Results from a general circulation model, *in* Sharpton, V. L., and Ward, P. D., eds., Global catastrophes in Earth history: An interdisciplinary conference on impacts, volcanism, and mass mortality: Geological Society of America Special Paper 247, 263–270.
12. CROFT, S.K. 1981. The modification stage of basin formation: Conditions of ring formation. *In* Multi-ring basins, P.H. Schultz & R.B. Merrill, Eds.: 227–257. Pergamon Press. Elmsford, New York.
13. EMILIANI, C., E.B. KRAUS, & E.M. SHOEMAKER. 1981. Sudden death at the end of the Mesozoic. Earth and Planetary Science Letters. **55:** 317–334.
14. GAULT, D.E. & J.A. WEDEKIND. 1978. Experimental studies of oblique impact. *In* Proceedings of the Lunar and Planetary Science Conference, 9th. 3843–3875.
15. GILMOUR, I., W.S. WOLBACH, & E. ANDERS. 1990. Early environmental effects of the terminal Cretaceous impact. *In* Global catastrophes in Earth history: An interdisciplinary conference on impacts, volcanism, and mass mortality, V.L. Sharpton & P.D. Ward, Eds.: 383–390. Geological Society of America Special Paper 247.
16. GRIEVE, R.A.F., P.B. ROBERTSON, & M.R. DENCE. 1981. Constraints on the formation of ring impact structures based on terrestrial data. *In* Multi-ring basins, P.H. Schultz & R.B. Merrill, Eds.: 37–57. Pergamon Press. Elmsford, New York.
17. HEAD, J.W. 1977. Origin of outer rings in lunar multi-ringed basins: Evidence from morphology and ring spacing. *In* Impact and Explosion Cratering, D.J. Roddy, R.O. Pepin, & R.B. Merrill, Eds.: 563–573. Pergamon Press. Elmsford, New York.
18. HILDEBRAND, A.R., G.T. PENFIELD, D.A. KRING, M. PILKINGTON, Z.A. CAMARGO, S.B. JACOBSEN, & W.V. BOYNTON. 1991. Chicxulub crater: A possible Cretaceous/Tertiary boundary impact crater on the Yucatán peninsula, Mexico. Geology. **19:** 867–871.
19. HÖRZ, F., & G.S. BANHOLZER. 1980. Deep-seated target materials in the continuous deposits of the Ries Crater, Germany. *In* Proceedings of the Conference on the Lunar Highlands Crust, J.J. Papike & R.B. Merrill, Eds.: 211–231. Pergamon Press. New York.
20. HÖRZ, F., R. OSTERTAG, & D.A. RAINEY. 1983. Bunte Breccia of the Ries: Continuous deposits of large impact craters. Reviews of Geophysics and Space Physics. **21:** 1667–1725.
21. JESSBERGER, E.K., A. CHRISTOFORIDIS, & J. KISSEL. 1988. Aspects of the major element composition of Halley's dust. Nature. **332:** 691–695.
22. IZETT, G.A., G.B. DALRYMPLE, & L.W. SNEE. 1991, $^{40}Ar/^{39}Ar$ age of Cretaceous-Tertiary boundary tektites from Haiti. Science. **252:** 1539–1542.
23. IZETT, G.A. 1991. Tektites in Cretaceous-Tertiary boundary rocks on Haiti and their bearing on the Alvarez extinction hypothesis. J. of Geophysical Research. **96:** 20,879–20,905.
24. KISSEL, J. *ET AL.* 1986. Composition of comet Halley dust particles from Vega observations. Nature. **321:** 280–282.

25. KOEBERL, C. & H. SIGURDSSON. 1992. Geochemistry of impact glasses from the K/T boundary in Haiti: Relation to smectites and a new type of glass. Geochimica et Cosmochimica Acta. **56:** 2113–2129.
26. KOEBERL, C., V.L. SHARPTON, B.C. SCHURAYTZ, S.B. SHIREY, J.D. BLUM, & L.E. MARÍN. 1994. Evidence for a meteoritic component in impact melt rock from the Chicxulub structure. Geochimica et Cosmochimica Acta. **58:**1679–1684.
27. KRING, D.A. & W.V. BOYNTON. 1992. Petrogenesis of an augite-bearing melt rock in the Chicxulub structure and its relationship to K/T impact spherules in Haiti. Nature. **358:** 141–144.
28. KROGH, T.E., S.L. KAMO, & B.F. BOHOR. 1993a. Fingerprinting the K/T impact site and determining the time of impact by U-Pb dating of single shocked zircons from distal ejecta. Earth and Planetary Science Letters. **119:** 425–429.
29. KROGH, T.E., S.L. KAMO, V.L. SHARPTON, L.E. MARÍN, & A.R. HILDEBRAND. 1993b. U-Pb ages of single shocked zircons linking distal K/T ejecta to the Chicxulub crater. Nature. **366:** 731–734.
30. KYTE, F.T., J. SMIT, & J.T. WASSON. 1985. Siderophile interelement variations in the Cretaceous-Tertiary boundary sediments from Caravaca. Earth and Planetary Science Letters. **73:** 183–195.
31. LOPEZ-RAMOS, E. 1979. Geologia de Mexico, 2nd edit. University of Mexico. Mexico City, Mexico. 453 p.
32. MACDOUGALL, J.D. 1988. Seawater strontium isotopes, acid rain, and the Cretaceous-Tertiary boundary. Science. **239:** 485-487.
33. MARSHALL, R.G. 1974. Petrology of subsurface Mesozoic rocks of the Yucatan Platform, Mexico: Master's thesis, University of New Orleans. New Orleans, Louisiana.
34. MELOSH, H.J. 1989, Impact cratering: A geological process. Oxford University Press, New York. 245 p.
35. MELOSH, H.J. 1995. Around and around we go. Nature. **376:** 386–387.
36. MELOSH, H.J. & W.B. McKinnon. 1978. The mechanics of ringed basin formation. Geophysical Research Letters. **5:** 985–988.
37. PENFIELD, G.T. & Z.A. CAMARGO. 1981. Definition of a major igneous zone in the central Yucatan platform with aeromagnetics and gravity (abstract). Society of Exploration Geophysicists Technical Program, Abstracts and Biographies. **51:** 37.
38. PIKE, R.J. & P.D. SPUDIS. 1987. Basin-ring spacing on the Moon, Mercury, and Mars. Earth, Moon, and Planets. **39:** 129–194.
39. PILKINGTON, M. & R.A.F. GRIEVE. 1992. The geophysical signature of terrestrial impact craters. Reviews of Geophysics. **30:** 161–181.
40. PILKINGTON, M., A.R. HILDEBRAND, & C. ORTIZ-ALEMAN. 1994. Gravity and magnetic field modeling and structure of the Chicxulub Crater. Mexico. J. Geophysical Research. **99:** 13,147–13,162.
41. POLLACK, J. B., O.B. TOON, T.P. ACKERMAN, C.P. MCKAY, & R.P. TURCO. 1983. Environmental effects of an impact-generated dust cloud: Implications for the Cretaceous-Tertiary extinctions: Science. **219:** p. 247–249.
42. POPE, K.O., K.H. BAINES, A.C. OCAMPO, & B.A. IVANOV. 1994. Impact winter and the Cretaceous/Tertiary extinctions: Results of a Chicxulub asteroid impact model. Earth and Planetary Science Letters. **128:** 719–725.
43. PRINN, R.G. & B. FEGLEY. 1987. Bolide impacts, acid rain, and biospheric traumas at the Cretaceous-Tertiary boundary. Earth and Planetary Science Letters. **83:** 1–15.
44. RAUP, D. 1989. The case for extraterrestrial causes of extinction. Philosophical Transactions of the Royal Society of London. . **325:** 421–435.
45. SCHABER, G.G., *ET AL.* 1992. Geology and distribution of impact craters on Venus: What are they telling us? J. Geophysical Research. **97:** 13,257–13,301.

46. SCHMIDT, R.M. & K.R. HOUSEN. 1987. Some recent advances in scaling of impact and explosion cratering. International Journal of Impact Engineering. **5:** 543–560.
47. SCHULTZ, P.H. 1976. Moon morphology. University of Texas Press. Austin, Texas. 926 p.
48. SCHURAYTZ, B.C., V.L. SHARPTON, & L.E. MARÍN. 1994, Petrology of impact-melt rocks at the Chicxulub multiring basin, Yucatán, Mexico. Geology. **22:** 868–872.
49. SHARPTON, V.L., B.C. SCHURAYTZ, K. BURKE, A.V. MARALI, & G. RYDER. 1990. Detritus in K/T boundary clays pf western North America; Evidence against a single oceanic impact. *In* Global catastrophes in Earth history; An interdisciplinary conference on impacts, volcanism, and mass mortality, V.L. Sharpton & P.D. Ward, Eds.: 349–357. Geological Society of America Special Paper 247.
50. SHARPTON, V.L., G.B. DALRYMPLE, L.E. MARÍN, G. RYDER, B.C. SCHURAYTZ, & J. URRUTIA-FUCUGAUCHI. 1992. New links between the Chicxulub impact structure and the Cretaceous/Tertiary boundary. Nature. **359:** 819–821.
51. SHARPTON, V.L., K. BURKE, A. CAMARGO-ZANOGUERA, S.A. HALL, S. LEE, L.E. MARÍN, G. SUÁREZ-REYNOSO, J.M. QUEZADA-MUÑETON, P.D. SPUDIS, & J. URRUTIA-FUCUGAUCHI. 1993. Chicxulub multiring impact basin: Size and other characteristics derived from gravity analysis. Science. **261:** 1564–1567.
52. SHARPTON, V.L., L.E. MARÍN, J.L. CARNEY, S. LEE, G. RYDER, B.C. SCHURAYTZ, P. SIKORA, & P.D. SPUDIS. 1996. A model of the Chicxulub impact basin based on evaluation of geophysical data, well logs, and drill core samples. Geological Society of America Special Paper 307, 55–74.
53. SHOEMAKER, E.M., R.F. WOLFE, & C.S. SHOEMAKER. 1990. Asteroid and comet flux in the neighborhood of Earth. *In* Global catastrophes in Earth history: An interdisciplinary conference on impacts, volcanism, and mass mortality, V.L. Sharpton & P.D. Ward, Eds.: 155–170. Geological Society of America Special Paper 247.
54. SIGURDSSON, H., S. D'HONDT, M.A. ARTHUR, T.J. BRALOWER, J.C. ZACHOS, M. VAN FOSSEN, & E.T. CHANNELL. 1991. Glass from the Cretaceous/Tertiary boundary in Haiti. Nature. **349:** 482–487.
55. SIGURDSSON, H., PH. BONTÉ, L. TURPIN, M. CHAUSSIDON, N. METRICH, M. STEINBERG, PH. PRADEL, & S. D'HONDT. 1991. Geochemical constraints on source region of Cretaceous/Tertiary impact glasses. Nature. **353:** 839–842.
56. SMIT, J., A. MONTANARI, SWINBOURNE, W. ALVAREZ, A.R. HILDEBRAND, S. MARGOLIS, P. CLAEYS, W. LOWRIE, & F. ASARO. 1992. Tektite-bearing deep-water clastic unit at the Cretaceous-Tertiary boundary in northeastern Mexico. Geology. **20:** 99–103.
57. SPUDIS, P.D. 1993. The geology of multi-ring impact basins. Cambridge University Press. Cambridge, England.
58. SWISHER, C.C., III, J.M. GRAJALES-NISHIMURA, A. MONTANARI, S.V. MARGOLIS, P. CLAEYS, W. ALVAREZ, P. RENNE, E. CEDILLO-PARDO, F.J-M.R. MAURRASSE, G.H. CURTIS, J. SMIT, AND M.O. MCWILLIAMS. 1992. Coeval $^{40}Ar/^{39}Ar$ ages of 65.0 million years ago from Chicxulub crater melt rock and Cretaceous-Tertiary boundary tektites. Science. **257**:. 954–958.
59. TOON, O.B., J.B. POLLACK, T.P. ACKERMAN, R.P. TURCO, C.P. MCKAY, & M.S. LIU. 1982. Evolution of an impact-generated dust cloud and its effects on the atmosphere. *In* Geological implications of impacts of large asteroids and comets on the Earth, L.T. Silver & P.H. Schultz, Eds.: 187–200. Geological Society of America Special Paper 190.
60. TURCO, R., O.B. TOON, C. PARK, R. WHITTEN, J.B. POLLACK, AND P. NOERDLINGER. 1981. Tunguska meteor fall of 1908; Effects on stratospheric ozone. Science. **214:** 19–23.

61. URRUTIA-FUCUGAUCHI, J., L. MARÍN, & V.L. SHARPTON. 1994. Reverse polarity magnetized melt rocks from the Cretaceous/Tertiary Chicxulub structure, Yucatan peninsula, Mexico. Tectonophysics. **237:** 105–112.
62. VICKERY, A. M. & H.J. MELOSH. 1990. Atmospheric erosion and impactor retention in large impacts, with application to mass extinctions. *In* Global catastrophes in Earth history: An interdisciplinary conference on impacts, volcanism, and mass mortality, V.L. Sharpton & P.D. Ward, Eds.: 289–300. Geological Society of America Special Paper 247.
63. WARD, W.C., G. KELLER, W. STINNESBECK, & T. ADATTE. 1995. Yucatan subsurface stratigraphy: Implications and constraints for the Chicxulub impact. Geology. **23:** 873–876.
64. WEISSMAN, P.R. 1988. Terrestrial impact rates for long and short-period comets. *In* Geological implications of impacts of large asteroids and comets on the Earth, L.T. Silver & P.H. Schultz, Eds.: 15–24. Geological Society of America Special Paper 190.

Tsunami Produced by the Impacts of Small Asteroids

JACK G. HILLS AND CHARLES L. MADER

Theoretical Division
Los Alamos National Laboratory
Los Alamos, New Mexico 87545

ABSTRACT: The fragmentation of a small asteroid in the atmosphere greatly increases its cross section for aerodynamic braking, so ground impact damage (craters, earthquakes, and tsunami) from a stone asteroid is nearly negligible if it is less than 200 meters in diameter. A larger one impacts the ground at nearly its velocity at the top of the atmosphere producing considerable impact damage. The protection offered by Earth's atmosphere is insidious in that smaller, more frequent impactors such as Tunguska only produce air blast damage and leave no long-term scars on the Earth's surface, while objects 2.5 times larger than it, which hit every few thousand years, cause coherent destruction over many thousands of kilometers of coast. Smaller impactors give no qualitative warning of the enormous destruction wrought when an asteroid larger than the threshold diameter of 200 meters hits an ocean. A water wave generated by an impactor has a long range because it is two-dimensional, so its height falls off inversely with distance from the impact. When the wave strikes a continental shelf, its speed decreases and its height increases to produce tsunamis. The average runup in height between a deep-water wave and its tsunami is more than an order of magnitude. Tsunamis produce most of the damage from asteroids with diameters between 200 meters and 1 km. An impact anywhere in the Atlantic by an asteroid 400 meters in diameter would devastate the coasts on both sides of the ocean by tsunami over 100 meters high. An asteroid 5 km in diameter hitting in mid Atlantic would produce tsunami that would inundate the entire upper East Coast of the United States to the Appalachian Mountains. Studies of ocean sediments may be used to determine when coastal areas have been hit by tsunamis in the past. Tsunami debris has been found to be associated with the Cretaceous-Tertiary impact and should be detectable for smaller impacts.

I. INTRODUCTION

Tsunami may be the most serious consequence of asteroid impacts unless the asteroid is massive enough to produce global, catastrophic changes in the atmosphere, as apparently occurred after the impact responsible for the Cretaceous-Tertiary extinction. Just as on land, much of the kinetic energy of an asteroid that impacts the ocean goes into the formation of a crater, but the crater is not stable. The outward propagation of its rim and its refilling produces a series of waves that propagate outward away from the impact [8].

In this paper we are primarily concerned with the impacts of small (compared to the depth of the ocean) asteroids that produce waves with amplitudes less than the depth of the ocean. Such deep-water waves do not dampen significantly until they run into shallows where they steepen into breakers and increase in height to form tsunami [13]. The average tsunami runup, the height of the tsunami in units of the deep-water wave that produced it, is about an order of magnitude.

The height of a deep-water wave only decreases inversely with the distance from its origin, so it can cause serious problems far from the impact. This results from the wave being inherently two-dimensional. The intensity of a three-dimensional disturbance such as an airburst or an earthquake falls off as the inverse square of the distance, so such a disturbance is far more localized than water waves.

There are many anecdotal illustrations of the long-range nature of tsunami; e.g., the earthquake in Chile in 1960 produced deep-water waves that traveled 150 degrees (over 17,000 km) around the Earth to produce tsunami in Japan that were from 1–5 meters high (average about 2 meters) and killed at least 114 people with another 90 people missing [19]. (It is estimated that the full amplitude of the deep-water wave before hitting Japan was 40 cm, so the maximum height above normal sea level was 20 cm, and it had a period of 60 minutes [11]). This implies an average tsunami runup of 10-fold and a maximum of 25-fold). In the Hawaiian Islands, at 10,600 km from the epicenter, the maximum runup was 15 meters. The major damage was in Hilo harbor where the maximum tsunami height was over 10 meters and 61 people were killed [6]. The average tsunami runup in Hawaii is about 40 fold. We shall see that asteroid impacts can produce tsunami vastly larger than the 1960 tsunami and in regions, such as the Atlantic, where coastal areas are poorly prepared for them.

II. IMPACTS INTO DEEP OCEANS

To determine heights of tsunami produced by an asteroid or comet, we first determine its kinetic energy at its impact with the ocean. Figure 1 (from the work of Hills and Goda [9]) shows the height in the atmosphere at which half the kinetic energy of a stony meteoroid is dissipated. We note that asteroids with radii exceeding 100 meters hit the ground with most of their original kinetic energy. The straight-line portion on the left side of the figure is for asteroids that do not fragment. Fragmentation can enormously increase the effective radius of smaller meteoroids and their rate of energy dissipation in the atmosphere. If we extend the straight line portion of the figure to sea level, we note that if it were not for fragmentation, asteroids with radii larger than 10 meters would be able to penetrate to sea level with most of their kinetic energy. The increased energy-dissipation cross section due to fragmentation causes stony asteroids with radii between 10 and 100 meters to dissipate most of their energy in the atmosphere rather than on impact with the ground.

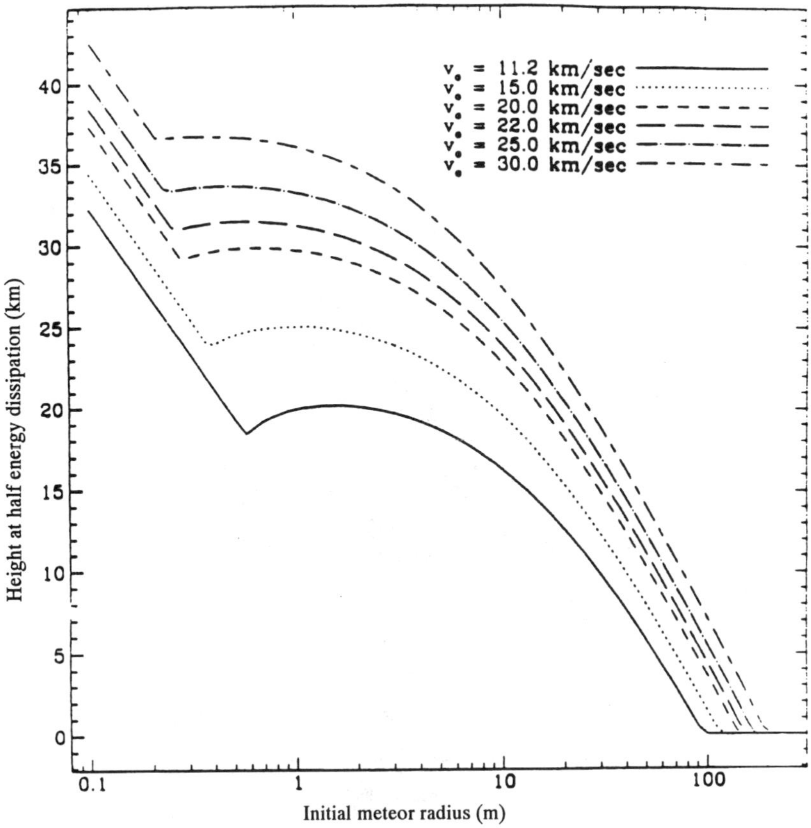

FIGURE 1. Soft Stone. The height (in km) in the atmosphere at which half the initial energy of the impactor has been absorbed. This is for soft (common) stony asteroids. It is given as a function of the radius of the impactor for various impact velocities (in km/s).

We use the impact energy at sea level to find the height of the deep-water wave. An empirical analysis of experiments with underwater nuclear explosives shows that the full height of a deep-water wave at a distance *r* from the underwater detonation of energy *Y* is given by [7]

$$h_W = 40{,}500\,ft\ \frac{(Y/kton)^{0.54}}{r/ft} = 6.5\left(\frac{Y}{gigaton}\right)^{0.54}\left(\frac{1000km}{r}\right). \quad (1)$$

This result is not sensitive to the depth at which the explosion occurs. The height, *h*, of the water wave above the ocean is half the full height of the wave, so $h = 3.3$ meters at 1000 km from a 1 gigaton = 4.2×10^{25} ergs explosion.

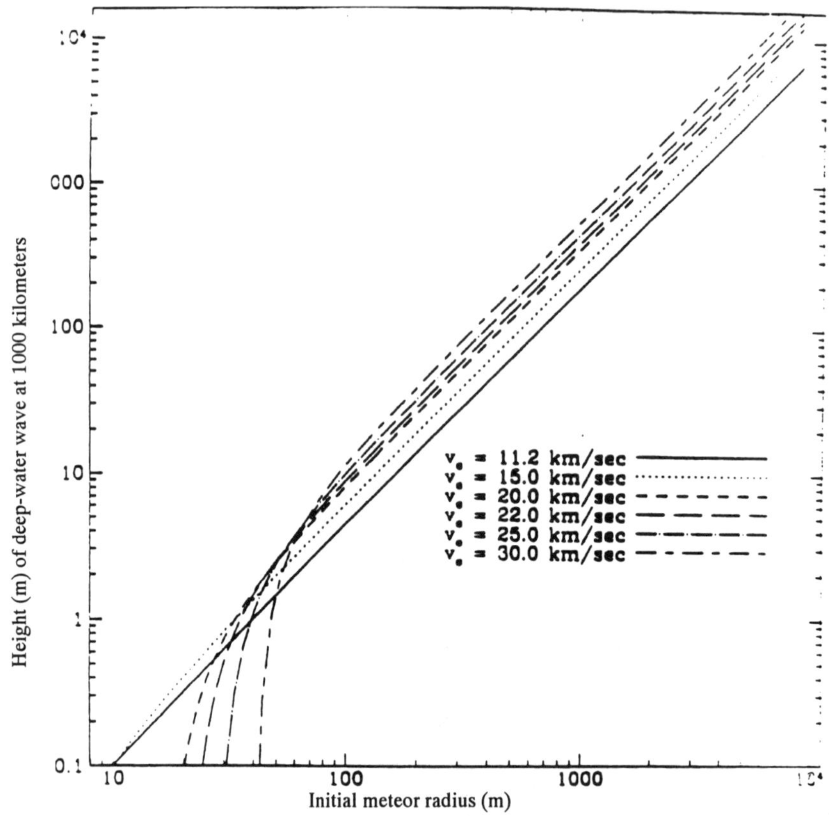

FIGURE 2. Iron. The full height (meters) of a deep-water wave 1000 km from the impact of a nickel-iron asteroid. The height is given as a function of impactor radius for various impact velocities. The height of the wave above mean ocean level is half the full height shown. This wave increases in height by over an order of magnitude to produce tsunami when it runs into a continental shelf.

A more recent analysis of Pacific test explosions in deep water with yields between 1 kiloton and 5 megatons and of modeled nuclear explosions of up to 100 megatons, shows a similar equation for the h above the ocean level. One of us (Mader) finds that

$$h = \frac{1}{2} h_W = 4.5\, meters \quad \left(\frac{Y}{gigaton}\right)^{1/2} \left(\frac{1000km}{r}\right). \tag{2}$$

The values given by this equation for $R > 100$ meters are in satisfactory agreement with those given by Eq. (1), considering the large extrapolation beyond the experimental points.

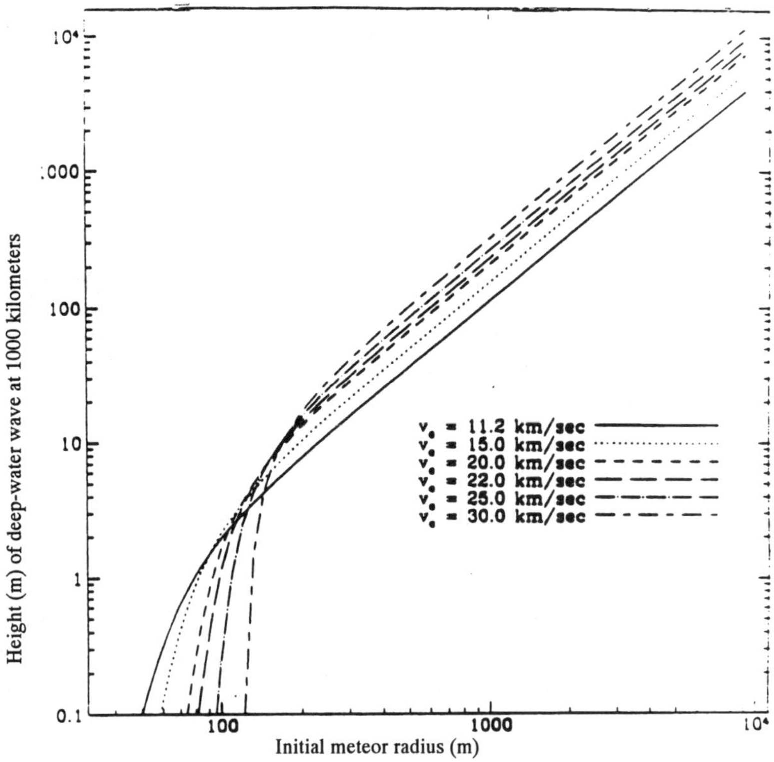

FIGURE 3. Soft stone. The full height (meters) of a deep-water wave 1000 km from the impact point of a stony asteroid. This height is given as a function of impactor radius for various impact velocities.

Hills and Goda [9] found the ground impact energies of comets, stony asteroids, and iron asteroids as a function of size and impact velocity taking into account the increase in their aerodynamic cross sections due to fragmentation. Figures 2 and 3 show the full height, H_W (twice the height h above sea level) of a deep-water wave 1000 km from the impact point for nickel-iron and stony meteorites, respectively, as a function of impactor radius for various impact velocities. The heights were obtained by putting the ground impact energies Y found by Hills and Goda [9] into Eq. (1).

We note that the wave heights for stony asteroids less than 100 meters in radius are significantly less than they would be without aerodynamic dissipation. This is also true of iron asteroids with radii less than 40 meters. The smaller cutoff radius for irons is due to their greater strength, which causes them to fragment less easily than stones.

For stones with radii $R > 100$ meters, which suffer no significant energy dissipation in the atmosphere, the deep-water wave height ($h = h_w/2$) above mean sea level at distance r (based on the heights determined by Eq. (1)) is given by

$$h = 7.8\,meters\;\left[\left(\frac{R}{203meters}\right)^3\left(\frac{V}{20km}\right)^2\left(\frac{\rho_M}{3g/cm^3}\right)\right]^{0.54}\left(\frac{1000km}{r}\right). \qquad (3)$$

Here a stony asteroid with a radius of 203 meters and a velocity of 20 km/s has an impact energy of 5 gigatons. An asteroid of this size, or larger, impacts Earth about every 10^4 years.

Asteroids of sufficient size produce craters that exceed the ocean depth. In these cases, Eqs. (1)–(3) and Figs. 2–3 are no longer valid. We shall discuss such impacts in the next section.

III. IMPACTS INTO SHALLOW SEAS

The average ocean depth is 4–5 km. If the depth of the impact crater exceeds the local ocean depth d, we can no longer use Eq. (1) to compute the height of the deep-water wave far from impact. It is known from nuclear weapon tests that an explosion in shallow water (e.g., Pacific test Bikini Baker) deposits less mechanical energy into the water than one in deep water [7]. Glasstone and Dolan find that the full height of the deep-water wave at distance r from the explosion is given by

$$h_w = 1450\,meters\;\left(\frac{d}{r}\right)\left(\frac{Y}{gigaton}\right)^{0.25}, \qquad (4)$$

where d is the depth of the water and Y is the yield. We note that the wave height increases less rapidly with yield than it does for waves generated in deep water, but there remains an inverse relationship between height and distance from the source. If we let $d = 5$ km, the average depth of the ocean, we find that Eqs. (1) and (4) give the same full height of $h_w = 8.1$ m at $r = 1000$ km for a yield of $Y = 1.5$ gigaton, which corresponds to a stony asteroid with a diameter of 272 meters and an impact velocity of 20 km/s.

Schmidt and Holsapple [18] found that the depth of a crater in water is about 12 times the impactor diameter. This suggests that where the impactor diameter significantly exceeds 8% of the depth, it is better to use Eq. (4) than Eq. (1) to determine the terminal height of the deep-water wave far from the impact. In the ocean, where $d = 5$ km, we should use Eq. (4) if the impactor diameter much exceeds 400 meters.

Hydrodynamic simulations by Nemchinov and associates (as given in [10]) of craters produced by asteroids with diameters comparable to the ocean depth suggest that the wave height falls between the values given by Eqs. (1) and (4), as expected. The fine structure that develops in these hydrodynamic simulations does not allow them to be run to times when the crater has collapsed into a series of outward propagating waves. The calculations must be stopped while the crater is still forming. From energy considerations, the present authors (Hills and Mader) estimate that the diameter of the water crater when it stops growing is typically about 25 to 30 times that of the asteroid that produced it.

IV. TSUNAMI

As the deep-water wave goes into a shoal, its speed decreases and its front increases in sharpness and amplitude until it breaks. This wave travels inland from the coast with decreasing speed and height above sea level. We shall first use analytic models to estimate the effect of the tsunami along a typical coast line, emphasizing the U.S. East Coast. We then use a full numerical model to determine the damage from the impact of a large asteroid into the mid Atlantic.

A. Analytic Model

We noted earlier that the 1960 Chile tsunami produced coastal runups in Japan that averaged 10-fold but reached about 25-fold in the Northern Islands. These values are fairly typical. On Hawaii, the average runup is about 40, but it can be less in areas with gradual continental shelves, such as off Florida. As an example, a stony asteroid with a radius of 200 meters (diameter 400 meters) that drops anywhere in the mid Atlantic will produce deep-water waves that are at least $h = 4$ meters high when they reach both the European and North American coasts. When it encounters land, this wave steepens into a tsunami with an average height of 40 meters (if it follows the Japanese runups) that hits both sides of the Atlantic nearly simultaneously.

Tsunami Flood Plane

When the tsunami impacts the shore, the maximum distance, X_{max}, to which it surges inland depends on the maximum depth of the water at the shoreline, the runup height h_o, the slope of the shore away from the coast, and the roughness of the ground that the water moves across (cf. [14]). If there is a flat coastal plane on which the flood depth h has a maximum value h_o, the depth at a distance X inland is given by

$$\frac{h}{h_o} = \left[1 + \left(\frac{X}{X_{\max}}\right)\right]^{4/3}, \tag{5}$$

where the maximum inward distance to which the water flows, scales as

$$X_{\max} = \frac{h_o^{4/3}}{n^2}A = Bh_o^{4/3}, \tag{6}$$

where n is the Manning roughness number of the terrain over which the water surges and A is a constant [5]. Here n varies from about 0.015 for very smooth terrain (e.g., mud flats and ice) to 0.070 for very rough coast areas (dense brush and trees and coarse lava formations). Developed areas typically have $n = 0.030 - 0.035$. For $n = 0.03$ and $h_o = 15$ meters (50 feet), $X_{\max} =$ 2.5 km (8000 feet) [5]. Putting this scaling factor into Eq. (5), we find that

$$X_{\max} = 1.4\ \text{km}\ \left(\frac{h_o}{10\text{meters}}\right)^{4/3}. \tag{7}$$

We note that in a developed area with a Manning roughness number of $n = 0.03$, a 40-meter tsunami would travel inland about 9 km, a 100-meter one would travel about 30 km, and a 200-meter ones would go 76 km. For croplands or grazing land with a Manning number approaching 0.015, the corresponding figures are 4 times larger. While there may be some difficulties in extrapolating Eq. (7) to these large values, it is clear that tsunami of these magnitudes would cause unprecedented damage to low-lying areas in North America such as Long Island.

The damage caused by tsunami results principally from the impact of the debris carried by the moving water. There is much debris in developed areas. This debris acts as a battering ram that effectively scours away the area impacted by the tsunami flood; e.g., in the 1960 Hilo, Hawaii inundation caused by the earthquake in Chile, the steel pipes supporting some of the parking meters in the city were bent to the ground by the ramming of debris carried by the flood. The higher the tsunami flood, the larger its mean flow velocity, and the more effective the ramming.

Because a disproportionate fraction of human resources are close to the coasts, tsunami are probably the most deadly manifestations of asteroid impacts apart from the very large Cretaceous-Tertiary type superkillers.

B. Numerical Model

While analytic models can approximate the general effect of a tsunami, detailed numerical models are needed to determine the runup and inundation along any particular coast. The height and direction of the deep-water wave along the coast may depend on reflections from nearby land masses as well as on the magnitude, distance, and direction of the impact. The runup depends on the height and direction of the wave and on the topology of the coast.

We did a numerical simulation of a tsunami along the East Coast of the United States produced by an asteroid falling into the mid Atlantic. It was modeled with the 2-dimensional Swan hydrodynamic code with 1 km spatial resolution. Figure 4 shows the position of the impact. We considered an initial perturbation in which the (square) crater was 150 km across with a depth equal to that of the ocean. We estimate that the formation of this crater requires an asteroid about 5 km in diameter, so this impactor is a little larger than the parent object of Comet Shoemaker-Levy 9 that impacted Jupiter in 1994 but an order of magnitude less massive than the impactor responsible for the Cretaceous-Tertiary extinction.

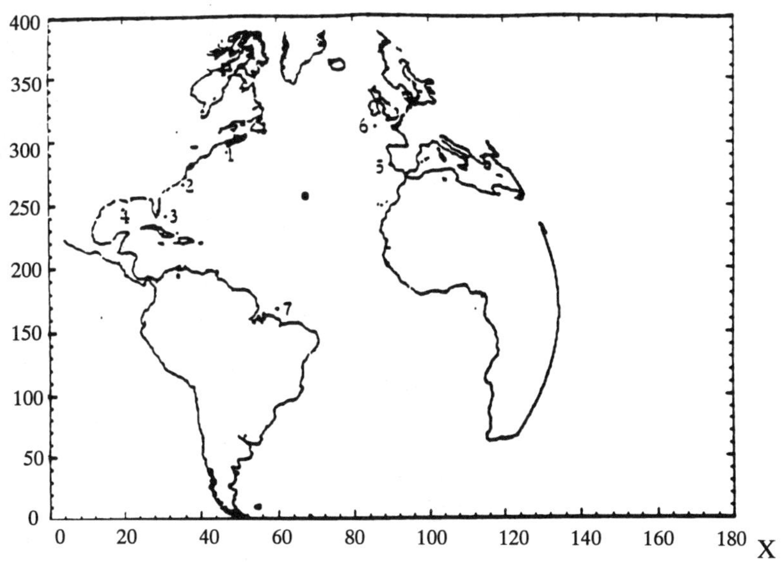

FIGURE 4. The position in the Atlantic of the 150 km diameter crater produced by an impactor. Also shown are seven locations off the continental shelves at which we determined the heights of the deep-water waves in the computer simulation.

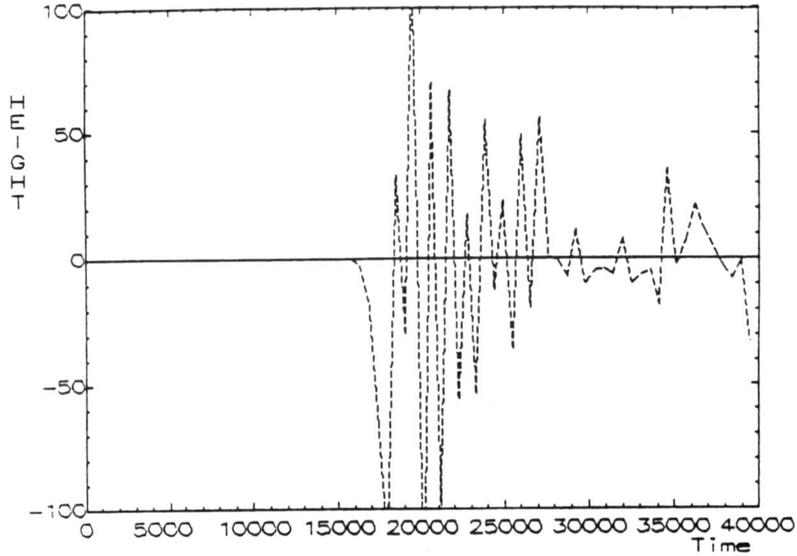

FIGURE 5. The wave height at Location 2 off the East Coast of the United States. This is still in deep water before any significant tsunami enhancement. We note the large number of separate waves that hit the coast over a period more than 6 hours.

Figure 4 shows seven representative locations just outside the continental shelf where the heights of the deep-water wave were determined. Table 1 shows characteristics of the series of deep-water waves that passed through each of these seven positions. The second column gives the maximum drop (in meters) in the level of the ocean as the deep-water waves passed by while the next column gives the maximum increase (in meters) in the level of the ocean. We see that off the central East Coast of the United States (Position 2) the maximum wave-height and fall-off are each 100 meters before the continental runup. Figure 5 shows the height of the wave as a function of time at Location 2. We note the large number of waves that will inundate the shore over a time of over six hours. The final column gives the period of the wave in seconds. These periods are similar to those measured in the 1960 Chilean tsunami. Shorter period deep-water waves disperse without energy dissipation, due to differences in their velocity with period, until their periods lengthen to these values.

Figure 6 shows the East Coast of the United of the United States before and during the tsunami surge. The coast is hit very hard by the surge. The wave travels inwards to the foothills of the Appalachian mountains in the upper two-thirds of the United States including surges of more than 200 km across Delaware, Maryland, and Virginia. Delaware, Long Island, and all of

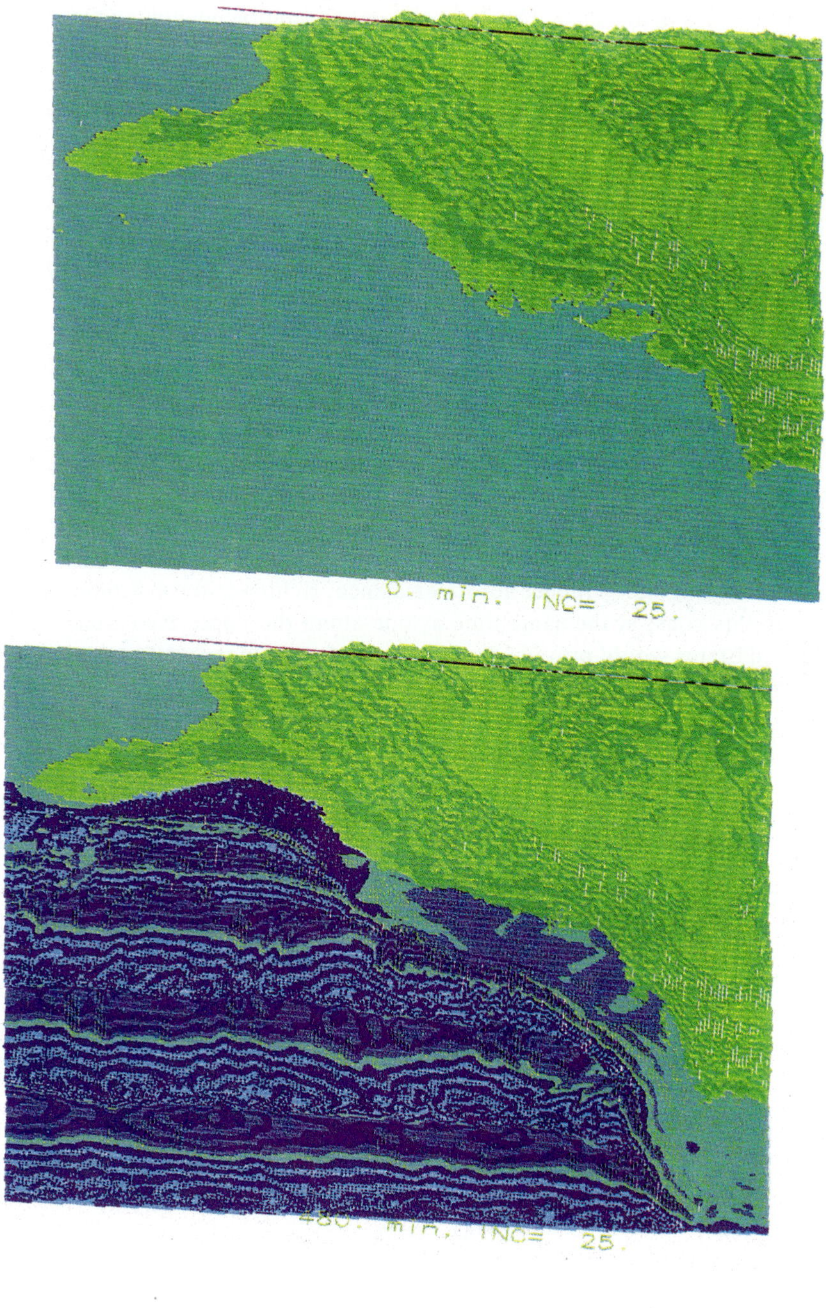

FIGURE 6. The East Coast of the United States before and during the tsunami produced by an asteroid 5 km in diameter hitting in the middle of the Atlantic Ocean.

Maryland below the Piedmont Plateau are completely inundated as are all coastal cities in this area. The damage would be unprecedented in human history.

There are surprises. The Florida coast is largely protected except for the Miami-West Palm Beach area, by a gradual continental shelf that reflects most of the tsunami energy back into the Atlantic. Inland areas of Florida are safe despite its low elevation. The enhanced damage in the Miami area compared to rest of Florida points to the particular danger to seaports from tsunami. Seaports are particularly valued if they have a deep offshore channel in otherwise shallow coastal waters. This channel can support a much more energetic tsunami than can the rest of the coast.

The tsunami causes much less damage to Europe than it does to North America because of a large continental shelf off most of the European coast. An exception is the Portugal-Spain peninsula, which has almost no continental shelf. The tsunami wraps itself around the peninsula up to the foothills of the mountains. The particular vulnerability of this region may have been forewarned by the tsunami produced by the Lisbon earthquake of two centuries ago. The lost city of Atlantis was allegedly along this section of the Atlantic coast before it was swallowed suddenly by the sea, although there is no archaeological evidence for its existence. Evidence for a strong tsunami along this coast at the same time as one along the upper East Coast of the United States would provide strong support for a major event in the Atlantic of the type expected from an asteroid impact.

TABLE 1. Deep-water Wave Characteristics from Atlantic Impactor

Location	Min	Max	Period
East Coast (Location 1)	–100	95	1000
East Coast (Location 2)	–100	100	1000
Florida (Location 3)	–95	30	1000
Gulf of Mexico (Location 4)	–5		
Portugal (Location 5)	–85	50	1200
England (Location 6)	–36	100	1500
Brazil (Location 7)	–100	50	1500

V. OBSERVATIONAL EVIDENCE FOR TSUNAMI FROM IMPACTORS

Very large tsunami have occurred. Deposits of unconsolidated corals hundreds of meters above sea level on the Hawaiian Islands of Lanai, Hawaii,

Oahu, Molokai, and Maui provide evidence of giant tsunami [12]. On Lanai they are found as high as 326 meters above sea level. A tsunami of similar height occurred in a fiord in Alaska in historical times. These occurrences show that there is no physics limiting large-scale tsunami at least 300 meters high. A tsunami at least 50–100 meters high appeared along the Texas coast after the Cretaceous-Tertiary impact [3].

Most searches for tsunami in the geological record have been done in the past few years, so it is likely that new evidence for them will appear at an increasingly rapid rate. It may be especially profitable to search for tsunami produced by impacts along the Atlantic coast which is less prone to earthquake-induced tsunami than is the Pacific. Geological (and perhaps archaeological) evidence for large tsunamis along the coasts of the major oceans (due to their large impact cross sections) may be the best counters for impacts of moderately large (R = 100–1000 m) asteroids.

VI. CONCLUSIONS

The atmosphere is ineffective in preventing impact damage to the ground when the diameter of a stony asteroid exceeds 200 meters. For iron meteorites that impact at greater than 20 km/s, the critical diameter is about 40–60 meters. These properties cause a threshold effect whereby stony asteroids less than 200 meters in diameter produce no significant ground (ocean) damage (but those larger than 60 meters in diameter can cause significant damage from airbursts [9]), while those larger than this value can cause catastrophic tsunami.

The growth of the height of the deep-water wave with increasing impact energy slows considerably when the crater depth becomes comparable to the depth of the ocean. This occurs at an impact energy of a few gigatons at a typical ocean depth. The probability is a few times 10^{-4} per year that an asteroid of sufficient size will impact an ocean on the Earth to produce tsunami with average heights exceeding 100 meters along the entire coast of the ocean.

REFERENCES

1. AHRENS, T.J. & J.D. O'KEEFE. 1983. Impact of an asteroid or comet in the ocean and extinction of terrestrial life. J. Geophysical Research Supp. **88:** A799–A806.
2. AHRENS, T.J. & J.D. O'KEEFE. 1987. Impact on the Earth, ocean, and atmosphere. Int. J. Impact Eng. **5:** 13–32.
3. BOURGEOIS, J., T.A. HANSEN, P.L. WIBERG, & E.G. KAUFFMAN. 1988. A tsunami deposit at the Cretaceous-Tertiary Boundary in Texas. Science. **241**: 567–570.
4. BRADDOCK, R.D. 1970. Tsunami propogation over large distances. *In* Tsunamis in the Pacific Ocean, W.M. Adams, Ed. : 285–303. East-West Center Press. Honolulu.

5. BRETSCHNEIDER, C.L. & P.G. WYBRO. 1977. Tsunami inundation prediction. *In* Proceedings of the Fifteenth Coastal Engineering Conference, Vol. 1: 1006–1024. Am. Soc. of Civil Engineers. New York.
6. COX, D.C. 1961. Effects of the May 1960 Tsunami in Hawaii and Other Polynesian Islands. *In* Proceedings of the Tsunami Meetings Associated with the Tenth Pacific Science Congress, D.C. Cox, Ed.: 87–95. University of Hawaii. Honolulu.
7. GLASSTONE, S. & P.J. DOLAN 1977. The Effects of Nuclear Weapons, 3rd Edit. U.S. Government Printing Office. Washington, D.C.
8. GAULT, D.E. & C.P. SONNETT. 1982. Laboratory simulations of pelagic asteroidal impact: Atmospheric injection, benthic topography, and the surface wave radiation Field. Geol. Soc Am Special Paper. **190:** 69–92.
9. HILLS, J.G. & M.P. GODA. 1993. The fragmentation of small asteroids in the atmosphere. Astronomical J. **105:** 1114–1144.
10. HILLS, J.G., I.V. NEMCHINOV, S.P. POPOV, & A.V. TETEREV. 1994. Tsunami generated by small asteroid impacts. *In* Hazards due to Comets & Asteroids: 779–790. Univ. Arizona Press. Tucson, Arizona.
11. IIDA, K. & Y. OHTA. 1961. On the heights of tsunami associated with distant and near earthquakes. *In* Proceedings of the Tsunami Meetings Associated with the Tenth Pacific Science Congress, D.C. Cox, Ed.: 105– 123. University of Hawaii. Honolulu.
12. JOHNSON, C. & D. KING. 1993. Can a Landslide Generate a 1000-ft Tsunami in Hawaii. Preprint.
13. MADER, C.L. 1988. Numerical Modeling of Water Waves. Univ. of California Press. Berkeley.
14. MADER, C.L. 1991. Modeling Hilo, Hawaii tsunami inundations. Science of Tsunami Hazards. **9:** 85–94.
15. MELOSH, H.J. 1989. Impact Cratering: A Geological Process. Oxford Univ. Press. Clarendon Press. Oxford. 246.
16. NEKRASOV, A.V. 1970. Transformations of tsunamis on the continental shelf. *In* Tsunamis in the Pacific Ocean, W.M. Adams, Ed.: 337–350. East-West Center Press. Honolulu.
17. RODDY, D.J., S.H.SHUSTER, M. ROSENBLATT, L.B. GRANT, P.J. HASSIG, & K.N. KREYENHAGEN. 1987. Computer simulations of large asteroid impacts into oceanic and continental sites-preliminary results on atmospheric, cratering, and ejecta dynamics. Int. J. Impact Eng. **5:** 123–135.
18. SCHMIDT, R.M. & K.A. HOLSAPPLE. 1982. Estimates of crater size for large-body impacts: Gravitational scaling results. Geol. Soc. of Amer. Spec. Paper. **190:** 93–101.
19. TAKASHASI, R. 1961. A summary report on the Chileantsunami of May 24, 1960 as observed along the coast of Japan. *In* Proceedings of the Tsunami Meetings Associated with the Tenth Pacific Science Congress, D.C. Cox, Ed.: 77–86. University of Hawaii. Honolulu.

On Kill Curves and Sampling Protocols: Studying the Relationships between Impact and Extinction

PETER D. WARD

University of Washington
Geology Department
Seattle, Washington 98195

The pioneering efforts of Raup (1990) have suggested that a relationship exists between crater diameter and percentage of organisms killed as a result of meteor or comet impact with the Earth. The new data (coming from study of the Manson and Chicxulub craters) suggest that the nature of target rock may be a factor nearly as important as impacter size, and that other aspects of the target, including its latitude, the atmospheric and climate conditions characterizing the Earth, as well as the stage of biological evolution and community development at the time of impact are factors which all must be factored into any new kill curve. It may be that no single "curve" is appropriate, but that a family of curves may be necessary to model the biological effects of large impacts.

We propose that a new protocol be developed to better constrain and understand the relationship between impact and extinction. Rather than searching known mass extinction boundaries for evidence of impact (an exercise which up to now has demonstrated that only the Chicxulub crater can be unambiguously related to a mass extinction of planetary scale), we propose that four known craters be investigated to see if they are temporally correlated with extinction at any detectable level. We suggest that Kara, Popigai, Manson, and Manicouagan Craters be investigated in the following way. First, what is their age? The Manson lesson is that the first step in understanding the relationship between impact and extinction is through reliable age dating. Second, can distal components of the impact ejecta (spherules, shocked quartz, and mineral signatures) be located from sedimentary record? Third, once identified, do these signatures coincide with paleontological or geochemical markers of extinction in either the synoptic literature, or from actual outcrops (or deep sea cores).

Cretaceous Habitats Before the End: The Record from 75,000,000-Year-Old Rocks in the San Juan Basin of New Mexico

DONALD L. WOLBERG

The Academy of Natural Sciences
1900 Benjamin Franklin Parkway
Philadelphia, Pennsylvania 19103

Because of the interest in extinctions, the end of the Cretaceous Period, and with it the end of the Mesozoic Era, some 65,000,000 years ago, has received a great deal of attention. In some ways, this focus has resulted in a biased view of the world of the Cretaceous; the Cretaceous spans a very long period of time, almost 80 myr, a span of time greater than that from the end of the Cretaceous to now. The Early Cretaceous record indicates a very different biota from that of the latest Cretaceous. A better balance of the richness of Cretaceous environments before massive extinctions occurred, and after the modernizing of Early Cretaceous floras and faunas, can be gleaned from the record at a handful of very special fossil localities. One such locality is known as the Fossil Forest of the San Juan Basin, northwestern New Mexico, an area of about 2500 acres of badland exposures developed in the Fruitland Formation. The rocks of the Fossil Forest consist of about 40 meters of exposed sandstones and siltstones, mudstones, and extensive coals. The area takes its name from the presence of several hundred in situ tree stumps and logs at five stratigraphic horizons, as well as three interbedded horizons with leaves; in all at least eight distinct forests are present. The trees represented include cypress, pines, redwoods, and palms. The leaf flora is more diverse and includes ferns, monocots, and dicots. Also contained within the rocks are freshwater invertebrates that includes the first record of insects in the Southwest.

During deposition of the Fruitland Formation in the area, the Fossil Forest was situated very near sea level; the eastern shoreline of the vast inland sea that occupied North America from the arctic to the Gulf of Mexico. The area was periodically inundated as shown by drowned forests and floated logs, mineral data, the presence of marine sharks, bivalves, and an ammonite. Mineral, sedimentological, and pollen data indicates multiple climatic shifts from warm, humid to warm, arid conditions. Analyses of fluid inclusions in resins found within sediments indicates that an atmosphere richer in oxygen than is now the case; 24%–30% as compared to the current 21% level. The area consisted of sandy flats near the mouth of a meandering, slowly moving river system in and near which amiids, gars, sharks and rays, turtles, crocodilians, other dinosaurs as well as freshwater bivalves and gastropods were present in abundance. Palm groves occupied better drained areas while plant material from a diverse flora accumulated in oxbows and poorly drained segments of the drainage system. Sharks and rays could be found in the sea and estuaries. Gars and paddlefish were in the freshwater reaches of the streams and rivers. Amphibians, lizards, crocodiles, and snakes are known, while turtles were present in abundance. A diverse mammalian fauna lived in the trees and undergrowth, and were largely nocturnal in habit. The mammals included insect eating animals as well as herbivores.

Dinosaurs are known from throughout the area; their bones and teeth are among the most common of fossils found. Dinosaur trackways indicate that these animals frequented areas at the shoreline; perhaps beaches provided easier access to different feeding areas than travel through dense forest. Dinosaur nesting areas with and without eggs are known. The dinosaur fauna from the Fossil Period includes several kinds of hadrosaurs, ankylosaurs, ceratopsians, coleurosaurs, and carnosaurs. Several different hadrosaurine dinosaurs, including *Edmontosaurus* and *Parasaurolophus* were present in abundance. A diversity of ceratopsians coexisted in the region and included *Monoclonius*, *Pentaceratops*, and *Torosaurus*. *Pentaceratops* is only known from the San Juan Basin region, indicating a certain degree of provinciality. Hadrosaurine dinosaurs nested near the sea as well as between drainages. Previous interpretations of nesting sites elsewhere hold that dinosaurs preferred upland habitats for nesting, away from the sea. The Fossil Forest discoveries indicate that nesting occurred very near sea level and require modification of this view. It may well be that eggs, nests, and infants are preserved in the rock record because of the proximity of fluctuating water levels.

Paleobiological Implications of Mass Extinctions

DAVID JABLONSKI

Department of Geophysical Sciences
University of Chicago
5734 South Ellis Avenue
Chicago, Illinois 60637

Survivorship patterns, or selectivity, might reveal much about both causes and long-term effects of extinction events. However, selectivity is manifest, if at all, at lower taxonomic levels (families and genera), biogeographically, or ecologically — and the relevant parameters are largely absent from synoptic databases. A global analysis of end-Cretaceous extinction in marine bivalves found that tropical settings outside of carbonate platforms suffered no greater losses than did extratropical faunas. Many of the other supposed selectivities (based largely on intuition or on extrapolation from background extinction) are also absent or more complex than expected for end-Cretaceous bivalves, including body size, position or breadth of distribution on the continental shelf, burrowing ability, and even feeding strategy. Only broad geographic distribution is significantly correlated with genus survival.

The evolutionary impact of mass extinctions is an active area of research. Many of the biotic replacements once thought to represent competitive victories over inferior lineages now appear to have been mediated by major extinction events, even though most species extinctions in the fossil record (probably >90%) occur outside of the five major extinction events. Mass extinctions have such profound biological consequences because they bite deep into standing diversity and disrupt background selection regimes, not because they account for most species terminations. Traits that favor survival during mass extinctions need have little correlation with those that enhance survival and diversification during background times, so mass extinctions can have unpredictable and lasting evolutionary effects. Not only do mass extinctions remove taxa and adaptations well-suited to the background regimes that represent the great bulk of geologic time, they create ecological and evolutionary opportunities by removing incumbent, dominant taxa and thus enable other groups to diversify in the aftermath of the extinction event. On the other hand, mass extinctions do not completely reset the evolutionary clock: many major evolutionary and ecological trends transcend even the Big Five events (e.g., the modernization of marine communities and the rise of flowering plants). We need to understand survivors and rebounds as well as victims and ecosystem collapse.

Asteroid Impacts and Mass Extinctions — No Cause for Concern

GERTA KELLER

Department of Geosciences
Princeton University
Princeton, New Jersey 08540

Asteroid impacts on Earth are relatively frequent with over a dozen craters known over the last 65 million years. With the exception of the largest at the Cretaceous/Tertiary boundary, none are associated with significant biotic extinctions. In addition, none of the major mass extinction episodes during the Paleozoic and Mesozoic eras can be directly linked to the effects of an asteroid impact. In fact, all major mass extinctions are marked by long-term declines in biotic diversity associated with climatic and environmental changes. These changes generally predate a bolide impact with no cause-and-effect relationship apparent.

This is also the case with the most famous mass extinction at the K/T boundary that is often attributed to a large bolide impact that may have hit Yucatan. This K/T boundary impact is popularly claimed to have caused the sudden mass extinction of most life on Earth from microorganisms to dinosaurs. The facts are different. There was a long-term decline during the last 2–3 myr of the Maastrichtian, with many organismal groups, such as inoceramids and rudistids, and ammonites declining long before the K/T boundary event. Climate cooled significantly during this time, reaching maximum cooling, and the sea level dropped by 70–100 m about 300 kyr before the K/T boundary. Between 100–200 kyr before the K/T boundary, climate warmed by 3–4°C, accompanied by a rise in sea level. This greenhouse warming was probably caused by the tremendous volcanic Deccan Trap eruptions in India. These climatic changes significantly stressed marine organisms, causing major population turnovers and some species extinctions. The greenhouse warming was terminated by rapid cooling of 2–3°C during the last 50–100 kyr of the Maastrichtian and followed by a sudden event, a bolide impact, at the K/T boundary.

The biotic effects of this impact were selective and limited to low latitudes. Among marine microorganisms, the planktic foraminifera were hardest hit, with 35 to 50% of 65 species extinct. All of the extinct species were highly specialized tropical-subtropical forms that were already highly stressed and endangered species during the preceding environmental changes. Their combined total abundance was less than 15% of the entire population. All cosmopolitan and environmentally tolerant species survived into the lower Paleocene. Among other fossil groups, no major extinctions occurred among ostracods, radiolarians, diatoms, and dinoflagellates, which comprise the bulk of the biomass and the base of the marine food chain. In high latitudes no mass extinctions occurred in any of the fossil groups including planktic foraminifera. Instead, a gradual long-term turnover is observed. This lack of a major sudden negative effect is also apparent in the absence of a $\delta^{13}C$ shift in high latitudes.

These data indicate that climatic and sea level changes during the last 300 kyr of the Maastrichtian were the major cause for declining biotic diversity, but that a bolide impact may have caused the extinction of the already endangered high-risk tropical and subtropical species.

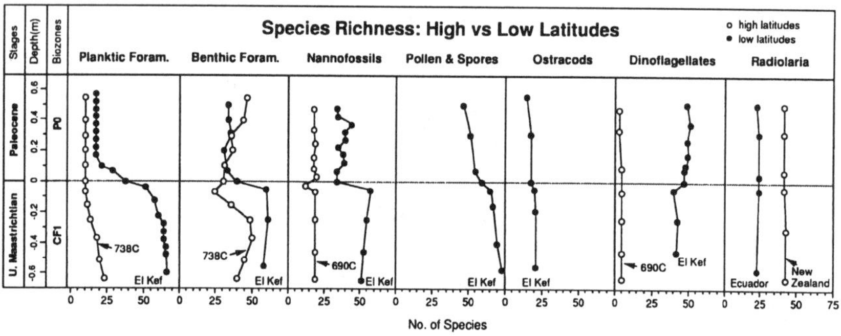

FIGURE 1. Species richness patterns across the K/T boundary in marine microfossils of high and low latitudes. Note that the major changes are apparent only in planktic foraminifera and nannofossils and only in low latitudes. The benthic foramininiferal species decline is a temporary emigration of species, most of which reappear in the Danian.

Environmental Perturbations Caused by the Impacts of Asteroids and Comets

OWEN B. TOON,[a] KEVIN ZAHNLE,[a] DAVID MORRISON,[b]
RICHARD P. TURCO,[c] AND CURT COVEY[d]

[a]*NASA AMES*
MS 245-3
Moffett Field, California 94035

[b]*NASA AMES Research Center*
NASA AMES
MS N245-1
Moffett Field, California 94035

[c]*University of California*
Los Angeles, California 90095
and
[d]*Lawrence Livermore Laboratory*
Livermore, California 94550

We review the major mechanisms proposed to cause extinctions at the Cretaceous-Tertiary geological boundary following an asteroid impact. We then discuss how the proposed extinction mechanisms may relate to the impact of asteroids or comets in general. We discuss the limitations of these mechanisms in terms of the spatial scale that may be affected, and the time scale over which the effects may last. Our goal is to provide relatively simple prescriptions for evaluating the importance of colliding objects having a range of energies and compositions. We also identify the many uncertainties concerning the environmental effects of impacts. We conclude that, for impact energies below about 10^4 Mts (megatons of TNT equivalent) — i.e., impact frequencies less than one in 6×10^4 yr, corresponding to comets and asteroids with diameters smaller than about 400 m and 650 m, respectively — blast damage, earthquakes, and fires should be important on a scale of 10^4 or 10^5 km^2, which corresponds to the area damaged in many natural disasters of recent history. However, tsunami could be more damaging, flooding a kilometer of coastal plane over entire ocean basins. In the energy range of 10^4 to 10^5 Mts (intervals up to 3×10^5 yr; comets and asteroids with sizes up to 800 m and 1.5 km, respectively) water vapor injections and ozone loss become significant on the global scale. If the submicrometer dust injection fraction from the pulverized target material is much higher than is presently thought to be most likely, then dust injection could also be important in this energy range. This energy range is a conservative lower limit where damage might occur beyond the experience of human history. The energy range from 10^5 to 10^6 Mts (intervals up to 2×10^6 yr; comets and asteroids up to 1.8 and 3 km diameter) is transitional between regional and global effects. The dust lifted in this energy range, the sulfur released from within impacting asteroids, and the soot from fires started by comets can produce climatologically significant global optical depths on the order of 10. Moreover, the ejecta plumes of these impacts may produce enough NO to destroy the ozone shield. Between 10^6 and 10^7 Mts (intervals up to 1.5×10^7 years; comet and asteroid diameters up to 4 and 6.5 km, respectively) dust and sulfate levels would be high enough to reduce light levels below those necessary for photosynthesis. Ballistic ejecta reentering the atmosphere as shooting stars would set fires over regions exceeding 10^7 km^2 and the resulting smoke will reduce light levels even further. At energies beyond 10^7 Mts blast and earthquake damage reach the regional scale (10^6 km^2). Tsunami cresting to 100 m and flooding 20 km inland could sweep the coastal zones of one of the world's ocean basins. Fires would be set globally. Light levels may drop so low from the smoke, dust, and sulfate that vision is not possible. At energies approaching 10^9 Mts the ocean surface waters may be acidified globally by sulfur from the interiors of comets and asteroids. The Cretaceous-Tertiary impact, in particular, involved evaporate substrates that very likely generated a dense widespread sulfate aerosol layer with consequent climatic effects. The combination of all of these physical effects would surely represent a devastating stress on the global biosphere.

A Unified Theory of Impact Crises and Mass Extinctions: Quantitative Tests

MICHAEL R. RAMPINO,[a,b] BRUCE M. HAGGERTY,[a] AND [c]THOMAS C. PAGANO

[a]*Earth and Environmental Science Program*
New York University
26 Stuyvesant Street
New York, New York 10003

NASA
[b]*Goddard Institute for Space Studies*
2880 Broadway
New York, New York 10025

and

[c]*Department of Physics*
New York University
New York, New York 10003

ABSTRACT: Several quantitative tests of a general hypothesis linking impacts of large asteroids and comets with mass extinctions of life are possible based on astronomical data, impact dynamics, and geological information. The waiting times of large-body impacts on the Earth derived from the flux of Earth-crossing asteroids and comets, and the estimated size of impacts capable of causing large-scale environmental disasters, predict that impacts of objects ≥ 5 km in diameter (≥ 10^7 Mt TNT equivalent) could be sufficient to explain the record of ~25 extinction pulses in the last 540 Myr, with the 5 recorded major mass extinctions related to impacts of the largest objects of ≥ 10 km in diameter (≥ 10^8 Mt events). Smaller impacts (~10^6 Mt), with significant regional environmental effects, could be responsible for the lesser boundaries in the geologic record.

Tests of the "kill curve" relationship for impact-induced extinctions based on new data on extinction intensities, and several well-dated large impact craters, also suggest that major mass extinctions require large impacts, and that a step in the kill curve may exist at impacts that produce craters of ~100 km diameter, smaller impacts being capable of only relatively weak extinction pulses. Single impact craters less than ~60 km in diameter should not be associated with detectable global extinction pulses (although they may explain stage and zone boundaries marked by lesser faunal turnover), but multiple impacts in that size range may produce significant stepped extinction pulses.

Statistical tests of the last occurrences of species at mass-extinction boundaries are generally consistent with predictions for abrupt or stepped extinctions, and several boundaries are known to show "catastrophic" signatures of environmental disasters and biomass crash, impoverished postextinction fauna and flora dominated by stress-tolerant and opportunistic species, and gradual ecological recovery and radiation of new taxa. Isotopic and other geochemical signatures are also generally consistent with the expected after-effects of catastrophic impacts. Seven of the recognized extinction pulses seem to be associated with concurrent (in some cases multiple) stratigraphic impact markers (e.g., layers with high iridium, shocked minerals, microtektites), and/or large, dated impact craters. Other less well-studied crisis intervals show elevated iridium, but well below that of the K/T spike, which might be explained by low-Ir impactors, ejecta blowoff, or sedimentary reworking and dilution of impact signatures.

The best explanation for a possible periodic component of ~30 Myr in mass extinctions and clusters of impacts is the pulselike modulation of the comet flux associated with the solar system's periodic passage through the plane of the Milky Way Galaxy. The quantitative agreement between paleontologic and astronomical data suggests an important underlying unification of the processes involved.

INTRODUCTION: A WORKING HYPOTHESIS

The recent collision of the fragments of Comet Shoemaker-Levy 9 with Jupiter focused attention on the problem of large-body impacts on the planets. Had these comet fragments hit the Earth, the results would surely have been mass devastation. The impact of a large comet or asteroid ~10 km in diameter at the Cretaceous/Tertiary (K/T) mass extinction boundary (65 Myr) is now well-established, and the removal of the old conceptual block of gradualism and growing evidence for large-body impacts in Earth history are pointing to a new paradigm in the Earth sciences in which "catastrophes" play an important part. Paleontologists recognize five major mass extinctions, equal to or greater in extinction magnitude than the K/T extinction, and about 20 other clearly identifiable peaks of extinction above the background during the past ~540 Myr [102]–[104] (Fig. 1). These extinctions involved both marine life and nonmarine organisms [72], [25], [53], [7] suggesting global environmental disasters as a cause. Although the method of "multiple working hypotheses" has a long history in geology, the appearance of a particularly strong hypothesis, such as the impact hypothesis, prompts us to consider its efficacy as a general explanation for mass extinctions of life.

Studies relating to the K/T event have come to the conclusion that the impact of an asteroid or comet ≥ 10 km in diameter (releasing ≥ 10^{24} J, 10^8 Mt TNT) would cause a global catastrophe of enormous proportions (e.g., [16], [117]). Such a large impact is predicted to result in major global environmental disasters primarily related to production of dense clouds of fine ejecta, and surface fires and smoke clouds from intense heating caused by reentry of ejecta into the Earth's atmosphere [56], [16]. Atmospheric shock waves from a large impact are calculated to create large amounts of nitric oxides, possibly producing nitric acid rain and NO_X in the stratosphere that would rapidly destroy the ozone layer. Other disastrous environmental effects of large impacts have been proposed that are "target-sensitive." These include enhanced greenhouse effect from water vapor injected into the atmosphere in an ocean impact [22], or from CO_2 released by impact into carbonate rocks [63], and the possible creation of large amounts of H_2SO_4 aerosols, and thus a marked cooling and acid rain, if the target rocks contained deposits of $CaSO_4$ [109], [70].

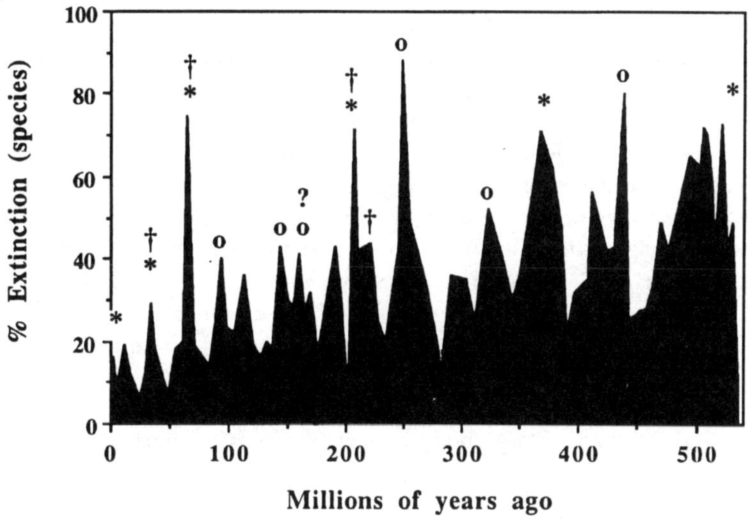

FIGURE 1. Percent extinction of marine species per geologic stage (or substage) during the Phanerozoic. Culled genus-level data [103], [104], converted to species level using reverse rarefaction techniques [85]. The following maxima are recognizable in Sepkoski's data. (Note that the formal end of stage in Myr, as dated by recently published time scales, may not be exactly coincident with extinctions): Pliocene (2.3), Mid-Miocene (11.2), Upper Eocene (34), Maastrichtian (65), Cenomanian (93.5), Aptian (112), Tithonian (144), Callovian (159), Pliensbachian (190), Norian (206), Carnian (221), Tatarian (248), Guadelupian (250), Stephanian (290), Serpukhovian (322), Famennian (363), Frasnian (367), Eifelian (380), Ludlovian (411), Ashgillian (439), Llanvirnian (469), Tremadocian (493), Trempeleauean (505), Franconian (508), Botomian (520), and possibly the Proterozoic/Cambrian (~540).

The earliest (540–493 Myr) extinction peaks may be anomalously high as a result of the relatively poor record of diversity: diagnostic stratigraphic evidence of impact (asterisks), possible stratigraphic evidence of impacts (open circles), and large, dated impact craters (crosses) (see text).

Astronomical observations of Earth-crossing asteroids and comets allow estimates of the expected mean times between collisions of objects of various sizes with the planet [107], [16]. On the large-diameter end of the size spectrum, the astronomical data and cratering records of the inner planets predict that in an ~100 Myr period, the Earth should be hit by several asteroids and comets larger than a few km in diameter (these release ~10^{23} J or 10^7 Mt TNT equivalent) and perhaps one ≥ 10 km diameter (≥ 10^{24} J, 10^8 Mt) object (Fig. 2) [107], [16].

Calculations suggest that the atmospheric dust and smoke, and the reentry heating effects of a ≥ 10^8 Mt impact would cause a considerably more severe global mass extinction event than an ~10^7 Mt impact [56], [117], but that a 10^7 Mt impact could cause environmental changes leading to a lesser, but still significant extinction event. With these considerations in mind, we can utilize the astronomical observations to predict a record of impact-induced extinction pulses during the Phanerozoic Eon (the last 540 Myr). The expected impact history predicts ~5 major mass extinctions, and about 25 ± 5 less severe pulses of extinction. This agrees surprisingly well with the record of extinctions of the last 540 Myr, which shows the 5 major and ~20 minor extinction pulses (Fig. 1). The agreement of these two independent data sets, one derived from paleontology and the other from astronomy, is impressive. The result suggests that the record of extinction events could be explained by comet and asteroid impacts. Further tests of a general relationship between extinctions and large-body impacts are possible with available data.

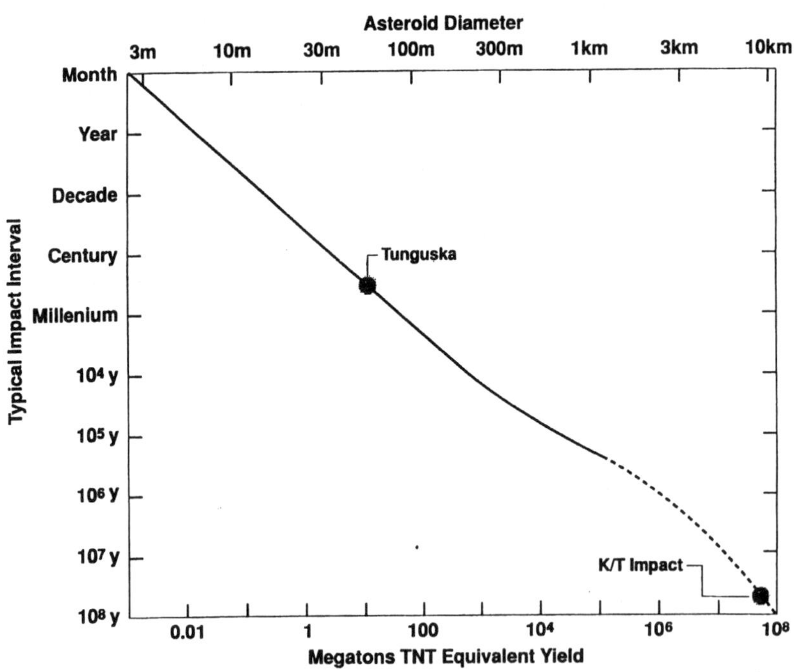

FIGURE 2. Estimated frequency of impacts on the Earth from the present population of comets and asteroids, and evidence from impact craters. The megaton TNT equivalents of impact energy are shown (after [16]).

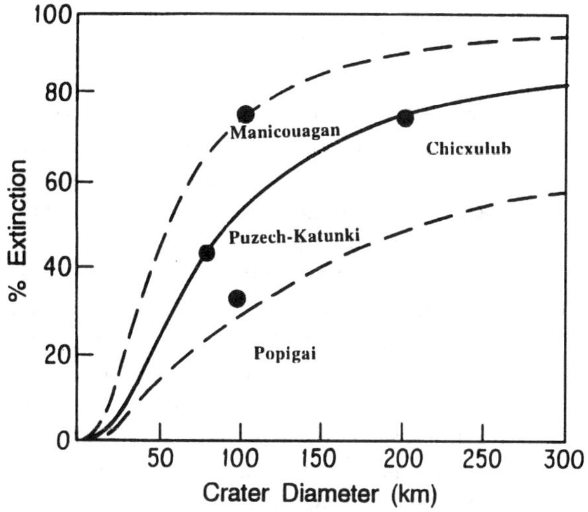

FIGURE 3. Kill curve for Phanerozoic marine species, with estimated error bars, plotted against estimated size of impact craters associated with extinctions of various kill magnitudes, assuming that the two are actually related [89], [90]. Estimated sizes of the largest craters with dates overlapping mass extinctions (from Table 1), are plotted against the species kill magnitude of those extinctions.

THE KILL CURVE: A QUANTITATIVE TEST

Natural phenomena (e.g., earthquakes, volcanic eruptions, floods) commonly show a variation in intensity and frequency in which lower intensity events are more common than high intensity events [35]. One can represent this variation by noting the average time interval between events of a given magnitude. Extinctions of life may be viewed this way, and Raup [89]–[91] constructed a species "kill curve" using extinction data for the last 540 Myr by determining the mean waiting times between extinctions of various severity. If the impact of a large asteroid or comet can produce extinctions, then a first-order relationship should exist between the impactor size and the number of species killed. Ward [133] recently pointed out that this quantitative kill curve concept represents, "one of the most powerful to emerge from the entire extinction debate."

Raup [89] combined the extinction data with information on the waiting times of terrestrial impact craters of various sizes to construct a possible theoretical relationship between mass extinctions and impact cratering (Fig. 3). He proposed that the relationship would follow a "growth curve" form, where small events might do relatively little damage, whereas large events would render the system so unstable as to cause mass extinctions, reaching a saturation for very large impacts, where the remaining few percent (%) of living species would be difficult to kill off. The form of the equation used is:

$$\% \text{ genera killed } = \frac{(2.7\ln D - 4.1)^5}{[e^{0.5} + (2.7\ln D - 4.1)^5]}, \tag{1}$$

where D = crater diameter in km. These data were transformed from % genera killed to % species killed using techniques developed by Raup [85].

Thus, Raup developed one of the first quantitatively testable relationships pertaining to the causes of mass extinction. Using the known rate of impacts, and the hypothesized relationship between impacts and mass extinctions, Raup [88] performed Monte Carlo simulations to determine the appearance of the extinction record supposing that impacts are sufficient to explain the entire record of the past 250 Myr. The results of these simulations compared well with the actual extinction record based on several criteria, including the severity of extinction events, the spacing of geologic stages, and the general level of background extinction. These calculations support the idea that the impact of a ~5 km diameter asteroid would be sufficient to cause extinction pulses clearly recognizable above the background rate.

Although the severity of impact-induced mass extinctions would probably depend on a number of variables (e.g., ambient climate conditions, susceptibility of fauna and flora, site of impact, etc.), the size and energy of the impactor is likely to be the most important factor. Recently, new data have become available as to the size and ages of impact craters [29], [30] and the severity of the mass extinctions [103], [104]. The extinction data generally show higher percentages of extinction per stage and substage than previous data. The theoretical curve can be compared with data representing specific large (> 80 km diameter) known impact craters with relatively well-defined ages that overlap the ages of mass-extinction boundaries (when the full dating uncertainties in both are taken into consideration) (Table 1). No other large, well-dated Phanerozoic craters are known. The observed points in Fig. 3 agree with the predicted curve within the envelope of error permitted by the geologic data, supporting at least a first-order relationship.

TABLE 1. Large Impact Craters [30], [139], [129] and Correlative (?) Extinctions [103], [104, data]. In cases where several impact craters have similar ages, we plotted the largest crater.

Name	~Diameter(km)	Age (Myr)	Extinction	% Genera	% Species
Puzech-Katunki	80	220 ± 10	Carnian	20	42
Popigai	100	35.7 ± 0.8	Late Eocene	12	30
Manicouagan[a]	100	212 ± 2	Late Triassic	47	76
Chicxulub	200?	65.2 ± 0.4	K/T	47	76
Duolon (?)	~100	~136	J/K	20	42
Taihu (?)	~70	~365??	Late Devon	41	70

[a]Manicouagan might correlate with the Carnian extinction event.

The shape of the species kill curve predicts that for single craters smaller than ~60 km in diameter there will be no associated extinction pulse that stands above the ~20–25% background level of % species extinction (e.g., [88], [39]), and indeed there are well-dated craters in this size range that apparently do not correlate with the local extinction maxima in Sepkoski's data, but may correlate well with geologic stage boundaries that are based on faunal turnover [84], [116].

Step Function in the Kill Curve?

The points in Fig. 3 could be interpreted as indicating a possible step-up in the kill curve at approximate crater sizes of ≥ 100 km diameter (~10^7 Mt of energy release), suggesting some kind of threshold effect(s). Raup [86] provided some evidence for an extinction threshold based on computer simulations of asteroids of different sizes hitting the Earth and producing "lethal radii" within which all life dies out. Using a definition of a mass extinction as elimination of at least ~20% of marine genera, a lethal radius of ≥ 10^4 km is required, or about 50% of the surface of the Earth. Anything less would be lost in the background of genera extinctions, but could cause enough faunal turnover to be represented in the geologic time scale as a stage or a zone boundary [88], [116]).

Using the calculations of Toon et al. [117], we estimate that the threshold impact energy required to cause global-scale wildfires through intense heating from ballistic ejecta re-entering the atmosphere is ~7×10^7 Mt (rise of atmospheric temperature above the ignition point over 50% of the Earth's surface). Their graphs also suggest that an impact releasing ~1.5×10^8 Mt could reduce global atmospheric transmission below the limit of photosynthesis for several months (optical depths ≥ 10 over ~50% of the Earth's surface), although other workers have suggested that smaller impacts could produce such conditions [28]. Model calculations predict that land-surface temperatures would decrease by ~15°C in less than a week under such conditions [69], [21].

Using the impactor energy/crater size relationship of Shoemaker et al. [107], the thresholds listed above from the work of Toon et al. [117] would be equivalent to crater diameters of ~135 (global wildfires) and ~167 km (global darkness), respectively. This is somewhat higher than the threshold of ~100 km diameter suggested by the available data in Fig. 3, but probably within the errors of the parameters and calculations involved.

Regional Effects of Smaller (~10^6 Mt) Impacts: The Manson Impact

Despite the growing acceptance of the connection between the end-Cretaceous mass extinction and the ~10^8 Mt impact that produced the ~200 km Chicxulub structure, the geologic community still lacks a general appreciation of the pervasive effects of smaller and more numerous impacts. Impacts involving energies of ~10^6 Mt should occur about every million years, or ~600 such impact during the Phanerozoic. About 30% of these impacts (~200) would have hit the continents, which means that North America (representing about 15% of total continental area) would have suffered about 30 such impacts, or approximately one every 20 million years.

A good example of the regional environmental effects of such an impact is the 35 km diameter Manson impact structure in northwest Iowa, which has been recently well-dated by $^{40}Ar/^{39}Ar$ in sanidine breccia clasts as having formed 73.8 ± 0.3 Ma [38]. A search for Manson distal ejecta in Late Cretaceous marine rocks to the west led to the discovery of shocked minerals in the Crow Creek member of the Pierre Shale in southeastern South Dakota, which has been interpreted as a possible tsunami deposit [38], [114].

Using the formula of Shoemaker et al. [107], the ~35 km diameter Manson crater is estimated to have resulted from an ~10^6 Mt impact, having a mean waiting time of a few Myr (Fig. 2). Adushkin and Nemchinov [1] give a scaling law for radii of severe shock devastation (~4–5 psi overpressure) around impacts as $R_s = 8E_k^{1/3}$, where R_s is in km, and E_k is the kinetic energy of the impactor in megatons (see also [16]). For the very large energies involved in large impacts, this might be an underestimate of destructive area, as the dynamic pressure, which defines blast damage due to drag force associated with strong winds that accompany passage of the blast wave, also increases, with wind velocities ≥ 260 km/hr.

The estimated scaling law for fire ignition is $R_f = 3E_r^{1/2}$, where R_f is in km and E_r is the thermal radiation energy in megatons (the ratio of thermal radiation to kinetic energy for an impact is ~20%). For large impacts, the shock wave deteriorates into an acoustical wave, and hence the shock wave would not blow out the impact-induced fires at large distances from the impact. According to [1], for an ~10^6 Mt impact event such as Manson, the radius of severe devastation would be ~800 km, whereas the radius of fire ignition would be ~1350 km, although recent work suggests smaller radii of destruction [149]. The radius of severe seismic effects of a 10^6 Mt impact (and the resulting $M = 9$ earthquake) is estimated at about 1,000 km [1].

The time of the Manson impact was associated with the disappearance of nine genera of dinosaurs and the rapid evolution of other dinosaur taxa in western North America [110], and major changes in marine vertebrates and mammals in the same region [98]. Catastrophic dinosaur bone beds, some containing entire herds of Hadrosaurids, occur in Campanian rocks dated at about 74 Myr in the Two Medicine Fm. in Montana, ~1,400 km from Manson [97], [123], [122].

In the San Juan Basin of New Mexico, at the Late Campanian "fossil forest" site (dated at between ~73.5 and 74.5 Myr), ~1,300 km from the Manson impact site, trees were killed and buried in place by some sudden event [95], possibly fire and subsequent floods related to the Manson impact. The overlying sediments contain a profusion of fossil mammals, possibly postdisaster opportunists.

The cumulative effects of multiple smaller impacts may also be important in producing some of the minor extinction peaks seen in Fig. 1. For example, during the time interval of the Jurassic/Cretaceous (J/K) boundary (~45% species extinction, ~142 to 144 Myr), at least two impact craters in the 20 to 40 km diameter size range are known, the Mjølnir structure (~40 km) in the Barents Sea and the Gosses Bluff structure (~22 km) in Australia. These impacts each represent ~10^6 Mt explosions that may be expected to have caused significant regional disasters, and some possible global environmental effects. The ejecta layer from the Mjølnir structure can be traced regionally, and is composed of a basal mud-flake conglomerate fining upward into sandy claystones containing shocked quartz grains and an Ir anomaly of up to ~1ppb with chondritic siderophile element ratios [24]. The ejecta lies ~3 m below the occurrence of ammonites of the basal Cretaceous stage of the Boreal realm, and hence is close to the Jurassic/Cretaceous boundary.

EVIDENCE OF LARGE IMPACTS

Impact Ejecta at or Near Extinction Boundaries

The discovery of the K/T impact layer prompted the search for impact signatures at other geological boundaries (see [43], [64]). Among the materials considered diagnostic of impact are shocked minerals (including shocked quartz, stishovite, shocked zircons, etc.), impact glass (microtektites/tektites), microspherules with structures indicating high-temperature origin, and Ni-rich spinels. Shocked minerals and tektite glass are quite rare in the geologic record (e.g., [100]), but these materials have been reported in stratigraphic horizons, regionally to globally, near the times of six recorded extinction pulses (Late Pliocene, Late Eocene, Cretaceous/Tertiary, Jurassic/Cretaceous, Triassic/Jurassic, and Late Devonian) (Table 2). A recent preliminary report of shocked quartz from the Permian/Triassic boundary in Antarctica and Australia [94] is in agreement with an earlier report of shocked quartz and stishovite at the P/Tr boundary in China [148].

Other kinds of material related to large impacts are coarse near-source ejecta deposits, mostly ballistic debris flows and mass flow deposits of various kinds, and tsunami and related wave-induced deposits. These deposits have been recognized regionally for the K/T impact at Chicxulub (see [52], [112], [61], and for a few other impact events (e.g., [38], [134], [77], but the vast majority of such impact-related sediments remains unrecognized in the geologic record or misinterpreted as tillites, tectonic megabreccias, incised valley fills, turbidites, or other kinds of deposits (e.g., [73]).

TABLE 2. Stratigraphic Evidence of Impact Debris at/or near Extinction Boundaries

Age	Evidence	References
Pliocene (2.3 Myr)	Microkrystites, microtektites	[43], [49], [67]
Late Eocene (~36 Myr)	Microtektites (multiple), tektites, microspherules, shocked quartz	[58], [20], [57]
Cretaceous/Tertiary (65 Myr)	Microtektites, tektites, shocked minerals, stishovite, Ni-rich spinels	[2], [3]
Jurassic/Cretaceous (~142 Myr)	Shocked quartz	[24]
Triassic/Jurassic (~206 Myr)	Shocked quartz (multiple?)	[10]
Late Devonian (~368–365 Myr)	Microtektites (multiple)	[17], [18], [125]

Iridium Anomalies

In the search for impact-related iridium anomalies, the K/T boundary has been used as the standard of reference [3]. An iridium anomaly in the ppb to tens of ppb range (hundreds of times the normal background iridium in sediments derived mostly from normal meteoroid ablation) is globally well documented at the K/T boundary. Criteria for impact signature has thus consisted of high (ppb) levels of iridium, and chondritic or near-chondritic element ratios. Kyte [43] argued, however, that the very high Ir concentrations at the K/T boundary anomaly should not be considered typical of stratigraphic iridium anomalies from impacts. For example, the K/T impact at Chicxulub may have been anomalous in several respects. It hit unusual target rock consisting of a carbonate and evaporite platform, and the impact may have been at an unusually low angle [101].

In fact, the search for iridium has resulted in reports of elevated iridium levels (commonly ≥ 10 times background values) at or near a number of extinction boundaries (Table 3). The Ir levels are generally significantly weaker (100s of ppt) than the K/T iridium anomaly, and the anomalies are commonly associated with nonchondritic element abundance patterns [43], [64], [126], [131]. This has led to the general conclusion that the Ir peaks are probably unrelated to impact processes [64], although Kyte [43] listed six such sedimentary horizons as "possible" impact horizons (the Cenomanian/Turonian, Callovian/Oxfordian, Early-Middle Jurassic, Permian/Triassic, Late Devonian, and Proterozoic/Cambrian). Studies also suggest that elevated iridium levels are uncommon in the geologic record away from boundaries [45], [5], [99].

TABLE 3. Elevated Iridium Reported at Geologic Boundaries (Results widely considered questionable marked by ?)

Locality	Ir (ppt)	References
	Pliocene (2.3 Myr)	
Core EL13-3, southeast Pacific	up to 5,000	[43]
	Late Eocene (36 Myr)	
Various localities	~4,000	[27]
	≤ 330	[59]
	Cretaceous/Tertiary (65 Myr)	
Worldwide	1,000s ppt	[3]
	Cenomanian/Turonian (92 Myr)	
Western US, Colombia, S.A., Western, Europe	≤ 560	see [64], [65]
	Jurassic/Cretaceous (~142 Myr)	
Central Siberia	ave. 7,800	[147]
Barents Sea	1,000	[24]
	Callovian/Oxfordian (~163 Myr)	
Spain and Poland	1,000–2,400?	[12]
	Middle-/Late Jurassic (~165 Myr)	
Italy	3,200	[96]
	Triassic/Jurassic (~206 Myr)	
Europe	≤ 400	see [54]
	Permian/Triassic (~250 Myr)	
Changxing, China	8,000?	see [64]
Meishan, China	600 ± 400?	[140]
Meishan, China	2,000?	[141]
Nammal, Pakistan	366	[140]
Bolzano, Italy	230	[140]
San Antonio, Italy	3,000?	[11]
Tesero, Italy	135	[62]
Casera Federata, Italy	100–145	—
Butterloch, Austria	90–95	—
Carnic Alps, Austria	165	[6]
	230	—
Lalung, India	73	[9]
	114	—
	Mississippian/Pennsylvanian (~320 Myr)	
Texas	380	see [64]
	Devonian/Carboniferous (~360 Myr)	
Oklahoma	560	see [64]
S. China	150	[15]
	Late Devonian (~368–365 Myr)	
China	230	[131]
Australia	300	see [64]
Europe	75–160	see [64]
	Ordovician/Silurian (~438 Myr)	
Anticosti Island, Canada	58	see [64]
Scotland	≤ 250	see [64]
China	≤ 230	[127], [128], [132]
	Proterozoic/Cambrian (~540 Myr)	
China	2,900	see [64], [140]

It is also noteworthy that diagnostic evidence of impacts (shocked minerals and/or microtektites) has been reported at some boundaries associated with relatively weak iridium anomalies — the Late Eocene (microtektites, microspherules, shocked quartz), Triassic/Jurassic (shocked quartz), Late Devonian (microtektites), and possibly the Late Permian (shocked quartz?) extinction boundaries. Microtektite layers in the Late Cenozoic are commonly associated with only background levels of Ir, or mild Ir anomalies [42].

Iridium and Impactor Size

Melosh [55] recently proposed that iridium concentrations at the Cretaceous/Tertiary (K/T) boundary provide the best way to resolve the debate surrounding the size of the Chicxulub impact crater [32]. Two factors, however (1) the great range in the iridium content of different types of extraterrestrial bodies, and (2) the tendency for large impacts to lose ejecta by atmospheric recoil, make it problematic to accurately estimate impact crater size from stratigraphic Ir anomalies, and raise a larger question regarding the size of Ir anomalies expected from catastrophic impacts.

Iridium concentration among known meteorites varies from $\sim 5 \times 10^{-3}$ ppb (achondrites and eucrites) to 10^5 ppb (some irons) [66]. Assuming a simple model for iridium distribution in impact ejecta, the mass M (in g) of the impactor is given by:

$$M = \sigma A / f_1 f_2, \tag{2}$$

where σ is the surface density of Ir (measured at Gubbio to be 8×10^{-9} g/cm^2), A is the area of the Earth, f_1 is the iridium mass fraction of the impactor, and f_2 is the estimated fraction of the impactor lofted to stratospheric height (following Alvarez et al. [2], we set $f_2 = 0.2$ for all impacts). Alvarez et al. [2] argued for a carbonaceous chondritic (CI) composition for the K/T impactor ($\sim 0.5 \times 10^{-3}$ ppb Ir), but others have suggested that element-abundance patterns at the boundary can be used to argue for various impactors ranging from metal-sulfide objects to comets (e.g., [45]). Although some workers believe that most large craters are produced by comets, impactor compositions of all major types have been tentatively identified using data from impact melt rocks [30].

Alvarez et al. [2] calculated a K/T impactor mass of $\sim 3.4 \times 10^{17}$ g. Within the range of iridium concentrations for meteorites given above, however, and based on (1), the mass of the impactor producing an iridium anomaly like that found at the K/T boundary could have been as low as 2×10^{15} g (iron). If the impactor was a comet (with a nominal composition of 50%

chondritic and 50% ice), the distributed Ir anomaly suggests a body with a mass of ~5 × 10^{17} g.

The conversion from projectile mass to crater size was computed using the formula from Shoemaker et al. [107]:

$$D = 0.074c_f(g_e/g)^{1/6}\,(W\rho_a/\rho_t)^{1/3.4}, \tag{3}$$

where D is the crater diameter, c_f the crater collapse factor (1.3 for craters larger than 4 km diameter on Earth), g_e the gravitational acceleration at the surface of the Earth, g the acceleration at the surface of the body on which the crater is formed (in this case, the Earth), W is the kinetic energy of the impacting body in kilotons TNT equivalent, ρ_a is the density of the impactor (ranging from 1.8 g/cm^3 for a comet to 7.3 g/cm^3 for an iron body), and ρ_t is the density of the target rock (for which we used 2.7 g/cm^3).

Typical impact velocities for asteroids are ~20 km/s vs ~60 km/s for long-period comets. The ranges of possible values for impactor velocity and iridium content imply a lower bound on the diameter of the crater of ~22 km (for an iron asteroid). A carbonaceous chondrite asteroid would produce a crater ~94 km in diameter, whereas for an impactor of cometary composition and velocity, a crater diameter of ~150 km is predicted.

Vickery and Melosh [124] performed similar calculations, based on chondritic and cometary Ir concentrations, and with a model incorporating iridium loss from the Earth due to atmospheric recoil. They estimated a lower diameter of 37 km, and an upper limit of 160 km for the K/T impact crater. Results differ because of assumed f_2 values from 0.25 to 1.0, differences in crater scaling, and their atmospheric blow-off impact model predicts that for impacts producing craters larger than ~180 km, too much Ir would be lost from the Earth to account for the K/T boundary Ir spike.

These results suggest that the uncertainties in impactor composition and velocity, and loss of ejecta by atmospheric recoil, are far too large to allow accurate estimates of crater diameter from stratigraphic iridium anomalies. Impacts of ice-rich long-period comets, capable of causing severe environmental disturbances, will produce significantly lower amounts of iridium (possibly as little as 1/100 Ir per Mt) than asteroid impacts [120], and thus might produce only relatively weak Ir spikes in the geologic record. Furthermore, the primary carrier of the iridium (e.g., microspherules) may be inhomogeneously mixed in the boundary layer [141], and bioturbation, reworking, and diagenesis of sediments might dilute Ir concentrations [54], [128]. Hence, geologic boundaries that show relatively small (100s of ppt) iridium peaks (Table 3) should be carefully reexamined for other evidence of possible impacts.

CHARACTERISTICS OF CATASTROPHIC EXTINCTIONS

Mass Extinctions: Sudden vs. Gradual?

There has been much debate about the nature of the fossil record at mass extinction boundaries such as the K/T (e.g., [110], [105]). The primary biostratigraphic questions regarding mass extinctions involve discriminating simultaneous (sudden) events from gradual or stepwise extinctions, and determining whether a stratigraphic gap between the last occurrence of a group of fossils and a major boundary actually indicates that the organisms died out before the boundary event [87]. A basic problem in using biostratigraphic last occurrences to infer patterns of extinction, however, is that last occurrences almost always underestimate the time of extinction. The geologic record is almost always undersampled. Signor and Lipps [108], pointed out that the vagaries of preservation and sampling could cause truncations of species ranges below the true level of extinction, and thus tend to make abrupt extinctions appear gradual.

Recent tests of the effects of sampling on faunal occurrence of vertebrates and invertebrates at the K/T boundary (e.g., [87], [105], [50], [133] and at other boundaries [14], [74] have concluded that sections which seem to show gradual extinction could be explained by the Signor-Lipps effect and other sampling problems. Detailed studies utilizing graphic correlation techniques suggest that 95% of the faunal extinctions at the end of the Permian Period (~250 Myr) occurred in an interval of time <18,000 years ([142]; although results have been disputed, D.H. Erwin, personal communication, 1996). Several extinction boundaries are known to show evidence of stepwise extinctions and of multiple impacts (e.g., the Eocene/Oligocene transition [41], [59], [57], the Late Devonian [53]), suggesting that pulses of impactors may be involved.

THE K/T BOUNDARY

Characteristics of the K/T boundary include: (1) a globally synchronous or near-synchronous mass mortality level or levels, marked by a negative shift in carbon-isotope ratios $\delta^{13}C$ in marine carbonates, indicating a biomass loss and drop in productivity (the Strangelove ocean), (2) a proliferation of opportunistic species, followed by recovery and radiation of surviving biota, (3) a marked negative shift in oxygen-isotope ratios $\delta^{18}O$, suggesting a brief global warming, (4) a positive shift in sulfur-isotope ratios $\delta^{34}S$, suggesting a trend toward anoxia in ocean waters, and (5) a sharp reduction in marine biogenic $CaCO_3$.

Isotopic perturbations. The K/T boundary is marked by a worldwide negative anomaly of $\delta^{13}C$ in marine carbonates lasting for several hundred thousand years [111]. The boundary is also commonly marked by a drastic reduction in $CaCO_3$ deposition in the deep sea, reflecting primarily the mass extinction of calcareous nannoplankton and planktonic foraminifera. Biogenic pelagic $CaCO_3$ deposition was depressed for up to 1 Myr after the boundary event in some places.

A mass mortality event would have significant effects on the cycling of carbon and other elements in the ocean [35], [13]. One predicted effect would be an abrupt negative shift in $\delta^{13}C$ of several per mil in marine carbonates and the organic fraction [48], [54]. The negative shift in $\delta^{13}C$ indicates a virtual cessation of global marine productivity, the so-called Strangelove-ocean effect, accompanied by a homogenization of the $\delta^{13}C$ content of surface and deep waters [35], [13]. The loss of calcium carbonate secreting organisms is predicted to cause an "alkalinity crisis" in the oceans, that might lead to large-scale inorganic precipitation of carbonates [13]. A mass mortality event may be coincident with the mass extinctions, but it should be emphasized that the severity and relative timing of the "final" extinction of various organisms may depend on many factors, both biological (including population size, distribution, habitat, and ecology), and geological (including the state of climate, sea level, and ocean chemistry).

Although oxygen-isotope studies across the K/T boundary have given somewhat conflicting results (see review in [13]), in a number of high-resolution K/T boundary sections, a marked negative shift in $\delta^{18}O$, beginning at the boundary and lasting at least several thousand years, has been interpreted as representing a rapid warming of up to ~8°C [111]. On the other hand, very short-term impact-induced cooling (impact winter), or decadal cooling from sulfate aerosols [70] might help to trigger longer intervals of decreased temperatures, such as ice ages or more severe periods within ice ages, through the positive feedback effects of increased snow and ice cover [46]. In that case, a marked positive oxygen-isotope anomaly, indicating cooling and growth of ice sheets, might occur at about the same time as some mass extinctions.

A sharp positive sulfur-isotope excursion reported at some K/T boundary sections suggests a depletion of oxygen in ocean waters in at least some areas at the time [40], which is consistent with organic-rich and pyrite-rich sediments in some K/T boundary sections. These conditions may have been a result, rather than a cause, of the extinctions.

Ecological disasters and biomass crash. Plankton extinction in the oceans seems to have been abrupt and catastrophic (e.g., [111], [71]), with proliferation and episodic blooms of opportunistic species (e.g., acritarchs, dinoflagellates), and possibly bacteria [33] just after the K/T boundary.

TABLE 4. Evidence of Environmental Disturbances at Mass Extinction Boundaries (References: [54], [78]–[80], [34], [130])

Extinction Event	$-\delta^{13}C$	$\delta^{18}O$	$+\delta^{34}S$	Ecological disturbances or biomass crash
Late Pliocene		X		X
Late Eocene		X		
[a]Cretaceous/Tertiary	X	X	X	X
Cenomanian/Turonian	X?	X		
Jurassic/Cretaceous	X	X?		
[a]Triassic/Jurassic	X?			
[a]Permian/Triassic	X	X	X	X
Devon./Carbon.	X	X		X
[a]Late Devonian	X	X	X	X
[a]Late Ordovician	X?	X	X	X
Late Precambrian	X	X	X	X?

[a]Five major mass extinctions.

On land, plant fossils and palynological data from several sites reveal evidence for a profound ecological disturbance at the K/T boundary. Late Cretaceous palynomorph assemblages commonly show severe reductions in abundance and diversity at the boundary, followed by a thin barren interval, and sudden rise in spore abundance — the so-called fern spike [138], and increase in fungal spores [9].

Table 4 summarizes the information on occurrences of isotope anomalies and evidence of severe ecological disturbances in the sea (e.g., plankton crash, proliferation of opportunistic species) and/or on land (e.g., fern spikes, fungal spore spikes, mass mortality) at extinction boundaries in the geologic record.

IMPACTS AND EXTINCTIONS: CORRELATION AND POSSIBLE PERIODICITY

Extinctions and Impacts

Seven of the ~25 extinction peaks in Fig. 1 (Pliocene, Late Eocene, K/T, J/K, Tr/Jr, Late Triassic, Late Devonian; and possibly the Proterozoic/Cambrian boundary) are associated with large impact craters and/or some form of stratigraphic evidence of impacts (e.g., large Ir anomalies, shocked quartz, and/or microtektites) (Tables 1 and 2). At least six more are associated with "possible" evidence of impact consisting of lower concentrations of Ir, but still anomalous with respect to background values (Table 3).

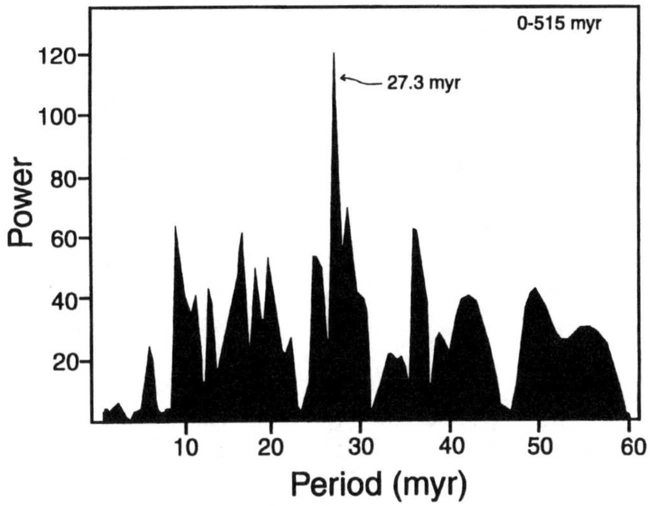

FIGURE 4. Fourier power spectrum for extinction events from Fig. 1 (0–515 Myr), computed as described in Rampino and Caldeira [75], [76].

The chances of hitting a target time series consisting of 24 extinction events (with the age error bars in this case taken as the difference in ages given by recent published time scales), with a time series consisting of the seven times of diagnostic stratigraphic evidence of impacts is $< 10^{-7}$, whereas the accidental correlation of all 13 "diagnostic" and "possible" indicators of impact with the mass extinction record is of even lower probability. Studies thus far suggest that iridium anomalies above background and shocked quartz are very rare in the geologic record (e.g., [45], [5], [100], [99]), suggesting that correlation of mass-extinction events and evidence of large impacts is very unlikely to be an accidental occurrence.

The discovery of periodic components in the record of mass extinctions would be strong evidence for a common cause and pacemaker. The mean occurrence rate of the 24 extinction pulses shown in Fig. 1 is one every ~23 Myr. Raup and Sepkoski [92], [93] reported a statistically significant 26.4 Myr periodicity in extinction time series for the last 250 Myr. Periods of ~26 to 31 Myr have been derived using various subsets of extinction events (family and genus levels), different geologic time scales, and various methods of time-series analysis, and Rampino and Stothers [83] reported detection of an ~29 Myr component in the record of extinctions of nonmarine vertebrates. The interpretation of these results has been a subject of considerable debate (e.g., see [79] for a discussion). The detection of a periodic component in extinctions does not imply either that the record contains a strict periodicity, or that every extinction event follow a 26 to 30 Myr cycle; the extinction record might be a mixture of periodic and random events.

One of the problems in studies of periodicity has been the shortness of the record, and the reported apparent lack of evidence of a statistically significant periodicity in extinction events prior to 250 Myr ago. Fourier analysis of an extended record of 21 extinctions going back to ~515 Myr, more than doubling the length of the series, resulted in a spectrum with the highest peak at 27.3 Myr (Fig. 4). A measure of significance was obtained by generating 1,000 pseudo-data sets, each containing the same number of randomly dated pseudo-events over the same time interval. For this analysis, the probability of generating higher spectral power at 27.3 Myr was ~2%, but in further tests the probability of generating higher spectral power at any period between 10 and 65 Myr fell below the 95% confidence level.

We have also tested the robustness of the ~27 Myr peak by performing Fourier analyses on a series of truncated extinction time series starting from 0 to 540 Myr, and subtracting one extinction at a time back to 253 Myr. As evidence for the robustness of the ~27 Myr periodicity in the extinction time series, a stable peak between 26.5 and 27.3 Myr remained the dominant feature in the spectra of the truncated extinction data sets [80].

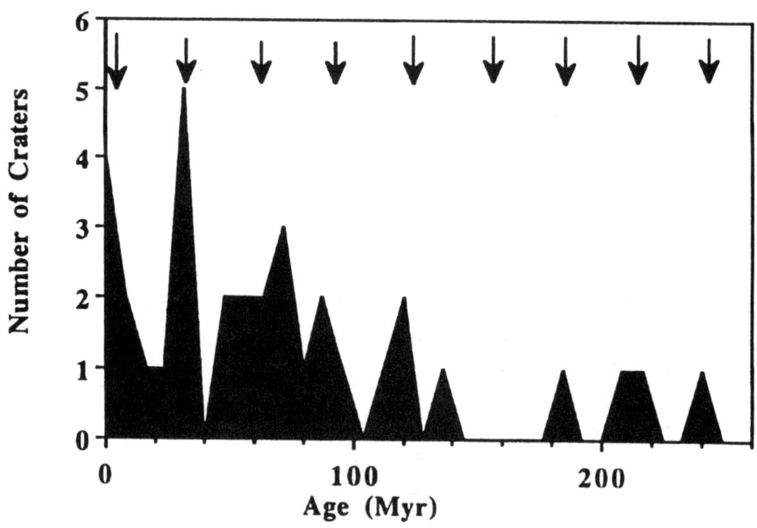

FIGURE 5. Distribution of the ages of the 33 best-dated impact craters (diameters ≥ 5 km) of the last 250 Myr (data from [30], [29]). Smoothed histogram of crater ages in 8 Myr bins. Arrows indicate approximate "best-fit" 30-Myr cycle.

The Record of Impact Craters

The impact-crater record has been interpreted as showing pulses at ~0, 35, 65, and 95 Myr ago [37], [106]. A major recurrent question has been: Does the terrestrial record of impact craters contain a periodic component similar to that found in the mass extinctions? Time-series analyses of terrestrial impact craters have provided some evidence of a possible 28 to 32 million year periodicity in impacts [81]–[83], [4], [106], [143]–[146], and have been interpreted as suggesting that the impact record is a mixture of periodic and random events [118], [26]. Several studies have concluded that, considering the problems in dating, the observed differences in the periodicity detected in impact craters and mass extinctions are to be expected [115], [26]. The ~30 Myr period is readily detected when an extended set of ~40–65 dated impact craters over the past 250 to 600 Myr is used, but the signal can become much weaker when smaller subsets of the best-dated craters are analyzed [143], [30].

In Fig. 5, we show the distribution of the ages of the 33 best-dated craters (≥ 5 km in diameter) of the last 250 Myr, and a "best-fit" periodic component of ~30 Myr. Critics have argued, however, that the number of well-dated craters is still too small to extract a consistent statistically significant periodicity (e.g., [30], [31]), and they suggest that consideration of periodicity and its implications be deferred until a better impact record is available.

THE SHIVA HYPOTHESIS: GALACTIC PACEMAKER FOR PERIODIC MASS EXTINCTIONS

If a general connection between impacts and mass extinctions exists, then the periodic component may be related to a similar period in galactic dynamics. This possible extraterrestrial cycle in Earth history has been named the "Shiva Hypothesis" after the Hindu god of destruction and renewal (see [79]). In the galactic model, increased numbers of long-period comets (comet showers) occur during the periodic passage of the solar system through the central plane of the Milky Way Galaxy [81], [19], [51]. The time between crossings of the galactic plane during the solar system's up-and-down oscillations through the galactic disk is estimated at ~30 ± 3 Myr, and the most recent plane-crossing occurred in the last few million years, in agreement with the record of mass extinctions and impact pulses. Models suggested for time modulation of the infalling Oort cloud comets include gravitational perturbations of the Oort comet cloud by adiabatically varying galactic tides [60], [51] and/or by near encounters with massive objects concentrated in the plane [81], [82], although some components of these models may be problematic (e.g., [135]).

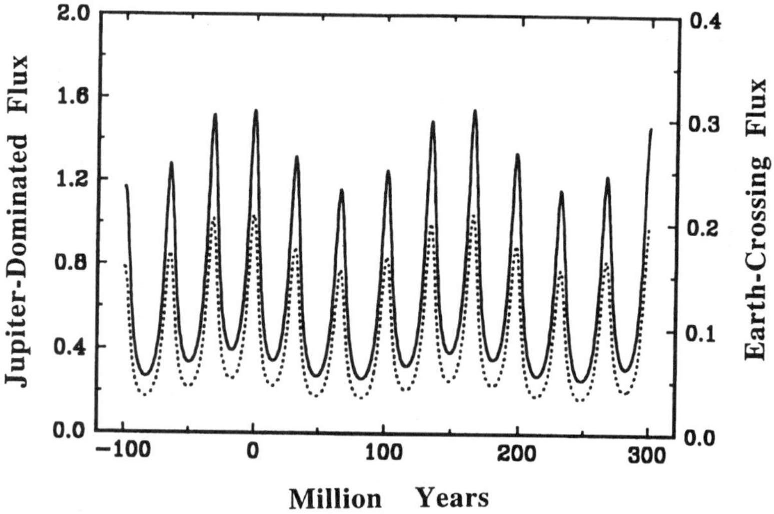

FIGURE 6. Time dependence of the Jupiter-dominated current (comets per year with $q < 5.8$ AU) from the outer Oort Cloud due to the adiabatically changing Galactic tide, based on a Galactic model (solid line). The Earth-crossing rate is also shown (dotted line). Both results are scaled to 0.2 Earth-crossing comets per year at present ($t = 0$). The model was run for an interval ranging from 100 Myr in the future to 300 Myr in the past (after [51]).

Recent calculations by Matese et al. [51] suggest that time modulation of the flux of new Jupiter-dominated Oort cloud comets could come from gravitational perturbations of the comet cloud by adiabatically varying galactic tides during the in-and-out of plane oscillation. They find that in a galactic model in which half of the disk matter is compact, the peak-to-trough comet flux variation should be ~4 to 1, with a half-width at half-maximum of ~4 Myr (Fig. 6). For the parameters chosen, Matese et al. [51] found that the most recent times of peak comet flux were ~31, 65, and 98 Myr ago, and that the cycle interval varied from 29.5 to 34.2 Myr over a 350 Myr run of the model. Valtonen et al. [119] note a small phase shift from the time of maximum tides to the time of maximum bombardment, suggesting that we are still approaching the maximum bombardment episode resulting from the latest Galactic plane crossing.

Pulses of impactors would explain the stepped nature of some extinction events [37], [113], clusters of similar-age craters, and multiple impact layers [59], [53]. Alternative models for producing periodic comet showers involving a companion star to the Sun [136], [23] or a tenth planet [137] have also been suggested, but these are somewhat *ad hoc* and have significant dynamical problems [135], [47], [121].

CONCLUSIONS

The impact theory of mass extinctions makes several predictions that can be tested with available data. The numbers and dynamics of Earth-crossing asteroids and comets predict that impactors greater than a few km in diameter should collide with the Earth on average every few tens of millions of years. Large impacts, releasing $\geq 10^7$ Mt TNT equivalent of explosive energy are calculated to produce global-scale disasters involving climatic cooling, reduced photosynthesis, and other effects, with very large impact events releasing $\geq 10^8$ Mt TNT producing additional severe effects such as complete loss of photosynthesis, "impact winter," and a global heat pulse and resulting wildfires from re-entering ejecta. These environmental effects have been predicted to lead to mass mortality and subsequent extinctions, and might also destabilize the climatic system, leading to longer term environmental changes. The extinction record of the last 540 Myr derived from paleontological studies shows 5 major pulses and about 20 minor pulses, in excellent agreement with the predictions from astronomical considerations of about 5 impacts with $\geq 10^8$ Mt and 25 ± 5 impacts of 10^7–10^8 Mt yields. Smaller impacts (~10^6 Mt events), singly or multiple, can cause regional devastation and may lead to other stage and epoch boundaries in the geologic record.

Extinction events, which may occur in steps, are commonly associated with negative shifts in carbon isotopes in marine sediments suggesting biomass crashes, abrupt shifts in oxygen isotopes indicating sudden climatic changes, and sometimes with positive sulfur-isotope anomalies suggesting lowered oxygenation of ocean waters. In an increasing number of cases, the extinction boundaries are found to be marked by die-off of marine plankton, decimation of terrestrial plant communities, impoverished postextinction ecosystems, and proliferation of opportunistic survivor species, with gradual recovery.

Seven pulses of extinction in the last 540 Myr seem to be associated with evidence of single or multiple impacts — layers containing high siderophile trace-element anomalies (especially iridium), and/or shocked minerals, tektites/microtektites, and impact craters with ages overlapping the boundaries. Some other boundaries are associated with elevated iridium concentrations above background that might be related to impacts of Ir-depleted objects, possibly comets, which may provide most of the largest impactors.

The record of mass extinctions and large impacts show correlations, and both may contain a mixture of stochastic events and a periodic component of ~26 to 30 Myr, which implies a common pacemaker. Such a pacemaker may be provided by the periodic modulation of the number of infalling Oort Cloud comets produced by encounters with interstellar clouds and with the changes in the Galactic tidal force as the solar system oscillates up and down through the disk-shaped Galaxy (½-period of ~30 ± 3 Myr), with maxima at times of Galactic plane crossing. The resulting theory represents a unification of terrestrial biological evolution and solar system interactions within the large-scale dynamics of the Galaxy.

ACKNOWLEDGMENTS

We thank R.A.F. Grieve, G. Izett, J.J. Matese, D.M. Raup, J.J. Sepkoski, Jr., P.M. Sheehan, J. Smit, M.B. Steiner, R.B. Stothers, M. Valtonen, K. Wang, P. Ward, and S. Yabushita for data, discussions, and criticism.

REFERENCES

1. ADUSHKIN, V.V. & I.V. NEMCHIMOV. 1994. Consequences of impacts of cosmic bodies on the surface of the Earth. *In* Hazards due to Asteroids and Comets, T. Gehrels, Ed.: 721–778. Univ. of Arizona Press. Tuscon, Arizona.
2. ALVAREZ, L.W., W. ALVAREZ, F. ASARO, & H.V. MICHEL. 1980. Extraterrestrial cause of the Cretaceous/Tertiary extinction: Experimental results and theoretical interpretation. Science. **208:** 1095–1108.
3. ALVAREZ, W. 1986. Toward a theory of impact crises. Eos, Trans. American Geophysical Union. **67:** 649–658.
4. ALVAREZ, W. & R.A. MULLER. 1984, Evidence from crater ages for periodic impacts on the Earth. Nature. **308:** 718–720.
5. ALVAREZ, W., F. ASARO, & A. MONTANARI. 1990. Iridium profile for 10 million years across the Cretaceous-Tertiary boundary at Gubbio (Italy). Science. **250:** 1700–1702.
6. ATTREP, M., JR., C.J. ORTH, & L.R. QUINTANA. 1991. The Permian-Triassic of the Gartnerkofel-1 Core (Carnic Alps, Austria): Geochemistry of common and trace elements II—INAA and RNAA. Abhandlungen der Geologischen Bundesanstalt. **45:** 123–137.
7. BENTON, M.J. 1995. Diversification and extinction in the history of life. Science. **268:** 52–58.
8. BHANDARI, N., M. GUPTA, J. PANDEY, & P.N. SHUKLA. 1994. Chemical profiles in K/T boundary section of Meghalaya, India: cometary, asteroidal of volcanic. Chemical Geology. **113:** 45–60.
9. BHANDARI, N., P.N. SHUKLA, & R.J. AZMI. 1992. Positive europium anomaly at the Permo-Triassic boundary, Spiti, India. Geophysical Research Letters. **19:** 1531–1534.
10. BICE, D.M., C.R. NEWTON, S. MCCAULEY, P.W. REINERS, & C.A. MCROBERTS. 1992. Shocked quartz at the Triassic-Jurassic boundary in Italy. Science. **259:** 443–446.
11. BRANDNER, R. 1988. The Permian-Triassic boundary section in the Dolomites (Southern Alps, Italy), San Antonio section. Ber. Geologischen Bundesanstalt. **15:** 49–56.
12. BROCHWICZ,-LEWINSKI, W., A. GASIEWICZ, G. MELENDEZ, L. SIQUEIROS, S. SUFFCZYNSKI, K. SZATKOWSKI, R. TARKOWSKI, & M. ZBIK. 1985. A possible Middle/Upper Jurassic boundary event. Abstracts, 27th International Geological Congress.
13. CALDEIRA, K. & M.R. RAMPINO. 1993. Aftermath of the end-Cretaceous mass extinction: Possible stabilization of the carbon cycle and climate. Paleoceanography. **8:** 515–525.

14. CASIER, J-G., F. LETHIERS, & P. CLAEYS. 1996. Ostracod evidence for an abrupt mass extinction at the Frasnian/Famennian boundary (Devils Gate, Nevada, USA). Compte Rendus Académie des Sciences, Paris. **322:** 415–422.
15. CHAI, Z.-F., X.-Y. MAO, S.-L. MA, S.-L. BAI, C.J. ORTH, Y.-Q. ZHOU, & J.-G. MA. 1987. Geochemical anomaly of the Devonian-Carboniferous boundary at Huangmao, Guangxi, China: Abstracts, International Geological Correlation Project 199, Rare Events in Geology, p. 29.
16. CHAPMAN, C.R. & D. MORRISON. 1994. Impacts on the Earth by asteroids and comets: Assessing the hazard. Nature. **367:** 33–40.
17. CLAEYS, P., J-G. CASIER, & S.V. MARGOLIS. 1992, Microtektites and mass extinctions: Evidence for a Late Devonian asteroid impact: Science. **257:** 1102–1104.
18. CLAEYS, P., F.T. KYTE, & J.-G. CASIER. 1994. Frasnian-Famennian boundary: Mass extinctions, anoxic oceans, microtektite layers, but not much iridium? *In* Papers Presented to New Developments Regarding the KT Event and Other Catastrophies in Earth History, LPI Contribution. No. 825: 22–24. Lunar and Planetary Inst. Houston, Texas.
19. CLUBE, S.V.M. & W.M. NAPIER. 1984. Comet capture from molecular clouds: A dynamical constraint on star and planet formation. Monthly Notices of the Royal Astronomical Society. **208:** 575–588.
20. CLYMER, A.K., D.M. BICE, & A. MONTANARI. 1996. Shocked quartz from the late Eocene: Impact evidence from Massignano, Italy. Geology. **24:** 483–486.
21. COVEY, C., S.J. GHAN, J.J. WALTON, & P.R. WEISSMAN. 1990. Global environmental effects of impact-generated aerosols; Results from a general circulation model. Geological Society of America Special Paper 247: 263–270.
22. CROFT, S.K. 1982. A first-order estimate of shock heating and vaporization in oceanic impacts: Geological Society of America Special Paper 190: 143–151.
23. DAVIS, M., P. HUT, & R.A. MULLER. 1984. Extinction of species by periodic comet showers. Nature. **308:** 715–717.
24. DYPVIK, H., S.T. GUDLAUGSSON, F. TSIKALAS, M. ATTREP, JR., R.E. FERRELL, JR., D.H. KRINSLEY, A. MØRK, J.I. FALEIDE, & J. NAGY. 1996. Mjølnir structure: An impact crater in the Barents Sea. Geology. **24:** 779–782.
25. ERWIN, D.H. 1993. The Great Paleozoic Crisis. Columbia Univ. Press. New York, NY.
26. FOGG, M.J. 1989. The relevance of the background impact flux to cyclic impact/mass extinction hypothesis. Icarus. **79:** 382–395.
27. GANAPATHY, R. 1982, Evidence for a major meteorite impact on the Earth 34 million years ago: Implications for Eocene extinctions. Science. **216:** 885–886.
28. GERSTL, S.A. & A. ZARDECKI. 1982, Reduction of photosynthetically active radiation under extreme stratospheric aerosol loads. Geological Society of America Special Paper 190: 201–210.
29. GRIEVE, R.A.F. 1996. Chesapeake Bay and other terminal Eocene impacts: Meteoritics and Planetary Science. **31:** 166–167.
30. GRIEVE, R.A.F. & L.J. PESONEN. 1996. Terrestrial impact craters: Their spatial and temporal distribution and impacting bodies. Earth, Moon, and Planets. **72:** 357–376.
31. GRIEVE, R.A.F. & E.M. SHOEMAKER. 1994. The record of past impacts on earth. *In* Hazards due to Asteroids and Comets, T. Gehrels, Ed.: 417–462. Univ. of Arizona Press. Tucson, Arizona.
32. HILDEBRAND, A.R., M. PILKINGTON, M. CONNORS, C. ORTIZ-ALEMAN, & R.E. CHAVEZ. 1995. Size of Chicxulub impact structure. Nature. **376:** 415–417.
33. HOLLANDER, D.J., J.A. MCKENZIE, & K.J. HSU. 1993. Carbon isotope evidence for unusual plankton blooms and fluctuations of surface water CO_2 in "Strangelove Ocean" after terminal Cretaceous event. Palaeogeography, Palaeoclimatology, Palaeoecology. **104:** 229–237.

34. HOLSER, W.T., M. MAGARITZ, & R.L. RIPPERDAN. 1995. Global isotopic events. *In* Global Events and Event Stratigraphy in the Phanerozoic, O.H. Walliser, Ed.: 63–88. Springer. Berlin, Germany.

35. HSÜ, K.J. & J.A. MCKENZIE. 1985. A "Strangelove" ocean in earliest Tertiary. *In* Carbon Cycle and Atmospheric CO_2, Natural Variations Archean to Present, E.T. Sundquist & W.S. Broecker, Eds.: 487–492. Geophysical Monograph 32. Amer. Geophys. Union. Washington, D.C.

36. HSÜ, K.J. & J.A. MCKENZIE. 1990. Carbon-isotope anomalies at era boundaries; Global catastrophes and their ultimate cause: Geological Society of America Special Paper 247: 61–70.

37. HUT, P., W. ALVAREZ, W.P. ELDER, T. HANSEN, E.G. KAUFFMAN, G. KELLER, E.M. SHOEMAKER, & P.R. WEISSMAN. 1987. Comet showers as a cause of mass extinctions. Nature. **329:** 118–126.

38. IZETT, G.A., W.A. COBBAN, J.D. OBRADOVICH, & M.J. KUNK. 1993. The Manson impact structure: $^{40}Ar/^{39}Ar$ age and its distal impact ejecta in the Pierre Shale in southeastern South Dakota. Science. **262:** 729–732.

39. JANSA, L.F., M.-P.AUBRY, & F.M. GRADSTEIN. 1990. Comets and extinctions; cause and effect? Geological Society of America Special Paper 247: 223–232.

40. KAJIWARA, Y.& K. KAIHO. 1992. Oceanic anoxia at the Cretaceous/Tertiary boundary supported by the sulfur isotope record. Palaeogeography, Palaeoclimatology, Palaeoecology. **99:** 151–162.

41. KELLER, G. 1986. Stepwise mass extinctions and impact events: Late Eocene to early Oligocene. Marine Micropaleontology. **10:** 267–293.

42. KOEBERL, C. 1993. Extraterrestrial component associated with Australasian microtektites in a core from ODP Site 758B. Earth and Planetary Science Letters. **119:** 453–458.

43. KYTE, F.T. 1988. The extraterrestrial component in marine sediments: Description and interpretation. Paleoceanography. **3:** 235–247.

44. KYTE, F.T. & J.T. WASSON. 1982. Geochemical constraints on the nature of large accretionary events. Geological Society of America Special Paper 190: 235–242.

45. KYTE, F.T. & J.T. WASSON. 1986. Accretion rate of extraterrestrial matter: Iridium in marine sediments deposited 33–67 Ma ago. Science. **232:** 1225–1229.

46. KYTE, F.T., Z. ZHOU, & J.T. WASSON. 1988. New evidence on the size and possible effects of a Late Pliocene asteroid impact. Science. **241:** 63–65.

47. LATHAM, D.W., T. MAZEH, R.J. DAVIS, R.P. STEFANIK, & H. ABT. 1991. A search for wide binaries at the North Galactic Pole using precise radial velocities. The Astronomical J. **101:**. 625–636.

48. MAGARITZ, M. 1989, ^{13}C Minima follow extinction events: A clue to faunal radiation: Geology. **17:** 337–340.

49. MARGOLIS, S.V., P. CLAEYS, & F.T. KYTE. 1991. Microtektites, microkrystites, and spinels from a Late Pliocene asteroid impact in the Southern Ocean. Science. **251:** 1594–1597.

50. MARSHALL, C.R. 1995. Distinguishing between sudden and gradual extinctions in the fossil record: Predicting the position of the Cretaceous-Tertiary iridium anomaly using the ammonite fossil record on Seymour Island, Antarctica. Geology. **23:** 731–734.

51. MATESE, J.J., P.G. WHITMAN, K.A. INNANEN, & M.J. VALTONEN. 1995. Periodic modulation of the Oort Cloud comet flux by the adiabatically changing galactic tide. Icarus. **116:** 255–268.

52. MAURRASSE, F.J-M.R. & G. SEN. 1991. Impacts, tsunamis, and the Haitian Cretaceous-Tertiary boundary layer. Science. **252:** 1690–1693.

53. McGHEE, G.R., JR. 1996. The Late Devonian Mass Extinction. Columbia Univ. Press. New York, NY. 303 p.

54. MCLAREN, D.J. & W.D. GOODFELLOW. 1990. Geological and biological consequences of giant impacts. Annual Review of Earth and Planetary Science. **18:** 123–171.

55. MELOSH, H.J. 1995. Around and around we go. Nature. **376:** 386–387.

56. MELOSH, H.J., N.M. SCHNEIDER, K.J. ZAHNLE, & D. LATHAM. 1990. Ignition of global wildfires at the Cretaceous/Tertiary boundary. Nature. **343:**. 251–254.

57. MILLER, K.G., W.A. BERGGREN, J. ZHANG, & A.A. PALMER-JULSON. 1991. Biostratigraphy and isotope stratigraphy of Upper Eocene microtektites at Site 612: How many impacts? Palaios. **6:** 17–38.

58. MONTANARI, A. 1990. Geochronology of the terminal Eocene impacts; An update. Geological Society of America Special Paper 247: 607–616.

59. MONTANARI, A., F. ASARO, H.V. MICHEL, & J.P. KENNETT. 1993. Iridium anomalies of Late Eocene age at Massignano (Italy), and ODP Site 689B (Maud Rise, Antarctica). Palaios. **8:** 420–438.

60. NAPIER, W.M. 1989, Terrestrial catastrophism and galactic cycles. *In* Catastrophism and Evolution, Astronomical Foundations, S.V.M. Clube, Ed.: 133–167. Cambridge Univ. Press. Cambridge, England.

61. OCAMPO, A.C., K.O. POPE, & A.G. FISCHER. 1996. Ejecta blanket of the Chicxulub crater from Albion Island, Belize. Geological Society of America Special Paper 307: 75–88.

62. ODDONE, W. & R. VANNUCCI. 1988. PGE and REE geochemistry at the B-W boundary in the Carnian and Dolomite Alps (Italy). Memorie della Societá Geologica Italiana. **34:** 121–128.

63. O'KEEFE, J.D.,& T.J. AHRENS. 1989. Impact production of CO_2 by the Cretaceous/Tertiary extinction bolide and the resultant heating of the Earth. Nature. **338:** 247–249.

64. ORTH, C.J. 1989. Geochemistry of the bio-event horizons. *In* Mass Extinctions: Processes and Evidence, S.K. Donovan, Ed.: 37–72. Columbia Univ. Press. New York, NY.

65. ORTH, C.J., M. ATTREP, JR., L.R. QUINTANA, W.P. ELDER, E.G. KAUFFMAN, R. DINER, & T. VILLAMIL. 1993. Elemental abundance anomalies in the late Cenomanian extinction interval: A search for the source(s). Earth and Planetary Science Letters. **117:** 189–204.

66. PALME, H. 1982. Identification of projectiles of large terrestrial impact craters and some implications for the interpretation of Ir-rich Cretaceous/Tertiary boundary layer. Geological Society of America Special Paper 190: 223–233.

67. PENG, H. 1994. An extraterrestrial event at the Tertiary-Quaternary boundary: Papers Presented to New Developments Regarding the KT Event and Other Catastrophies in Earth History, LPI Contribution No. 825: 88–89. Lunar and Planetary Inst. Houston, Texas.

68. POAG, C.W., D.S. POWARS, L.J. POPPE, & R.B. MIXON. 1994. Meteoroid mayhem in Ole Virginny: Source of the North American tektite strewn field. Geology. **22:** 691–694.

69. POLLACK, J.B., O.B. TOON, T.P. ACKERMAN, C.P. MCKAY, & R.P. TURCO. 1983. Environmental effects of an impact-generated dust cloud: Implications for the Cretaceous-Tertiary extinctions. Science. **219:** 287–290.

70. POPE, K., K.H. BAINES, A.C. OCAMPO, & B.A. IVANOV. 1994. Impact winter and the Cretaceous/Tertiary extinctions: Results of a Chicxulub asteroid impact model. Earth and Planetary Science Letters. **128:** 719–725.

71. POSPICHAL, J.J. 1994. Calcareous nannofossils at the K-T boundary, El Kef: No evidence for stepwise, gradual, or sequential extinctions. Geology. **22:** 99–102.

72. RAMPINO, M.R. 1988, Are marine and nonmarine extinctions correlated? Eos, Transactions of the American Geophysical Union. **69:** 889, 895.

73. RAMPINO, M.R. 1994. Tillites, diamictites, and ballistic ejecta of large impacts. J. Geology. **102:** 439–456.

74. RAMPINO, M.R. & A.C. ADLER. 1996. Abrupt latest Permian mass extinction: Results of a test of the Signor-Lipps effect. Geology. Submitted.

75. RAMPINO, M.R. & K. CALDEIRA. 1992. Episodes of terrestrial geologic activity during the past 260 million years: A quantitative approach. Celestial Mechanics and Dynamical Astronomy. **54:** 143–159.

76. RAMPINO, M.R. & K. CALDEIRA. 1993. Major episodes of geologic change: Correlations, time structure and possible causes: Earth and Planetary Science Letters. **114:** 215–227.

77. RAMPINO, M.R., K. ERNSTSON, A.G. FISCHER, D.T. KING, JR., A.C. OCAMPO, & K.O. POPE. 1996. Characteristics of clasts in K/T debris-flow diamictites in Belize compared with other known proximal ejecta deposits: Geological Society of America Abstracts with Programs. **28:** A-182.

78. RAMPINO, M.R. & B.M. HAGGERTY. 1994. Extraterrestrial impacts and mass extinctions of life. *In* Hazards due to Asteroids and Comets, T. Gehrels, Ed.: 827–857. Univ. of Arizona Press. Tucson, Arizona.

79. RAMPINO, M.R. & B.M. HAGGERTY. 1996a. The "Shiva hypothesis": Impact crises, mass extinctions, and the Galaxy. Earth, Moon and Planets. **72:** 441–460.

80. RAMPINO, M.R. & B.M. HAGGERTY. 1996b. Impact crises and mass extinctions: A working hypothesis. Geological Society of America Special Paper 307: 11–30.

81. RAMPINO, M.R. & R.B. STOTHERS. 1984a. Terrestrial mass extinctions, cometary impacts and the Sun's motion perpendicular to the galactic plane. Nature. **308:** 709–712.

82. RAMPINO, M.R. & R.B. STOTHERS. 1984b. Geological rhythms and cometary impacts. Science. **226:** 1427–1430.

83. RAMPINO, M.R. & R.B. STOTHERS. 1986, Geologic periodicities and the Galaxy, in Smoluchowski, R., Bahcall, J.N., and Matthews, M.S., eds., The Galaxy and the Solar System: Tucson, Univ. of Arizona Press, p. 241–259.

84. RAMPINO, M.R. & R.B. STOTHERS. 1987. Episodic nature of the Cenozoic marine record. Paleoceanography. **2:** 255–258.

85. RAUP, D.M. 1979. Size of the Permo-Triassic bottleneck and its evolutionary implications. Science. **206:** 217–218.

86. RAUP, D.M. 1982: Biogeographic extinction: A feasibility test: Geological Society of America Special Paper 190: 277–281.

87. RAUP, D.M. 1989. The case for extraterrestrial causes of extinction: Philosophical Transactions of the Royal Society of London. **B325:** 421–435.

88. RAUP, D.M. 1990. Impact as a general cause of extinction; a feasibility test. Geological Society of America Special Paper 247: 27–32.

89. RAUP, D.M. 1991a. A kill curve for Phanerozoic marine species. Paleobiology. **17:** 37–48.

90. RAUP, D.M. 1991b. Extinction: Bad Genes or Bad Luck? Norton. New York, NY.

91. RAUP, D.M. 1992. Large-body impact and extinction in the Phanerozoic. Paleobiology. **18:** 80–88.

92. RAUP, D.M. & J.J. SEPKOSKI, JR. 1984. Periodicity of extinctions in the geologic past. Proceedings of the National Academy of Sciences USA. **81:** 801–805.

93. RAUP, D.M. & J.J. SEPKOSKI, JR. 1986. Periodic extinctions of families and genera. Science. **231:** 833–836.

94. RETALLACK, G.J., A. SEYEDOLALI, W.T. HOLSER, D. KRINSLEY, AND E.S. KRULL. 1996. Shocked quartz at the Permian-Triassic boundary in Australia and Antarctica: Geological Society of America Abstracts with Programs. **28:** A-368.

95. RIGBY, J.K., JR., & D.L. WOLBERG. 1987. The Therian mammalian fauna (Campanian) of Quarry 1, Fossil Forest study area, San Juan Basin, New Mexico. Geological Society of America Special Paper 209: 51–79.
96. ROCCIA, R., D BOCLET, , P. BONTE, A. CASTELLARIN, & C. JEHANNO. 1986. An iridium anomaly in the Middle-Lower Jurassic of the Venetian region, northern Italy. J. Geophysical Research. **91:** E259–E262.
97. ROGERS, R.R. 1990. Taphonomy of three dinosaur bonebeds in the Upper Cretaceous Two Medicine Formation of Northwestern Montana: Evidence for drought-related mortality. Palaios. **5:** 394–413.
98. RUSSELL, D.A. 1993. Manson crater extinctions. Science. **262:** 1956.
99. SCHMITZ, B. & F. ASARO. 1996. Iridium geochemistry of volcanic ash layers from early Eocene rifting of the northeastern North Atlantic and some other Phanerozoic events. Geological Society of America Bulletin. **108:** 489–504.
100. SCHMITZ, B., L. JEPPSSON, & J. EKVALL. 1994. A search for shocked quartz grains and impact ejecta in early Silurian sediments on Gotland, Sweden. Geological Magazine. **131:** 361–167.
101. SCHULTZ, P.H. & D.E. GAULT. 1990. Prolonged global catastrophes from oblique impacts. Geological Society of America Special Paper 247: 239–261.
102. SEPKOSKI, J.J., JR. 1992. A compendium of fossil marine animal families, 2nd edit. Milwaukee Public Museum Contributions to Biology and Geology 83.
103. SEPKOSKI, J.J., JR. 1994. Extinction and the fossil record. Geotimes. **39:** (3) 15–17.
104. SEPKOSKI, J.J., JR. 1995. Patterns of Phanerozoic extinction: A perspective from global data bases. *In* Global Events and Event Stratigraphy in the Phanerozoic, O.H. Walliser, Ed.: 35–51. Springer. Berlin, Germany.
105. SHEEHAN, P.M., D.E. FASTOVSKY, R.G. HOFFMANN, C.B. BERGHAUS, & D.L. GABRIEL. 1991. Sudden extinction of the dinosaurs: Latest Cretaceous, Upper Great Plains, U.S.A. Science. **254:** 835–839.
106. SHOEMAKER, E.M. & R.F. WOLFE. 1986. Mass extinctions, crater ages, and comet showers. *In* The Galaxy and the Solar System, R. Smoluchowski, J.N. Bahcall, & M.S. Matthews, Eds.: 338–386. Univ. of Arizona Press. Tucson, Arizona.
107. SHOEMAKER, E.M., R.F. WOLFE, & C.S. SHOEMAKER. 1990. Asteroid and comet flux in the neighborhood of Earth. Geological Society of America Special Paper 247: 155–170.
108. SIGNOR, P.W., III, & J.H. LIPPS. 1982. Sampling bias, gradual extinction patterns and catastrophes in the fossil record. Geological Society of America Special Paper 190: 291–296.
109. SIGURDSSON, H., S. D'HONDT, & S. CAREY. 1992. The impact of the Cretaceous/Tertiary bolide on evaporite terrane and generation of major sulfuric acid aerosol: Earth and Planetary Science Letters. **109:**. 543–559.
110. SLOAN, R.E., J.K. RIGBY, L.H. VAN VALEN, & D. GABRIEL. 1986. Gradual dinosaur extinction and simultaneous ungulate radiation in the Hell Creek Formation. Science. **232:** 629–632.
111. SMIT, J. 1990. Meteorite impact, extinctions and the Cretaceous-Tertiary boundary. Geologie en Mijnbouw. **69:** 187–204.
112. SMIT, J., TH.B. ROEP, W. ALVAREZ, A. MONTANARI, P. CLAEYS, J.M. GRAJALES-NISHIMURA, & J. BERMUDEZ. 1996. Coarse-grained clastic sandstone complex at the K/T boundary around the Gulf of Mexico: Deposition by tsunami waves induced by the Chicxulub impact? Geological Society of America Special Paper 307: 151–182.
113. STANLEY, S.M. & X. YANG. 1994. Two extinction events in the Late Permian. Science. **266:** 1340–1344.

114. STEINER, M.B. & E.M. SHOEMAKER. 1996. An hypothesized Manson impact tsunami: Paleomagnetic and stratigraphic evidence in the Crow Creek Member, Pierre Shale. Geological Society of America Special Paper 302: 419–432.
115. STOTHERS, R.B. 1989. Structure and dating errors in the geologic time scale and periodicity of mass extinctions. Geophysical Research Letters. **16:** 119–122.
116. STOTHERS, R.B. 1993. Impact craters at geologic stage boundaries. Geophysical Research Letters. **20:** 887–891.
117. TOON, O.B., K. ZAHNLE, R.P. TURCO, & C. COVEY. 1994. Environmental perturbations caused by asteroid impacts. *In* Hazards due to Asteroids and Comets, T. Gehrels, Ed. : 791–826. Univ. of Arizona Press. Tucson, Arizona.
118. TREFIL, J.S.& D.M. RAUP. 1987. Numerical simulations and the problem of periodicity in the cratering record. Earth and Planetary Science Letters. **82:** 159–164.
119. VALTONEN, M.J., J.Q. ZHENG, J.J. MATESE, & P.G. WHITMAN. 1995. Near-Earth populations of comets from the Oort Cloud and their impacts with planets. Earth, Moon, and Planets. **71:** 219–223.
120. VAN DEN BERGH, S. 1994. Astronomical catastrophes in earth history. Publications of the Astronomical Society of the Pacific. **106:** 689–695.
121. VANDERVOORT, P.O. & E.A. SATHER. 1993. On the resonant orbit of a solar companion star in the gravitational field of the Galaxy. Icarus. **105:** 26–47.
122. VARRICCHIO, D.J. 1995. Taphonomy of Jack's Birthday Site, a diverse dinosaur bonebed from the Upper Cretaceous Two Medicine Formation of Montana. Palaeogeography, Palaeoclimatology, Palaeoecology. **114:** 297–323.
123. VARRICCHIO, D.J. & J.R. HORNER. 1993. Hadrosaurid and lambeosaurid bone beds from the Upper Cretaceous Two Medicine Formation of Montana: Taphonomic and biologic implications. Canadian J. Earth Sciences. **30:** 997–1006.
124. VICKERY, A.M., & H.J. MELOSH. 1990. Atmospheric erosion and impactor retention in large impacts, with application to mass extinctions. Geological Society of America Special Paper 247: 289–300.
125. WANG, K. 1992. Glassy microspherules (microtektites) from an Upper Devonian limestone. Science. **256:** 1547–1550.
126. WANG, K., M. ATTREP, JR., & C.J. ORTH. 1993a. Global iridium anomaly, mass extinction, and redox change at the Devonian Carboniferous boundary. Geology. **21:** 1071–1074.
127. WANG, K., B.D.E. CHATTERTON, M. ATTREP, JR., & C.J. ORTH. 1992. Iridium abundance maxima in the latest Ordovician mass extinction horizon, Yangtze Basin, China, Terrestrial or extraterrestrial. Geology. **20:** 39–42.
128. WANG, K., B.D.E. CHATTERTON, M. ATTREP, JR., & C.J. ORTH. 1993b, Late Ordovician mass extinction in the Selwyn Basin, northwestern Canada: geochemical, sedimentological, and paleontological evidence: Canadian J. Earth Sciences. **30:** 1870–1880.
129. WANG, K., H.H.J. GELDSETZER, & B.D.E. CHATTERTON. 1994. A Late Devonian extraterrestrial impact and extinction in eastern Gondwana: Geochemical, sedimentological, and faunal evidence. Geological Society of America Special Paper 293:111–120.
130. WANG, K., W. GOODFELLOW, D. MCLAREN, H. GELDSETZER, & A. SWEET. 1996. Attributes of Phanerozoic mass extinctions. Unpublished.
131. WANG, K., C.J. ORTH, M. ATTREP, JR., B.D.E. CHATTERTON, H. HOU, & H.H.J. GELDSETZER. 1991. Geochemical evidence for a catastrophic biotic event at the Frasnian/Famennian boundary in south China. Geology. **19:** 776–779.
132. WANG, K., C.J. ORTH, M. ATTREP, JR., B.D.E. CHATTERTON, X. WANG, & J-J. LI. 1993c. The great latest Ordovician extinction on the South China Plate: Chemos-

tratigraphic studies of the Ordovician-Silurian boundary interval on the Yangtze Platform. Palaeogeography, Palaeoclimatology, Palaeoecology. **104:** 61–79.
133. WARD, P.D. 1996. After the fall: Lessons and directions from the K/T debate. Palaios. **10:** 530–538.
134. WARME, J.E. & C.A. SANDBERG. 1996. Alamo megabreccia: Record of Late Devonian impact in southern Nevada. GSA Today. **6:** 1–7.
135. WEISSMAN, P.R. 1986. The Oort Cloud and the Galaxy: Dynamical interactions. *In* The Galaxy and the Solar System, R. Smoluchowski, J.N. Bahcall, & M.S. Matthews, Eds.: 204–237. Univ. of Arizona Press. Tucson, Arizona.
136. WHITMIRE, D.P. & A.A. JACKSON. 1984. Are periodic mass extinctions driven by a distant solar companion. Nature. **308:**. 713–715.
137. WHITMIRE, D.P. & J.J. MATESE. 1985. Periodic comet showers and Planet X. Nature. **313:** 36–38.
138. WOLFE, J.A. & G.R. UPCHURCH. 1986. Vegetation, climatic and floral changes at the Cretaceous-Tertiary boundary. Nature. **324:** 148–152.
139. WU, S. 1987. The Duolun impact crater, China: Abstracts, International Geological Correlation Project 199 Meeting, Mar. 3–4, 1987, Beijing, China.
140. XU, D., Z. YAN, ET AL. 1989. Astrogeological Events in China. Van Nostrand Reinhold. New York, NY.
141. XU, D. & Z. YAN. 1993. Carbon isotope and iridium event markers near the Permian/Triassic boundary in the Meishan section, Zhejiang Province, China. Palaeogeography, Palaeoclimatology, Palaeoecology. **104:** 171–176.
142. XU, G. 1991. Stratigraphical time-correlation and mass extinction event near Permian-Triassic boundary in South China. J. Chinese University of Geosciences. **2:** 36–46.
143. YABUSHITA, S. 1991. A statistical test for periodicity hypothesis in the crater formation rate. Monthly Notices of the Royal Astronomical Society. **250:** 481–485.
144. YABUSHITA, S. 1992. Periodicity in the crater formation rate and implications for astronomical modeling. *In* Dynamics and Evolution of Minor Bodies with Galactic and Geological Implications, S.V.M. Clube, S. Yabushita, & J. Henrard, Eds.: 161–168. Kluwer. Dordrecht, The Netherlands.
145. YABUSHITA, S. 1996a. Are cratering and probably related geological records periodic? Earth, Moon, and Planets. **72:** 343–356.
146. YABUSHITA, S. 1996b. Statistical tests of a periodicity hypothesis for crater formation rate-II. Monthly Notices of the Royal Astronomical Society. **279:** 727–732.
147. ZHAKAROV, V.A., A.S. LAPUKHOV, & O.V. SHENFIL. 1993. Iridium anomaly at the Jurassic-Cretaceous boundary in northern Siberia. Russian J. Geology and Geophysics. **34:** 83–90.
148. ZHOU, Y.Q. & C.F. CHAI. 1991. The discovery of shocked quartz and stishovite in the Permian/Triassic boundary clay of Huangshi, China. Meteoritics. **26:** 413.
149. TOON, O.B., K. ZAHNE, D. MORRISON, R.P. TURCO, & C. COVER. 1997. Environmental pertubations caused by impacts of asteroids and comets. Reviews of Geophysics. **35:** 41–78.

NEO Mission Dynamics and Advanced Space Propulsion

P.M. SFORZA[a] AND J.L. REMO[b]

[a]*Polytechnic University*
6 Metrotech Center
Brooklyn, New York 11201
and
[b]*Quantametrics*
1 Brackenwood Path, Head of the Harbor
St. James, New York 11780

ABSTRACT: An analysis identifying opportunities for extending space propulsion capabilities to ranges around 1 AU from the Earth is presented to suggest an effective and practical response to critical interaction missions to Near Earth Objects (NEOs). Consideration is focused on timely interception of NEOs approaching Earth well beyond cislunar range during a time period out to about 2020. It is shown such missions might be best accomplished by means of nuclear thermal or nuclear electric propulsion. Considering the development effort expended thus far on such systems it is recommended that completing this effort is of paramount importance for ensuring future deep-space performance capability.

INTRODUCTION

Near-Earth object (NEO) missions may require a spacecraft with the capability to rendezvous at great distances (1 AU) from the Earth, within a relatively short amount of time (on the order of a year), with a large payload, and with a capability for extensive in-flight trajectory corrections. This includes the capability to rendezvous away from the ecliptic, although the interception should be planned so that this could be avoided, if at all possible. While all of the above requirements will present a challenge to the most advanced contemporary spacecraft, the last two requirements essentially preclude the launching of a standard chemical rocket with a short burn time on a ballistic trajectory with only a minimal (vernier) thrust correction, usually on an upper stage. Add to these requirements the possible necessity of engaging the target a large distance from the ecliptic, and the demands on the spacecraft propulsion system can become formidable. To understand why the spacecraft must have these capabilities, one must understand something of the nature of NEO orbits. We briefly describe some of the NEO orbital characteristics and their prediction relevant to spacecraft interactions.

PREDICTING NEO ORBITS

It may initially appear that the problem of predicting orbits in general and close approaches and even possible Earth collisions with NEOs, given the best available initial conditions, is straightforward. Theoretically, at least, for certain classes of asteroids and comets with reasonably stable orbits, this may very well be the case if the (forward integration) computations include the use of general relativistic equations of motions incorporating the planetary perturbations and related tidal effects, as well as the effects of the Earth and the Moon. An excellent review on this topic as it relates to predicting close approaches of asteroids and comets to the Earth is given by Yeomans and Chodas [20]. In principle, the computational approach should be ideally suited to asteroid motion, but even for asteroids there is a need for well-determined orbits. Those orbits which are best determined are those where radar observations are utilized [9]. For these radar observed orbits relativistic equations can sometimes provide additional accuracy. However, an additional set of problems arise when dealing with both long- and short-period comets. For the reasonably accurate computation of those short-period comets which are active, and thereby subject to nongravitational interactions such as outgassing effects, special computational considerations must be taken into account. Even then, one may not be able reliably to obtain the desired precision necessary to predict a close approach sufficiently well to compute whether there is a need for countermeasure. Also, if it is perceived that a threat exists, there may be limitations on the precision of a countermeasure solution. Indeed, as comets and asteroids undergo close planetary encounters, their positional uncertainty will increase. Such uncertainties will place an additional burden on intercept missions.

The orbital uncertainty in predicting close approaches exists for both the long-period (which may necessarily have a short observational data interval) and short-period comets for several reasons which include first and foremost nongravitational interactions. An example of this causing a problem arose in the planned postperihelion observations of the recent Hyakutake comet with the Hubble Space Telescope (HST). Second is the offsetting of the true center of mass and the center of light. This effect is to be regarded as less important than the effect of nongravitational forces. Third is errors in the star catalogs used for reducing the positions of both comets and asteroids. To some extent, the difficulty in detection could be remedied considerably by improving the magnitude sensitivity (to mag 22/24) and extending the NEO search region in ecliptic longitude. However the other sources of uncertainty are not so easily overcome, and mission planners must take them into account.

LEAD TIME

The actual estimation of the warning time for a long-period comet inpact is arbitrary. However, extracting from Marsden and Steel [6], an estimate of the number of days between "final" opposition and pre-perihelic impact is about 250, as applied to high (retrograde) inclinations with perihelia at 1 AU. Another set of solutions are applicable to low inclinations, ranging from 0 at perihelion distance 0.5 AU to about 75 days at 1.0 AU. Since these intervals are far too short for realistic mission considerations, one may give for the lead time for these low inclination comets that time since the previous opposition, which is in the 425–500 day range. This assumes, of course, that the comet is bright enough to have been seen 500 days before impact! With respect to high inclination observations, the comet may also be detected at its previous opposition, at about 595 days (rather than 250 days) before impact, again provided it was bright enough. Problems may arise, however with inclinations in the range of 60–120°. For a follow-up on these lead times and sky coverage issues, one is referred to Marsden's paper in this proceedings and Marsden and Steel [6]. We may now infer that such uncertainties in orbit prediction and lead time as discussed above will place additional burdens on NEO interaction missions that must be launched on relatively short notice.

NEO INTERCEPTION

Given the above uncertainties in precisely predicting NEO orbits, great demands will be placed on a reliable interception vehicle. It is precisely because of the fact that certain NEO orbits cannot be precisely predicted and warning times may be short, that an interception vehicle must be capable of large sustained speeds and extended range. In particular, the spacecraft must first be capable of being launched within a short period of time in order to maximize the interval for transit to interception. Second, the spacecraft must be able to both continually track the NEO and correct its own trajectory at several stages of the mission in order to take into account deviations from a theoretically computed NEO orbit. Third, the spacecraft vehicle must also be capable of reaching an extended range on the order of 1 AU in relatively short time, carrying a payload adequate to carry out a range of interaction missions. This range must also include an out-of-the-ecliptic capability, if necessary, in order to reconnoiter or otherwise interact with the target NEO.

MISSION DESCRIPTION

The approach developed by Moeckel [7] provides a simple and convenient means to analyze deep-space missions for the purposes of comparing a variety of candidate advanced propulsion concepts. The simplification arises from considering the missions to occur in gravity-free space. Moeckel shows that this approach yields good estimates of trip time for interstellar and planetary missions as long as escape from, or descent to, low orbits about the major planets is not required. Three types of missions are considered: (a) flyby, (b) rendezvous, and (c) round trip. These missions are idealized as one-dimensional excursions between the Earth and a body a fixed distance R_o away, i.e., there is no relative motion between the two. The trajectories may then be readily analyzed in R, t space, where t is time.

The major performance index for propulsion systems is the specific impulse, defined by

$$I = F/g_o\dot{m},$$

where

$$F = \dot{m}V_e$$

is the thrust produced, V_e is the (constant) propellant exhaust velocity, $\dot{m}$ is the propellant mass flow rate, and g_o is the acceleration at the Earth's surface. Thus

$$I = V_e/g_o,$$

and therefore, the speed to which the propellant can be accelerated defines the efficiency of propellant use.

Propulsion systems which accelerate the propellant by direct conversion of thermal energy in the heated propellant to kinetic energy by means of a nozzle are limited by the temperature capability of the thrust chamber. They are called Type I systems and are characterized by high thrust and high propellant flow rates. The most common example of this type is the chemical rocket, while the most advanced such system within practical grasp in the near term is the nuclear thermal propulsion (NTP) system. Propulsion systems which generate electric power and use it to electromagnetically accelerate charged propellant particles to high velocities thereby escape the thermal limitations of Type I systems. These Type II systems are instead limited by the specific mass parameter α', the mass of the propulsion system required per unit electrical output. They are characterized by low thrust and very low propellant flow rates. The most common Type II systems use solar energy while advanced systems on the near-term horizon use fission energy to generate electricity and are called nuclear electric propulsion (NEP) systems. Sforza and Remo [16] compared the performance of these two very different types of propulsion systems for the NEO intercept problem. Type I propulsion systems were considered to be impulsive thrusters where acceleration to coast speed is very rapid, while Type II propulsion systems were considered to provide slow but continuous acceleration with a coast, at final velocity, of relatively short duration.

The results obtained indicate about equal flyby performance at around 2.5 AU from Earth for the two types of propulsion systems. Within that range the Type I propulsion systems seem to have superior performance. For rendezvous missions the range of superior Type I performance falls to about 1.5 AU. However, for the return missions the Type II systems appear to have the advantage beyond about 0.5 AU from Earth. Therefore, for scientific and mitigation missions, one-way trips in which the spacecraft must be considered expendable, advanced Type I propulsion systems, specifically NTP($I \simeq$ 1000 s) appear to command overwhelming consideration for near-term (25 years) development of flight articles. On the other hand, for later commercial missions, in which the spacecraft is reusable, advanced Type II systems, specifically NEP ($\alpha' \approx 10$ kg/kw) are prime candidates for development. Progress achieved in building a Type I nuclear thermal propulsion system should aid in the development of a Type II nuclear electric propulsion system, which makes the suggested approach doubly attractive.

MISSION SPECIFICATION

Critical missions are considered to be those which involve interception of a NEO closing with Earth along a trajectory which admits the possibility of a close encounter. We may extend the field-free approach originated by Moeckel [7] in order to assist in the comparison of propulsion systems which might be best suited to meet this emergency within a near-term scenario, say the period up to 2020. The results of such an analysis will also be valuable in treating similar, but less serious situations, where the NEO is approaching the Earth in a nonthreatening fashion. In this case, the NEO's approach permits a less demanding effort for visiting, and particularly returning from, such a body.

RANGE AND TIME TO INTERCEPT, NTP SYSTEMS

The range to intercept for NTP systems may be obtained kinematically by applying appropriate velocity changes ΔV_i necessary to follow the trajectories shown in Fig. 1. The results for the ratio of intercept range R_i to initial range R_o at spacecraft launch are as follows:

$$\text{flyby:} \quad R_i/R_o = (1 - V'/\Delta V_t)^{-1},$$

$$\text{rendezous:} \quad R_i/R_o = (1 + V'/\Delta V_t)/(1 - V'/\Delta V_t), \tag{1}$$

$$\text{return:} \quad R_i/R_o = (1 - 4V'/\Delta V_t)^{-1}.$$

In the same fashion, the time required to intercept, in years, is given by

$$\begin{aligned} \text{flyby:} \quad & T_i = (R_o/0.211V')[(V'/\Delta V_t)/(1 - V'/\Delta V_t)], \\ \text{rendezous:} \quad & T_i = 2(R_o/0.211V')[(V'/\Delta V_t)/(1 - V'/\Delta V_t)], \\ \text{return:} \quad & T_i = 4(R_o/0.211V')[(V'/\Delta V_t)/(1 - V'/\Delta V_t)], \end{aligned} \tag{2}$$

where R_o is in AU and V' and ΔV_t are in km/s. Note that for the case of a NEO closing with the Earth, its velocity $V' = dR/dt \prec 0$. Thus, from Eq. (1), any flyby mission will result in an intercept, no matter how close to Earth, while a rendezvous mission can only succeed (i.e., $R_i/R_o \succ 0$) if $|V'| \prec \Delta V_t$, and a return mission can only succeed (i.e., for a stay time > 0) if $|V'| \prec \Delta V_t/4$.

Moeckel [7] shows that for a Type I vehicle made up of N identical stages the total velocity change achievable is

$$\Delta V_t = g_o N I \log_e[(1 + k)/(\mu_{pl}^{1/N} + k + \mu_{ps})], \tag{3}$$

where k is the ratio of tankage mass to propellant mass. For the ideal case where $k = 0$ and $\mu_{ps} = 0$, Eq. (3) becomes

$$\Delta V_t = -g_o I \log_e(M_{pl}/M_o). \tag{4}$$

This relation is depicted in Fig. 2. It is clear from Fig. 2 and Eq. (1) that for high closing NEO velocities, V′, it may be impossible to achieve rendezvous or return missions to the NEO for a given payload fraction and specific impulse because of the limitation on velocity change achievable given in Eq. (4).

For identical stages, the payload ratio is given by

$$\mu_{pl} = (M_{n+1}/M_n)^N, \tag{5}$$

so that for a reasonable stage mass ratio, say $M_{n+1}/M_n = 0.1$ the mass fraction of payload becomes $\mu_{pl} = 10^{-N}$. For an initial spacecraft mass of around 20 Mg, which could be placed in LEO by a Titan IV or a Shuttle Orbiter, the payload deliverable to the NEO is as follows:

$$\begin{aligned} M_{pl} &= 2000 \text{ kg}, \quad N = 1, \\ M_{pl} &= 200 \text{ kg}, \quad N = 2, \\ M_{pl} &= 2 \text{ kg}, \quad N = 3. \end{aligned} \tag{6}$$

Therefore, to ensure substantial payload capability for the limited initial mass placeable in low-Earth orbit (IMLEO) it seems that multiple stage vehicles do not warrant further consideration for the critical missions contemplated here and a single stage to intercept vehicle (SSTI) is indicated. This result of minimizing the number of stages is in keeping with the requirements for maintaining high reliability in the spacecraft to be used in such relatively long-duration missions [15].

Rendezvous mitigation missions are considered to provide the capability for timely surgical interventions for precisely controlled attempts at orbit management, such as those suggested by Remo and Sforza [12]. The ratio of range at rendezvous to initial range R_i/R_o as given by Eq. (1) is shown as a function of NEO closing velocity V' in Fig. 3 for $\mu_{pl} = 0.1$ and various values of specific impulse. It is evident that for rendezvous to occur far from Earth, it is desirable to have high specific impulse. For this case the maximum NEO closing speed which can be accommodated is

$$V'_{\max} = 0.0225\, I$$

and this corresponds to 11.25, 22.5, and 45 km/s for I = 500, 1000, and 2000 s, respectively. This demonstrates the specific impulse limitations on flexibility for mitigation missions of the rendezvous type. Note that the time required to intercept can also be inferred from Fig. 3, since it is given by

$$T_i = (R_o/0.211V')(1 - R_i/R_o), \tag{7}$$

where T_i is in years. Thus, low values of specific impulse require long times to achieve rendezvous and the possibility for repeated attempts at mitigation likewise diminishes.

If more direct intervention is considered acceptable it is possible to consider flyby missions as true interceptions; here the consequences of limited specific impulse are less severe and an intercept at $R_i \succ 0$ is always possible. Such a situation is shown in Fig. 4 where the ratio of intercept range to initial range

$$R_i/R_o = (1 + V'/\Delta V_t)^{-1} \tag{8}$$

is plotted as a function of a NEO closing velocity V' for $\mu_{pl} = .1$ and various values of specific impulse. Once again, the time of intercept is given by the right-hand side of Eq. (3) and is clearly less demanding than in the case of rendezvous.

In order for the stay time at the NEO to be nonnegative, the total velocity change required for a return mission must be more than 4 times the NEO closing speed. Therefore, the employment of reusable return vehicles for the mitigation mission is likely only for very small closing speeds. Indeed the maximum NEO velocity that can be negotiated for a return mission is given by the inequality

$$V' \prec 5.64 \times 10^{-3}\, I, \tag{9}$$

where V' is in km/s and the results of Eq. (4) have been used. It would appear that return missions do not possess sufficient operational latitude in terms of NEO closing speed to consider them strongly for mitigation missions.

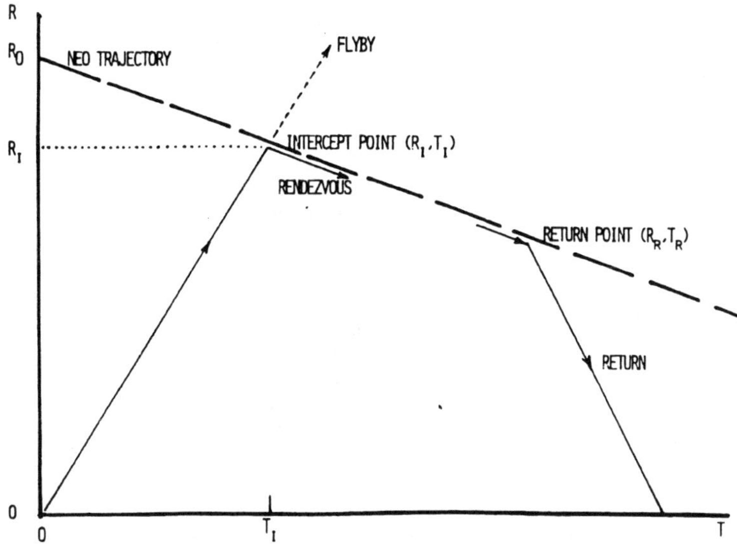

FIGURE 1. Schematic diagram of trajectories for missions with Type I propulsion systems.

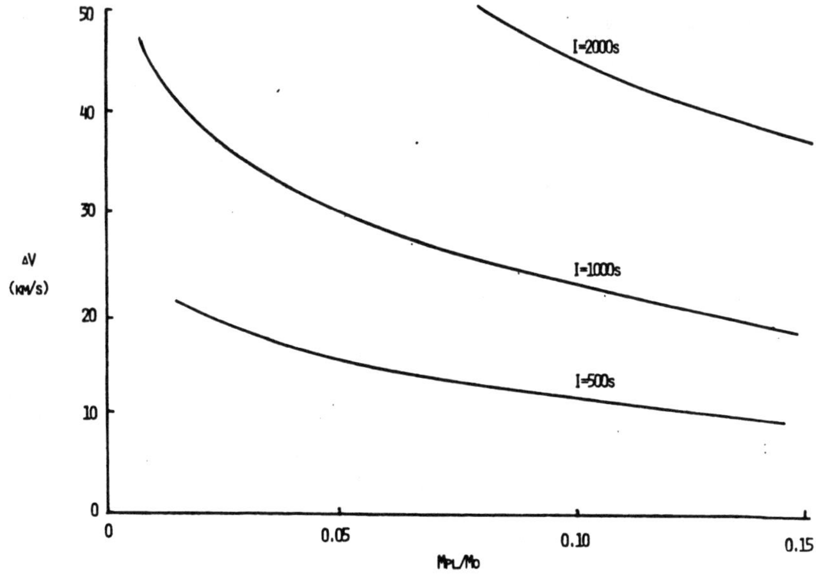

FIGURE 2. Total velocity increment as a function of payload ratio for Type I propulsion systems.

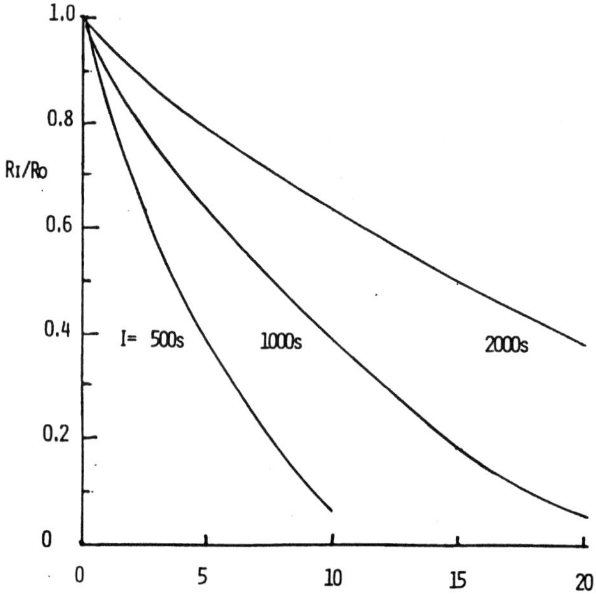

FIGURE 3. Ratio of intercept range to initial range for a Type I rendezvous mission with a NEO closing at V' for various values of specific impulse and $\mu_{pl} = 0.1$.

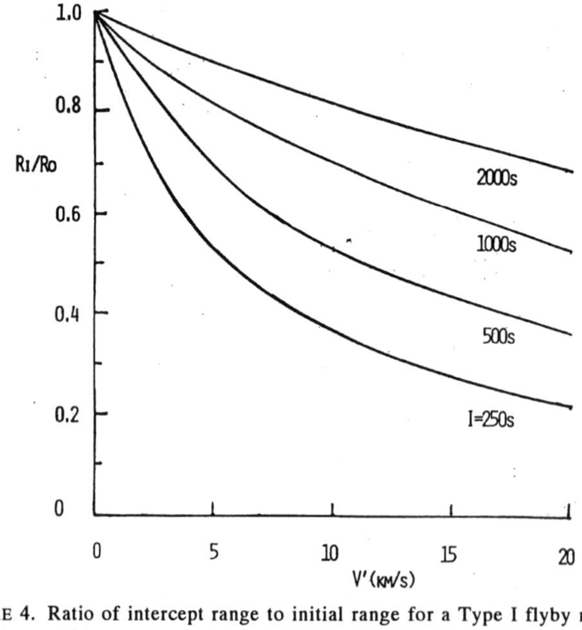

FIGURE 4. Ratio of intercept range to initial range for a Type I flyby mission to a NEO closing at V' for various values of specific impulse and $\mu_{pl} = 0.1$.

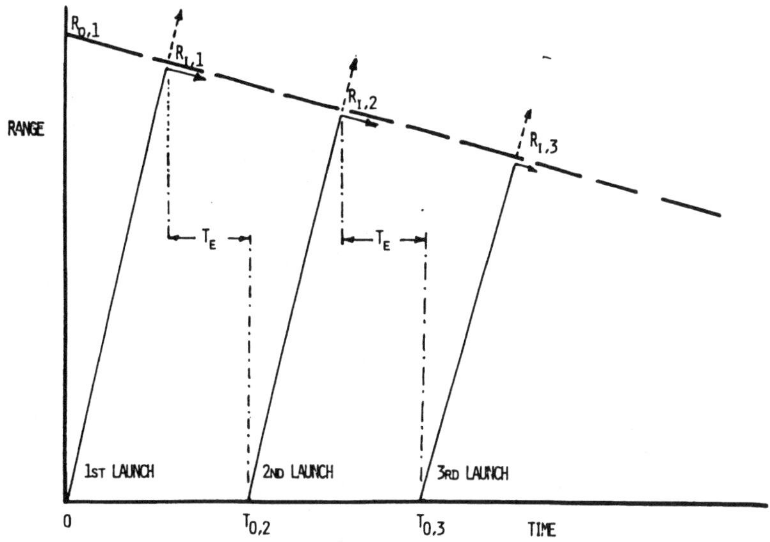

FIGURE 5. Schematic diagram of a typical multiple launch mission. The quantity T_e denotes the evaluation time required before launching another spacecraft.

NTP SEQUENTIAL MITIGATION MISSIONS

After a spacecraft rendezvous with a NEO carries out its mitigation mission, the effects produced must be assessed and, if it was not successful, another spacecraft must be launched. This procedure is illustrated in Fig. 5, where the time interval denoted by T_e represents the necessary time for evaluation and decision to launch again. The ratio of the range at intercept by the nth spacecraft to that at the time of the initial launch is given by

$$R_{i,n}/R_{0,1} = (R_{i,1}/R_{0,1})^n \prod_{j=1}^{n} (1 + 0.211 V' T_e / R_{0,j}), \tag{10}$$

where the ratio $R_{i,1}/R_{0,1}$ is given, for the flyby and rendezvous missions considered here, by Eq. (1). In general then

$$R_{i,n}/R_{0,1} \prec (R_{i,1}/R_{0,1})^n \tag{11}$$

and the inequality is slight if the evaluation time T_e is small enough such that

$$\frac{.211 V' T_e}{R_{0,i}} \lll 1. \qquad (12)$$

For evaluation times on the order of days, $T_e = 0(10^{-2}\ \text{yr})$, the deleterious effects on the interception range of the nth spacecraft become important when

$$R_{0,i}/V' = 0(10^{-2})$$

which, for $V' = 0(10\ \text{km/s})$, indicates that success in mitigation becomes problematical if intercept is not accomplished by the time the NEO is at around 0.1 AU. This speaks for the desirability of achieving first intercept at large ranges with spacecraft powered by propulsion systems with high specific impulse. There is obviously a systems optimization problem embedded here, since interception at large ranges presupposes sufficiently accurate tracking information to determine that a given NEO can be determined to be a threat at those ranges. This suggests the likelihood of generating an inordinate number of false alarms to be dealt with by means of expensive interdiction resources. On the other hand, reliance upon existing, limited performance, propulsion system engenders risks if multiple missions become required for NEOs of high (10 km/s or more) closing speed. Furthermore, if one sacrifices high specific impulse for flyby (direct intervention) mitigation missions instead of (surgical) rendezvous mitigation missions, one obtains, in return, the risks of producing a fragmented NEO closing with the Earth.

APPLICABILITY OF NEP SYSTEMS.

It has been pointed out previously that in the range of $0.5 \prec R_i \prec 1.5$ AU Type I propulsion systems have generally superior time performance for flyby and rendezvous missions. In addition, the relative unsuitability of return missions for critical NEO interventions which has been demonstrated here appears to indicate that Type II propulsion systems, which have better performance in the return mode, represents a less desirable alternative. Finally, although much study has been devoted to nuclear electric systems [8], and specific masses in the range of $0.1 \prec \alpha' \prec 10$ kg/kw appear to be achievable, no large-scale hardware experience has been demonstrated. On the other hand, large-scale Type I propulsion systems have generated values of I = 825 s for solid core reactors and $I \simeq 1000$ s seem to be achievable for solid particle bed reactor systems. Therefore, for the time extending out to 2020, the Type I propulsion system bears the most promise for providing realistic capability for NEO mitigation missions. The likelihood of extending the Type I system performance to higher levels of I = 2000 s requires the development of gas core fission reactors. Such reactors represent a new technology which requires substantial basic research, while Type II nuclear electric systems can exploit the solid core reactor experience. The ability to drive the specific mass to $\alpha' = 0(1)$ would provide propulsion performance on a par with Type I systems with $I = 2000$ s. Therefore, the near-term outlook demands a doubling of performance from the best Type I chemical systems (i.e., up to $I \simeq 1000$ s) and a further doubling of performance to the equivalent of $I \simeq 2000$ s. The latter may be best achieved by development of a Type I gas core fission rocket or a Type II nuclear electric system with $\alpha' \simeq 1$kg/kw.

DISCUSSION

An example of a planetary spacecraft mission that can operate at distances of 2.2 AU from the Sun (1.2 AU from Earth) is the (chemical) vehicle used for the Near-Earth Asteroid Rendezvous (NEAR) mission. However, this mission will take almost 3 years to reach its objective, the asteroid Eros, since a deep-space maneuver ($\Delta V = 279$ m/s) after a flyby of (253) Mathilde more than 16 months after launch is required, followed by an Earth swingby almost 2 years after launch [2]. Arrival at Eros will occur about one year after the Earth swingby. Such a mission provides a good example of how limited capabilities both in the launch vehicle and payload can still permit valuable scientific explorations of space. However, if reconnaissance and mitigation must be carried out within a limited time frame and with large payloads, chemical rockets are of little value; NTP vehicles are needed for large payload and rapid transit to a target.

The analysis presented here provides a first-order computation of the propulsion requirements for timely and deep-space missions, such as may be required for NEO threat mitigation. This work indicates that Type I nuclear thermal propulsion (NTP) systems currently provide the optimal platform for critical rendezvous and interception missions around 1 AU from the Earth. For such systems, the basic principles and analyses of NTP and its application to "near term" design configurations suitable to NEO missions are discussed by Venetoklis et al. [17]. In addition, operational issues germane to the analysis presented in this paper — the role of NTP in space exploration and NEO mitigation — are outlined by Willoughby et al. [19]. From somewhat slightly different perspectives, both of these studies indicate the need for nuclear propulsion systems for deep-space operations.

The ability to respond to an extraordinary NEO incident as described above strongly indicates the use and feasibility of nuclear propulsion of one type or another. It is exactly this propulsion mode for which government support, whether NASA, DOE, or DOD, has been severely curtailed. It seems particularly important to encourage research and development in this area of propulsion with the aim of demonstrating and testing hardware. Such a program would seem to provide an additional motivation for a presence in space as represented by the Space Station and proposals for a Moon base.

CONCLUSIONS

An analysis of propulsion system performance in missions designed to reconnoiter and mitigate the threat of NEOs closing with the Earth at planetary speeds has been presented. The present level of uncertainty concerning the material and structural nature of asteroids and comets and their response to the effects of orbital management makes the development of appropriate mitigation measures highly speculative. For any anticipated deflective posture, prudence dictates that the response be complete at large distances from the Earth so that the possibility of additional such responses, if required, may be carried out with reserve interceptor craft.

The one-dimensional field-free approach due to Moeckel [7] was used here to broadly characterize the capabilities of different propulsion systems in intercepting NEOs at a fixed distance from the Earth in the range of 0.5 to 1.5 AU. That method was extended in the present paper to include the case where the NEO is closing with the Earth. Results showed that flyby and rendezvous mitigation missions are realizable while return missions have insufficient operational latitude to be effective. For an initial mass equal to that of a Titan IV payload in LEO, it is demonstrated that only a single stage interceptor is feasible for the mitigation mission. The limitation of the specific impulse of a chemical propulsion system has been shown to be a serious obstacle for such missions. For the time frame extending out to the year 2020, the logical choice for a replacement is the solid core nuclear thermal rocket, with an effective doubling of the specific impulse. Further increase in performance is most likely to be found in serious development of a solid core nuclear electric rocket, exploiting the experience gained in building the thermal rocket.

Development times for advanced spacecraft are long, and many of the richly diverse propulsion systems that have been studied are far from proven. The important factor is to break out of the apparently endless concept phase and into flight test in order to gain valuable experience. A staged design and development effort on a Type I and a Type II experimental space vehicle for the NEO mission should be initiated to capitalize on efforts restarted in the 80's on advanced propulsion systems like nuclear thermal and nuclear electric, before the effects of their recent restopping in the 90's ends in loss of the skill base.

ACKNOWLEDGMENT

We are especially indebted to B. Marsden for clarifying issues regarding estimation of the lead time for NEO impact.

REFERENCES

1. ALVAREZ, L.W., W. ALVAREZ, F. ASARO, & H.V. MICHEL. 1980. Extraterrestrial Cause for the Cretaceous-Tertiary Extinction. Science. **208:** 1095–1108.
2. FARQUHAR, R. & J. VEVERKA. 1995. Romancing the Stone: The Near-Earth Asteroid Rendezvou. The Planetary Report. **XV:** (5) 8–11.
3. GEHRELS, T., Ed. 1994. Hazards Due to Comets and Asteroids. University of Arizona Press. Tuscon, Arizona
4. HELIN, E.F.& E.M. SHOEMAKER. 1979. Palomar Planet-Crossing Asteroid survey 1973–1978. Icarus. **40:** 321–328.
5. LAU, C.O. & N.D. HULKOWER. 1985. On the Accessibility of Near-Earth Asteroids. AAS/AIAA Astrodynamics Specialist Conference, AAS85-352, Colorado, August 12–15. (Also in Astrodynamics 1985, Volume 51, B. Kaufman et al., Eds. Univelt Inc. San Diego, California.
6. MARSDEN, B.G. & D.I. STEEL. 1994. Warning times and impact probabilities for long-period comets. *In* Hazards Due to Comets and Asteroids, T. Gehrels, Ed.: 221–239. University of Arizona Press. Tuscon, Arizona.
7. MOECKEL, W.E. 1972. Comparison of Advanced Propulsion Concepts for Deep Space Exploration. J. Spacecraft. **9:** (12) 863–868.
8. MYRABO, L.N. 1984. Space Nuclear Multi-Mode Reactors. *In* Orbit-Raising and Maneuvering Propulsion, L.H. Caveny, Ed.: Vol. 89: 477–494. Progress in Aeronautics and Astronautics. AIAA. New York, NY.
9. OSTRO, S. The Role of Groundbased Radar in Near-Earth Object Hazard Identification and Mitigation. *In* Hazards Due to Comets and Asteroids, T. Gehrels, Ed.: 259–282. University of Arizona Press. Tuscon, Arizona.
10. RABINOWITZ, D. 1994. The Size and Shape of the Near-Earth Asteroid Belt. Icarus. **111:**. 364–377.
11. RABINOWITZ, D., E. BOWELL, G. SHOEMAKER, & K. MUINONEN. 1994. The Population of Earth-Crossing Asteroids. *In* Hazards Due to Comets and Asteroids, T. Gehrels, Ed.: 285–312. University of Arizona Press. Tuscon, Arizona.
12. REMO, J.L. & P.M. SFORZA. 1995. Subsurface Momentum Coupling Analysis for Near-Earth-Object Orbital Management. Acta Astronautica. **35:** (1) 27–33.
13. REMO, J.L. 1994. Classifying and Modeling NEO Material Properties and Interactions. *In* Hazards Due to Comets and Asteroids, T. Gehrels, Ed.: 551–596. University of Arizona Press. Tuscon, Arizona.
14. SAUER, C.G., JR. 1991. Current Trajectory Options for a Comet Nucleus Sample Return Missions. AAS91-474: 1055–1069.
15. SFORZA, P.M., M. SHOOMAN, & D. PELACCIO. 1993. A Safety and Reliability Analysis for Space Nuclear Thermal Propulsion Systems. Acta Astronautica. **30:** 67–83.
16. SFORZA, P.M. & J.L. REMO. 1997. Propulsion Options for Missions to Near Earth Objects. Acta Astronautica. In press.
17. VENETOKLIS, P., E. GUSTAFSON, E. MAISE, & J. POWELL. 1994. Applications of Nuclear Propulsion to NEO Interceptors. *In* Hazards Due to Comets and Asteroids, T. Gehrels, Ed.: 1089–1110. University of Arizona Press. Tuscon, Arizona.
18. WALLACE, R.A., W.H. BLUME, N.D. HULKOWER, & C.L. YEN. 1985. Interplanetary Missions for the Late Twentieth Century. J. Spacecraft. **22:** (3) 316–324.
19. WILLOUGHBY, A.J., M.L. MCGUIRE, S.K. BOROWSKI, & S.D. HOWE. 1994. The Role of Nuclear Thermal Propulsion in Mitigating Earth-Threatening Asteroids.*In* Hazards Due to Comets and Asteroids, T. Gehrels, Ed.: 1073–1088. University of Arizona Press. Tuscon, Arizona.

20. YEOMANS, D.K. & P.W.CHODAS. Predicting Close Approaches of Asteroids and Comets to Earth. *In* Hazards Due to Comets and Asteroids, T. Gehrels, Ed.: 1241–258. University of Arizona Press. Tuscon, Arizona.

High-Performance Ultra-light Nuclear Rockets for Near-Earth Objects Interaction Missions[a]

JAMES POWELL, GEORGE MAISE, HANS LUDEWIG, AND MICHAEL TODOSOW

Department of Advanced Technology
Brookhaven National Laboratory
Upton, New York 11973-5000

ABSTRACT: The performance capabilities and technology features of ultra compact nuclear thermal rockets based on very high power density (30 Megawatts per liter) fuel elements are described. Nuclear rockets appear particularly attractive for carrying out missions to investigate or intercept near-Earth objects (NEOs) that potentially could impact on the Earth. Many of these NEO threats, whether asteroids or comets, have extremely high closing velocities, i.e., tens of kilometers per second relative to the Earth. Nuclear rockets using hydrogen propellant enable flight velocities 2 to 3 times those achievable with chemical rockets, allowing interaction with a potential NEO threat at a much shorter time, and at much greater range. Two versions of an ultra compact nuclear rocket based on very high heat transfer rates are described: the PBR (Particle Bed Reactor), which has undergone substantial hardware development effort, and MITEE (MIniature ReacTor EnginE) which is a design derivative of the PBR. Nominal performance capabilities for the PBR are: thermal power $\simeq$ 1000 MW thrust $\simeq$ 45,000 lbsf, and weight $\simeq$ 500 kg. For MITEE, nominal capabilities are: thermal power 100 MW; thrust $\simeq$ 4500 lbsf, and weight $\simeq$ 50 kg. Development of operational PBR/MITEE systems would enable spacecraft launched from LEO (low-Earth orbit) to investigate intercept NEO's at a range of ~100 million kilometers in times of ~30 days.

INTRODUCTION

Major development efforts on nuclear thermal rocket propulsion have been carried out in the US and former Soviet Union since the 1950's, with a cumulative total expenditure of well over ten billion dollars (in current dollars). Although nuclear thermal rockets have not proceeded to the operational, or even the flight test stage, there has been extensive ground testing of nuclear rocket systems.

Nuclear thermal rockets offer the potential for a major increase in performance as compared to existing chemical rockets, which have essentially matured technologically with only marginal improvement likely. Using pure hydrogen propellant, nuclear rockets can achieve specific impulses (lbs of thrust per lb of mass per second) of ~1000 seconds, as compared to only ~425 seconds for H_2/O_2 rockets and ~250 seconds for solid propellant rockets.

[a]This work was perfomed under the auspices of the U.S. Department of Energy.

The higher specific impulse of nuclear rockets offers the potential for much greater operational velocities. For equivalent engine thrust/weight ratios, and the same mass ratios and payload weights, a nuclear rocket would achieve 2 to 3 times the ΔV (velocity increment) achievable with chemical rockets. This translates into much higher terminal velocities (a factor of 2 to 3 higher), or much greater payloads for a given terminal velocity, or some combination of the two.

Development efforts on nuclear rockets in the US initially focussed on the NERVA [1] System. NERVA was based on the use of uranium containing graphite or carbide fuel rods located in channels through a graphite moderator core. Hydrogen flowed axially down the multiple channels, exiting from the bottom of the core at ~2750 K. A number of NERVA type reactors were ground tested at the Nevada Test Site, at powers up to 5000 MW(th).

Because of its large size fuel elements, and the long axial flow path, the fuel element power density in NERVA reactors was constrained to be relatively low, on the order of 2 to 3 MW per liter. This constraint resulted in low thrust to weight ratios, on the order of 5 to 1, for NERVA engines. This feature combined with the large relatively heavy reactors dictated by the choice of moderator (graphite or zirconium hydride plus graphite), severely limited the potential usefulness of NERVA.

Development work on NERVA ceased in 1972 when no practical mission could be identified. Research and development on NERVA type systems continued in the Soviet Union [2] until the early '90s, when it too was canceled. The Soviet version used twisted ribbon type uranium-zirconium-niobium carbide fuel rods stacked along an axial flow channel inside a graphite moderated core. Fuel elements were tested at temperatures up to 3000 K for periods on the order of one hour. Performance appeared satisfactory.

The Soviet type NERVA had the same low power density and heavy weight limitations of the US NERVA, however. Such nuclear rockets, while potential candidates for heavy lift type missions, e.g., manned journeys to Mars, do not appear promising for lightweight, high velocity applications, such as unmanned planetary scientific probes, or interaction missions with NEOs.

Work on a much higher performance nuclear rocket concept [3], the Rotating Bed Reactor (RBR), was initiated at Brookhaven National Laboratory in the 1960s. The RBR nuclear rocket used small diameter (~400 micron) coated HTGR type fuel particles instead of large solid fuel rods. The particles were directly cooled by the flowing hydrogen propellant. Their much smaller size, and the consequent much greater heat transfer area per unit volume of fuel (~100 cm^2 per cubic centimeter of fuel bed) enabled much greater power densities than NERVA — on the order of 30 Megawatts per liter.

In the RBR, the fuel particles were held by centrifugal force inside a porous rotating cylindrical basket (~1000 rpm) through which the inlet hydrogen passed. After passing through the fuel bed, the hot H_2 exited through an outlet nozzle at the bottom of the rotating basket. An external moderator surrounding the annular fuel downscattered the released fast neutrons to thermal energies, where they then diffused back into, and were absorbed by the fuel bed.

The RBR development was at an early stage when the US nuclear rocket program stopped in 1972, with only cold flow, nonnuclear testing having been carried out.

Major US R&D on nuclear rockets resumed in 1987 on the Particle Bed Reactor (PBR), a derivative of the RBR. In the PBR [6], [4], [5] the elements consist of fuel particles that are placed into annular packed beds held between two porous cylindrical "frits." Hydrogen coolant flows radially inwards through the outer "cold" frit at cryogenic temperature, then through the thin bed of fuel particles, and finally exits through the hot frit at ~3000 K. The hot hydrogen then flows out through the central hot channel to the rocket exit chamber and nozzle.

The PBR program was directed toward defense applications. Used as an upper stage on a standard booster in place of a conventional hydrogen/oxygen stage, it would lift into low-Earth orbit (LEO) 3 to 4 times the payload that the conventional upper stage could.

Development of the PBR proceeded successfully until 1993, when the program was terminated due to the ending of the Cold War and the PBR mission. Major technical advancements to that point included the development of fuel particles capable of operation at 3000 K, along with hot and cold frits and other reactor hardware. Full size PBR fuel elements were thermally hydraulically tested at 2500 K; these tests demonstrated the capability for 30 MW per liter power density in transient blow down experiments. Nuclear critical performance, including critical mass and 3D power distributions, was verified in a zero power critical assembly of the PBR. Tests of a PBR fuel element were also carried out in an operating reactor. The PBR nuclear engine appears very attractive for planetary exploration and NEO interaction missions, because of its low weight (~500 kg) and high specific impulse (~1000 seconds).

The PBR design and hardware development was guided by the goal of a high thrust (45,000 lbsf) engine for heavy lift launch applications. Planetary and NEO mission launched from LEO orbits, however, do not require high thrust capability. Removing this constraint enables further major reductions in engine weight, since core size is no longer determined by the requirement for high thermal powers (~1000 MW for the PBR) but only by the requirement that the core be critical. This allows a considerably smaller, lighter reactor.

A new concept for an ultra compact nuclear engine is described in this paper, termed the MITEE (MIniature Reactor Engine) nuclear rocket. MITEE is a derivative of the PBR, achieving comparable power densities, on the order of 30 MW per liter of fuel element. However, its total power is much lower than the PBR (~100 MW compared to ~1000 MW for the PBR). Its total weight is also much lower (~50 kg compared to ~500 kg for the PBR).

The MITEE engine appears very attractive for accelerating small light payloads from Earth orbit to ultra high velocities, i.e., several tens of kilometers per second. At a velocity of 35 kilometers per second for example, the payload could travel 100 million kilometers in a month. Probes could reach the outer planets in a few years, rather than decades, and interaction missions with asteroids and comets could occur at long ranges, i.e., ~1 AU, instead of close-in to Earth.

Figure 1 illustrates an overview of nuclear rocket projected capabilities. The MITEE engine would appear to be the end of the line for nuclear rockets, since the lower limit to its weight will be determined by criticality considerations, and it would use the lightest possible materials of construction.

The remainder of this paper describes the PBR and MITEE engines in greater detail, together with their technology issues and requirements.

THE PBR ENGINE

Figure 2 shows the basic construction of the PBR reactor [4]. The small diameter (~400 microns) fuel particles are contained between two porous thin cylindrical tubes, an outer "cold frit," and an inner "hot frit". Cold (~100 K) H_2 coolant flows radially inward through the annual packed particle bed; after exiting from the hot frit, the hot (~3000 K) H_2 turns and flows out through the central flow channel to the outlet chamber and exhaust nozzle.

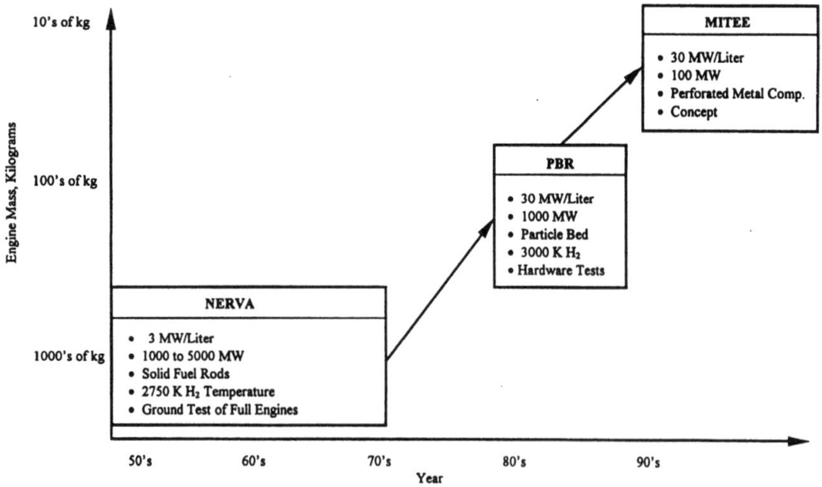

FIGURE 1. Overview of nuclear rocket capabilities.

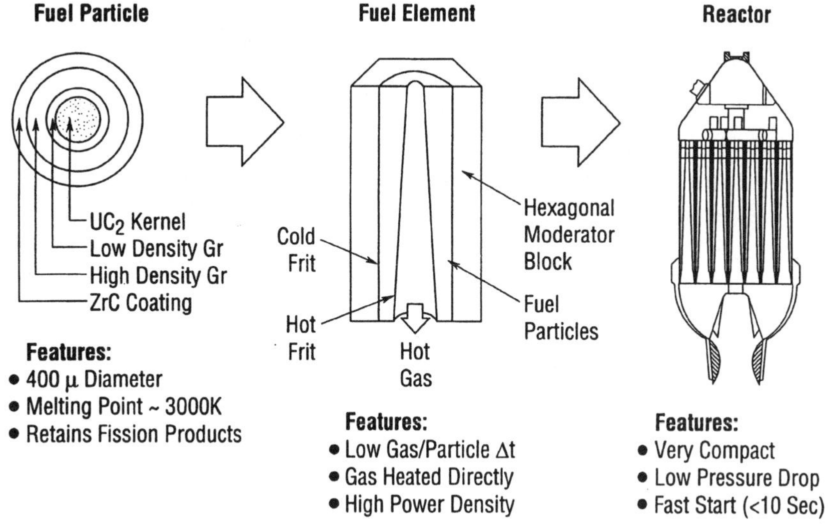

FIGURE 2. The particle bed reactor.

THE PBR ENGINE

Figure 2 shows the basic construction of the PBR reactor [4]. The small diameter (~400 microns) fuel particles are contained between two porous thin cylindrical tubes, an outer "cold frit," and an inner "hot frit". Cold (~100 K) H_2 coolant flows radially inward through the annual packed particle bed; after exiting from the hot frit, the hot (~3000 K) H_2 turns and flows out through the central flow channel to the outlet chamber and exhaust nozzle.

The annular fuel elements are arranged in a surrounding matrix of moderator made of a beryllium structure with hydrogenous inserts (polyethylene or ^{7}LiH). The full reactor core typically contains 19 or 37 fuel elements, depending on the desired power level. The complete PBR engine (Fig. 3) includes the reactor, exhaust plenum chamber, nozzle, turbo pump, turbo pump exhaust, structure, and thrust vector controls.

Detailed PBR engine designs were carried out for a wide range of parameter space, including power level, fuel element diameters, fuel element pitches, moderator composition bed power density, number of fuel elements. The neutronic analysis utilized the explicit geometry Monte Carlo (MCNP) code with point cross sections to accurately model the all aspects of the reactor in 3 dimensions. Table 1 lists the design parameters and features for a typical 1000 MW design.

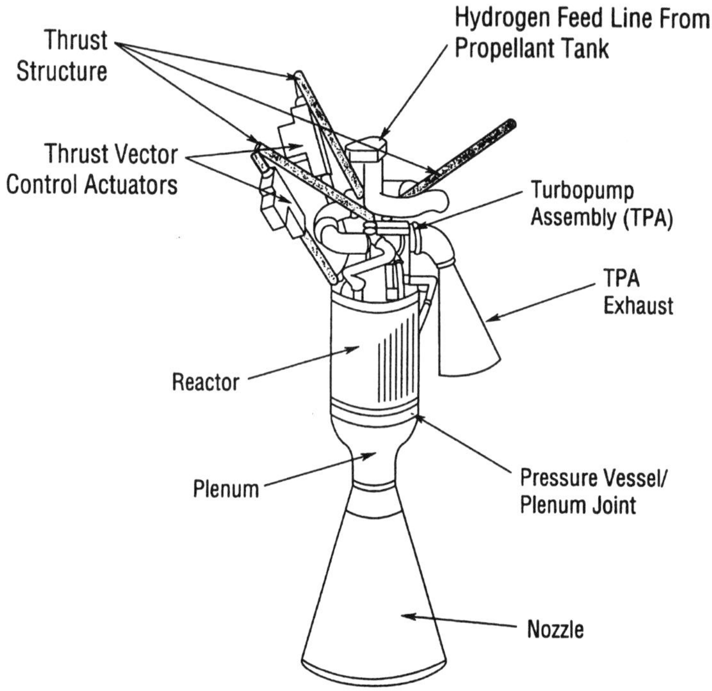

FIGURE 3. PBR engine description.

TABLE 1. Design Features and Parameters for a 1000 Megawatt PBR Nuclear Engine

Power level	= 1000 MW
Mixed mean outlet temperature	= 3000 K
Average fuel bed power density	= 40 MW/Liter
H_2 chamber pressure	= 70 Atm
Number of fuel elements	= 19
K_{eff}	= 1.18 (criticality constant)
Specific impulse	= 950 seconds
Fuel element pitch/diameter ratio	= 1.5
Frit mass	= 29 kg (aluminum cold frit, coated carbon-carbon hot frit)
Fuel bed	= 56 kg
Moderator	= 159 kg (beryllium - polyethylene)
Grids	= 100 kg
Core total	= 345 kg
Pressure vessel	= 45 kg (carbon - carbon)
Balance of plant	= 647 kg (turbo-pump, thrust vector control, nozzle, propellant management system, and instrumentation)
Thrust to weight ratio	= 31

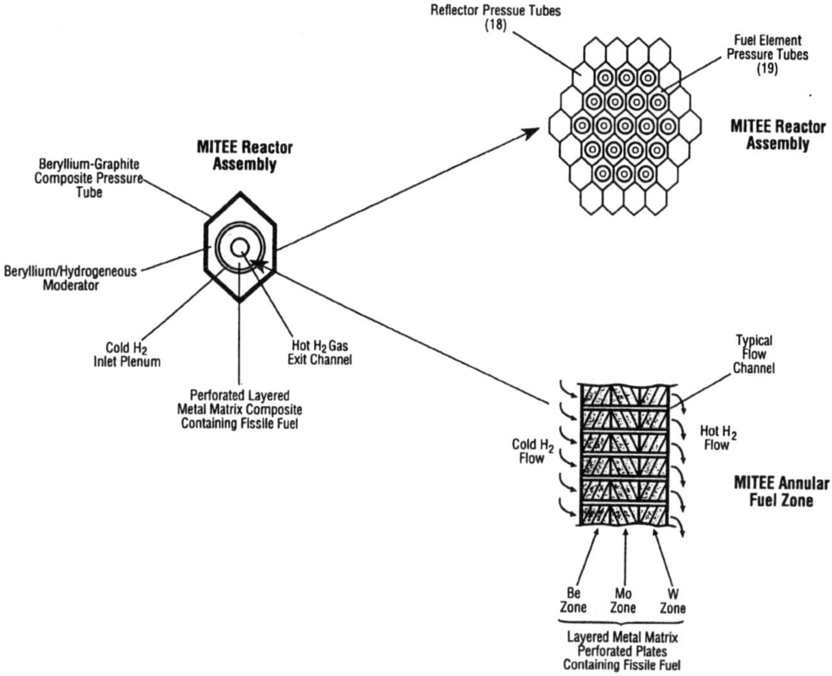

FIGURE 4. The MITEE (MIniature ReacTor EnginE) concept.

The time-dependent neutronic behavior of PBR engines was also modeled in detail, including transient power response during fast startup, control requirements, and temperature stability. Three-dimension power distributions were obtained for all points in the fuel elements and moderator.

The neutronic analyses were validated by carrying out a series of critical experiments in a zero power version of the PBR. The experimental results for K_{eff}, temperature coefficient, 3D power distribution, and other reactor parameters agreed very closely with analytical predictions.

The power density performance of the PBR was experimentally validated by an extensive series of transient blowdown tests on hot PBR fuel elements (the elements did not contain fissile fuel). The elements were first heated to high temperature (~2500 K) and then subjected to a full flow blowdown with cold inlet high pressure H_2. Hydrogen heating rates equivalent to particle bed power densities of 30 Megawatts per liter were measured during the transient blowdown phase. Individual single elements were tested, along with a multi-element assembly (7 elements) that was representative of the central portion of 19 element PBR core.

Major accomplishments in the PBR materials development program included the fabrication and testing of coated fissile fuel particles capable of operating in 3000 K high pressure hydrogen for long periods; high strength coated graphite and carbon-carbon hot frits that could operate in high pressure 3000 K hydrogen for long periods; and porous aluminum cold frits that enabled precise control of local H_2 inlet flow, so that local H_2 flow rates could be optimally matched to the 3D axial, azimuthal, and radial variations in nuclear power generation rates.

Along with the materials development program, a PBR hardware component development effort was also carried out. This included a lightweight, high temperature H_2 turbo pump that utilized a small portion of the hot outlet H_2 flow for drive power; a coated high temperature carbon-carbon nozzle; and a lightweight, high strength carbon-carbon pressure vessel.

At the close of the PBR program, the first of a planned series of PBR nuclear fuel elements was undergoing tests in the Sandia ACRR reactor. This test (NET-1) demonstrated the generation of hot H_2 coolant from a particle bed fuel element. Following completion of the NET tests, it was then planned to operate a 7 element test reactor (PIPET) at the Nevada Test Site. Following the successful demonstration of full power density (30 MW per liter), full temperature (3000 K) PBR fuel elements in PIPET, it was planned to carry out the ground test of a complete PBR engine. All testing was to be carried out in a fully contained system, with zero release of fission products to the environment.

While the overall objective of developing an operational PBR engine was not achieved, due to termination of funding in 1993, the program was very successful. Over its 6-year period, it resolved the major technical unknowns that were issues at the start of the program, and developed the materials, the neutronic and mechanical design, and many of the components required for the operational engine. If the program had continued, it appears likely that the target date for testing a flight capable engine in the late 90's would have been met.

THE MITEE ENGINE

Toward the close of the PBR project, Brookhaven National Laboratory (BNL) was investigating modifications to further reduce the size and weight of the PBR rocket. These included incorporating fissile fuel into the hot (the afterburner concept) and cold frits, a hydride spike (inside the hot gas channel), the "sock" reactor (nested frits containing fuel), and moderator with fuel inclusions. These approaches substantially reduced size and weight. However, due to the termination of the PBR program, there was insufficient time to fully investigate their true potential and to develop self-consistent reactor designs.

Based on these earlier studies, a new concept for a ultra compact and lightweight nuclear rocket is now proposed, termed MITEE. Figure 4 illustrates the basic MITEE concept. In contrast to the PBR, which utilized a relatively large single pressure vessel, the MITEE core consists of a set of hexagonal pressure tubes, each containing an outer shell of moderator and an inner cylindrical fuel element. The H_2 coolant flow is radially inwards through the cylindrical fuel element, with cold H_2 entering the outer surface at low temperature ~100 K) and exiting the inner surface at high temperature (~3000 K).

Unlike the PBR, where the fissile fuel was contained in small individual fuel particles (~400 microns in diameter), in MITEE, the fuel is contained as fibers or particles in metal thin plates of perforated matrix composites that form the multilayered annular cylindrical fuel element. As in the PBR, there is a central hot gas channel, down which the hot H_2 flows to exit the reactor.

In contrast to the PBR, however, the MITEE hot gas channels do not exit into a common large hot gas plenum and a single throat/nozzle unit. Instead, each pressure tube has its own individual nozzle (Fig. 5) which exhausts to space. The combined thrust from the assembly of nozzles then provides the total engine thrust. This arrangement results in a much simpler and lighter reactor.

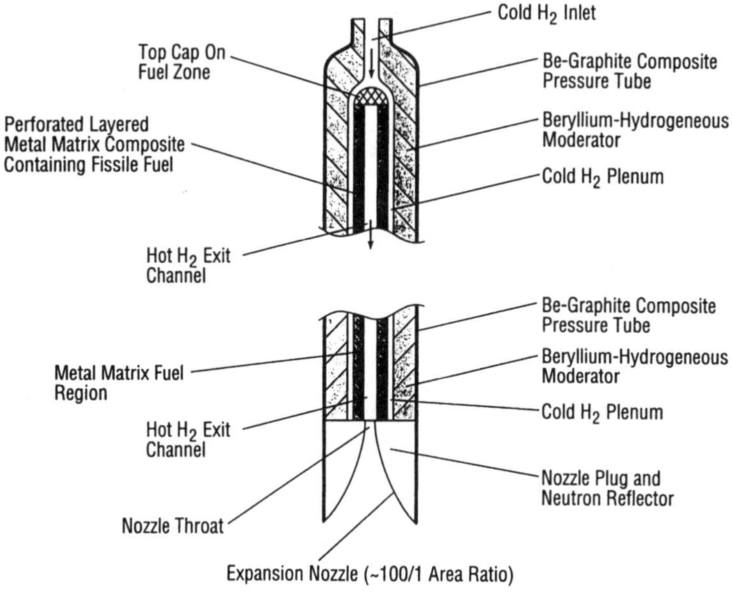

FIGURE 5. MITEE pressure tube and exit nozzle.

As shown in Fig. 4, the core is made up by assembling a number (probably either 19 or 37) of the pressure tube elements. The core is surrounded by a row of pressure tubes that contain only reflector material.

Table 2 compares the main features of the PBR and MITEE. H_2 exit temperature, specific impulse and power densities are simliar, with the principal differences between the two types of reactors being the nature of the core, the power and thrust level, and the overall weight.

Two versions of MITEE appear possible:

- MITEE-1 with a solid moderator (lithium hydride or polyethylene)
- MITEE-2 with liquid H2 moderator.

TABLE 2. Comparative Features of the PBR and MITEE Nuclear Engines

Feature	PBR	MITEE
Type of construction	Single pressure vessel	Assembly of multiple individual pressure tubes
Type of nozzle	Single nozzle	Multiple nozzles — one for each pressure tube
Fuel form	Coated particles (~400 μ diameter)	UC/C or UO_2 particles or fibers in thin metal matrix composite plates
Heat transfer to H_2	Packed bed of fuel particles	Perforated metal matrix plates
Weight	~500 kg	~50 kg
Power level	~1000 MW	~50 to 100 MW
Power density in fuel zone	~30 to 40 MW/Liter	~20 to 30 MW/Liter
Thrust, lbsf	~45,000	~3000 to 4000
H_2 exit temperature (mixed mean)	~3000 K	~3000 K

TABLE 3. Principal Features of MITEE-1 and MITEE-2 Moderators

	MITEE-1		MITEE-2
Feature	LiH Moderator	Polyethylene Moderator	Liquid H_2 Moderator
Moderator form	LiH particles in Be-Graphite metal matrix composite	Polyethylene melted into Be-Graphite honey-comb pressure tube	Liquid H_2 inside Be-Graphite pressure tube
Moderator maximum temperature	~700 K	~400 K	~30 K
Volume fraction of moderator	~0.8	~0.8	~0.9
H_2 density in mod material, cm^{-3}	6.2×10^{22}	8.4×10^{22}	4.2×10^{22}
Effective H_2 density, cm^{-3}	5.0×10^{22}	6.5×10^{22}	3.8×10^{22}
Effective mass density of mod/Be-G mixture, g/cm^3	1.02	1.14	0.24
Note: Be-Gr mass density = 1.8 g/cm^3			

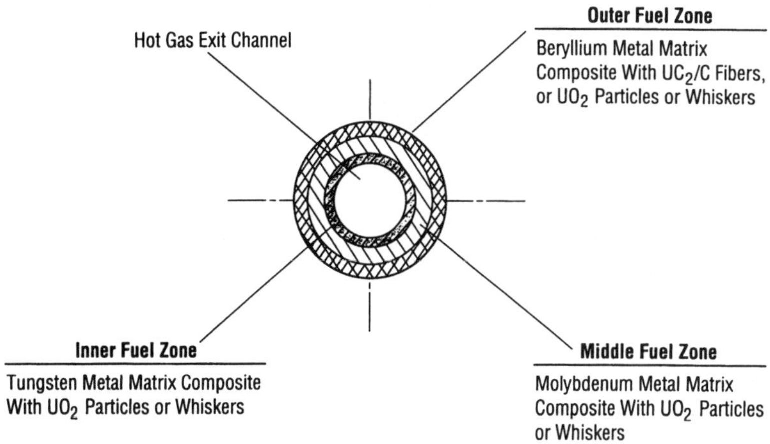

FIGURE 6. MITEE fuel region.

Table 3 compares the principal features of the moderator for the MITEE-1 and MITEE-2 designs. The liquid H_2 moderator has by far the lowest density; however, its core size will be substantially larger, as will be discussed shortly, which will tend to narrow the differences in the overall weights for the various reactor designs. Because of the low temperature, the effective thermal diffusion length in liquid H_2 will be significantly smaller than that for room temperature water (L ~ 2.5 cm). This will tend to require a larger number of pressure tubes (e.g., 37 for liquid H_2 compared to 19 for the solid moderators). To some extent, the enhanced absorption for low temperature neutrons in cold hydrogen will be compensated by the reduced scattering cross section of para-hydrogen, which will tend to increase the value of L and offset the increased absorption cross section.

The annual MITEE region, shown in Fig. 6, would have 3 zones: (1) an outer zone of a beryllium metal matrix composite with graphite fibers that were loaded with UC_2 or UO_2 particles or whiskers, (2) a middle zone of a molybdenum metal matrix composite with UO_2 particles or whiskers, and (3) an inner zone of a tungsten metal (separated tungsten –184, to reduce the effective neutron absorption cross section) matrix composite with UO_2 particles or whiskers.

Table 4 gives nominal properties and operating temperatures for the 3 fuel forms. To first order, each zone would have the same ^{235}U loading per unit volume. The magnitude of the actual loading would be determined by the fuel volume required to meet the desired power output, and the required critical mass. The maximum achievable fuel loading per unit volume probably will be on the order of 3 g/cm^3 (including the gas flow channels in the metal matrix). For a midrange reactor of 75 MW with a power density of 25 MW/Liter, this

TABLE 4. Nominal Properties and Operating Temperatures for Metal Matrix Fuel Forms

Parameter	Be Matrix	Fuel Form M_O Matrix	W Matrix (separated W^{184})
U-235 form	UC_2/C fibers or UO_2 particles or UO_2 whiskers	UO_2 particles or UO_2 whiskers	UO_2 particles or UO_2 whiskers
Matrix melting point, K	1557	2893	3643
Matrix maximum operating temperature, K	1200	2300	3000
Matrix density, g/cm^3 (100% density factor, U^{235})	1.8	10.2	19.3
Effective density of metal in fuel zone, g/cm^3 (20% holes and 20% U^{235} by volume in matrix)	1.08	6.12	11.58
Fraction of fuel region occupied by fuel form	0.40	0.37	0.23
Effective density of metal in fuel region, g/cm^3	0.43	2.26	2.66

corresponds to a maximum achievable M_{235} of 75/25 (3), or 9 kg, of uranium. Based on critical data, this appears to be more than will be required.

The metal matrix zones consist of multiple thin perforated plates (probably arranged as a spiral wrap for each of the 3 metal zones) through which the H_2 coolant flows. For weight estimation purposes, the plates are assumed to have 20% hole area. In addition, the metal matrix composite is assumed to be 80% metal and 20% U-235 fuel. Summing up the effective density factor in Table 4, the average metal density in the fuel region is equal to 0.4 + 2.26 + 2.66 = 5.35 g/cm^3. Thus, a fuel region volume of 3 liters would have a total metal weight of 3 × 5.35, or about 16 kg.

MITEE reactor weights can be estimated as follows. The critical core volume of homogeneous water moderated spheres is ~10 liters if an infinite water reflector is used (a 5 cm thick water reflector approximates an infinite reflector), with a corresponding critical mass of approximately 1 kg of U^{235}. The 10 liter core volume corresponds to a radius of $R_C^\bigstar = 13.4$ centimeters.

The critical core radius using alternate hydrogenous moderators (i.e., LiH, polyethylene, or liquid H_2) will scale as

$$R_C = R_C^\bigstar \left(\frac{n_{H20}}{\bar{n}_H}\right)\left(\frac{1}{f_{mod}}\right), \tag{1}$$

where

$\bar{n}_H$ = effective H atom density in moderator (cf. Table 3),

f_{mod} = volume fraction of moderator in core.

and the corresponding core volume is given by

$$V_C = V_C^{\star} \left(\frac{n_{H20}}{\bar{n}_H}\right)^3 \left(\frac{1}{f_{mod}}\right)^3 , \tag{2}$$

where the value of f_{mod} is given by

$$\begin{aligned} f_{mod} &= \frac{V_C - V_{\text{Metal Matrix}} - V_{\text{Hot Channel}}}{V_C} , \\ &= 1 - \frac{(V_{MM} + V_{HC})}{V_C} . \end{aligned} \tag{3}$$

For a nominal (75 MW) MITEE reactor, $V_{MM} \sim 3$ liters at a power density of 25 MW/Liter, and $V_{HC} \sim 1$ liter, so that

$$f_{mod} \cong 1 - \frac{4}{V_C} , \tag{4}$$

where V_C is given in liters.

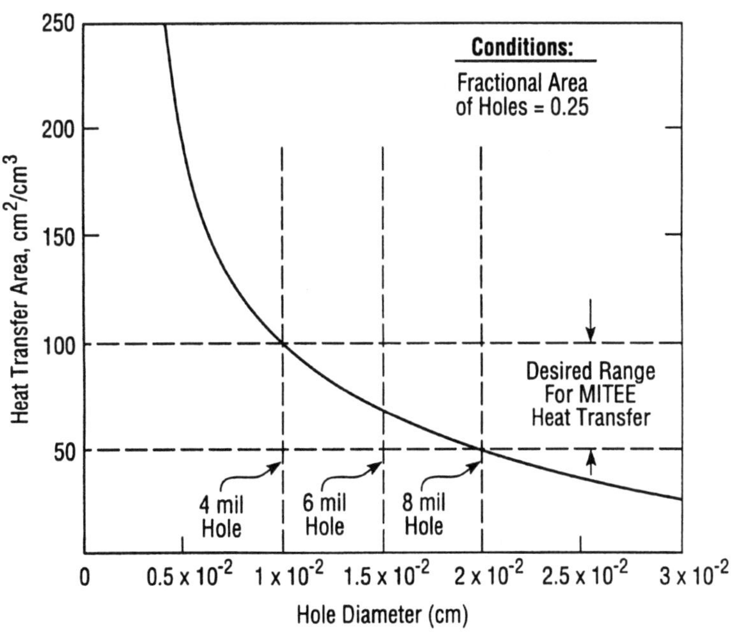

FIGURE 7. MITEE heat transfer area in metal matrix as a function of hole diameter.

TABLE 5. MITEE Reactor Dimensions and Weights as a Function of Moderator Type

	MITEE-1		MITEE-2
Parameter	LiH Mod	Poly E Mod	Liquid H_2 Mod
n_H, atoms/cm^3	5×10^{22}	6.5×10^{22}	3.8×10^{22}
Metal matrix volume, Liters	3	3	3
Hot channel volume, Liters	1	1	1
f_{mod}	0.85	0.77	0.92
R_C, centimeters	18.8	16.2	23.1
Core volume, Liters	27.5	17.5	51.1
Moderator volume, Liters	23.5	13.5	47.1
Moderator weight, kg	24.0	15.4	11.3
Metal matrix weight, kg	16	16	16
U_{235} weight, kg	~2	~2	~3
Total core weight, kg	42	33.4	30.3
Reflector thickness, cm	7.5	7.5	7.5
Reflector volume, Liters	48.4	37.9	68.3
Reflector effective density, g/cm^3 (Liquid H_2 + Be-G)	0.25	0.25	0.25
Reflector weight, kg	12.1	9.5	17.1
Total reactor weight, kg	54.1	43.9	47.4

Iterating, the values of f_{mod}, R_C, and V_C shown in Table 5 as a function of the choice of MITEE moderator are obtained. The critical mass of uranium is approximated by

$$M_{25} = M_{25}^{\star} x \left(\frac{R_C}{R_C^{\star}}\right)^2 \; ; \; M_{25}^{\star} \sim 1 \text{ kg}. \tag{5}$$

The critical mass depends on the square of R_C rather than the cube, since the lower moderator density partially offsets the larger core volume. (The relationship in Eq. (5) assumes that the ratio of hydrogen atoms to uranium atoms remains fixed.]

Moderator, fuel, metal matrix, reflector, and total weights are shown in Table 5 for the 3 MITEE designs. Interestingly, all three designs have approximately the same total weight, that is, about 50 kg. This is a result of the lower density liquid H_2 moderator having a larger core and heavier reflector, even through the density of the liquid H_2 moderator itself is much lower than the density of the solid moderators.

Given this situation, MITEE-1 with the solid lithium hydride moderator appears to be the best choice. It is more stable and easier to cool than polyethylene. By operating at a high moderator temperature, neutron absorption by hydrogen will be much less than with the liquid H_2 moderator, so that the core looks more like a homogeneous reactor than a heterogeneous one. It also enables the use of fewer fuel elements (e.g., 19 instead of 37), without having excessive neutron absorption in the moderator.

The heat transfer area in the perforated metal matrix sheets is calculated by taking the surface area of the holes through the sheet. This neglects the additional heat transfer from the surface of the sheet, and thus underestimates the actual heat transfer capability of the metal matrix fuel form. Figure 7 shows the heat transfer area as a function of hole diameter for the case where 25% of the sheet surface area is penetrated by holes. At a hole diameter of 1.5×10^{-2} cm (6 mils) the heat transfer area is 67 cm^2/cm^3, comparable to the surface area in PBR beds.

Figure 8 shows a proposed construction for the layered array of perforated sheets. The gas flow holes through the sheets are located in a grid of slightly depressed lines formed in the sheets. When the sheets are layered together, the raised zones prevent closure of the holes in the sheets. Gas exiting through the set of holes in one sheet has no problem in flowing to, and entering, the set of holes in the next sheet.

This flow arrangement helps to mix the gas flow between sheets, and reduces the chance of thermal instability. The first sheet in the multilayer stack (the "cold frit") will have smaller holes so as to distribute and match the H_2 flow to local variations in the radial, axial, and azimuthal nuclear power distributors.

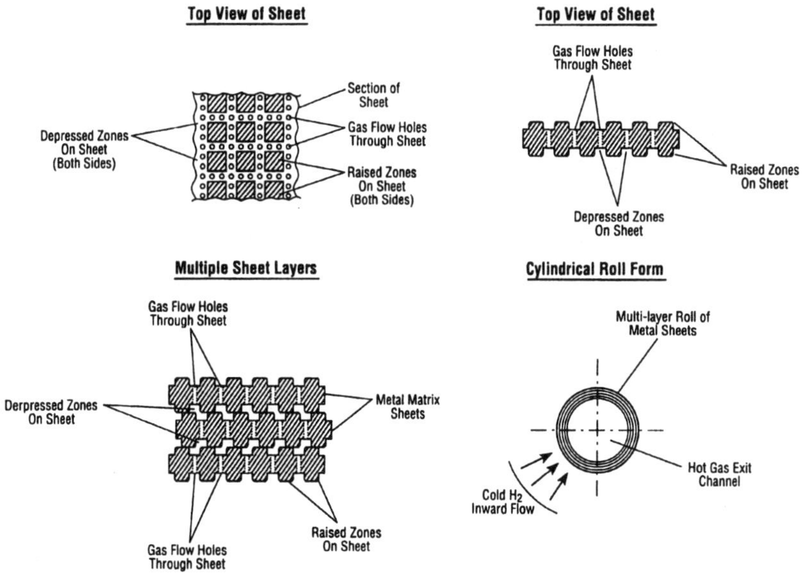

FIGURE 8. Construction of multiple matrix sheets.

TABLE 6. Nominal Design Parameters for MITEE

75 MW reactor power
3000 K exit temperature
25 MW/Liter average power density in metal matrix
19 fuel elements (hexagonal shape)
37 cm core diameter
Radius of hot gas channel = 0.685 cm
Total H_2 flow rate = 1.67 kg per second
Outer radius of metal matrix region in fuel element = 1.35 cm
Be matrix volume fraction = 0.40
Mo matrix volume fraction = 0.37
W matrix volume fraction = 0.23 (Tungsten-184)
Moderator volume fraction = 0.85
Core moderator composition 80% LiH (ρ = 0.82 g/cm^3) 20% graphite-Be (70% Be, 30% graphite)
Core moderator temperature = 700 K
Reflector composition 90% liquid H_2 (ρ = 0.07 g/cm^3) 10% graphite-Be (70% Be, 30% graphite)
Hole volume fraction in metal matrix = 0.25
Hole diameter = 1.5×10^{-2} cm (6 mil)
Exit temperature from metal matrix zone 1200 K from Be matrix 2300 K from Mo matrix 3000 K from W matrix

Table 6 gives nominal design parameters for the MITEE-1 engine with LiH moderator. The ^{235}U loading is not given, since it will have to be determined by MCNP calculations. Based on this initial examination, the MITEE-1 reactor will weigh approximately 50 kilograms. The weight of the turbo pump (~10 kg) assembly has not been included. Further optimization of the reactor is expected to reduce its weight somewhat, allowing the total engine weight to be about 50 kilograms.

UTILITY OF THE MITEE ENGINE FOR NEO INTERACTION MISSIONS

We now consider the quantitative advantages of the MITEE engine over a chemical rocket for a typical "terminal interception" of an NEO on collision course with Earth. Terminal interception, as its name implies, differs from what is termed "remote interdiction" where the orbit of a colliding

NEO is well known several periods prior to the collision, and deflection of the NEO can be effected at an early time. The Earth-crossing asteroids may eventually become subject to remote interdiction, but not long-period comets which may visit the Earth only once. At this writing, only a small fraction of even the "well-behaved" Earth-crossing asteroids have been found. Thus, for some time to come, both comets and asteroids will be subject to "terminal interception."

Considering now a case involving terminal interception, Fig. 9A shows an Earth-crossing NEO on collision course with Earth, represented in heliocentric coordinates. The orbit of the NEO is elliptic around the Sun. Transformation of the coordinates from heliocentric to geocentric, as shown in Fig. 9B, makes the problem very nearly one-dimensional, if we confine our attention to terminal interception, i.e., to a time scale of about one month. (Note: The one-dimensional analysis would certainly be inappropriate if we considered interaction time scales corresponding to the period of the NEO or the period of the Earth, i.e., one year.)

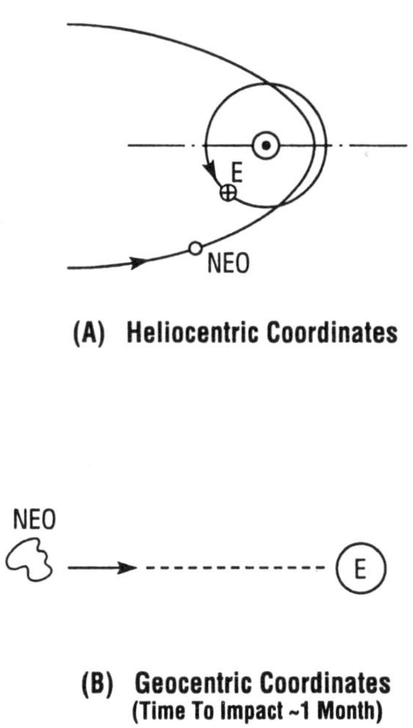

FIGURE 9. Coordinate systems for terminal intercept problem.

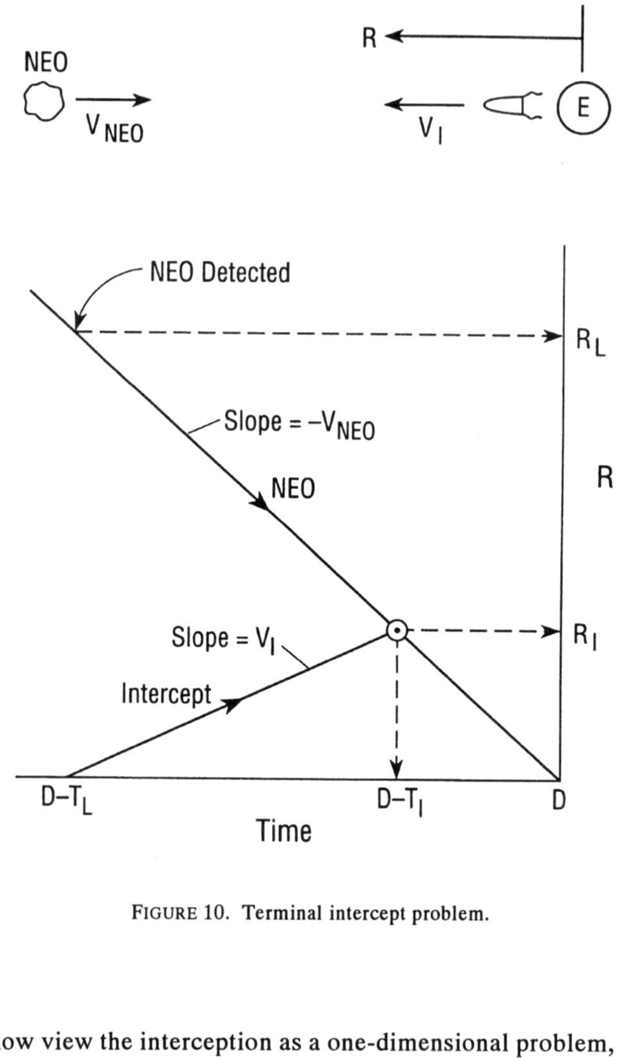

FIGURE 10. Terminal intercept problem.

If we now view the interception as a one-dimensional problem, we can illustrate the interception graphically on a time vs. radial-distance-from-Earth plot, as shown in Fig. 10. The abscissa is the time scale, represented in days prior to impact, with "D" as the day of predicted impact. The graph illustrates a case where an impacting NEO is discovered at a distance R_L from Earth on a date T_L days prior to impact. At the same time that the NEO is discovered, an interceptor is launched to interrogate/deflect/destroy the NEO. The slopes of the two lines, representing the motions of the NEO and the interceptor, correspond to their respective velocities. The interception occurs at T_I days before impact at a distance of R_I from the Earth.

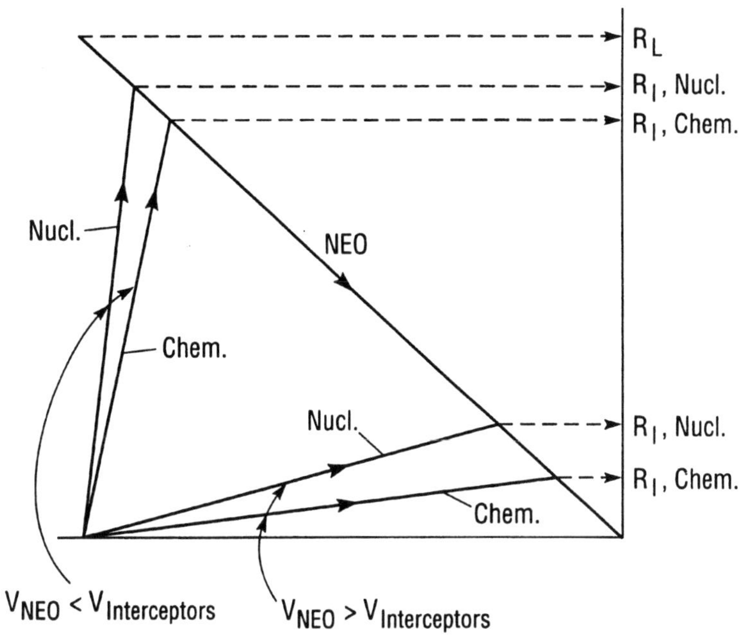

FIGURE 11. Comparison of chemical and nuclear powered rockets.

A graph such as Fig. 10 can be used to illustrate the advantages of a nuclear rocket such as MITEE. In Fig. 11 we show a comparison of the nuclear and the chemical rocket for two different situations: (1) The NEO velocity is moderate compared to the interceptor velocity, and (2) NEO velocity is larger than or comparable to the interceptor velocity. The first might correspond to an Earth-crossing asteroid, whereas the second corresponds to a very long-period comet. The latter typically have closing velocities of about 55 km/sec. The conclusions that can be drawn from Fig. 11 are:

1. Nuclear rocket is always better than chemical because interception occurs at a larger distance.
2. The advantage of the nuclear rocket is greatest when the NEO closes at a high velocity. These NEOs are likely to be long-period comets, which will probably be discovered only a short time prior to impact.

The analysis described above was used for a numerical example of the interception problem. Assume that a long-period comet is discovered to be on a collision course with Earth 30 days prior to impact. The comet has a closing velocity of 55 km/sec. Interception is considered via a chemical rocket and a nuclear rocket.

The interceptor powered by a chemical rocket has a terminal velocity of 7.5 km/sec. The interceptor reaches the comet 3.6 days prior to impact, when the comet is at a distance of 17,000,000 km from Earth. If the vehicle carried an explosive device it would have to impart a lateral velocity of 20.6 m/sec to the comet to miss the Earth.

For the interceptor powered by a nuclear rocket, the situation would be as follows. The terminal velocity of the interceptor is 15 km/sec. The interceptor reaches the comet in 6.4 days prior to impact, when the comet is at a distance of 30,000,000 km from Earth. If the vehicle carried an explosive device it would have to impart a lateral velocity of 11.5 m/sec to the comet to miss the Earth. The advantage of the nuclear rocket is clearly evident from this example.

CONCLUSIONS

Ultra compact nuclear engines offer the potential of achieving extremely high ΔVs of NEO intercept and rendezvous missions, using light-weight spacecraft. Total Vs of 30 to 40 km per second appear possible using an ultra-compact MITEE engine. The total weight of the MITEE engine is projected to be approximately 50 kilograms, with a thrust level of ~2000 kilograms force. The MITEE engine could accelerate a 50 kilogram payload to a flyby velocity of 35 km/sec based on a total spacecraft takeoff weight of 3 metric tons.

The MITEE engine is an advanced lightweight version of the PBR nuclear engine, for which an extensive development program was carried out during the period 1987 to 1993. During the program, the neutronic and thermal hydraulic performance of the PBR was experimentally validated, with demonstration of fuel element power densities of 30 Megawatt power densities. Hardware for the PBR — fuel particles, moderators, frits, etc. — were also developed and demonstrated.

The MITEE engine appears to be of excellent promise for NEO interaction missions, and would intercept and rendezvous with NEOs at much longer ranges than would be possible with chemical rockets.

ACKNOWLEDGMENTS

The authors express their deep appreciation and thanks to Ms. Barbara Roland for her efforts in preparing and typing the manuscript, and to Ms. Judith Otto for preparing the drawings.

REFERENCES

1. DURHAM, F.P. 1972. Nuclear Engine Definition Study Preliminary Report, Vol. 1, Engine Description. Los Alamos Laboratory, Informal Report, Los Alamos, NM, LA-5044-MS (September).
2. GOLDIN, A.V., ET AL. 1991. Development of Nuclear Rocket Engines in the USSR, presented by J.R. Welch. AIAA Propulsion Meeting, Sacramento, California.
3. HATCH, L.P., W.H. REGAN, & J.R. POWELL. 1960. Fluidized Beds for Rocket Propulsion. Nudeonics. **18:** (12) 102.
4. LUDEWIG, H., ET AL. 1993. Summary of Particle Bed Reactor Designs for the Space Nuclear Thermal Propulsion Program. BNL 52408.
5. LUDEWIG, H., J. POWELL, M. TODOSOW, G. MAISE, R. BARLETTA, & D.G. SCHWEITZER. 1996. Design of particle bed reactors for the space nuclear thermal propulsion program. Progress in Nuclear Energy. **30:** 1–65.
6. POWELL, J.R. & F.L. HORN. 1985. high power density reactors based on direct particle cooling. *In* Proc. Second Symposium on Space Nuclear Power Systems, Albequerque, New Mexico.

Near-Earth Resources

RICHARD GERTSCH,[a] JOHN L. REMO,[b] AND LESLIE SOUR GERTSCH[c]

[a]*Mining Engineering Department*
University of Missouri at Rolla
Rolla, Missouri 65401

[b]*Quantametrics, Inc.*
1 Brackenwood Path, Head of the Harbor
St. James, New York 11780
and
[c]*Mining Engineering Department & Center for Space Mining*
Colorado School of Mines
Golden, Colorado 80401

ABSTRACT: The technologies required to detect, track, categorize, and intercept objects in Earth-impacting orbits can also provide access to their rich storehouse of materials. Mitigation of a clear and present near-Earth object (NEO) threat to the Earth must provide the greatest assurance of success with the least risk to the planet. In some cases, mining a threatening NEO may become a viable alternative or supplement to a deflection or interception scenario. This converts the NEO threat to a near-Earth resource. NEOs can supply materials for a wide range of operations both in space and on Earth, as they are thought to contain large amounts of water, carbon, structural metals, industrial feedstocks of many types, and precious metals. This wealth has low overhead for utilization in space; some known NEOs would require lower transportation energy expenditure than lunar resources. Mining a NEO inherently requires, among other things, altering the mass distribution of the body during exposure, removal, and processing of the ore. These processes can be tailored to facilitate deflection of the body from Earth impact by altering its orbital characteristics. The advantage of NEO mining is that it can mitigate the threat — the primary effort — while converting it into resources for space exploration. This additional effort, within an appropriate time scale, allows sequential mitigation of the NEO in a controlled manner while providing the resources contained within the NEO for use either in space or on Earth. Both goals, to be successful, will require maximum utilization of all sources of knowledge. The two most important are an extensive reconnaissance of the target NEO and the long history of terrestrial mining practice. This paper discusses how current mining technology might be adapted to mine NEOs, whether threatening or not. It summarizes our knowledge of NEO composition, physical properties, and mining and processing methods, and points out areas where further research, especially physical testing in space, is vital. The great potential of NEO resources and the successful mitigation of NEO threats will be best realized if their utilization is considered from the earliest planning of Earth-NEO mitigations.

1. INTRODUCTION

Although near-Earth objects (NEOs) are potentially catastrophic to the Earth, they also provide opportunities to support space exploration and development. Asteroids and comets whose orbits approach Earth represent a huge storehouse of natural resources free of Earth's gravity well. For the exploration of space, both near and far, the utilization of such resources may mean the difference between success or failure of a space mission. Using space resources for space exploration is not a new idea. Others have examined the idea of mining asteroids simply for their wealth [7], [42], [32], [14]–[16], or for extraterrestrial manufacturing ([44] and others). Hartmann and Sokolov [22] discussed the policy aspects of this approach to impact hazard mitigation.

Mining is a risky undertaking, wherever it is performed. In addition to the constraints with which terrestrial miners and metallurgists are familiar, the largely unknown and hazardous environment of space will impose further difficulties. Conversely, the properties of space may offer opportunities that are unavailable or uneconomical on Earth. Only direct experience in space over a long term will reveal the possibilities.

2.1. Orebody Characterization

Before resource recovery can start, target NEOs must be identified, then understood to a degree that encourages success of both mining and processing. The goal of orebody characterization is to determine the quality, quantity, and availability of the ore. Detailed NEO characterization requires precursor missions to pin down the mineralogy, mechanical properties, and structure of any NEO orebody over the appropriate scale .

Characterization begins with remote sensing. Success of a NEO mining venture will require numerous physical samples on a scale sufficient for understanding the three-dimensional compositional and mechanical structure of the body. The theoretical bases of spectral remote sensing are well understood by the space community, as they have been the major means of exploring nonterrestrial bodies. But these techniques have not been well calibrated with physical samples of NEOs; meteorites serve as an analog to asteroid material, but do not provide a basis for a comprehensive understanding of asteroid structure and composition [52]. Mineralogical characterization of comets is even more speculative since comet materials are not available. In all cases, remote sensors reveal the composition of only the outermost layers of the body, with uncertainty increasing rapidly with depth. However, remote characterization of NEOs benefits from the much greater ratio of surface area to body volume than is the case on Earth.

Statistically well-designed physical sampling programs are vital to the success of NEO resource recovery. Rigorous geostatistical sampling theories have been developed for terrestrial use to account for degree of geologic variability, grain sizes of the material to be sampled, and size of the samples to be collected. For example, the more the orebody grade varies, the shorter the required sample interval. Further, the larger the largest material grain size, the larger the sample required to ensure adequate representation. The scales of all sampling are crucial, because the resulting orebody model intimately guides both mining and processing. Terrestrial underground hardrock mining can require samples at intervals as short as one or two meters. Terrestrial open pit mines commonly sample on the order of 10 meters, even after orebody homogeneity has been demonstrated. Bruce [3] and Brown [2] discuss terrestrial orebody characterization requirements. Any feature of the NEO orebody that affects resource extraction must be sampled at an interval that fully characterizes the feature (the final stages of terrestrial characterization programs sample in three dimensions by drilling grids of boreholes and collecting cores or cuttings samples on the order of two-meter intervals). Physical sampling is expensive and time-consuming, and thus a tempting target for cost-cutting; many terrestrial mining projects still suffer from problems due ultimately to poor sampling techniques or resolution. It is important to match the sampling scale to the scale of the geological ore-control structures on Earth; it will be imperative to do so on NEOs.

In short, characterization of a NEO orebody requires a thorough sampling of its interior. Given the ubiquitous terrestrial difficulties in obtaining representative information, the reconnaissance mission should in fact substantially over-sample the NEO because corrective mining options in space will be far more limited than on Earth.

Anticipated fundamental differences in mode of formation of NEOs from terrestrial orebodies ultimately will offer opportunities to refine NEO characterization methodologies. Potential characterization shortcuts cannot be counted on, however, until a significant base of experience is developed, and even then should be used with extreme caution. Sparse data gives only incomplete clues. Recall the lines of reasoning that postulated Martian canals, steamy Venusian swamps, and deadly deep Lunar dust. Contrast the Galileo probe data with the expected composition and behavior of Jupiter's atmosphere. In these cases, and more, careful logic foundered on data insufficient to support the conclusions drawn. When data is sparse or erroneous, non-unique solutions are not always recognized as such. Physically realistic conclusions must be constrained and verified with extensive physical measurements.

2.2. Mining

Unit operations are the basic steps needed to perform a function, in this case, mining (removing material of value from its source and delivering it to a processing plant). Processing is defined as removing the fraction of interest from the material, with optional chemical transformations. The means by which the unit operations are accomplished, and their relative importance, may or may not be very different from on Earth, depending on the physical processes involved.

Terrestrial mining consists of four unit operations:

(1) *fragmentation* breaks the material into movable size, then
(2) *excavation* (also called loading) moves it into a
(3) *transportation system* [13], while
(4) *ground support* prevents cave-ins and landslides. Ground support will take on new dimensions in zero gravity. Control of rock fragments in zero gravity is partially addressed under materials handling, but the integrity of the body being mined may also become a problem.

Depending on the mechanical characteristics and size of the orebody, the production rate, and the environment, several unit operations may be combined in a single machine, or two machines may be required for one operation (e.g., fragmentation by drilling and blasting). One unit operation may be easy, even trivial, while another may be very difficult. All, however, must be accomplished for mining to succeed.

2.2.1. Fragmentation

Fragmentation frees the ore from the surrounding rock mass and breaks it into fragments small enough for the later mining and processing steps. There are only two economic methods of fragmenting rock on Earth: mechanical attack, and drilling and blasting. The latter is by far the most common option, although mechanical methods generate fewer fine particles. Researchers study exotic fragmentation methods constantly, such as water jets, microwaves, thermal spelling, laser irradiation, and electrostatic arcing (see, e.g., [36]), that are uneconomical under most terrestrial conditions. However, in the vastly different environment of space, some of these techniques may prove to have significant advantages, while others, such as waterjets, may be less useful.

2.2.1.1. Drilling and Blasting

Rock fragmentation using conventional explosives is controlled by two considerations: powder factor and blast geometry. Powder factor is the amount of explosive required to fragment a unit mass or volume of rock [39]. Blast geometry determines how the energy is delivered in space and time. More importantly, the geometry of the sequenced explosions must break rock to an adequate free face or fragmentation will not occur properly [30]. Powder factor is commonly expressed as kilograms of explosive needed to fragment one cubic meter of rock. The powder factor for hard rock, whether crystalline such as basalt or sedimentary such as sandstone, is remarkably consistent: approximately 0.57 kg/m^3. This is the average needed to produce a distribution of rock fragment sizes smaller than a meter or so in diameter, a common terrestrial mining scale. Using a powder factor of $0.6\ kg/m^3$, a 100-meter diameter asteroid would need approximately 310,000 kg of conventional blasting agent, or an equivalent amount of unconventional explosive.

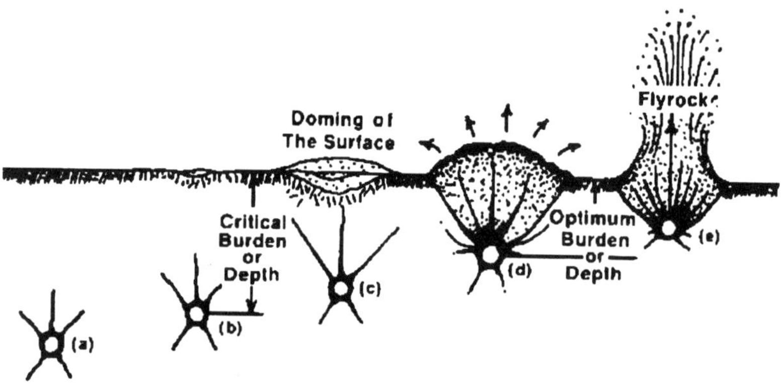

(A) Optimal Placement of an Explosive Charge in a Rock Mass. The rock must be able to break toward a nearby free face (exposed surface), and there must be an appropriate amount of explosive per unit volume of rock (i.e., the powder factor must be optimized for the rock). Too little burden (e) causes flyrock, too much burden (a) to (c) wastes energy and does not fragment the rock efficiently. If the free face is too far away (i.e., excessive burden, case (a)), the explosive will crush the rock locally but will not fragment the required volume. The spacing between the charges also must be optimized for efficient fragmentation.

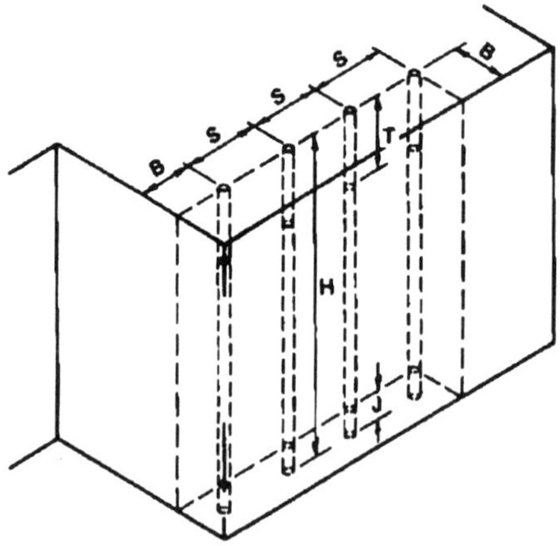

(B) Geometry of Blast Design in a Rock Mass. B = burden (distance from nearest free face), S = spacing (between blastholes), H = blasthole height, T = stemming (inert filler), and J = subdrilling (blasthole depth below grade). Each blasthole should contain the optimum explosive mass to fragment the volume of rock defined by the spacing, burden, and bench height ($H - J$).

FIGURE 1. Basic Blast Geometries and Definitions (after Morhard [39]).

Powder factor does not address the geometry of using this explosive. Poorly applied explosive energy can result in either too much or too little fragmentation (Fig. 1A), both undesirable. Blast design requires efficient use of an available free face, which is simply a rock surface toward which the explosive can break the rock; on Earth, the interface between rock and air. In space it is the interface between rock and vacuum. An explosive charge initiates a compression wave that expands to a free face, then reflects back from it in tension. This transient tensile regime, along with gas expansion and flexural rupture, causes the rock at the face to fracture. Fragmentation is impossible if no free face is reached before the compression wave dissipates [39].

Terrestrial mining practice distributes explosives spatially within holes drilled in the rock, hence the term drilling and blasting. Each blasthole fragments a rectangular column of rock (Fig. 1B). The holes are placed so that over time each hole uses the locally optimum powder factor, which is the same powder factor as for the entire rock mass. Spacing (the distance from hole to hole), burden (the distance from the nearest free face), and hole depth define the volume of breakable rock. Explosive energy and the rate of energy release are controlled by the type of explosive and the diameter of the drill hole.

The blastholes that make up a shot are fired at accurately timed intervals so that earlier-shot holes create new free faces for later holes to break toward. The sequence and timing intervals of the delays control the amount of rock fragmentation and the spatial disposition of the fragments (or blast cast). For example, on Earth, if a high degree of fragmentation is desired, the interval between blasts will be about four milliseconds per meter of burden, but if maximum throw is desired, the holes will be initiated simultaneously [30]. If the muck pile (muck is the miner's term for broken rock) is to be excavated by a loading tool (e.g., a front-end loader), then the pile is left against the wall to improve the crowding performance of the bucket.

Before conventional explosives can be placed effectively in the rock for fragmentation, a pattern of holes must be drilled into the rock mass. Figure 2 illustrates the forces involved in drilling a hole. The downhole pressure (also called weight-on-bit, thrust, and pulldown) and the torque required to rotate the bit against the rock at the hole bottom can be of significant magnitude. Table 1 lists general drilling requirements for terrestrial rocks. On Earth the reaction forces for these are derived mainly from the machine weight, with some use of small anchors as hold-downs. In the negligible gravity field of a NEO, this approach will not work. Perhaps other means for drilling holes will have to be developed, utilizing unfamiliar or novel techniques. However, conventional terrestrial techniques should work, if the hold-down problem can be solved.

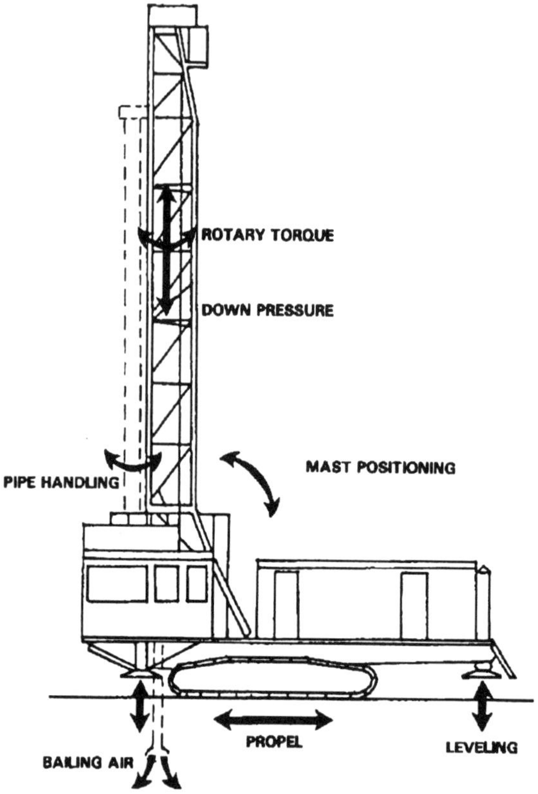

FIGURE 2. Forces on a Rock Drill. Thrust (down pressure) and rotary torque are the most important forces on a drill. Rock drills are used in two key terrestrial mining operations: drilling holes for explosive placement and drilling holes for rock bolts and anchors. Any drill requires a method to supply reaction forces for the drilling motions. Rotary drills (oil field technology) consume most of their drilling power in torque, while percussion drills (underground mining technology) consume most of it in thrust. Other functions must also be considered: bailing air removes rock cuttings; pipe handling is necessary when drilling long holes; and finally the drill must be positioned by mast, propel, and leveling motions (after Martin et al. [35]).

TABLE 1. Terrestrial Rotary Blasthole Drilling Requirements Vary with Rock Strength and Bit Diameter (up to 0.6 m)[a]

Type of rock	Rotary torque (*N*-m/cm bit diem)	Pulldown force (*N*/cm bit diameter)	Rotary speed (rpm)
soft (e.g., sedimentary < 100 MPa UCS[b])	400–1,000	2,600–4,400	70–100
hard (e.g., crystalline > 100 MPa UCS)	100–350	7,000–14,000	40–80

[a]These approximate values are for roller-cone technology (from [35]).
[b]UCS is uniaxial compressive strength.

The basic principles of blasting will remain the same in space, but implementation will have to be modified, perhaps drastically. This is due in large part to the absence of gravity assist in space. Blasting practice may also be affected by some of the differences in physical properties between terrestrial and extraterrestrial materials. Another significant factor to consider is the effect of force vectors produced during blasting on the orbital dynamics of the NEO. These effects can be used to modify the gross behavior of the body, but only if they are well controlled. Very careful blast design will be required to predict accurately the effects of each blast on the angular momenta of the NEO and the rock fragments. Blasting force vectors can be used in conjunction with other methods of altering NEO orbits, such as tether management and engine firing.

2.2.1.2. Mechanical Fragmentation

For some NEOs, such as those with low strength ore or whose orebodies are much smaller in volume than the surrounding host material, mechanical mining machines may be more efficient and easier to control than drilling and blasting. On Earth, machines presently are used to mine rocks ranging in hardness from soft clays to moderately hard sandstones and limestones, all with less than about 100 MPa uniaxial compressive strength (UCS). This approach has several advantages over explosive fragmentation: easier automation, better ground support, more flexibility, better selectivity, and continuous operation. Mechanical fragmentation takes place when the force applied by a mechanical cutter on a rock surface exceeds the local strength of the rock, forming a chip or drill cutting (Fig. 3). Some energy also is expended to crush the rock directly beneath the cutter. Since the energy required to form any rock fragment is proportional to the fragment's surface area, crushing is an energy-intensive process when compared to forming large chips. Rock drill bits and cutterheads are designed to reduce crushing and enhance chipping as much as possible. This goal would not change in space.

The highest rock cutting forces can be very high, on the order of several hundred thousand kilonewtons per cutter. On Earth, machine weight and/or stabilization devices provide the necessary reaction force. Since NEO gravity is negligible, cables or anchoring mechanisms (there are many types, as will be seen) would be needed to provide sufficient cutting forces.

Different mechanical cutter designs have been developed for various rock properties (Fig. 4). The abrasiveness of the rock controls the rate of wear of these cutters; abrasiveness is largely a function of the volume percentage of quartz and feldspar in the rock. In several meteorite classes, however, quartz (Mohs hardness ~7) is almost absent and feldspars (Mohs ~6) are only accessory minerals [52]. However, mafic minerals (e.g., pyroxene and olivine, Mohs ~5.5 to 6.5) are common. This means that cutter wear could be less than terrestrial experience, but still may be a factor to consider if unfamiliar wear mechanisms predominate in the space environment.

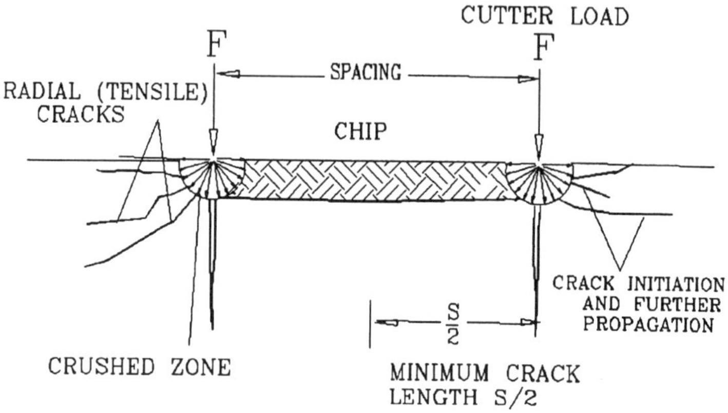

FIGURE 3. Fragmentation Action of Mechanical Rock Tools. When a rock cutting tool is forced into a rock mass, the material is first crushed under the load. As the load increases, the rock under the crushed zone begins to crack. When the cracks reach a free surface, a chip or cutting forms. The difference between chips and cuttings is mostly size: drills produce chips, larger excavators produce cuttings. Interaction of the radial cracks from one cut to another is important: cutting is most efficient when a crack forms one-half the distance to the adjacent cut (after Neil et al. [40]).

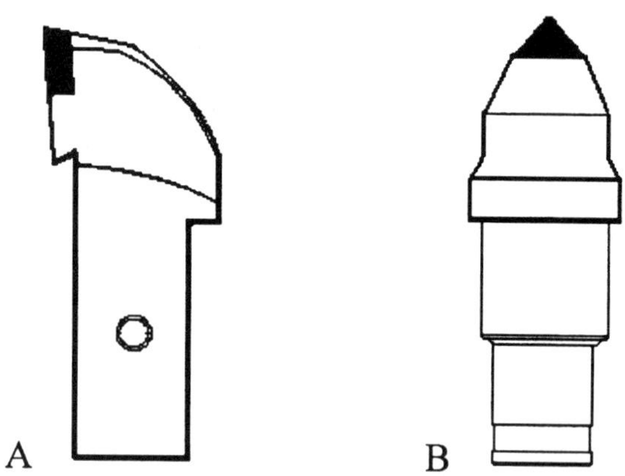

FIGURE 4. Typical Rock Cutting Tools. Drag bits (A) and point-attack cutters (B) are commonly used to fragment soft to medium-hard rock in terrestrial practice. Both feature a tungsten carbide tip brazed onto a steel body. The lower shank (2 to 5 cm width or diameter) fits into a tool holder on the excavator cutterhead. Cutterheads may contain over 50 cutting tools (see Fig. 7), arranged in a pattern that is most effectively designed from empirical cutting tests. Because of the heat and brittleness limitations of the carbide tip, neither of these tool designs is effective in hard crystalline rock. On the Earth, these cutters are used for fragmenting coal, evaporites, and softer sedimentary rocks.

Terrestrial mechanical excavators accomplish both fragmentation and excavation. However, this requires gravity assist; mechanical gathering arms, or other mechanical pickup devices near the working face, remove the fragmented rock as it drops from the face (see Fig. 7). The implementation will have to change in space, but rock fragment removal will be even more important in the near-absence of gravity.

A key feature of mechanical excavators is that the specific energy of cutting is a function of cutting geometry. Spacing between cutters and penetration of each cutter into the rock affect specific energy (SE). Since SE is the energy needed to fragment a unit volume of rock, the lower the energy the more efficient the fragmentation (Fig. 5). The optimal spacing and penetration are usually determined experimentally for each rock type encountered. Disc cutters usually require about 10 kW-hr/m^3 of rock; point attack cutters require about 30 kW-hr/m^3 (explosive fragmentation is one to two kW-hr/m^3).

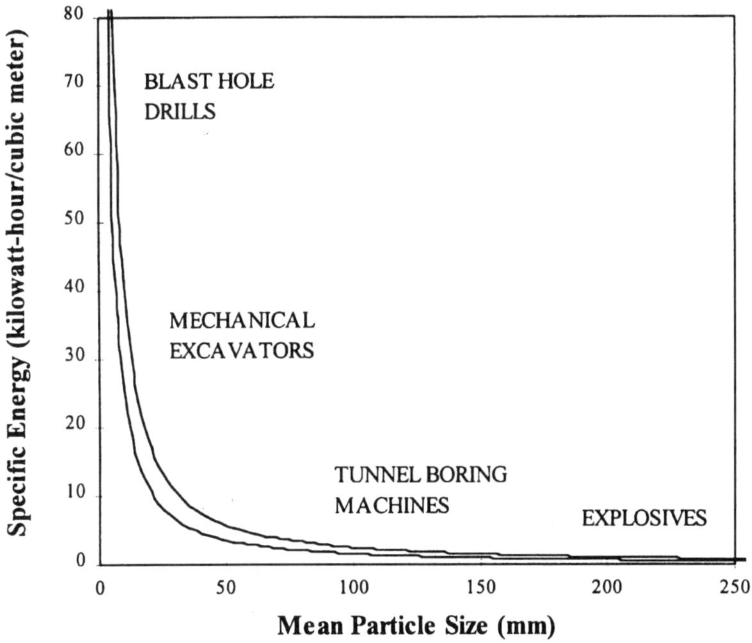

FIGURE 5. Specific Energy of Rock Fragmentation as a Function of Particle Size. Specific energy is partially controlled by the type of excavation device, because each device has a characteristic product size distribution associated with it. For example, a drill bit produces a smaller average chip than a tunnel boring machine, and both produce smaller average fragments than blasting. Specific energy also is a function of rock strength: stronger rock requires more fragmentation energy to produce the same product size (after Neil et al. [40]).

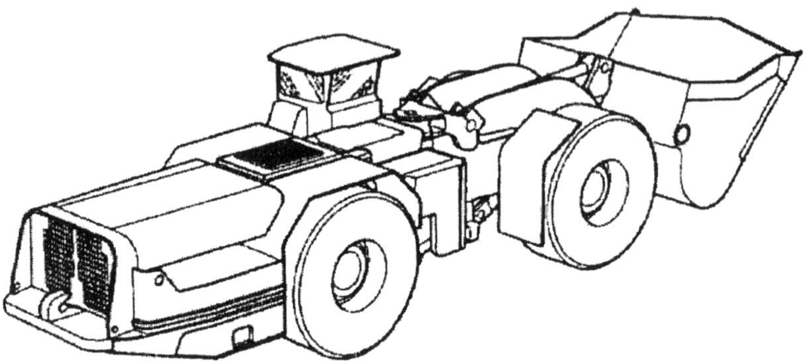

FIGURE 6. The Most Common Terrestrial Excavation System. This particular machine is designed for underground mining, and is called a load-haul-dump (LHD). The more common surface design is called a front end loader (FEL). These machines are very common in terrestrial mining and construction. Their major purpose is excavation: lifting fragmented material from its place of production and placing it in the transportation system. In microgravity, the excavation problem is entirely different than the terrestrial case, but is just as important. Drawing courtesy of Atlas Copco Wagner Inc.

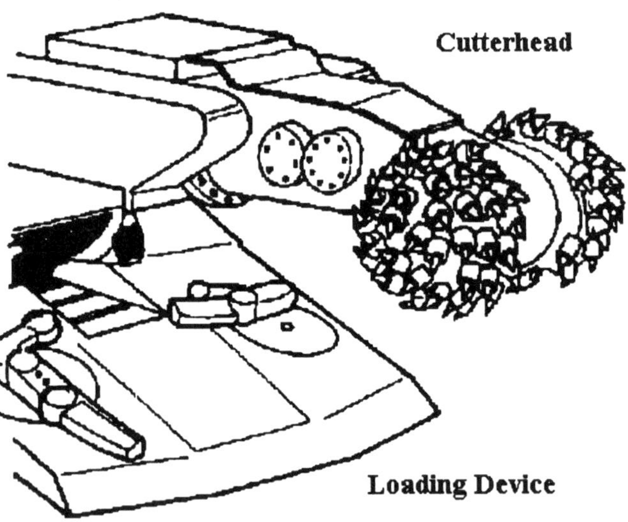

FIGURE 7. Mechanical Miner with Loading Device. As the cutterhead fragments rock, the cuttings drop onto the loading device, called an apron in this configuration. The moving arms on the apron gather the fragmented material and draw it onto a central conveying device for delivery to the transportation system. This process depends on gravity to function. In microgravity, a cloud of various sized cuttings will float around the cutterhead, so other collection means will be required (after Neil et al. [40]).

2.2.2. Excavation

Excavation removes the broken ore material from the place of fragmentation and places it in the materials handling system (Fig. 6). In most terrestrial cases, large equipment (front-end loaders, hydraulic excavators, or electric shovels) performs excavation, which in essence has the purpose of overcoming gravity. The blasted material is lifted from its place of formation and placed into the transportation system. Other machines such as scrapers, draglines, and bulldozers also find use as excavators. Some, for example scrapers and dozers, can also fragment. As noted, terrestrial mechanical mining equipment usually excavates rock as it is fragmented (Fig. 7).

Excavation is a necessary step on Earth because the constant gravity force requires it. The broken material must be lifted from where it is formed, which is a significant cost. In space, excavation may be more easily combined with fragmentation in a single step. This may be true even when fragmenting with chemical explosives. For example, controlled blasting could throw the muck directly into processing bags [23]. Blasting of waste rock could be directed along a different vector than valuable ore, a crude sorting technique often used in terrestrial mines within the constraints imposed by gravity. Finally, launching material away from a NEO with blasting could (by providing an orbital impulse) be a method to change the course of the NEO in a controlled manner.

2.2.3. Transportation

Once the ore has been fragmented and excavated, it is transported to a processing location, close or far away. On Earth, this is accomplished with trucks, conveyors, trains, or in slurry pipelines, in approximate order of popularity. In terrestrial gravity, transportation is an energy-consuming process; in freefall it takes on a new order of difficulty because the broken material must be held in place. A rotating NEO will increase the difficulty of fragment control due to centrifugal force.

In freefall, newly-formed rock fragments, whether mechanically or explosively formed, may tend to remain suspended in the vicinity of the working face long enough to obscure vision and reduce the efficiency of excavation. NEO mining operations could easily result in an ever-growing cloud of fragments and dust, even if most of the fragmented material is captured. This will cause navigation hazards in the vicinity of the operation.

Fragmented material from a NEO can be thought of as a flying muck pile. All transportation scenarios require maintaining control of the muck pile. The control could be sophisticated, such as separating ore and waste at this point. The fragmentation method influences the size distribution of the muck, which in turn influences its dynamic behavior in freefall. A wide distribution of fragment sizes, especially a high percentage of fines, will cause control problems. Huge mesh sacks or crude but robust boxcars (bins with lids) may be required to shepherd flying ore and waste piles along the required orbits. Even the best blast design will leave some fragments in unpredictable trajectories. Mechanical mining machines also generate difficult-to-control rock fragments.

Transport mechanisms come into play at other stages of the process as well, including moving the finished product to market. In all instances, the difficulties and advantages of moving large amounts of variably sized particles through interplanetary space can only be guessed at before actual experience is gained.

Another factor to be considered when planning transport is the time value of money. Low-energy orbits may tie up the material and the machinery to process it for long periods of time, thus lowering the return on investment.

2.2.4. Ground Support

In terrestrial mining and excavation, ground support refers to controlling rock in the presence of gravity for the purpose of maintaining rock slopes and rock openings. The most common example uses rock bolts to keep mine roofs from falling (Fig. 8). These bolts and their close relations, rock anchors, require holes drilled in the rock for placement. Mechanical friction or chemical adhesives then hold them in place (Fig. 9). The reactive force generated between the anchor and the wall of the borehole is key to the rock bolt's performance. Cone-shaped anchor holes are avoided; cylindrical and inverted-cone holes (rare) are more effective. Shaped explosive charges, even in series, tend to create cone-shaped holes in homogeneous material. If rock anchors are to be used extensively, drilling technologies must be developed that function robustly in microgravity. Using a penetrator as an anchor is very rare in terrestrial practice, but could be useful in NEO applications if fired from an orbiting device.

An alternative to point-applied rock anchors in NEOs is to wrap a cable sling around the body, either singly or as a network. This would spread applied loads, reducing the possibility of premature material failure of the NEO, and provide a number of hardpoints for attachment of tethers, equipment, and mining machinery. This concept is developed further by Gertsch et al. (this volume).

In space, rotating NEO orebodies can also suffer the consequences of centrifugal force. As mining progresses, the integrity of the body can decrease due to actions such as blasting, and enclosure within a restraint may become necessary. This, too, is discussed in the accompanying paper.

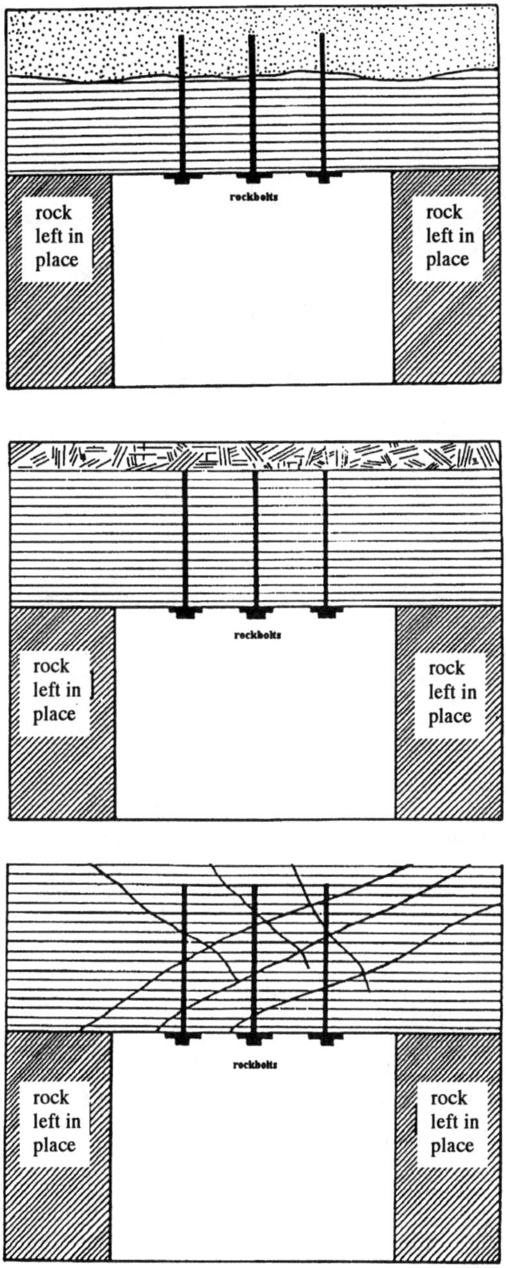

FIGURE 8. Rock Bolts in Mine Roof Support. Shown are the principle ways that rock bolts support underground mine roofs: (A) Bonding weak laminated material to stronger rock above that can support it. (B) Increasing the friction force acting between laminations. (C) Interlocking fractured rock (after Peng [47]).

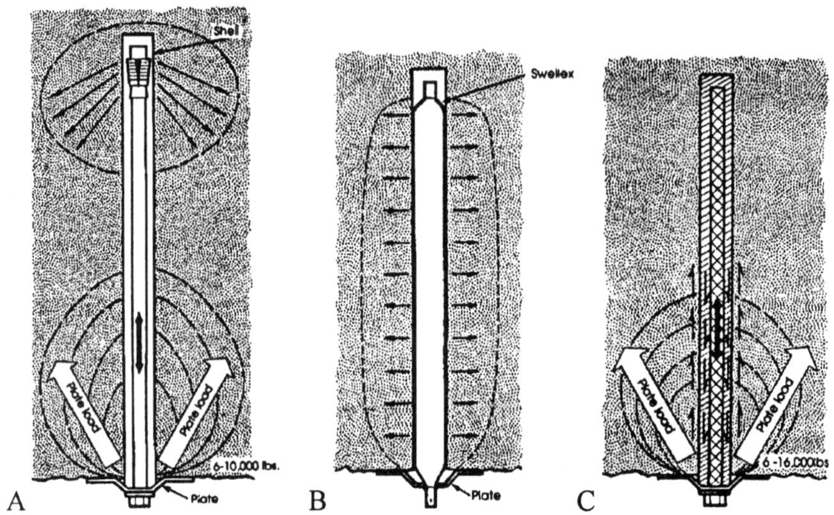

FIGURE 9. Basic Terrestrial Rock Bolt Types. These are starting points for designing methods to anchor to NEOs. All begin with a drill hole into which the bolt is placed. For roof control, each anchors an end plate that restrains the rock surface. Active systems (A and B) are installed under tension; they immediately place restraining load, generated from the anchor, on the rock surface. Passive systems (C) develop restraining load only as the rock deforms against the end plate. Illustrations after Scott [57].

(A) Expanding shell bolt — the anchoring end expands mechanically and places a point load on a very short portion of the drill hole. The anchor pressure can be very high, requiring strong rock to create a secure bond.

(B) Expansion bolt inflated in the drill hole — the entire length actively generates reactive load.

(C) Passive cemented bolt- eventually creates a load over its entire length as the rock deforms.

2.3. Processing

Terrestrial mineral processing consists of three unit operations:

(1) *comminution* reduces the size of the ore fragments to liberate the valuable fraction;
(2) *separation* removes either ore or waste fractions; and
(3) *materials handling* moves the material through the processing plant.

As in mining, processing unit operations are sensitive to mine size and production rate. Unlike mining, processing unit operations are sensitive to the chemical as well as the physical characteristics of the ore.

Perhaps the major feature of terrestrial mineral processing plants is the ubiquitous use of water. All three unit operations are usually wet and many will not work without the use of water. For example, material is moved through the process plant frequently in slurry pipelines, rather than conveyors or trucks. This will present significant problems when transferring processing technology to space.

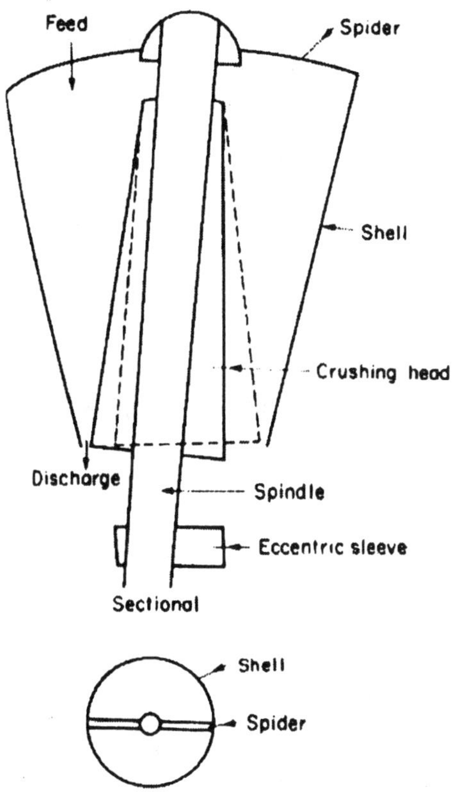

FIGURE 10. Gyratory Crusher. Gyratory crushers and jaw crushers (not shown) perform the first stage of comminution (fragment size reduction) in terrestrial mining and milling practice. They prepare the material for finer or secondary size reduction, which can be either crushing or grinding. The two types of primary crushers use the same fragmentation mechanism, alternately compressing and releasing the rock fragments. These designs rely on gravity to maintain the flow of material (after Wills [61]).

2.3.1. Comminution

When ore is delivered to the processing plant, it usually is too large or of an inappropriate size distribution for efficient processing. The valuable fraction may be locked up in unwanted matrix material, making immediate separation impossible or at best inefficient. Comminution, or crushing and grinding, liberates the valuable material from the waste by reducing the fragment size of the muck. The optimum size of the particles in the valuable fraction governs the amount of size reduction required. If the valuable fraction is micron sized, extensive grinding may be required. If it is millimeter sized, one-step crushing may suffice. The processing technique to be used to separate the waste from the valuable mineral also influences the amount of comminution required: low efficiency separation processes require more comminution.

According to Rittinger's Law [61], the energy consumed in size reduction is proportional to the new surface formed during the reduction process. However, energy consumed is also influenced by the reduction ratio (initial size compared to final size); the larger the ratio, the less efficient the reduction process. Consequently, a single step of crushing or grinding rarely exceeds a size reduction of six times the diameter. Further, size reduction of very small particles is less efficient [49]. Practical determination of the work necessary to achieve size reductions has been an area of intense research. Although a rigorous method of predicting grinding power requirements is not yet available, Bond's approach works well [27] and has wide industry acceptance:

$$W = 10W_i\left(\frac{1}{\sqrt{d_o}} - \frac{1}{\sqrt{d_I}}\right) \quad (1)$$

where

W = work of reduction in kW-hr,

W_i = Bond's work index in kW-hr/ton (ton equals 2000 pounds),

d_o = final size in microns, and

d_I = initial size in microns.

Bond's work index allows for both the hardness of the material and the mechanical efficiency of the grinding apparatus. Since only approximately 1% of the machine power is converted to new useful surface, it is remarkable that the method works as well as it does [27]. Typical values for Bond's work index, W_i, are given in Table 2.

TABLE 2. Typical Work Indices for Bond's Method[a]

Material	W_i, kW-hr/ton	Material	W_i, kW-hr/ton
All Materials	15.19	Feldspar	12.84
Basalt	22.45	Granite	15.83
Clay	7.81	Limestone	12.77
Coal	12.51	Mica	148.00
Dolomite	12.44	Quartz	14.05

[a]These values are more appropriate for grinding than crushing. However, they give a good initial estimate for both. Terrestrial practice usually starts with Bond's method, which is then refined through exhaustive testing of the material.

Fragmentation during mining begins the comminution process. In general, mechanical excavators deliver smaller and more uniformly sized material, with a smaller fraction of fines, than blasting does. Controlled blasting can in rare cases accomplish most of the required comminution, but blasting as a rule is more sensitive to variations in rock properties, whether structural or mineralogical, and consequently can deliver large variations in particle sizes.

Crushing is the next step in comminution after fragmentation. On Earth, massive primary crushers of either the jaw or gyratory type (Fig. 10) reduce run-of-mine ore to an appropriate size to feed the secondary crushers. Tramp metal, slimes, and severely oversized fragments are usually removed at this stage. Secondary crushers, lighter in construction, reduce the ore to a size suitable for grinding. Three designs are in common terrestrial use: cone, impact, and roll crushers (Fig. 11).

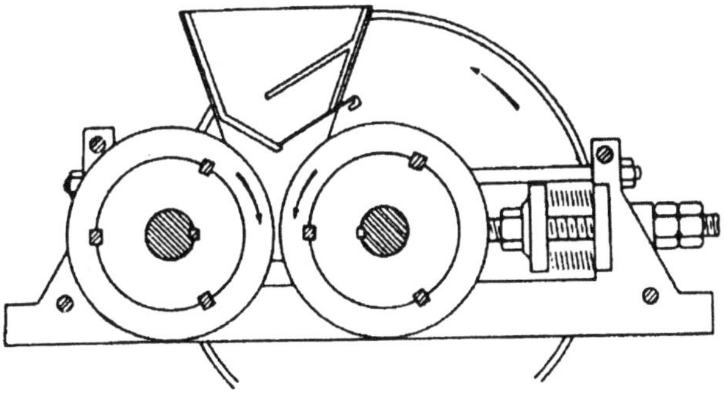

FIGURE 11. Roll Crusher. Roll, cone, and impact crushers perform secondary crushing after the larger primary crushers have started the comminution process. Roll crushers require very large rolls in relation to the size of the feed particles to achieve reasonable reduction ratios. Cone crushers, similar to but smaller than gyratory crushers, are increasing in popularity. Impact crushers, such as hammermills, create particles more suited for surface-binding applications such as aggregate (after Wills [61]).

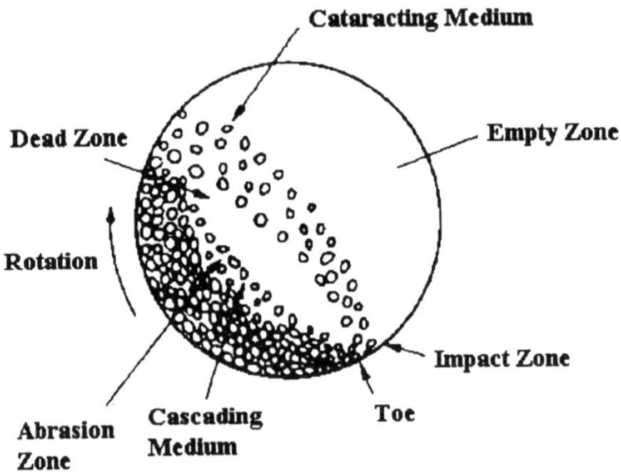

FIGURE 12. Grinding Action of Rod and Ball Mills. These machines can produce a very fine (very small diameter) product at high production rates. Usually the feed material has already been reduced in size by crushing. This action completes liberation of the mineral of value from the waste (after Wills [61]).

Grinding further reduces rock fragment size by a combination of impact and abrasion in tumbling mills (Fig. 12), large rotating cylindrical vessels designed to make maximum use of gravity. Grinding consumes more energy than any other aspect of mineral processing [61]. It cannot be avoided, however, because small particles are necessary for efficient separation, and crushing alone usually is insufficient.

The most effective means of achieving the proper size distribution of the mined material depend on the physical properties of the mineral constituents, their size, their distribution, and the shapes of the individual mineral grains within the rock. The more the grains of mineral and waste are interlocked, the less clean are the grains produced by comminution and the more difficult it will be to separate them efficiently. The optimum grain size is determined by careful bench- and pilot-scale testing, especially in unusual situations. This will be particularly true in space.

Common terrestrial crushing and grinding techniques are associated intimately with gravity. This is evident from the construction of many mills on steep hillsides; among other things, gravity holds the muck in the jaws of the crushing mechanism. Other means of providing the force necessary for fragment fracture are known, however, such as highspeed centrifugal impact grinding mills [33]. The availability of familiar forms of energy will be less in space; innovative energy sources may prove more practical once they have been studied and developed.

Many schemes have been proposed to extract minerals in situ by chemical attack, or leaching (e.g., carbonyl [32] and zone refining [43]). However, these scenarios are fraught with difficulty and their feasibility has not yet been demonstrated. If the material is porous enough, in situ leaching is very effective, as terrestrial cyanide precious metal leaching has demonstrated. However, preleaching comminution usually is required. One terrestrial example, heap leaching of gold ores, commonly requires crushing the ore rock to less than 25 mm passing (25 mm maximum particle size) [18]. In any case, leaching requires large amounts of fluids, and a head gradient (due usually to gravity, on Earth) to control the direction and speed of flow. Generally, it can be very difficult to control fluid pathways and maintain fluid boundaries in geologic materials, and microgravity adds another novel element to moving fluids through these materials.

2.3.2. Separation

Once the mineral of interest is liberated (i.e., optimum particle size is achieved), it is separated from the waste particles. This can be a series of several steps, each designed to recover a different mineral, size fraction, or other property. The process may take only one step, as in smelting (see Section 2.3.3). A wide variety of separation techniques are used in terrestrial mineral processing. These range from primitive gravity separation (e.g., panning) to sophisticated exploitation of differences in surface wettability of mineral grains (froth flotation). Almost all of them take place in water. Given the difficulty of finding and using water in space, alternative fluids may have to be found. The commonly used property-related separation methods that may or may not require a fluid medium include:

(1) *Density.* A common technique, on Earth this is called gravity separation. However, related approaches use centrifugal force (with or without gravity assist) in devices called spirals, sluices, cones, shaking tables, vanners, and jigs.

By narrowing, pinched sluices cause material streams to stratify as they descend in elevation; upon discharge the layers can be separated by simple flow splitters (Fig. 13). Shaking tables rely on friction to slow the fluid velocity near a fixed surface to create lateral displacement of entrained particles of different densities. Jigging, the oldest "gravity" concentration method known, separates minerals of different densities in a bed fluidized by a pulsating current. The pulsation frequency and the direction of fluid flow are adjusted to control the dilation of the particle mass so that denser particles are allowed to sink. Small dense particles penetrate the interstices of the bed and large dense particles fall in a condition approximated by hindered settling. Less dense particles of all sizes rise to the surface of the mass.

These methods are highly sensitive to the presence of fines and operate best with narrowly-sized material. The wider the range of densities, however, the wider the range of particle sizes that can be separated successfully [61].

Centrifugal concentrators have been developed recently for very fine particles, using accelerations from 5 to 50 G's [29], [59], [4].

(2) *Chemistry.* This includes differential reactions or distillations. Examples are leaching and the carbonyl process.

(3) *Surface physical properties.* The common terrestrial version, froth flotation, carries mineral particles on air bubbles in water using differential wetting. Unwanted minerals can be suppressed by appropriate choice of fluids.

(4) *Electrical and magnetic properties.* These have been proposed previously for extraterrestrial processing [1]. Magnetic pickups are widely used to remove tramp metal from processing streams to avoid equipment damage but these methods are otherwise rare in terrestrial practice.

(5) *Particle size.* The most effective particle size separations occur when the particle sizes or shapes of the mineral and the waste are different, but their densities are similar.

The relative efficiencies of these approaches will change in space, with currently inefficient methods possibly becoming methods of choice.

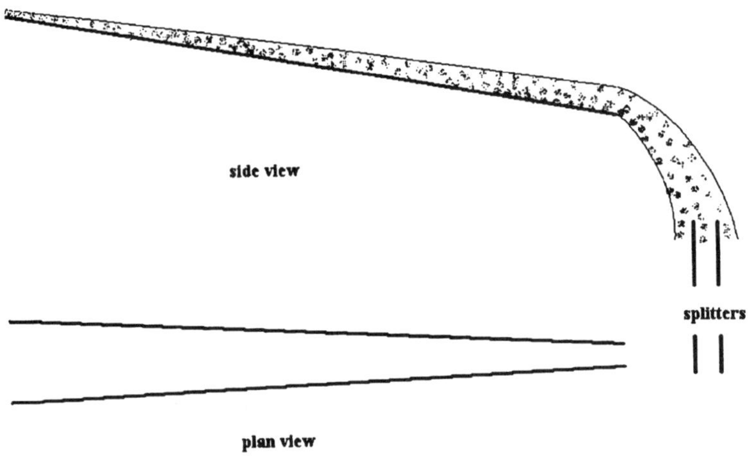

FIGURE 13. A Pinched Sluice. After liberation, the valuable mineral fraction must be separated from the particles of waste rock. In this approach to gravity separation, the flow gradient combined with narrowing sidewalls causes the particles to stratify. During projectile motion upon discharge from the sluice the resulting layers can be directed to separate bins by splitters (after Wills [61]).

2.3.3. Smelting

Smelting is a heat or flame oxidation or reduction process. Many terrestrial base metal ores, such as copper and zinc, are sulfide compounds, so sulfur oxides constitute a major by-product of smelting. Sulfur negatively affects terrestrial environmental quality. The gas products of smelting in space, on the other hand, can be condensed and captured easily in the low ambient temperature, and contained for use in other processes. This is very difficult on Earth.

In space, solar heat can be focused to perform many processes, including smelting, whose energy costs would be prohibitive on Earth. Smelting could take on new significance and be applied more broadly in space. A wide range of materials can be manipulated and refined by heat. Perhaps the simplest and easiest is removing water from comets and other bodies. At the other end of the spectrum, enough energy theoretically will turn any material into a plasma, which then could be passed through a magnetic field and separated into its constituent elements (this approach has yet to be demonstrated, however).

In any case, the huge amount of solar power available should not be underestimated. Energy-intensive processing may become more feasible than the terrestrial case. Stine [58], for example, calculated that a simple parabolic mylar mirror, 100 m diameter, in Earth orbit could vaporize several kilograms of copper per second.

2.3.4. Materials Handling

Process plant transportation (also called materials handling) has the same goal and many of the same features as in mining: material being processed must be moved about the plant. In addition to the materials handling systems discussed already, process materials handling systems use more conveyors and slurry pipeline systems. Since terrestrial separation processes are usually accomplished wet, slurry materials handling makes good economic sense there.

In space, the absence of free gases may be an advantage for materials handling in microgravity. Gas concentrations that would be considered vacuum on Earth still are dense enough to induce flow in confined areas in zero gravity, due to density gradients. Significant gas losses must be expected, however, from adhesion of gas molecules to material surfaces created in hard vacuum.

Although free fall at first glance appears to be an advantage for materials handling in mining, in reality it will cause many unforeseen difficulties. Essentially all of human experience has taken place in a significant gravity field. Moving fragmented materials in space will be similar to playing a three-dimensional billiard game with the added novelty of unbalanced rotating balls in a dust cloud. Further, the flow of material through bins and chutes is controlled by the Earth's gravity vector. Substitutes for processes now taken almost for granted (e.g., conveyor transport, ore sorting, etc.) will have to be developed; certainly no one substitute will serve for all.

Placing NEOs in bags for fragmentation and processing has long been proposed [42] to contain the billiards game. It could do double duty for fragmentation and processing. However, after bagging, getting the material out in a controlled flow in zero gravity may be a challenge.

Space vacuum also will present difficulties, some of them almost certainly unknown at this time. Friction, which is not completely understood in any case, will cause ore particles to behave very differently than terrestrial experience has encountered. Metallic surfaces in contact may tend to weld together, and particle flow may become problematical. One serious problem with process streams in vacuum is that movement may be required into and out of vacuum. Airlocks tend to inhibit and complicate continuous process streams, and favor batch processes.

3. CHALLENGES FOR NEO MINING AND PROCESSING

Terrestrial mining is a combination of brute strength and engineering finesse. The former is required by the constantly changing character of the rock, and the latter is required by the economics of business. These fundamental aspects won't change in space, but the means by which the unit operations are accomplished will. This section points out some of the attributes of terrestrial mining practice that will be essential for successful exploitation of NEOs and describes some of the problems that will arise in mining them.

3.1. Mining and Processing Machinery Design Principles

Terrestrial mining equipment is generally as simple in design as possible (for an in depth discussion of design principles, see [9]–[11]). They also tend to be strong and massive [35]. In general, mining machines should be difficult to break and easy to repair when broken. In the nearly inaccessible reaches of even near-Earth space, these principles must, in the main, be adhered to even more closely.

Mining equipment is built robustly because it must withstand impact forces much greater than the average loading force. The impacts occur frequently as the equipment comes in contact with the rock, sometimes quite violently. However, robust machinery usually means greater mass to be transported to the NEO, posing a logistics problem.

A machine attribute that will be important in NEO mining is flexibility, but flexible machines tend to be both massive and complex. Flexibility means the ability to deal successfully with new and challenging mining situations. Mining machinery, like mine planning, must allow changes in mining method, geometry, rate, etc., should conditions change, because those conditions cannot be completely characterized prior to mission deployment.

3.2. Rotation Control

Controlling the natural spin of the NEO will be made more difficult by the force vectors imparted by the mining process, which will alter the NEO's angular and linear momenta. Since NEOs are relatively small bodies, the effect could be significant. Mining-induced changes must be monitored and controlled to avoid instability of the main body or unwanted changes of the orbits of either the main body or any resulting fragment clouds. Mine planning will have to account for the effects of altering the distribution of mass of the NEO as well as the effects of blasting impulses.

3.3. Automation

Automation is in its infancy in terrestrial mining [12] because of the great variety and variability inherent in the mining process. Unlike artificial materials manufactured with precisely repeatable motions, mining consists of infinite near-repeats of highly variable patterns of movement and force application. These subtle but ubiquitous variations make automation of some mining processes very difficult. Mining frequently encounters novel situations which require flexible machine perception and operational response. Flexibility is extremely difficult to incorporate in automated operations. Controlling mining processes from nearby (line-of-sight) is difficult enough for well-defined and well-bounded tasks. The uncertainty inherent in truly remote mining increases that difficulty enormously. Human intelligence has long been an integral part of all mining operations, and is proving difficult to replace on the whole. It is, however, possible to automate some portions of the mining process [56]. Work in this area is accelerating [28].

For example, automation will have to account for the differences between cyclic and continuous processes, especially as material moves into and out of gaseous atmospheres. Some mining unit operations work only in a cyclic mode, such as drilling and blasting, where each step is performed in a logical sequence. Still other processes are inherently continuous, such as mechanical excavation as currently practiced on Earth. There can be continuity between the two modes in some cases. To take an example from agriculture, combine harvesters operate continuously only as long as their internal storage bins can be unloaded onto transport vehicles on the run at appropriate intervals; if that does not happen, the harvesting becomes a batch process (at significant loss of productivity). Continuous coal mining machines are similar to combine harvesters, except they have little internal storage. The machine can only operate when a truck is present to accept the mined material. One terrestrial solution is to connect the continuous miner directly to a conveyor material handling system, but this adds significant complexity. For use in space, significant internal muck storage could be added to a mechanical miner to make it truly continuous and minimize the difficulty of automation.

A completely automated and autonomous NEO mining system is not possible in the near future without significant expense and risk. The degrees of freedom are simply too great. However, many individual tasks are suitable for automation and/or tele-operation; a combination of these with human supervision of the less well-bounded tasks may be the optimal solution.

3.4. Long Mission Duration

Asteroid missions will average several years in length, rarely taking less than one year each way (Sforza and Remo, this volume; [31]). Minimum terrestrial mine life for an asteroid-sized orebody, once characterization is complete, is between five and ten years [21]. Given this, it seems very unlikely that mining will be a viable mitigation method for long-period comets, where six to nine months warning is the current best lead time. Not only will mining a NEO take several years to complete, but an ore body characterization mission of equal flight duration to the mining mission must be flown first. Every NEO mining profile will include the following phases, each of which will affect mission duration:

(1) Fly characterization mission;
(2) Characterize the orebody;
(3) Design mining method and fabricate equipment based on orebody characterization;
(4) Deliver equipment to NEO;
(5) De-spin and/or control NEO orbital motion;
(6) Mine until NEO threat is mitigated.

Long mission duration also allows little room for error; if a hammer is left behind, it cannot be sent in the next shipment. This means that the mass that must be lifted from Earth for a flexible, robust NEO mining plan will be great. Any reasonable means of reducing the amount of material that has to be brought up from the Earth's gravity well would increase the chance of success. Selectively mining and processing a select part of the NEO to prebuild or complete equipment on-site may help. For example, wear parts, cutting tools (pick or disc cutters), or basic structural components (beams, columns, cables, plates, etc.) could be fabricated on-site with relatively simple technologies [9]. Use of NEO material for life support also would reduce the mission mass requirements.

3.5. Health and Safety

Mission length also affects health and safety. Human participation in the mining process greatly increases its flexibility, but at the cost of complicating the health and safety and life support requirements. Extended stays in free fall are physically debilitating. In addition, the protective atmosphere of the Earth shields us from numerous hazards, among them solar radiation and particle impacts. Both are life-threatening in space, and conscious efforts will be necessary to replace the atmosphere's beneficial effects. The effect of reduced gravity, while not fatal in the short term, contributes to physiological deterioration.

The problems of living and working in space are dealt with at length elsewhere (e.g., [26]). In addition to these, however, mining itself is physically hazardous. The combination of dense objects and high forces with uncertain, constantly changing conditions has always made mining one of the more dangerous terrestrial occupations. However, the measures taken to protect operators from the even more inimical environment of space may result in a decrease in the relative hazard due solely to the mining process. In short, the dangers of the space environment may overshadow the dangers of mining. Neither should be underestimated.

Mission design can mitigate some of the problems of remoteness and human exposure to the hazards of space. Mining and processing equipment and other massive items could be sent ahead on a long, slow path that minimizes energy expenditure. The human operators then could leave later on a high-energy, more direct route that would deliver them to the NEO after robotic processors had slagged enough raw material to build a radiation shelter.

3.6. Reduction of Terrestrial Environmental Impact

Moving resource exploitation to space also removes much of its environmental impact from the Earth [45]. This reason alone may be enough for NEO resource exploitation to continue after mitigation of a threat NEO. The costs associated with rehabilitating the environment disturbed by mining on Earth could help balance the costs of mining off the Earth.

3.7. Energy Availability

The relative availability of different types of energy that can be used for mining and processing will be very different than on Earth. We are accustomed to relying on energy from the Sun, directly and indirectly, as well as on gravitational potential energy. In theory, energy can be extracted from any gradient: temperature, density, magnetic, etc. Initially more familiar energy sources will be exploited, such as solar, but the impetus will be to develop more efficient energy sources that now would seem exotic or farfetched.

3.8. Zero Gravity and Mining

Fragmentation and excavation will be complicated on NEOs by the absence of gravity. Machine weight alone can not provide sufficient reaction force without significant increases in machine inertia. All six degrees of machine motion (translation and rotation about the three coordinate axes) must be consciously controlled, and additional reaction force will be needed for drills and mechanical excavators to penetrate the rock. On Earth, this is done when necessary by tying the machine to the basal rock mass with anchors set in bored holes. Anchoring is a time-consuming operation, and a very difficult one in low gravity. Reaction engines/thrusters are much too compliant for mechanical rock drilling and excavation, besides requiring large amounts of reaction mass. Stiff, strong physical constraints are essential for breaking rock.

3.9. Zero Gravity and Mineral Processing

The absence of both a strong, unidirectional gravity vector and flowing fluids will require fundamental modification of most extractive metallurgical processes. The terrestrial gravity vector is used both to control material flow and to separate particles by density or size, and will have to be replaced in space operations. In some processes it can be simulated by centrifugal force [58] which can be generated by two bodies linked with a tether rotating about their common center of gravity. However, centrifugal force is not a complete analog of gravity. The acceleration gradient inherent in the phenomenon will have to be taken into account.

3.10. Fluid Flow

The lack of gravity in NEO processing may mitigate somewhat the scarcity of flowing fluids. Head gradients can be generated with very low gas density, since it is the gradients, rather than absolute fluid heads, that create flow. Studies need to be done to determine the minimum gas volumes needed to move the maximum fragmented material. Use of physical space will be very different than on Earth; remarkably sparse process streams may be the norm in space.

3.11. Vacuum

The behavior of bulk materials in vacuum before exposure to an atmosphere is subtly different from its behavior after exposure and return, because the particles originally had no adhered coating of gas molecules. The internal friction coefficient is expected to be significantly higher before any atmospheric exposure; other effects may become apparent during practice [41], [46].

Moving material in and out of vacuum will be a batch process. This may be difficult to incorporate into processes that on Earth are usually continuous. Steps that on Earth are performed wet may have to be entirely redesigned, or replaced by other technology. Moving wet process feed into/out of vacuum will require study to ascertain its feasibility.

3.12. Temperature Extremes

In some metallurgical processes, temperature gradients may replace gravity as a motivating factor. Thermal cycling could become a useful fragmentation mechanism for geologic materials (it is one of the oldest rock-breaking methods known). Confirmation of the possible utility of various methods requires multiple-scale physical experiments on Earth and in space. Thermal cycling effects must be minimized, however, in other parts of the mining process. In addition to low-temperature embrittlement, thermal cycling fatigues the metals in mining machinery.

3.13. In Situ Processing

In situ processing is attractive for processing of some NEOs, such as those with significant volatile components or those with available pathways for the working fluid to circulate. Because solution pathways can be a problem to construct, and containing the solution in the mining area can be difficult, in situ processing is uncommon in terrestrial practice. But where feasible, it is simple and inexpensive; it combines all unit operations into one step. In space either gas or liquid leachants may be feasible.

Carbonyl-based removal of metals has been proposed for low-percentage metal-containing NEOs that might also have porous fluid pathways, such as carbonaceous chondrites [32]. Other in situ schemes such as zone refining have also been proposed [43]. However, severe technical problems still remain to be addressed for both these scenarios. Further, in situ processing tends to remove small amounts of material, leaving a lot of waste mass in place, still heading for Earth. It may be most useful as one of a series of mining/processing steps custom-designed for a particular NEO; the remaining mass then may be more amenable to mechanical excavation techniques.

TABLE 3. Anticipated Near-Earth Asteroid Resources and their Corresponding Asteroid Types

Raw mmaterial	Corresponding asteriod types	Recovery processes	Composition	Utilization
hydrous silicates	C, B, G, W (wet M)	heating distillation electrolysis comminution density separation	water, oxygen, hydrogen, Si, SiO_2 hydrocarbons, Fe, Mg, Ca, Na, Cl, K, elemental carbon	industrial feedstocks, life support, cement, materials processing, propellant, polymers, processing, synthetics, oxidizing and reducing agents
carbon-rich anhydrous silicates	P, D, F	heating distillation density separation	same as above except for water and possibly hydrocarbons	see previous
silicate rock	S, Q, V	cutting comminution melting density separation, also reduction if interested in metals	rock, Si, Fe, Mg, Al, O, Ca, Na, etc. primarily as oxides	rock products, cement/ shielding, semiconductors (solar cells,micro-processor chips, diodes), optical components (fibers, reflectors, waveguides)
diamonds	S	separation by surface properties	C	cutting, grinding, electronics
nickel-iron (with silicates)	M, S, Q	melting, crushing, density separation, carbonyl, reduction and solution	Ni, Fe, kamacite, taenite, olivine, pyroxene	structural, electrical, & mechanical components, stainless steel foils, forms, cylinders, etc.
precious metals	M, S, Q	chemical refining, density separation	Ag, Au, Ga, Ge, Ir, Os, Pt, Rh	micro electronics, thin films, solar cells/concentrators, export

4. GENERAL CHARACTERISTICS OF NEAR-EARTH OBJECTS

A large volume of resources is contained by NEOs, which consist of both asteroids and comets. These bodies, although of different origins, share common orbital characteristics that bring them within close proximity to Earth's orbit. This relative closeness, the source of their potential hazard, also makes them ideal candidates for reconnaissance, exploration, and mining.

4.1. Asteroids

Asteroids are found mainly within the asteroid belt between the orbits of Mars and Jupiter, as predicted by Bode's Law, as well as in planetary-crossing orbits. They can be classified by orbital characteristics, spectra and albedo, and inferred physical properties first proposed by Remo [51] and later extended [52]. Based on the study of meteorites, computer models, and (solar) reflectance properties, asteroids are thought to have undergone heating episodes which selectively vaporized the lighter substances such as water, leaving mostly silica, carbon, sulfur, and metal (Fe, Ni, Mg). Actually, water-bearing asteroids (most of taxonomic types C, B, G) are the majority in the asteroid belt and a substantial minority (~10 to 15%, see Table 4) in the near-Earth population. Most asteroids are less than 1 kilometer in size. The physical structure of asteroids may range from a dense crystalline material to a loose aggregate.

Raw materials found in meteorites, means of recovery, and possible utilization are presented in Table 3. Orbital and spectral classifications for asteroids are given in Table 4. Details of the material properties of meteorites as asteroid analogs may be found in Remo [52]. The compositions of the various meteorite categories are given in Tables 5 and 6.

NEOs can be divided easily by Earth-based observation into three orbit-based groups: Atens, Apollos, and Amors, shown in Table 4. The characteristics of these groups are discussed elsewhere [37].

NEO asteroids also are classified according to their spectra, albedo, and probable surface composition. These superficial properties are inferred from spectrophotometry, broad-based photometry, polarimetry, reflectance spectroscopy, and radiometry. Such remotely obtained data will begin to refine the selection of promising target bodies for resource recovery.

The C spectral class includes more than 75% of all known asteroids and is regarded by many to be the source of most carbonaceous chondrite meteorites that fall to Earth. They are very dark, and thus certainly under-represented in our observational databases.

S-class bodies consist of metallic nickel and iron, mixed with iron-magnesian silicate rock in (apparently) all proportions from metal-rich to metal-poor. They are represented by stony iron meteorites (metal-rich igneous rocks), lodrandrites and acapulcoites (metal-bearing igneous rocks), ureilites (carbon-rich igneous rocks with diamond inclusions but largely devoid of metal), CV3 and CO3 carbonaceous chondrites (carbon-bearing anhydrous silicate rocks), and ordinary chondrites (anhydrous silicate rocks with 1% to 20% NiFe metal by weight). They constitute about 17% of all classified asteroids, but about 50% of NEOs.

TABLE 4. Orbit-based and spectral classifications of NEOs (from [37], [34])

Group	Total count	Min-Max inclination (°)	Min-Max eccentricity	Min-Max diameter (km)	Spectral class[a] C	D	M	Q	S	V	others
Atens	22	0–49	0.09–0.87	0 02–5 0	1		1		1		1
Apollos	166	0–68	0.05–0.92	0.006–12	3			1		5	3
Amors	170	0–52	0.03–0.71	0.02–41	1	1	1		9	3	6

[a]Insufficient information currently exists to classify most near-Earth asteroids.

The M class is mostly pure nickel-iron (Ni-Fe), although it includes bodies with substantial amounts of anhydrous silicates. Metallic bodies showing evidence of hydrated silicates are designated class W, for "wet M." The M and W classes have slightly lower average albedos than the S class; all three are brighter than the C class.

Less common, V-class bodies consist of material very similar to terrestrial basalt. They are anticipated to be largely devoid of free metal or volatiles. D-class bodies are reddish carbonaceous material that is typical of comets and Trojan asteroids. They and the P and F classes are believed to contain anhydrous silicates and, perhaps in protected interiors, water ice or gas hydrates (clathrates).

4.2. Comets

Comets are believed to originate in the outer edges of the solar system in a region called the Oort cloud which is thought to contain 10^{13} objects. The best known of these comets is Halley, with a period of 76 years. Coalescing far from the Sun at temperatures as low as 10–15 K in interstellar space, they contain a high proportion of water ice, in addition to other frozen "volatiles" and some complex hydrocarbons. They also can have relatively substantial fractions of carbon, silicate, and metal (Table 5). Comets belonging to the Kuiper belt tend to have shorter periods than those of the Oort cloud. However, both types of comet can be captured in tight solar orbits, and it is impossible to determine the origin of a particular orbit.

The comets that cross Earth's orbit have been disturbed by strong gravitational perturbations primarily from Jupiter; they have extremely elliptical orbits that may not lie within the plane of the solar system. Active comets maintain velocities in the vicinity of Earth much higher than those of most other NEOs. Very long-period comets not previously observed are thus correspondingly more difficult to detect with adequate lead times for economical resource exploitation.

TABLE 5. Some Near-Earth Comet Resources

Raw material	Recovery processes	Composition	Utilization
FeMg silicates (olivine), amorphous core silicate micro-structures, silicon carbide, possibly micro-diamonds	heating distillation	Si, Fe, Mg, O	shielding, cement, structural materials, microprocessors, electro-optic components
complex organic refractories	heating distillation	polycyclic-aromatic hydrocarbons	feedstocks for synthetics
volatiles	distillation cracking	H_2O, CO, CO_2, H_2CO, OCN^2, CH_3OH, NH_4^+, NH_3, CH_4, ...	life support, propellant, fertilizer, processing chemicals, polymers, explosives

It has been suggested that some Earth-approaching asteroids (perhaps 10% to 50%) are old cometary nuclei that have exhausted their near-surface inventory, but retain substantial ice inventory deep in their interior. The previous cometary activity of near Earth asteroid 1979VA lends credence to this model. The Apollo class of near-Earth asteroids is speculated to consist of comets that have been stripped of all their volatile components. What remains, the nucleus of an extinct comet, is thought to contain mostly hydrocarbons and some silicates. It is speculated that up to half of the nearest asteroids may be short-period comets. Furthermore, comets not close enough to the Sun to form comas are considered to be dormant; this type is anticipated to contain a higher proportion of volatile components than extinct comets do.

5. MECHANICAL CLASSIFICATION OF NEAR-EARTH OBJECTS

A material classification system for NEO hazard mitigation [51], [52] divides NEOs into four broad mechanical types: ice composites, friable rock, igneous rock, and nickel-iron (which includes stony-irons). This classification is also useful for planning resource recovery, as outlined in Table 6.

These material categories can be related to the different spectral classes (Table 7). Possible differences exist between the two classifications, however, due to "weathering" of the surfaces in the space environment. The consesequences could be analogous to terrestrial experience, where oxidized metallic ores (found in the upper, or weathered, portions of orebodies) re-

TABLE 6. NEO Mechanical Classes and their Physical Properties, with Possible Mining and Processing Methodologies[a]

Category	Approximate tensile strength	Exploration resolution	Terrestrial excavation technology	Available processing technology
0. ice composites	ice ~ 2 MPa to ? for composites	10 to 100 m	blasting, heating, distillation	phase separation in bags
1. friable rock	can be very low (~0.2 MPa) to 8 MPa	10 to 100 m	blasting, mechanical (ripping)	phase separation in bags, mechanical, chemical, magnetic
2. hard rock (crystalline)	8 to 33 MPa + (Farmer [6])	10 to 20 m	blasting, mechanical (disc cutters)	mechanical, chemical, magnetic
3a. metallic Ni-Fe (massive)	impervious to drill & blast	100 m	concurrent with processing	melting/smelting, carbonyl methods
3b. hard rock-metallic Ni-Fe	rock-metal boundary strength		various: rock-metal interface determines processing method	mechanical, chemical, magnetic; melting/smelting

[a]Ice composites can be very strong or very weak depending on the material in the ice.

TABLE 7. Relationship between NEO Mechanical Properties, Inferred Mineralogy, and Spectral Classes.

category	spectral class	number in main belt	inferred surface mineralogy	meteoritic analog
0. ice composites	comet observations possible P & D	many	Fe & Mg silicates (olivine), carbon, complex organic refractories, & volatiles	extinct or dormant comet, cosmic dust particles
1. friable rock	B C G	6 41 5	hydrated silicates + carbon/organics/ opaques	CII-CM2 assemblages and those from aqueous alteration &/or metamorphism of CI/CM precursors
	D F P W(wet M)	19 10 23 2	carbon/organic-rich anhydrous silicates	organic-rich cosmic dust grains & CI1/ CM2 + organics
	T	1	possibly similar to D and P	
2. hard rock	E	8	enstatite or other Fe-free silicates	enstatite achondrites
	Q	1	olivine+pyroxene +metal	ordinary chondrites
	R S	1	pyroxene +olivine	pyroxene-olivine chondrites ureilites, CV3/C03 chondrites
	V	1	pyroxene±feldspar	basaltic achondrites
3. metallic (Ni-Fe)	M	21	metal, metal+enstatite	irons, ensatite chondrites
	S	73	metal±olivine± proxene	pallasites w/accessory pyroxene
	T		metal	olivine-dominated stony-irons, iron-sulfides, carbides & silicates, primitive achondrites

quire metallurgical techniques very different from sulfide metallic ores (from the lower, unweathered portions). For centuries, processing technologies developed for the more accessible upper portions of terrestrial orebodies could not handle the much larger reserves of sulfide ores.

While the mechanical classes are more directly useful than orbital or spectral classifications for eventual NEO recovery, the present state of knowledge in all classes is suitable only for preliminary targeting for further exploration. The data currently available are much too crude for detailed mining or processing plans, even with the sparse samples provided by meteorites.

5.1. Ice Composites (Class 0)

This class corresponds generally to comets, both active and passive. The "dirty snowball" model of the structure of comet nuclei [60] implies random inclusions within a mixture of ices; extinct nuclei of this type may be more accurately classified as friable material (below). Other models are being proposed as well, such as the "birds nest" composite structure of Greenberg and Remo [19]. Comets are expected to be rich in water and volatiles, and to contain as well organic compounds and carbon. Cosmic dust particle compositions imply that anhydrous silicates (such as silicon carbide) are an important component of cometary materials.

5.2. Friable Rock (Class 1)

Friable material at first glance appears to be the easiest to fragment and excavate; it is material that breaks apart easily. Both its uniaxial compressive strength (UCS) and tensile strength are low, and its degree of natural fracturing may be high. Such materials are characterized by low bulk elastic modulus. This category includes volatile depleted and extinct comets, highly fractured asteroids and collections of loosely agglomerated material.

5.3. Hard Rock (Class 2)

This class is well known in terrestrial mining and processing. Terrestrial rock varies considerably in UCS, but has relatively low tensile strength, usually 4% to 10% of its UCS [6]. It is the low relative tensile strength that allows fragmentation. Most known near-Earth asteroids are thought to fall within this class.

5.4. Metallics (Class 3)

The very high strength of nickel-iron alloys has been established by Johnson and Remo [24], Remo and Johnson [54], Johnson, Remo, and Davis [25]; on Earth these alloys are similar to nickel-rich steel. Even when combined with rock, as in the stony-iron class of meteorites, this type of NEO represents a high grade source of structural metal. This category includes some S and most M spectral classes of asteroids. For mining and processing purposes, a special subcategory will be discussed later: rock and metal mixes.

6. MINING NEAR-EARTH OBJECTS: ACCESSING NEO RESOURCES

The unit operations required for mining and processing may have to be modified to fulfill the goals of NEO mining: hazard mitigation, with wealth generation secondary. This section discusses the basics of how terrestrial mining and processing technology might be applied to the four different NEO classes, and how the classes will affect mining technologies.

6.1. Ice Composites (Class 0)

Broadly speaking, ease of fragmentation depends largely on the tensile strength of any material, particularly if it is brittle. Ice excavation is less common on Earth than rock excavation. Blasting of ice for underground tunnels can require powder factors as high as 2.7 kg/m^3 [50]. Conversely, a powder factor as low as 0.1 kg/m^3 will crater one-meter-thick ice layers floating on water [20]. Blasting Arctic ice using mining scale fragmentation geometries has been successful at powder factors of 0.4 kg/m^3 [38]. In general, powder factors for pure ice should be less than for rock, but ice composites can be significantly stronger.

Pure water ice is plastic at low strain rates and brittle at high strain rates, but simple water ice NEOs are almost certainly rare. The material strength effects of mixing other cometary materials with water ice are unknown, as are the differences between terrestrial ice and the low-temperature, high-vacuum ice of space. Very low temperatures in the terrestrial atmosphere severely embrittle ice, causing very low tensile strength and thus easy fragmentation. The strength range of mixed ice composites, on the other hand, could be quite broad. The quasi-static and dynamic properties of ice are treated elsewhere in this volume (Furnish and Remo).

Inclusions of narrow size distribution within an ice matrix probably decrease the strength of the composite, particularly if the aspect ratio (width:length) of the inclusions is near unity [5]. Ice inclusions with more extreme aspect ratios (e.g., wood fibers) can make ice surprisingly strong, as illustrated by ice aircraft carrier research [48].

Water ice fragments tend to be abrasive, and included material may make them more so. Along with excavatability, abrasiveness controls the economics of equipment selection. Mining ice may be as hard on equipment as mining crystalline rock, even if it is not as strong.

6.2. Friable Rock (Class 1)

At first glance, the readily fractured nature of friable material would seem to be an advantage. However, it is highly unlikely that the natural fragment size distribution would be optimal for efficient extraction of the mineral of interest. Instead of reducing the number of comminution steps required prior to separation, friability may actually increase them. Any extra crushing and grinding will add costs and increase the time to mission completion.

On Earth, mechanical mining machines easily fragment friable materials and rival drilling and blasting in efficiency, so the choice between the two is one of site-specific economics. As mentioned above, in space these materials may be difficult to fragment and excavate in a controlled manner with either method. The problems caused by the non-gravity-controlled, three-dimensional dynamics of loose material aggregations in space may loom larger than with other material classes, where the fragmentation process is more directly controlled.

One method of mining friable bodies in space would surround them with a bag, as discussed in the accompanying paper (Gertsch et al., this volume). However, this would merely contain the deadly billiards game, not prevent it. Further, a less cohesive NEO would require a harness in addition to a bag to prevent the body from failing prematurely. At some point in the mining process the remainder of the NEO will become mechanically unsound and must be controlled.

Low mechanical strength does not always translate into ease of mining. Terrestrial clays, for example, are very weak but difficult to fragment because they are plastic and sticky. (It is hard to imagine a clay or claylike NEO, but the consequences would be catastrophic, if not previously planned for.) Further, material can be too easy to fragment, resulting in unwanted particle degradation during processing and transport.

Some NEOs are expected to be unconsolidated, and to possess little or no internal cohesion. The size distributions of their constituent particles are not completely known; some pieces may be too large for processing and therefore require further fragmentation. Some of the constituent chunks of material may themselves be very hard. This would be a difficult problem: hard-to-fragment overlarge chunks in the middle of a flying gravel pit. Fragmenting one chunk will affect the motion of others, possibly introducing a collective dynamic that would hinder mining and processing efforts. Careless handling could cause some to reach NEO escape velocity when not desired.

6.3. Hard Rock (Class 2)

Terrestrial rock is almost always blasted, but rock with UCS below 100 MPa is commonly fragmented mechanically [40]. If the NEO is massive (whole, sparsely fractured), exporting proven terrestrial techniques to mine it may be the best option. The physical properties of even relatively massive rock can vary dramatically on scales of a few meters, however, so detailed exploration remains paramount before realistic missions can be planned.

The strength of meteorites provides only an upper bound on the strength of their parent bodies. It is the unjointed, strong pieces that survive the fall to Earth. Real rock masses are riddled with faults, fractures, joints, cracks, mineral segregations, and other weaknesses that significantly influence the overall response to loading. NEO strength will be controlled likewise by the distribution of such discontinuities, rather than by the maximum strength of intact samples.

Successful blasting powder factors, near 0.6 kg/m^3, hold consistently over a wide range of terrestrial rock types (sedimentary, igneous, and metamorphic) and strengths (< 50 MPa UCS). Terrestrial rock, regardless of strength, will succumb to blasting. Since rock has a low tensile strength that is relatively constant from rock type to rock type, fragmentation of massive rock with explosives varies little in energy requirements. It is not clear at this point how dissimilar asteroid fragmentation properties may be from their terrestrial counterparts; empirical testing is required.

The same holds true for mechanical fragmentation and excavation. Mechanically speaking, any rock can be cut with the appropriate cutting tools. Economics, not machine strength, limit their terrestrial use. The rate of cutter wear controls the economic feasibility, and is related to the abrasiveness of the rock. Rock abrasiveness will be an important factor in NEO mining as well. If the rock of a target NEO is shown to have low abrasiveness by comprehensive premining rock characterization, then lower-mass cutters can be used. This would reduce mission mass.

6.4. Metallics (Class 3)

The composition of metallic bodies ranges from all metal to largely metal. Their mechanical strength and characteristics will vary greatly, depending on the metal fraction and its particle size. Because of this, if a mechanically significant amount of rock is included in the body, it is designated Class 3b rather than Class 3.

6.4.1. Nickel Iron

Blasting effectiveness in metallic materials is a function of tensile strength and other properties. On Earth, drilling and blasting of native metal ores is difficult due to their ductility. Mechanical excavation also is more difficult in metal than in rock. However, this may not be so important in space; metals tend to become more brittle and less ductile with low temperatures. In the conditions of space, metal exhibits a wider range of physical properties than terrestrially, and each body should be characterized carefully [24].

UCS measurements of Ni-Fe meteorite alloys under space temperatures already have been conducted [53], [54], [25]. This program must be expanded to other properties; UCS alone is a poor predictor of material excavatability [40].

6.4.2. Stony Iron (Class 3b)

This class combines the problems/advantages of metal and rock, while adding the difficulty of handling and processing these two disparate materials simultaneously. Each has very different mechanical properties. The relative amounts of these materials present and the degree of intermix will radically affect the orbital and mechanical response of these NEOs to mining. The conditions of formation of these materials also will affect their behavior; since these conditions are known poorly if at all, mining and processing questions will have to be answered by direct empirical testing.

Blasting may fragment the rock well while leaving the metal portions distorted but largely intact. The microscopic geometry of the rock-metal interface also may control the processing methodology. Fragmentation by whatever means will result in a bimodal distribution of particle sizes and particle properties that may lend themselves to specialized sorting/processing methods.

7. CONCLUSIONS

This paper discussed the necessary classification of near-Earth objects and the basics of the terrestrial mining methods that will have to be adapted to mine them. Mining, in addition to removing or otherwise mitigating a serious hazard to life on the Earth, also will provide access to a surprisingly large store of natural resources. Those resources will make feasible the human exploration and development of space.

Mining and processing NEOs to supply raw materials needed on Earth and in space could provide an elegant solution for mitigating the hazard posed by Earth-striking NEOs: a threat becomes a resource for humanity. NEO mining promises to be a difficult task, fraught with unknowns; each NEO is vastly different from any other in size, structure, composition, and dynamic properties. Resource recovery can be achieved, but must be done with care, foresight, adequate reconnaissance, and thorough planning. When the Earth is threatened, no expense will be spared to mitigate the threat, but this cannot be an excuse for wasteful and fanciful mining schemes. Some viable mining and processing methods, strategies, and scenarios are not really viable mitigation schemes. Successful NEO exploitation for 1,000,000 oz (31,100 kg) of gold worth $390,000,000 that leaves the Earth in danger is counterproductive. Mining must be secondary to and in support of protecting Earth's population.

Many of the technologies and methodologies required to realize the immense resource potential of NEOs can and should be derived from the terrestrial mining and processing industries. Terrestrial mining techniques represent a huge amount of effort that has already found practical and efficient ways to mine successfully under very difficult conditions. New technologies are not required; it will be more straightforward to adapt terrestrial processes to the novel conditions of space.

NEO resource recovery should not be limited to threatening objects. One feasibility measure for NEO resource recovery will be economic. For a threatening NEO, cost cannot be a measure, but mining a nonthreatening NEO must make economic and technical sense, and should be carried out in a manner such that the nonthreatening NEO cannot become a threat.

A successful NEO mining mission requires a precursor mission to characterize the NEO with sufficient detail to ensure success. Earth prospectors have long known they must accurately determine quality, quantity, and availability before embarking on a mining venture. Space will be no different. Geology, mineralogy, and mechanical properties must be determined beforehand, which means physical sampling of the NEO with the detail that only dense drilling can provide. Omitting this step invites disaster.

Time will be a particularly important consideration for the feasibility of mining as a threat mitigation scheme. Given that the NEO must be characterized by a precursor mission, and that mining itself can be a slow process, short-lead-time bodies probably cannot be mitigated fully by mining means. However, since many NEOs contain water, mining it for crew needs or propellant could shorten mission time by lowering the mass required for the nonmining mitigation mission.

While the tradeoffs between manned and automated missions are not clear, mining will probably require a manned solution. On the one hand, mining is a complex and difficult affair that taxes the best human-controlled machines. This could translate into a significant and expensive research and development effort to ensure the success of an automated mission. On the other hand, the difficulty of maintaining a manned presence in space for long periods should not be underestimated. Perhaps the best solution would send the equipment by long-period low-energy routes, while the crew travels quickly in high-energy vehicles (Sforza and Remo, this volume).

Theory predicts an easy time of moving material around in free fall, because the absence of gravity removes the single largest problem in moving heavy objects. However, the absence of gravity also removes the most familiar aspect of motion as we perceive it in our terrestrial experience. As a species, we are profoundly affected by and are familiar with gravity, and we have no intuitive experience with the alternative, from either the human or mechanical perspectives. The movement and control of geologic materials and machinery will be difficult to solve.

Despite the challenges that will accompany the effort, mining a NEO that threatens to extinguish life on Earth is a viable option that may make the difference between a failed and a successful mitigation.

The accompanying paper, "Mining Near-Earth Resources" (Gertsch et al., this volume), discusses in more detail the specific application of the mining methodologies discussed here to the four classes of NEOs. Several innovative approaches are presented.

ACKNOWLEDGMENTS

The reviewers of this paper, Mike Gaffey, Andy Cutler, and R. Larry Grayson, provided many constructive suggestions. Their assistance is very much appreciated.

REFERENCES

1. AGOSTO, W.N. 1985. Electrostatic concentration of lunar soil materials. *In* Lunar Bases and Space Activities of the 21st Century, W.W. Mendell, Ed.: 453–464. Lunar and Planetary Institute. Houston, Texas.
2. BROWN, R.L. 1990. Exploration and geology techniques: Overview of exploration, Surface Mining, Bruce Kennedy, Ed.: Chap 2.1, 15–21. Society for Mining Metallurgy and Exploration Inc. Littleton, Colorado.
3. BRUCE, C. 1982. Ore body evaluation. *In* Underground Mining Methods Handbook, W.A. Hustrulid, Ed.: 3–16. Society of Mining Engineers. AIME. New York, NY.
4. CHAN, B.S.K., R.H. MOZLEY, & G.J.C. CHILDS. 1991. Extended trials with the high tonnage Multi-Gravity Separator. Minerals Engineering. **4:** (3/4) 489.
5. COBLE, R.L. & W.D. KINGERY. 1963. Ice reinforcement. *In* Ice and Snow: Properties, Processes, and Applications, W.D. Kingery, Ed.: Chap. 12, 130–148.
6. FARMER, I.W. 1968. Engineering Properties of Rocks, E.& F.N. Spon Ltd. London, 180 pp.
7. GAFFEY, M. & T. MCCORD. 1977. Mining the asteroids. Mercury, Vol 6, No. 6, p 1.
8. GENTRY, D.W & T.J. O'NEIL, 1984. Mine Investment Analysis. Society of Mining Engineers of American Institute of Mining Metallurgy and Petroleum Engineers, New York.
9. GERTSCH, R.E. 1983. A method for mining lunar soil. *In* Space Manufacturing 1983, Proc. Space Studies Institute Conference on Space Manufacturing, Advances in the Astronautical Sciences, American Astronautical Society.
10. GERTSCH, R.E. & L.E. GERTSCH. 1990a. Lunar surface mines and mining machinery: Design criteria,. *In* Engineering, Construction, and Operations in Space, Proc Space 90, Albuquerque, New Mexico. ASCE. New York, NY.
11. GERTSCH, L.E. & R.E. GERTSCH. 1990b. Application of lunar criteria to three terrestrial mining methods. *In* Engineering, Construction, and Operations in Space, Proc Space 90, Albuquerque, NM. ASCE.
12. GERTSCH, R.E., L.E. GERTSCH, & L. OZDEMIR. 1991. Issues in extraterrestrial mine automation. *In* Proc. International Symposium on Mine Mechanization and Automation, Colorado School of Mines. Golden, Colorado.
13. GERTSCH, R.E. & L. OZDEMIR. 1991. Unit operations in lunar mining. *In* Space Manufacturing, Proc. Space Studies Institute. AIAA.

14. GERTSCH, R.E. 1992a. Exploring, mining and evaluating nonterrestrial resources. *In* Space Resources, Vol. 3, Materials, McKay M.F. et al. (eds), National Aeronautics and Space Administration. Washington, DC.
15. GERTSCH, R.E., 1992b. To build a mine: Prospect to product. Space Resources, Vol. 3, Materials, M.F. McKay et al., Eds. National Aeronautics and Space Administration. Washington, DC.
16. GERTSCH, R.E. 1992c. Asteroid mining, Space Resources, Vol. 3, Materials, M.F. McKay et al., Eds. National Aeronautics and Space Administration, Washington, DC.
17. GERTSCH, R.E. 1992d. Some Notes on Terrestrial Blasting Design and NEO Interception," Proc Near Earth Object Interception Workshop, National Aeronautics and Space Administration and Los Alarnos National Laboratory, LA-12476-C.
18. GERTSCH, R.E. 1995. Modular mill design for flexible heap leach operations. Mining Engineering, Vol. 47, No. 9. Society for Mining Metallurgy and Exploration. Littleton, Colorado.
19. GREENBURG, J.M. & J.L. REMO. 1994. A comet nucleus model. *In* Hazards due to Comets and Asteroids, T. Gehrels, Ed. Univ. Arizona Press, Tuscon, Arizona.
20. GUSTAFSSON, R. 1973. Swedish Blasting Technique. SPI. Gothenburg, Sweden.
21. HARTMAN, H.L., 1992. Elements of mining. *In* SME Mining Engineering Handbook, Vol. 1, H.L. Hartman et al., Eds.: Chap 1.2, 24–38. Society for Mining Metallurgy and Exploration Inc. Littleton, Colorado.
22. HARTMANN, W.K. & A. SOKOLOV. 1994. Evaluating space resources in the context of Earth impact hazards: Asteroid threat or asteroid opportunity? *In* Hazards Due to Comets and Asteroids, Tom Gehrels, Ed. 1213–1224.
23. ISU, 1990. International Asteroid Mission, Final Report, The International Space University, Summer Session,York University. Toronto, Canada.
24. JOHNSON, A.A. & J.L. REMO. 1974. A new interpretation of the mechanical properties of the Gibeon meteorite. J. Geophys. Res. **79:** 1142–1146.
25. JOHNSON, A.A., J.L. REMO, & R.R. DAVIS. 1978. Low temperature impact properties of the meteorite Hoba. J. Geophys. Res. **84**.
26. JOHNSON, R.D. & C. HOLBROW, Eds. 1977. Space Settlements, A Design Study. NASA SP413. Washington, DC.
27. KELLY, E.G. & D.J. SPOTTISWOOD. 1982. Introduction to Mineral Processing. John Wiley and Sons. New York, NY.
28. KING, R.H., 1992. Automation and robotics. SME Mining Engineering Handbook, Vol 2, H.L. Hartman et al., Eds.: Chap 22.2, 1908–1917. Society for Mining Metallurgy and Exploration Inc., Littleton, Colorado.
29. KNELSON, B. 1992. The Knelson Concentrator: Metamorphosis from crude beginning to sophisticated world wide acceptance. Minerals Engineering. **5:** (11/12).
30. KONYA, C.J. & E.J. WALTER. 1990. Surface Blast Design. Prentice Hall. Englewood Cliffs, New Jersey.
31. LAU, C.O. & N.D. HULKOWER. 1987. Accessibility of near-Earth asteroids. J. Guidance, Control, and Dynamics. **10:** (3). AIAA. New York, NY.
32. LEWIS, J.S. & R.A. LEWIS. 1987. Space Resources, Columbia Univ. Press. New York, NY.
33. LLOYD ET AL. 1982. Centrifugal grinding on a commercial scale.Engng. & Min J. **49**..
34. MARSDEN, B.G. & G.V. WILLIAMS, Eds. 1996. Minor Planet Circulars. International Astronomical Union, Smithsonian Astrophysical Observatory, January. Cambridge, Massachusetts.
35. MARTIN, J.W., T.J. MARTIN, T.P. BENNET, & K.M. MARTIN. 1982. Surface Mining Equipment, Martin Consultants Inc. Golden, Colorado. 455 pp.

36. MAURER, W.C. 1968. Novel Drilling Techniques. Pergamon Press. Elmsford, New York. 114 pp.
37. MCFADDEN, L.A., THOLEN, D.J., & VEDER, G.J. 1989. Physical properties of Aten, Apollo, and Amor asteroids. *In* Asteroids II, R.P. Binzel, T. Gehrels, & M. Matthews, Eds.: 442–467. Univ of Arizona Press. Tucson, Arizona.
38. MELLOR, M. & KOVACS, A. 1977. Destruction of ice islands with explosives. *Presented at the* Fourth International Conference on Port and Ocean Engineering, Memorial University of Newfoundland, St. John's. Newfoundland, Canada.
39. MORHARD, R.C., 1987. Explosives and Rock Blasting. Atlas Powder Company. Dallas, Texas.
40. NEIL, D.M., ROSTAMI, J., GERSTCH, R., & L. OZDEMIR. 1994. Production estimating techniques for underground mining using roadheaders. *Presented at the* Society for Mining Metallurgy and Exploration Engineers Annual Meeting, Denver, Colorado.
41. NICHOLSON, D.E., 1971. Gravity Flow of Powder in a Lunar Environment. 1. Property Determination of Simulated Lunar Soil, US Bureau of Mines, RI 7543.
42. O'LEARY, B., M. J. GAFFEY, D.J. ROSS, & R. SALKELD. 1979. Retrieval of asteroidal materials. *In* Space Resources and Space Settlements, NASA SP-428, John Billingham and William Gilbreath, Eds.: 173–189.
43. O'LEARY, B. 1983. Mining the Earth approaching asteroids for their precious and strategic metals. *In* Space Manufacturing 1983, Proc. Space Studies Institute Conference on Space Manufacturing, Advances in the Astronautical Sciences. American Astronautical Society.
44. O'NEILL, G.K. 1974. The colonization of space. Physics Today. **27:** (9) 3–40.
45. O'NEILL, G.K., 1977. The High Frontier, Human Colonies in Space, William Morrow and Co., Inc. New York, NY.
46. PARISEAU, W.G. 1971. Gravity Flow of Powder in a Lunar Environment. 2. Analysis of Flow Initiation, US Bureau of Mines RI 7577.
47. PENG, S. S., 1978. Coal Mine Ground Control. John Wiley & Sons. New York, NY.
48. PERUTZ, M.F. 1948. A description of the iceberg aircraft carrier and the bearing of the mechanical properties of frozen wood pulp upon some problems of glacier flow. J Glaciol. **1:** (3) 95.
49. PRASHER, C.L. 1987. Crushing and Grinding Process Handbook. John Wiley and Sons Ltd. New York. 474 pp.
50. RAUSCH, D.O. 1959. Studies of ice excavation. Colo. School of Mines Quarterly. **54:** (2).
51. REMO, J.L., 1993. Asteroid/meteorite analogs and material properties. Appendix to Workshop 4, NASA NEO Conf., LA-1247-C. Los Alamos, NM.
52. REMO, J.L. 1994. Classifying and modeling NEO material properties and interactions. *In* Hazards Due to Comets and Asteroids, T. Gehrels, Ed. Univ Arizona Press. Tuscon, Arizona.
53. REMO, J.L. & A.A. JOHNSON. 1974. The ductile-brittle transition in meteoric irons. Meteorites. **9:** 209–213.
54. REMO, J.L. & A.A. JOHNSON. 1975. A preliminary study of the ductile-brittle transition under impact conditions in the material from an octahedrite. J. Geophys. Res. **80:** 3744-3748.
55. REMO, J.L. 1995. Laser power beaming in space. Report of Quantametrics, Inc. St. James, NewYork.
56. SCHNAKENBERG, G.H. 1990. U.S. Bureau of Mines coal mining automation research. Preprint No. 90-87, SME Annual Meeting, February. Salt Lake City, Utah.
57. SCOTT, J. J. 1989. Roof bolting — A sophisticated art. Coal, Aug. Reprint.
58. STINE, G.H. 1979. The Third Industrial Revolution. Grosset and Dunlap Company. New York.

59. WELLS, J. & C. PATEL. 1991. Contemporary practices in gravity recovery installations in the Canadian gold mining industry. Minerals Engineering. **4:** (3/4) 399.
60. WHIPPLE, F.L. 1977. The constitution of cometary nuclei. Comets, Asteroids, and Meteorites, A. H. Delsemme, Ed.: 25–35. Univ of Toledo. Toledo, Ohio.
61. WILLS, B.A. 1992. Mineral Processing Technology: An Introduction to the Practical Aspects of Ore Treatment and Mineral Recovery, 5th edit. Pergamon Press, New York. 855 pp.

Mining Near-Earth Resources

RICHARD GERTSCH,[a] LESLIE SOUR GERTSCH,[b] AND JOHN L. REMO[c]

[a]*Mining Engineering Department*
University of Missouri at Rolla
Rolla, Missouri 65401

[b]*Mining Engineering Department & Center for Space Mining*
Colorado School of Mines
Golden, Colorado 80401
and
[c]*Quantametrics, Inc.*
1 Brackenwood Path, Head of the Harbor
St. James, New York 11780

ABSTRACT: Mining the potentially vast storehouse of natural resources contained within near-Earth objects (NEOs) could assist mitigation of the danger that a threatening object presents to life on Earth. Properly planned, NEO mining could provide a substantial basis for the exploration and development of space, in addition to providing important tools and opportunities for mitigating impact hazards.

The key to successful NEO mining is the appropriate modification of tried-and-true terrestrial mining techniques for the very unfamiliar and dangerous environment of space. Mining engineering is unlike traditional engineering disciplines in that, as a whole, it is less easily quantified; the practitioner must retain flexibility and resilience in the face of wide and unknown variations in material properties. This unique element of mining engineering will be an asset in space operations. Understanding of the basics of mining and processing will improve the planning and designing of NEO mitigation missions.

Space possesses many properties that are unique in human experience. Some of these can provide efficient new means of accomplishing old tasks. Others will prove to be obstacles that must be overcome. Mining and processing techniques that are inefficient curiosities on Earth may form the basis of powerful new methodologies in space. On the other hand, some vital processes will demand innovative engineering for success. This paper discusses the effects of the space environment on the application to NEO resources of mining methodologies developed over more than a thousand years on Earth. It will focus on the four mechanical classifications of NEOs described in Gertsch et al. (this volume).

1. INTRODUCTION

The value of certain NEO types can be immense; the in situ value of a nickel-rich ataxite has been estimated at over 2,200 1979 dollars per metric ton [16]. However, at present there is insufficient demand or supporting infrastructure to mine these materials, however rich, for terrestrial markets. Consequently their exploitation cannot be justified because their cost is prohibitive. In all terrestrial mining, the value of the ore must support the cost of extracting and processing it. The same holds in space. However, if a catastrophic hazard emerges, impact threats to the Earth must be mitigated regardless of expense. This paper applies terrestrial mining techniques to removing the danger presented by NEOs on Earth-impacting or -grazing orbits. Mining any NEO will involve surmounting many difficulties not encountered with terrestrial mining. For a threat NEO, the question becomes whether the difficulties of mining outweigh the difficulties associated with other mitigation techniques. If enough time is available, and if the NEO composition is appropriate, mining has a strong advantage over other methods: it sequentially diminishes the threat in a controlled manner without nuclear explosives and turns it into a resource for further extraterrestrial expansion. However, if the mitigation scenario is to deflect or intercept a NEO using a low energy, long-term mission, mining the NEO to supply the mitigation mission without further exploitation of the materials should also be considered.

On the basis of information gleaned from remote observational programs, sparse meteorite samples, and even sparser flyby missions, NEOs can be classified into four general groups [22], [23] as shown in Table 1.

The mechanical, physical, and chemical properties of each group are expected to be approximately consistent, but as more detailed exploration is completed, the details of this classification system almost certainly will change. In the meantime, it provides the means for general planning of an approach to mining and processing of NEOs.

Exploration of the potential orebody, that is, detailed characterization of all pertinent material properties (both ore and waste), is the extremely important first step in any mining program, regardless of location. Following the start of exploration (which continues throughout the life of the mine), mining consists of the unit operations fragmentation, excavation, and materials handling. The mined material is then passed on to the processing section, where unit operations consist of comminution, separation, and materials handling. Only then is the product ready for sale or use. These basics of the resource utilization process are discussed by Gertsch et al. (this volume) and in more detail by Hartman [8], Hustrulid [10], Kennedy [15], and Wills [30].

This paper is a starting point; it presents specific proposals that address many, but not all, of the issues that will confront the would-be asteroid miner. While specific mining and processing methods are proposed here, more importantly they are discussed in light of their effectiveness in meeting the challenges of mining the four mechanical classes of NEOs. Just as importantly, they illustrate the high degree of difficulty that will be encountered in mining and processing NEOs in the harsh and unfamiliar arena of space.

TABLE 1. NEO Mechanical Classes and Their Physical Properties, with Possible Mining and Processing Methodologies. Ice composites can be very strong or very weak depending on the material in the ice.

Category	Approximate tensile strength	Exploration resolution	Terrestrial excavation technology	Available processing technology
0. ice composites	ice ~2 MPa to ? for composites	10 to 100 m	blasting, heating, distillation	phase separation in bags
1. friable rock	can be very low (~0.2 MPa) to 8 MPa	10 to 100 m	blasting, mechanical (ripping)	phase separation in bags, mechanical, chemical, magnetic
2. hard rock (crystalline)	8 to 33 MPa (Farmer, 1968)	10 to 20 m	blasting, mechanical (disc cutters)	mechanical, chemical, magnetic
3. metallic Ni-Fe (massive)	impervious to drill & blast	100 m	concurrent with processing	melting/smelting, carbonyl methods
3b. hard rock-metallic Ni-Fe	rock-metal boundary strength	1 to 100 m	various: rock-metal interface determines processing method	mechanical, chemical, magnetic; melting/smelting

The bulk of the proposed methods are based on the concept of bagging the NEO, whether fully enclosing the NEO (e.g., [19]) or partially covering an area of the NEO surface [5], [11]. First proposed almost 20 years ago, the bag appears applicable to many of the different mechanical classes and to the many chemical and mechanical variations of each class. However, while the concept is simple, its execution will not be. The detailed operations needed to control and bag the NEO, and then manipulate the material contained in the bag, certainly will be complex; many aspects will possess small margin for error.

In general, NEO mining and processing methods based on bags will consist of the following steps. Not all of these proposed steps will be required, nor necessarily in this particular sequence, depending on each NEO encountered. Some steps will have to be performed simultaneously. In approximate order of initiation, they are:

(1) Anchoring to the NEO, and tether attachment;
(2) NEO motion control: Partial or complete de-spin and de-wobble;
(3) Body/fragment restraint placement;
(4) Operations platform placement;
(5) Bag placement;
(6) Auxiliary and support equipment placement;
(7) Mining operations;

(8) Processing operations;
(9) Transportation operations;
(10) Mitigation by orbital modification.

The following sections discuss these topics in more detail, grouped into the more general categories of NEO preparation and mining methods.

2. NEO PREPARATION

In terrestrial mining terminology, the preparation stage is termed mine development. This is when the orebody is defined in as much detail as possible, and the infrastructure necessary for a successful mining venture is put into place. Mine development includes assuring access to the minesite; constructing buildings, processing modules, etc.; and ensuring continuing supplies of whatever materials will be needed during the actual mining/milling.

Operating in space will increase the importance of the development stage by at least an order of magnitude. Not only will the working environment and the physical needs of the mining/milling process be unfamiliar, but assistance will be very far away. It is not possible to provide for every eventuality prior to the mission, so it becomes even more important than usual for flexibility and robustness to be incorporated into the mitigation plan at all levels. The following proposals approach these problems from the viewpoint of mining engineering, which has benefited from a long history of dealing with the unknown.

2.1. Anchoring and Tethering

Anchoring to the NEO is the start of de-spinning (if required), habitat placement, drilling, mining, processing, and material transportation. Before any operation can begin, the NEO must be under control and the machinery and people must be fixed firmly to it. Several anchor/tether methods will be required to accommodate the various NEO situations that will be encountered. These include both noninvasive slings and rock anchors emplaced within the body of the NEO:

(1) Single Sling (Fig. 1A): A sling looped around the NEO spreads its load over a larger area than even a series of point anchors, and is appropriate for moderately strong to very strong bodies. Any consolidated NEO is a candidate for a single sling, even bodies containing a predominant fracturing plane, if the sling can be oriented approximately normal to that plane. Slings are notable for not invading the integrity of the NEO.
(2) Multiple Sling (Fig. 1B): Multiple slings may be required for weak or unconsolidated NEOs, or bodies that are very large. The load is spread even more widely than with a single sling. At some point, the slings be-

come dense enough to become a net enclosing the NEO. Multiple cables wrapped one or more times apiece around the body are a simple type of multiple sling arrangement (Fig. 1D).

(3) Penetration Anchor: This method could be appropriate for easy-to-penetrate NEOs that retain high residual strength after penetration. Ice composites and friable rock NEOs are candidates. The penetrator resembles a projectile and is fired into the NEO (Fig. 1C). It is easy to place, but can exhibit relatively poor holding power; it does not withstand impulses well. The kinetic energy that allows deep penetrations also tends to fracture the rock mass and reduce the anchor's holding power. This may be a good preliminary anchoring technique, for example to hold a drilling platform that allows boreholes to be drilled for stronger anchors.

(4) Drill and Expand: For high tensile-strength NEOs, the drill-and-expand approach creates a very strong anchor if the local rock is very competent and well constrained. This requires a stable platform to generate reaction forces for the rock drill. A point anchor placed in the hole grips the walls of the hole with a mechanical expanding shoe. This type of anchor works best in massive, competent rock; remote examination of the borehole will be necessary to select an optimal placement location. Generally, the greater the depth of placement, the higher the confinement of the stressed rock volume, and the less opportunity for early pullout.

(5) Drill and Friction: For high tensile-strength NEOs, drill-and-friction also creates a very strong anchor if the local rock is very competent. Also requiring a drilled borehole, it is placed by hammering the friction element into the hole and trapping the anchor there; therefore, a reaction force is needed for this step as well.

(6) Drill and Glue: For moderate tensile-strength NEOs, drill-and-glue can create a strong anchor if the glue penetrates the local rock mass. The glue (usually epoxy resin on Earth) tends to hold weak or fragmented rock in place. If the body has poor cohesion and many weakness planes, the end result may be that it disintegrates and leaves behind a small glued-together ball of rock around the anchor. New glues probably will be required since vacuum and temperature will have profound effects on terrestrial glues and their ability to cure.

The point anchors rely on the tensile strength of the NEO, a less important parameter for the slings. Geologic materials are always weak in tensile strength, with the lone exception of pure metallic bodies unweakened by rock inclusions. The drill-and-place systems (expand, friction, or glue) are mechanically similar to each other, but their placement methods vary, allowing some flexibility in their application. The most likely scenarios will include several types of anchor systems. These may be combined with slings during the later stages of processing as the NEO loses its cohesiveness.

TABLE 2. Rotational Kinetic Energy (joules) for Homogeneous Granite Spheres of Various Sizes and Rotational Periods

Period (hr)	Diameter(m) 100	500	1,000	5,000	10,000
5	70e+03	219e+06	7e+09	22e+12	702e+12
12	12e+03	38e+06	1e+09	4e+12	122e+12
24	3e+03	10e+06	305e+06	952e+09	30e+12
120	122e+00	381e+03	12e+06	38e+09	1e+12
240	30e+00	95e+03	3e+06	10e+09	305e+09

2.2. NEO Motion Control and Rotational Energy Utilization

If motion control is necessary, it should begin soon after placement of the tether. It is not clear that de-spin will always be necessary or desirable, because of the possibility of using the rotational energy of the NEO for transportation and mining assist. A 100 meter diameter granite sphere rotating once in five hours contains 70,000 joules of kinetic energy. A larger or faster rotating body (Table 2) may contain enough energy from its rotation alone to sling mined and processed material to places of utilization. Also, the slinging action would impart an impulse to the NEO which could assist moving it from the threat orbit if planned properly. Several other ways to utilize rotational energy are discussed below.Wobble (if present) is a more difficult problem than spin. NEO wobble does not lend itself to extracting energy, because the motion is more complex and there is even less energy to extract than contained in spin.

2.2.1. Utilization/Removal of NEO Rotational Energy

(1) De-spin and De-wobble: The material of the NEO could be selectively removed, altering its mass distribution and providing feed material for a mass driver oriented to reduce the spin of the NEO. This would be very similar to the transportation and orbit modification described below.

(2) Transportation: Extract the rotational energy to move material and personnel. This has long been proposed as a transportation method in deep space (e.g., see [14], [1, Sections III & IV], [21]). However, the usually low rotation rates (periods on the order of days) give relatively small amounts of energy to extract (Table 2).

(3) Orbit Modification: Proposals abound to use the mass of the asteroid to move itself (e.g., [18], [14]). Utilizing rotational energy for other purposes will change the orbit of the NEO. Whether or not that orbit change

will be sufficient to avoid Earth collision depends on the proportion of usable reaction mass to unusable mass contained by the body. Using mass-removal to alter a NEO threat orbit could be complementary to transporting material to other locations.

(4) Mining Assist: Spin energy might assist mining in several ways. Terrestrial systems utilize gravity to enhance materials handling wherever feasible. For example, the rotating cutterhead of a roadheader (see Fig. 7 in Gertsch et al., this volume) breaks rock fragments from the working face (rock surface being excavated) and lets them fall onto an excavation device which then removes them from the area.

In free fall, at least two spin-assisted centrifugal material-removal scenarios are possible. The first utilizes the mining machine and the near proximity of NEO-synchronous orbit as a materials transportation method [5]. Since NEO-synchronous orbit is quite close to the surface of the NEO, it may be relatively easy to fling the fragmented material above it. If there is a material guide that rotates with the NEO (e.g., a pipeline), the material begins to fall away from the NEO, directed into a partial bag or perhaps into a processing module. If the NEO were several years away from impact, this technique alone could mitigate its hazard.

The second mining method based on centrifugal material removal would fragment the NEO into two processing bags, with half the mass going into each bag. The spin is utilized in splitting the material into two portions, and then in processing each portion. The NEO would be split in two and rubblized by blasting: the explosive charges would fragment the rock mass, while throwing it into two opposite processing containers. This is similar to a terrestrial mining method known as caving that relies on blasting or gravity to fragment a rock mass that has been undercut. Ideally, the material fragments into appropriately-sized pieces during the fragmentation process. Determining whether a particular rock mass will yield to this technique is difficult, and the consequences of a wrong prediction can be severe. Control and secondary fragmentation of oversized boulders (a possible result) may be difficult in space.

(5) Process Control: Instead of simply tossing material away from the NEO, spin can also be used to let the material fall away from it in controlled stages. Again, the material must be in a bag or pipeline and must be flung past geosynchronous orbit. At each stage, the material is subjected to a different processing method. After a fraction is removed at one stage, the remaining material drops to the next stage.

(6) Artificial Gravity for Crew Habitat: While a habitat at the end of a rotating tether would experience a centripetal force that could simulate the effect of gravity, the tether length required to generate 0.5 G while rotating at one revolution per day is over 900 million meters, which is impractical. Artificial gravity might be better generated in other ways for

crew habitats, such as the familiar cylinder rotating at a faster rate than the NEO could withstand.

2.2.2. Partial or Complete De-Spin

After a tether is connected to the anchoring system, motion control becomes one of the next tasks for the anchor and tether system. However, despin will not necessarily precede restraint placement, as restraints may be needed to control an unconsolidated NEO during the maneuver. Before despin takes place the connection to the NEO can be strengthened by allowing the spinning NEO to take up several turns of the tether (Fig. 1D). This is known as a capstan, and it gives a very strong friction connection to the NEO. However it is accomplished, motion control begins when a sufficient connection has been assured.

As mentioned, de-wobble with a tether is more difficult than de-spin. To reduce wobble quickly, a thruster hard-mounted to the NEO would work more efficiently, but this is more complicated than connecting and deploying a tether. Fast wobble reduction with a tether will require application of high forces to the tether, which would be complicated by the relatively high compliance inherent in a tether system.

2.3. Restraints

A NEO restraint is a network of cable and/or structures that binds the body together and controls it while it is under attack by the mining machinery. It is a device to prevent catastrophic failures, such as splitting or slabbing. Even when a NEO experiences engineered fragmentation, a wide range of uncontrolled failure types is possible. This might occur, for example, after solar energy distills the volatile material from a friable body surrounded by a bag. A mechanically weakened and unstable geologic assemblage would remain, and could break apart from the stresses imposed by additional mining. Restraint additional to that provided by the anchor system may be necessary to prevent the remnants from tumbling and perhaps destroying the processing bag.

The restraint system also could provide locations for processing plants, control centers, workshops, and other supplementary systems that don't require artificial gravity (see operating platforms, below). One use of the restraint system would have it provide reaction forces for drilling, fragmentation, and excavation. In an extreme case, the mesh created by the restraint system could be fine enough to hold the smallest fragments created by the mining machinery in place.

NEO restraint can be accomplished in several ways, which can be combined as the situation requires:

(1) Anchor and Tether Wrap: One or more anchored tethers would be wound around the NEO, preventing major breakups but having little effect on

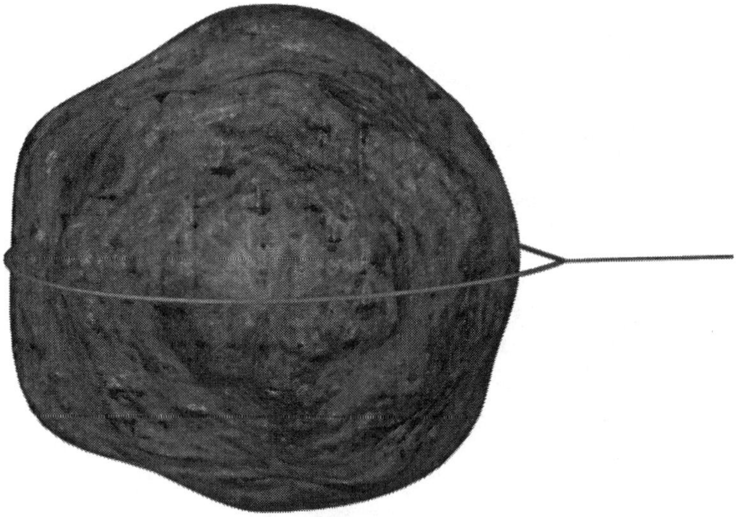

1A. Attachment to the NEO can start with a single sling that acts like a lariat. This simple system is quite strong if the "slip knot" does not kink the cable.

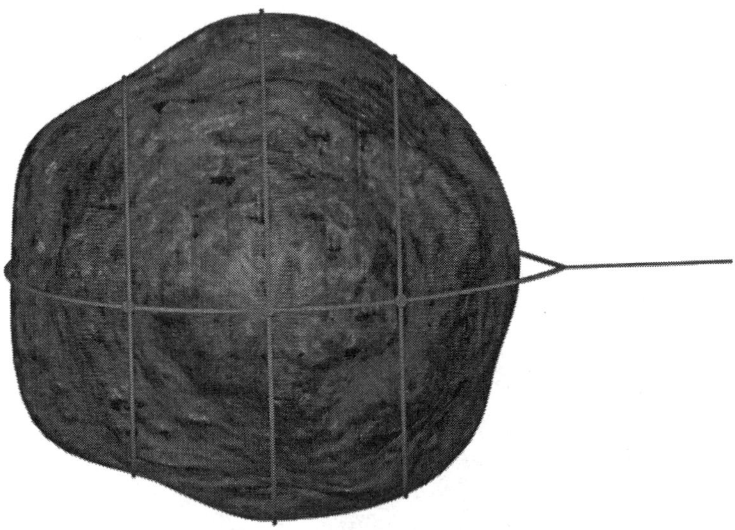

1B. After a single base loop is placed, it can be the basis for large scale weaving of a cable network around the NEO. The network then becomes a working platform.

FIGURE 1. Some Methods for Securing a Tether to a NEO. Slings should be a very effective method to secure tethers to fairly large NEOs, at least approximately 300 meters in diameter. The limiting factor may be the compliance of the cable, rather than its tensile strength. Cable stretch would cause the reaction forces of a mining tool, such as a drill or mechanical cutter-head, to fluctuate, reducing efficiency and increasing destructive mechanical vibrations. Point anchors may be useful for creating the first toehold on the NEO, but may have to be supplemented with slings for reliability.

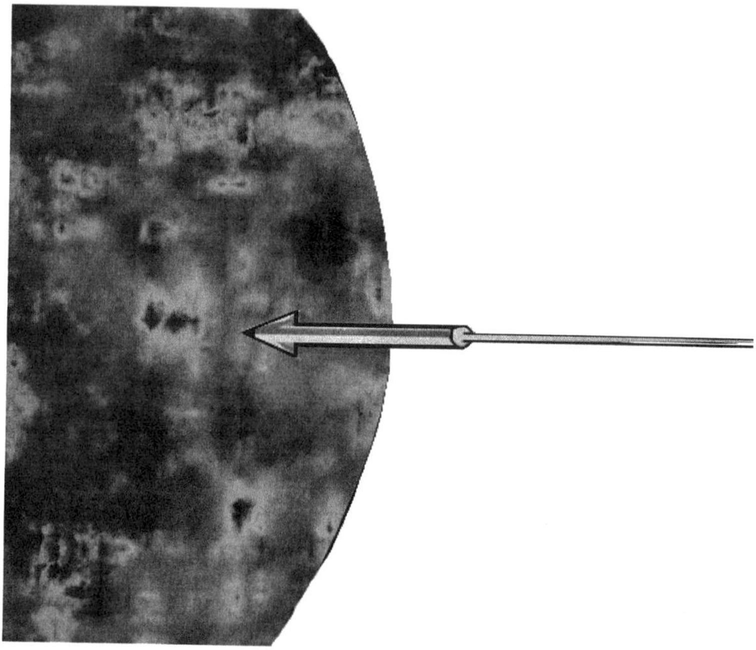

FIGURE 1C. Penetrators are point anchors that can be fired into the rock instead of being placed in a predrilled hole. They work well in fractured or soft material, but easily penetrated material often does not grip anchors well. Consequently, penetrator anchors can not withstand impulses of any magnitude. They already are being used on Earth, as are the rock bolts described by Gertsch et al. (this volume).

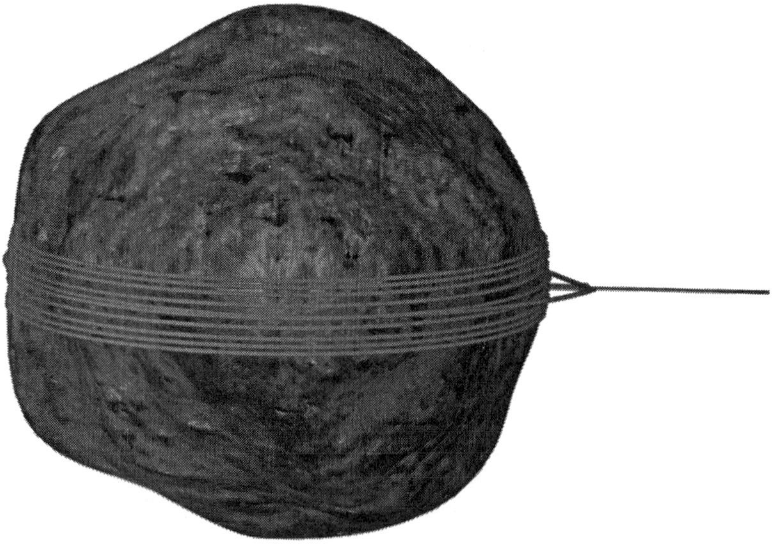

FIGURE 1D. The capstan wrap – several loops of cable around the NEO – overcomes the anchor grippability problem by using the friction of the cable against the NEO for extra holding ability.

smaller material failures. The tethers could be fastened together where they cross each other, strengthening the framework.

(2) Cable Cage: A continuation of the tether wrap, the cable cage would weave a network of cables, possibly of several graduated sizes, in and around a sparse network of tethers wrapped around the body. A cable cage could be made as coarse or as fine as needed, subject to time and material constraints.

(3) Rigid Cage: This option would require the most material, and would be made more feasible by the availability of locally manufactured structural elements. For pure restraint purposes, it would be needed only for weak, unconsolidated NEOs. However, a rigid cage would most easily provide the reaction forces needed by mining and drilling machinery working on any type of NEO. The drawbacks of rigid cages include their relative complexity, expense, and high mass requirements. Even with a premining mission to create structural members from on-site materials (reducing launch mass requirements), assembly would not be a trivial matter. Custom fitting of framework members would require close human supervision.

(4) Bag: Bagging may be required for poorly consolidated NEOs or NEOs requiring fragmentation for processing. This approach is discussed in Section 2.5.

The level of difficulty involved in these NEO restraint methods increases from the tether wrap to the rigid cage concept. Bagging, while a continuation of the trend toward total enclosure of the NEO, would require less mass and possibly less labor than the other methods. A weak NEO nevertheless may require the more difficult approaches or a combination of them.

2.4. Operating Platforms

The simplest operating platform is a cable wrapped once around the NEO [5]. A single cable actually can fulfill several functions: a tether to de-spin the NEO, a basic restraint system, a force reaction device, or simply a place to attach things: drilling platform, explosives loading platform, machinery base, habitat anchor, and spacecraft dock, for example. Like the mining and processing unit operations discussed in the accompanying paper (Gertsch et al., this volume), the unit operations of tethering, restraining, and providing reaction forces can be accomplished by separate or combined systems, depending on the particular situation encountered.

In the general case, operating platforms will be set up after emplacement of any tethers or restraints, since platform construction will need a staging area. Secure working platforms from which to attack the NEO also may be required; their design will depend on the mining method. The requirements of both the mining and the milling (processing) systems will have to be addressed at this stage. A well-designed tether-restraint-platform system will enhance the safety and efficiency of moving equipment and personnel.

The layout of the mining platforms generally will reflect the mining sequence dictated by the distribution of ore, or by the needs of the mitigation plan. Mining machines will work beneath the platforms, gripping and pushing against their frameworks to achieve the fracturing, crowding, and loading forces necessary to fragment and excavate the rock (Fig. 2). If partial (not covering the entire NEO), the reaction frames must be large enough that mining can take place under them for some time before they must be dismantled and/or moved to continue mining elsewhere. Upon completion of the mining pass for which the framework was designed, it then would have to be dismantled and removed or modified for the next pass. At the present point in the planning process it is impossible to predict the cost-benefit ratio of continually modifying the framework. It is likely that such a labor-intensive approach would be feasible for a NEO containing isolated pockets of high-value ore, and infeasible for a NEO containing more disseminated ore, or one consisting entirely of desired product (e.g., ice). The more imminent the impact hazard with the Earth, the more likely this approach would be adopted, if it also resulted in a time savings.

As mining progresses deeper, especially in the isolated-pocket case, the need for a reaction force system would diminish — it would be replaced by the NEO itself. At this point the terrestrial system of hydraulic grippers to anchor tunnel boring machines to the walls of the tunnels they have just excavated could be applied almost directly. The framework then could be recycled and its material/components used as needed elsewhere.

2.5. Bagging

After the NEO motion is under control and restrained, a bag or bags could be placed either around the entire body or covering any portion of its surface. Bags will have to be lightweight, robust, and flexible in the cold vacuum of interplanetary space. The demands of the mining and processing to be accomplished within their confines also will dictate their material properties closely.

Bagging a NEO for mining and processing is one engineering response to the unfamiliar uncertainties of working in space; for example, with no gravity, no atmosphere, and significant temperature variations due to sunlight vs. shadow [4]. Enclosing all or part of the working volume will mitigate some of these drawbacks. Specifically, there are three major and complementary reasons to bag an asteroid: to contain material fragments, to provide a processing vessel, and to provide an operating platform for mining machinery.

Bagging is feasible for a very wide range of NEO sizes, chemical properties, and mechanical properties. To determine the appropriate bagging scenario, the following factors should be considered:

FIGURE 2A. Reaction Forces Could be Derived from the Cable Sling Network. These artist's conceptions show a mechanical mining machine (A, from Gertsch [5]) and a blasthole drill (B) gripping cables wrapped around a NEO. The former would mechanically fragment, excavate, and begin transportation of rock fragments by launching them toward a processing module, and the latter would drill blastholes for explosives. Although not shown here, provision would have to be made to accommodate the uneven surface of the NEO while maintaining consistent reaction force vectors.

FIGURE 2B

(1) Size: There will almost certainly be an upper bound on NEO size for complete bagging, a limit that will change with NEO mechanical properties. Stronger bodies need less restraint. At the extreme large sizes, when fragment control is still necessary, partial bagging may work [11], but restraint of the NEO remnants will become necessary, anyway, above some level of mass reduction from mining.
(2) Strength: NEO cohesion may be so low that the body will tend to disintegrate uncontrollably during fragmentation. The effect could be global and/or local. Large or pervasive weakness planes (called joints or faults) can cause a NEO to split into several fragments [6]. Higher tensile strength would increase the NEO's resistance to fragmentation, thus requiring more explosive or mechanical energy input [2], [29]. However, high fragmentation energy can create a large fraction of fine material (dust), which may create particle containment problems. High fragmentation energy will also stress the bag and the restraint system, because the fragments tend to be expelled at higher velocities. The problem of high energy fragments is compounded by the tendency of explosives to expel material out of the opening at the end of the drill hole. Blastholes in space probably will require a high strength seal (called stemming in terrestrial practice [17]).
(3) Chemistry: In addition to its effect on NEO material strength, chemistry will affect the mining process and the subsequent bagging method. A metallic Ni-Fe NEO [12], [13], [24], [25] will be attacked very differently than an ice composite NEO [3], [7]; the bagging, restraining, and operating platform designs will change accordingly.
(4) Waste Fraction: In every orebody there is an unusable portion. Since this fraction has no economic use, it must be handled as little as possible. Waste from the NEO will remain in the mining and processing bag after the useable fraction has been exploited. The size distribution and mechanical properties of this fraction will affect bag design and deployment. Although of little intrinsic value, waste disposal nevertheless is of importance: a chunk of mining waste heading mistakenly for Earth is not a desirable outcome.

3. NEO MINING METHODS

This section discusses applications of the mining and processing techniques described in the accompanying paper by Gertsch et al. (this volume) to the broad mechanical NEO classifications outlined in Table 1, once the NEO has been prepared as discussed above. As in terrestrial mining engineering, the mitigation/mining plan for a NEO will be controlled by the properties of the NEO and their distribution throughout the body.

3.1. Mining Ice Composites and Noncohesive Friable Rock

A NEO containing a large volatile fraction could be a simple matter to mine and process. At the least, ice composites and high water-fraction friable rock NEOs will in general have lower mechanical strength than stony materials, with the fragmentation difficulty increasing for hard rock and metallic NEOs [3], [12], [24]. The basic mining scheme would place the NEO within an impervious and structurally robust bag (Fig. 3). Solar energy focused on the exterior of the bag would first melt the water-ice (or free the water of hydration of a friable rock NEO) and then turn the water into steam. After being forced from the bag, the mineral-laden water or steam would be processed. A portion could be diverted into a jet for motion control and/or orbit modification. Even if not part of the main processing scheme, the steam jet technique could serve as a backup or emergency orbit control mechanism.

If the water-ice fraction is large enough, a controlled fluid jet from the NEO would be sufficient to de-spin and de-wobble the NEO (if required) and possibly even to change the ΔV sufficiently to move it from the threatening orbit. Further, as water is drawn from the NEO by melting and vaporizing, the water could become an effective transport and/or processing medium for the extraction of other valuable materials. If the water mass flow is inadequate to move or even significantly de-spin the NEO, the scheme still could effectively collect water and other volatile materials for use by the mining crew or the material transport needs.

While there will certainly be variations forced by the composition of an actual NEO, the mining and processing scenario will follow the same basic procedure. First, a tether must be anchored to it. This would form the base for enclosing the body to whatever degree is required with a restraining system, after whatever amount of initial de-spin is needed. The final preparation step would enclose the NEO with an impervious bag. The bag must be strong enough to withstand the dynamic forces that will be induced when the water is melted and removed, and the remainder of the NEO breaks up. Since the NEO will lose significant mass during the process, the anchor will eventually lose its structural integrity and disengage from the NEO. The tether, then, must be securely attached to the bag such that they become a single structural unit. At this point the bag itself may become the restraint for the NEO remnants. The tether and bag combination then must withstand the forces necessary to de-spin the NEO and move it from the threatening orbit. In addition, the tether will hold the pipeline for water and volatile transport along its length.

De-spin, orbit modification, and material processing would begin as inter-related operations following tether and bag placement. Several infrastructure modules will be necessary for these steps: solar reflector(s) for process energy, pipeline(s), processing plant(s), and fluid jet engine(s). The bag itself can be used as a fractional distillation device, with other process-

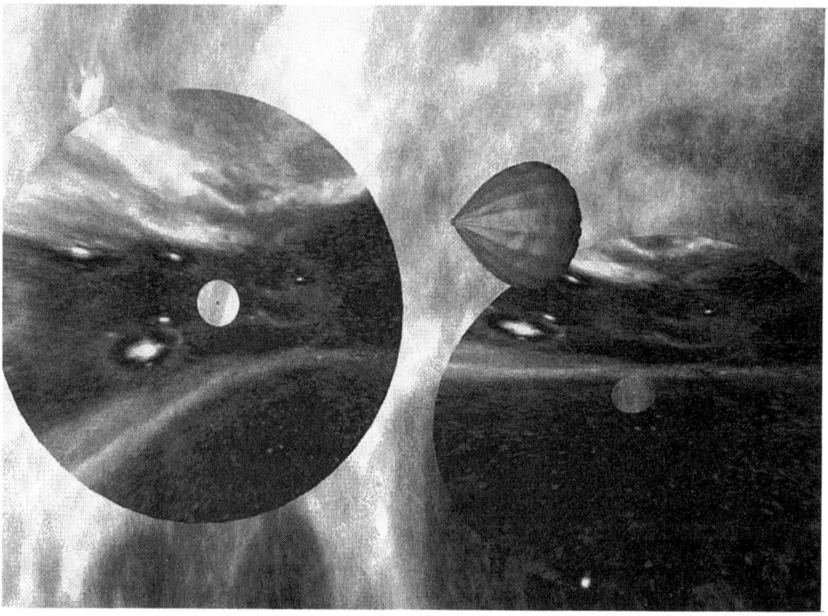

3A. Solar energy directed onto the bag with one or more reflectors vaporizes the volatile fraction.

3B. Vapor pressure starts the transport of volatile material through the piping on the tether toward the processing station. The vaporized material could be used also in a steam jet to divert the NEO from a threat orbit. In the meantime, solid material liberated by the release of the volatiles is contained by the bag. This artist's conception shows the processing module (foreground) linked to the bagged NEO (background) with a steam jet engine at upper right.

FIGURE 3. Mining NEOs with a Large Fraction of Volatile Material. The entire NEO is placed within an impervious bag, after being trussed with a network of slings tied into a main tether.

ing plants placed along the pipeline from the bag to the end of the tether. Very large lightweight solar reflectors can direct heat energy as needed anywhere along the processing plant stream [28]. Their energy will drive the processing modules and the orbit-modifying fluid jet, placed probably at the far end of the tether. Habitat and support modules can be attached at various places on the harness and tether. A way to move people and materials along the bag and tether would complete the infrastructure.

Processing and utilization of the raw materials would begin once the necessary modules are in place. The solar reflectors would heat the bag and begin driving off the water, initiating fractional distillation in the bag as the first separation step of the various minerals. The water can be turned into steam or removed as water depending upon the chemical mix of the asteroid. As the water is removed, other minerals can be extracted as required. Some minerals could use the water as a transport medium, others could be removed separately in their liquid or vapor phases. As the mineral-laden water enters the pipeline, mineral separation modules along its length would use solar energy to extract fractions of interest as the fluid (liquid or vapor) travels outward toward the end of the tether. As material is extracted from the fluid stream, the remaining fraction will be heated to higher temperature. By the time it reaches the jet engine at the end of the tether, the temperature of the fluid (probably steam) must be as high as possible. It will then be jetted in a controlled manner either to de-spin the NEO or to modify its orbit.

There can be enormous variation on how this mining and processing plant will be configured. The simplest scenario produces no product except for orbit modification; all the volatile matter is consumed by the jet. The next level of complexity might use NEO material to supply the crew and machinery of the orbit modification mission. The most complex scenario would supply necessary materials for the last two cases, but also generate products for use in space or on the Earth.

3.2. Mining Cohesive Friable Rock and Hard Rock

NEOs requiring artificial fragmentation to some degree constitute another mining and processing class. This includes both friable rock and hard crystalline rock (Classes 1 and 2), but some metal and metal-rock NEOs (Classes 3 and 3b) may be effectively treated with this approach.

3.2.1. Whole-NEO Rubblization

In a manner similar to the bag and boil method described above, whole-NEO rubblization would process the entire body as a unit (Fig. 4). Rubblization is a terrestrial mining term for completely fragmenting a rock mass to an approximately uniform particle size in situ, using a series of timed explosive blasts similar to the bench round shown in Gertsch et al. (this volume, Fig. 1B). The purpose of rubblization is to liberate the valuable minerals locked in the rock matrix without having to excavate and transport the material. After liberation, mineral separation would be the next step, for example metal removal via some leaching scheme, but only if the leachate can come into effective contact with the metal particles. One major area of concern is the remaining waste fraction. If the metal constitutes only a small percent of the NEO mass, a sizable yet unusable portion of the NEO would remain in the threatening orbit. The waste fraction might be transported out of orbit using the rotational energy of the NEO, and the impulse caused by the mass removal might also assist in modifying the orbit. However, due to the long rotational periods (see Section 2.2) many NEOs will not have sufficient rotational energy.

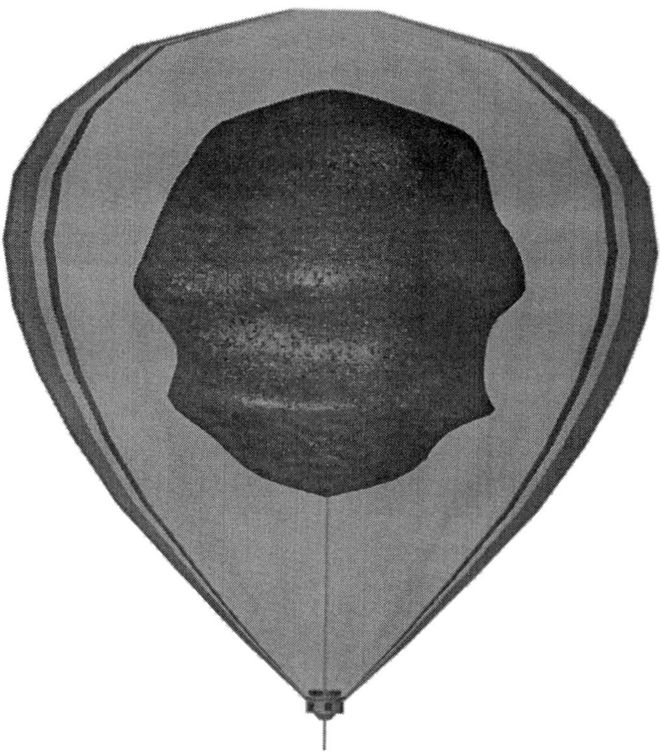

FIGURE 4. Whole-NEO Rubblization. This would be a similar process to the boiling bag scenario shown in Fig. 3. In this case, however, the NEO is fragmented mechanically or explosively inside the bag. This approach relies on the fragmentation process producing particles that are of the appropriate size distribution and mean diameter to liberate the ore mineral without recourse to further comminution. Processing would be completed by chemical or thermal methods, within the bag and/or in processing modules nearby to which material streams can be directed.

This mining method starts with anchoring and wrapping a set of cables around the NEO. A rigid cage restraint system may be required in some cases. Either system must provide reaction forces for drilling of enough blastholes to distribute the explosive sufficiently throughout the body. Controlled rubblization requires careful dissemination of the explosive energy in three-dimensional space. Depending on the processing method planned, spin reduction can be introduced either before or after rubblization. If the NEO is left spinning until after rubblization, the rotational energy could assist in processing and transporting valuable material. Once the NEO is completely bagged, the rubblization blast can occur.

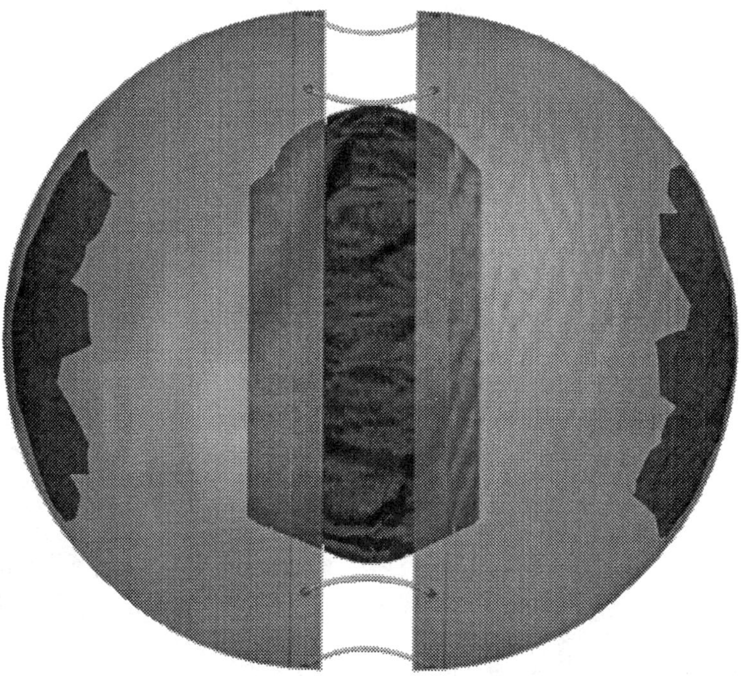

FIGURE 5. Whole NEO Processing Using a Rubblize-and-Split Method. After placement of anchors, the NEO is trussed with cables (not shown). Blastholes are drilled into the NEO using the cable mesh as a working platform while the NEO is being surrounded by two impervious bags. The blast rubblizes the entire NEO with an appropriate timing sequence that fragments it layer by layer. As the blast develops, it throws each rubblized half into its own processing bag. The two bags are allowed to travel to the end of the tether while continuing to rotate about their center of mass. The bags draw tight like a purse string. The rotating masses at the end of the tether are processed using the rotation to assist size or density separation. Material (processed or unprocessed) can be metered out the far side of either bag.

One method of conserving and utilizing the rotational energy would split the NEO into two nearly-equal masses during the rubblization blast or blasts (Fig. 5). The two bagged masses would be left rotating at opposite ends of a tether about the system center of mass, roughly halfway between the two bags. Similar in action to a jig (see Gertsch et al., this volume, Section 2.3.2), the separation process would begin by winding the tether around a central spindle, pulling the two bagged halves of material closer together. The wound cable would then be released, whereupon the two bagged masses would fall apart from each other. Before the tether snapped taut, the material would experience a dilation due to the freefall and the gases trapped in the bag. When the tether did snap taut, the material would experience a sudden, brief deceleration. The centrifugal force combined with the explosive gas products would tend to separate materials of different particle sizes and/or densities. This technique is based on intermittent acceleration and is part of a family of separation techniques known in terrestrial processing as gravity concentration [30]. Given the low rotational energies of some NEOs, spin may have to be increased to accomplish this type of separation efficiently.

After the initial separation in the bag, processing would continue by controlled extraction of material from the bag. Again centrifugal force would transport the material to subsequent processing stages. After processing, rotational energy would move the useful product to its next point of use, and move the waste material to a nonthreatening orbit. Details of the postgravity separation processing stages depend on the geology and chemistry of the NEO.

FIGURE 6. Partial-Bag Mining Scenario. This is similar to the technique shown in Fig. 4, but would be more appropriate for large NEOs, differentiated bodies with pockets of high-value ore, or threat bodies that can be divided easily into manageable subbodies; in short, NEOs that would be difficult to simply mine away in a single step. Either explosives or mechanical mining machines would be used to fragment and excavate the material. The requisite processing modules are not shown (from [5]).

The advantage of this apparently ambitious mining and processing method is the huge capacity possible, while keeping manipulation of the NEO to a minimum. If this approach proves feasible, human labor requirements should be lower than for other methods.

3.2.2. Partial and Sequential Fragmentation

Rubblizing a NEO in one step is a task that should not be undertaken lightly. The more common mining experience will most likely involve sequential fragmentation of the NEO. For example, Gertsch [5] proposed a system that anchors a bag over only a section of the NEO (Fig. 6). A mechanical mining machine similar to that shown in Fig. 2A would fragment and excavate the material, throwing the muck (broken rock) into the bag. An alternative but still sequential approach would capture explosively fragmented material in a bag and then transport the ore, still within the bag, to the processing module.

Sequential mining is at the heart of terrestrial mining practice. It allows modifications of the mining and processing methodologies to reflect changing rock conditions and to incorporate technical improvements developed during the life of the mine. It is a flexible, robust approach to any mining project. Given the variety of NEO compositions and the selection of possible mining methods already used terrestrially, it is difficult to catalog here all the conceivable variations of sequential mining of NEOs. The interested reader is referred to Hartman [8], Hustrulid [10], and Kennedy [15] for pertinent material.

3.2.3. Mass Wasting

Essentially this method would simply modify the orbit of the NEO by rapidly removing material from the NEO using controlled explosive blasts. While not a true mining method in itself, mass wasting can at least accomplish one product: threat removal.

To achieve mass wasting through the use of high explosives in space, a stable, anchored platform must be established so the NEO can be drilled and then blasted out into space (for a discussion of explosives emplacement in NEOs see [5]). Control of the size and orientation of the blasts would make them act as reaction engines, moving the NEO. Very many blasts may be needed, removing a great deal of material, to alter a threat orbit sufficiently, depending on the size of the asteroid.

3.3. Mining Metal and Metal-Rock

A NEO composed of nickel and iron can provide a high mass density, structurally sound, and easily manipulated feed for a linear accelerator rocket known as a mass driver [19]. Although simple in concept, this process contains some inherent difficulties. Melting the nickel-iron will require large amounts of energy. Then the metal must be drawn from the three-dimensional molten puddle, transported, formed, and delivered to the mass driver. Smelting in a reduced gravity environment places additional burdens on the processing. Although this approach would be difficult, it would not be impossible. Space-based research may develop alternative methods that will increase the efficiency of the process.

4. MARKETS FOR NEO RESOURCES

At its heart, mining is defined by economics, just like any other business. An orebody is nothing more than a mineralized mass that can be mined at a profit. This means that the quantity of ore shrinks and grows in response to market forces. If there is no market, there are no orebodies. The question may be asked: what is the price of saving the Earth? The answer is that there cannot be a cost limit. If mining merely adds expense to the mitigation process, and if there is no market for the materials, then it makes little sense to add mining/processing components to an already complex problem. On the other hand, if mining safely enhances mitigation, then it may be the appropriate choice even if the market for the mined material is weak.

Markets exist where and when there are customers for the product. For a threat NEO, this principle can be modified somewhat. The NEO must be deflected, and an enormous amount of capital expenditure will be associated with mitigating the threat. If that capital expense can be shifted to income-generating opportunities without causing any safety degradation, the effort is worth considering. If the lives of say a third world poverty society will be enriched by mining a threat NEO (with no increase of the threat), then the possibility should reasonably be explored. Two market areas are possible for NEO-derived materials: terrestrial and extraterrestrial.

Although it is difficult to predict just when, eventually a NEO will have to be deflected to prevent collision with Earth or human space structures. Using the material from a catastrophic threat for constructive economic purposes will be getting two benefits for the price of one. As pointed out by Hartmann and Sokolov [9], profit motive outlasts fear in the minds of the public, from whom all funding — public and private — arises.

4.1. Terrestrial Markets

Most of the likely markets on Earth already exist, but will change and grow as a result of NEO exploitation. New ones will be born. Unless carefully managed, the vastness of the resources potentially available from NEOs could destabilize terrestrial markets. The effect of New World gold and silver on European currencies in the sixteenth century provides an illustrative case study. A million-ton NEO with a barely measurable gold fraction would overwhelm the modern terrestrial gold market if the NEO could be mined economically. Speculators in copper, gold, and silver futures could have a field day in space.

4.2. Extraterrestrial Markets

Extraterrestrial markets may resemble those of frontiers such as Australia and the American West. The high cost of lifting raw materials, especially water and volatiles, up from Earth's gravity well will be analogous to the high cost of eggs, flour, and other consumables in nineteenth century mining camps. The surest way to profit in those situations was to sell supplies to the miners, rather than dig for a rich strike yourself. Extraterrestrial markets will serve all human activity in space, encompassing structural materials, shielding materials, feedstocks for secondary and tertiary markets, and consumables.

4.2.1. NEO Interception/Deflection Missions

Manned NEO interception and deflection missions need mass for propellants, life support, energy, radiation shielding, and possibly the manufacture of spare parts. The threat NEO can supply many of these needs. Particularly useful are NEOs with high water fractions, such as those in category 0.

4.2.2. Space Exploration

Human activity in space relies on safe transportation through extremely long distances. Transportation in turn requires propellant of one sort or another, structural elements, life support materials, and radiation shielding. It has long been postulated that all of these can be supplied eventually from NEOs [20]. Their successful long-term utilization will depend on the existence of a mature extraction and processing industry in space.

4.2.3. High-Value and Specialty Materials

High value and specialty items, such as custom-manufactured materials that take advantage of the unique properties of space, will find niches in both markets. For example, items such as ultra-pure materials, alloys impossible to manufacture on Earth, and large well-ordered crystals are proposed products, but their economic viability is still unproven. Other advantages of space-based manufacturing certainly will be realized once experience is gained there. The still relatively unknown requirements of life and manufacture in space also will create demand for products created in the space environment.

4.2.4. Space Manufacturing and Agriculture

A mature space manufacturing system must be supplied from space [20]. NEOs can supply resources for a wide variety of space activity, including near-Earth operations and exploration of the solar system and deep space.

Solar power satellites have long been proposed as the next effort toward a manned presence in space [20], [27]. However, there is no clearly identified path to this eventuality. Perhaps a more modest approach using laser power beaming for communications [26] is a more logical scenario. Eventually there will be a manufacturing industry in space, part of which will manufacture a wide range of products and part of which will construct and outfit deep space expeditions.

Water is supremely important to terrestrial industrial processes. As experience in space industry grows, old methodologies will be altered and new ones will be developed. Eventually water will be required relatively less for manufacturing and more for agriculture. It always will be required for human life support, if not in the quantities to which Earth dwellers are accustomed. Recycling and reuse can only minimize, not avoid, losses due to inherent inefficiencies. There will be a constant, minimum level of need for water in all extraterrestrial activities. The level will fluctuate as changes occur in processing and use methodologies, but will never be zero. At least part of this demand could be supplied by the ice composite class of NEOs.

Carbon and volatiles will be needed in a wide variety of industrial processes, while organics are necessary for agriculture and the manufacture of synthetics. Precious metals have a wide number of specific uses which will expand if the supply is increased. They will find use as high performance materials in space and on Earth. The former is due to the relatively higher degree of remote sensing and machine operation that will be required in space. Also, solar collectors will require metals for structures and coatings.

Structural metals generally are so massive that their lifting from Earth's gravity well will remain expensive. Manufacture at the top of the well will eliminate that cost. It also will allow manufacturing process refinements, and finished material properties, that are not practical on Earth.

4.3. Tourism

Space tourism will combine the two markets of Earth and space, in that the demand is essentially terrestrial, but many raw materials would come most economically from space sources. Tourism is closely tied to entertainment; both are high-return terrestrial markets that can be expected to expand using the novelties provided by the extra-terrestrial environment. Perhaps a space journey will be the 21st century equivalent of 19th century exotic expeditions to remote corners of the Earth.

5. CONCLUSIONS

The extremely large amount of valuable natural resources represented by even the known population of NEOs increases the attractiveness of mining them, independently of the need for mitigating their impact hazard to the Earth. Mining as a component of impact hazard mitigation affords challenging opportunities for adapting mining and processing techniques developed over hundreds of generations of terrestrial experience. The fundamentally unfamiliar environment of space will offer both advantages and disadvantages: processes infeasible in terrestrial practice may become powerful new tools in space, yet the necessary goals may become difficult to achieve through familiar terrestrial means.

The target NEO, once selected, will first have to be anchored to a tether. This will assist in the subsequent infrastructure development for hazard mitigation, mining, and processing. Depending on the properties of the NEO, the infrastructure may include additional tethers, secondary and tertiary anchoring, bagging, operational platforms, and mining machines and processing plants. Hazard mitigation, the most important goal for handling a threat NEO, will drive the design of the mining/processing sequence in concert with the material properties and structure of the NEO. Once the body has been surrounded with a processing bag, several options become available for utilizing the materials contained within it.

Ice composites can be mined almost completely by surrounding them with a robust bag and extracting the considerable volatile portion through heating. Friable bodies will likely break up easily, especially once their volatile component has been removed. Neither NEO type will be easy to anchor to, and probably will have to be restrained with a multiple sling setup. Bags for processing this type of NEO will have to withstand large temperature changes and some internal pressure, as well as being lightweight, robust, and flexible.

Hard rock and metallic bodies of varying stone-metal ratios will be stronger, and therefore more likely to withstand the stress concentrations imposed by point anchor systems. They can be processed either by mechanical mining, or by application of chemical high explosives. Again, enclosure within bags, perhaps reinforced with additional restraint systems, is a likely approach for a wide range of processing scenarios.

All NEOs can be expected to have spin (and some may have wobble) components in their motions. The dynamics of each body will have to be studied carefully to find the optimal usage of this energy. For many of them, the rotational energy can be utilized in the fragmentation, excavation, comminution, separation, and materials handling steps.

The mechanical classification of NEOs suggested by Remo [23] has provided the basis in this paper for a discussion of how these bodies could be mined, with an emphasis on methodologies for changing the orbital characteristics of the NEO. A permanent human presence in space could result from the effort to destroy or deflect a NEO that is discovered to be on an intercept course with our planet. Once that has been accomplished, the potential markets for materials mined and processed in space may be perceived as strong enough that NEO mining will continue with nonthreatening NEOs. Fear may drive the initial effort, but the profit motive will keep it going.

ACKNOWLEDGMENTS

The reviewers of this paper, Mike Gaffey, Andy Cutler, and R. Larry Grayson, provided many constructive suggestions. Their assistance is very much appreciated. In addition, the efforts of Andy Schmucker in creating most of the illustrations are gratefully acknowledged.

REFERENCES

1. BILLINGHAM, J., W. GILBREATH, & B. O'LEARY, EDS., 1979. Space Resources and Space Settlements, NASA SP-428, Washington DC.
2. FURNISH, M.D., G.T. GRAY III, & J. REMO. 1993. Dynamical behavior of octahedrite from the Henbury Meteorite. *In* High Pressure Science and Technology–1993, S.C. Schmidt, J.W. Shaner, G.A. Samura and M. Ross, Eds. AIP Press. New York, NY.
3. FURNISH, M.D., M.B. BOSLOUGH, G.T. GRAY III, & J.L. REMO. 1995. Dynamical Properties Measurements for Asteroid, Comet, and Meteorite Material Applicable to Impact Modeling and Mitigation Calculations. *In* Proceedings of 1994 Hypervelocity Impact Symposium. Int. J. Impact Engrg.
4. GERTSCH, R.E. & L.E. GERTSCH. 1990a. Lunar surface mines and mining machinery: Design criteria. *In* Engineering, Construction, and Operations in Space, Proceedings of Space 90, Albuquerque, New Mexico. ASCE. New York, NY.
5. GERTSCH, R.E. 1992c. Asteroid mining. *In* Space Resources, Vol. 3, Materials, McKay M.F. et al., Eds. National Aeronautics and Space Administration. Washington, DC.
6. GERTSCH, R.E. 1992d. Some Notes on Terrestrial Blasting Design and NEO Interception. *In* Proceedings of Near Earth Object Interception Workshop. National Aeronautics and Space Administration and Los Alamos National Laboratory, LA-12476-C.
7. GREENBERG, J.M. & J.L. REMO. 1994. A comet nucleus model. *In* Hazards due to Comets and Asteroids, T. Gehrels, Ed. Univ. Arizona Press. Tuscon, Arizona.
8. HARTMAN, H.L., Sen. Ed. 1992. SME Mining Engineering Handbook. Society for Mining Metallurgy and Exploration Inc. Littleton, Colorado.
9. HARTMANN, W. K. & A. SOKOLOV. 1994. Evaluating space resources in the context of Earth impact hazards: Asteroid threat or asteroid opportunity? *In* Hazards Due to Comets and Asteroids, Tom Gehrels, Ed.:1213–1224.
10. HUSTRULID, W.A., Ed. 1982. Underground Mining Methods Handbook. Society of Mining Engineers, AIME. New York, NY.
11. ISU. 1990. International Asteroid Mission, Final Report. The International Space University, Summer Session, York University. Toronto, Canada.
12. JOHNSON, A.A. & J.L. REMO. 1974. A new interpretation of the mechanical properties of the Gibeon meteorite. J. Geophys. Res. **79:** 1142–1146.
13. JOHNSON, A.A., J. L. REMO, & R. R. DAVIS. 1978. Low temperature impact properties of the meteorite Hoba. J. Geophys. Res. **84**.

14. JOHNSON, R.D. & C. HOLBROW, Eds. 1977. Space Settlements, A Design Study, NASA SP-413, Washington, DC.
15. KENNEDY, B., Ed. 1990. Surface Mining. Society for Mining Metallurgy and Exploration Inc. Littleton, Colorado.
16. KUCH, D.L. 1979. Near-Earth terrestrial resources. Space Manufacturing, 1979. Proc. Space Studies Institute. AIAA. New York, NY.
17. MORHARD, R.C., 1987. Explosives and Rock Blasting. Atlas Powder Company. Dallas, Texas.
18. O'LEARY, B. 1977. Mass driver retrieval of Earth-approaching asteroids. Proceedings of the 3rd Princeton Conference on Space Manufacturing Facilities. AIAA. New York, NY.
19. O'LEARY, B., M.J. GAFFEY, D.J. ROSS, & R. SALKELD. 1979. Retrieval of asteroidal materials, J. Billingham & W. Gilbreath, Eds.: 173–189.Space Resources and Space Settlements, NASA SP-428.
20. O'NEILL, G.K. 1977. The High Frontier, Human Colonies in Space. William Morrow and Co., Inc. New York, NY.
21. PIGNOLET, G. 1979. Retrieving asteroids for space materials. XXX Congress, International Astronautical Federation, IAF-79-F-22. Pergamon Press. Elmsford, New York.
22. REMO, J.L. 1993. Asteroid/meteorite analogs and material properties. Appendix to Workshop 4, NASA NEO Conference, LA-1247-C, Los Alamos, NM.
23. REMO, J.L. 1994. Classifying and modeling NEO material properties and interactions. *In* Hazards Due to Comets and Asteroids, T. Gehrels, Ed. Univ. Arizona Press. Tuscon, Arizona.
24. REMO, J.L. & A.A. JOHNSON. 1974. The ductile-brittle transition in meteoric irons. Meteorites. **9:** 209–213.
25. REMO, J.L. & A.A. JOHNSON. 1975. A preliminary study of the ductile-brittle transition under impact conditions in the material from an octahedrite. J. Geophys. Res. **80:** 3744–3748.
26. REMO, J.L. 1995. Laser power beaming in space. Report of Quantametrics, Inc. St. James, New York.
27. ROBERTSON, D.F. 1995. Reach out and touch the stars. Analog, mid-Dec.: 62–73
28. STINE, G.H. 1979. The Third Industrial Revolution. Grosset and Dunlap Company, New York, NY.
29. TEDESCHI, W.J., J.L. REMO, J.F. SCHULTZE, & R.P. YOUNG, JR. 1995. Experimental hypervelocity impact effects on simulated planetesimal materials. Proceedings of the 1994 Hypervelocity Impact Symposium. Int. J. Impact Engrg.
30. WILLS, B.A. 1992. Mineral Processing Technology: An Introduction to the Practical Aspects of Ore Treatment and Mineral Recovery, 5th edit. Pergamon Press. New York. 855 pp.

Several Aspects of Space Protection of the Earth: Conceptual Development

V.A. SIMONENKO
Institute for Technical Physics
Chelyabinsk, Russia

A set of general demands to a system for space protection of the Earth (SPE) with wide opportunities is presented. It includes categorization of near-Earth objects (NEO) on lead time, type of NEOs' orbits, character, speed of approach to the Earth, and NEO material properties. The possible structure of the SPE system on the basis of existing or near-future technologies is discussed.

Depending on quantitative values of corresponding defining parameters of NEO's orbital movement and internal properties, various demands for the means of action upon threatened objects are discussed. For the most probable methods of action (kinetic projectiles and nuclear explosions), possible regimes of applications are presented and analyzed. The essential aspects for future studies of these regimes, possible results, and their consequences for the Earth and space environment are illuminated.

The demands for the launch and ejection subsystems, for delivery, guidance, and terminal navigation components, are also discussed.

Special attention is devoted to desirable directions for modern and near-future experimental and theoretical research, in particular, for material properties of NEOs, physical and mechanical processes for various regimes of action upon them, and for a more reliable description of the consequences of deflecting or shattering actions. Possible experimental programs for direct study of NEOs properties as a whole are discussed.

It becomes advisable to develop coordinated research on the general scope of the SPE problems on an international basis for better progress and more efficient use of limited scientific and technical resources.

Optimal Detection of Near-Earth Object Threats

GREGORY H. CANAVAN

Los Alamos National Laboratory
Los Alamos, New Mexico 87545

ABSTRACT: Detection of near-Earth Objects (NEOs) has concentrated on search at long range. Improved technology and search strategies could significantly increase search rates and decrease survey times.

INTRODUCTION

Earlier detection work concentrated on searches near opposition [1], where NEOs are at ranges of several AU (astronomical unit, the distance from the Sun to the Earth). This note discusses improvements in ground-based search sensors and strategies that could enable efficient search over a much wider area of the sky, which would provide much more rapid surveys and earlier warning of potential impacts [2]. The requirements for detection are derived from signal and background considerations. Optimal detection uses exposure times that are much shorter than those typically used for detection at maximum range, which permits the sensor to scan the whole fresh sky each month to maximize the opportunities for detection.

For large NEOs, it is possible to use $\approx 1\ m^2$ telescopes with large focal plane arrays to produce ranges of a few AU and detection rates capable of supporting decadal searches. As array sizes are increased, the detection rate increases as the square root of the number of detectors. Very large focal planes would eventually fill the telescope's flat field. To avoid overfilling it, it would be necessary to increase aperture and focal plane size together. In both limits, large ground-based systems have strongly diminishing marginal returns due to limitations from atmospheric turbulence.

OPTIMAL GROUND-BASED DETECTION

The preferred search strategies and technologies for the detection of NEOs has undergone a significant shift in the last few years — away from an earlier emphasis on the use of very large telescopes at maximum range toward the use of modest telescopes with very large charge coupled device (CCD) focal planes that cover the fresh sky completely each month. The reasons for this shift for large NEOs can be discussed simply and also provides a useful background for discussion of the difficulties involved in detecting smaller objects. For low signal-to-noise ratio (SNR) detection, which necessitates long integration time t, the signal, S_{req}, required to achieve a given SNR is approximately

$$S_{req} = \sqrt{SNR^2 B/Qt}, \tag{1}$$

where Q is the sensor quantum efficiency and $SNR \approx 4$ is the limiting ratio for detection. The background is $B \approx B''A\theta^2$, where A is its aperture, θ its pixel diameter, and $B'' \approx 0.63/\text{m}^2\text{-arcsec}^2$. For ground-based sensors, θ is limited by atmospheric seeing to a few arcsec, so $B \propto A$, and $S_{req} \propto \sqrt{A/t}$. The signal received by a sensor of aperture A from an illuminated object of diameter D at range r is

$$S_{rec} = JAD^2/r^2(1 + r)^2, \tag{2}$$

where J is proportional to the solar illumination and r is in AU (1 Astronomical Unit $\approx$ 150 million km). Equating S_{req} to S_{rec} determines the sensor parameters required for detection

$$\sqrt{SNR^2 B''\theta^2/AQt} = JD^2/r^2(1 + r)^2, \tag{3}$$

from which the area-exposure product required scales as $At \propto [r^2(1 + r)^2/D^2]^2$. If the sensor uses an exposure time t and views a solid angle w, it covers the sky at angular rate w/t. It searches out to range r in a volume of space that contains a density of NEOs $N \approx K/D^n$, where K and $n \approx 2$ (for $0.1 < D < 2$ km) are constants. Thus, it should detect NEOs at a rate

$$R \approx (w/t)\int dr\, r^2 N = (w/t)\int dr\ \ r^2 K/D^2$$

$$= (w/t)\sqrt{AQt/SNR^2 B''\theta^2}\, J \int dr\, r^2 K/r^2(1 + r)^2$$

$$\approx (w/t)\sqrt{AQt/SNR^2 B''\theta^2}\, JK. \tag{4}$$

The rate R scales primarily on $(w/t)\sqrt{At}$, where (w/t) is the rate of covering sky, and $\sqrt{At}$ is the depth to which it is covered. For the sensor to cover the whole fresh sky each month requires that it attain a search rate $W' \approx 100\ \text{deg}^2/\text{hr}$. To achieve that rate with a sensor with a field of view (FOV) w requires an exposure time no longer than

$$t \approx w/W'. \tag{5}$$

Shorter exposure times would cover parts of the sky twice, which would not produce additional discoveries. Longer exposures would leave parts of the sky unexamined. Thus, $t \approx w/W'$ is the optimum time for maximizing the *rate* of detection. It is generally much shorter than the time for maximizing the *range* of detection, which is the usual metric. The parameters Q, SNR, B'', and θ^2 are relatively inflexible, so R scales as

$$R_{\text{opt}} \propto (w/t)\sqrt{At} \propto W'\sqrt{Aw/W'} \propto \sqrt{Aw/W'}\,. \tag{6}$$

This result shows that the optimum detection rate scales simply as the square root of the aperture area and search rate. However, it is shown below that for practical systems, its scaling is simpler still.

LARGE DETECTOR ARRAYS

For NEOs larger than a kilometer, it is possible to use telescopes a meter or so in diameter with large charge-coupled device (CCD) focal planes of a few million detectors to achieve ranges of a few AU, which produces detection rates adequate to support search times of a few decades. It is useful to examine the optimal sensors for this purpose. A sensor of f-number F (= focal length/primary diameter) and a CCD focal plane array with N detectors of pitch d has a geometric solid angle FOV $w \approx N(d/F)^2/A$; thus, from Eq. (6), its optimized search rate scales as

$$R_{\text{opt}} \propto \sqrt{A[N(d/F^2)/A]} \propto (d/F)\sqrt{N}\,. \tag{7}$$

In this equation, F cannot be decreased much below the current value of $F \approx 2$ without degrading optical quality, and d decreases only with technological development. Thus, increasing the number of detectors N appears to be the only variable with significant potential for increasing detection rate. Indeed, at the current level of technology, R_{opt} is simply proportional to the square root of N. The optimal detection rate is independent of A, because for given N, the additional light gathering that results from increasing A is just offset by the reduction of FOV it produces. Increasing A would just provide more flat space in the focal plane to put additional detectors — if more space is needed. To the extent that space is available in the focal planes of existing telescopes, the most effective way to improve ground-based sensors is to improve their focal planes — not build bigger telescopes.

For very large focal planes, there are limits. For given A, as N increases, w increases as N/A. At some value of N, the maximum usable field of view, w_{max}, will be reached. Thereafter, A will have to be increased in proportion to N to maintain image quality. In that limit,

$$R_{opt} \propto \sqrt{A w_{max}} \propto \sqrt{A}\,, \tag{8}$$

and aperture again has utility. The transition between these two limits occurs where $w_{max} \approx N(d/F)^2/A$, or

$$N/A = w_{max}(d/F)^2. \tag{9}$$

For nominal parameters this occurs at $N/A \approx (2\ \text{deg}/57.3\ \text{deg/rad})^2(2/25\ \text{micron})^2 \approx 8 \times 10^6/\text{m}^2$. For $a < 8 \times 10^6$ array, a 1 m^2 aperture sensor would thus follow the $\sqrt{N}$ scaling of Eq. (7). For significantly larger arrays, it would follow the $\sqrt{A}$ scaling of Eq. (8). As CCD technology improves and d decreases, the transitional value of N/A will increase as $1/d^2$. When the sensor is designed against the limit of Eq. (9), $R_{opt} \propto \sqrt{AN(d/F^2)/A}$. This is formally the same scaling as that in Eq. (7). The difference is that in Eq. (7), w is free to vary, so A cancels out and N is the dominant variable, while in Eq. (9), w is fixed, so N must be used to offset any change in A, so A is the dominant variable.

In these two limits the scaling of R_{opt} is $\sqrt{N}$ or $\sqrt{A}$. Since costs increase roughly as N or A, the return per unit investment falls as $1/\sqrt{N}$ or $1/\sqrt{A}$ in these two limits, which means that large ground-based systems have strongly diminishing marginal returns. Tracing back through the analysis shows that both of these limitations stem from the inability of ground-based sensors to achieve pixel sizes smaller than those dictated by atmospheric turbulence. That limitation could be overcome by space-based systems, which are discussed elsewhere [3].

With the results above, it would be possible to estimate the absolute detection rates from various sensors. However, the uncertainties in the NEO distribution and signatures and the uncertainties in the performance of sensors that are only designed parametrically as yet make that less accurate than simply normalizing on the empirical detection rates of current sensors. Given that is done, the principal results of the analysis above are the $\sqrt{N}$ or $\sqrt{A}$ limits and the transition between them at $N/A = w_{max}(F/d)^2$, which do not need numerical confirmation.

SUMMARY

Preferred search strategies and technologies for NEO detection have undergone significant shifts in the last few years. Earlier work on detection concentrated on searches near opposition, where ranges are longest. Improvements in ground-based search sensors and strategies make search faster and warning more timely. The requirements for detection follow immediately from signal and background considerations. Integrating over space gives an approximate evaluation of the detection rate. Optimal detection uses exposure times that are proportional to sensor FOV, and much shorter than those typically used for detection at maximum range. That permits the sensor to scan the whole fresh sky each month to maximize the opportunities for detection.

For large NEOs, it is possible to use $\approx 1\ m^2$ telescopes with megapixel CCD focal planes to produce ranges of a few AU and detection rates capable of supporting decadal searches. As array sizes are increased, the detection rate initially increases as the square root of the number of detectors. The aperture is initially unimportant, because the additional light gathering it produces is just offset by the corresponding reduction in FOV. Very large focal planes would eventually fill the telescope's flat field. To avoid going outside that area, it would be necessary to allow its aperture to vary with detection rate. In both of these limits, large ground-based systems have strongly diminishing marginal returns, which result from the limitations from atmospheric turbulence.

REFERENCES

1. SHOEMAKER, E. 1995. Report of the Near-Earth Object Survey Working Group. Washington, D.C., NASA Hq., June 1995.
2. CANAVAN, G. 1994. Cost and benefit of near-Earth object detection and interception. *In* Hazards Due to Comets and Asteroids, T. Gehrels, Ed. Univ. of Arizona Press. Tucson, Arizona.
3. CANAVAN, G. 1996. Optimal detection of short-warning near-Earth object threats. *In* Proceedings Space 96, Fifth International Conference on Space, Albuquerque, New Mexico.

Technology for the Detection of Near-Earth Objects

G.H. STOKES AND D.F. KOSTISHACK

[a]*MIT Lincoln Laboratory*
244 Wood Street
Lexington, Massachusetts 02173

The detection and cataloging of near-Earth asteroids and comets is of considerable current interest in the international scientific and policy making communities. Electro-optical technology developed by MIT Lincoln Laboratory for satellite tracking applications is well suited to the detection and tracking of faint near-Earth objects. This technology includes state-of-the-art, large-format CCDs with noise and readout performances which considerably exceed those of CCDs employed in current NEO search systems. The performance of the Lincoln Laboratory CCDs and surveillance techniques enables the construction of a NEO search system using 1-meter class telescopes while retaining the performance of the Spaceguard system [1] which proposes 2.5-meter class telescopes. Use of smaller telescopes results in considerable system construction cost savings.

The performance of the new CCDs and the associated support system has been demonstrated as part of ongoing, routine device testing and field operations. In 1993, field measurements demonstrated the performance of the CCD technology by using a 1024 by 1024 pixel format CCD mounted on a 31-inch telescope to conduct a limited search for asteroids. Field test efforts in 1995 will employ the next generation CCD, with a 2560 by 1960 pixel format, to validate the technology for use in an NEO search.

REFERENCE

1. MORRISON, D., Chair. 1992. The Spaceguard Study: Report of the NASA International Near-Earth-Object Detection Workshop, Washington D.C., NASA (23 January 1992).

A Near-Earth Object Hazard Index

RICHARD P. BINZEL

Department of Earth, Atmospheric, and Planetary Sciences
Massachusetts Institute of Technology
Cambridge, Massachusetts 02139

ABSTRACT: With the advent of expanded near-Earth object (NEO) surveys, the number of known close approaches to Earth will increase dramatically. For many of these objects, uncertainties in their preliminary orbits will result in calculations yielding nonzero probabilities of collision over the next century. An *index* system is proposed which places the hazard posed by any given close approach into an easily understandable context that allows simple and efficient communication between astronomers and the public.

INTRODUCTION

Collisions of asteroids and comets with the Earth is a topic so provocative and so prone to sensationalism that great care must be taken to assess and publicly communicate the realistic hazard (or nonhazard) posed by close approaches of currently known and newly-discovered objects. Low-probability events having high consequences are, by their very nature, not within the realm of common experiences. To meet this communications challenge, a simple *index* is proposed which:

(1) Sets a standard for what information should be communicated regarding the hazard posed by close approaches of near-Earth objects.
(2) Provides a lexicon for efficient public communication of collision probabilities and potential consequences for close encounters.
(3) Provides a context for public understanding of the relative threat posed by any individual close encounter event.

Forethought on communicating the potential hazards posed by close encounters of near-Earth objects is necessary because the advent of expanded NEO surveys will dramatically increase the number of known objects and correspondingly increase the number of predictable close approach events. Forecasting these events over the course of one century, a reasonable horizon for practical concern, will be difficult due to uncertainties associated with preliminary orbit determinations. A critical consequence of these uncertainties is that for many newly-discovered NEOs, it will not be possible initially to preclude a nonzero collision probability for many close approach events occurring within the next century. For such events, we have a responsibility to assess as accurately and objectively as possible the collision probability and the potential consequences of an impact.

THE HAZARD INDEX

It is desirable that any proposed system for public communication should: (a) be understandable at many different levels; (b) provide a context for understanding the full range of the problem; and (c) have a rigorous basis in scientific calculation. The *index* proposed here is based on calculated values of an object's collision probability and impact kinetic energy. The proposed *index* translates the collision probability into a scale ranging from 0 to 5, which encompasses a full range of probabilistic outcomes over 8 orders of magnitude. At its most simple level, the *index* is understandable as "0 is good and 5 is bad." By establishing a scale bounded by upper and lower values, one can immediately achieve a sense of context for any event. For example, a close approach event reported to have "an *index* value of 1 out of a 5 point scale" is likely to be appropriately disregarded as not being a matter of practical concern.

At the next level of sophistication, the *index* uses descriptive terms to indicate the collision likelihood. For the *index* categories 0 to 5, respectively, these likelihoods are described as "none," "extremely improbable" "very improbable," "unlikely," "possible," and "certain." For likelihoods which are "unlikely" or greater, responsible communication would also include the date of closest approach. Finally, the most sophisticated level of communication would report the value of the collision probability, or perhaps the range of values derived from differing assumptions or different modeling algorithms.

The attributes of the proposed *hazard index* are most easily examined in Table 1, where the level of sophistication in the information increases from left to right. Table 1 details the collision probability ranges which are logarithmically associated with each category. Because each category spans two orders of magnitude, the *index* is robust with regard to calculation uncertainties.

TABLE 1. Description of Near-Earth Object Hazard Index

Hazard Index Category	Likelihood of Collision	Probability Range for the Category (P = 100 yr cumulative probability)
0	None	$P < 10^{-8}$
1	Extremely Improbable	$10^{-8} \leq P < 10^{-6}$
2	Very Improbable	$10^{-6} \leq P < 10^{-4}$
3	Unlikely. Closest approach on date X	$10^{-4} \leq P < 10^{-2}$
4	Possible on date X	$0.01 \leq P < 0.99$
5	Certain on date X	$P \geq 0.99$

TABLE 2. Description of Potential Collision Outcomes

Collision Outcome Category	Likely Consequences of Collision	Natural Frequency of Occurrence	Kinetic Energy Range for the Category (KE in megatons TNT)
Global	Serious global effects. Severe regional damage.	Once per 100,000 years, or less often.	$KE > 10^5$
Regional	Possible regional damage. Severe local impact damage.	Between once per 100,000 years and once per 1000 years.	$10^5 \geq KE > 100$
Local	Possible local impact damage.	Between once per 1000 years and once per 50 years.	$100 \geq KE > 1.0$
No consequence	No damage likely. Probable atmospheric dissipation.	Once per 50 years, or more often.	$KE \leq 1.0$

Collision probability is only one dimension in communicating the hazard (or nonhazard) of an NEO close approach event. The second dimension is the potential outcome of an actual collision, for which the most simple parametrization is by the impact kinetic energy. To incorporate this second dimension, the proposed *index* translates the range of impact outcomes, spanning 7 orders of magnitude in kinetic energy, into descriptive terms ranging from "no consequence" to possible "local," "regional," or "global" effects. As is the case for the collision probability dimension, by establishing categories which bound the full range of possibilities, the context of a particular event can be more easily understood.

Table 2 presents the increasing levels of sophistication for describing the potential outcomes of collision events. The first step is to give a more detailed description of what constitutes, for example, a "local" event. An additional important step, which further puts an event into context, is to communicate the natural frequency of occurrence of events within each category. Finally, the highest level of sophistication for this categorization is to present the range of calculated estimates for the actual kinetic energy value. Because each category spans three orders of magnitude, they are also robust with regard to very large uncertainties in the kinetic energy calculations.

PRACTICAL APPLICATION OF THE HAZARD INDEX

For nearly all newly-discovered NEOs, application of the *hazard index* will be immediate and trivial. Most preliminary orbits will clearly show that the nodes occur far from the Earth's orbit, resulting in an immediate and final determination of zero for the *index* value. Unfortunately, it will not be the case that all new discoveries will be immediately assignable to category zero even when relatively good preliminary orbits are in hand. These objects are the focus of discussion within this paper.

For objects which are not immediately assigned to category zero, their *index* values will not be constant. Rather, their calculated *index* values will vary with time until they reach either a value of 0 (no collision) or 5 (certain collision). The binary outcome is a consequence of the problem being deterministic: any object in question either will not or will strike the Earth within the next century. The rational expectation for such an object is that its *index* value will stepwise decrease to zero as the measured arc length and precision of the orbit increases. For most objects, such a definitive determination is likely to be possible during the course of a single apparition. If the astrometric coverage during the discovery apparition is poor, or if a future encounter is particularly close, astrometric measurements during two or more apparitions may be necessary to yield a definitive determination of the *index* value. During the interval when the determination is uncertain, the *hazard index* will serve to provide a context for public communication of the likelihood that an object's collision probability will prove to be zero.

While the proposed *hazard index* utilizes a timescale of one century for achieving a deterministic outcome of whether or not an object is likely to collide with the Earth, this timescale is not appropriate for all objects. For example, many short-period comets have orbital periods longer than 100 years. In such cases, the *hazard index* can be based on the probabilistic outcome during the next perihelion passage. Other examples for employing longer timescales would be for objects with very well-determined orbits which are benign during the next century, but which make predictable close approaches with significantly nonzero collision probabilities at later dates. In such cases, prudence dictates that their *index* values should correspond to the hazard posed by close encounter events occurring more than one century ahead from the present.

A TWO-DIMENSIONAL VIEW OF THE HAZARD SPACE

An ideal way to assess the portent of a close encounter event may be to examine it within the two-dimensional "hazard space" defined by the collision probability and kinetic energy. An example of this hazard space is shown in Fig. 1. In the figure, the most seriously threatening events are those which fall toward the upper right corner, denoting high-probability collisions having large kinetic energies. The level of threat decreases for events which fall further to the right or toward the bottom of the diagram. Annual meteor showers contain objects for which the collision probability is certain, but they have kinetic energies too low to be of any practical consequence. These events fall in the bottom-right corner of the diagram.

Within the two-dimensional plot of the hazard space, all currently known close approaches by discovered NEOs fall along the left-most edge and are events of no consequence. The rational expectation for newly-discovered objects, which are not immediately placed at the left edge, is that their position will evolve toward that side as the measured arc lengths and precision of their orbits increase. Newly-discovered objects which are placed in the upper right, or whose plotted positions evolve in that direction, are the most urgent candidates for high precision orbit measurements. Those falling furthest toward the upper right are the best candidates for consideration of mitigation strategies. However, evaluating the need for mitigation requires the consideration of numerous additional dimensions beyond collision probability and kinetic energy. Some examples of these additional dimensions are the composition of the potential impactor and the potential impact location.

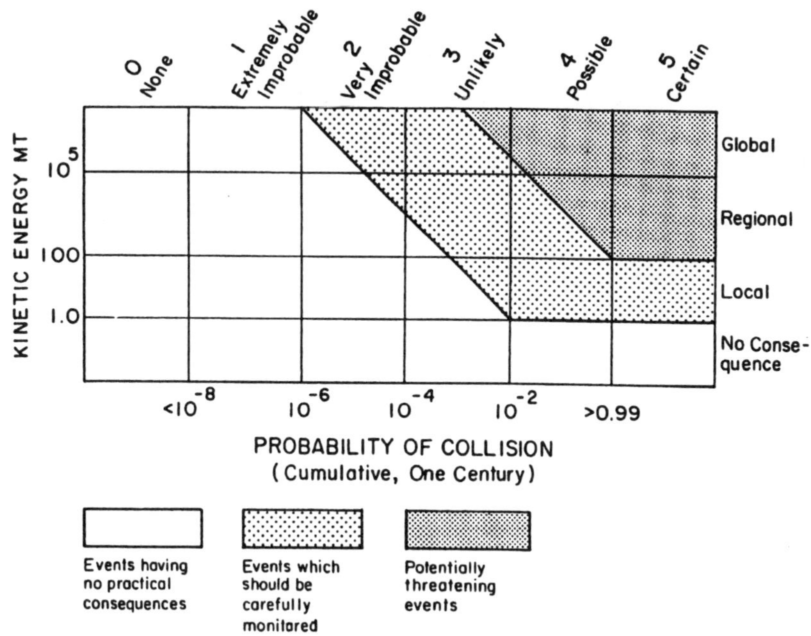

FIGURE 1. A two-dimensional view of the hazard space arising from close approaches of near-Earth objects (NEOs). Individual close approach events may be plotted within this figure based on their collision probability (increasing toward the right) and their kinetic energy (as measured in megatons of TNT, increasing toward the top). The boxes denote the various categories for the proposed *hazard index*. Collisional probabilities are labeled along the bottom axis and their corresponding *hazard index* categories, ranging from 0 "none" to 5 "certain," are labeled along the top. Values for the kinetic energies are labeled along the left axis and categories labeling their potential consequences are labeled on the right. In general, the relative threat posed by any individual event increases toward the upper-right corner, corresponding to higher probability and higher energy collisions. The shading in the figure represents a subjective assessment of the threat posed by a wide range of events. All close-encounter events for all currently known NEOs fall along the far left edge of the diagram within hazard category zero.

MATHEMATICAL DESCRIPTION OF THE HAZARD INDEX

The proposed *hazard index* is based on the calculated probability of collision by an object making one or more close approaches to Earth during the next 100 years. Because an object in an orbit which makes repeated close approaches may be more hazardous than an object which makes a single close approach, the *hazard index* may be best parametrized by using the cumulative collision probability over 100 years. Let $P_1, P_2, P_3, \ldots, P_n$ equal the discrete collision probabilities for the first, second, third, ..., nth close approach events occurring for an object during the next century. The cumulative collision probability over one century, P, can then be written as:

$$P = 1 - (1 - P_1)(1 - P_2)(1 - P_3) \ldots (1 - P_n). \tag{1}$$

Within Eq. (1), any individual value for $P_1, P_2, P_3, \ldots$, or P_n which is substantially larger than the others will dominate in computing the value for P. For objects with orbital periods greater than 100 years, we *assign* P to be equal to the discrete collision probability at the next perihelion passage. The valid range for the values of $P_1, P_2, P_3, \ldots, P_n$ is the interval 0 to 1.

Once we have computed a value for P, the *hazard index* value, H, is defined by:

$$\begin{aligned} H &= 0 \quad \text{for } P < 10^{-8}, \\ H &= 5 + 0.5 \operatorname{Log}(P) \quad \text{for } 10^{-8} \le P \le 1, \end{aligned} \tag{2}$$

where the valid range for H is the interval 0 to 5. For $1 \le H < 3$, it may be most reasonable to report just the integer value (e.g., "1" or "2"). For $H \ge 3$, the date of closest approach (corresponding to the highest discrete probability event) should also be reported. Furthermore, if an object has two or more close approach events having discrete probabilities $> 10^{-4}$ (individually corresponding to $H \ge 3$) , it may be of interest to compute and report separate *hazard index* values and approach dates for each event.

CONCLUSIONS

Forethought now should be applied to the problem of how to communicate the hazard posed by NEO close approach events. A uniform system with a common lexicon will be far more effective in communicating with and educating the public than unorganized and haphazard reporting. Clear, careful, and consistent communication will be the most important elements in abating the potential for over sensationalism which threatens the credibility of the field. An *index* system is proposed here which puts the hazard (or nonhazard) of NEO close approaches into an easily understandable context for simple and effective public communication.

ACKNOWLEDGMENTS

The development of this *index* has benefited greatly from from discussions with numerous colleagues, most notably: Clark R. Chapman (Planetary Science Institute), David Morrison (NASA Ames), Alan W. Harris (Jet Propulsion Laboratory), and J. Kelly Beatty (Sky & Telescope).

Momentum Coupling to NEOs

B.P. SHAFER,[a] M.D. GARCIA,[a] R.A. MANAGAN,[b] J.L. REMO,[c] C.E. ROSENKILDE,[b] R.J. SCAMMON,[a] C.M. SNELL,[a] AND R.F. STELLINGWERF[a]

[a]*Los Alamos National Laboratory*
Los Alamos, New Mexico 87545

[b]*Lawrence Livermore National Laboratory*
Liver more, California 94550
and
[c]*Quantametrics Inc.*
1 Brackenwood Path,Head of the Harbor
St. James, New York 11780

ABSTRACT: We considered the coupling of momentum to near-Earth objects (NEOs) for the purposes of collision avoidance or collision damage mitigation through trajectory modification. The momentum sources considered were pulsed lasers, kinetic energy impactors, and nuclear (or conventional) explosives, detonated above or below the surface of the NEO by the nonnuclear and nuclear momentum sources listed above. We did *not* address the system trade-offs or the environmental implications of momentum source delivery.

INTRODUCTION

Momentum coupling to Near-Earth Objects (NEOs) was first reported by Hammerling and Remo [3], [4] for interactions with x-rays and neutrons. Shafer et al. [6] presented the first overall review of the coupling of energy and momentum to asteroids and comets. Specifically, this study focused on the impulses delivered to a NEO by pulsed lasers, kinetic energy impactors, and nuclear or conventional explosives detonated above or below the surface of the NEO. The results presented here include results from the previous study in addition to new analytical and experimental results for laser and kinetic energy coupling to NEOs. In this study we present assessments of the efficacy of a variety of nonnuclear and nuclear energy sources in diverting Earth-threatening comets or asteroids through momentum coupling. The energy sources and specific energies of those sources are shown in Fig. 1. The specific methods of applying the nonnuclear or nuclear energy sources are illustrated in Fig. 2.

Energy Sources
- Pulsed lasers
- Impact kinetic energy
- Nuclear (or conventional) explosives in several delivery or emplacement modes, i.e.,
    - Standoff (flyby)
    - Shallow burial (surface penetrator)
    - Deep burial (mining required)

Specific Energies
- High explosive 6 MJ kg^{-1}
- Kinetic energy (10 km s^{-1}) 50 MJ kg^{-1}
- Nuclear explosive 4×10^6 MJ kg^{-1}

FIGURE 1. Energy sources and specific energies.

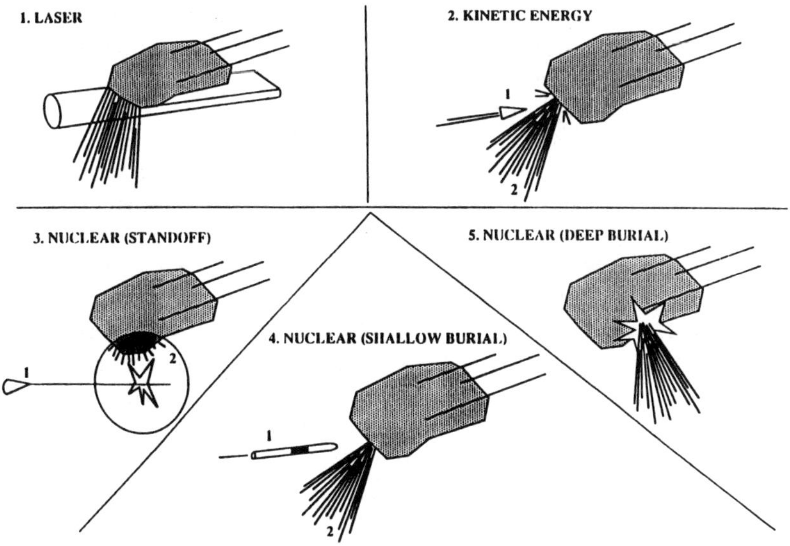

FIGURE 2. Energy delivery mechanisms. (Figures 2–4, 6, 7, and 12–19 are adapted from [6] and reprinted with permission from The University of Arizona Press.

Impulse coupling coefficient "C"

Common usage (mixed units):

$C = I/E =$ (dyne s cm^{-2})/(J cm^{-2}), in terms of specific impulse and specific energy.

$C = I/E =$ dyne s/J, in terms of total impulse and total energy.

Proposed change:

$C = I/E =$ dyne s/J = (Ns/J) $\times 10^{-5}$.

Define $V_{REF} \equiv 10^5$ m s^{-1} = 100 km s^{-1}.

Then define a nondimensional impulse $I^* \equiv C \times V_{REF} = I \times V_{REF}/E$.

Numerical values of C (in old units) and I^* are *identical.*

FIGURE 3. Impulse coupling coefficients.

IMPULSE COUPLING FOR NONNUCLEAR AND NUCLEAR ENERGY SOURCES

We will examine in detail momentum coupling from four of the five energy sources shown in Fig. 2. Common measures of momentum (or impulse) coupling coefficients are listed in Fig. 3, along with a new definition which we employ in our study. In particular, we define the impulse coupling coefficient I^*, which we use throughout our study.

Pulsed-laser Impulse Coupling

Experimental pulsed-laser impulse coupling coefficients are shown in Fig. 4 for low and high atomic-number elements. The figure presents measurements and least-squares fits to the data for six elements. I^* shows little variation with atomic number and lies in the range 5–10 (an exception is beryllium with a value near 3). These coefficients result from laser-induced surface blowoff into vacuum.

The vacuum boundary condition provides the *least* coupling and, consequently, the lowest values for I^*. If the laser beam were to pass through a transparent medium in contact with the exposed surface, the coupling coefficient would increase dramatically. Coupling coefficients greater than 1000 (enhancements of greater than 200) are inferred from velocity-time records obtained by Trott [8] and Remo, Farnsworth, and Trott [1] for meteoritic materials in contact with lithium fluoride. Even as little as one atmosphere of air in contact with the surface can increase the coupling coefficient as much as 5 or 10.

The implications for asteroids and comets are as follows. For asteroids, we have a solid surface in contact with the hard vacuum of space. Therefore, the coupling coefficients of Fig. 4 are applicable. However, for comets, the laser must penetrate the diffuse material of the coma before impinging upon the surface of the nucleus. There is a strong possibility that the coma particles can produce an effect similar to that produced by a transparent material in contact with the comet nucleus. Therefore, the coma might produce an enhancement of impulse coupling to the nucleus. The answer to this question will require additional experimentation and analysis.

Kinetic Energy Impulse Coupling

Hypervelocity impactors typically produce impact craters that are much larger than the impactors. The material ejected from the crater produces additional target momentum which adds to the initial impactor momentum, thereby enhancing the kinetic energy (KE) impulse coupling. Our earlier results in Shafer et al. [6] included a number of models for KE impulse coupling. The results obtained by previous investigators were included with the results of a dimensional analysis that indicated a quadratic dependence of I^* upon impactor velocity (see [6] for details). This implies that I^* approaches a constant as impactor velocity increases, rather than I^* decreasing with increasing impactor velocity as indicated by the other KE impulse coupling models (Fig. 5).

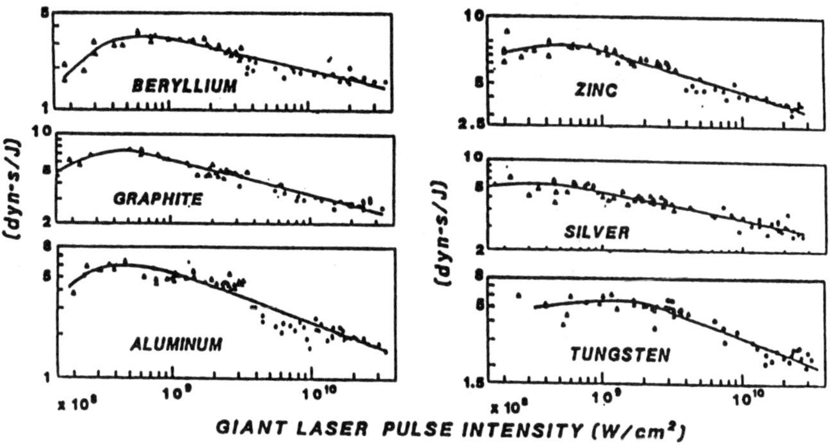

FIGURE 4. Pulsed laser impulse coupling; max I^* = 5–10.

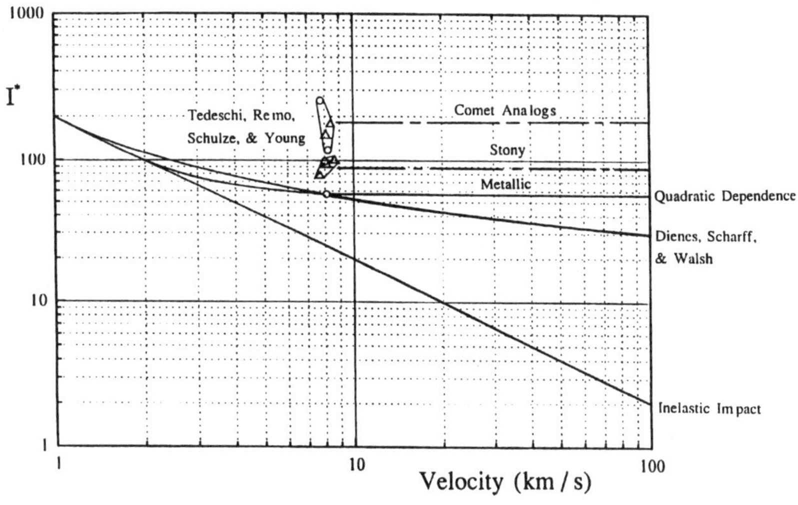

FIGURE 5. Kinetic energy impulse coupling.

In the previous work, the models were calibrated using hypervelocity data for annealed aluminum targets. Target momentum is known to be inversely proportional to target strength, i.e., the greater the target strength, the less the momentum produced by a given set of impact conditions. For the present work, we were fortunate to have impact crater results for meteorites and comet surrogate materials from the experiments of Tedeschi, Remo, Schultze, and Young [7]. These results are shown in Fig. 5, and indicate as we would expect, increasing KE impulse coupling I^* with decreasing target strength.

X-ray and Neutron Impulse Coupling

We next consider the coupling of x-rays and neutrons generated by nuclear explosives to NEOs. The first situation we consider is shown in Fig. 2 as "nuclear (standoff)" and denotes a surface irradiation of the comet or asteroid. Closed form approximations are used to estimate both x-ray and neutron coupling.

We begin with a parametric study of the coupling of blackbody x-radiation to a surrogate material, pure silica. Shown in Fig. 6 is the total specific impulse as a function of x-ray fluence for blackbody (BB) temperatures that vary from 0.1 keV to 100 keV. For the higher BB temperatures, a prominent impulse threshold effect is evident. This effect is demonstrated even more clearly in Fig. 7, in which the impulse coupling coefficient, I^*, is shown as a function of total fluence for BB temperatures that vary from 5 keV to 100 keV. Variations of total impulse with NEO composition and sublimation energy for two BB temperatures, 0.1 and 5 keV, are shown in Fig. 8. Four materials; silica, carbonaceous and ordinary chondrites, and nickel-iron, are compared. Three values of sublimation energy; nominal, ½ times nominal, and 1½ times nominal are compared for silica. A comparison of impulse coupling coefficients for the four materials is shown in Fig. 9. Total impulses for line and BB spectra are compared in Fig. 10 for the four materials. The corresponding impulse coupling coefficients are shown in Fig. 11. These figures illustrate the dependencies of total impulses and impulse coupling coefficients upon x-ray spectra and NEO material properties for a wide range of fluence values.

We next consider impulses produced by neutron deposition. Three spectra were compared, the fission and thermonuclear spectra presented in Glasstone and Dolan [2] and monoenergetic 14 MeV neutrons. Impulse coupling coefficients for the three spectra are shown in Fig. 12. In the final figure in this section (Fig. 13), x-ray and neutron coupling coefficients are compared. Hydrocode results, denoted V and E (for volume deposition and edge deposition, respectively), confirmed the closed form approximations for neutron impulse coupling. Neutrons were found to produce a clear coupling advantage over x-rays because of the superior penetration of neutrons as compared to x-rays.

Finally, we will consider the coupling of buried nuclear explosives to asteroids and comets. Subsurface energy deposition can provide the most efficient energy coupling of all the methods shown in Fig. 2, provided the depth of burial is optimized. In other words, with the burial depth too shallow, the overburden volume is too small and less than optimal mass is ejected which results in less than optimal momentum imparted to the NEO. On the other hand, with the burial too deep, the overburden volume is too large and the explosion may be completely contained, and no net momentum will be imparted to the NEO.

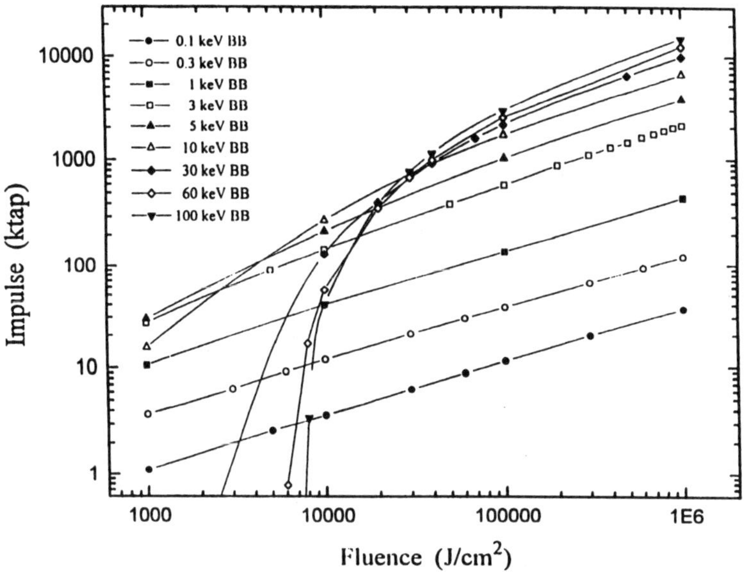

FIGURE 6. X-ray impulse as a function of fluence.

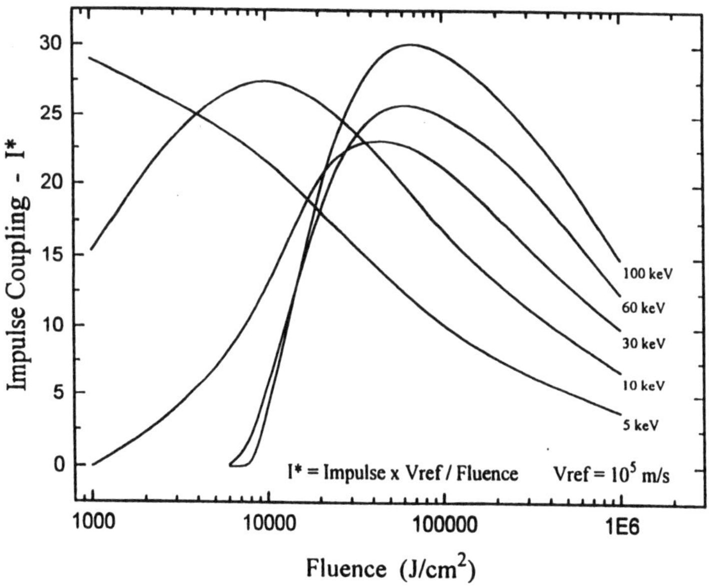

FIGURE 7. Nondimensional x-ray impulse coupling as a function of fluence.

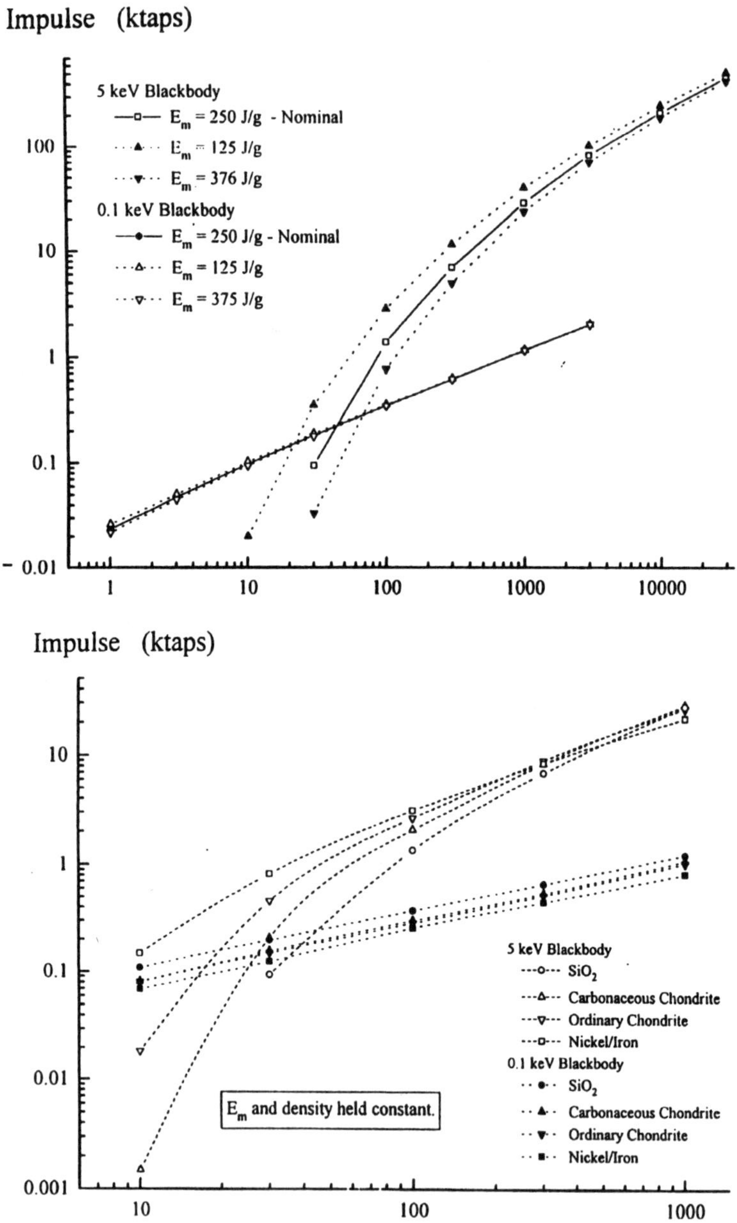

FIGURE 8. Variations of total impulse with NEO composition and sublimation energy.

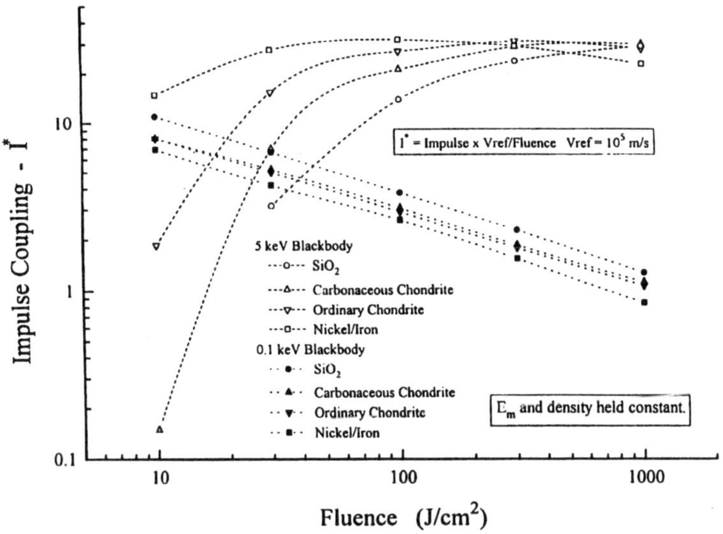

FIGURE 9. Variation of impulse coupling with NEO composition.

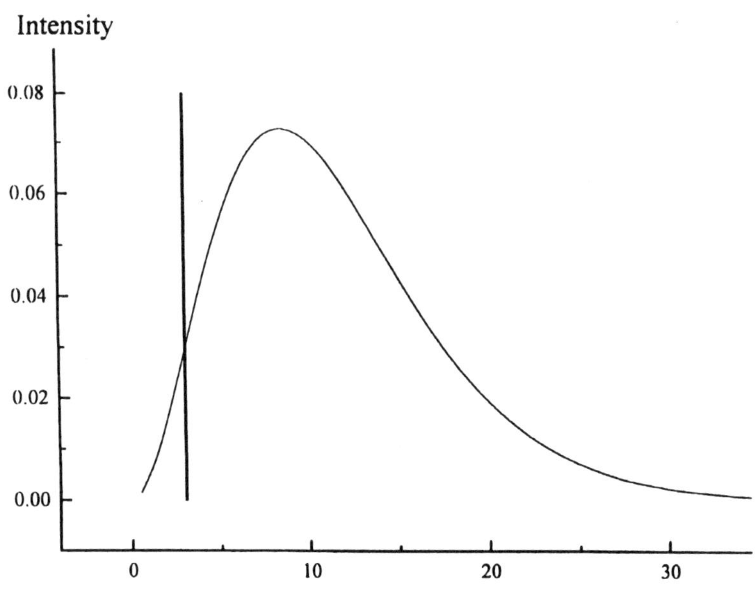

FIGURE 10. Line and blackbody spectra.

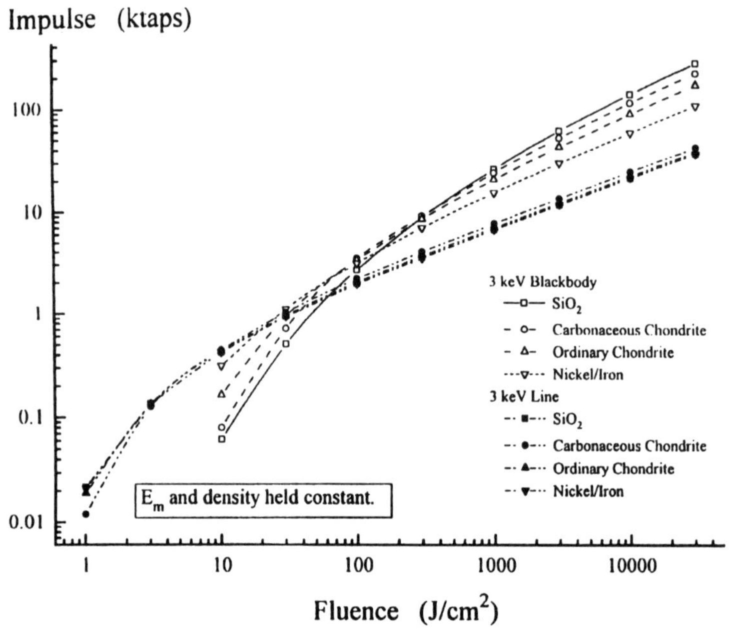

FIGURE 10 (*continued*). Comparisons of total impulse for line and blackbody spectra.

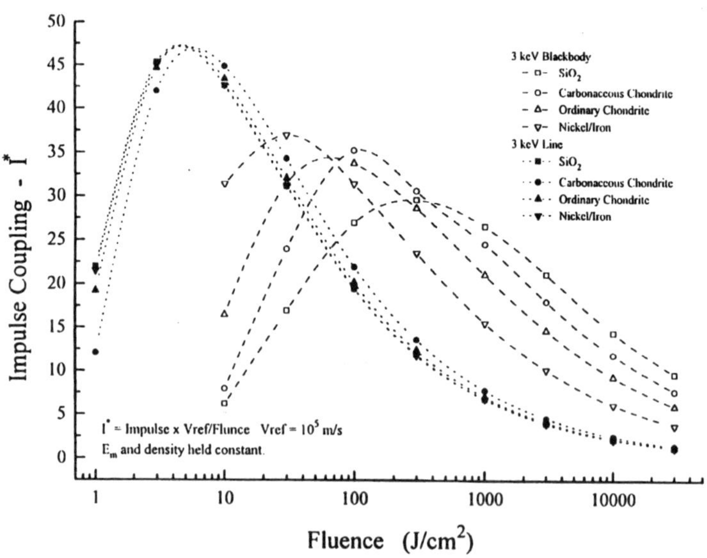

FIGURE 11. Comparisons of impulse coupling for line and blackbody spectra.

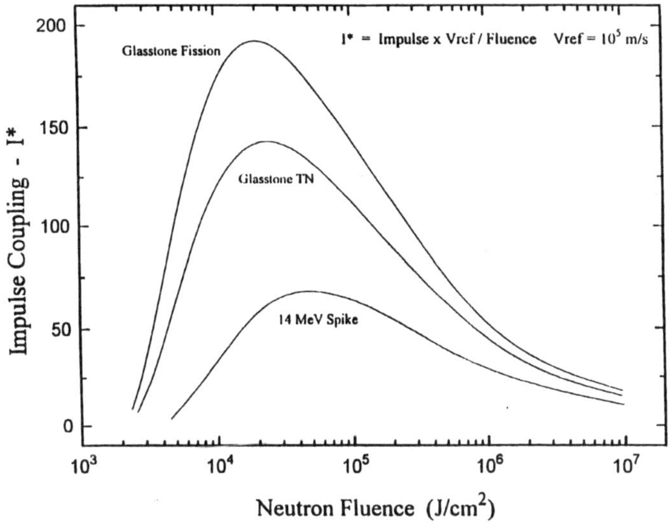

FIGURE 12. Neutron impulse coupling as a function of fluence and spectrum.

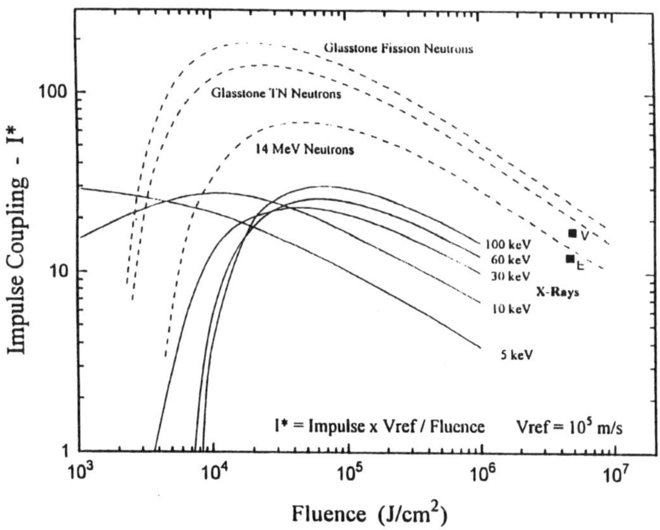

FIGURE 13. X-ray and neutron coupling as functions of fluence and spectra.

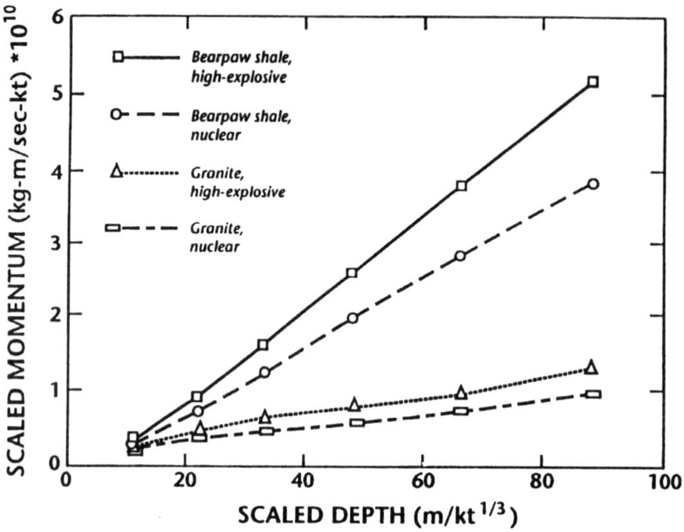

FIGURE 14. Coupling of buried chemical and nuclear explosives to shale and granite.

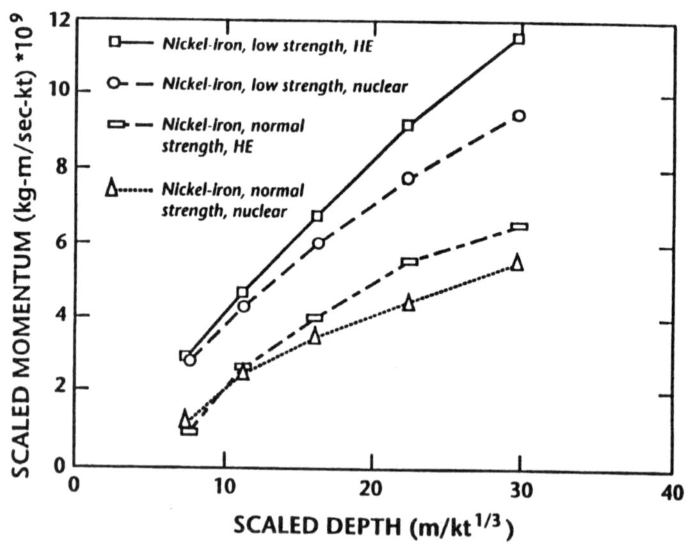

FIGURE 15. Coupling of buried chemical and nuclear explosives to nickel-iron.

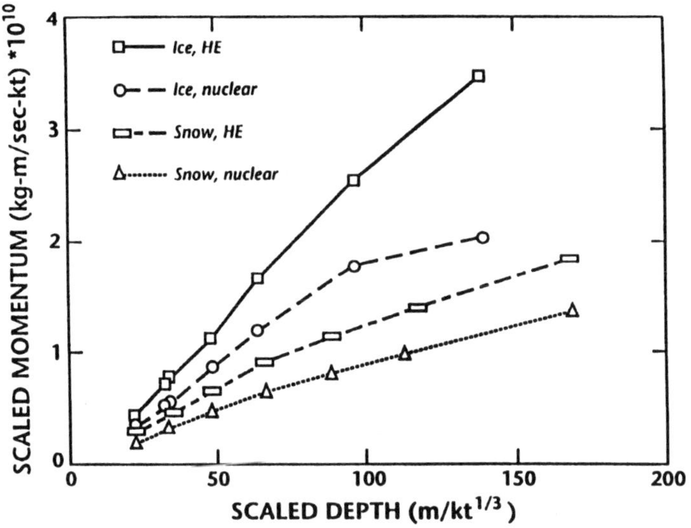

FIGURE 16. Coupling of buried chemical and nuclear explosives to ice and snow.

Shown in the next series of figures are scaled momentum as a function of scaled depth of burial for conventional and nuclear explosives, and for several surrogate materials for asteroids and comets. The materials are shale and granite to simulate weak and strong chondrites, respectively. Nickel-iron represents a metallic asteroid. Finally, high- and low-density comets are simulated by ice and snow, respectively. For the figures that follow, units of scaled momentum are kg m/s per kiloton of explosive energy, and units of scaled depth of burial are meters divided by the cube root of the explosive energy. Shown in Fig. 14 are the scaled coupling relationships for shale and granite subjected to conventional and nuclear explosives. Figure 15 illustrates the scaled coupling relationships for normal and low-strength nickel-iron materials. Finally, in Fig. 16 are found the scaled coupling relationships for ice and snow. Note that in all cases, higher material strengths resulted in lower momenta because higher strengths result in more explosive energy required to fracture the overburden and less energy available to accelerate the overburden. Also note that per unit of total energy, conventional explosives couple more efficiently than nuclear explosives. However, we see from Fig. 1, that nuclear explosives have specific energies nearly a million times greater than conventional explosives, which more than compensates for the slight decrease in coupling efficiency. We note from the results above, the practical applicability of the NEO materials classification given by Remo [5] for the coupling of momentum to NEOs. The NEO material categories 0, 1, 2, and 3 appear valid for kinetic energy, nuclear, and laser interactions.

		I^*
Pulsed lasers		2–10
Kinetic energy	10 km s^{-1}	50–300 (*25–50*)
	100 km s^{-1}	30–300 (*10–20*)
Standoff x-rays	5 keV	4–28
	10 keV	7–27
	30 keV	10–23
	60 keV	12–26
	100 keV	15–30

FIGURE 17. Summary of specific impulse coupling results.

	I^*
Standoff neutrons 14 MeV	10–70
Thermonuclear	14–140
Fission	16–190
Deep burial bearpaw shale	60–900
Hard granite	40–220
Nickel-iron	20–130
Ice	80–480
Snow	50–320

FIGURE 18. Summary of specific impulse coupling results (*continued*).

For surface loading with photons and neutrons, and $\phi > \phi\ (I^*_{max})$:

$$I^* \propto 1/\sqrt{E}$$

$$I \propto \sqrt{E}$$

For kinetic energy loading

$$I^* \propto E^{-a}\ (a = 0\text{–}0.13)$$

$$I \propto E^{b}\ (b = 0.87\text{–}1.0)$$

FIGURE 19. General impulse coupling results.

SUMMARY AND CONCLUSIONS

Our coupling results are summarized in Figs. 17 and 18. For kinetic energy coupling, the results in parentheses are earlier derived results prior to the reporting of the experimental results of Tedeschi, Remo, Schultze, and Young [7]. The results to the left of the parentheses are the new results of Tedeschi et al. The new results show a nearly *ten-fold* increase in KE impulse coupling coefficient. This large coupling coefficient for KE implies that the shallow burial option depicted in Fig. 2 may be a very viable option for NEO diversion. Presented in Fig. 19 are general impulse coupling dependencies for surface loading with photons and neutrons, and the coupling dependencies for KE loading. Finally, we identify the following research needs:

(1) Impulse coupling *confirmation* experiments for very intense fluxes of photons and neutrons.
(2) Extend kinetic energy impulse coupling experiments to ultra-high velocities in order to *confirm* proposed velocity dependencies.
(3) *Implicit* in (1) and (2) is the urgent need for NEO material property studies.
(4) Explore in much greater detail the issues involved in the "slow" versus the "fast" release of nuclear energy, i.e., nuclear propulsion versus nuclear explosion.
(5) Explore in much greater detail the shallow burial option for NEO diversion.

REFERENCES

1. REMO, J.L., A. FARNSWORTH, & W.M. TROTT. 1997. Laser interaction with meteoritic materials. In preparation.
2. GLASSTONE, S., & P.J. DOLAN. 1977. The Effects of Nuclear Weapons , 3rd edit.: 304-365. Washington, D.C.: U.S. Dept. of Defense and the Energy Research and Development Administration.
3. HAMMERLING, P., & J.L. REMO. 1993. NEO interaction with X-ray and neutron radiation. *In* Proceedings of the Near-Earth Object Interception Workshop, G.H. Canavan, J.C. Solem, & J.D.G. Rather, Eds.: 186–193. Los Alamos Laboratory. Los Alamos, New Mexico.
4. HAMMERLING, P., & J.L. REMO. 1995. NEO interaction with nuclear radiation. Acta Astronautica. **36:** (6) 337–346.
5. REMO, J.L. 1994. Classifying and modeling NEO material properties and interactions. *In* Hazards Due to Comets and Asteroids, T. Gehrels, Ed.: 551–596. University of Arizona Press. Tucson, Arizona.
6. SHAFER, B.P., M.D. GARCIA, R.A. MANAGAN, J.L. REMO, C.E. ROSENKILDE, R.J. SCAMMON, C.M. SNELL, & R.F. STELLINGWERF. 1994. The Coupling of Energy to Asteroids and Comets. *In* Hazards Due to Comets and Asteroids, T. Gehrels, Ed.: 955–1012. University of Arizona Press. Tucson, Arizona.
7. TEDESCHI, W.J., J.L. REMO, J.F. SCHULTZE, & R.P. YOUNG, JR. 1995. Experimental hypervelocity impact effects on simulated planetesimal materials. Int. J. Impact Engineering. **17:** (1–6) 837–848.
8. TROTT, W.M. 1995. Private communication to J.L. Remo. Sandia National Laboratories. Albuquerque, New Mexico.

Ice Issues, Porosity, and Snow Experiments for Dynamic NEO and Comet Modeling

M.D. FURNISH[a] AND J.L. REMO[b]

[a]*Instrumentation Development Department*
Sandia National Laboratories
Albuquerque, New Mexico 87185
and
[b]*Quantametrics, Inc.*
1 Brackenwood Path, Head of Harbor
St. James, New York 11780

ABSTRACT: Category 0 NEO bodies, such as many comets, appear to be comprised largely of porous ice. Quasi-static and dynamic properties of ice are expected to control the corresponding properties of Category 0 NEO material. These properties depend on temperature, strain rate, and possibly sample history. Corrections must be made for porosity and the presence of impurities such as silicates. These corrections reflect the added complexity of energy partitioning in a porous, inhomogeneous material as well as added $P \cdot \Delta V$ work during compression. A study of the dymanic (shock) properties of natural snow over the stress range 0–4 GPa is described. Comparison of the results of this study with computational modeling suggests that existing ice models with introduced porosity are applicable in some regimes, while water models with introduced porosity provide better fidelity in other regimes.

OVERALL DESCRIPTION OF COMETS

Previous work regarding the material properties of near-Earth objects (NEOs) [1] introduced a classification scheme for NEOs based on material properties. The purpose of establishing such a classification scheme is the reduction of uncertainty in planning strategies for orbit-modifying interceptions as well as for possible resource exploitation. In this classification scheme, NEO materials are divided into four categories, 0, 1, 2, and 3, based primarily on structural, thermal, and other properties. NEO materials included in groups 1, 2, and 3 are described as analogs to classes of meteorite materials, while comet analogs are related to group 0. The assumption is that meteorites are the fragments of asteroids and that near-Earth asteroid (NEA) material properties are essentially those of meteorites, with a few exceptions. This assumption is supported by the fact that orbits of recovered meteorites can be extrapolated to asteroid orbits. The direct material properties association of meteorites with NEAs is based on the extensive ground-based observations of asteroid (reflectance) spectral characteristics from the infrared through the visible, which suggests matches with meteorite mineral spectra. Radar studies and satellite (Galileo) flybys of Gaspra and Ida provided a close-up look at overall asteroid morphology which is consistent with an association between asteroids and meteorites.

The fact that meteorites are generally available for laboratory analysis has provided researchers with an opportunity to carry out extensive analyses on meteorite chemical composition as well as on mechanical, thermal, electrical, magnetic, and optical properties. Extrapolating these properties to NEAs may be uncertain in terms of the exact mineral composition, but their mechanical and thermal characteristics can be constrained within the three NEO classes which are briefly described:

1. NEO 1: Composed of the structurally weakest (friable) meteorites and resemble primitive asteroid classes.
2. NEO 2: Composed of metamorphic and igneous materials.
3. NEO 3: Composed of materials resembling metallic meteorites.

One of the purposes of this work is to extend the detailed discussion of NEO materials to the NEO 0 category.

SOME CHARACTERISTICS OF COMET NUCLEI

The comet mass and solid components are contained within the nucleus, which can range in size from 1 to 20 km. The refractory solids and frozen volatile materials contained within the nucleus (mostly ice, silicates, and various organic materials) are considered by many to be representative of the cosmochemical record of the primordial solar nebula. For this reason alone, comets are worthy of being studied. From the perspective of developing a reasonable NEO mitigation policy, the size, structure, integrity, and chemical composition are critical parameters.

Early spectroscopy of comets revealed the presence of gaseous carbon. Additional atoms, molecules, and ions found in climates include H_2, O_2, CN, CH, OH, H_2O^+, CN^+, CH^+, OH^+, $N_2{}^+$, CO^+, and $CO_2{}^+$. More complex organic molecules recently discovered include CH_3CN, H_2CO, and HCO. A large variety of molecules has been discovered in comets. Their origin and role in determining comet structure is the subject of another paper in this volume (Greenberg and Remo). The large variety of molecules and processes of accretion in interstellar space can provide a broad range of morphologies for comets. Most models suggest a weak compaction of grains and dust held together by water ice [2] similar to the dirty snowball model first suggested by Whipple. All of these models imply a low mass density (<1.0 gm/cm^3). A recent model suggests a "birds nest" type of structure [3].

In 1986 five spacecraft probed Halleys comet. The closest approach, by the probe Giotto, was to within 600 km. Data from these missions showed the gas in the comet head to be about 80% (vol.%) water vapor. The dust grains, found to be rich in C, H, O, and N, elementally resemble carbon-rich

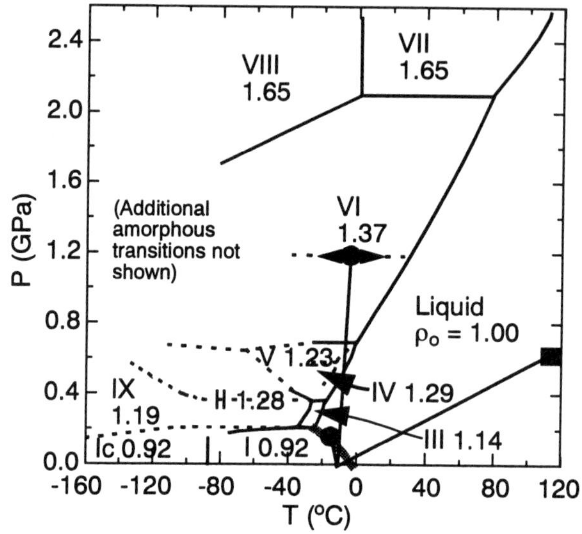

FIGURE 1. Ice phase diagram (quasi-static; adapted from Kamb [8] and Durham et al. [9]). Circles show quasi-static results of Larson [10]; square shows snow Hugoniot point (see discussion in text). Densities given are in gm/cm^3 and are for the lower stability limit of each phase.

meteorite materials. Since the largest component of comets is ice, it is reasonable to initiate our studies of the dynamic properties of the (NEO 0) comet material analogs with experiments on snow, ice, and related low-density composite analog materials.

COMET TEMPERATURE RANGE

Comet evolution and material properties are strongly affected by the amount of radiation the comet receives from the Sun. The primary effect of this radiative heating is to cause amorphous water (water ice) to undergo a crystalline phase change at about 140K. The secondary effect is to induce the sublimation of volatile ices at the surface of the comet nucleus. The phase diagram in Fig.1 indicates that there are at least nine ice phases within the temperature range from –160°C to 12°C (113 K to 285 K). The physical properties of ice and, therefore comets, will depend strongly on the comet temperature. Short-period comets are thought to experience a temperature range of 190 K to 240 K. Table 1 (adapted from Mumma et al. [2]) shows critical (solar insolation driven) temperatures encountered by a comet in its orbit from the Oort cloud to perihelion. At each of the temperatures listed, the comet properties change.

The most important temperature ranges for experimentation on comet analog (NEO 0) materials may be determined in view of the critical temperatures in Table 1 and the associated effects of the phase diagram in Fig. 1.

Modeling interactions with Class 0 NEOs therefore requires some knowledge of the orbital position of the NEO at the time of interaction. The NEO surface temperature is strongly affected by orbital position, while the interior temperature range also depends on the size and thermal conductivity of the NEO. The thermal conductivity appears to be strongly phase dependent, complicating the problem. As an example, the results of a thermal modeling of a hypothetical comet 1 km in radius [7] are presented in Table 2.

Comparing Fig. 1 with Table 2 it is clear that the thermal and mechanical properties of the comet will vary considerably throughout its orbit. The required mitigation, or resource exploitation method, will depend on the position of the comet in its orbit at the time of interaction. Most likely, an interception would occur near perihelion (distance from the Sun ≤ 1 AU) given constraints of launch vehicles and advance warning capabilities.

TABLE 1. Important Temperatures for Modeling the Thermal and Mechanical Properties of Comets

Temperature (K)	Associated Physical Characteristics
5–6	Ambient Oort cloud temperature
25–60	Comet temperature of formation based on the measurement of volatile abundances
27	Organic (yellow) material explodes on rapid heating (seconds), or slow warming as a chain reaction in radical chemistry [4]
40	Heavy cracking, whitening of mixed water ice [5]
70	40 m/s grain impact produces temperature spikes [4], [6]
125–150	Exothermic phase transition to crystalline form occurs in amorphous water ice. Excess trapped gases are released
160–170	Water ice sublimation rates become significant

TABLE 2. Internal Temperatures for a Hypothetical Comet of 1 km Radius at Different Orbital Positions

Orbital Position	Temperature Range (K)
Aphelion	125 – 145
3 AU inbound	235 – 245
3 AU outbound	170 – 145
Perihelion	Up to 240

CHOOSING ANALOGUES FOR COMETLIKE MATERIAL

The NEO 0 category is primarily composed of dormant or extinct comet nuclei, which are very poorly understood because:

1. Spectra are available from only a limited number of active comets at or near perihelion during their active phases. These spectra are generally those of the comet coma, from which the properties of the comet nucleus must be inferred.
2. Only one comet, Halley, has been visited by flyby missions.
3. No sizeable amount of unmodified comet material has ever been recovered for laboratory analysis. Meteoroids of probable cometary origin may represent highly fractionated and devolatilized components of depleted comets.

In spite of the above uncertainties, comets are generally regarded to have a density in the range of 0.2 to 1.5 gm/cm^3, low tensile strength and a composition dominated by water ice. Variable fractions of silicates and metals, as well as volatile and refractory organics (CHON) appear to be present (Greenberg and Remo, this volume). It is important to understand the material properties of comets because their volatile and structurally weak nature affects their dynamics by introducing nongravitational forces (jetting of volatile gasses) which can significantly affect their orbital trajectory. The tensile strength of the comet (and in some cases the compressive and shear strengths as well) strongly influence the Roche limit, behavior of the comet upon atmospheric entry, and cratering dynamics.

From the above, it is clear that there are significant difficulties inherent in determining material properties of NEO 0 category bodies (more so than with typical asteroids, for example). Nonetheless, it is possible that comets may pose a greater hazard to the Earth because of their probability of incidence. If mitigation measures are to be planned, these material properties must be understood.

Comets are generally believed to be composed of low-density material made primarily of ice with silicate grains and organics (see Greenberg and Remo, in this volume). Emphasizing the icy characterization of comets, here we report on some recent experimental work to deduce properties of highly porous ice, snow, and other underdense materials. The primary objective of this work is to experimentally characterize the dynamic properties (such as Hugoniot data) of ice composites and to assess the validity of related equation-of-state (EOS) models.

Application of the present work to problems of NEO modeling requires extrapolation to natural conditions. It is understood that the temperatures and pressures at which these experiments were carried out do not represent conditions in interplanetary space. The materials studied here are ideal natural and synthesized laboratory materials at atmospheric pressure and relatively high temperature, i.e. natural snow, lightweight concrete, and snow-matching grout. These materials may or may not be applicable to comet properties in general. Ideally, terrestrially recovered comet materials could be synthesized and dynamic experiments performed on these analogs at temperatures and pressures similar to those encountered in interplanetary space (e.g., see Tables 1 and 2). Nonetheless, until comet material is sampled in situ and/or recovered for laboratory analysis and accurate analogs synthesized, the materials used in this study represent a valid and helpful approach.

DYNAMIC PROPERTIES

In a study concerned with either the effects of a possible planetary impact of a NEO or with deflection strategies, the following material properties are required:

1. Hugoniot: The locus of points in pressure/density space accessible by shocking a material; this includes the effects of dynamic compressibility as well as material strength and pertinent phase transitions.
2. Release properties: The pressure/density path followed by the material during stress release following shock compression. This property strongly affects shock-attenuation rates.
3. Material strength information under various strain rate scenarios, varying from quasi-static (strain rate $\leq 10^{-3}$ sec^{-1}) to shock loading (strain rate 10^{5} to 10^{6} sec^{-1})
4. Shock vaporization (critical point, vapor dome).

Similar information is required for calculating momentum transfer and object breakup in mitigation scenarios.

The dynamic properties of a Category 0 NEO are likely to be those of ice, with perturbations for porosity and the presence of other materials. The ideal approach to assessing the dynamic material properties of a Category 0 NEO is to assess properties of ice, and then introduce impurities and porosity in a controlled manner. The literature contains significant information to aid in this effort, although difficulties remain.

SHOCK INTERACTIONS WITH POROUS MATERIALS

NEO Context

The passage of a shock wave through an inhomogeneous and porous material, such as a cometlike ice composite structure, provides several modes of energy dissipation. The energy partition from the shock-compression process includes the following mechanisms:

1. friction,
2. fracture initiation and growth,
3. defect introduction,
4. void collapse,
5. free surface spallation,
6. microkinetic effects,
7. melting and vaporization,
8. radiation effects, and,
9. reaction bonding energy.

While all of the above mechanisms dissipate energy, the voids provide ideal low-density regions within which large amounts of shock-wave energy can be initially trapped and then dissipated. The medium around the cavity will be subject to all of the above effects to a much greater degree than will the surrounding nonporous medium. Surface material around the pores will be accelerated into the voids, introducing fractures and undergoing kinetic interactions which will release thermal energy. The heating and melting within the pores will provide the conditions for chemical reactions to be initiated among certain molecules, as well as for melting and vaporization of material. Also, the (transient) high energy of the pores can provide conditions for the quasi-adiabatic expansion of the pores, thereby introducing a localized density decrease. In terms of the usual density increase associated with a compressive wave, this is an anomalous effect. As well, following release to ambient pressure, the shocked materials may have substantially different mechanical and thermal properties than did the initial materials. The degree of change in general will depend on the amplitude and duration of the loading.

Physical models of comets [2], [3] tend to treat comets as porous solids. Since comets are seen to outgas at perihelion, it is likely that selective vaporization increases porosity. Hence, one would expect that a high-density shock wave traversing a comet will introduce chemical and mechanical alteration in addition to (the desired) momentum transfer. How well we understand the energy partitioning of a shock wave as it traverses the comet depends on how well we can model the EOS of the comet and the response of the EOS to shock loading.

Approximation of Energy Partition in a Porous Material

The energy coupled into a material by a shock wave is partitioned into irreversible and reversible components, where the irreversible component is manifested as residual heating after release. For a porous material the irreversible component is especially important because of its relatively large magnitude.

The work done by a shock wave in irreversibly compressing a void is:

$$W = 0.5 \times f P\,(V_i - V_f) = 0.5 \times f P\,V_i\,(1 - \delta),$$

where the volume fraction remaining after shock compression to pressure P is

$$\delta = V_f / V_i$$

and f is the fraction of work converted strictly into high temperature thermal modes. An estimate of f can be obtained through the use of carefully placed thermal sensors.

The energy required to heat, melt, and vaporize a unit of comet material is:

$$E_{h,l,v} = C_{ps}\,((T_l - T_i) + h_l + C_{pl}\,((T_v - T_l) + h_v,$$

where C_{ps} and C_{pl} are the specific heats of the solid and liquid phases, T_i, T_l, and T_v are the initial, melting, and vaporization temperatures, and h_l and h_v are the specific heats of vaporization.

The ratio $W/E_{h,l,v}$ will give an approximation of the upper limit of the fraction of material that can be vaporized by a given shock wave. In a porous material, compressional heating will be heterogenously distributed, giving rise to localized melting and vaporization.

Theoretically, it is possible to analyze the postshock material structure and temperature to infer the partitioning of energy. In practice the most practical way of inferring the energy partitioning is through measurement of the hysteresis of the stress-strain loop describing the material response to passage of a compressional wave. The present studies provide information contributing to this goal for several materials of interest.

This partitioning is important to calculations of NEO mitigation strategy effectiveness insasmuch as energy partitioned into heating (irreversible deformation and compression) is not available for changing the NEO velocity or causing macroscopic material ejection.

PROPERTIES OF INTEREST FOR PURE ICE

Quasi-static high pressure studies of ice have shown the existence of at least nine ice phases (plus water), as summarized in Fig. 1. A denser set of low pressure phases appears to exist at low pressures and temperatures below 38–63 K; these are typed by the Kuiper-belt comets which are presumed to have a density of approximately 1.07 gm/cm^3 (vs. 0.94 for Oort belt comets) [11]. Early structural descriptions of the individual phases are described by Kamb [8].

In general, brittle behavior is exhibited at higher strain rates ($>10^{-4}$ s^{-1}), lower temperatures (< 195 K) and lower pressures (< 200 MPa) [9]. For temperatures above 195 K (–78°C), behavior is dominantly ductile. As well, ductile strengths decrease greatly near the ice I ↔ II boundary. The above is greatly simplified (e.g., a change in mechanism at 243 K is inferred as well), and the reader is referred to Heard et al.'s paper [9] for more details.

Various shock-wave studies have been conducted on ice, covering a wide range of stress conditions. Gaffney [12] summarizes earlier work, concluding the following:

1. The elastic limit (HEL) is at 0.1–0.2 GPa mean stress, 1.5–3 GPa axial stress, and is manifested by a precursor traveling at 3.7 km/s (slight orientation dependence).
2. A phase transition to Ice VI occurs over the span 5–6.9 GPa.
3. Melting is complete by 10 GPa (a result depending somewhat on initial temperature, not noted here).
4. The Hugoniot is well represented above 8 GPa by a particle velocity–shock velocity relation:

$$U_S = 1.32 + 1.68U_P - 0.035\, U_P^2\ (r^2 = 0.994) \quad \text{or}$$

$$U_S = 1.79 + 1.42U_P\ (r^2 = 0.993).$$

Larson et al. [13] performed experiments which showed a much lower melting transition stress. For initial temperatures of –10°C, shocks propagating at 0.2–1.5 GPa in ice showed strong attenuation, which they attributed to a hysteretic melting transition.

A practical question arising during the process of setting up a modeling of an event is what equation of state to apply. Experiments overseen by Gaffney (personal communication, 1992) give a pressure-density plot agreeing more closely with those given by the ANEOS and SESAME tabular EOS for water than the corresponding EOSs for ice. These EOSs are applicable to a wide variety of wavecodes presently in circulation.

PROPERTIES OF HIGHLY POROUS ICE (SNOW) AND OTHER UNDERDENSE MATERIALS

Background

Category 0 NEOs may have densities as low as 0.2, requiring substantial average porosities. For these materials, the dynamical studies are generally of primary interest inasmuch as the quasi-static studies quickly become studies of the corresponding fully-compressed materials. Various shock-wave studies have been conducted on natural and artificial snow and other porous materials.

Experimental measurements of dynamic properties of these "underdense" materials pose three primary difficulties. First, the samples are perishable; they can melt or sublimate during storage, preparation, and testing. Second, they are brittle and crushable; they cannot withstand such treatment as traditional machining or launch in a gun system. Third, with increasing porosity the calculated Hugoniot density becomes rapidly more sensitive to errors in wave time-of-arrival measurements. Carefully chosen simulants eliminate preservation (perishability) difficulties, but the other difficulties remain.

There are several experimental constraints of special interest for impact testing of these materials. Sample sizes depend on configuration used (e.g., explosive tile drivers or gun impact). For gun impact tests, for example, samples generally are 4–10 mm thick and 6–7 cm diameter (2–2.2 cm diameter for impact velocities of 2.5–7 km/sec, where a two-stage light-gas gun must be used). Hugoniot stresses available depend on the exact configuration and material tested. While for iron-nickel materials, Hugoniot stresses to 55 GPa may be achieved with the larger diameter samples, and 265 GPa, with the smaller samples. For underdense materials, final stresses are highly sample-dependent. For water ice impacted by tantalum at 2.3 km/sec, for example, a Hugoniot stress of about 10 GPa is produced. With increasing porosity, this stress decreases.

Let us briefly summarize key dynamic studies which have been performed on snow. Bakanova et al. [14] conducted several very-high stress tests on natural snow of densities of 0.915, 0.60, and 0.35 gm/cm^3, using high-explosive drivers and electrical contact time measurement methods. Maximum stresses measured Hugoniot states up to 50.3, 35.4, and 22.2 GPa, respectively. Johnson et al. [15], [16] achieved low stresses (to 0.04 GPa) in snow of various densities with an eight-inch gas gun, instrumenting the sample with embedded carbon gauges. They found that (1) the Rankine-Hugoniot jump conditions, although not strictly applicable in this stress region, were not far in error (< 10%), and (2) there was little temperature or stress history dependence of the Hugoniot state of the sample. Studies by Solie et al. [17] and Erlich and Curran [18] were recently concluded. Solie et al. [17] detonated sheet explosives at the surface of a snowpack (relatively dry; $\rho_0 \approx 0.25$ gm/cm^3) instrumented with carbon and PVDF gauges. They observed extremely high attenuation. Experimental difficulties included major corrections for the impedance mismatch between the gauges and the snow (giving large error bars for their inferred in situ states; also a problem for Johnson et al. [15], [16]) and rapid attenuation (88% in 0.2 m for shock velocity). Adding a sheet to prevent entry of the blast gasses into the snow increased the stress observed at 10 cm by a factor of two. Input stresses ranged to 0.9 GPa. Erlich and Curran [18] combined yttrium stress gauges and particle velocity gauges to measure the response of artificial snow samples and highly porous grout samples (intended to be of identical material to those used in the present study) to a shock introduced by dilute explosives tiles (DET). Again, rapid attenuation was observed; for example, stresses decreased from 0.1 GPa to 0.025 GPa in 6 cm.

These studies taken together suggest that natural snow gives rapid attenuation where impulsive loading provides stresses of ~0.1 GPa and that the Rankine-Hugoniot wave equations are of restricted value except at very high stress levels (≥ 0.1 GPa) or where a nonattenuating wave is indicated.

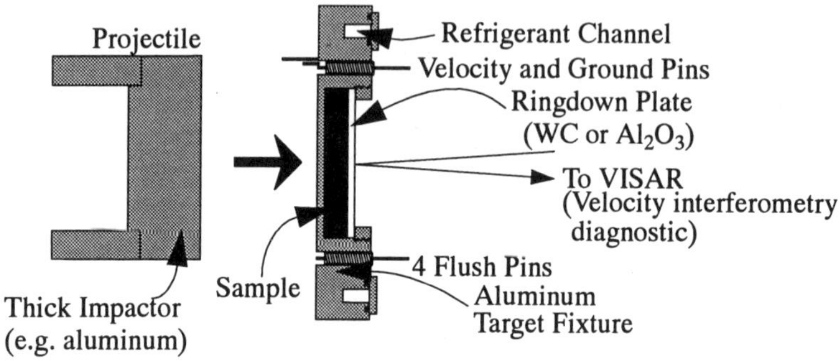

FIGURE 2. Experimental configuration used for Hugoniot/reshock/release experiments on natural snow and simulants. Projectile was 89 mm or 102 mm

Experiments to Assess EOS Validity

Method

A recent set of experiments by Furnish et al. [19], [20] made use of a 4-inch gas gun and a 3.5″ powder gun to assess the Hugoniot, reshock, and release properties of natural snow (initial density of 0.5 gm/cm^3) and of two potential simulants (possibly similar dynamic properties), specifically a lightweight concrete (initial density of 0.7 gm/cm^3) and a mixture of foam beads and Portland cement (initial density of 0.22 gm/cm^3). This set of experiments provided Hugoniot data over an intermediate stress range (0.3–3.5 GPa) which allows an assessment of several extant equations of state (EOSs). This stress range is critical for correct modeling of many NEO scenarios.

The experimental and data reduction methods are described in detail elsewhere [19], [20], and will be briefly summarized here. The configuration is shown in Fig. 2. Hugoniot states are determined from wave travel time through the sample, which is deduced from the time interval between initial impact and wave arrival time at the rear surface of the ringdown plate. Reshock states and release paths are derived from the initial waveform rise rate and the later-time shape of the waveform. The release is accomplished by a ringdown of the high-impedance ringdown plate.

For highly distended materials, the loading wave may be taken as a single wave. Multiwave structures do not normally occur except at extremely low stress levels because the volume change is dominated by porosity collapse.

This configuration bears close analogies to the ringdown configuration described by Chhabildas and Miller [21] for measuring the release behavior of crystalline quartz. In their configuration, however, the sample was launched into the ringdown plate. The release measured was from a simple

Hugoniot state. Their configuration, however, requires that the sample withstand exposure to vacuum.

In practice, this configuration normally did not deliver the sharply stepped waveform expected from a ringdown design; rather, the waveform was generally a sloped ramp with indistinct steps. Hugoniot states were deduced using the waveform toe time-of-arrival, with reshocked states and release paths calculated by treating the ringdown plate as rigid.

Results Compared with EOS Predictions

The primary intent of this section is to compare the results of Furnish et al. [19], [20] with equations of state of interest. The reader interested in the detailed results is referred to the earlier articles.

Pressure-density plots of the Hugoniot results are shown in Fig. 3. As can be seen, the Hugoniot values tend toward the expected fully-compressed densities of the respective solids. Comparing the snow Hugoniot to the higher-pressure ice Hugoniot of Gaffney (above), the role of the porosity in adding heating becomes apparent. The negative slope in the plot of the grout Hugoniot may indicate unsteady wave propagation in this material (also suggested by the long toe observed in the grout waveforms).

A comparison of theoretical EOS predictions with waveform and Hugoniot results for snow is made in Fig. 4. Two EOSs are considered, both of the SESAME form (SESAME 7151 — a tabular ice model — and SESAME 7150 — a water model). As used in this case, both include irreversible pore collapse, a strength of 100 bars, a starting temperature of 0.023 eV (–6°C), an uncrushed-state sound speed of 300 m/s, a transition modulus ($V \cdot \delta P/\delta V$) of 0.1 GPa, and an initial specific energy of -3.3377×10^9 dynes/gm. The results shown in Fig. 4 indicate that the porous water model provides a more appropriate representation of snow for Hugoniot stresses in the stress range 0.1–1.0 GPa than does porous ice. As well, it matches the older Hugoniot data [12]. A test run (not shown) confirmed that the effect of varying the initial specific energy between –3.3377E9 dynes/gm and 0 (for SESAME EOS 7151) was negligible.

The higher-stress snow tests do not lie on the porous water curve in Fig. 4; they lie closer to the porous ice curve. Correcting for the increased compressional heating shifts the water curve on the left side of Fig. 4 as shown. The appearance is still that the porous ice curve is more appropriate for the three higher-stress snow tests (Hugoniot stresses of 1.3–3.5 GPa).

It was also found that these tests are insensitive to strength changes in the 1–4 MPa range, as well as to the other explicit input parameters for the equation of state.

Hence, or stress levels to 1 GPa, good agreement is found between the SESAME EOS for water (7150), the Mie-Grüneisen EOS and the experimental data. At higher stress levels, the SESAME EOS for ice (7151) provides better agreement.

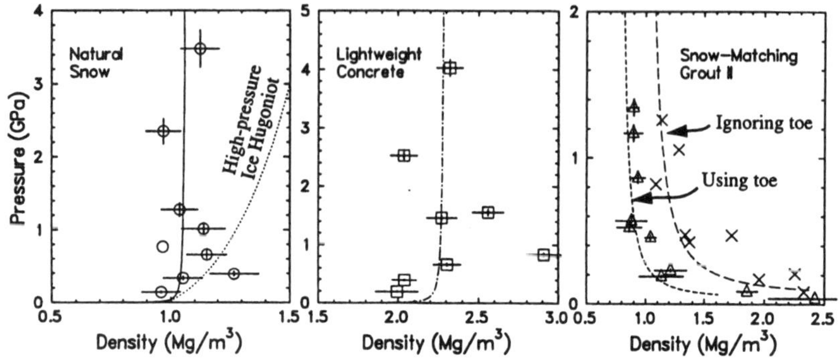

FIGURE 3. Hugoniot data and fits in pressure-density space. The "snow-matching grout II" waveforms showed low-amplitude precursors; calculations were made including and disregarding these "toes."

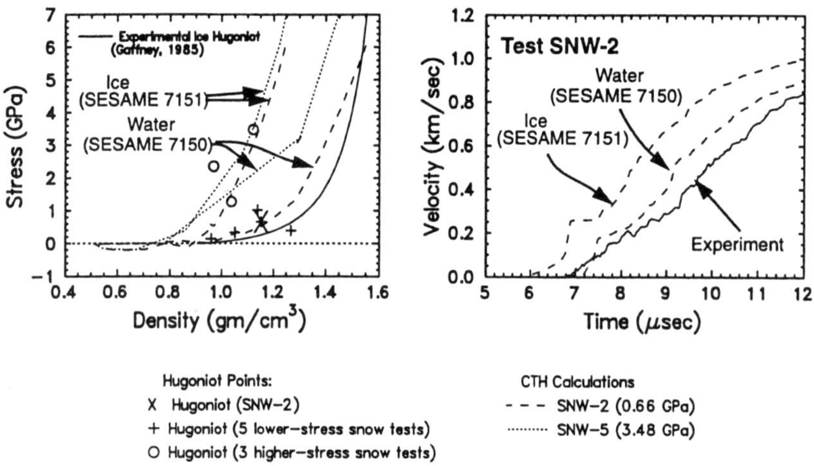

FIGURE 4. CTH model (1-dimensional) and experimental wave profiles for experiment SNW-2 (natural snow, 900 m/s impact velocity), Eos varied. Time = 0 impact. Thermal effects shift pressure/density path at higher Hugoniot stress, as illustrated for conditions of test SNW-5.

PROPERTIES OF FROZEN SOILS AND SILICATE/ICE MIXTURES

A broad base of literature pertinent to impact events exists for soils with an ice component, including quasi-static high-pressure characterizations of strength and compressibility [22], [23], nondestructive testing [24], [25], shock-wave testing [10], [26], [27], and Earth penetrator tests [28], [29]. Selected trends are outlined below. Portions of this information will be applicable directly to important NEO problems.

The dominant effect observed in all of these experiments is a strong temperature dependence of the strength and moduli of such materials over the range between –10°C and 0°C. This is largely due to a depression of the freezing point by partial adsorption of water onto grains [10], [24], [25]. The effect is strong for clays, but negligible for coarse sand. Figure 5 exhibits this temperature dependence for Hanover silt. The strong temperature dependence appears to continue to significantly lower temperatures. For certain NEO scenarios this effect should be considered.

The quasi-static strength of ice is generally increased by the addition of any amounts of sand. Adding 60 wt% (approximately 30 vol%) sand to ice increases its maximum strength by approximately 30% [23], while a sand mixture saturated with ice has approximately four times the strength of ice alone ($(\sigma_1 - \sigma_3)_{max}$ = 8 MPa for ice [23]). For a given concentration of sand, peak strength increases with confining pressure until the hydrostatic pressure reaches 25–40 MPa, then decreases toward or past the unconfined values.

Dynamic moduli and damping ratios in a frozen silt do not seem strongly affected by confining pressure [24]. However, the moduli increase with frequency and decrease with temperature and strain amplitude, while the damping ratio acts in the opposite way.

Details of the shock-loading process in soils are strongly dependent on whether the soil is frozen/thawed and saturated/unsaturated. Nakano and Froula [26] tested samples of Ottawa sand to arrive at the following conclusions. Undersaturated soils generally are compressible enough that a single sharp shock is observed regardless of whether the water is frozen (although shock impedances vary). Saturated soils may propagate a single shock when thawed, but freezing introduces a strength effect sufficient to give rise to a precursor whose amplitude (axial stress ~0.3 GPa) is consistent with that for pure ice.

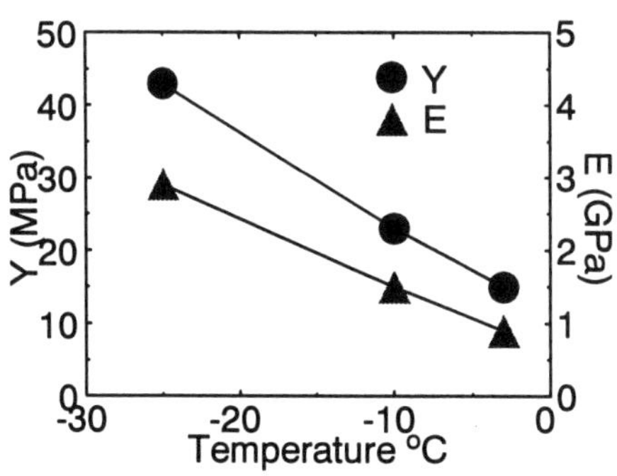

FIGURE 5. Temperature dependence of Young's modulus E and yield strength Y ($(\sigma_1 - \sigma_3)_{max}$). Data from Aitken et al. [28].

Sharp shock waves in frozen soils (starting temperatures around –10°C) are subject to strong attenuation due to hysteresis in the ice melting transition (occurring at 0.1–0.2 GPa), by contrast with the relatively nonhysteretic melting behavior of pure ice [10]. Addition of porosity will increase this attenuation [27], while we presume lowering the starting temperature so that the melting transition does not occur will decrease attenuation. An inspection of Fig. 1 suggests that the Hugoniot of fully-saturated sandy soils with an initial temperature below ~–30°C will pass through entirely solid phases of ice to stresses significant above 2.5 GPa. The shock-melting data of Schroeder and McMaster [30] constrains this melting point to be greater than 25 GPa. Porous soils will have a markedly greater shock heating (analogous to the snow Hugoniot point in Fig. 1); materials shocked from much lower initial temperatures may undergo melting at relatively low stresses.

A set of early Hugoniot/release tests [31] was conducted to ~40 GPa on a variety of frozen soils (each with saturations chosen at increments from dry to fully saturated). Results were dominated by crush-up behavior, with some evidence for the quartz-stishovite transition at the higher end. The origin of kinks in the shock velocity–particle velocity curves for the partially and fully saturated curves near 15 GPa pressure is unclear. Unfortunately no comparison with unfrozen material was made in this study. Ice Hugoniot points are in reasonable agreement with the later fit of Gaffney [12].

Soil penetration tests, both with small-scale [28] and large-scale [29] penetrators, show a strong dependence of penetration depth and axial acceleration on the frozen/thawed state of the soil, although studies to date have not resolved their dependence on other parameters such as saturation and initial temperature.

CONCLUSIONS

Category 0 NEOs are likely to be comprised largely of water ice; however, variables such as porosity and included heavier materials may substantially affect their response to impact. A moderate temperature dependence of the dynamical properties of ice is enhanced by the addition of silicates. Strength is often increased by the addition of silicates and reduced by porosity and increased temperature. At temperatures greater than –30°C, modest shock pressures (0.2–0.5 GPa) are sufficient to cause partial ice melting because of the negative Clapeyron curve bounding Ice I and water. At lower initial temperatures (probably common to many NEOs) the material must be very porous to experience the melting transition under a modest pressure shock wave. The question of which published equation of state for water or ice is most appropriate to use is still somewhat unsettled (more experiments are suggested), although the suggestion of this work is that a water equation of state is best for shocking snows to stresses of 0.1–0.2 GPa, with an ice equation-of-state for higher stresses.

REFERENCES

1. REMO, J. 1994. Classifying and modeling NEO material properties and interactions. *In* Hazards due to Comets and Asteroids, T. Gehrels, Ed.: 551–596. University of Arizona Press. Tuscon, Arizona.
2. MUMMA, M.J., P.R. WEISMAN, & S.A. STERN. 1993. Comets and the origin of the solar system. *In* Protostars and Planets III, E.H. Levy & J.I. Lumine, Eds.: 1177–1252. University of Arizona Press, Tuscon, Arizona.
3. GREENBERG, J.M. & J.L. REMO. 1993. A current working model of comet nucleus and implications of NEO interactions. *In* Protostars and Planets III, E.H. Levy & J.I. Lumine, Eds.: 635–647. University of Arizona Press, Tuscon, Arizona.
4. GREENBERG, J.M. 1986. Predicting that Comet Halley is dark. Nature. **321:** 385–387.
5. BAR-NUN, ET AL. 1985. (ref?)
6. D'HENDECOURT, L.B. 1980. Laboratory evidence for molecule ejection at low temperatures. Internal report. Laboratory for Astrophysics, Univ. of Leiden.
7. HERMAN, G. & P.R. WEISSMAN. 1987. Internal temperature of cometary nuclei. Icarus. **69:** 314–328.
8. KAMB, BARCLAY. Structural studies of the high-pressure forms of ice. Trans. Amer. Cryst. Assoc. **5:** 61–79.
9. DURHAM, W.B., H.C. HEARD, & S.H. KIRBY. 1983. Experimental deformation of polycrystalline H_2O ice at high pressure and low temperature: Preliminary results. J. Geophys. Res. **88:** B377–B392.
10. LARSON, D.B. 1978. Explosive energy coupling in ice and frozen soils. *In* Proceedings of 3rd International Conference on Permafrost, Edmonton, Alberta, 1978: 807–812.
11. JENNISKENS, P. & D.F. BLAKE. 1994. Structural transitions in amorphous water ice and astrophysical implications. Science. **265:** 753–756.
12. GAFFNEY, E.S. 1985. Hugoniot of water ice. *In* Ices in the Solar System, J. Klinger et al. Eds.: 119–148. D. Reidel.
13. LARSON, D.B., G.D. BEARSON, & J.R. TAYLOR. 1973. Shock-wave studies of ice and two frozen soils. *In* Permafrost; North American Contribution, Second International Conference; Physics, Physical Chemistry, and Mechanics of Frozen Ground and Ice. 318–325.
14. BAKANOVA, A.A., V.N. ZUBAREV, YU. N. SUTULOV, & R.F. TRUNIN. 1975. Thermodynamic properties of water at high pressures and temperatures. Sov. Phys. JETP. **41:** 544–548.
15. JOHNSON, J.B., J.A. BROWN, E.S. GAFFNEY, G.L. BLAISDELL, & D.J. SOLIE. 1992. Shock response of snow—analysis of experimental methods and constitutive model development. CRREL Report 92-12.
16. JOHNSON, J.B., J.A. BROWN, E.S. GAFFNEY, G.L. BLAISDELL, M. STURM, & S.A. BARRETT. 1992. Shock wave studies of snow. *In* Shock Compression of Condensed Matter 1991, S.C. Schmidt, R.D. Dick, J.W. Forbes and D.G. Tasker, Eds.: 107–110. Elsevier. New York.
17. SOLIE, D.J., J.B. JOHNSON, & S.J. BARRETT. 1994. The response of natural snow to explosive shock waves. *In* High Pressure Science and Technology—1993, S.C. Schmidt, J.W. Shaner, G.A. Samara, & M. Ross, Eds.: 1139–1142. AIP Press. New York.

18. ERLICH, D.C. & D.C. CURRAN. 1994. An experimental technique for studying shock propagation in large-scale samples of snow and other highly-porous materials. *In* High Pressure Science and Technology—1993, S.C. Schmidt, J.W. Shaner, G.A. Samara, & M. Ross, Eds.: 1135–1138. AIP Press. New York.
19. FURNISH, M.D., M.B. BOSLOUGH, G.T. GRAY, JR., & J.L. REMO. 1995. Dynamical properties measurements for asteroid, comet and meteorite material applicable to impact modeling and mitigation calculations. Int. J. Impact Engrg. **17:** 341–352.
20. FURNISH, M.D. & M.B. BOSLOUGH. 1995. Measuring Hugoniot, reshock and release properties of natural snow and simulants. Sandia National Laboratories Report, SAND92-0985.
21. CHHABILDAS, L.C. & J.M. MILLER. 1985. Release-adiabat measurements in crystalline quartz, Sandia Laboratories Report, SAND85-1092. (See also L.C. Chhabildas & D.E. Grady. 1984. Dynamic material response of quartz at high strain rates. Mat. Res. Soc. Symp. Proc. **22:** (Pt. III) 147–150. Earlier credit for ringdown technique due to P. C. Lysne, R. R. Boade, C. M. Percival, & O. E. Jones. 1969. J. Appl. Phys. **40:** 3786.
22. CHAMBERLAIN, E. 1973. Mechanical properties of frozen ground under high pressure. *In* Permafrost; North American Contribution, Second International Conference; Physics, Physical Chemistry, and Mechanics of Frozen Ground and Ice: 295–305.
23. JONES, S.K. & V.R. PARAMESWARAN. 1983. Deformation behavior of frozen sand-ice materials under triaxial compression. *In* Permafrost, Proceedings of the Fourth International Conference. National Academy Press: 560–565.
24. VINSON, T.S., C.R. WILSON, & P. BOLANDER. 1983. Dynamic properties of naturally frozen silt. *In* Permafrost, Proceedings of the Fourth International Conference. National Academy Press: 1315–1320.
25. VINSON, T.S. 1978. Parameter effects on dynamic properties of frozen soils. J. Geotech. Eng'rg Div. **GT10:** 1289–306.
26. NAKANO, Y. & N. FROULA. 1973. Sound and shock transmission in frozen soils. *In* Permafrost; North American Contribution, Second International Conference; Physics, Physical Chemistry, and Mechanics of Frozen Ground and Ice: 359–369.
27. LARSON, D.B. 1976. Explosive energy coupling in ice and frozen soils. LLNL report UCRL-78962.
28. AITKEN, G.W., G.W. SWINZOW, & D.R. FARRELL. 1976. Projectile and fragment penetration in snow and frozen soil. CRREL report.
29. DOCKERY, H.A., ET AL. 1987. Alaskan frozen soil impact tests of the B83-C/S and strategic Earth penetrator. LLNL report UCRL-53813.
30. SCHROEDER, R.C. & W.H. MCMASTER. 1973. Shock compression freezing and melting of water and ice. J. Appl. Phys. **44:** 2591–2594.
31. ANDERSON, G.D. 1968. The equation of state of ice and composite frozen soil material. Research Report 257 for U. S. Army Cold Regions Research and Engineering Laboratory.

Magnetospheric Effects as a New Aspect of the Asteroid Impact Problem: Necessity and Possibilities of Laboratory Simulation Experiments

YURI P. ZAKHAROV, SERGEI A. NIKITIN, ARNOLD G. PONOMARENKO, AND SHIGEYUKI MINAMI

Institute of Laser Physics
630090 Novosibirsk, Russia

This paper discusses the possible consequences to the Earth's magnetosphere, when due to too short an advanced warning, attempts at mitigation of a near-Earth object (NEO) must be made in close proximity to the Earth. The energy E_o, an explosive plasma release during impact may be compared with the kinetic energy E_k of the NEO and with the energy, E_e ($E_e \sim E_k$), needed for NEO deflection by a strong (protective force) explosive, at distances close to the scale of the magnetosphere. If the energy, E_m, of the Earth's dipole field latter is relatively small ($E_m \ll E_o$ for a NEO size ~1 km), global or even catastrophic disturbances could occur. These ecologically important magnetospheric aspects of the NEO impact problem have been discussed recently; particularly in the context of the comet SL-9/Jupiter impact. In the latter case, the effect on Jupiter's magnetosphere of the "NEO" explosions was very small ($x = E_o/E_m \sim 0.001$, where E_m is the "outer" magnetic energy of the planetary dipole field) and the corresponding model of its "fireball" development could be simulated numerically in "zero" approximation, with the assumption of an undisturbed magnetospheric media as a whole. However, in general, and, in the rather probable case of NEO impacts with values $x \sim 1$, the development of such 3D, nonstationary MHD or PIC-models without additional supporting laboratory data are very questionable at this time. Such information can be obtained from new kinds of simulation experiments with the laboratory magnetosphere, the so-called "terrella."

In this paper we analyze the possibilities of such simulated terrella experiments as they relate to the program of laboratory simulation of nonstationary magnetospheric processes at the KI-1 facility (Institute of Laser Physics) as well as recent data from it concerning the dynamics of a laser-produced plasma cloud, expanding into a dipole magnetic field in a large-scale chamber (diameter 1.2 m × 5 m) of KI-1. We also discuss the program and results of terrella experiments on the simulation of the stationary magnetospheric phenomena at the OCU facility (Osaka City University) as well as related data on plasma plume release.

The design and preliminary data of simulation experiments on the global magnetospheric disturbances are discussed for the cases of intra- and extra-magnetospheric explosions with energy in a range of values $x \sim 0.01$–1.

A conclusion is drawn about the necessity to develop a wide international collaboration in this field, including not only laboratory, but also numerical simulations of the nonstationary and high-energy magnetospheric phenomena caused by various NEO-explosions in near-Earth space.

ACKNOWLEDGMENT

The authors thank Peter Hammerling for useful discussions.

Laboratory Planetary Physics

P. HAMMERLING AND J.L. REMO

Quantametrics, Inc.
1 Brackenwood Path, Head of the Harbor
St. James, New York 11780

ABSTRACT: This paper describes experimental approaches to determine how the microstructure of near-Earth objects (NEOs), responds to high pressure and loading conditions. Meteorites are used as analog materials and are illuminated by high-energy lasers or x-rays to bring them to the required conditions.

I. INTRODUCTION

The current interest in near-Earth objects (NEOs) has underscored the lack of detailed knowledge of not only their material properties in space, but also their interactions in a planetary environment, which, due to their velocities, occur at very high energy densities [1], [2]. In order to understand how these objects respond under high pressure and temperature conditions, laboratory simulation experiments can be performed without resorting to large-scale energy detonations. In this research we describe how laboratory experiments on NEO analog materials can be achieved using high-power lasers to create the high pressures, temperatures, and related effect which can then be used to deduce the NEO equations-of-state (EOS), x-ray opacities and impulse response. The Earth's mantle contains approximately 2/3 of the Earth's mass and is composed among other elements of iron and magnesium bearing silicates with pressures in the tens of kilobars. At its center the Earth's core is thought to be mainly iron at temperatures of around 7000 K with pressures of several megabars and a density in excess of around 12 g/cm^3.

High power lasers have long been used to create high pressures in the laboratory using materials such as Al, Cu, etc., and various plastics [3]. With the advent of the use of *hohlraums* that convert the laser light into soft x-rays it is also possible to measure opacities and the effects of x-ray driven ablation [4], [5]. In both situations, pressures ranging from Kbar to Mbar and even Gbar scales can be produced, albeit on very small samples and on time scales of the order of nanoseconds. Nonetheless, such experiments, using sophisticated diagnostics, can obtain data in regions not as easily or conveniently reached by conventional methods such as diamond anvils or impact experiments, and more unconventionally as in underground nuclear tests. These ideas are inspired by experiments that have been done at LLNL, LANL, and other groups world wide in the context of Inertial Confinement

Fusion (ICF) [3]–[13] and represent a synthesis of the existing methods as applied to meteoritic material. Intense femto-second lasers are a new tool which can be used to explore the behavior of matter under extreme conditions [14].

Figure 1 illustrates a generic *hohlraum* target consisting of a right circular cylinder made of a high-Z material such as gold. The laser beams impinge through small holes at the ends and the sample to be irradiated is mounted on the side as shown in the sketch [15]. The inner wall, due to absorption of the incident laser energy, is heated to a high temperature and thus giving rise to a black-body radiation field that is in equilibrium with the wall. By suitable placement of the incident beams a very uniform radiation field can be generated around a target in the center of the *hohlraum*, as in the ICF application, or across the sample in the wall for the application considered here (Fig. 1). The radiation produced is in the soft x-ray region and is absorbed by the target through a radiative diffusion process. The cavity temperature achieved depends on the energy input, cavity dimensions, and the losses into the walls, target, and entrance holes. Using present facilities, with incident laser energies in the tens of kilojoules, cavity temperatures in the range of several hundred electron volts can be achieved with soft x-ray fluences of up to a few times 10^{14} W/cm^2 on target.

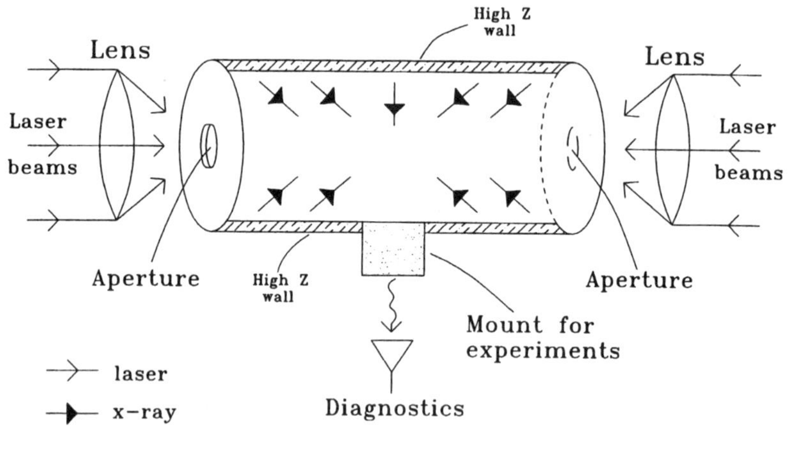

FIGURE 1. This illustrates a generic *hohlraum* target consisting of a right circular cylinder made of a high-Z material such as gold. The laser beams impinge through small holes at the ends and the sample to be irradiated is mounted on the side as shown in the sketch.

The targets used in such laboratory experiments may be made of meteorite materials as surrogates for most asteroidlike NEOs. There is no such convenient surrogate available in sufficient quantity for comet analog experimentation. However, comet compositions may be inferred from a combination of observational and laboratory data and suitable targets synthesized for experimentation. The information derived from these experiments could also be of use in understanding some of the processes that occur in planetary formation. Experiments on iron and some alloys could also be applicable to studies relating to modeling of the Earth's core. Additional experiments can be carried out on porous materials which can assist both in modeling the compaction of planetary materials as well as providing important data for industrial processing.

II. FEASIBILITY OF EOS, OPACITY, AND IMPULSE EXPERIMENTS

With respect to the NEOs, the regime of interest can be determined by the following simple argument. Velocity changes, ΔV, of the order of several cm/sec have been estimated as sufficient to deflect a large asteroid from collision with the Earth. For deflection by x-ray absorption, the ablation pressure, P_a, is related to the desired velocity change by:

$$\Delta V = (P_a \Delta t)/m, \tag{1}$$

where m is the areal mass and Δt the time over which the impulse is applied.

Assuming a spherical target and that the pressure is applied over only half its area, and taking Δt to be a microsecond, a density, ρ, of 5 g/cm^3, a radius of 1 km, with

$$m = \rho R/3, \tag{2}$$

$$\Delta V(\text{cm/sec}) = 3P_a\ (\text{Mbar}). \tag{3}$$

Thus, ablation pressures of the order of Mbars (1bar = 10^6 dynes/cm^2) will be required to deflect a large asteroid.

Assuming the *hohlraum* radiation is equivalent to that of a black-body at temperature T, the ablation pressure resulting from the absorption of soft x-rays can be related to the radiation temperature by a scaling law derived from a similarity solution of the hydrodynamic equations [16]:

$$P \propto T^x. \tag{4}$$

The value of the exponent depends on the assumed density and temperature dependence of the opacity and EOS.

For example, Lindl [4] quotes:

$$P(\text{Mbar}) \approx 3T^{3.5}, \tag{5}$$

where T is in units of 100 eV.

A similar relation has been found in laser plasma interaction experiments [17], [18]:

$$P(\text{Mbar}) \approx 15\, I^{2/3}\, \lambda^{-2/3}, \tag{6}$$

where I is the laser intensity in units of 10^{14} W/cm^2 and λ the laser wavelength expressed in microns.

Cavity temperatures in the range of one to several hundred eV and Mbar pressures are now being achieved using the LLNL NOVA laser, the Phébus laser in France, as well as in laboratories in Germany and Japan. Mbar experiments using soft x-rays or direct laser illumination are thus well within the current state of the art and are now routinely performed. As distinguished from experiments performed on simpler materials, those to be done using meteoritic samples must deal with them in their complexity of composition. The class of objects range from stony meteorites composed of mostly low-Z elements to the iron-nickel meteorites. Since the materials are heterogeneous, sample preparation and characterization will be important. Comets, which consist mostly of ice in various phases, present a special problem which needs to be discussed separately.

III. SHOCK HUGONIOT AND EOS DETERMINATION

A. Theory and Methodology

An equation of state is a thermodynamic relationship between quantities such as the pressure, p, specific energy, ε, and specific volume, V, of a material. For the application at hand, one requires EOS data ranging from that for cold matter to around half an electron volt.

The comportment of the shocked material is often plotted in a (P,V) diagram, the curves $p_H(V_1, p_0, V_0)$ are called Hugoniots. Using the laws of conservation of energy, momentum, and mass across a shock wave one can derive the following relations [19]:

$$\begin{aligned} V_0/V &= D/(D-u), \\ p &= Du/V_0, \\ p &= (D^2/V_0)(1 - V/V_0), \\ \varepsilon - \varepsilon_0 &= p(V_0 - V)/2, \end{aligned} \tag{7}$$

where D denotes the speed of the shock, ε denotes the specific energy, u is the particle velocity in the undisturbed medium, and $p_0 = 0$ has been assumed in the last equation. Quantities without subscripts refer to conditions behind the shock front. It is seen that a measurement of two of the variables suffices to determine the others, e.g., from D and u, p, and ε can be found from the above relations. Ultimately, what we wish to establish are Mie-Grüneisen types of equations of state for various classes of meteorites:

$$p - p_0 = [\Gamma(V)/V](E - E_0). \tag{8}$$

The subscript refers to the initial condition. E is the internal energy and $\Gamma(V)$ the Grüneisen coefficient. It is often a good approximation to set [20]

$$\Gamma(V)/V \cong \Gamma(V_0)/V_0.$$

For certain materials, experiments show that there is a linear relation between the shock and the particle speeds:

$$D = A + Bu, \tag{9}$$

in which event, the Hugoniot curve, using the third relation from Eq. (7) becomes [19]:

$$p_H = \frac{A^2(V_0 - V)}{(B-1)^2 V^2 \left[\frac{B}{B-1} - \frac{V_0}{V}\right]^2}. \tag{10}$$

Equation (4) is also written as [21]

$$D = C_0 + Su + Qu^2, \tag{11}$$

where the coefficients are calculated from a least-squares fit to the data, thus C_0 should not be interpreted as the sound speed.

B. Experimental Methods

In general there are three techniques that have been used to help establish Hugoniot curves, the "free surface" method, the impedance matching method, and the collision or impact method [19].

Free Surface Technique

In the first approach, a shock is launched into the target from one side and the velocity of the rear, free surface is measured together with the shock speed. Since the free-surface velocity is approximately twice the material velocity, the two necessary parameters are determined. This technique is valid if the shock strength is not too high and if the material is nonporous. The speed of the rear surface can be measured using laser-Doppler interferometry. If both the incident and reflected light are normal to the moving surface, then there will be a wavelength shift between the incident and reflected light given by

$$\Delta\lambda/\lambda = -2V/c. \tag{12}$$

The VISAR (velocity interferometry system for any reflector) technique uses a Michelson interferometer [22]. Alternatively, a Fabry-Pérot interferometer can be used [23]. The incident and reflected light can be transported using optical fibres. Using multiple fibres or a slit one can obtain information from more points or from a line. Besides the speed, one also can measure the rear surface temperature using time-resolved infrared pyrometry. For example, with three fibres coupled to IR optics and three filters spanning the 2–5 μm region together with HgCdTe detectors, one can determine the temperature as well as the emissivities.

Collision Method

The collision method creates a shock by high-speed impact of a "flyer" onto a target. If the flyer and target materials are identical, then the material velocity is half the flyer speed. Again the shock speed must be measured in the target material. For nonidentical materials, the flyer Hugoniot must be known in order to deduce the particle velocity from the data. Crucial to the use of this technique is the ability to measure the flyer velocity, that the flyer's integrity be maintained and that it impact normally to the target's surface. Figure 2 shows one way of implementing this technique [12].

Impedance Matching

The impedance matching technique sends a shock through two materials in contact, the Hugoniot of one of them being known. The shock data can be deconvolved to give the material velocity; measurement of the shock speed in the target material then determines the Hugoniot. Figure 3 sketches how this has been implemented in one laboratory [5].

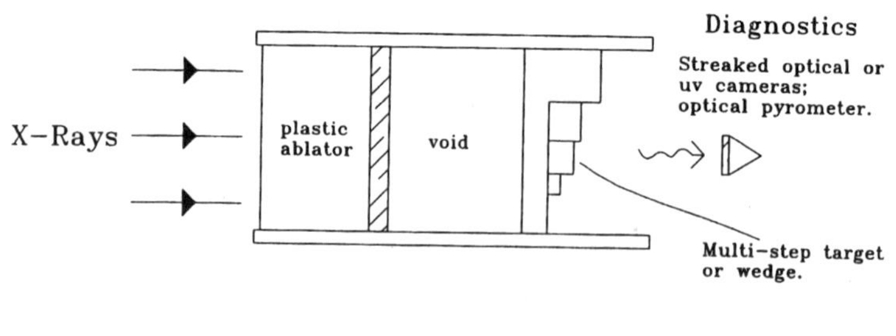

FIGURE 2. Flyer foil technique for measuring Hugoniot curves adapted to an x-ray drive from a hohlraum source. Experiments at LLNL on Au foil targets using gold flyer foils resulted in a dynamic pressure of 0.75 Gbar, inferred from the shock speed.

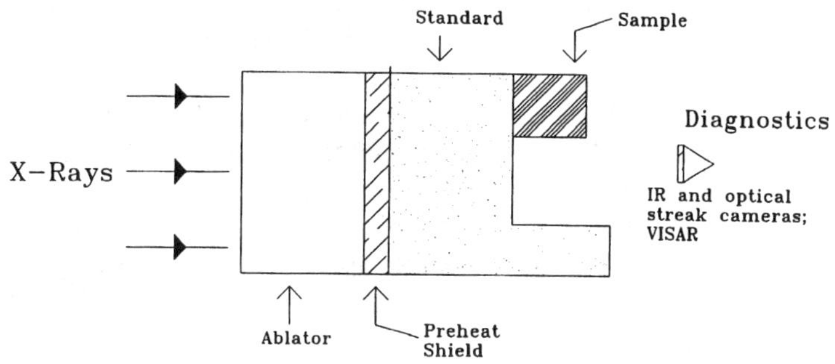

FIGURE 3. Impedance matching technique for measuring Hugoniot EOS as used at AWRE with a hohlraum source. Al was the standard material. Pressure experiments were in the range of 2–10 Mb in Al and 2–19 Mb in Cu.

Knowing $p(V)$ and $\varepsilon(V)$, the cold compression curve $p_C(V)$ can be determined [19].

If the target material heats up appreciably during the experiment the contribution from the electron pressure can be appreciable. In general this is given by:

$$p_e = \frac{2}{3}\varepsilon_e/V, \tag{13}$$

where ε_e is the electron thermal energy per unit mass and V is the specific volume.

In practice [12], [13], on measuring the shock speed, the pressure and target material density are obtained from EOS tables. The most commonly used tables are the SESAME tables [24], developed at the Los Alamos National Laboratory. A general purpose equation of state that is adapted for use in hydrodynamic simulations has been developed by researchers at the Lawrence Livermore National Laboratory [25].

All of the above experimental methods have been used in plasma physics experiments. Below we discuss their adaptation to the meteoritic application.

Laser beams can be used to induce shock waves in experimental samples. However, precision measurements require shocks which are spatially uniform over the sample area. This uniformity is difficult to obtain using intense laser light due to the latter's intensity fluctuations. This difficulty eventually can be overcome by using spatial or temporal beam smoothing techniques [26]–[28]. Planar shocks can be generated using the thermal x-rays generated in laser heated cavities or *hohlraums* as mentioned in the Introduction and sketched in Fig. 1. Figure 2 shows how the flyer foil technique can be adapted to the *hohlraum* capsule [15]. A wedge-shaped, single- or multistep target also could be used. Figure 3 gives a target configuration, used at the AWE facility in Aldermaston, that implements the impedance matching technique [5]. The standard material is one such as Al whose properties are well known.

The use of meteorite samples, for example chondritic or Fe-Ni, may present some practical difficulties for homogeneous target fabrication: the target linear dimensions being about 500 μm with thickness ranging from a few to some tens of microns. This limitation, however, should make it easier to experiment on cyogenically cooled samples which are more appropriate to outer-space conditions.

IV. IMPULSE MEASUREMENTS

The impulse transmitted to the target by thermal x-ray irradiation can be measured by use of piezo-electric detectors, either quartz or plastics such as PVDF, as shown in Fig. 4 [29]–[31]. We have previously discussed, using a simple model, the impulse produced by absorption of x-rays in NEO model materials [32]–[35]. The model predicts that the specific impulse, I, varies with absorbed x-ray fluence, F_0, according to the nondimensional relation:

$$I^* = 2\sqrt{2}\{\sqrt{F_0^\star - 1} - \tan^{-1}[\sqrt{F_0^\star - 1}]\}, \tag{14}$$

where the scaled variables are defined by

$$\begin{aligned} I &= (\sqrt{\varepsilon_\nu}/\mu) I^* \\ F_0{}^\star &= F_0/F_{min} \geq 1 \\ F_{min} &= \varepsilon_\nu/\mu \end{aligned} \tag{15}$$

In the above, ε_ν denotes the vaporization energy per unit mass and μ (cm^2/gm) the x-ray opacity of the material under consideration.

Asymptotically, when $F_0{}^\star > 1$

$$I = 2\sqrt{2F_0/\mu}. \tag{16}$$

This predicted behavior has been seen in preliminary experiments on Al performed at the Ecole Polytechnique [29] using a quartz piezo detector and x-rays generated from a back-lighter source. The latter used L-band emission at around 1.4 kev from a copper target illuminated by 1.06 μm light from a Nd-glass laser. The x-ray intensity on target ranged from 10^{10}–10^{11} W/cm^2 (10–100 J/cm^2). The recorded impulse was made approximately at the onset of the liquid-vapor transition rather at vaporization [36]. Figure 4 illustrates the principle of this method.

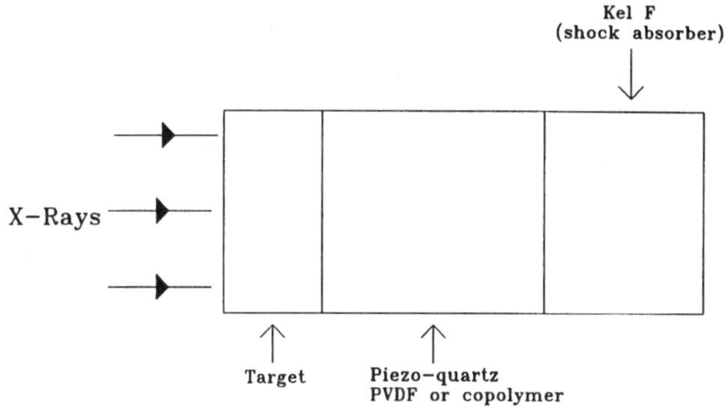

FIGURE 4. Impulse measurements can also be made with a hohlraum x-ray source as indicated in the sketch. This is an adaptation of techniques developed at the University of Poitier and performed using the Nd-glass laser at the LULI facility at the Ecole Polytechnique. Preliminary experiments using x-rays emitted from the rear of a laser heated target have also been done at LULI.

This type of experiment could be done with meteoritic samples using higher laser energies and/or more sensitive detectors. The response of this material to a black-body radiation source could be measured using a *hohlraum* configuration of the type already described above. The analysis of such an experiment probably will require a more sophisticated treatment such as a radiation-hydrodynamic code which also requires, as an input, the opacity of the material. This is discussed in more detail in Section VII.

V. RELATION TO PLANETARY FORMATION AND IMPACT

Planetesimal accumulation and subsequent fragmentation have been studied in the context of understanding the observed mass distribution of the asteroids as discussed in [37]–[39], and references therein. In trying to generalize from laboratory impact experiments to collisions involving bodies of asteroidal dimensions, i.e., from targets of the order of 1–10 cm to those of a km range, it is imperative to have some type of scaling law that is based on experimental data. A recent paper [38] summarizes the work to date on the scaling of Q^*, the specific energy required to fragment a body, with the size of the object. Two main approaches have been used, dimensional analysis [40] and a hydrocode [41]. The current hydrocode results give a lower value for the fraction, f_{KE}, of the projectile's kinetic energy, that ends up as kinetic energy of fragments of the target, than for the dimensional analysis model, and are not in as good agreement with the asteroid number vs. diameter data [38].

The high pressure generated by the impact sends a strong shock into the target leading to its eventual disruption and fragmentation. A key input to a hydrocode are the constitutive equations that describe the material behavior via an equation of state (EOS). At high pressures, in the hydrodynamic regime, the linear relation between pressure and volume changes and the volume varies less rapidly. It is customary to rely on empirical equations of state in this region. For example, Melosh et al. [41], in the hydrocode alluded to above, use a Murnaghan form for the isotherm relating the pressure to the density:

$$P = \frac{B_0}{n}[(\rho/\rho_0)^n - 1], \tag{17}$$

where B_0 is the bulk modulus of the material, n a constant adjusted to the material, and ρ_0 the density at standard conditions.

However, as discussed by Livits et al. [42], in their numerical study of the effect of the EOS on shock waves produced by impulsive loading, the flow structure and the shock formation are EOS dependent. Hence, it is important to use as accurate an EOS as possible in the hydrocode description.

A first step in this program to acquire laboratory data relevant to modeling the impact of meteoritic and comet analog matter with the Earth or in the planetesmal context has been taken in a recent series of experiments [43], [44]. Impact and stress-strain experiments were performed on iron-nickel and stony meteoritic material, as well as on snow and snow as analogs of cometary material. The experiments proposed above would complement and extend this data to a higher pressure and temperature regime as well as directly simulate the response to nuclear x-ray irradiation.

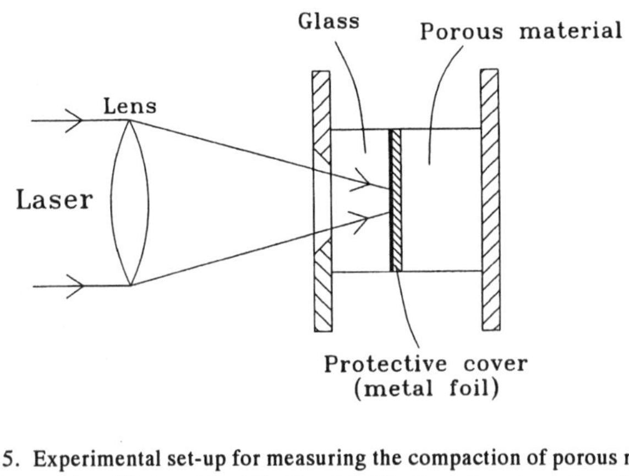

FIGURE 5. Experimental set-up for measuring the compaction of porous materials by an intense laser pulse. (University of Poitier, LULI). Experiments on low-density foam targets driven by soft x-rays from a back-lighter source have been performed at the Rutherford Appleton Laboratory and using a hohlraum source at LLNL.

VI. POROUS MATERIALS

The behavior of porous materials under high stress can be studied by an adaptation of the techniques discussed above. Figure 5 shows an example of a target holder that can be used with an intense laser source or attached to the wall of a cylindrical hohlraum to study the response to high-temperature black-body radiation. Target porosity is essential to take into account in studying porous NEO analogs, particularly comets in the context of impact experiments, collisional disruption, and shock-wave effects [45]. At the high pressures we are interested in, the Mie-Grüneisen equation can be adapted to account for porosity [20]. The Hugoniot equation for a porous material, $P_H^{\star}(V)$ can be obtained using the Rankine-Hugoniot equations and the Mie Grüneisen equation of state referred to the Hugoniot, $P_H(V)$, of the compact material [20]:

$$P_H^{\star}(V) = P_H(V)\frac{1-\frac{\Gamma}{2}(V_0/(V-1))}{1-\frac{\Gamma}{2}(\alpha V_0/(V-1))}, \tag{18}$$

$$\alpha \equiv V_0^{\star}/V.$$

In Eq. (18), V_0 and $\Gamma(V)$ are the initial specific volume and the Grüneisen coefficient of the compact material and α is the porosity. Equation (18) does not appear in its familiar form since we have not used the usual assumption that $\Gamma(V)/V = \Gamma_0/V_0$ = const, relating the Grüneisen coefficient and specific volume to their values at standard conditions. Highly porous materials under high-pressure loading can exhibit, due to thermal pressure, anomalous Hugoniot curves where the final density is lower than the initial density [19], and hence it was thought better to allow for the possible temperature dependence of Γ. The transition to the anomalous region occurs when $V_0{}^*/V = 1 + 2/\Gamma$.

Another formulation treats the Grüneisen parameter as a function of the energy, rather than the volume and appears to give a better description of the response of porous materials to shock waves [46]. The data for porous ice in Oh and Persson [46] cover the particle velocity range from about 1–8 km/s, while that for natural snow and snow match grout in Furnish et al. [43] cover the range ~0.5–2.5 km/s. In the region where they overlap these two independent experiments yield similar results. These new results also exhibit the anomalous behavior mentioned above.

An elegant method that can characterize porous media before and after impact has been demonstrated recently for preshock characterization, namely the application of synchrotron generated x-radiation to computer microtomography of the material [47]. The resulting tomograms in digital format can be used in other computer descriptions that relate the micro to the macro description.

The response of porous materials to an impulsive load should also differ from the original compact material due to dissipation, random directional unloading of the shock, and the possibility of an anomalous Hugoniot. Porous materials are known to be poor transmitters of shock waves [45].

VII. OPACITY MEASUREMENTS

As can be seen from the simple expressions above, Eqs. (14)–(16), a critical input to understanding the impulse response to an x-ray radiation source is knowledge of the opacity of the material being studied. In general this parameter is a function of the composition of the material, the wavelength of the radiation, and the temperature of the sample.

In the diffusion approximation in one dimension, the radiation flux, S, obeys the equation:

$$S = -K\frac{\partial T}{\partial x}, \tag{19}$$

where T is the temperature.

In particular, we can express the diffusion constant in terms of the Rosseland mean-free-path, l_R, by [19]

$$K = \frac{16}{3}\sigma T^3 l_R, \tag{20}$$

where σ is the Stefan-Boltzmann constant. The Rosseland mean-free-path is related to the opacity by

$$\rho\mu = l_R^{-1}. \tag{21}$$

An upper bound to the opacity is given by the Bernstein-Dyson maximum opacity theorm [48],

$$(\rho l_R)^{-1} = (4.57 \times 10^5(Z/A)(R_y/kT)\ \mathrm{cm^2/gm}. \tag{22}$$

Z/A is the ratio of the nuclear charge to the atomic number of the material, R_y, the Rydberg constant and k Boltzmann's constant.

Pakula and Sigel [16] have solved the one-dimensional hydrodynamic equations including radiative conduction as above with K and l_R obeying power laws, where ρ is the density:

$$K = K_0 \frac{T^n}{\rho^m},$$

$$l_R = A \frac{T^j}{\rho^m}, \tag{23}$$

$$n = j + 3.$$

The ablation pressure then has the form, where t is the time:

$$P_A \propto T^{\frac{j+3+m}{m+1}} t^{-1/m+1}. \tag{24}$$

The predicted ablation pressure thus depends, in this model, on the form of the opacity. Using the Bernstein-Dyson, hydrogenic plasma, and cold-matter opacities, respectively, we have:

$$P_A \propto T^{5/2} t^{-1/2},$$

$$P_A \propto T^{17/6} t^{-1/3}, \tag{25}$$

$$P_A \propto T^{7/2} t^{-1/2}.$$

The last relation in Eq. (25) can be recognized as that given in Eq. (5).

In practice, one usually uses tabulated opacity tables [49]–[51]. Very useful approximate results for Planck and Rosseland mean opacities have been given by Tsakiris and Eidman [52]. In particular, for a variety of elements, they can parameterize their results in the form of Eq. (20).

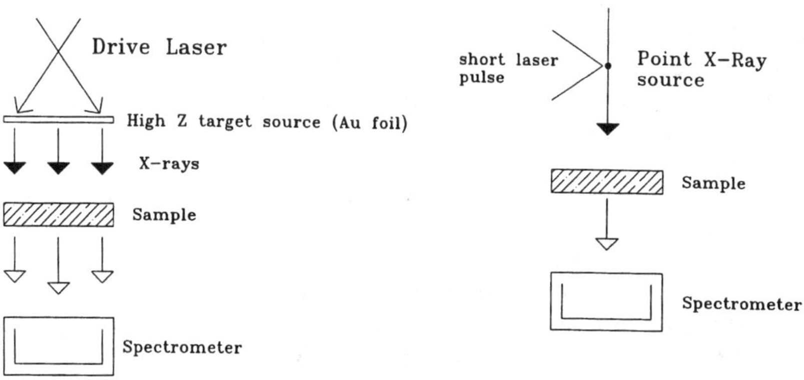

(a) Broad-band source (b) Mono-energetic source

FIGURE 6. Opacity measurement using back-lighter technique (AWRE, LLNL).

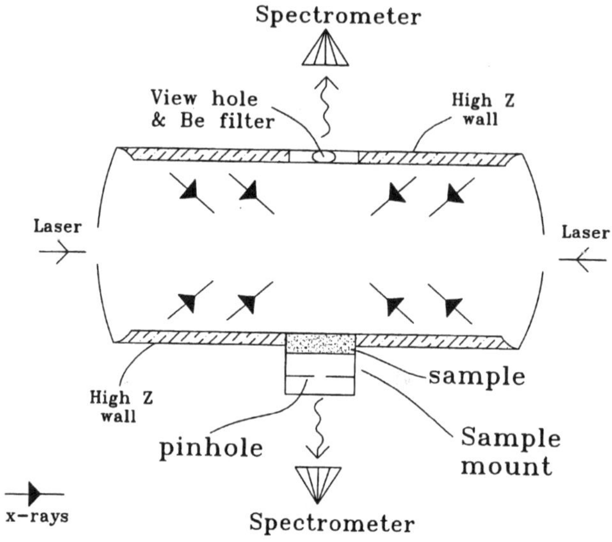

FIGURE 7. Opacity measurement using a hohlraum source.

One can use such data together with a constitutive model to get an estimate for the order of magnitude of the opacities of meteoric and comet material. If the tabular data can be parameterized by a power law, the ablation pressure can be estimated using the above results. The actual data required for meteorite and comet analogs for NEOs can be obtained in several ways.

The first uses a back-lighter pulsed laser beam focused on a foil which emits x-rays in the band of interest. These x-rays then traverse the sample and are measured by a time-resolved spectrometer [53]. The opacity of the target is obtained by comparing the spectrum of the x-rays that traverse it to that of the source. The method just outlined applies to cold matter. Alternatively, if one is interested in the opacity under conditions of higher pressure and temperature, one can confine the sample by tamping it with a surrounding plastic and then heat it with x-rays produced by an intense laser beam incident on another foil and use the back-lighter beam as a probe [53]. A typical set-up is shown in Fig. 6.

Clearly, a *hohlraum* configuration, Fig. 7, also could be used to illuminate a sample with black-body radiation to give an integrated response [54], [55]. Such an experiment yields both the Planck and Rosseland mean-free-paths which can then be compared to the opacity models. This approach can give a laboratory simulation of the response to be expected from a NEO subjected to nuclear x-radiation, and can be used to verify the simple scaling laws. As mentioned already, a full numerical simulation of the expected response of matter to x-ray heating, ablation, and shock-wave propagation requires a radiation-hydrodynamic code based as much as possible on measured opacities, supplemented by the tables referred to above.

CONCLUSIONS

Hohlraum techniques provide methods of measuring the opacity and radiation coupling of NEO analog materials without having to utilize large x-ray simulation facilities and, more importantly, not having to resort to the actual detonation of nuclear devices. A considerable amount of data can be obtained on various classes of NEO analog materials by utilizing laser-driven *hohlraums*. Such *hohlraum* experiments can be carried out in a systematic and cost-effective matter, and can provide the required data base for understanding NEO and related protoplanetary and planetesimal materials interaction.

Experiments using direct illumination using high-power lasers can supplement the hohlraum results may be helpful to those carrying out modeling on:

1. Planetesimal and planetary accretion models.
2. Comet interaction models with planetary atmospheres (e.g., SL-9)
3. Meteorite interactions in planetary atmospheres, planetary impacts, and asteroid collisions;
4. NEO orbital momentum change;
5. Nonplanetary modeling on industrial materials.

Additional applications of the above techniques include simulating x-radiation in astrophysical environments such as stellar atmospheres, and with the advent of very powerful (~MJ) lasers, stellar interiors. These and other topics of interest to astrophysics are discussed by Rose [57] and Goldstein et al. [58]. However, we should emphasize that many experiments relevant to planetary physics conditions can be done in smaller facilities. For example, pressures of 5–10 Mbar have been achieved in *hohlraum* experiments on thin foil targets using laser energies of around 1 kilojoule [13]. Although the emphasis in this paper has been on *hohlraum* sources, it is possible to obtain good EOS data and generate high-pressure shocks using direct laser illumination of smoothed beams as has been done in preliminary experiments at the LULI facility in France [59], [60]. Finally, two recent experiments using x-radiation generated by a *hohlraum* should be mentioned. In one, radiation was transferred through a cylindrical tube [61], and in the second, the Rosseland mean opacity of iron was measured and compared with various opacity models [62]. As already mentioned, all these new techniques and data complement that obtained in other types of facilities. This latter information is summarized in two recent reviews [63], [64].

We would like to point out that the availability of *hohlraum* and backlighter methodologies mitigates the necessity to carry out large scale x-ray interaction experiments on NEOs in space in order to gain a base line of data. Such experiments, without substantial preliminary preparation could precipitate a hitherto nonexistent hazard. If, on the other hand, an unambiguous NEO threat appears, *hohlraum* data can provide a first-order information base, which, used in conjunction with reconnaissance and other available information, could aid in the design and implementation of an effective threat response.

REFERENCES

1. REMO, J.L. 1993. Classification and modeling dynamic processes and interactions of NEO materials. *In* Hazards Due Meteorites and Comets, T. Gehrels, Ed. University of Arizona Press. Tucson, Arizona.
2. REMO, J.L. 1992. Asteroid/meteorite analogs and materials properties. *In* Proceedings of the Near Earth Object Interception Workshop, Jan. 14–16 1992, G.H.Canavan, J.C.Solem, & J.D.G. Rather, Eds.: 163–174. Los Alamos National Laboratory, Los Alamos, New Mexico. LA-12476-C.
3. WOOD, L.L., R.N. KEELER, & J.H. NUCKOLLS. 1977. Concerning the generation of very high pressures for EOS studies with ultra-high power laser pulses. Presented at the 6th AIRAPT International High Pressure Conference at the University of Colorado at Boulder, July 25–29, 1977. UCRL-79610.
4. LINDL, J. 1994. Two decades of progress towards ICF ignition and burn. *In* Proceedings of the 23rd ECLIM. St. John's College, University of Oxford, Sept. 19–23, 1994; 1995. Physics of Plasmas. **2:** 3933.
5. EVANS, A.M. , N.J. FREEMAN, & S.D. ROTHMAN. 1994. Feasibility of Hugoniot EOS measurements on Helen. *In* Proceedings of the 23rd ECLIM. St. John's College, University of Oxford, Sept. 19–23, 1994.
6. VEESER, L.R. & J.C. SOLEM. 1978. Phys. Rev. Lett. **40:** 1391.
7. TRAINOR, R.J., ET AL. 1979. Phys. Rev. Lett. **42:** 1154.
8. COTTET, F., ET AL. 1984. Phys. Rev. Lett. **52:** 1884; 1985. Appl. Phys. Lett. **47:** 67.
9. SCHMALTZ, R.F. & J. MEYER-TER-VEYN. 1985. Phys. Fluids. **28:** 932.
10. a. FABBRO, R., B.FARRAL, J.VIRMONT, H. PEPIN, F. COTTET, & J.P.ROMAIN. 1986. Laser and particle beams. **4:** 413.
10. b. FABBRO, R. & B. FERRAL. 1994. Laser and particle beams. **9:** 397.
11. HAMMEL, B.A., ET AL. 1993. Phys. Fluids B. **5:** 2259.
12. CAUBLE, R., ET AL. 1993. Phys. Rev. Lett. **70:** 2102; 1994. J. Quant. Spectrosc. Radiat. Transfer. **51:** 433.
13. LÖWER, TH., ET AL. 1994. Phys. Rev. Lett. **72:** 3186.
14. DOWNER, M.C., ET AL. 1993. Properties of condensed matter under planetary interior conditions measured by femtosecond laser spectroscopy. *In* Laser Interactions with Atoms, Solids, and Plasmas, R.M. More, Ed. Plenum Publishing Corp. New York, NY.
15. KAUFFMAN, R.L., ET. AL. 1994. Phys. Rev. Lett. **73:** 2320.
16. PAKULA, R. & R. SIGEL. 1985. Phys. Fluids. **28:** 232.
17. CUNNINGHAM, R.F., ET AL. 1983. Phys. Fluids. **26:** 2011.
18. MAYER, B. & G. THIELL. 1984. Phys. Fluids. **27:** 302.

19. ZEL'DOVICH, YA.B. & YU.P. RAIZER. 1966. Physics of Shock Waves and High-Temperature Phenomena. Academic Press. New York, NY.
20. MEYERS, M. 1994. Dynamic Behavior of Materials. John Wiley. New York, NY.
21. MARSH, S.P., ED. 1980. LASL Shock Hugoniot Data. University of California Press. Berkeley, California.
22. BARKER, L.M. & R.E. HOLLENBACH. 1972. J. Appl. Phys. **43:** 4669.
23. MCMILLAN, C.F., ET AL. 1988. Rev. Sci. Instrum. **59:** 1.
24. HOOLIGAN, K.S., Ed. 1984. T-4 Handbook of Material Properties Data Bases: Vol.1c, Equations of State. LANL Report No. LA-10160-MS,UC-34, Nov. 1984 (unpublished).
25. MORE, R.M., ET AL. 1988. Phys. Fluids. **31:** 3059.
26. KATO, Y., ET AL. 1984. Phys. Rev. Lett. **53:** 1057.
27. LEHMBERG, R.H. & S.P. OBENSCHAIN. 1983. Opt. Comm. **46:** 27.
28. SKUPSKY, S., ET AL. 1989. J. Appl. Phys. **66:** 3456.
29. GAUTHIER, J.C., ET AL. 1993. Impulse generated by a laser-produced x-ray source on a metallic layer. *In* Applications of Laser Plasma Radiation, M.C. Richardson, Ed.: 220. SPIE , vol. 2015.
30. ROMAIN, J.P., ET AL. 1994. Measurements of laser induced shock pressures using PVDF gauges. *In* Shockwaves in Condensed Matter Conference. AIP Press. New York, NY.
31. JOHNSON, D.E., ET AL.1994. PVDF measurement of soft x-ray induced shock and filter debris impulse. *In* Shockwaves in Condensed Matter Conference: 1911–1914. AIP Press. New York, NY.
32. HAMMERLING, P. & J.L. REMO. 1995. NEO Interaction with Nuclear Radiation. Acta Astronautica. **36:** 337–346.
33. BLOEMBERGEN, N., Chairman. 1987. Report to the American Physical Society by the Study Group on Science and Technology of Directed Energy Weapons. Rev. Mod. Phys. **59.**
34. LAWRENCE, R.J. 1988. An Effective Properties Model for Pulsed Radiation Interactions. Sandia National Laboratories Report, SAND88-0245, April, 1988.
35. LAWRENCE, R.J. 1991. The Equivalence of Simple Models for Radiation-Induced Impulse. Sandia National Laboratory Report, SAND 91-1171C9.
36. GAUTHIER, J.C. Private communication.
37. BARGE, P. & R. PELLAT. 1991. Icarus. **93:** 270; 1993. Icarus. **104:** 79.
38. DAVIS, D.R., E.V. RYAN, & P. FARINELLA. 1994. Planet. Space Sci. **42:** 599.
39. FUJIHARA, A., P. CERRRONI, D. DAVIS, E. RYAN, M. DI MARTINO, K. HOLSAPPLE, & K. HOUSEN. 1989. Experiments and scaling laws for catastrophic collisions. *In* Asteroids II, R.P. Binzel, T. Gehrels, and M.S. Mathews, Eds. University of Arizona Press. Tucson, Arizona.
40. HOUSEN, K. & K. HOLSAPPLE. 1990. Icarus. **84:** 226.
41. MELOSH, H.J., E.V. RYAN, & E. ASPHAUG. 1992. J. Geophys. Res. **97:** 14735.
42. LIFITS, S.A., S.I. ANISIMOV, & J. MEYER-TER-VEHN. 1991. Z. Naturforsch. **47A:** 45.
43. FURNISH, M.D., ET AL. Dynamical properties measurements for asteroid, comet and meteorite material application to impact modeling and mitigation calculations. 1994. Presented at the 1994 HyperVelocity Impact Symposium (HVIS), Santa Fe, New Mexico, Oct. 16–20,1994, paper 160F.
44. TEDESCHI, W.J., J.L. REMO, J.F. SCHULTZE, & R.P. YOUNG, JR. 1994. Experimental Hypervelocity Impact Effects on Simulated Planetesimal Materials, presented at the 1994 HyperVelocity Impact Symposium (HVIS), Santa Fe, NM, Oct. 16-20,1994, paper 072; and 1995: Int. J. Impact Engineering, **17:** 1–6, 837–848.
45. LOVE, S.G., F. Horz, & D.E. Brownlee. 1993. Icarus. **105:** 216.
46. OH, K.-H. & P.-A. PERSSON. 1989. J. Appl. Phys. **65:** 3852.

47. SPANNE, P., ET.AL. 1994. Phys. Rev. Lett. **73:** 2001.
48. BERNSTEIN, J. & F. J. DYSON. 1959. General Atomic Report No. GA 848.
49. SESAME. 1983. Report on the Los Alamos equation of state Library, T-4 Group. Los Alamos National Laboratory Report No. LALP-83-4. Los Alamos, New Mexico.
50. WEISS, A., J.J. KEADY, & N.H. MAGEE. Atomic Data Nucl. Data. **45:** 209.
51. ROGERS, F.J. & C.A. IGLESIAS. 1992. Astrophys. J. Suppl. Ser. **79:** 507.
52. TSAKIRIS, G.D. & K. EIDMAN. 1987. J. Quant. Spectrosc. Radiat. Transfer. **38:** 353.
53. LEE, R.W., ET.AL. Verification of OPAL Code Predictions for Conditions of Astrophysical Interest. Lawrence Livermore National Laboratory ICF Quarterly Report, **3** (2), Jan.–Mar. 1993. UCRL-LR-105821-93-2.
54. SPRINGER, P.T., ET AL. 1994. J. Quant. Spectrosc. Radiat. Transfer. **51:** 371.
55. WHITE, V.J.L., ET AL. 1994. Phys. Rev. E. **49:** R4803.
56. AFSHAR-RAD, T., ET AL. 1994. Phys. Rev. Lett. **73:** 74.
57. ROSE, S.J. 1991. Laser and Particle Beams. **9:** 869.
58. GOLDSTEIN, W.H., ET AL. 1994. Laboratory high-energy astrophysics on lasers. *In* Laser Interaction with Matter, S.J. Rose, Ed.: 235–238. Proceedings of the 23rd European Conference, St. John's College. University of Oxford, Sept. 19–23, 1994.
59. KOENIG, M., ET AL. 1994. Experimental test of the relative coherence of equation of states of different materials compressed by shock waves. *In* Laser Interaction with Matter, S.J. Rose, Ed.: 309–312. Proceedings of the 23rd European Conference, St. John's College. University of Oxford, Sept. 19–23, 1994.
60. KOENIG, M., ET AL. 1994. High pressure shock generation in optically smoothed laser-plasmas. *In* Laser Interaction with Matter, S.J. Rose, Ed.: 313–316. Proceedings of the 23rd European Conference, St. John's College. University of Oxford, Sept. 19–23, 1994.
61. PERRY, T.S., ET AL. 1994. Experimental techniques to measure thermal radiation heat transfer. J. Quant. Spectrosc. Radiat. Transfer. **51:** 273–280.
62. SPRINGER, P.T., ET AL. 1994. Spectroscopic measurements of Rosseland mean opacity. J. Quant. Spectrosc. Radiat. Transfer. **51:** 371–377.
63. AVORIN, E.N., B.K. VODOLAGA, V. SIMONENKO, & V.E. FORTOV. 1993. Intense shock waves and extreme states of matter. Sov. Phys. Uspekhi. **36:** 337.
64. FORTOV, V.E. 1995. Shock compressed ideal plasmas. *In* Elementary Processes in Dense Plasmas: 389–402. Proceedings of the Oji International Seminar, Tomakomai, Japan, June 27–July 1, 1994. Addison-Wesley Publ. Co. Reading, MA.

A Proposal to the United Nations Regarding the International Discovery Programs of Near-Earth Asteroids

TOM GEHRELS

Space Science Building
University of Arizona
Tucson, Arizona 85721

ABSTRACT: The largest near-Earth asteroids are dangerous to all people. It seems appropriate that the United Nations undertake an administrative role in supporting the discovery programs. With a relatively small amount of support we can have the problem largely under control by the year 2020.

The odds of our international society getting obliterated by an asteroid impact in the next 50 years has been estimated at 6,000 to 1. That probability and its frightening consequences, as well as the methods for discovery and mitigation, are described in a textbook by 120 authors in 46 chapters, edited by Gehrels [1].

The first step in mitigation is to discover with astronomical telescopes the estimated 1,700 suspects that are larger than 1 kilometer in diameter, the ones energetic enough to cause demise on a worldwide scale.

Table 1 shows the programs that are presently in operation or will be in operation soon. At the Palomar-Schmidt Telescope, the programs were discontinued in 1994, but Eleanor Helin has recently taken it up again. The Spacewatch telescopes are of the University of Arizona [2], [3]. This is the first professional program that uses charge-coupled devices (CCDs) instead of photographic materials. Amateurs, however, also use CCDs and telescopes as large as one meter; about two or three near-Earth asteroids are presently being discovered by amateur astronomers. At the Anglo-Australian Observatory, photographic plates are used at the 1.2-meter Schmidt Telescope, while a CCD system is being installed at another telescope. At the Côte d'Azur Observatory in Southern France a CCD system should come on-line soon for part-time usage of their 1.0 m Schmidt Telescope. At the Lowell Observatory, Flagstaff, Arizona, the 0.6 m Schmidt Telescope is being prepared for operation with a CCD system. A group at the Jet Propulsion Laboratory is preparing a CCD system to be used on a 1.0 m telescope of the U.S. Air Force. The Spacewatch team is building a 1.8 m telescope which might begin to come on-line by 1997.

TABLE 1. Near-Earth Detection Programs

Group	Telescope	Operational
Palomar	0.46 meter	1973-94
Spacewatch	O.9 m	1990-present
Anglo-Australian	1.2 m	1990-present
Amateurs	< 1.0 m	1994-present
Côte d'Azur	1.0 m	1996?
Lowell Observatory	0.6 m	1996?
Jet Propulsion Laboratory	1.0 m	1996?
Spacewatch	1.8 m	1997?

However, the question marks in Table 1 indicate uncertainties due to shortage of funding. At the present rate of development, I estimate that it will take at least 50 years to near-completion of discovery for the 1,700 suspects. Fifty more years of uncertainty whether or not we might be eliminated by a rock!

I propose that the United Nations take an interest in this first, most urgent mitigation phase of finding the 1,700 suspects. My proposal is to use *existing* telescopes. It also relies on existing teams, the people who have an established willingness to do the fund raising and observing.

A one-kilometer asteroid dwells in the magnitude range 19–21 most of its time in elliptical orbit about the Sun. Telescopes in the 1-meter range are therefore suitable; I see opportunities at the following.

The team at the Anglo-Australian Observatory has already been mentioned. They are effectively working on this problem at a variety of telescopes; they need additional support on the order of $0.25 M/year to bring their CCD system on-line. The 1-meter Schmidt of the European Southern Observatory could be made available part-time, and European asteroid observers are readily found; about the same level of support ought to stimulate their fund raising and bring this telescope on-line for this work. These two programs in the Southern Hemisphere would be the proper counterpart to those in the Northern Hemisphere.

Even some of the programs on the telescopes being put into operation, mentioned above, need additional help. The discovery programs at the Côte d'Azur Observatory, the Lowell Observatory, and at the University of Arizona are barely surviving. A challenge will have to be met of handling hundreds of discoveries per year, far more than has ever been processed before.

For astrometric and photometric follow-up, one or two large telescopes appear available, part-time, with a small group of observers who can work with various large telescopes on Mauna Kea, Hawaii; additional support, again of the same magnitude, could expand their work to the needed level.

The relative smallness of the fields and the vagaries of the weather require detailed scheduling and processing of the discoveries and their follow-up. I propose that this task would be requested from the Minor Planet Center of the International Astronomical Union in Cambridge, Massachusetts. It is exquisitely organized and has been productive over the years in keeping track of discoveries and follow-up observations for hundreds of thousands of comet and asteroid observations. The Center would, however, need additional financial support, again of the same magnitude.

For the seven programs identified above, the United Nations' prestige and support would provide stimulus and seed money with which to raise their needed additional financing. The endorsement by the UN would also facilitate getting the telescope assignments. The case needs to be made that the near-Earth asteroid problem is urgent in view of the serious consequences, and that the UN represents all people on Earth for their protection from a unique hazard.

I therefore propose that the UN help acquire the funding and oversee the execution of this proposal. Agencies in various countries have to be asked to support this initiative as well as NASA and the Air Force of the United States. The orbital periods of the hazardous 1,700 are between three and six years. A first support plan might be undertaken for a five-year period. After review, the support might then be renewed for another five years. Because of the UN prestige and seed-money aspects, the infusion may triple the present effectiveness, such that near-completion of the discoveries may be reached by the year 2006 for an expenditure of not more than \$1.75 M/year, a total of \$20 M. This is less than half a penny per inhabitant of the Earth, each one of whom would suffer terribly — and possibly die — of the effects of a collision with a large asteroid.

NOTE ADDED IN PROOF (APRIL 14, 1997): The state of near-Earth detection programs (Table 1) is fluctuating. The Anglo-Australian program was closed by the Australian government, and the crew dispersed, on January 1, 1997, but rumor has it that the government is reconsidering. Additional groups have joined us in Italy, China, and New Mexico. I am now estimating that, with support as outlined in this article, we may have the problem of the 1,700 globally dangerous objects largely under control by the year 2020 (see *Scientific American*, **274:** 34–39, 1996.)

REFERENCES

1. GEHRELS, T., ED. 1994. Hazards Due to Comets and Asteroids . Univ. Ariz. Press. Tucson, Arizona. 1300 pp.
2. GEHRELS, T. 1991. Scanning with Charge-Coupled Devices. Space Sci. Rev. **58:** 347–375.
3. SCOTTI, J.V. 1993. Computer-Aided Near-Earth Asteroid Detection: a Spacewatch Perspective, in Asteroids, Comets and Meteors 1993, A. Milani, M. diMartino, & D. Cellino, Eds. :**160:** 17–30.

International Efforts toward the Spaceguard System

A. CARUSI

Consiglio Nazionale delle Ricerche
Instituto di Astrofisica Spaziale Reparto Di Planetologia
Viale dell'Universitá 11
00185 Rome, Italy

ABSTRACT: Near-Earth Objects are responsible for the cratering of the Earth-Moon system. Depending on the size of the impacting body, the effects of these events on the terrestrial environment range from nothing at all to catastrophes which may pose a very serious threat to life on Earth. Our knowledge of the population of bodies responsible for Earth cratering is currently very poor, because only a minority of them has been discovered and because the parameters and the possible evolution of their orbits are very uncertain. In the last few years several initiatives have been started in order to plan alert systems. A global observational network, the Spaceguard System, is being studied at an international level. Consciousness of the threat has also driven actions at a political level that are, however, still very far from conveying to these researches the resources that would be necessary. In this paper I give a short account of the activity that the International Astronomical Union has initiated by creating the Working Group on near-Earth objects.

INTRODUCTION

The formation of craters on the solid bodies in the solar system is an ubiquitous and perennial process. After 30 years of space exploration, it is clear that this process has affected in the past — and will be affecting in the future — all objects: from small, rocky asteroids to planets and satellites, including the Earth.

In the last few decades scientific research has reached a satisfactory understanding of how craters are formed, whereas the frequency of these events according to the type and size of the impacting objects is still not completely understood. Furthermore, the most recent studies on the occurrences of impacts on the Earth's surface are continuously unveiling the close connection between cosmic impacts and the development of life on our home planet. They indicate how impacts have contributed, during the geologic eras, to the global evolution of the biosphere through dramatic and repeated destructive episodes, sometimes leading to mass extinctions of living organisms and species.

It is a widespread opinion, among planetary scientists, that impact events may represent a serious and continuous threat to human civilization, having as their likely consequences the potential to trigger large-scale and long-standing ecological catastrophes that may cause the collapse of social infrastructures, especially in less developed countries.

Modern scientific and technological researches are able to contrast the threat posed by cosmic impacts to that of other major natural disasters such as volcanic eruptions, floods and earthquakes. However, such researches — whose importance for the protection of the environment has not yet been fully recognized — are still poorly funded and not well coordinated at an international level; catastrophic impacts do not recognize political boundaries, and the global character of the hazard needs a corresponding international response.

POSSIBLE ACTIONS

Although the probability of an impact at the highest levels of risk is extremely small (of the order of once every 300,000 to 30 million years, depending on the size), its likely consequences are so high that it seems at least prudent to reach the certainty that no asteroid or comet in that diameter range (over 1 km) is currently on an orbit which could lead it to a collision with the Earth.

Unfortunately this certainty is at present impossible to obtain. As already said, we have discovered only a minority of the potentially hazardous objects, and for many of them our current knowledge of their motion is very poor. In recent years several studies have been conducted aimed at designing an observational strategy capable of dramatically increasing the rate of discovery (see [1], [2]). It is generally believed that a network of dedicated telescopes, equiped with modern CCDs (charge-coupled devices), in operation for several decades, would discover some 95% of the objects larger than 1 km, i.e., those which could pose a global threat (the so-called "Spaceguard System"). However, there are two sets of problems related to this plan. The first one is scientific in character and related to the discovery-confirmation process; the second one is mainly of political nature and involves the current situation of observers and observatories.

As a matter of fact, discovery is not sufficient. In order to be able to compute the orbit of an NEO with sufficient accuracy to guarantee that it will not collide with our planet during, say, the next couple of centuries, we need a great number of observations, spread over a rather long period of time. This activity is called follow-up, and is of crucial importance. It is not at all sure that the discovery network could, at the same time, make sufficiently reliable follow-up; on the contrary, it is probably more useful that a number of different telescopes be dedicated to follow-up observations, once the objects have been identified.

There are currently only three centers making regular observations aimed at discovering NEOs. Two of them are located in the USA, and one is in Australia. All of these projects have big funding and personnel problems. A project like the proposed Spaceguard System, on the contrary, should count on a relatively large number of observing stations (six to ten, for discovery and follow-up) and certainly involve many scientists. An important point, in this respect, is first the geographical distribution of observatories, which should be evenly located in the northern and southern hemisphere. A second point is that the activity at all these stations should be highly coordinated, possibly on an international basis. A third point involves international agreements to allow observers to move from one place to another, without requiring an often impossible financial effort from the home institutions. The organizational aspects of the problem are as much important as the scientific and technical ones, because a network like the one being studied could work efficiently only if these problematic aspects are solved with the concurrence of many nations.

It is not the purpose of this paper to discuss the matter of active protection from cosmic impacts. It is necessary to mention, however, that any attempt to catalogue unknown objects and investigate their physical nature would greatly benefit from observations from space, both using space observatories in Earth orbit and by close examination of selected targets by space probes.

At present, the major space agencies are studying missions to minor bodies, such as the NEAR (NASA) or the Rosetta (ESA) missions. A great deal of new knowledge will certainly be produced by these missions, especially concerning the physical nature of the objects visited.

Moreover, it is possible to design low-cost satellites to be launched into orbit around the Earth that could perform the type of observations not possible from ground, either because of the screening effect of the atmosphere (especially in the infrared region of the spectrum), or because the objects to be observed are located too close to the Sun.

POLITICAL INITIATIVES

Several initiatives have already been taken at a more political and organizational level, especially in the USA, to deepen our current knowledge of the NEO hazard and to study plans for intervention. The U.S. Congress, in 1990, directed NASA to establish two Working Groups in order to assess detection problems (Detection WG, chaired by David Morrison) and interception and active defense problems (Interception WG, chaired by J. Rahe and J. Rather). The two Working Groups have produced two reports, published by NASA in 1992 [1], [3]. Subsequently, in 1994, a new directive of the U.S. Congress led to the formation of the NASA NEO Survey Working Group, led by Eugene Shoemaker. The aim of this committee was to meet a goal of discovering all potentially hazardous objects within ten years. The Shoemaker committee produced its report in June 1995 [4].

In 1991, the International Astronomical Union (IAU) created the *Working Group on Near-Earth Objects*, chaired by Andrea Carusi, charging it with the specific task of raising international consciousness to the problem, as well as to assist the development of national and international initiatives. The WGNEO has been reappointed in 1994 and has recently organized an International Workshop with the aim of providing a detailed plan for a Spaceguard System. This plan should form the basis for a future international cooperative effort in the field.

In early 1993, the European Science Foundation (ESF) initiated a new Scientific Network, entitled "Impact Cratering and Evolution of Planet Earth." This Network, especially active in the geological and paleontological fields, was intended to allow the exchange of scientists and expertise in different fields, ranging from the study of terrestrial cratering records to the investigation of mass extinction phenomena in the past. The Network has been active for three years and will probably be renewed as an ESF Research Program in the future.

Major international scientific meetings dealing with NEOs and their impact threat have been held in Flagstaff (USA), San Juan Capistrano (USA), St. Petersburg (Russia), Liege (Belgium), Tucson (USA), Erice (Italy), Belgirate (Italy), Sagamihara (Japan), New York (USA), Vulcano (Italy). In 1996 the major opportunities will be Capri (Italy), Versailles (France), and St. Petersburg (Russia).

THE SPACEGUARD FOUNDATION

At the meeting organized by the WGNEO in Vulcano, in September 1995, the NEO community discussed at length all aspects related to the set up of a surveillance system, largely based on the proposal of the Morrison Report and its subsequent refinements. However, no detailed plan of intervention could ever be put into existence without (i) a source of money, and (ii) an organization to take care of the coordination of researches at world level. The WGNEO has therefore decided to set up an international organization, provisionally called *The Spaceguard Foundation*, in charge to collect support and contributions and to assure coordination.

This decision has been very timely. As a matter of fact, on March 20, 1996, the Parliamentary Assembly of the Council of Europe unanimously approved a Resolution on the protection of the Earth from cosmic impacts. The Resolution invites the member states and the European Space Agency to urge and support the development of the Spaceguard Foundation, and to provide support for an international program aimed at the discovery and follow-up of NEOs. This is an important achievement because, for the first time, an international organization has considered the problem of NEO hazards worth an effort extended to many nations. It would be of great relevance that the United Nations also take a complementary approach.

The Spaceguard Foundation will be a private institution, international in character and based in Europe. Its principal aims will be: to support (financially and with personnel) the existing researches, to help the start of new initiatives, to coordinate the activity of all participating centers, to collect, process, and manage the resulting data, to maintain scientific and political contacts, and to participate, when and where appropriate, in space projects.

It is hoped that this initiative will be received favorably by the scientific community at large, and by the decisional centers who may provide financial support. The next four or five years will be crucial to the establishment, with the help of the Foundation, of the Spaceguard System.

REFERENCES

1. MORRISON, D., ED. 1992. The Spaceguard Survey. Report of the NASA International Near-Earth Objects Detection Workshop, NASA, Office of Space Science and Application.
2. GEHRELS, T., ED. 1994. Hazards due to Comets and Asteroids. University of Arizona Press. Tucson, Arizona.
3. RATHER, J.D.G. & J.H. RAHE, J.H., EDS. 1992. Report of the NASA NearEarth Objects Interception Workshop, NASA, Office of Space Science and Application.
4. SHOEMAKER, E.M., ED. 1995. Report of the Near-Earth Objects Survey Working Group, NASA, Office of Space Science and Application.

Space Debris Issue at the United Nations[a]

PETR LÁLA

Deputy to the Director
United Nations Office for Outer Space Affairs, Rm E-0970
P.O.Box 500, A-1400
Vienna, Austria

ABSTRACT: The space debris issue has been discussed by the United Nations Committee on the Peaceful Uses of Outer Space (COPUOS) for many years. In recognition of its importance, at the 1994 session of the Scientific and Technical Subcommittee of COPUOS, it became a separate agenda item. The Committee agreed that international cooperation is needed to evolve appropriate and affordable strategies to minimize the potential impact of space debris on future space missions. This paper summarizes discussions held during the last two sessions of the Subcommittee, including *inter alia* space debris measurement techniques, mathematical modelling of the debris environment, characterizing the space debris environment, and spacecraft design measures to protect against space debris. The focus of the 1995 session was on the subject of acquisition and understanding of data on the characteristics of the space debris environment, which is closely related to the observations of NEO.

1. INTRODUCTION

The focal point of United Nations activities in the field of peaceful uses of outer space is the Committee on the Peaceful Uses of Outer Space (COPUOS), set up in 1959 by the General Assembly. Since its establishment, the Committee and its two subcommittees — the Scientific and Technical Subcommittee and the Legal Subcommittee — have addressed such issues as benefits from space activities, the definition and delimitation of outer space and the use of the geostationary orbit, implications of remote sensing, space sciences, space-based communications, navigation, and meteorological systems, as well as use of nuclear power sources in outer space, space debris, and spin-off benefits of space technology. All three bodies are serviced by the Office for Outer Space Affairs at Vienna (Austria). The Committee consists of 61 Member States and a number of international organizations (Committee on Space Research (COSPAR), European Space Agency (ESA), International Academy of Astronautics (IAA), International Astronautical Federation (IAF), International Astronomical Union (IAU), International Law Association (ILA), INMARSAT, INTELSAT etc.) have observer status with it. The most significant developments undertaken by the Committee have been the consideration, negotiation, and conclusion of five

[a]The views, interpretations, and opinions presented in this paper do not necessarily reflect the position of the United Nations.

international treaties as well as formulation and adoption by the General Assembly of four sets of legal principles concerning space activities [1].

2. SESSIONS OF THE SCIENTIFIC AND TECHNICAL SUBCOMMITTEE OF COPUOS

The item on space debris was included into agenda of the Scientific and Technical Subcommittee at its February 1994 session [2]. The Subcommittee expressed its satisfaction at having the subject of space debris as a separate agenda item after many years of discussion in various international forums, including the Subcommittee and the Committee. It agreed that consideration of space debris was important and that international cooperation was needed to evolve appropriate and affordable strategies to minimize the potential impact of space debris on future space missions. At its 32nd session in February 1995, the Subcommittee continued its consideration of this agenda item, this time on a priority basis [3].

In order to have a common understanding of the term "space debris," the following explanation was proposed: Space debris are all man-made objects in Earth orbit or reentering the dense layers of the atmosphere that are non-functional with no reasonable expectation of their being able to assume or resume their intended functions or any other functions for which they are or can be authorized, including their fragments and parts.

The Subcommittee agreed that it was important to have a firm scientific and technical basis for future action on the complex attributes of space debris and that the Subcommittee should, *inter alia*, focus on understanding aspects of research related to space debris, including debris measurement techniques; mathematical modelling of the debris environment, characterizing the space debris environment; and measures to mitigate the risks of space debris, including spacecraft design measures to protect against space debris. In order to advance in its consideration of space debris, the following work plan was adopted by the Subcommittee:

1996: Measurements of space debris, understanding of data, and effects of this environment on space systems.

Measurements of space debris comprise all processes by which information on the near-Earth particulate environment is gained through ground- and space-based sensors. The effect (impact of particulates and resulting damage) of this environment on space systems should be described.

1997: Modelling of space debris environment and risk assessment.

A space debris model is a mathematical description of the current and future distribution in space of debris as a function of its size and other physical parameters. Aspects to be addressed are:

- Analysis of fragmentation models
- Short- and long-term evolution of the space debris population
- Comparison of models.

The various methods for collision risk assessment should be critically reviewed.

1998: Space debris mitigation measures.

Mitigation comprises reduction of the space debris population growth and protection against particulate impact. Measures for the reduction of space debris growth include methods for debris prevention and removal. Protection against space debris includes:

- Physical protection with shielding
- Protection through collision avoidance.

Each session should review the current operational debris mitigation practices and consider future mitigation methods with regard to cost efficiency.

The Subcommittee noted that a certain amount of research on space debris had already been undertaken in some countries, which had allowed for a better understanding of the sources of debris, the areas in near-Earth orbit that were reaching high levels of space debris density, the probabilities and effects of collisions and the necessity to minimize the creation of space debris [4]. The Subcommittee encouraged Member States and relevant international organizations to provide information on practices that they had adopted and that had proven effective in minimizing the creation of space debris. The Subcommittee heard scientific and technical presentations on the subject of space debris by experts from different Member States and observer organizations [5], [6]. The most interesting information is summarized in the next parts of this paper.

3. MEASUREMENTS OF SPACE DEBRIS

Objects moving in the near-Earth space are tracked and catalogued by the United States Space Command Space Surveillance System. This system operates more then two dozen radar and optical facilities to monitor near-Earth space and maintains a catalogue of orbital elements of all tracked objects. Recently, these unique facilities are offered also for monitoring and cataloguing of the near-Earth objects [7]. The minimum diameter of observable objects is about 10 cm for low-Earth orbits (LEO) and 1 m for geostationary orbits (GSO). The catalogued objects represent about 99% of the total on-orbit mass (roughly 3,000 metric tons), but only 5 to 20 per cent of the debris larger than 1 cm. The number of debris in the 1–10 cm range is estimated to be between 35,000 and 150,000, and the population of particles larger than 1 mm is probably more than 1,000 times the catalogued population.

In the past year, 89 successful launches into outer space have been carried out (compared to 79 in 1993). The total number of trackable objects in space is steadily increasing (7,860 at the end of 1994). More than 40% are fragments of rocket upper stages and spacecraft as a result of explosion. As of the end of 1994, 124 breakups are known. Only about 6% of the catalogued objects are operational spacecraft. In 1994, six fragmentations have been observed:

- 7 February 1994, Titan IIG stage (international designation 1994-004B), orbit 220 × 254 km, 745 fragments were registered, but decayed by now;
- 27 April 1994, Ariane 4 third stage (1991-015C), transfer orbit, 9 fragments;
- 7May 1994, Proton kickstage (1991-010D), orbit 225 × 21805 km, 39 fragments;
- 6–8 November 1994, Proton stage (1992-047H), orbit 479 × 19033 km, 36 fragments;
- November 1994, USA 40 satellite (1989-061A), 62 fragments;
- 26 December 1994, Russian Rokot stage (1994-085B), 1865 × 2185 km, 30 fragments.

A twenty-four hour measurement campaign in LEO with facilities located in Europe was carried out on 13–14 December 1994. The radar stations involved were in Wachtberg Werthhoven (Germany) and Fylingdales (United Kingdom), optical in Herstmonceux (United Kingdom), and Zimmerwald (Switzerland). The purpose of the campaign was to observe a specific volume in space, detect all objects of a minimum size passing through that volume, and compare results with predictions by a space debris model. Due to unfavorable weather conditions, the optical site at Zimmerwald could not make any observations, while Herstmonceux had for few hours clear sky for observations.

The German radar searched in a volume centered at a range of 786 ± 75 km with a beam width of 0.5 degree, for which ESA's MASTER reference model predicts 84 detonations of objects larger than 2 cm. Preliminary analysis of a limited amount of data within 19 hours resulted in 26 detections of such objects. Similar campaigns carried out with the Haystack radar (United States) show that current space debris models overestimate the number of debris at lower altitude (300–500 km), while at higher altitude (800–1000 km) the number is underestimated.

The Synoptic Monitoring of Orbital Debris (SYNMOD) project, carried out by the Institute for Space Science and Technology (ISST Europe) in Grasse Magagnose in France, in collaboration with several laboratories and institutes, is supervised by European and American organizations. The French space agency CNES has been involved in two basic projects aimed at geostationary (GSO) and transfer (GTO) orbit population. The observations are conducted at the Calern observatory which uses a telescope fitted with the CCD camera. An observation campaign lasting three months is planned in order to verify whether the observation site and the measuring devices are really suitable. Another project, called Rosace, is intended to improve the determination of geostationary satellite orbits from ground-based observations. The system combines a Newton telescope and a charge coupled device (CCD) matrix. The point is to measure the angular separation between the image of the satellite trajectory and stars with precisely known positions. The timing is provided through the global positioning system (GPS) receiver. Measurement accuracy is expected to be 1 arcsec and 1 ms in time. Intergration of the prototype used for this experiment will start in 1995.

Europe is making a significant contribution to the expansion of knowledge of small size particles with ESA's European Retrievable Carrier (EURECA), the returned solar array panel of the Hubble Space Telescope (HST) and other spacecraft. The spacecraft surfaces are covered with a large number of impacts of micrometeoroids and debris. The size of individual impact craters and holes range from 20 microns to several millimeters. A preliminary analysis of EURECA shows that the number of large impacts was greater than expected. The larger impact crater diameter is 6.4 mm, resulting from an object of 0.5 to 1 mm diameter. With 3.6 years on orbit at the time of the repair mission, the HST had about 4 times the space exposure time of EURECA. First visual inspection of the retrieved panel revealed a large number of impacts, with the largest hole diameter of the order of 2–3 mm. Results from the EURECA and HST panel impact analyses will be used to validate current reference flux models for small size meteoroids and space debris.

4. MODELLING OF SPACE DEBRIS ENVIRONMENT

Assessing the present and future risk to all space objects by the space debris environment requires the use of models. The ESA MASTER Model (Meteoroids And Space Debris Terrestrial Reference Model) [8] has been developed by the Institute for Space Flight Technology and Nuclear Reactor Technology of the Technical University of Braunschweig, Germany (IFR/TUBS) under an ESA contract. The scope of the model covers man-made and natural objects larger than 0.1 mm in altitudes below 2,000 km. Analytical break-up models are used for the individual generation of the untrackable debris population for the known on-orbit fragmentations. The observable part of the population is considered by the implementation of ground-based radar measurements. For each generated population member, information is retained on its origin, mass, size and orbit. The population can be propagated to a reference epoch, at which a 3D (radius, right ascension, declination) spatial density distribution and the resulting flux to any circular or elliptic user defined target mission can be established. Flux information is resolved with respect to the impactor source, its mass, the impact velocity, impact azimuth angle, impact elevation angle, and the position of the impact occurrence along the target's orbit.

The historical orbital fragmentations have been simulated at IFR/TUBS and the corresponding object flux has been established. The flux resulting from these simulations is smaller than that observed by deterministic sources, e.g., space returned hardware (LDEF) and is adapted to the measurements for application purposes. Currently ongoing studies will extend the model up to the GEO region and will provide a CPU time efficient, user friendly, and highly accurate analyst and engineering application to be distributed via CD ROM. The model will be applicable on both work station and PC environments and will include several features for spacecraft designers and operators as well as for ground based observatory operators. It will enable a broad user community to carry out investigations on space debris, which have previously been reserved to debris analysts with powerful computer hardware.

Recent studies suggest that, in the crowded orbital regimes of LEO, the number of collision partners will reach the critical level required to sustain collisional cascading within the next 10 to 15 years. This is the process whereby the breakup of one large object can produce many hundreds of fragments, each capable of colliding with, and damaging other objects. These secondary events may then produce more fragments, the process can be repeated and the orbital regime can soon become a particulate equivalent of the Van Allen radiation belts, a hostile and inaccessible environment for normal operations. Some studies conclude that it is the larger objects which play the critical role in the cascade phenomenon. If the number of these objects in orbit can be reduced or minimized, the onset of the cascading will be avoided and fragmentation events will remain localized.

The average relative velocity between any particular spacecraft orbiting near-Earth space and orbital debris objects ranges from about 10 km/sec for low inclination spacecraft orbits, to about 13 km/sec for near polar orbits. Due to their great kinetic energy, particles larger than 0.1 mm may cause damage or degradation of a spacecraft and 1 cm fragment has sufficient energy to disturb significantly any satellite's operation; and debris in the 1–10 cm size range (too small to be sensed by ground systems), are large enough to cause a catastrophic damage to many satellites.

Based on existing meteoroid and debris models, the flux of particles near a space vehicle can be modeled. ESA analyses [9] show that the current population of debris does not represent an immediate and excessive danger. For example, during a five-year period, the chance of the ERS satellite to collide with a 1 cm particle, or larger, amounts up to 1 to 2%. However, the risk of collision with debris is steadily growing. Of most concern are the long-term prospects in low-Earth and geostationary orbits, the two most heavily-used and endangered regions in space. Theoretical models of debris generation caused by the disintegration of objects or by collision between objects also are performed in other countries, e.g., in India. These studies are aimed at better understanding of the untracked debris environment and predicting collision probability for operational spacecraft. In spite of different initial assumptions, their conclusions are in general agreement.

E. SPACE DEBRIS MITIGATION MEASURES

Several strategies, including shielding of the critical components of a spacecraft, can be adopted for protecting the spacecraft from the consequences of orbital debris impacts. However, the only efficient way to avoid collisions of the spacecraft with orbital debris greater than 1 cm in size is the maneuver of changing the spacecraft's orbit. To achieve this, the spacecraft must be maneuverable and predicted future positions of the spacecraft and space debris must be known with sufficient accuracy. The accuracy requirements of predictions depend on the spacecraft and its maneuverability. Predictions must be sufficiently accurate to avoid false alarms and to provide high confidence of predicted collisions.

Space debris pose a significant danger for such missions as the international Space Station, the Hubble Space Telescope, and other large satellites [10]. The Space Station's large cross section and the anticipated length of its mission magnify significantly the danger of a catastrophic failure. To provide an accurate warning of possible collisions, it is essential to maintain a catalogue of debris orbital elements sets and to calculate precise predicted positions of a large number of objects. The Naval Space Surveillance Center plans to maintain a catalog of 25,000 objects larger than 1 cm in support of the International Space Station. Since the Space Station maneuverability will be limited to a few maneuvers per year and at a distance of a few hundred meters, the accuracy of the predicted positions must be very high.

The International Academy of Astronautics (IAA) initiated a study to be performed by experts from its Committee on Safety, Rescue, and Quality. The objective were to evaluate the need and urgency for action and to indicate ways to reduce the hazards. This document, which has been approved in October 1993 as an official IAA Position Paper [11], recommends for immediate application the following actions:

- No deliberate breakups of spacecraft which produce debris in long-lived orbits;
- Minimization of mission-related debris;
- Safing procedures for all rocket bodies and spacecraft which remain in orbit after completion of their mission;
- Selection of transfer orbit parameters to ensure the rapid decay of transfer stages;
- Reorbiting of geostationary satellites at end-of-life (minimum altitude increase 300–400 km);
- Separated Apogee Boost Motors (ABM) used for geostationary satellites should be inserted into a disposal orbit at least 300 km above the geostationary orbit;
- Upper stages used to move geostationary satellites from GTO to GSO should be inserted into a disposal orbit at least 300 km above the geostationary orbit and freed of residual propellant.

As space debris is a global problem which only can be solved by a joint effort, discussions and cooperation among space agencies and related organizations are of paramount importance. Regular coordination meetings have been held between ESA and NASA since 1987. In 1993, the Inter-Agency Space Debris Coordination Committee (IADC) has been created. Its founding members are ESA, entities from Japan, NASA, and the Russian Space Agency (RKA). The primary purpose of the IADC is to exchange information on space debris research, to facilitate opportunities for cooperation in space debris research, to review the progress of ongoing cooperative activities and to identify debris mitigation options. In particular, the need was recognized to implement a common database of the space debris environment which will be supported by the surveillance networks and observation facilities of all four participants.

To mitigate the problem of space debris even in the absence of any binding international agreement, space agencies, and several satellite users have already adopted some voluntary measures. For example, the French CNES together with Arianespace started the residual fuel venting from the upper stage of Ariane launcher in case of launches into low altitude circular orbits (launchings of satellites SPOT 2, ERS 1, and Topex/Poseidon) to prevent explosions. Since the flight 59 (in September 1993), fuel venting of the third stage is carried out routinely regardless of the type of the target orbit. The following ESA geostationary satellites have been reorbited so far: OTS 2 (now orbiting 318 km above GSO), GEOS 2 (260 km), Meteosat 2 (334 km), ECS 2 (335 km), and Olympus 1 (due to a failure, this satellite has been left at orbit 213 km below the GSO).

The Intelsat organization has adopted the following practice to minimize the creation of space debris:

- At the end of their operational lifetimes, Intelsat will boost its communication satellites into an orbit at least 150 km above the geostationary arc; the intended increase in orbit will be 300 km for the Intelsat VI and all later series satellites;
- Intelsat will discourage manufacturers of its spacecraft from using designs that jettison spacecraft parts, especially near GSO; e.g., solid rocket motor casings and solar array cable wraps will stay attached to the spacecraft.

Similarly, the United Kingdom recognizes the unique nature of the geosynchronous altitude and the need to preserve this global resource for future development and exploitation. Consequently, the Skynet family of geosynchronous communications satellites controlled by the United Kingdom have the following operational requirements:

- For all satellites which are currently in orbit, a fuel budget is allocated which is capable of performing a tri-impulse manoeuvre to a circular orbit with a minimum altitude of 150 km above the geostationary ring at the end of operational life; and
- Design requirements for future series of satellites specify a capability to achieve a minimum altitude of 500 km above the geostationary ring using a similar tri-impulse manoeuvre at the end of the operational life.

For all cases, in order to eliminate the potential for explosion, appropriate operational procedures shall be established to passivate all energetic subsystems when the satellite has been placed in a graveyard orbit.

5. CONCLUSION

The question of space debris is in fact only a part of a more general problem of protection and preservation of the outer space environment. In recognition of its importance, it is now a priority agenda item of the Scientific and Technical Subcommittee of the United Nations Committee on the Peaceful Uses of Outer Space (COPUOS). It is expected that after careful deliberations, sufficient technical background information will be available for the Legal Subcommittee to develop legal norms for space debris mitigation. Since the question of NEO hazards is now becoming a subject of increasing interest and discussions in the scientific community, some Member States of the United Nations might take the initiative to bring this problem to the attention of COPUOS. If this is the case, scientific and technical problems should be addressed first in order to assess the situation realistically.

REFERENCES

1. UNITED NATIONS TREATIES AND PRINCIPLES ON OUTER SPACE. 1994. Document A/AC.105/572, United Nations.
2. REPORT OF THE SCIENTIFIC AND TECHNICAL SUBCOMMITTEE ON THE WORK OF ITS THIRTY-FIRST SESSION. 1994. Document A/AC.105/571, March 10. United Nations.
3. REPORT OF THE SCIENTIFIC AND TECHNICAL SUBCOMMITTEE ON THE WORK OF ITS THIRTY-SECOND SESSION. 1995. Document A/AC.105/605, February 24. United Nations.
4. NATIONAL RESEARCH ON SPACE DEBRIS; SAFETY OF NUCLEAR-POWERED SATELLITES; PROBLEMS OF COLLISIONS OF NUCLEAR-POWERED SOURCES WITH SPACE DEBRIS. 1995. Document A/AC.105/593 and Adds.1–4, February 24. United Nations
5. SCIENTIFIC AND TECHNICAL PRESENTATIONS TO THE SCIENTIFIC AND TECHNICAL SUBCOMMITTEE AT ITS THIRTY-FIRST SESSION. 1994. Document A/AC.105/574, May 12. United Nations.
6. SCIENTIFIC AND TECHNICAL PRESENTATIONS TO THE SCIENTIFIC AND TECHNICAL SUBCOMMITTEE AT ITS THIRTY-SECOND SESSION. 1995. Document A/AC.105/606, April 17. United Nations.
7. DORRAH, J.H. 1997. The Development of Wide Field of View Optical Search Systems for Meteoroid and Satellite Tracking. *In* Proceedings of the International Conference on Near-Earth Objects, 1995.
8. 1992. Proceedings of the Conference Stand der Space Debris — Forschung in Deutschland und Internationale Regulierungsfragen, Bonn,Germany. 9.12..
9. 1993. Proceedings of the First European Conference on Space Debris, Organized by the European Space Agency, Darmstadt, Germany, 5–7 April.
10. 1995. Orbital Debris: A Technical Assessment, National Academy Press, Washington, D.C.
11. 1995. Position Paper on Orbital Debris, International Academy of Astronautics. Paris.

United Nations Contributions to the Worldwide Development of Astronomy[a]

HANS J. HAUBOLD

United Nations Office for Outer Space Affairs
Vienna International Centre
1400 Vienna, Austria

ABSTRACT: The United Nations in cooperation with the European Space Agency organized a unique series of annual Workshops on Basic Space Science since 1991 for the benefit of astronomers and space scientists in Asia and the Pacific, Latin America and the Caribbean, Africa, and Western Asia. The merits and accomplishments of these Workshops and their regional follow-up projects are reviewed in this paper.

BASIC SPACE SCIENCE = PLANETARY EXPLORATION + ASTRONOMY

With the establishment of the United Nations Committee on the Peaceful Uses of Outer Space (COPUOS) [1] in 1959, the Office for Outer Space Affairs [1] of the United Nations Secretariat was assigned the responsibility for implementing decisions of the Committee, its Legal Subcommittee, and its Scientific and Technical Subcommittee related to the promotion of international cooperation in outer space matters. Among the primary tasks carried out by the Committee is the development of international treaties, conventions, or legal principles governing the activities of Member States in the peaceful exploration and use of outer space and the provision of technical assistance and information on space technology and applications to interested Member States.

Following the decision of COPUOS to take upon itself the promotion of international cooperation in space science and technology, the United Nations Program on Space Applications [1] was established in 1969 with the objective, *inter alia*, to provide scientists from developing countries with educational programs in remote sensing, satellite meteorology, satellite communication, and basic space science. The activities of this program are subject to the annual review and approval of COPUOS. Because of the increasing number of developing countries that are actively engaged in space science research, the United Nations and COPUOS has over the past several years begun to place an increased emphasis on promoting education and re-

[a]The views expressed in this paper are those of the author, writing in his personal capacity.

search in space science and technology, particularly in the two fields which are on the agenda of COPUOS, planetary exploration and astronomy. These two fields together comprise what has come to be known as basic space science in the international space community.

Of the 185 countries that are Member States of the United Nations, nearly 100 have professional or amateur astronomical organizations. Only about 60 of these countries, however, are sufficiently involved in astronomy to belong to the International Astronomical Union (IAU). Only about 20 countries, representing 15 percent of the world's population, have access to the full range of astronomical facilities and information. This does not include most of the Eastern European, Baltic, and former countries of the Soviet Union, whose fragile economies keep them from achieving their full potential, despite the excellence of their astronomical heritage and education (Percy [2]).

UN AND ESA INITIATED WORKSHOPS ON BASIC SPACE SCIENCE

In 1991, as part of the United Nations Program on Space Applications, the United Nations, in cooperation with the European Space Agency (ESA), initiated the organization of annual Workshops on Basic Space Science (Jasentuliyana [3]) for developing countries. These workshops were designed to be held as a unique series in each of the following major regions of the world: Asia and the Pacific, Latin America and the Caribbean, Africa, Western Asia, and Europe. This subdivision by region follows, *inter alia*, United Nations principles to assess the relevance of space activities to worldwide economic and social development.

To date, United Nations/ESA Workshops on Basic Space Science have been held in India (1991), Costa Rica and Colombia (1992), Nigeria (1993), and Egypt (1994). The four workshops were attended by 300 invited participants from 50 countries and 15 national and international organizations. An assessment of the achievements of this series of Workshops will be made during a final workshop to be held in Europe in 1996. In 1992, the Planetary Society joined the effort to organize the UN/ESA Workshops in order to provide better coverage of planetary exploration in the program of the workshops. Between 1991 and 1994 the workshops were also coorganized by the German Space Agency (DARA), the National Aeronautics and Space Administration (NASA), the International Centre for Theoretical Physics (ICTP), and the Austrian Space Agency (ASA)(Haubold et al. [4]).

The scientific topics covered by the workshops depended on the interest of local organizers and the research activities undertaken in the respective region. The topics were chosen from among the following fields:

- International cooperation in basic space science;
- Education for space science;
- Solar terrestrial interaction;
- Planetary science;
- Space astronomy and astrophysics;
- Cosmology; and
- Data bases and on-line data in astronomy.

The workshops were hosted by the government of the respective country in which they were held; government representatives were actively involved in the organizational and scientific preparations for the workshops which helped ensure close interaction between the government and the local scientific community. This ultimately proved crucial to facilitating achievement of the workshops' objectives. A vital part of the program of each workshop were the Working Group sessions which provided all participants a common platform from which to make critical observations and recommendations addressing the development of astronomy in their specific region. These observations and recommendations are available in the published Workshop Proceedings [3]–[9] and are also contained in United Nations reports [10] on each workshop. These reports are available for distribution to governments (particularly through COPUOS and its Scientific and Technical Subcommittee), national, regional, and international organizations as well as concerned scientific institutions and universities. This collection of observations and recommendations constitutes a unique international framework for the development of astronomy covering four major regions (and almost all developing countries) of the world. To prepare the workshops on a regional basis, the United Nations invited astronomers to submit studies on the current status of astronomy in the regions which will be published in a United Nations booklet in the near future.

FOLLOW-UP PROJECTS BENEFIT ASTRONOMY

The increasing costs for big-science projects and a changing international research atmosphere have fostered the trend in the scientific community to focus on the development of large internationally-funded facilities, which may take advantage of the climatological and geographic attributes of any country in the world, particularly in developing countries. On the other hand it has been strongly emphasized that the establishment of networks of existing small scientific facilities should be high on the agenda of strengthening international cooperation, particularly in geomagnetic studies, electrojet current measurements, solar photometry, astrometry, galactic mapping, and optical astronomy.

This fact also addresses the concept of international, electronically-linked observing programs — similar to the "Whole Earth Telescope" project — the value of which have been become more evident through successfully observing runs involving tens of globally distributed telescopes in recent years. These programs could be expanded to include more active participation of developing countries at relatively low cost to them.

In addition to the common direct benefits of an international workshop, the UN/ESA Workshops have generated a number of follow-up projects which will be implemented on a long term basis.

Astronomical Telescope for Sri Lanka [11]

The government of Japan took the initiative to support the establishment of national astronomical observatories in the Asia and Pacific region through the provision of moderate-sized astronomical research telescopes or planetaria. Over the past several years, Singapore received a Mitaka-kokhi 40 cm reflector for its science centre of public education, and Malaysia started operating a Minolta Planetarium at its space science education centre. Through the Japanese Cultural Grant Aid Program, Thailand was able to install a Goto 45cm reflector at the Department of Physics of the Chulalongkorn University Bangkok, and the Bosscha Observatory (Hidayat [12]) in Lembang, Indonesia, is using its Goto 45cm reflector for astronomical research. As a result of the UN/ESA Workshops, Sri Lanka received a 45 cm astronomical telescope which will be installed at the Arthur C. Clarke Centre for Modern Technologies at Katubedda Moratuwa, Sri Lanka. These moderate-size optical telescopes set up at appropriate locations on Earth have contributed significantly to astronomical research. The 45cm reflecting telescope from Goto, for example, is equipped with a photometer, a spectrograph, and a photographic camera. Although the telescope was designed primarily for photometric observational studies of variable stars, it also allows for the observation of comets and asteroids. This telescope also offers the possibility of studying interstellar, interplanetary, and atmospheric dust. A network of telescopes of this type throughout a region or worldwide could form an even more powerful tool for other types of astronomical research (Warner [9], Budding [9]). Moreover, it could foster regional and international cooperation in astronomical research as in the case of the "Spacewatch" program.

Colombia [6] and Honduras [7]: Taking the Initiative in Central America

An international cooperative program for the Galactic Emission Project (GEM) formed by Brazil, Colombia, Italy, and the United States is now operating a radio telescope in Colombia in order to survey the galactic emission at long wavelengths (de Amici et al. [6]). The GEM project is funded by the United States National Science Foundation and by COLCIENCIAS in

Colombia. Because of its equatorial latitude and the presence of high peaks in the Andean mountain range, Colombia is an ideal location for astronomical and radio astronomical research. Access to the northern and southern celestial hemispheres with the same instrumentation opens a window of opportunity for research in astronomy. The plan for the establishment of an Astronomical Observatory in Honduras (Pineda de Carias [7]) also promises to be an important project for the scientific development of the region and was discussed extensively during the UN/ESA Workshops on Basic Space Science.

A Large Astronomical Facility for Africa? [7]

A proposal for an Inter-African Astronomical Observatory and Science Park on the Gamsberg in Namibia is currently being supported by the United Nations as a result of the UN/ESA Workshops. Because of its unique geographic location, southern Africa can make an immense contribution to astronomy. Certain time-critical phenomena and 24-hour coverage can be obtained only through astronomical observatories in continents (excluding Antarctica) south of the equator. The Gamsberg has been identified as one of the most suitable sites for an observatory in southern Africa. It is a table mountain 120km south-west of Windhoek above the Namib desert at an altitude of 2350m above sea level. It experiences a large number of cloudless nights, a dark sky, excellent atmospheric transparency and low humidity. Comparative test measurements have demonstrated that the above qualities are just as good as those of well-known astronomical sites in Chile. The mountain top is owned by the Max-Planck-Society of Germany and a small astronomical station was established there in the 1970s (Elsaesser [7]). Besides astronomy, the mountain is of considerable interest to other scientific disciplines such as cosmic ray physics, atmospheric research and meteorology, biology, geology, etc. The huge plateau of about 250 hectares offers enough space for various independent installations. The Max-Planck-Institute for Astronomy at Heidelberg, Germany, is trying to initiate the development of a new scientific centre on the Gamsberg. This can only be achieved, however, with international collaboration and support in kind and cash. South Africa has expressed an interest in operating the astronomical observatory on behalf of the international community. The ideal solution would be an Inter-African Science Park. The Government of Namibia, as well as the recently established Windhoek University, have expressed support for this project. If established, this facility could become an important focus in the development of basic space science in African countries (Okeke and Onuora [7]). If the facility is established with a viable infrastructure, it is possible that it would be attractive to northern hemisphere countries as well, especially those wishing to establish facilities in the southern hemisphere.

FIGURE 1. The parabolic reflector antenna of the Galactic Emission Mapping Project in Columbia (courtesy of S. Torres, who is proudly presenting the project in the figure).

Kottamia is the Largest Observatory in Western Asia [8], [9]

The Kottamia Observatory, housing a 74-inch reflecting telescope, is located in the north-eastern Egyptian desert, 80km from Helwan, Egypt, on the Cairo-Suez road. Because the Kottamia Observatory telescope facility could supply major observational capabilities for basic space science in the region, its importance cannot be underestimated. Hence, to safeguard the importance of this telescope facility (the largest in the Western Asia region), further development would be needed (Mikhail [8], [9]). Mindful of this, deliberations of the UN/ESA Workshops noted with great satisfaction that there is a willingness to cooperate in this effort and further stressed that a collaborative effort to bring about the modernization of the Kottamia Observatory represented one of the most important collaborative efforts in Western Asia. It is the intention of the Government of Egypt to open the Kottamia Observatory to scientists from the entire region in order to allow access to the important observational opportunities for basic space science. The plans of astronomers from other countries, both in Western Asia and Africa to share in the future development of the Kottamia Observatory will greatly contribute to the future development of basic space science in the region and will enhance significantly the opportunities for cooperation among astronomers in the region.

FIGURE 2. The Gamsberg in Nambia seen from the northeast (courtesy of H. Elsaesser.)

FIGURE 3. The Kottamia Observatory in the Egyptian desert south of Cairo (courtesy of M.Y. Tawadros).

A WORLDWIDE UN NETWORK OF TELESCOPES FOR OBSERVING NEAR-EARTH OBJECTS? [13]

The recent impact of the Shoemaker-Levy 9 comet on Jupiter has raised new concerns about the possibility of a near-Earth object hitting the Earth. Pursuing an understanding of the Earth's interactions with near-Earth objects has become an issue of global importance. In an effort to provide a scientific basis for future cooperative international research and space exploration, The Explorers Club and the United Nations Office for Outer Space Affairs organized an International Conference on near-Earth objects (this volume). Leading researchers in the fields of astronomy, planetary science, astrophysics, paleontology, and astronautics assembled from around the world to present their ideas on various related topic. The program of the conference included interdisciplinary aspects of near-Earth objects from the point of few of natural science. Among the topics addressed by the conference was the establishment of adequate observational facilities for near-Earth objects in both the northern and southern hemispheres. As a first step, these facilities should be improvements of existing astronomical telescopes, including those in developing countries. Observation programs could be coordinated with activities of amateur astronomy groups and organized on an international level which may lead to the establishment of a network of moderate-sized astronomical telescopes as discussed in the UN/ESA Workshops on Basic Space Science. Taking into account the fact that the primary objective of this conference was to interpret the threat posed by near-Earth objects within the cooperative framework of existing guidelines of international space law and international space programs, it was suggested that the United Nations should take the lead in establishing a worldwide network of telescopes for the observation of near-Earth objects.

CONCLUSIONS

The observations and recommendations made during the past four UN/ESA Workshops on Basic Space Science can be summarized in the following topics which need to be addressed urgently on a regional and international level: (i) promotion of the advancement and dissemination of the knowledge of basic space science and its application to human welfare (Prasad et al. [7], Bennett [6], Bennett and Morrow [6]), (ii) provision of on-line database and e-mail services, (iii) provision of abstracting and indexing services in basic space science, (iv) dissemination of reliable information on basic space science to the public, (v) collection and analysis of the statistics on the profession and on education in basic space science, (vi) encouragement of the documentation and study of the history and philosophy of basic space science, and (vii) cooperation with organizations on educational projects at all levels.

Among the above topics, the electronic networking of scientific institutions may have the most immediate impact on the situation in developing countries. There exist extensive data archives in space science readily available at virtually no cost to any astronomer which has established access to the Internet. Space astronomy missions such as IUE, HST, COBE, ROSAT, IRAS, etc., have made their data archives publicly accessible through electronic networks (Albrecht and Pirenne [9], Wamsteker and Griffin [9]). These archives are available to astronomers in any country on Earth as long at they have access to the Internet. These electronic networks also allow immediate access to electronic mail channels and electronic publications (Astrophysical Data System), solving the traditional problem of isolation and obsolete libraries in many developing countries. The combined efforts of individual astronomers and the support of governments and international organizations could realize the concept of a "Global Village" in terms of education and research worldwide.

ACKNOWLEDGMENTS

Special thanks to Barbara Haubold for her skill, efficiency, and hard work of typing and preparing the papers for all the Workshops. Bruce Schoenfeld's support in editing this paper is appreciated.

REFERENCES

1. CHIPMAN, R., ED. 1982. The World in Space. Prentice-Hall, Inc. Englewood Cliffs, New Jersey.
2. PERCY, J.R. 1995. Mercury. **24:** (2) 15–18.
3. HAUBOLD, H.J. & R.K. KHANNA, EDS. 1992. Basic Space Science. American Institute of Physics Conference Proceedings, Vol. 245. AIP. New York.
4. HAUBOLD, H.J., A.C. OCAMPO, S. TORRES, & W. WAMSTEKER. 1995. ESA Bulletin, February, No.81: 18–21.
5. FERNANDEZ, W. & H.J. HAUBOLD, EDS. 1993. Basic Space Science. Earth, Moon, and Planets 63:93-170.
6. HAUBOLD, H.J. & S. TORRESS, EDS. 1994. Basic Space Science. Astrophysics and Space Science. **214:** 1–260.
7. HAUBOLD, H.J. & L.I.ONUORA, EDS. 1995. Basic Space Science. American Institute of Physics Conference Proceedings, Vol. 320. AIP. New York, NY.
8. HAUBOLD, H.J. & J.S.MIKHAIL, EDS. 1995. Basic Space Science. Earth, Moon, and Planets. **70:** 1–229.
9. HAUBOLD, H.J. & J.S. MIKHAIL, EDS. 1996. Basic Space Science. Astrophysics and Space Science. **228:**. 1–405.

10. Reports on the UN/ESA Workshops and the status of the follow-up projects are available on the World Wide Web at the electronic address\par ftp://ecf.hq.eso.org/pub/un/un-homepage.html.
11. GEHRELS, T. 1988. On the Glassy Sea: An Astronomer's Journey. American Institute of Physics, New York.
12. HIDAYAT, B., ET AL. 1992. *In* Evolution of Stars and Galactic Structure, K.Isida & B.Hidayat, Eds. National Astronomical Observatory. Tokyo.
13. MORRISON, D., ED. 1992. The Spaceguard Survey: Report of the NASA International Near-Earth Object Detection Workshop. National Aeronautics and Space Administration, Washington.

Index of Contributors